BOTANY
An Introduction
to Plant Biology

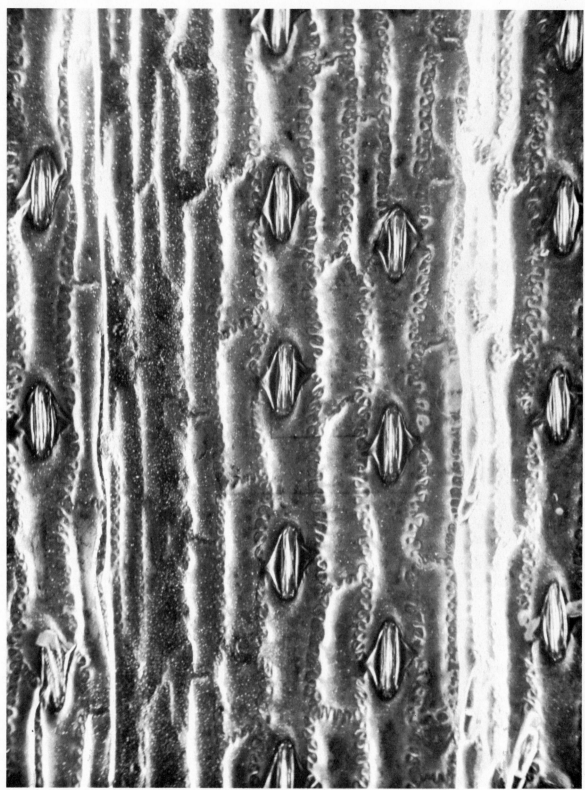

Stereoscan micrograph of the lower surface of a fresh corn leaf. Corn (*Zea mays*) is a C_4 plant (see Chap. 13). The shiny strip is a raised vein which is reflecting light. Note the intricately interlocking cell walls and the stomata arranged in files.

BOTANY

An Introduction to Plant Biology

Fifth Edition

T. Elliot Weier
C. Ralph Stocking
Michael G. Barbour

University of California
Davis, California

JOHN WILEY & SONS

New York   London   Sydney   Toronto

Line drawings by Alice B. Addicott with
the assistance of Jean A. McKinnon.

Library of Congress Cataloging in Publication Data:
Weier, Thomas Elliot, 1903–
    Botany, an introduction to plant biology.

    First-3d ed. by W. W. Robbins and others.
    1. Botany. I. Stocking, C. Ralph, 1913–   joint
author. II. Barbour, Michael G., joint author.
III. Robbins, Wilfred William, 1884–1952. Botany, an
introduction to plant science. IV. Title.

QK47.W44 1974   581   73-18035
ISBN 0-471-92468-7
Printed in the United States of America
10 9 8 7

Preface

The objectives of this book remain unchanged in the fifth edition: to acquaint students with all the fields of modern botany and with the great diversity of plant life that inhabits our biosphere. This book is still an introductory text. It is for beginning biology students at the college level, and has been written with them particularly in mind. Major changes in this edition include the following:

1. We have expanded our treatment of molecular biology in Chapters 6 and 25. The modern interpretation of how DNA is translated into protein is described with new text and original illustrations. Dr. Robert M. Thornton of the Botany Department at Davis gave us valuable assistance.

2. Chapters 13 and 14, Photosynthesis and Respiration, were largely rewritten to include recent advances (for example, in the light reactions of photosynthesis, photorespiration, and the C-4 pathway of photosynthesis) and to break the metabolic pathways down into understandable, bite-size pieces. However, our emphasis remains on energy transformation and the overall role of respiration and photosynthesis in plants—and not on chemical terminology.

3. Chapter 20, Plant Growth and Development, was completely rewritten by Dr. Bruce A. Bonner, also of the Botany Department at Davis. His research and teaching interests in photoreceptor systems and in hormones enabled him to write a chapter of exceptional clarity and authority, with carefully drawn conclusions that take the reader up to the limits of current knowledge. We are indebted to him for his contribution to the text.

4. Chapter 18, Plant Ecology, was expanded with a section on fire ecology and the use of fire as a vegetation management tool. Examples are drawn from both the east and west coasts of North America. Dr. Harold H. Biswell, Professor of Forestry at the University of California, Berkeley, critically reviewed this section and was also helpful in supplying illustrations.

5. Chapters on the bacteria, fungi, and viruses (21, 23, 24, 25) have been brought up-to-date thanks to the great help of members of the Plant Pathology Department at Davis. Dr. Joseph M. Ogawa made contributions to the text and added material on the Rickettsias and Myxoplasmas, plant disease agents whose importance has only recently been discovered. Dr. Edward E. Butler was of particular help to our artist and in providing stereoscan micrographs of fungi. With Dr. Robert J. Shepherd's

help, the viral reproductive cycle was expanded in level of detail, and now follows closely new material in Chapter 6 on molecular biology.

6. The final chapter, Evolution, has been completely rewritten. Two-thirds of the chapter is devoted to the fossil record of the plant kingdom, beginning with microfossils exquisitely preserved in chert 1–3 billion years old. Professor J. William Schopf of the Geology Department, University of California, Los Angeles, was most helpful in providing illustrations of these microfossils. A recent book by Professor Harlan Banks (*Evolution and Plants of the Past*, Wadsworth, Belmont, Ca., 1970) stimulated us to expand our treatment of the rest of the geologic record. Dr. Ernest M. Gifford, Jr., of the Botany Department at Davis gave our artist considerable help in maintaining a high level of accuracy. The uneven rate of evolution is emphasized, and changes in the atmosphere or the climate are explored as probable causes for bursts of evolutionary change. New material on introgressive hybridization, r and K selection strategies, and electrophoresis has been added to Chapters 19 and 30.

7. Many new illustrations are scattered throughout the book. Four 4-color plates have been added, making a total of 18 full-color pages. Topics in the new color plates include: osmosis, morphology of etiolated seedlings, symptoms of some viral and bacterial diseases, liverworts, and vegetation changes in North America over the past 60 million years. The plates are scattered throughout the book, close to appropriate text pages for easy reference. There are numerous new scanning electron microscope pictures, especially in the chapters on stem, leaf, flower, seed, fungi, and bryophytes. Many new drawings accompany the major revisions summarized above, but in addition an effort was made in every chapter to redo figures from earlier editions that we considered substandard. In Chapter 5, for example, three-dimensional drawings replace the old stick diagrams of sugar and protein molecules.

As in the last edition, special praise goes to the artist, Alice Baldwin Addicott, who has taken such care with the many new drawings and diagrams she prepared. The majority of the nonphotographic illustrations in the book are now original with her, and they bear her signature (ABA). She pursued the truth relentlessly, through microscope, library, and consultation with campus authorities, fortunately not taking our word alone because we were sometimes wrong. If readers learn from this book, much of the material they retain must be due to her good work.

The close and frequently exciting interchanges with colleagues have been a constant source of stimulation and pleasure during the preparation of this edition. Their response to requests for illustrations and advice has been excellent. Other faculty members, graduate students, and staff in our own department have also been most helpful. Drs. Norma Lang, Richard Falk, Terrance Murphy, and David Bayer were frequently consulted and gave freely of their time and knowledge. Russell Chapman was helpful in the preparation of certain new illustrations in the algae chapter. We are particularly indebted to Dan Hess for the hours he spent on the scanning electron microscope. Sonia Cook, Harold Drever, Walter Russell, and Dorothy Brandon were always ready to provide plant material for the pen or camera.

We owe a special debt to our reviewers: Dr. Joe Arditti, University of California, Irvine, who critically reviewed the previous edition; and Dr. Rex L. Lowe of Bowling Green State University and Dr. Jerry D. Davis of the University of Wisconsin, La Cross, who reviewed the manuscript for this fifth edition.

Since we have received the important and willing assistance of many colleagues in the preparation of this edition, the book in a sense expresses the points of view of a wide range of botanists. However, we are responsible for its accuracy of statement, its organization, and its philosophy. Writing this text has been difficult but exciting for us. We hope that some of that excitement will be felt by the reader.

T. Elliot Weier
C. Ralph Stocking
Davis, California   *Michael G. Barbour*

Contents

List of Color Plates

1

Introduction

An educated person is one who has a fair knowledge and an appreciative understanding of the environment in which he lives, and of the part man plays in modifying it. Our environment is most complex. It embraces everything that surrounds us, both living and nonliving. It includes the air we breathe, the food we eat, the water we drink; it includes temperature, rainfall, light, humidity of the atmosphere, wind, soil, and atmospheric pressure; it involves the varied plant and animal life, both wild and domesticated; and it **includes, as well as plants and animals, the human beings with whom we are associated, their racial characteristics,** and **their social** and **economic systems.** All elements of the environment affect our lives, our health, our well-being, our happiness, our freedom, our philosophy, and our religious views.

This environment, in its totality, is called the **biosphere.** Our current interest lies in plant life, and this first chapter briefly discusses some of the roles that plants play in helping to stabilize the biosphere. It also points out ways in which botanists study plants and their interrelationships with other components of the biosphere. Chapter 2 sketches some of the similarities and diversities among plants, as well as briefly relating the botanists' efforts to arrange the similarities into a system that will bring order to the diversity. This is classification and it draws upon knowledge from many fields of botany: anatomy, plant biochemistry, heredity, and morphology. This effort, which is the central theme of Chapter 2, results in a more or less satisfying separation of plants into groups that give evidences of a genetic relationship through the sharing of a common ancestor. All details of anatomy, cytology, and physiology used here will be discussed in more depth in subsequent chapters, and are of importance here only insofar as they serve to clarify the groupings of plants. Do not try to master them at this point.

The Biosphere

Our environment constitutes the biosphere of the planet Earth. It is a thin layer of water, soil, and atmosphere which supports life as we know it. The soil and water teem with microorganisms, but relatively few large organisms inhabit the soil. On the other hand, many species, including some of the largest forms, inhabit the water. But it is the narrow band of atmosphere about the land which supports the organisms of our immediate concern.

The temperature range within our solar system is from nearly absolute zero to 10^7°C, and the wavelengths of radiant energy range from cosmic rays of 10^{-13}–10^{-9} cm to long radio waves of 10^4–10^5 cm. Life requires a temperature between 0 and 50°C, with the optimum around 30°C. The human eye is sensitive to radiant energy from the sun between 500 and 700 nm in wavelength (5 to 7×10^{-5} cm).* Plants, the sole converters of radiant energy, use mainly blue and red light for this purpose. Their growth responds to blue light and other wavelengths in a variety of ways (Chap. 20). Thus, life is tolerant to a very narrow temperature band and responds to an equally narrow band of radiant energy.

The solar radiation reaching the earth's surface is modified by the composition of the atmosphere, which is a mixture of 21% oxygen, 78% nitrogen, 0.03% carbon dioxide, and less than 0.006% of three other gases. At sea level, the column of atmosphere above 1 in.^{2} weighs 15 lb and will support a column of water 32 ft high (see Fig. 5.14). There is abundant life in the biosphere from sea level to an elevation varying from about 2.5 miles at the equator to about 1 mile in the northern Canadian Rockies. In the oceans, abundant life is found from sea level to a depth of 100 ft.

But the biosphere has not always been constituted as described. Life probably originated in an atmosphere devoid of oxygen. Thus, the first living organisms were anaerobic forms and probably never advanced much beyond simple filaments or masses of cells (see Fig. 2.5A). They may have been subjected to large amounts of ultraviolet radiation. With the advent of photosynthesis (Chap. 13), there arose organisms able to utilize the radiant energy of sunlight and to split water into hydrogen and oxygen. The hydrogen was used to reduce carbon dioxide to a carbohydrate (Chap. 13), thus permitting storage of radiant energy as chemical energy for future use. The oxygen escaped to the atmosphere, and ozone (O_3), formed in the atmosphere from oxygen, filtered out much of the ultraviolet radiation. Early primitive plants played a dominant part in establishing oxygen in the atmosphere, and present-day plants are essential for its maintenance there.

It may not forever remain so. Man's ability to utilize the enormous amounts of fossil energy stored in eons past, together with his utilization of nuclear energy, is bringing about changes in the biosphere. We already know that these changes are presently harmful to some forms of life, including man himself. We do not know the permanent long-term

*For a table of metric and English equivalents, see p. 671.

changes that may benefit or harm man, and man's ability to bring about such changes in the biosphere poses a very important problem for contemporary biologists, planners, and legislators.

Certain Biological Universals

If we assume that the original anaerobic organisms were similar in appearance and function to those of today (see Fig. 21.12), and if the first aerobic and photosynthetic forms are also similar to those of today (see Fig. 2.5A), then certain general principles about life emerge. These principles may be called **biological universals.** First, *all life is cellular* (Fig. 2.5). Since viruses cannot live without host cells, we shall not consider them true living organisms. Second, the use and transfer of energy within all cells involves the same compound, known as **adenosine triphosphate,** or **ATP** for short (Chap. 5). Third, some steps in respiration are identical in all aerobic organisms and in some anaerobic forms (Chap. 14). Fourth, some steps in respiration in all organisms and in all cells are associated with membranes (Chap. 4). Fifth, the green pigment **chlorophyll a** is required in the conversion of the radiant energy of sunlight to food in all photosynthetic plants; also, ATP is involved in respiration and photosynthesis, and the light-trapping steps in photosynthesis are always associated with membranes (Chap. 4). And sixth, genetic information which carries inheritance from one generation to the next is stored in the same compound in all organisms. This is **deoxyribose nucleic acid,** or simply **DNA** (Chap. 5). In other words, if the present-day bacteria and single-celled algae (both anaerobic and aerobic forms), which we consider to be primitive organisms, are similar to their ancestors of 400 million years ago, certain important basic processes of life become apparent: Respiration, photosynthesis, and inheritance are basically the same in all organisms and have been so from the beginning. If this assumption is false, we have no starting point, and an understanding of past life will be much more difficult to attain.

Differences between Organisms of Anaerobic and Aerobic Time

We have just assumed that present-day anaerobic forms closely resemble the forms living in anaerobic time. Are there any major differences between the organisms of anaerobic and aerobic time? There seem to be. In present-day aerobic organisms, the processes of respiration, photosynthesis, and inheritance are confined to their own special compartments within the cell: mitochondria, chloroplasts, and nuclei, respectively (see Figs. 2.5*B*, 4.2, 4.3, 6.2). But in the primitive forms, these three activities, as well as others, occupy communal space in a single noncompartmented cell (Fig. 2.5*A*). Since the more primitive forms lack a true nucleus, they are known as the **prokaryotes.** The forms with a true nucleus are the **eukaryotes.** All organisms are converters of energy, but only plants convert the radiant energy of sunlight into food. Other organisms consume this food. This is a less scientific but more descriptive way of saying that there are two kinds of kingdoms, Animal (consumers) and Plant (producers). *Man is the greatest consumer of all.*

Plants and Plant Scientists

In the plant kingdom only the blue-green algae and bacteria are prokaryotes and only the fungi, together with a few isolated species, lack chlorophyll. The metabolic processes of all plants, with the possible exception of the fungi and bacteria, are similar; they are all producers!

In all of life on earth, plants are the only producers. All consumers, particularly man, are dependent upon plants for food, fiber, wood, energy, and oxygen. A knowledge of plants, their habitats, structure, metabolism, and inheritance is thus the basic foundation for man's survival.

What are the problems that concern the plant scientists in their efforts to accumulate significant knowledge of the plant life of the biosphere? Plants must first be named. Early botanists, such as the Greek, Theophrastus (300 BC), were largely concerned with naming plants. In fact, many current scientific names for plants are the same as, or similar to, those given to the same plants by Greek or Roman botanists. Concomitant with naming plants came an increase in the numbers of known plants. Plant explorers accompanied the expansion of empires. As a result, botanic gardens and herbaria, storehouses of botanic information, were established in Sweden, England, and France. In order to retrieve information there must exist some system of classification. Modern methods of classifying plants date from the Swedish botanist, Linnaeus (1707–1778). New species of plants are still being described, but the standing of a plant scientist no longer depends upon the numbers of plants he can recognize.

Some other important problems claiming the attention of plant scientists are as follows: (*a*) What are the principles underlying the great diversity of plants? (*b*) What are the mechanisms that control the precise patterns of growth? (*c*) Since plants are the only producers, what kind of plants, growing under what conditions of temperature, soil, moisture, and nutrients will produce most abundantly? (*d*) How are plants related to their environment, and will the changes in the biosphere brought about by man eventually prove harmful to man? These problems can only be studied with experiments.

It is through experimentation that we hope to derive answers. Experimentation asks questions about plants that can be answered by manipulating plants under specific conditions designed to answer the question. In 1770, the Dutch physician van Ingenhousz asked the question, "Will a glowing sliver burst into flame in an atmosphere in which plants have been confined in the dark?" The answer was, "No." Will it burst into flame in an atmosphere in which plants have been confined in the light? Now the answer is, "Yes." What is the difference between the two atmospheres?

Asking the right question is frequently the most difficult step. Generally, it should be a simple question, hopefully with not more than two possible answers. The procedure must be within the realm of available experimental technique. The experimental system set up for study of the question must be under reasonably good control. The fewer the variables there are in the system, the better. And, of considerable significance, the answer to the right questions must lead to a greater understanding of the whole problem and, in addition, to other questions.

Plant Taxonomists

Taxonomists wish to complete an inventory of the earth's plant resources; they want to categorize, for easy reference, the sum total of existing plant variation. As the inventory becomes complete (and it is far from complete today), they hope it will be possible to organize this diversity into some evolutionary scheme. What sort of patterns exist, and do these patterns mean genetic relationship and evolution from common ancestors? Which plants or plant characteristics are the most primitive, that is, imitating early, now-extinct species; and which characteristics are advanced, that is, indicating a recent time of evolution? Taxonomists use morphology, anatomy, biochemistry, hybridization, and chromosome studies to arrive at tentative conclusions about the **phylogenetic** (evolutionary) relationships between species.

Taxonomy is basic to botany, for all botanists use taxonomic information in their research. The physiologist, for example, must know the particular species he works with. Since species differ in their physiology, his results would be meaningless if the name of his study organism was not known; other workers could neither corroborate nor extend his findings. The anatomist, developmental morphologist, and cytologist all face the same necessity of knowing the species with which they work.

Plant Ecologists

Plant ecologists want to understand how plants are adapted to their environment. What are the environmental forces that regulate plant growth and what kinds of responses can plants make to them? Living organisms and their environment are interdependent in complicated, often unexpected ways. Can man learn to imitate this natural balance or to manipulate it for his own ends? Is it possible, with the vegetation we have, to meet several pressing needs at once: those of agriculture, industry, recreation, conservation? What is the wisest, most efficient way to use our plant resources? Have we changed our environment dangerously by pollution? What level of pollution can be safely tolerated? Are plants adapting to pollution?

The ecologist also wishes to inventory the earth's plant resources. Unlike the taxonomist, he is interested in categorizing groupings of species (communities) rather than individual species. How is it that certain species repeatedly group together to the exclusion of all other species? What is the geologic history of these communities, and what are their paths of migration through time?

Ecologists are often generalists, utilizing information from taxonomy, physiology, anatomy, geology, and meteorology to search for answers to questions like those posed above.

Plant Morphologists

The great diversity of plant forms is readily apparent. There are seaweeds, mosses, ferns, pine trees, corn plants, palms, and dandelions. Obviously, corn bears seeds (Chap. 16), but do ferns, mosses, and seaweeds also bear seeds? If they do not, how do they propagate (Chaps. 22, 26, 27)? Both pines and palms have single straight trunks, but the palm has large divided leaves, found at the top only, and the pine has many lateral branches clad with needles. Why? (Chap. 17.) Botanists interested in plant form are known as morphologists. Their interests lie in the diversity of plant form and the manner in which cells are arranged to give this form. All plants require water and food; are water and food conducted and transported in cells of similar form in plants of the ocean and plants of the desert? How do the reproductive structures of plants of the ocean, marsh, and desert differ (Chaps. 22, 27)? Are the cellular details

of the stems of mosses, ferns, pine trees, and palms alike (Chaps. 7, 27)?

The morphologists are also interested in how form develops in such a precise fashion. When a fertilized egg cell in the flower of a pea plant starts to grow, it follows a pathway similar to that followed by fertilized eggs of all other seed plants. Eventually a pea seed develops identical to all other pea seeds but different from every other kind of seed. When the seed is planted, it germinates and grows just as do the seeds of all other species of plants, but the pea seed always produces pea plants, and only pea plants. Cells divide in precise planes and enlarge in precise locations within the developing seed at regular rates. Differentiation of cells to form tissues also occurs in a regular fashion. The morphologist has learned how to modify some of these steps. He would like to know more of the manner in which they are controlled and directed within the plant itself. A better knowledge of these things will give us a deeper insight into the developmental process of all cells.

Plant Physiologists

While it is true that the basic processes of respiration and photosynthesis are similar in all plants, the end products of the synthetic processes in plants are as variable as there are different species of plants. Why? There are hard wheats and soft wheats, and the flour made from them is different. Some wheat seeds need to be held at a low temperature before they will germinate. Some lettuce seeds will not germinate until they have been briefly exposed to red light. Again, why? The answer to these and similar questions lie in the realm of plant physiology.

Morphologist and physiologist frequently unite in searching for aspects of form that have a chemical basis. A vegetative shoot of *Chrysanthemum* will not flower as long as it receives more than 10 hr of daylight. Soon after the vegetative shoot receives a day-night cycle of only 9 hr of daylight and 15 hr of darkness, flower buds form. What are the morphological changes that take place in the *Chrysanthemum* shoot as it changes from the vegetative to the flowering condition? What chemical substance, i.e., hormone, is induced by the change in light period and how does it bring about the change in the growth pattern?

One of the major goals of the plant physiologist is to understand all the biochemical and physical processes that are the basis of life, and that are responsible for the great diversity of life form and activities. The more the physiologist knows about life reactions, the more he is able to regulate and control these activities and so shape plant growth and development. Is it possible to predict or even regulate when a desirable plant will flower? Can fruit without seeds be produced? Why are some plants bushy-shaped dwarfs and others of the same species tall and vine-like?

Cytologists

Cytologists, or cellular biologists, as they are now more popularly known, are concerned with many problems at the cellular level. It might seem that since the basic phenomena of inheritance, photosynthesis, and respiration are essentially identical in all organisms and these three processes are confined to specific organelles within the cell, that understanding them would be easy. We do know that the genetic code present in the nucleus directs the synthesis of specific proteins. But we do not know how protein is actually synthesized in the cell. We are not sure of the physical arrangement of DNA in a chromosome. We would like to know the precise relationship between the steps in photosynthesis and respiration and the molecules of the membranes associated with these processes. Respiration and photosynthesis supply energy for cellular synthesis.

Protein synthesis probably takes place in all parts of the cell, even in the mitochondria and chloroplasts which supply energy for synthesis elsewhere in the cell.

Since protein synthesis is coded by DNA (Chap. 6), and mitochondria and chloroplasts have their own DNA (p. 39), what is the relationship between the nuclear or chromosomal DNA and mitochondrial and chloroplastic DNA in the synthesis of the important array of enzymes characteristic of these organelles? What other organelles may be involved? How? Questions are being answered slowly. The cell physiologist and the biochemist find themselves seeking solutions to similar questions but seeking answers by different approaches.

Conclusions

Since plants (except fungi and some bacteria) are producers and animals (including man) are consumers, we must know about plants. Universities must continue to educate people knowledgeable in the ways of plants, just as they train specialists in space, medicine, law, and social theory. For without plants there would be no life.

2

Classification— The Relationship of Plants

Plants are (*a*) those organisms supplied with chlorophyll, capable of trapping the radiant energy of sunlight, transforming it into stored potential energy while at the same time releasing oxygen to the atmosphere; and (*b*) an assemblage of organisms lacking chlorophyll but morphologically very similar to certain groups of the lower plants supplied with chlorophyll.

Over 550,000 species of plants clothe the earth and inhabit most of the waters to a depth of approximately 100 ft. What is a **species?** How are species arranged or classified, to bring order to this vast assemblage of plants? A classification serves one practical purpose: the storage of information in a retrievable manner. Most classifications have a common basis; similar things are grouped together. Retrievable information about Americans is kept in computer centers in Washington. Books are kept in libraries. Plants are kept in herbaria and botanic gardens. *A species is a convenient unit of information.*

Retrieval of information is made easy in a library by placing like information together on shelves and giving it a numerical designation. All botanical literature in a library using the Library of Congress classification is found under the letters QK. Medical literature is shelved under R. This is accompanied by an elaborate card index system in which extensive cross-references allow the sought-after information to be found in a number of ways. One type of information about kinds of plants may be retrieved by the use of books called floras. In these floras, plant names and descriptions are arranged so that with enough information about a given plant specimen, unknown plants may be named and placed in the proper group by a skilled person.

A Natural Classification

Any classification then, is the placing of similar things together. The problem in biological classification is the meaning of the word "similar." Are plants living in the same habitat similar? (Fig. 2.1; see Color Plate 1.) They must certainly have similar nutritional requirements. The Swedish botanist Linnaeus (Chap. 19) divided all flowering plants into 24 classes based on the number and arrangement of stamens in each flower (Chap. 15). This worked surprisingly well. All plants currently grouped together as members of the pea family have 10 stamens (see Fig. 2.4). All buttercups and their relatives have many stamens, but so do some roses (Fig. 2.2*A*), mallows (Fig. 2.2*B*), and camellias (Fig. 2.2*C*). Consideration of other charac-

Figure 2.2 There are 35 different families having flowers with many stamens. Three are shown, *A, Rosa; B, Hibiscus; C, Camellia.*

teristics today allows us to separate plants bearing flowers with many stamens into at least 35 different categories or families. What do modern taxonomists mean by similar?

Plants are placed in categories thought to have a close genetic relationship. Thus, in the most homogeneous populations, the plants are not only very similar in many aspects of their morphology but they *interbreed freely* (Figs. 2.3*A, B*; see Color Plate 1). Their genetic constitutions are similar, even identical; they are very closely related. Such a population of closely related interbreeding individuals constitutes a **species.** The most basic requirement of a biological classification is that it show a true genetic relationship. Several different species may have some characteristics in common and an occasional mating may take place between members of two different but similar species (Fig. 2.3*C*). Such similar species are grouped together in **genera** (*s.* genus), which are thus groups of genetically related species. A comparison of some genera will show that they have traits in common (Fig. 2.3*D*). Genera with similar traits are grouped in **families,** which are therefore also genetically related, but less closely than genera. Plants in the same family have common characteristics that indicate they may all have evolved from the same ancestor. For instance, members of the pea family have flower parts in fives, pods like pea or bean pods, and dissected leaves like pea or locust leaves. Many members of the family have flowers resembling the sweet pea, with ten stamens, nine of them grown together (Fig. 2.4).

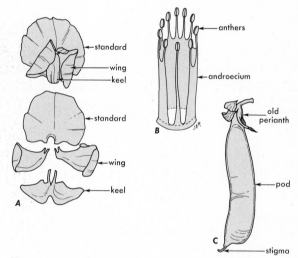

Figure 2.4 Diagram showing traits that are important in the classification of the sweet pea (*Lathyrus odoratus* L.). *A,* flower, exploded view; *B,* 10 stamens, 9 attached together, 1 free; *C,* pod with seeds.

Families are grouped together in **orders,** and orders may be grouped in **classes.** At the top level it is customary to divide all plants into **divisions,** or **phyla.** Differences in photosynthetic pigments (Chap. 22), manner of leaf development (Chap. 27), structure of the conducting, or vascular, tissues (Chap. 27), and methods of reproduction (Chap. 22) are the most important elements in separating plants into divisions.

Figure 2.1 Cranberry, club moss, and a mushroom living in close association. They have similar environmental requirements. Are they similar?

Figure 2.3 *A* and *B*, two clones of irises (*Iris germanica*). These two clones freely interbreed and thus constitute a species. Notice, however, that the flower colors of the two clones are different. *C*, a single population of Dutch iris (*Iris xiphium*). These plants interbreed freely, but only very rarely do they cross with members of the populations shown in *A* and *B*. *C* is a second species in the genus *Iris*. *D*, a population of *Freesia*. These plants have some characteristics similar to those of *Iris*. Both *Iris* and *Freesia* are genera in the family Iridaceae.

A *B*

C *D*

A

B *C*

D

Color Plate 2   **Thallophytes**

Figure 2.7 Pigments, hence color, are of importance in distinguishing divisions in the thallophytes. *A, Amanita muscaria,* or the fly mushroom, is a Basidiomycete; it lacks chlorophyll but does have a bright red pigment in the cap. *B, Ulva,* or sea lettuce, of the division Chlorophyta has both chlorophylls *a* and *b*. Because its thallus is only two cells in thickness, the green color shows best through overlapping folds of the blade. *C,* the Phaeophyta, represented by *Postelsia,* or the sea palm, have chlorophylls *a* and *c* and a brown pigment. *D,* the Rhodophyta have only chlorophyll *a* and a red pigment which shows well in the blades of *Iridophycus. E,* the blade of *Porphyra* are two cells thick; because of the thinness of the blades and a frequently poor development of the red pigment, this species of Rhodophyta is not as red as other members of the division. *F,* a narrow strip of intertidal zone shows the green alga *Ulva,* a brown alga resembling *Fucus,* and the red alga *Gigartina.* (All with the assistance of David Brown.)

E *F*

A classification depicting genetic relationships is said to be a **natural** or **phylogenetic classification.** Grouping plants by habitats (such as desert or marsh plants), or by the number of the stamens or color of their flowers has limited usefulness and is said to be an **artificial classification.** A true natural classification is impossible to arrive at and there have been several major attempts, some of which are discussed in later chapters.

Thus, the word similar, in a natural classification, means organisms having genetic constitutions indicating relationship. This system is exemplified in the complete classification of the sweet pea given below. Note the characteristic ending for each category.

Sweet Pea

	Name	Ending
Species designation	*odoratus*	—
Genus	*Lathyrus*	—
Family	Leguminosae	aceae[a]
Order	Rosales	ales
Subclass	Dicotyldeonae	ae
Class	Angiospermae	ae
Division	Anthophyta	phyta

[a] Taxonomists have agreed that all names of families should end in "aceae" (from the Latin suffix *aceae, aceous,* meaning "of the nature of"). But in this case, they have retained an old name which is very popular. (Fabaceae is an alternative name for the pea family which conforms with the rule.) However, see other plant families such as Rosaceae, "of the nature of a rose;" Liliaceae, "of the nature of a lily;" and the like.

Some Convenient Groups of Plants

Broad groups of plants may be made that include several divisions. While these categories are convenient, they do not have taxonomic importance. For instance, all plants can be divided in two groups: those with true nuclei (the **eukaryotes,** Fig. 2.5B) and those without true nuclei (the **prokaryotes,** Fig. 2.5A). A large group of primitive plants, mainly aquatic and filamentous, may be set apart from other plants by a very distinctive characteristic: the sexual cells (eggs, sperms, or gametes) are not protected by a jacket of vegetative or sterile cells (Fig. 2.5C). These plants may be referred to collectively as the **thallophytes.** The thallophytes constitute 8 algal, 2 fungal, and a single bacterial division. The only unifying characteristic of all the remaining plants is that their reproductive cells are protected by a jacket of sterile cells (Fig. 2.5D). A small group of these plants

lack vascular conducting tissue. They are the **liverworts** and **mosses** (bryophytes) (see Fig. 2.8). All other plants have developed a specialized conducting tissue, and are known as **vascular plants.** The more advanced vascular plants bear seeds and are frequently referred to as the **seed plants** (Fig. 2.5F). These are rather precisely separated into **gymnosperms** (see Fig. 2.13), whose seeds are borne naked on a cone scale, and **angiosperms** (see Fig. 2.15), whose seeds are protected within an ovary.

Plant Divisions

We have seen that plants constituting a species are very closely related. And we have said that all members of a family arose, in the primitive past, from a common ancestor. Do all members of a division have a common ancestor? Are divisions related? Do all plants have a common ancestor? A partial answer comes from the fossil record.

Plants of past eons have left their remains, or marks, on geologic strata. These remains or marks are **fossils** (Fig. 2.6). Fossils are found in sedimentary rocks, in volcanic ash, and in quantity in coal. By comparing the morphology of present-day plants with the morphology of fossils from past geologic periods, it is possible to establish broad lines of relationship between the divisions. Unfortunately, fossils are few in number, do not reveal the entire plant, and are preserved only in certain habitats. Therefore, the fossil record must be used with caution. This is why agreement on the evolutionary relationship (phylogeny) of plant divisions is difficult to achieve.

Divisions of Thallophytes

Thallophytes, a very large group of primitive plants, do not have a jacket of sterile protecting cells around their reproductive cells (Fig. 2.5C). Those without chlorophyll are the **bacteria** and **fungi** (Fig. 2.7A; see Color Plate 2); those with chlorophyll are **algae** (Figs. 2.5E, 2.7B, C, D).

Algal Divisions. The algae may be separated further into groups on the basis of photosynthetic pigments. Some of these groups have division rank. A large group of algae contain chlorophylls a and b. These algae are easily divided into two divisions. One of these, the **Euglenophyta,** consists of single-celled forms (**unicells**) that lack a cell wall. The other division is a large one, the green algae (**Chlorophyta,** Fig. 2.7B). Because higher plants also contain chlorophylls a and b, it is thought that primitive Chlorophyta are ancestors of the higher plants (Chap. 22). Chlorophyll b is lacking in all other algae.

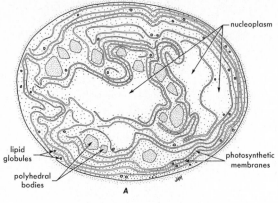

nucleoplasm

lipid globules

polyhedral bodies

photosynthetic membranes

A

Figure 2.5 Diagrams to show differences between groups of plants; *A*, a prokaryotic cell (blue-green alga); *B*, a eukaryotic cell (green alga, *Chlorella*); *C*, a group of reproductive cells lacking a protective jacket (the brown alga, *Ectocarpus*); *D*, a group of reproductive cells provided with a protective jacket (the liverwort, *Riccia*); *E*, a thallophyte (the single-celled green alga, *Acetabularia*); *F*, a seed-bearing vascular plant (bean, *Phaseolus vulgaris*).

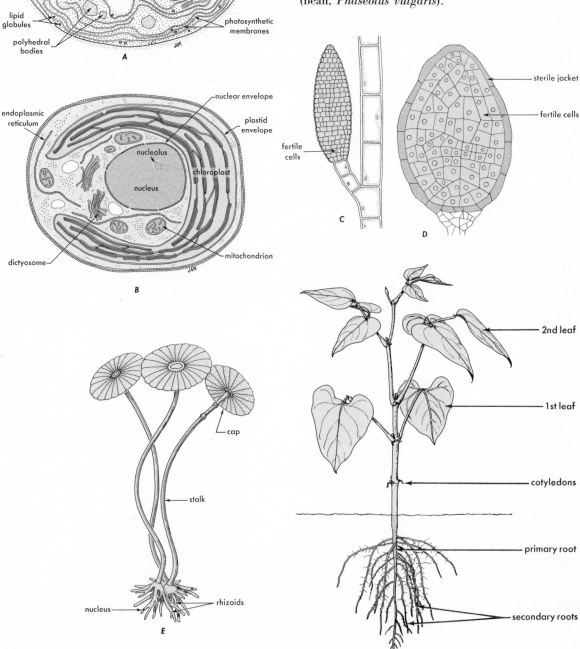

nuclear envelope

plastid envelope

endoplasmic reticulum

nucleolus

chloroplast

nucleus

dictyosome

mitochondrion

B

sterile jacket

fertile cells

fertile cells

C

D

cap

stalk

nucleus

rhizoids

E

2nd leaf

1st leaf

cotyledons

primary root

secondary roots

F

Figure 2.6 An imprint of a leaf of *Pecopteris*, a common plant growing during the Pennsylvanian geologic period about 300 million years ago. (Courtesy of Field Museum of Natural History.)

A second large algal group is known to contain only chlorophyll a together with accessory water-soluble pigments, one blue and the other red (Chap. 22). This assemblage of forms is divided into two divisions. One division, the blue-green algae (**Cyanophyta,** Fig. 2.5A), are prokaryotes and hence lack true nuclei. The other division, the red algae (**Rhodophyta,** Fig. 2.7D), is composed of eukaryotes that inhabit the coastal waters of the oceans in the tropic and temperate zones. While both divisions contain only chlorophyll a and similar accessory pigments, their structure is so different that any relationship between the two divisions must be very distant (Chap. 22).

A third group definitely contains chlorophylls a and c plus certain other distinguishing plastid pigments which separate them into three divisions. One division comprises the seaweeds—kelps—of ocean coasts. These are brown algae, or **Phaeophyta** (Fig. 2.7C). A second division in this group is the **Chrysophyta,** consisting of a small number of golden brown algae. The diatoms, or Bacillariophyta, comprise a third large division in this group. The diatoms are quite small, but anyone who has looked at a fresh slide of sea water under the microscope has probably seen them, for they are extremely common unicells and filaments. Their cell walls are impregnated with silica, and are so constructed that the two halves fit together like the top and bottom of a petri dish. The third division having chlorophylls a and c is comprised of small cells, many of which lack cell walls. It is known as the **Pyrrophyta.** These algae are not common, but upon occasion a sudden increase in their population to vast numbers causes "red tides," poisoning many miles of ocean coastlines.

The algae in a fourth group were thought to contain only chlorophyll a, and it was separated from the Cyanophyta and Rhodophyta because the water-soluble pigments were not present. This group consists of only a single relatively small division, the **Xanthophyta.** Recent work indicates that chlorophyll c is present, in at least many species, and that the Xanthophyta may best be characterized by certain xanthophyll plastid pigments.

Fungal Divisions. One thallophyte division lacks chlorophyll; thus, the plants cannot make their own food and are therefore called **heterotrophs.** A primitive subdivision of fungi, the **Myxomycotina,** are the slime molds which, in their vegetative stage, are nothing but naked masses of protoplasm. An advanced group, the **Eumycotina,** are true fungi (Fig. 2.7A). Some of them very closely resemble the algae, and one group is known as the algal fungi. The thallophyte divisions are summarized in Table 2.1.

Classification of Land Plants

There is general agreement that the land plants take their origins from the green algae, and a particular genus, *Fritschiella* (see Fig. 22.5C), has been selected as a possible starting point. One group of land plants lacking a vascular system forms a natural division of liverworts and mosses, the Bryophyta. All other land plants have conducting, or vascular, tissue. General agreement is lacking as to their natural relationship.

TABLE 2.1   Summary of Thallophyte Divisions

Number	Division	Common name	Chlorophylls	Description
1	Cyanophyta	Blue-green algae	*a*	Prokaryotic cells
2	Pyrrophyta		*a, c*	Mostly unicells
3	Chrysophyta	Golden algae	*a, c*	Some unicells
4	Bacillariophyta	Diatoms		Many unicells, silica in walls
5	Phaeophyta	Brown algae	*a, c*	Marine forms, some very large
6	Rhodophyta	Red algae	*a*	Mostly marine forms, many with a feathery plant body
7	Xanthophyta	Golden-green algae	*a, c*	Many unicells, some simple filaments
8	Euglenophyta	Euglena	*a, b*	Motile unicells without cell walls
9	Chlorophyta	Green algae	*a, b*	Mostly rather small forms of great diversity
10	Schizomycophyta	Bacteria	—	Prokaryotic cells, a few with bacterial chlorophyll
11	Mycota	Slime molds	—	Amorphous masses of proto-plasm in vegetative stage
		True fungi	—	Mostly filamentous forms of considerable diversity

All vascular plants have a system for conducting water and nutrients throughout the plant. They all have roots, stems, and leaves (Fig. 2.5*F*), except for some primitive living genera which lack roots (see Fig. 2.9). Also, they have all developed a cuticle on their aerial portions which serves as protection against desiccation. The classification problem facing us is: Did all these characteristics, which distinguish vascular plants from all other plants, arise just once in the long evolutionary history of plants? If so, all plants living today have evolved from the same vascular ancestral form. This theory of plant evolution assumes a **monophyletic origin** of vascular plants; all such plants would then belong to one division: the Tracheophyta (Table 2.2).

It is equally conceivable that there were numerous steps in the evolution of the characteristics of the vascular plants, and that the cuticle, vascular tissue, roots, stems, and leaves may have arisen in several different periods and localities. The plants living today, then, would not have a common vascular ancestral form. There would be several common ancestral forms, at least one for each suggested division. This is known as a **polyphyletic origin.**

It therefore becomes difficult to present a system for the classification of the vascular plants in an introductory botany course. Two such systems are compared in Table 2.2. The first column summarizes a monophyletic system. Column 2 shows a polyphyletic system used in a textbook of plant morphology, which is more similar to the system we shall follow in this book (column 3). There is fairly general agreement among botanists on the main subgroups

of vascular plants; the differences in these columns lie in the number of divisions. According to column 1 there is a single division for all vascular plants; thus implying a single ancestor. According to column 2, there are ten vascular plant divisions, implying at least ten ancestors.

The fourth column gives the common names for these groups of plants. While these names refer as precisely to a group of plants as do the scientific names, the common names do not suggest relationships between the plant groups they designate.

Divisions of Land Plants

The mosses and liverworts (Bryophyta, Fig. 2.8, Chap. 26) are easily separated from all other land plants. Instead of true roots, they have rhizoids that act as absorbing organs. They also have small leafy stalks which produce sperm and egg cells. Frequently, in mosses, a brown upright stalk bearing a capsule is seen growing from the low leafy stalk; this is the offspring of sperm and egg fusion. The mosses, which lack vascular tissue, are thought to have arisen from the Chlorophyta but not to have given rise to any other land plants. In other words, the mosses we see today may be much as they were when plants first invaded the land.

The most primitive living vascular plants are known as the **psilophytes,** and there are only two living genera, *Psilotum* and *Tmesipteris*. *Psilotum* is easily cultivated and is fairly common as a potted plant in botany greenhouses (Fig. 2.9). The name *Psilotum* means naked, or smooth, and refers to the

Figure 2.8 The hair cap moss, *Polytrichum*, is a representative of the Bryophyta.

Figure 2.9 A potted plant of *Psilotum nudum*, a living species closely resembling a fossil form found in sedimentary rocks laid down 400 million years ago.

simple dichotomously branching stem, with small scale-like leaves and with rhizoids rather than roots (Chap. 27). *Psilotum* appears very similar in structure to a fossil (*Rhynia*) from Scotland found in sedimentary rocks that were deposited 400 million years ago. Two other very ancient groups of plants, the **club mosses** or **lycopods** (Lycophyta), and the **horsetails**

or *Equisetum* (Sphenophyta), once formed a dominant earth flora with many tree forms. We shall consider three of the five genera of club mosses, and a single genus of horsetails, *Equisetum*. Both are ancient plants, being easily related to fossils found in rocks nearly 400 million years old.

Equisetum has hollow, jointed stems which have

TABLE 2.2 Comparison of Systems of Classifying the Major Divisions of the Land Plants

Monophyletic (H. P. Banks, 1968)	Polyphyletic (Scagel et al., 1967)	Used in this text	Common names
Bryophyta	Bryophyta	Bryophyta	Bryophytes; mosses and liverworts
Tracheophyta			Vascular plants
	Psilophyta	Psilophyta	Psilophytes
Rhyniophytina	*Rhynia*	*Rhynia*	*Rhynia*
Zosterophyllophytina			
Psilophytina	*Psilotum*	*Psilotum*	*Psilotum*
Lycophytina	Lycophyta	Lycophyta	Lycopods, club mosses
Sphenophytina	Arthrophyta	Sphenophyta[a]	*Equisetum*, horsetails
Trimerophytina			
Pterophytina			
Cladoxylopsida			
Filicopsida	Pterophyta	Pterophyta	Ferns
Coenopteropsida			
Progymnospermopsida	Pteridospermophyta	Pteridospermophyta	Seed ferns
Cycadopsida	Cycadophyta	Cycadophyta	Cycads
	Ginkgophyta	Ginkgophyta	*Ginkgo*
Coniferopsida	Coniferophyta	Coniferophyta	Conifers
Gnetopsida	Gnetophyta	Gnetophyta	Gnetum
Angiospermopsida	Anthophyta	Anthophyta	Angiosperms

[a]Sphenophyta is an older name for this division; it is retained here because of confusion arising from similarity of the new proposed name, Arthrophyta, with Anthophyta and the animal phylum Arthropoda.

an encircling sheath of small leaves at each joint. Spores, or reproductive cells, are borne in specialized cones at the summit of some of the upright stalks (Fig. 2.10).

The club mosses (Lycophyta) more closely resemble the higher plants (Fig. 2.11). One genus is known as ground pine and has been used extensively for Christmas wreaths. The club mosses are separated from the other groups of vascular plants by their distinctive reproductive structures; their vascular system; the single veins in their small leaves; and the union of stem and leaf without any break in the vascular tissue of the stem (Chap. 27).

The living **ferns** (Pterophyta) are generally small plants and are characterized by their highly dissected, characteristic leaves, which unroll at the tip as they open. Seeds are not formed by any fern (Fig. 2.12).

The remaining six divisions of plants (Table 2.2) are seed plants. The first, the seed ferns, are now extinct. Four of these divisions have seeds borne upon scales in a cone (Fig. 2.13) or as naked seeds on short stems; these are known collectively as the **gymnosperms.** The remaining division of seed plants has seeds enclosed in a seed case or ovary (Fig. 2.14); these are the **angiosperms** (Anthophyta, Fig. 2.15).

Two of the divisions of gymnosperms, *Ginkgo* (Ginkgophyta, see Fig. 28.8) and the cycads (Cycadophyta, see Figs. 28.5, 28.6) are the most primitive of the living seed plants. *Ginkgo biloba* is the only species in its division. It grows to be a large tree, has been cultivated for many centuries in Chinese temple gardens, and is currently finding much favor in western countries. It is extinct in the wild. *Ginkgo* has fan-shaped leaves (see Fig. 10.5*B*) with many veins radiating from the petiole and a break or gap in the vascular cylinder where the vascular tissue from the leaf joins the stem. Naked seeds are borne on the end of short stems (see Fig. 28.8).

The cycads (Cycadophyta) comprise eleven genera, two of which are extinct. Living cycads range in size from small forms to large trees resembling palms (See Fig. 28.6).

The most conspicuous division of gymnosperms are the **conifers** (Coniferophyta, Fig. 2.16). Together with the angiosperms (Anthophyta), they dominate the present flora. The needle or scale leaves of the conifers have but one or two veins, and in some

Figure 2.10 Top, left: Horsetails, *Equisetum telmateia* L.; this is the only genus in the division Sphenophyta.

Figure 2.11 Bottom, left: There are five genera in the Lycophyta. This is *Lycopodium obscurum* L., or running pine.

Figure 2.12 The fern, *Polypodium vulgare*, representing the Pterophyta.

Figure 2.13 An open cone of the pinon pine, *Pinus monophylla*, showing two seeds in place upon a scale. A gymnosperm.

Figure 2.14 A fully matured ovary of a bell pepper, *Capsicum frutescens var grossum*, showing seeds within the matured ovary. An angiosperm.

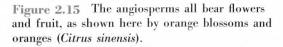

Figure 2.15 The angiosperms all bear flowers and fruit, as shown here by orange blossoms and oranges (*Citrus sinensis*).

Figure 2.16 *Pinus jeffreyi* representing the Coniferophyta. Ponderosa pine (*P. ponderosa*), the most abundant western states pine, looks very similar.

forms there does not appear to be a break in the vascular cylinder where the vascular tissue from the leaf joins with the vascular cylinder of the stem. In contrast, leaves of the angiosperms (Anthophyta) have many veins and there is always a gap at the junction of leaf vascular tissue with the stem vascular tissue. We may divide the angiosperms into two subclasses. The seeds of one have two seed-leaves, or **cotyledons,** as seen in the peanut or bean. These are the Dicotyledonae. The seeds of the other group (onions, lilies) have a single seed-leaf. These are the Monocotyledonae.

The association between the divisions is shown by the key on the next page.

The Binomial

The species designation of sweet pea is *odoratus* and it belongs to the genus *Lathyrus* in the Leguminosae family. To call it just *odoratus* would not be very

meaningful, for it does not show relationship with anything else. To call it simply *Lathyrus* would not separate it from other species in the genus *Lathryus*. The scientific name of every species is a binomial consisting of two words: the genus plus the species (*Lathyrus odoratus*).

Frequently, the genus name refers to the ancient name of the same or a similar plant. *Lathyrus* is the Greek word for pea. An effort is usually made to make the species name descriptive. *Odoratus* is the Latin word designating that something is sweet-smelling. So, altogether the Greek genus and Latin species translate to sweet-smelling pea. There are also black peas, *L. niger;* cultivated peas, *L. sativus;* beach peas, *L. maritimus;* splendid peas, *L. splendens;* and so on. Since an effort is made to keep the scientific names of plants and animals in the correct Latin or Greek form, they are really foreign terms and as such are always written in *italics*. As the species designations are generally descriptive, they may be used in conjunction with other genera. We have, for instance, black pepper, *Piper nigrum;* black mustard, *Brassica nigra;* and black walnut, *Juglans nigra*. There is sweet alyssum, *Alyssum odoratum,* and cultivated oats, *Avena sativa*. A genus name, however, is used for only one group of species and cannot be used for any other group, plant or animal.

The Swedish botanist, Linnaeus (Fig. 2.17) was largely responsible for establishing the binomial system, not only for plants but for all living organisms. In fact, the full name of the sweet pea is *Lathyrus ordoratus* L. The cultivated pea should likewise be *Lathyrus sativus* L. The L. after these two names indicates that they were first described by Linnaeus. The large numbers of plant names followed by the letter L. are a living tribute to the greatness of Linnaeus' taxonomic contributions. And today when the sweet pea binomial appears in the literature, every plant scientist, no matter what language he speaks, knows that reference is made to a specific pea with certain characteristics.*

It is not uncommon, particularly in cultivated

*Realizing that many species of plants have several common names, thus giving rise to confusion, the American Joint Committee on Horticultural Nomenclature has prepared a book entitled *Standardized Plant Names*, which in the second edition is described as "a revised and enlarged listing of approved scientific and common names of plants and plant products in American commerce or use." In the preface to the second edition the statement is made that "the purpose of *Standardized Plant Names* is to bring intelligent order out of the chaos in names of plants and plant products existing the world over. Such standardization, supported by adequate authority, will not only promote satisfactory understanding between those who sell and those who buy, but will also improve the multifarious relations, scientific, educational and social, into which the advancing plant consciousness of America has grown."

	Divisions	Common names
A. Plants without roots, stems or leaves, and no protective jacket of vegetative cells around the reproductive cells		thallophytes
B. Plants lacking chlorophyll		
C. Prokaryotes	Schizomycophyta	bacteria
C. Eukaryotes		fungi
D. Vegetative stage of naked cytoplasm	Myxomycota	slime molds
D. Vegetative stage with cell wall, mostly filamentous	Eumycota	fungi
B. Plants with chlorophyll		algae
C. Chlorophyll a alone		
D. Prokaryotes	Cyanophyta	blue-green algae
D. Eukaryotes		
E. With water-soluble blue and red pigments	Rhodophyta	red algae
E. With plastid pigment, xanthophyll	Xanthophyta	
C. Chlorophylls a and c		
D. Large seaweeds with a cell wall of cellulose	Phaeophyta	brown algae
D. Small, mostly unicells, cell wall of cellulose	Chrysophyta	golden algae
D. Small, mostly unicells, cell wall of silica	Bacillariophyta	diatoms
D. Small unicells generally lacking a cell wall	Pyrrophyta	
C. Chlorophylls a and b		
D. Unicells without a cell wall	Euglenophyta	*Euglena*
D. Greatly diversified forms with cell walls	Chlorophyta	green algae
A. Protective jacket of vegetative cells present		
B. Plants lacking vascular tissue	Bryophyta	liverworts, mosses
B. Plants with vascular tissue		vascular plants
C. Plants without seeds		
D. Roots absent	Psilophyta	*Psilotum*
D. Roots present		
E. Small leaves, no gap at union with stem, leaves do not unroll as they open		
F. Stems not jointed	Lycophyta	club mosses
F. Stems jointed	Sphenophyta	horsetails
E. Leaves well-developed, gap at union with stem, leaves unroll as they open	Pterophyta	ferns
C. Plants with seeds		seed plants
D. Seeds not covered		gymnosperms
E. Seeds on short stems	Ginkgophyta	*Ginkgo*
E. Seeds in cones		
F. Trees palm-like	Cycadophyta	cycads
F. Trees conical	Coniferophyta	conifers
D. Seeds covered	Anthophyta	angiosperms
E. Two seed leaves	Dicotyledonae	dicotyledons
E. One seed leaf	Monocotyledonae	monocotyledons

plants, to have races develop within a species. Indeed, in the many years devoted to breeding sweet peas, such races have arisen. One has a condensed, bushy type of growth. Bailey gives this the name *Lathyrus odoratus nanellus* Bailey or dwarf sweet pea. This is a trinomial.

Several thousand new species of plants are discovered every year. The plant specimens may have been collected in parts of the world previously unexplored by botanists, although new species are being discovered every year in areas traversed many times by botanists. These specimens ultimately fall into the hands of a systematic botanist who is interested in and has given special attention to the group (family or genus) to which the plant in question belongs. This botanist sets out to identify the species. In so doing, he compares it with allied forms, of which he has herbarium specimens or published descriptions. He may find that the plant has characters similar to those of a species heretofore described; or he may find that the plant has characters so different from those of any known species that he concludes that it is a new species (**species novum, sp. nov.**). In other words, it is a species that has never

The Binomial   17

Figure 2.17 Linnaeus (1707–1778), the Swedish botanist who initiated the binomial system of nomenclature. (Courtesy of the Library of The New York Botanical Garden.)

been described. Accordingly, he describes this new species and gives it a name. The description is published in one of the many recognized botanical journals, and the specimen or specimens he used in making the description are properly labeled and placed in one or more of the herbaria of the world. This specimen is known as the **type specimen.**

Herbaria and Floras

There are about 550,000 species of plants, divided as we have just indicated. No attempt has been made to describe all of these plants in one treatise. It requires, for instance, ten large volumes to describe the vascular plants in Engler and Prantl's current *Genera Plantarum*. For general convenience, books describing regional floras are in common use. We have, for instance, for the United States:

Leroy Abrams (1940–1960). *Illustrated Flora of the Pacific States.* 4 vols. Stanford University Press. Stanford, Ca.

L. H. Bailey (1949). *Manual of Cultivated Plants.* Macmillan Co., New York, 1116 pp.

H. A. Gleason (1952). *Illustrated Flora of the Northeastern United States and Adjacent Canada.* 3 vols. Lancaster Press, Lancaster, Pa.

A. S. Hitchcock (1950). *Manual of the Grasses of the United States.* 2nd ed. U.S. Government Printing Office, Washington, D.C., 1051 pp.

J. K. Small (1933). *Manual of the Southeastern Flora.* University of North Carolina Press, Chapel Hill, N.C., 1554 pp.

We have noted that through the use of such manuals one could, with sufficient knowledge and skill, identify any unknown plant. This is sometimes difficult, even for an expert. For the agriculturist, cattle rancher, conservationist, tomato grower, or orchidist struggling with forage plants, poisonous plants, or weeds, the task of identifying plants is likely to be difficult and time-consuming.

Specimen plants, stored in herbaria, make identification of unknown plants easier and more definite. Such specimens have been identified by specialists; sometimes they must be sent long distances to these storehouses of information about plant classification. Here, unknown plants may be accurately identified. Herbaria are generally associated with universities, botanic gardens, and certain government agencies. Several of the larger herbaria are listed below.

Herbarium	Number of specimens
Royal Botanic Gardens, Kew, England	6,500,000
Komarov Botanical Institute, Leningrad, USSR	4,500,000
Conservatoire et Jardin Botaniques de Geneve, Switzerland	4,000,000
U.S. National Herbarium, Washington, D.C.	3,000,000
New York Botanical Garden, New York	3,000,000
Missouri Botanical Garden, St. Louis	2,200,000
National Herbarium, Melbourne, Australia	1,500,000

Botanic Gardens

Botanic gardens are of particular interest because they store the plants themselves (Fig. 2.18). Furthermore, these gardens are not confined to the dissemination of knowledge about plants, but are also concerned with dissemination of the plants themselves. Some of the great botanic gardens of today, like the Kew Gardens in England, have a long history of distributing valuable crop and ornamental species to

Figure 2.18 Botanic gardens are storehouses and disseminators of information. *A*, the extensive collection of monocotyledons at the Brooklyn Botanic Garden is displayed in a large and attractive border planting. *B*, The Childrens' Garden at the Brooklyn Botanic Garden gives practical information about plants to the children of the neighborhood. (*A*, courtesy of Brooklyn Botanic Garden; *B*, courtesy of Mr. Bertram Wiener.)

areas far from their native habitats. Sometimes the species actually grow better in their new habitats because insect or fungal parasites are absent. In other cases, the garden grows many individual plants but selects only the most robust ones for transplanting.

Collecting and distributing valuable plants has sometimes involved questionable activities such as smuggling. For example, by the mid-nineteenth century, rubber had become a valuable, desired plant product. The best rubber trees (*Hevea brasiliensis*) were native to Brazil, and the Brazilian government definitely wished to maintain its monopoly. Export of rubber tree cuttings or seeds was strictly forbidden.

The India Office of England, however, was determined to initiate rubber plantations in the English colony of Ceylon. In 1876 it secretly commissioned an English rubber worker in Brazil named H. A. Wickham to smuggle out rubber tree seeds, agreeing to pay him $50 per thousand. Wickham proved to be very efficient and soon collected some 70,000 seeds, packing them in covered baskets which he labeled simply, "Botanical specimens for Her Majesty's Gardens at Kew." He eventually found a willing ship captain with available space and slipped the seeds past the customs office at Belem without incident.

Once in England, he turned the seed over to botanists at Kew Gardens, who carefully germinated

Figure 2.19 Firestone rubber plantation near Robertsville, Liberia. (Courtesy of Dr. Georg Gerster–Rapho Guillumette.)

National Museum of Natural History, Paris, France (1635)

Royal Botanic Garden, Edinburgh, Scotland (1670)

University Botanic Gardens, Cambridge, England (1762)

Botanic Garden of Harvard University, Cambridge, Mass. (1907)

Botanical Garden of Rio de Janeiro, Brazil (1808)

Botanic Gardens of New South Wales, Sydney, Australia (1816)

Government Botanic Gardens, Buitenzorg, Java (1817)

U.S. Botanic Garden, Washington, D.C. (1820)

Royal Botanic Gardens, Kew, England (1841)

Missouri Botanical Garden, St. Louis, Mo. (1859)

The New York Botanical Garden, New York City (1895)

Brooklyn Botanic Garden, Brooklyn, N.Y. (1910) (Fig. 2.18)

These gardens have plantings of species from all parts of the world, arranged by plant families or according to habitats. They contain outdoor plantings and often greenhouses controlled to grow tropical plants. Their reference libraries may contain thousands of volumes of books, pamphlets, and photographs, and the herbaria may be extensive. For example, the New York Botanical Garden, with 40 acres, has a reference library of 43,500 volumes and thousands of pamphlets; an herbarium of approximately 3,000,000 specimens; and a systematic plantation, arboretum, fruticetum, rose gardens, and rock gardens. The Missouri Botanical Garden, with 75 acres, has a library of 50,000 volumes, an herbarium of 2,200,000 specimens, 6500 species growing under glass, and 7500 species growing out-of-doors. The famed Royal Botanic Gardens, at Kew, with 288 acres, has a reference library of 44,000 volumes, an herbarium of 6,500,000 specimens, an arboretum and fruticetum together of 7000 species and varieties, 13,000 species growing under glass, and 8000 herbaceous species growing out-of-doors.

Botanical Journals

We have previously mentioned the storing of knowledge about plants in a retrievable form. But knowledge must be accumulated before it can be stored. Today, the greatest increase in our knowledge of plants and animals comes from the biology departments of many universities. Since most of this work receives support from public funds, it must be available to all who desire to know about it. As a consequence, this new information is published in some 2000 scientific journals. A number of examples of such journals is given in Table 2.3.

and nurtured 2400 seedlings from the 70,000 seeds. Most of these plants were shipped within a year to plantations which had been made ready in Ceylon. Wickham had his picture taken in a Ceylon plantation in 1912, when it was producing a princely $18,000 worth of rubber per acre per year! Not surprisingly, he was knighted in 1920. Today, 95% of the world's rubber comes from the Far East. Other plantations have been started in Africa (Fig. 2.19).

Botanic gardens serve as a definite link in the distribution of knowledge about plants to the general public. In addition to actual displays of growing plants, informative lectures and booklets dealing with plants are available to the interested and advanced amateur.

Among the oldest and best-known botanic gardens in the world (with the dates of establishment in parentheses) are the following:

Royal Botanic Garden of Padua, Padua, Italy (1545)

TABLE 2.3 Representative Journals Publishing the Results of Botanical Research

Library of Congress call number	Journal	Published by	Country of origin	Date established
	I. General			
QK 1 A55	*American Journal of Botany*	Botanical Society of America	U.S.	1914
QK 1 A65	*Annals of Botany*	Oxford University Press	England	1887
QK 1 C35	*Canadian Journal of Botany*	National Research Council, Canada	Canada	1922
QK 1 D4	*Deutsche botanische Gesellschaft, Berichte*	Deutsche botanische Gesellschaft	West Germany	1882
QK 1 F5	*Flora*	Gustav Fischer Verlag	East Germany	1817
QK 1 J3	*Japanese Journal of Botany*	National Research Council of Japan	Japan	1923
QK 1 L45	*Linnean Society of London, Journal of Botany*	Linnean Society of London	England	1856
QK 1 T6	*Bulletin Torrey Botanical Club*	Torrey Botanical Club	U.S.	1870
QK 1 E3	*Economic Botany*	Society for Economic Botany	U.S.	1947
QK 1 R8	*Doklady Botanical Science*	Academy of Sciences of the USSR	Russia	1921
	II. Cells and Cellular Biology			
QK 1 R38	*Revue de Cytologie et de Biologie Végétales*	Centre National de la Recherche Scientific	France	1937
QH 301 C5	*Chromosoma*	Springer-Verlag	Germany	1939
QH 581 A1J6	*Journal of Cell Biology*	Rockefeller University Press	U.S.	1955
QH 591 P6	*Protoplasma*	Springer-Verlag	Austria	1927
QH 301 C9	*Cytologia*	Wada-Kunkokae Foundation	Japan	1929
	III. Taxonomy			
QK 1 R5	*Rhodora*	New England Botanical Club	U.S.	1899
QK 1 B675	*Brittonia*	American Society of Plant Taxonomists	U.S.	1931
QK 1 P465	*Phyton*	Fundacion Romulo Raggio	Argentina	1948
	IV. Ecology			
QH 301 J6	*Journal of Ecology*	British Ecological Society and Blackwell Scientific Publications	England	1913
QH 301 E34	*Ecology*	Ecological Society of America	U.S.	1920
QK 901 A1V4	*Vegetatio*	Association of International de Phytosociologie	Netherlands	1948

(continued)

TABLE 2.3 (*Continued*)

Library of Congress call number	Journal	Published by	Country of origin	Date established
V. Inheritance and Evolution				
QH 301 G4	*Genetics*	Genetics Society of America	U.S.	1916
QH 301 H4	*Hereditas*	Nordick Genetikerförening	Sweden	1920
QH 301 E9	*Evolution*	Society for the Study of Evolution	U.S.	1946
VI. Plant Physiology				
QK 1 P48	*Plant and Cell Physiology*	Japanese Society of Plant Physiologists	Japan	1959
QK 1 P5	*Plant Physiology*	American Society of Plant Physiologists	U.S.	1926
QK 1 P452	*Physiologie Végétale*	Gauthier-Villars	France	1963
VII. Morphology, Including Algae and Fungi				
QK 600 M9	*Mycologia*	Mycological Society of America	U.S.	1909
QK 564 J6	*Journal of Phycology*	Phycological Society of America	U.S.	1965
QK 564 P46	*Phycologia*	International Phycological Society	U.S.	1961
QK 534 B7	*Bryologist*	American Bryological Society	U.S.	1898

3

The Plant Body of Seed Plants

In this chapter we shall give attention to the general structures and functions of a familiar seed plant, the common garden bean. Any one of many thousands of seed plants would serve our purposes as well. It is recognized that seed plants are very complex organisms, as compared with unicellular, filamentous, and plate-like plant bodies or even multicellular plants like the mushrooms. Other types of plants have different types of plant bodies. They will be discussed in some detail after the structure and function of the plant body of seed plants is understood. See Color Plates 12, 13, 14, 15, 16, and 17 for other plant body types.

The bean seed is borne in a pod on a "mother plant." When removed from the pod under suitable growing conditions, the bean seed germinates and a young plant (seedling) emerges (Fig. 3.1D) having roots, stems, and leaves. The plant *grows,* extends its roots into the soil and its stems with their leaves into the atmosphere, and after a number of weeks attains adult size. Flowers are formed, then pods "set," and in the pods are seeds ("beans"). The bean plant has then completed a **life cycle.** Many of the physiological processes involved are complex and require a knowledge of chemistry and physics in order to be understood. A bean plant is an individual—it has a birth, a young life, an adult life, a period of reproduction, and then it dies. It has **organs,** just as does the human being, which perform various functions necessary to maintain its life. These organs—**roots, stems, leaves, flower parts**—are composed of **tissues,** groups of cells that carry on different activities (Figs. 3.1E, F, 3.2). The cells are the small compartments that make up tissues; they possess living material **(protoplasm)** and hence are the *living units of plant structure and function* (see Fig. 4.2). The plant body is composed of cells and materials elaborated by cells. All the different kinds of work the plant does are really the work of its cells.

A plant body like that of the bean (Fig. 3.1D) possesses all the organs and tissues necessary to maintain life. The stem (a) supports the foliage leaves and flowers, (b) conducts water and mineral salts from the roots, (c) conducts plant foods from tissues where they are manufactured to tissues where they are needed for growth or where they are stored for future use, and (d) may store foods. The roots (a) anchor the plant in the soil, (b) absorb water and mineral salts from the soil, and (c) may store food. The leaves (a) manufacture foods and (b) give off water in the form of water vapor. These activities involve both chemical and physical processes, which will be discussed in later chapters.

23

One cannot have a clear understanding of the structures and functions of a plant without a knowledge of its anatomy, such as is revealed by the use of a compound microscope. All-important at the very beginning is an understanding of cells—the units of plant structure and function. Chapters 4–6 deal with the cell.

The gross development of the seed plant body is shown in Figs. 3.1A–D. After a bean seed is soaked in water for several hours, the outer skin or seed coat is easily removed. The body exposed is the **embryo** (Figs. 3.1A, B). The term "embryo" as used in biology refers to a living organism in the early stages of development. In most plants and the higher animals, the embryo is more or less dependent upon the maternal parent for its food. The bean embryo is a very young plant. There are two large **cotyledons** or seed leaves, several young **foliage leaves,** a **stem** (composed of **epicotyl** and **hypocotyl**), and a **rudimentary root (radicle).** The two cotyledons contain stored food, chiefly starch. Although the young foliage leaves are small, their veins can be easily seen with a lens. Figs. 3.1A, B show the bean embryo spread open so that the relationship of the different parts is more easily observed.

The embryo plant body has two regions of importance to future growth—the **shoot tip** and the **root tip** (Figs. 3.1E, F). The shoot tip includes (a) at the very end of the young plant, between the rudimentary leaves, a dome-shaped mass of undifferentiated cells constituting the shoot apex and (b) cells directly below the apex that are enlarging and in early stages of elongation and differentiation. The root tip, at the lower end of the young plant, is composed of similar tissues: (a) the root apex of undifferentiated cells, (b) the elongating and differentiating cells above the root apex, and (c) a root cap protecting the root apex. The undifferentiated cells that make up these two tips are capable of **division, differentiation,** and **growth. Meristematic cells** are cells capable of active cell division with a resulting increase in the number of cells. Until the death of the plant, there are always meristematic tissues at the tips of stems and roots.

As long as they are alive, all multicellular plants contain localized regions of meristematic tissue, enabling them to grow and differentiate new stems and roots. Indeed, we shall see that a continual growth of new roots is essential for continued absorption of water and nutrients. Plants are thus said to have open growth, and it has been demonstrated that many fully differentiated individual plant cells contain a full genetic complement and are able to give rise, through mitosis, to new individuals. On the other hand, animals do not have meristematic tissue;

only the undifferentiated cells of embryos are able to divide and to differentiate into specialized cells. With but few exceptions, mature animal tissue can only produce new cells similar to those of their own tissue type. Moreover, while well-fed animals may increase somewhat in girth, they do not *continually grow* in girth or in length as does a tree or vine.

During the early stages in the development of the seedling bean plant, the cells are nourished by food (chiefly starch) stored in the cotyledons. Before this food moves from the cotyledons to other parts of the plant, it is changed to sugar. Sugar, unlike starch, is soluble in the water of cells, and in solution it moves from the cotyledons to all other cells of the embryo, especially to those actively growing cells of the shoot tip and root tip. As soon as the cotyledons are raised into the light by the lengthening of the shoot, they become green. Then they begin the manufacture of food. By this time, the roots are well-established in the soil and are absorbing water and mineral salts. The foliage leaves that were present in the seed expand; they, too, engage in the process of food-making. As food moves from the cotyledons, they shrivel and after a time disappear altogether. The seedling, with its roots absorbing water and mineral salts from the soil, and with its green leaves manufacturing food, is now an *independent plant*. The plant, as shown in Fig. 3.1D, is equipped with all the organs and tissues necessary to carry on the processes that are required to keep it alive and growing. At this stage the structural framework of the adult plant is established. It is an axis with meristematic tissue at both ends (Fig. 3.2). The lower portion of the axis anchors the plant in the soil and absorbs water and mineral nutrients. The upper end of the axis produces leaves for photosynthesis, and flowers and fruits for reproductive purposes. The central portion of the axis supports the photosynthetic and reproductive parts in the light and air of the biosphere and serves as a communication pathway between the absorbing and synthetic ends of the axis.

The axis may branch. Branch roots arise from certain meristematic cells inside the main root (Fig. 3.2A). Branch shoots originate in undifferentiated cells found in an angle formed by a young leaf at the point of its attachment to the stem. This is the leaf **axis** (Fig. 3.2A). Leaves originate from the undifferentiated cell or meristematic cells at the upper end of the shoot or shoot apex (Figs. 3.1, 3.2A). These undeveloped leaves are known as **leaf primordia.**

A diagram of a seed plant (Fig. 3.2) will serve to summarize the foregoing discussion and illustrate the structural organization of a seed plant. The plant body is made up of two parts: the shoot and the root.

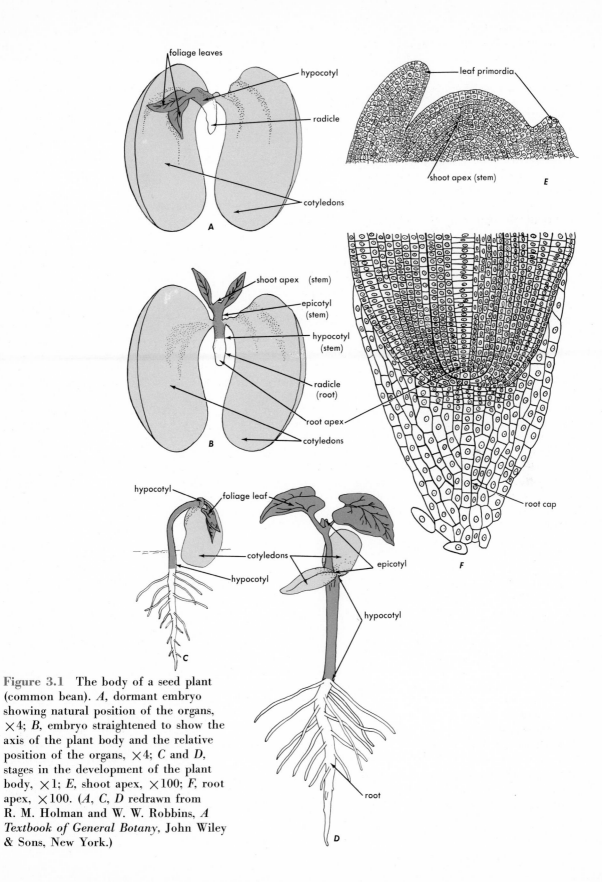

Figure 3.1 The body of a seed plant (common bean). *A*, dormant embryo showing natural position of the organs, ×4; *B*, embryo straightened to show the axis of the plant body and the relative position of the organs, ×4; *C* and *D*, stages in the development of the plant body, ×1; *E*, shoot apex, ×100; *F*, root apex, ×100. (*A, C, D* redrawn from R. M. Holman and W. W. Robbins, *A Textbook of General Botany*, John Wiley & Sons, New York.)

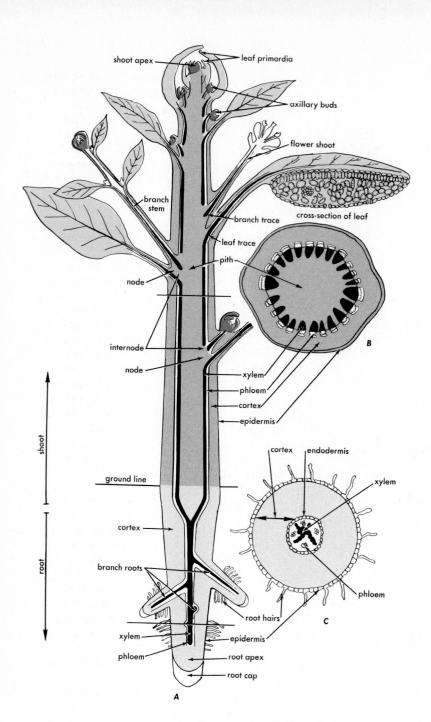

shoot apex — leaf primordia

axillary buds

flower shoot

branch stem

branch trace

cross-section of leaf

leaf trace

pith

node

B

internode

node

xylem

phloem

cortex

epidermis

shoot

cortex endodermis

xylem

ground line

cortex

phloem

root

C

branch roots

root hairs

xylem — epidermis

phloem — root apex

root cap

A

Figure 3.2 *A*, the principal organs and tissues of the body of a seed plant; *B*, cross-section of stem; *C*, cross-section of root. (*A*, redrawn from R. M. Holman and W. W. Robbins, *A Textbook of General Botany*, John Wiley & Sons, New York.)

The shoot is composed of two kinds of organs—stems and leaves. In the seedling there are two kinds of leaves, the cotyledons or seed leaves, which are temporary in most species, and the foliage leaves, which are relatively longer-lasting. The stem, in contrast to the root, is divided into **nodes** and **internodes,** regions that alternate throughout the length of the stem. Nodes are the slightly enlarged portions where leaves and buds arise and where branches originate from the buds. An internode is the region between two successive nodes. As we shall see later, the anatomy of nodes and internodes differs. In Fig. 3.2, structures and tissues are shown which will be discussed in later chapters.

4

The Metabolic Plant Cell

Galen, the last of the great Greek doctors, who lived in Asia Minor during the second century AD, thought that all tissues, such as spleen, brain, and kidney of animals, leaves, stems, and roots of plants, were a "sensible element, of similar parts all through, simple and uncompounded." Others before him had thought that animal tissues were simply coagulated "juices" seeping through the walls of the intestine. No one dreamt that these tissues and their plant counterparts had a fabulous and complicated structure, a structure that could not be seen until the invention of a microscope, which Zacharias Jansen, a spectacle maker of Holland, accomplished in 1590. This instrument was later improved by Robert Hooke, an Englishman, who was not only interested in optics but was also an architect and an experimenter with flying machines. Hooke, who lived from 1636 to 1703, examined all sorts of natural objects with his improved microscope. Among these were thin slices of cork (the dead outer bark of an oak). Fig. 4.1 is an illustration of the cork tissue as Hooke saw it under his microscope. This figure was published in 1664 in an article entitled *Micrographia, or Some Physiological Descriptions of Minute Bodies Made by Magnifying Glasses.* The term **cell** was first used by Hooke to denote in cork "little boxes or cells distinct from one another . . . that perfectly enclosed air." He estimated that a cubic inch of cork would contain about 1259 million such cells.

Because of the prominence of cell walls in plant tissue, the cell early became the unit of structure and of life for botanists. However, during these early years, zoologists considered the tissues as the true centers of life of the body. There were supposed to be 21 different tissues, able to change within themselves, depending upon the organ in which they were located. Life was thought to reside in these tissues. However, numerous observations upon protozoa and animal tissues led to a gradual accumulation of evidence that cells also existed in animals. In 1838 the German zoologist Theodor Schwann and the Belgian botanist Matthias Jacob Schleiden collaborated in a paper entitled, *Microscope Investigations on the Similarity of the Structure and Growth in Animals and Plants.*

This paper established on a firm basis *the theory* that the cell is a basic unit of structure *in both plants and animals.* However, another 30 years of research were necessary for the general acceptance of such a new idea and for the establishment of *the fact* that *all organisms are composed of cells,* that cells are indeed the basic units or building blocks of life. Cells are maintained by the division of cells. Thus, meriste-

27

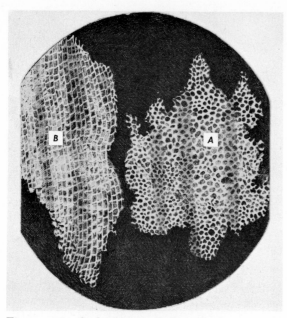

Figure 4.1 Cork tissue as Robert Hooke observed it under his microscope. (From Hooke, *Micrographia* (1664), The Council of the Royal Society of London for Improving Natural Knowledge.)

matic cells of shoot apices (which we have just discussed) have a direct line of descent going back to the beginning and continuing into the future. In animals, such a continuity of cells occurs only in the germ plasm. Cells have existed as long as life and will exist as long as life exists.

Cell Concept

We have already noted the existence of two types of cells, the prokaryotic cells which characterize more primitive organisms, and the eukaryotic cells of all other animals and plants (Figs. 2.5*A*, *B*). A detailed discussion of the prokaryotic cell will be reserved for the chapters discussing the more primitive plants. The following discussion is concerned mainly with the structure of the eukaryotic plant cell (Fig. 4.2).

All eukaryotic cells separate many of the metabolic processes of the cell into discrete membrane-bound, organized, protoplasmic particles called **organelles.** Each organelle has a characteristic structure and performs the same function wherever it occurs. The organelles are: the **nuclei** (Fig. 4.3), the **mitochondria** (Figs. 4.3, 4.8), the **endoplasmic reticulum** (Figs. 4.3, 4.7), the **dictyosomes** (Figs. 4.3, 4.11), the **micro-**

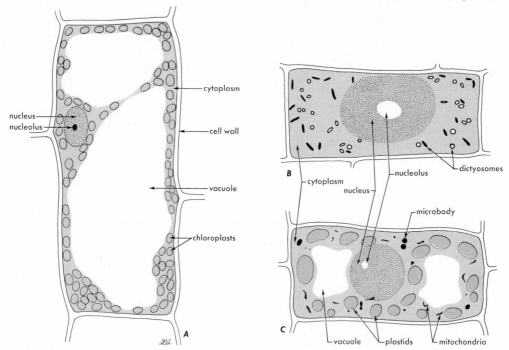

Figure 4.2 Plant cells. *A*, diagram of a living cell from an *Elodea* leaf. When viewed on the light microscope, organelles in living cells, other than chloroplasts, lack contrast. The nucleus and nucleolus are generally visible. Sometimes one of the categories of microbodies may be seen. Mitochondria are seen only with perfect lighting conditions, ×700. *B*, a cell from a root tip of *Vicia faba* heavily stained with osmium to show the dictyosomes, ×500. *C*, a cell from a young leaf of *Elodea*, fixed to show plastids and mitochondria, ×500. (*B* after R. Bowen, *Zeitschr. Zelf. Mikr. Anat.* **9**, 1.)

bodies (Figs. 4.8*B*, 4.13), and the **plastids** (Figs. 4.9, 4.10). While the **ribosome** is a small particle rather than a membrane-bound organelle, it does have a specific niche in cell metabolism, and it is convenient to include it in this list.

In the nuclei, nuclear division, and the associated ribosomes, cells have a means of storing genetic information and a system of distributing it to direct cellular development. In the mitochondria, cells have a mechanism for delivering useful energy for cellular activities. In the endoplasmic reticulum, dictyosomes, and possibly the microbodies, there is a mechanism for the synthesis and probably degradation of cellular metabolites. The organelles are bathed in ground **cytoplasm** through which diffusion of materials occurs, where certain phases of metabolism take place, and where the activities of the organelles are coordinated.

Eukaryotic plant and animal cells differ in several basic ways: (*a*) Plant cells specialized to carry out photosynthesis are provided with one to several organelles, the **chloroplast** (Figs. 4.2, 4.3), in which reside most of the enzymes, energy compounds, and intermediate substances required for photosynthesis. (*b*) With few exceptions, plant cells are provided with a **cell wall** composed of **cellulose** and secondarily deposited materials which gives greater rigidity and lessened mobility to plants. (*c*) Most plant cells are provided with a large central aqueous **vacuole** (Figs. 4.2, 4.19). (*d*) Synthesis of materials occurs at a much higher rate in some animal cells, for example gland cells, where the organelles involved in synthesis (endoplasmic reticulum and dictyosomes) are more numerous.

Considering for the moment only the morphology of plant cells, we may ask four questions. (*a*) What is the structure of the meristematic cell? (*b*) What changes occur during cell division? (*c*) What changes take place during differentiation? (*d*) How is the specialized cell particularly fitted to perform its part in the overall economy of the whole plant? These questions will demand our attention in the pages that follow. We shall first consider the structure of a cell from a young bean leaf, first as it appears under the light microscope at about 1000× magnification, and then as it appears under the electron microscope at from 3000–60,000× magnification.

Cells become specialized to perform definite functions. This results in different specialized tissues and organs. In all plants and animals, cells have been organized into comparable tissues to perform comparable functions. For instance, upon reaching a certain state of complexity, most green plants— thallophytes, mosses, or vascular plants—develop flattened blades similar to leaves for carrying on photosynthesis (see Figs. 22.7, 26.12, 27.7). All multi-

cellular plants and animals may start life as a single cell, the fertilized egg cell (see Fig. 16.21). The subsequent lineage of cells derived from the zygote differentiates into several distinct tissues, each specialized to perform a different function. Most plants retain, for relatively long periods, undifferentiated cells capable of becoming specialized cells. Indeed, it seems that most all plant cells carry full potency to return to the undifferentiated state and to again divide and give rise to a line of cells able to form a complete plant or a complete functioning tissue (Chap. 20).

Units of Measurement

At this point a consideration of a few measurements is in order (Table 4.1 and p. 671). We have already stated that Robert Hooke calculated that there would be 1259 million cells in a cubic inch of cork. This figure is based on his estimate of about 1100 cork cells along a line 1 inch long. One thousand cells along 1 inch would give us about 40 cells for the length of 1 mm, so a single cork cell is $\frac{1}{40}$ mm in diameter. It now becomes convenient to use a smaller unit of measurement, the **micron**, symbolized by the Greek letter mu, μ. There are $1000\,\mu$ in 1 mm, and a cell $\frac{1}{40}$ mm in diameter would be $25\,\mu$ in diameter. This is about correct for a cork cell, although many plant cells are larger.

Under the best conditions one can detect particles $0.25\,\mu$ in diameter under the light microscope. This is about one-half the wavelength of blue light, and shorter wavelengths are needed to see smaller particles. Just as a large ocean wave is not modified as it passes by a small post, particles smaller than one-half the wavelength of light do not disturb the advancing wave front of light, and therefore the particle cannot be seen.

An electron beam has a wave motion with a distance of $0.0000005\,\mu$ from crest to crest. Such small distances call for a still shorter unit of measurement. This is the **nanometer,** abbreviated as nm.* There are 1000 nm in 1 μ. With the short wavelengths of a beam of electrons, a resolution of distances separating atoms is theoretically possible. But the difficulties in making perfect magnetic lenses and in the nature of biological material itself means that lines or particles in cells closer than 2 nm can not be distinctly separated from each other regardless of the magnification obtained. However, the best figure obtainable with the light microscope is $0.25\,\mu$, or 250 nm, so the electron microscopes allow more than a 100-fold increase in magnification.

---

*An angstrom (A), the unit formerly used, was 1/10,000 μ. There are 10 A in 1 nm.

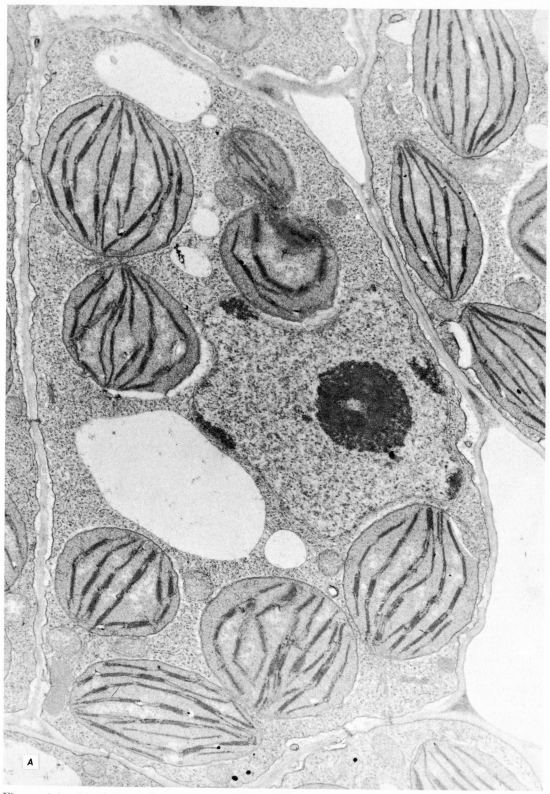

Figure 4.3 A cell from a leaf of a bean seedling (*Phaseolus vulgaris*). *A* is an electron micrograph; *B* is a diagram of *A*. Chloroplasts, nucleus, and vacuoles are same size as in *A*; other organelles are slightly enlarged or otherwise accentuated. Ribosomes not drawn. Original magnification of *A* is ×10,000.

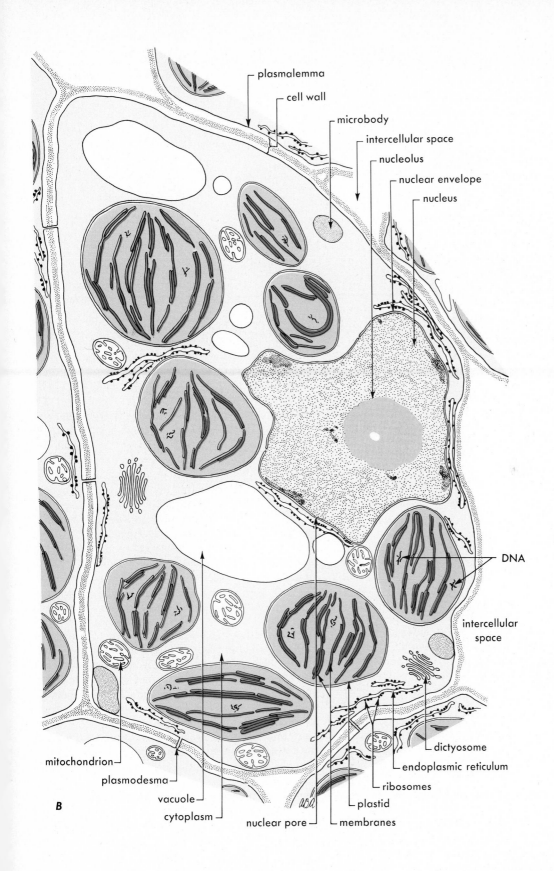

plasmalemma

cell wall

microbody

intercellular space

nucleolus

nuclear envelope

nucleus

DNA

intercellular space

dictyosome

endoplasmic reticulum

mitochondrion

plasmodesma

ribosomes

vacuole

plastid

cytoplasm

nuclear pore

membranes

B

31

TABLE 4.1   Some Dimensions[a]

	Microns	Nanometers
1 inch	25,000	2,500,000
1 mm	1,000	1,000,000
1 μ		1,000
Resolution of unaided eye	100–300	
Resolution of light microscope	0.25	250
Practical resolution of electron microscope for biological material		2
Wavelength of blue light	0.50	500
Wavelength of red light	0.70	700
Wavelength of electron beam 75 KV (KV is 1000 volts)		0.05
Diameter of cork cell	25	
Volume of cork cell	15,625 μ^3	
Diameter of leaf cell	200–400	
Diameter of vessel in xylem	500	
Diameter of *Scenedesmus*	10	
Length of *Scenedesmus*	30	

[a] One foot magnified 100,000× is nearly 19 miles; 1 inch magnified so each angstrom in it measures 1 inch in length appears about 46,000 miles long.

Since the unaided eye cannot see particles closer together than 0.2 mm or 200 μ, a magnification of ×100,000 would be needed to visualize a particle 2 nm in diameter. Because of the complexity of the cell and the relative thickness of the tissue section (25× the 2 nm particle), direct magnifications of ×100,000 are not satisfactory. So for higher magnifications, photographic negatives are generally made at from 40,000–50,000 magnification, and the additional magnification is obtained in an enlarged photographic print. Most biological work with the electron microscope is done at lower magnifications.

Technique

Living cells may be observed for many hours on the light microscope by maintaining them in a proper medium. Because the parts of the protoplast have an index of refraction close to that of water (the protoplasm bends light rays to the same degree that water does), many details of cell structure are not apparent in living cells. Enhancement of detail is attained by studying living cells with phase contrast optics or with polarized light. Still more detail is obtained by killing cells in various mixtures of alcohol, acetic acid, chromic acid, formaldehyde, or osmium tetroxide. The latter two compounds are of particular interest, because in killing cells they do not coagulate the proteins. The killed tissue is dehydrated, embedded in paraffin, and cut into sections from 5 to 20 μ in thickness. When this tissue is mounted on a glass slide and properly stained, much detail becomes apparent.

Plants, or plant tissues, may be examined with a beam of electrons very much as they are examined with a light beam. However, electrons are not visible to the eye, and some device, such as activation of phosphors on a fluorescent screen, must be used to enable us to see the beam of electrons. There are two types of electron microscopes just as there are two types of light microscopes: transmission microscopes and reflecting microscopes (stereoscan microscopes). In the first type, the beam passes through the specimen as in the compound microscope. In the second case, the vehicle is reflected photons or electrons as with the stereodissecting scope.

For transmission electron microscopy, small pieces of leaf or other tissue are killed in glutaraldehyde, stained with osmium tetroxide, and dehydrated and embedded in a plastic that must be held at 70°C for 24 hr to harden it properly. Sections of tissue and plastic, about 50 nm thick, are then placed in the electron microscope in a high vacuum. A beam of electrons is passed through the sections, and an image is formed on a fluorescent screen. Magnetic fields bend the beam of electrons just as glass bends light rays. Thus, the lenses in the electron microscope are precisely formed magnetic fields. The dark masses seen on the screen of the transmission electron microscope will not be images of organic matter but shadows of the high-atomic-weight metal, osmium, which has become preferentially bound to some components of the protoplasm. It is not currently known why or how osmium is bound to cellular constituents. Actually, an electron microscope is a highly refined television set tuned onto a closed circuit (Fig. 4.4). The contrast in the image on the electron microscope screen is low and details can be studied best on photographic prints prepared to enhance contrast.

A recent development in technique makes it possible to examine the ultrastructure of frozen cells

Figure 4.4 An electron microscope (Hitachi, HU-12) which gives magnifications from 1000 to 500,000 diameters. (Courtesy of Perkin-Elmer Corp.)

shadowed or coated with a thin layer of gold. Thus prepared, the tissue has normal appearance, does not further deteriorate, and has a better reflecting surface in the electron beam.

The Plant Cell: Structure and Function

Cells may be considered from the viewpoint of their (a) structure (Chap. 4), (b) activity (Chap. 6), or (c) chemical organization (Chap. 5). Knowledge of the interrelationships of all three approaches is necessary for an understanding of the cell as a whole. Within a cell, certain activities are confined to definite structures or organelles. Photosynthesis takes place in chloroplasts, respiration is confined to mitochondria, and the nuclei store the genetic information.

Cells are dynamic living units maintained in equilibrium with their surroundings only through the expenditure of energy. Cells are continually renewing their substance. Cells divide. To keep pace with the increasing number of cells, the mass of protoplasm must also increase. Cells also accumulate materials to be used in their own metabolism. For instance, leaf cells synthesize sugar from simpler substances, and store it in large amounts as starch.

It is sometimes convenient to think of cells in terms of the structure and chemical nature of the organelles—such as cholorplasts— which lie within the cell. Among other things, chloroplasts are formed of proteins, lipids, and pigments. Some of the lipids are phospholipids containing phosphorus, while other lipids contain sulfur and have a very high surface activity. The most common pigments are chlorophylls a, b, and c, and a variety of **carotenoids** (substances resembling, or related to, carotene, p. 63). We know least about the proteins. A great deal is known about the chemistry of plastids, mitochondria, and nuclei. Less is known about the other organelles.

The products, or accumulations, of protoplasmic activity (water, sugar, cellulose, hormones, and similar items) are found in the dynamic cooperating unit of protoplasm, called the **protoplast** (Figs. 4.2A, 4.3). The protoplast secretes about itself a cell wall of cellulose, which frequently later receives a deposition of secondary products (Fig. 4.21). All of the protoplast, plus the cell wall, is the **plant cell.**

The activities of whole cells may be studied. Their ability, through the expenditure of energy, to accumulate materials in greater concentration than occurs in the surrounding media is a subject of considerable interest. The organelles may be isolated from cells, and their activities or chemical characteristics

without chemical fixation. Cells or isolated organelles are rapidly frozen at the temperature of liquid nitrogen. They are placed under a vacuum and the frozen suspension is kept at $-147°C$. The top of the frozen drop is chipped away and some of the top ice is allowed to evaporate. This leaves about $\frac{1}{3}$ nm of frozen tissue extending up out of the frozen block. A very thin layer of carbon and platinum is now deposited over the tissue and ice surface. This layer of carbon and platinum, actually the impression of the exposed tissue, may now be placed on a grid and observed with the electron microscope (Fig. 4.18). The technique is referred to as **freeze-etch.** Bacteria may be quickly frozen to $-147°C$ and upon thawing still survive, so bacterial cells examined by the freeze-etch technique are still alive. This is as close to living cells as the electron microscopist is likely to get.

For viewing with the stereoscan electron microscope, an object is mounted on a small steel plate and placed in a vacuum in the path of a beam of electrons. Living tissue may be examined directly. However, it rapidly desiccates in the vacuum and does not provide a good reflecting surface for the electrons. So plant parts to be studied with the stereoscan microscope are usually killed, dehydrated carefully, with a final drying in liquid CO_2, and then

may be studied in test tubes. Both isolated mitochondria and chloroplasts will carry on many of their normal cellular activities for a short period *in vitro*.

Whatever the approach to the study of cells, it is important to remember that energy is required to maintain a living cell as an individual entity. The maintenance of a steady living state, or equilibrium, through the expenditure of energy is known as **homeostasis.** Disruption of the source of the necessary energy must result in the death of the cell.

The parts of the cell are listed in Table 4.2. Also included are the dates when the parts were first observed by the light microscopist and by the electron microscopist, the approximate dimensions of the parts, and a notation as to their function. Dates for first observation with the light microscope are in **boldface type.**

Light Microscopy

The Living Cell

In a living leaf of *Elodea* or a young bean plant (*Phaseolus vulgaris*), the chloroplasts are the most prominent organelles. They are embedded in the clear hyaline cytoplasm which is confined to the periphery of the cell, or to thin strands crossing the clear central vacuole (Fig. 4.2A). The central vacuole is largely water with dissolved salts, various organic

TABLE 4.2 The Parts of the Cell[a]

Cell part	Dimension	Function
Plant cell, **1880**		
Primary cell wall, **1665**	2–5 μ	Protection, strength
Cellulose microfibrils, 1948	10–25 nm (indefinite length)	Mechanical strength
Cellulose molecules, 1922 (x ray), 1949	0.834 × 0.8 nm	Strength
Amorphous matrix		
Middle lamella	2 μ	Adhesion between cell walls
Protoplast, **1846**	0.025–2 mm	Homeostatic unit
Protoplasm, **1840**		Synthetic and developmental reactions under direction of nucleus
Cytoplasm, **1882,** 1957		
Chloroplast, **1702,** 1953	2–20 μ	Photosynthesis
Chromoplast, **1900,** 1958	2–10 μ	Accumulation of carotenes and similar pigments
Amyloplast, **1884,** 1955	2–25 μ	Starch storage
Leucoplast, **1883,** 1957	2–10 μ	?
Mitochondrion, **1897,** 1947	2 × 2–10 μ	Respiration
Dictyosomes, **1927,** 1956	3 μ	Enzyme synthesis
Endoplasmic reticulum, 1957	17 nm (indefinite length)	Protein synthesis
Ribosomes, 1955	20 nm	Protein synthesis
Sphaerosomes, **1919,** 1967	2 μ	Lipid synthesis (?)
Microbodies, 1965	0.1–2.0 μ	Compartmentalization of various enzymes
Microfibrils, 1959	28 nm (indefinite length)	Various functions
Tonoplast, **1877,** 1960	8 nm	Regulation of exchange between vacuole and cytoplasm
Plasmalemma, **1880**(?), 1954	8 nm	Regulation of exchange between cytoplasm and external solution
Crystals, 1963	10 μ	Unclear, possibly protein storage
Plasmodesmata, **1879,** 1957	2 μ	Protoplasmic bridge between cells
Nucleus, **1831**	5–30 μ	Contains genetic information necessary for normal cell development and activity
Nuclear envelope, **1907,** 1955	25 nm	Separates nucleoplasm from cytoplasm
Nucleoplasm, **1879**		
Chromatin, **1880**		
Nucleoproteins, **1869**		

solutes, and sometimes water-soluble pigments. The interfaces of the cytoplasm–vacuole and the cytoplasm–cell wall are characterized by special cytoplasmic membranes that exercise some control over the passage of substances into and out of the cytoplasm. These membranes are not visible with the light microscope, but, because of their activity, their existence has long been recognized. The membrane separating cytoplasm and vacuole is known as the **tonoplast;** that bounding the outer surface of the cytoplasm is the **plasmalemma.** The cytoplasm and plasmalemma may be removed, leaving the tonoplast as a sac enclosing the vacuole. A single nucleus (Fig.

4.2A) is embedded in the cytoplasm and may be seen flattened against one of the walls. In the living cell it generally appears structureless, although with careful focusing, one to four small denser bodies, the **nucleoli** (nucleolus s.), can be seen (Fig. 4.2A). There is also an envelope around the nucleus. This may be punctured and a clear fluid, the **nucleoplasm,** will be able to escape. In addition, small spherical particles about 0.54 μ in diameter may be seen in the living cytoplasm. They belong to a class of particles called **microbodies** (Fig. 4.8B).

The cytoplasm in a leaf cell, especially when warmed by the light of a microscope, will show

TABLE 4.2 (*Continued*)

Cell part	Dimension	Function
Nucleic acids, **1889**		
DNA, **1924**		
DNA helix, 1953	1.8 nm (indefinite length)	Carries genetic code
Unit fibers, 1963	12.5 nm (indefinite length)	Encompasses DNA helix and nucleoproteins
Nucleolus, **1882,** 1958 (animal, 1952)	1–5 μ	RNA synthesis
RNA		Transferal of information from DNA to cytoplasm
Ribosomes, 1956	15 nm	Protein synthesis
Mitosis, **1882,** 1960		Replication of DNA and distribution to daughter cells
Interphase, **1830**		Replication of DNA
Chromosome, **1888,** 1955	2–200 μ	Vehicle carrying DNA helix in replication and distribution
Kinetochore		Region of chromosome to which spindle fibers are attached
Centromere **1925**		
Chromatid, **1900**	1–10 μ	One-half of a chromosome
Prophase		Stages of mitosis
Metaphase, **1884**		
Anaphase, **1884**		
Telophase, **1894**		
Spindle, **1883,** 1960		Cytoplasmic structure involved in moving chromosomes
Fibers, **1881,** 1960	Various indefinite lengths	During mitosis
Microtubules, 1960	28 nm (indefinite length)	
Cytokinesis, **1891,** 1960		Division of cytoplasm to daughter cells, frequently unequal
Equatorial plate, **1875,** 1958	Various	First stage in separation of daughter cells
Vacuole, **1835,** 1957		Various functions important in water economy of cell
Water		
Inorganic salts		
Various organic solutes		
Crystals		

[a] Date in **boldface type** refers to the year first reported with the light microscope; date in lightface type indicates the year first seen with the electron microscope.

active streaming, carrying both the chloroplasts and the microbodies with it. With time, the velocity increases up to 5–10 mm per min. This rapid flowing of the cytoplasm is a dramatic indication of its plasticity, and should be recalled when studying microscope slides and electron micrographs, for it is sometimes difficult to reconcile the activity of flowing cytoplasm with the static and frequently complicated images seen in electron micrographs.

Fixed and Stained Cells

Cells fixed and stained show considerably more detail. The cell wall is evident and the cellulose wall of one protoplast is frequently seen separated from the cellulose wall of an adjacent protoplast by an intervening substance. This is the **middle lamella** (Fig. 4.21), which is largely pectin and formed as a final step in cell division (p. 82). Chloroplasts, when present, are evident. After proper fixation and staining, numerous small filamentous bodies, perhaps 2–4 μ long and 0.54 μ in diameter, are seen in the background cytoplasm. These are **mitochondria,** which are typically seen in the light microscopes as simple, darkly staining granules (Fig. 4.2C). Microbodies and dictyosomes are apparent after special fixing and staining. The **nuclear membrane** and the nucleolus are both much more distinct after fixation and staining. Within the nucleus, there are generally clumps of stained material called **chromatin** (Fig. 4.2C). The **nucleoplasm,** however, does not stain.

Electron Microscopy

A cell from a young bean leaf (*Phaseolus vulgaris*) is shown in Fig. 4.3. Since the cells at this stage have not elongated and contain all the organelles, they are satisfactory examples of plant cell structure as seen by the electron microscope. Fig. 4.3A illustrates the relative size and abundance of the organelles just mentioned. Fig. 4.3B is a diagram of this cell which should be used to definitely locate the organelles in the micrograph. The chloroplasts are the most abundant and conspicuous oganelles, since only the nucleus is larger. Mitochondria and dictyosomes are small, and the dictyosomes occur in relatively small numbers. There are not many membranes of endoplasmic reticulum and they are masked by the abundant ribosomes. In the diagram, mitochondria, dictyosomes, and the endoplasmic reticulum are enlarged. The free ribosomes are not indicated. The tonoplast is accentuated.

Membranes

Perhaps the most notable thing about cells at this magnification is their **compartmentalization** into **membrane-bound organelles.** The protoplasm itself is bounded externally by the plasmalemma (Figs. 4.3, 4.5, 4.18) and internally by the tonoplast (Figs. 4.3, 4.5). The plasmalemma separates the protoplast from the external cell wall or other nonliving systems. Within the protoplast, the tonoplast separates the protoplasm from the vacuole. In any organized living system, high-energy regions must be separated from low-energy regions by membranes.

Note that each of these membranes consists of a light line sandwiched between two dark lines. The approximate widths of the three membrane components are 2, 3.5, 2 nm, respectively. Note that the plasmalemma is more heavily stained and the two darker components may not have identical widths. These membranes are known as **unit membranes.** They are thought to be composed of a central bi-

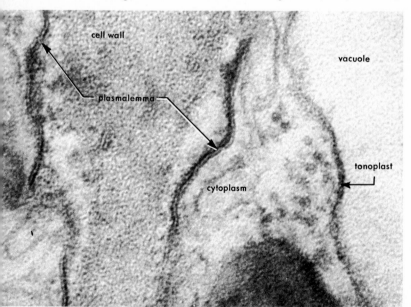

Figure 4.5 A cell wall bounded on each side by a thin peripheral layer of cytoplasm. Bordering portions of the vacuoles of each protoplast are present. The tonoplasts separate cytoplasm and vacuole. The plasmalemma bounds the protoplast, separating it from the cell wall. The plasmalemma stains heavier than the tonoplast. The unit membrane structure is evident in portions of all three membranes. Magnification ×100,000.

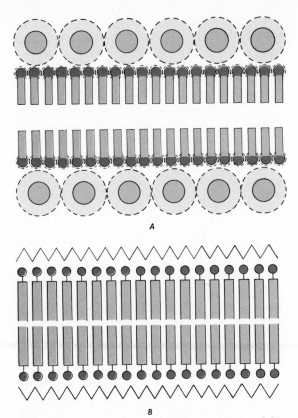

Figure 4.6 The unit membrane. *A*, the model proposed by Davson and Danielli in 1935. It shows a bimolecular layer of lipid bounded by a monomolecular layer of globular proteins. *B*, the model proposed by Robertson in 1959. Here the bounding monomolecular layer is fully stretched protein.

molecular leaflet of lipid sandwiched between two monomolecular layers of fully extended protein (Fig. 4.6*B*).

A drop of oil or other lipid may be spread on water to form a monomolecular film. (The colors seen when oil is spread on water are indicative of very thin oil films, some of which may be only a single molecule in thickness.) A force of 9 dynes is required to break such a lipid film. Natural oils spread on water (Harvey and Danielli used natural oils from salmon eggs) are disrupted by forces as low as 2 dynes. Protein was found to be associated with the salmon egg oil and to account for the lessened strength of the film formed when this oil was floated on water. Later experiments demonstrated that a membrane consisting of a bimolecular layer of lipid sandwiched between monomolecular layers of protein could account for many of the properties of cellular membranes. The Davson–Danielli model of a cellular membrane is given in Fig. 4.6*A*. Note that

these workers proposed the protein layers to be in the more natural form of globular proteins. The idea that the protein is fully extended (Fig. 4.6*B*) was proposed by Robertson because, in studying the membranes of the rather inert myelin sheath protecting nerve fibers, he saw that there was not space enough for the globular proteins. The difference in the chemical activity of globular proteins and fully stretched proteins is significant and would greatly influence the activity of the membranes. It was proposed early in the initial days of electron microscopy, and generally accepted, that all cellular membranes were tripartite membranes composed of a bimolecular leaflet of lipid sandwiched between two monomolecular layers of fully stretched protein, similar to that of the myelin sheath of a nerve cell (Fig. 4.6*B*). The problem of membrane structure has great significance because many important cellular reactions are membrane-bound. Electron transport activity in both photosynthesis (p. 215) and respiration (p. 244) is bound to membranes. The binding of enzymes to membranes relegates the enzymes to a particular spot in a team of cooperating enzymes.

Ribosomes

The ground cytoplasm, after glutaraldehyde fixation and osmium staining, is moderately electron-transparent and has a soft gray granular appearance. In it always appears at least one type of densely stained granule, roughly angular and from 17 to 20 nm in diameter (Fig. 4.3*A*). The enzyme **RNAase** specifically degrades RNA. Treating tissue with RNAase results in the disappearance of these granules, with no change in the other organelles. This is good evidence that RNA is present in these particulates, which are called **ribosomes.** Ribosomes are present in plastids and mitochondria, as well as in the cytoplasm. They appear to be lacking in nuclei (Figs. 4.3*A*, 4.16). Some micrographs show them attached to the cytoplasmic side of the nuclear envelope. If they do occur within the nucleus, they are probably confined to the nucleolus (Fig. 4.16*C*). Ribosomes are frequently associated with the cytoplasmic membrane system, the endoplasmic reticulum (Figs. 4.3, 4.7*A*). Under certain circumstances they appear in groups, frequently as helical aggregations. These arrays are designated as **polyribosomes** or simply **polysomes** (Fig. 4.7*B*).

Organelles

Endoplasmic Reticulum. In cross-section, the endoplasmic reticulum is represented by profiles of two parallel membranes separated by a narrow light

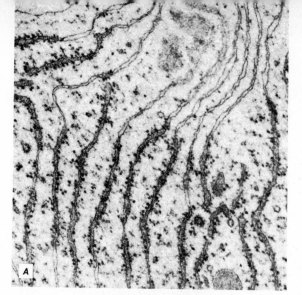

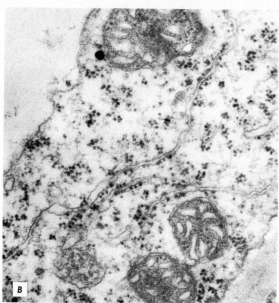

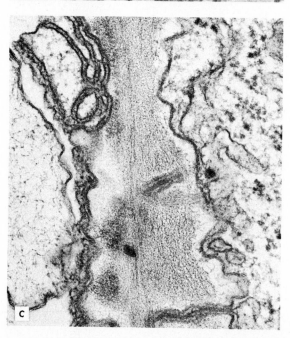

Figure 4.7 Endoplasmic reticulum. *A*, endoplasmic reticulum in an apical cell of the alga *Chara*. Both rough and smooth endoplasmic reticulum are present and are continuous with each other. A microbody is also visible, ×15,000. *B*, endoplasmic reticulum in a meristematic cell from a root tip of wheat. Both rough and smooth endoplasmic reticulum may be seen. There are also free ribosomes and some in helical aggregations that form polyribosomes, ×30,000. *C*, endoplasmic reticulum in a sieve element and companion cell of curcurbit phloem. The endoplasmic reticulum may be seen in the plasmodesmata that pass through the cell walls connecting the two cells, ×60,000. (*A*, courtesy of J. B. Pantastico; *B*, courtesy of P. Bartels; *C*, courtesy of K. Esau and J. Cronshaw, *J. Ultrastruct. Res.* **23**, 1. Copyright Academic Press.)

space about 4 nm wide (Figs. 4.7, 4.18). These profiles are actually cross-sections through extensive flattened vesicles. Note that these membranes again form a closed system; their ends are never open to the ground cytoplasm. The endoplasmic reticulum frequently has ribosomes appressed to its outer, or cytoplasmic, surface. This is known as **rough endoplasmic reticulum. Smooth endoplasmic reticulum** lacks ribosomes and is not as common as rough endoplasmic reticulum (Fig. 4.7A). There is considerable evidence suggesting that protein synthesis may be associated with the rough endoplasmic reticulum. For example, endoplasmic reticulum is most highly developed in cells active in protein synthesis, such as the pancreas, and the glandular hairs of *Drosera* leaves (see Figs. 10.16B, C). It is poorly developed in mature leaf cells (Fig. 4.3A). The endoplasmic reticulum is associated with **plasmodesmata,** or cytoplasmic connections from cell to cell across cell walls (see Chap. 7).

As exemplified by the endoplasmic reticulum, membrane systems within the protoplast, with the single exception of microtubules, form closed systems. Thus, membranes form sacs or vesicles. The vesicles may be roughly spherical or they may be flattened. Flattened vesicular forms may be in stacks or may occur singly, irregularly, or regularly, distributed throughout the cytoplasm and organelles.

Mitochondria. The mitochondrial envelope separates internal mitochondrial space from the ground cytoplasm (Figs. 4.8, 4.18). The mitochondrial envelope is double and each component is a typical unit membrane. The outer component forms a continuous barrier around the mitochondrion. The internal component connects with a series of internal membranes, the **cristae** (Fig. 4.8). The cristae form a closed system, separating an internal space from the **mito-**

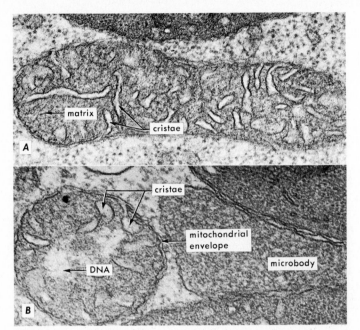
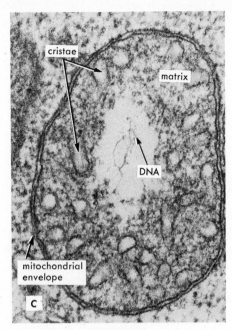

Figure 4.8 Mitochondria. *A*, mitochondrion from a young leaf of a safflower (*Carthamus tintorius*) seedling, ×60,000. *B*, mitochondrion and a microbody from a 7-day-old seedling of bean (*Phaseolus vulgaris*). The mitochrondrion is spherical; the association between a crista and the inner component of the envelope is distinctly seen. The light areas have fine fibrils in them, probably of DNA. The double envelope of the mitochondrion may be compared with the single membrane bounding the microbody, ×70,000. *C*, mitochondrion in a cortical cell of the brown alga *Egregia menziesii*; the double envelope, cristae, and DNA strands are apparent, ×75,000. (*A*, courtesy of A. H. P. Engelbrecht; *C*, courtesy of T. and A. A. Bisalputra, *J. Cell Biol.* **33**, 511.)

chondrial matrix. Note that the cristae are continuous with the internal space of the envelope. The form and amount of the cristae varies. They are numerous in mitochondria from animal cells and generally are parallel to each other and at right angles to the long axis of the mitochondrion. Cristae in plant mitochondria frequently are sparse, irregularly arranged, and not always seemingly connected with the inner membrane of the envelope.

Plastids. The chloroplasts, the plastids which contain chlorophyll, are prominent organelles (Fig. 4.9). The chloroplasts are bounded by an envelope of two membranes, just as are the mitochondria. The plastids have a granular background material, the **stroma.** Particulates having the properties of cytoplasmic ribosomes may be seen in the stroma. They are from 17 to 20 nm in diameter. Clear areas are also present, and in favorable sections, such clear areas containing an array of slender fibrils may be seen. These fibrils are 5 nm in diameter, which is the diameter of the DNA fibrils in other locations. The fibrils are absent from sections treated to remove DNA. Presumably, then they represent chloroplastic DNA. They are

more apparent in plastids from young leaves, as shown in Fig. 4.3.

Plastid Membranes. Some plastid membranes occur in cylindrical stacks (Fig. 4.9) called **grana.** The grana are connected at irregular intervals by **frets** or **lamellae** (thylakoids).

Various Kinds of Plastids. Plastids respond readily to changes in age or environmental conditions. A seedling of bean or barley just emerging from the ground may have yellow first leaves. The inner leaves of lettuce or cabbage are yellowish. Both leaves will become green in the light, yet the plastids in the two types of leaves have a very different appearance. The plastids (**etioplasts**) in the leaves of the etiolated bean seedling have an elaborate crystalline-like structure generally referred to as a **prolamellar body** (Fig. 4.10*B*). The plastids in the lettuce leaf are small and have only a few single lamellae (Fig. 4.10*A*). These plastids, since they will become normal plastids in light, may be called **proplastids.** Other colorless plastids, **leucoplasts,** are found in the epidermal cells of leaves, in onion, in storage tissue of apples and in other white tissues. Some of these plastids

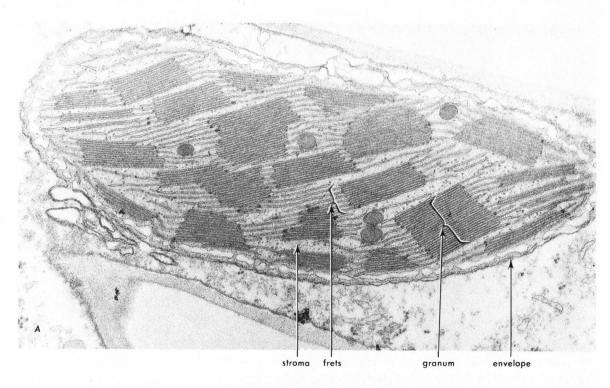

stroma   frets   granum   envelope

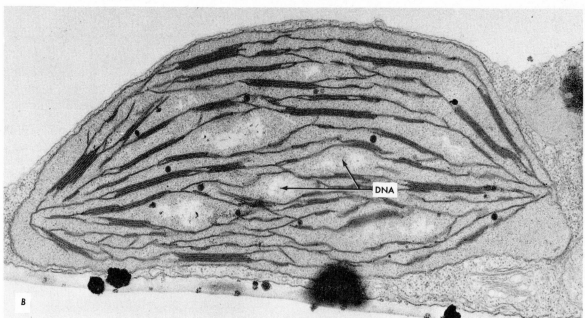

DNA

Figure 4.9 Chloroplast from a corn (*Zea mays*) leaf, ×25,000. (Courtesy of L. K. Shumway.)

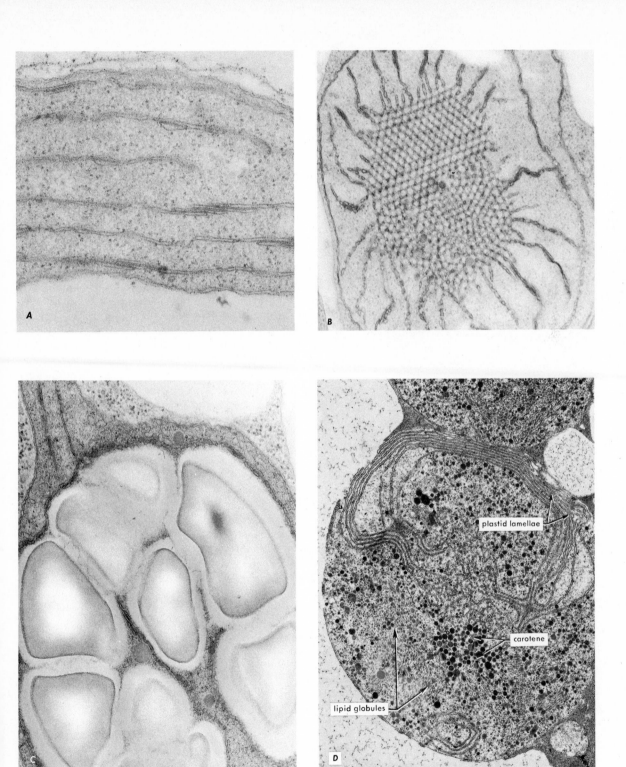

Figure 4.10 Four types of plastids. *A*, a small leucoplast from an inner white leaf of endive (*Cichorium endiva*), ×20,000; *B*, an etioplast from a yellow leaf of a dark-grown bean seedling (*Phaseolus vulgaris*), ×20,000; *C*, an amyloplast from a bean seedling (*Phaseolus vulgaris*), ×20,000; *D*, a chromoplast from a mature red pepper (*Capsicum*), ×21,000. (*D*, courtesy of A. R. Spurr and W. H. Harris, *Am. J. Bot.* **44**, 1210).

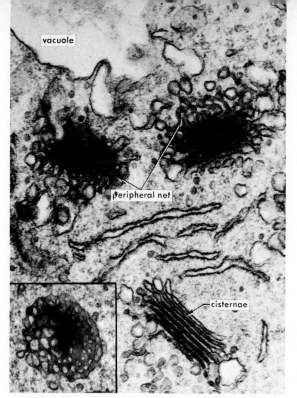

vacuole

peripheral net

cisternae

Figure 4.11   Four dictyosomes in a cell from the
root tip of a water plant (*Hydrocharis*). One
dictyosome has been sectioned at right angles to its
cisternae, and another one almost parallel to its
cisternae. Note the peripheral net around this
second dictyosome. The two other dictyosomes
have been sectioned obliquely to the cisternae and
some material appears to be passing from them to
the vacuole, ×15,000. (Courtesy of E. G. Cutter.)

store large amounts of starch and may be called
amyloplasts (Fig. 4.10C).

As fruits ripen on trees and leaves prepare to fall
in the end of summer, they change from green to
red, orange, or yellow. This occurs because there is
a destruction of chlorophyll, accompanied by an
accumulation of yellow or red pigments known as
the **carotenoids.** The plastids with a dominance of
the red and yellow pigments are **chromoplasts.** A
chromoplast from red pepper is shown in Fig. 4.10D.
Plants grown in solutions deficient in mineral nutrients show characteristic changes in plastid structure.

Carotenoids.   As just noted, there are two categories
of plastid pigments, the chlorophylls and the carotenoids. The function of the chlorophylls is to trap
light energy. The distribution of three of the chlorophylls (*a, b, c*) has already been considered. There
are many types of carotenoids, and specific groupings may be constant throughout different divisions
or genera. Their chemistry is briefly discussed in
Chapter 5. It is thought that they may be active in
photosynthesis through the trapping of various
wavelengths of light and passing the energy on to
chlorophyll.

Cytoplasmic Inheritance.   Both mitochondria (Fig.
4.8C) and chloroplasts (Fig. 4.3) have a supply of
DNA and RNA. Does this mean that they have their
own store of genetic information and that they are
able to synthesize their own proteins? The case is
not so clear for mitochondria, but evidence to be
presented in Chapter 17 (on inheritance) definitely
demonstrates that chloroplast traits may be passed
on from one generation to the next independently
of the nucleus.

Dictyosomes.   Dictyosomes are composed of from
5 to 15 circular flattened vesicles, or **cisternae,**
aligned in stacks (Figs. 4.11, 4.18). In cross-sections,
many small vesicles appear at the margins of the
cisternae. However, face views reveal that the margins of each cisterna form a coarse net with true
vesicles only at the extreme outer regions (Fig. 4.11).
The peripheral vesicles, in most cases, contain
granules of secretory material elaborated in the dictyosome cisternae, possibly in collaboration with the
endoplasmic reticulum. There is considerable evidence that the precursors of cell wall material are
synthesized in the dictyosomes. The stack of cisternae appears to be polarized; that is, there is a
forming face and a disappearing face. The latter is
the concave face which is somehow used up in the
formation of the vesicles.

There are many dictyosomes in plant cells, and
collectively they may be referred to as the **golgi
apparatus,** or golgi zone. This name is derived from
the Italian neurologist, C. Golgi, who first described
them in nerve cells in 1898. In animal cells, the golgi
apparatus is generally associated with other cell
constituents to form a complex region active in the
elaboration of secretory products.

The dictyosomes are about $2\,\mu$ in diameter and
about $0.5\,\mu$ in width. They are thus large enough to
be seen with the light microsope. They probably
were seen in 1925 by Robert Bowen, who called
them osmiophilic platelets (Fig. 4.2B) because they
combined with OsO_4. He was sure that they were
the plant counterpart of animal dictyosomes.

Microtubules.   There remains a membranous system
which is not closed to the ground cytoplasm: the
microtubules (Fig. 4.12). Microtubules are of indefinite length, are 28 nm in diameter, and are provided
with a pore 8 nm in diameter. The function of the
microtubules is not definitely known. They occur in
the spindle during mitosis (see Fig. 6.5). They also
occur in the cilia, or flagellae, of motile cells of all
eukaryotes. A function of motility has been ascribed
to them. In some conducting cells they are wavy, and
it is thought that they may be active in moving
material through conducting cells. They may have a
structural function, and they may conduct stimuli.
In many elongating cells, the microtubules are par-

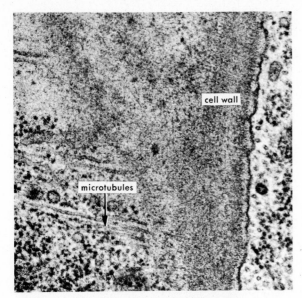

cell wall

microtubules

Figure 4.12 Microtubules. A cell from a root of *Juniperus*. Note the microtubules at right angles to the cell wall, ×50,000. (Courtesy of M. C. Ledbetter and K. R. Porter, *J. Cell Biol.* **19**, 139.)

allel with each other and lie at right angles to the long dimension of the cell wall (Fig. 4.12).

Microbodies. Preparations of a number of plant and animal tissues show a variety of spherical organelles that vary from about mitochondrial size to very much smaller. These organelles can be distinguished morphologically from mitochondria, since they have a single outer membrane rather than a double membrane envelope that is characteristic of the mitochondria. Cisternae and cristae are absent from microbodies, and the central area frequently appears rather dense under the electron microscope and may contain a variety of crystals (Figs. 4.8*B*, 4.13, 4.18).

Microbodies not only are morphologically different from other cellular structures, but also can be separated from them by centrifuging a mixture of ground tissue containing all organelles on a density gradient of sucrose solution.

Considerable research is being carried out at the present time to determine the exact chemical and physiological significance of these organelles. Evidence is accumulating that many different enzymes may be found in single membrane-bound spherical organelles in both plant and animal cells. To some extent, the types of enzymes found depend upon the tissue and kind of cells being studied. Thus, it appears that there is not just one kind of microbody but that the term "microbody" as defined above includes a fairly large number of different kinds of organelles that contain different enzymes and perform different functions in the cell.

The beginning student can appreciate some of the

contradictions in terminology when he realizes that microbodies isolated from leaves and containing a complex of oxidative enzymes have been called peroxisomes; those isolated from castor bean seeds and containing other oxidative enzymes have been called glyoxysomes; while some microbodies observed in plant cells have been given the name sphaerosomes. Liver and various other animal tissues have yielded microbodies called lysosomes (high in hydrolytic enzymes) and peroxisomes (containing certain oxidative enzymes).

Crystals. Other particulates are found in the protoplasm. Among the more widespread of these are crystals, sometimes of protein. They have been observed in the nucleus, cytoplasm, mitochondria,

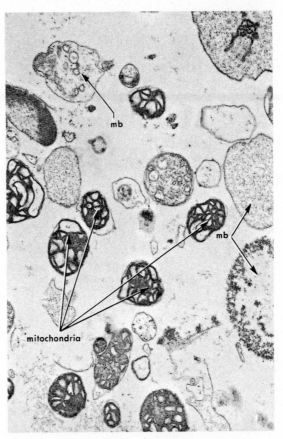

mb

mb

mitochondria

Figure 4.13 Microbodies. Particulates from squash cotyledons, as well as other plant and animal parts, may be isolated and concentrated so that their ultrastructure may be studied with the electron microscope. Mitochondria are easily recognized. Also observed are a number of other particulates, bounded by a single membrane and varying in structural details. These are microbodies, designated by **mb** on the figure. The single membranes of some have been ruptured by the processing techniques, ×5000. (Courtesy of John Lott.)

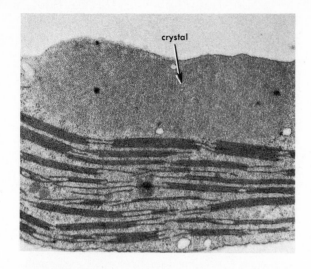

crystal

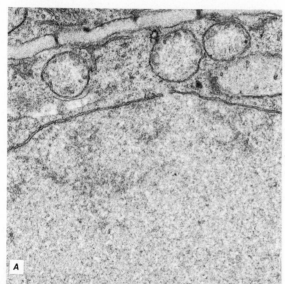

A

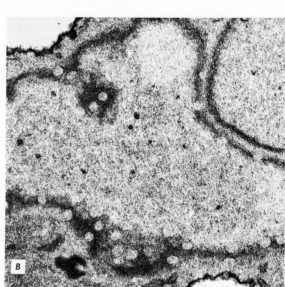

B

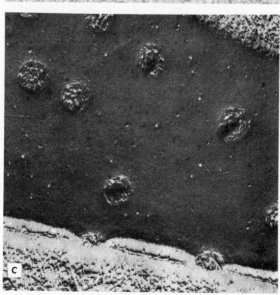

C

Figure 4.14 Left: Crystals within the cytoplasm. Crystals, probably of protein, in a chloroplast isolated from a leaf of bean (*Vicia faba*). Similar ones are frequently seen, not only in chloroplasts, but also in mitochondria and in the cytoplasm, ×60,000. (Courtesy of L. K. Shumway, *Protoplasma* **76**, 182. Copyright Springer-Verlag.)

Figure 4.15 The nuclear envelope. *A*, a section of a nucleus from a meristematic cell of the shoot apex of *Chenopodium album*. The section is at right angles to the envelope, which is distinctly double and shows definite interruptions. *B*, a section parallel to the nuclear envelope from a cell in a squash cotyledon. The pores are very distinct, ×20,000. *C*, the surface of the nuclear envelope as it appears in a freeze-etch preparation, ×30,000. (*A*, courtesy of E. M. Gifford, Jr., and K. Stewart; *B*, courtesy of J. Lott; *C*, courtesy of D. Branton.)

and chloroplast. In isolated chloroplasts (Fig. 4.14), they are generally noted when the isolation medium has a high molar concentration.

Nucleus

Structure. In general, the complex details of nuclear structure and organization have been worked out with the light microscope, using fixatives of high acidity (lower than pH6). Such fixatives, which coagulate proteins, are of little use in studying cellular detail at a resolution level of 5 nm. Consequently, with but three exceptions—(a) the nuclear envelope (Fig. 4.15), (b) the nuclear **unit fibers** (Fig. 4.16B), and (c) a significant stage of meiosis (Chap. 17)—the electron microscope has yielded a paucity of information about the nucleus in contrast to the immense amount of information derived from electron microscope studies on the cytoplasm and cytoplasmic organelles. New information about the nucleus and nuclear activity has come mainly from two sources: (a) studies of the growth of organisms in which nuclear control of differentiation has been subjected to experimental techniques and (b) biochemical studies of the nucleus and of inheritance. These investigations have furnished us with an insight into nuclear activity and structure that is far beyond the details of the cytoplasmic organization we have just discussed.

Electron micrographs of nuclei show few details not already recognized by the light microscopist (Figs. 4.3, 4.18). There are one or more nucleoli and some denser areas embedded in a granular nucleoplasm. The denser areas may represent sections of chromosomes, but such a relationship has not yet been definitely demonstrated. High magnification sometimes suggests fibrils in sectioned nuclei (Fig. 4.16A), and fibrils are certainly present in isolated nuclei (Fig. 4.16B) which have been freeze-dried.

The nuclear envelope is double and is provided with pores, which are shown in cross-section in Fig. 4.15A. Fig. 4.15B shows a tangential cut through a small portion of the envelope of a nucleus in a cell of a squash cotyledon. The pores, in face view, are very evident. They are also evident in the surface of the nuclei when the freeze-etch technique is used for preparation (Fig. 4.18). Although the pores appear as distinct holes, and there is some evidence that rather large molecules can pass through them, not all workers are agreed that they do serve as passageways.

Figure 4.16 The nucleoplasm. *A*, portion of a section of a nucleus from a leaf of tobacco (*Nicotiana rustica*); fibers and aggregations of fibers are present. Under well-controlled processing procedures, the diameters of such fibers approximate 12.5 nm. *B*, fibers, about 22.5 nm in diameter, are a constant feature of the nucleoplasm obtained from isolated nuclei of barley (*Hordeum vulgare*) endosperm processed by the critical freezing point method. *C*, a view of a section of a nucleolus from the root tip of the water plant, *Hydrocharis*. Note the aggregation of polysome-like particles and the absence of a limiting membrane between the nucleolus and the nucleoplasm. All magnifications ×30,000. (*B*, courtesy of S. L. Wolfe, *J. Cell Biol.* **37**, 610; *C*, courtesy of E. G. Cutter.)

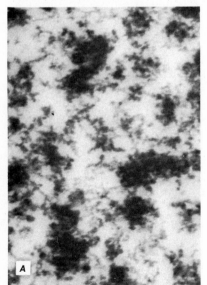
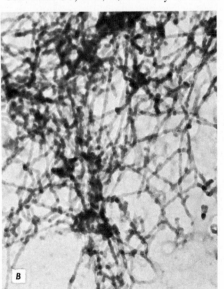
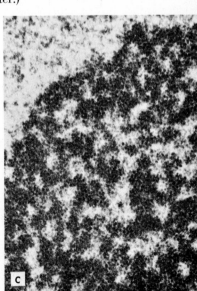

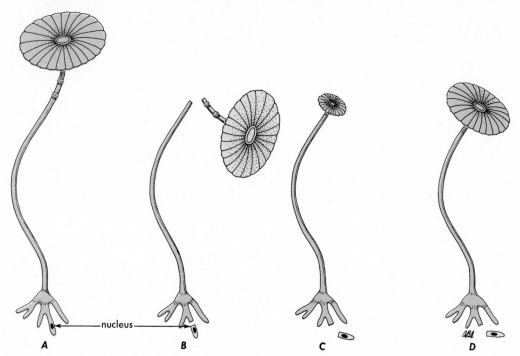

Figure 4.17 Regeneration of a new cap by an enucleated cell of the green alga *Acetabularia*. *A*, a cell of *Acetabularia*; note nucleus at tip of rhizoid. *B*, nucleus and cap have been cut off. *C*, new cap grows. *D*, growth of new cap completed; it may be smaller than the original cap.

Function. While it is well-established that genetic information is stored within, and distributed from, the nucleus, the details of how this occurs are not all clear. Cells deprived of nuclei may continue normal activities for as long as 3 months. The green alga, *Acetabularia* (Figs. 2.5E, 4.17) is a single-celled plant 1–1.5 in. tall. It shows considerable differentiation for a single cell. It has rhizoids for attachment, an upright stalk, and an expanded cap which differs from species to species. The nucleus is situated at the bottom of one of the rhizoidal branches (Fig. 4.17A). It may easily be removed by cutting off the tip of the rhizoid containing it (Fig. 4.17B). Such an enucleated *Acetabularia* plant has been kept alive in the light for 7 months, although *normal synthetic activities* ceased after 2 weeks. When the cap and a portion of the stalk are removed from an enucleated cell (Fig. 4.17B), a new cap will be regenerated. The shorter the piece of stalk that is removed, the better the regeneration. It appears that materials capable of inducing the formation of a new cap are released by the nucleus and accumulate in the upper portion of the stalk. Here they are readily available for the renewal of a cap, should the original cap be damaged or lost. These experiments would seem to indicate that the presence of a nucleus is not imme-

diately necessary for the normal day-to-day metabolic activities of *Acetabularia*. There must have been sufficient genetic information present in one form or another, in the cytoplasm, to provide for the immediate regeneration of the new cap. It can be shown in *Acetabularia* that this material is used up. There must be a continual renewal of the cytoplasmic informational materials. The nucleus is also necessary for cell division. We shall see later on (Chap. 6) that growth and differentiation do not occur without cell division. In the periods between cell divisions, the nucleus is actively preparing for a subsequent division.

This line of reasoning leads to an enigma—the lack of a nucleus in the food-conducting cells of the angiosperms, the sieve tube members of the phloem (p. 115). Mature and actively conducting angiosperm sieve tubes lack a nucleus. Such cells normally function for periods not longer than 3 years, but in palm trees they must function for 200 years! The only other enucleated fuctioning cell is the mammalian erythrocyte, and its life span is about 3 months.

Fig. 4.18 shows the freeze-etch appearance of many of the organelles we have just discussed. The surface view of the nuclear envelope with pores, both in face view and sectioned, is particularly striking. Three dictyosomes are shown: a face view, an oblique view, and a cross-section. The peripheral net in the face view is very distinct. Two vesicles are adjacent to the peripheral net and one appears to be associated with the endoplasmic reticulum. A microbody appears close to the sectioned dictyosome. A cell wall separates the cells, and the upper cell contains mitochondria and vacuoles.

Nonprotoplasmic Portions of the Cell

All the organelles and particulates so far discussed have been an integral part of the protoplasm. Energy has been expended within the protoplasm to maintain its state of homeostasis. The cell also contains regions outside the protoplasm which are maintained in a state of equilibrium just as are organelles. The vacuoles and crystals within the protoplast, as well as the cell walls bounding the exterior of the protoplast, are such extraprotoplasmic regions.

Vacuoles are definitely a part of the protoplast (Figs. 4.3, 4.19). We shall see later (Chap. 5) that vacuoles have a higher energy potential (solute concentration) than the solutions outside the cells and that energy expended by the protoplasm is required to maintain this higher energy potential.

Certain of the materials of the vacuole may be absorbed passively from solutions surrounding cells. This absorption requires no energy. Other substances, like inorganic salts, are actively accumulated; energy must be expended to take them into the vacuole and to retain them within the vacuole in a higher concentration than that outside the protoplast. Other materials, like sugars and pigments, are synthesized within the cytoplasm and discharged into the vacuole. These materials present an interesting problem in that they easily pass through the tonoplast to reach the vacuole. They also pass easily from one cell to another, but they do not pass out of the cell into external solutions.

In many young cells, the cytoplasm occupies much of the space in the cell. Small vacuoles are, however, present (Fig. 4.19A). As the cell grows larger, the small vacuoles within the cytoplasm increase in size, coalesce, and become fewer in number (Fig. 4.19B, C). Finally, when the cell has attained its mature size, only a few large vacuoles, or even only one, may remain (Fig. 4.19D). The protoplasmic contents (nucleus and cytoplasm) of the cell lie compressed against the cell wall. The nucleus may occupy a position near the center of the cell, in which event it is connected with the cytoplasm around the cell wall by strands of cytoplasm (Fig. 4.19D).

Vacuoles do not contain pure water, but rather a highly dilute solution of many substances. This aqueous solution in the cell is termed **cell sap.** Among the substances dissolved in the water, and thus constituents of cell sap, are the following: (a) atmospheric gases, including nitrogen, oxygen, and carbon dioxide; (b) inorganic salts, such as nitrates, sulfates, phosphates, and chlorides of potassium, sodium, calcium, iron, and magnesium; (c) organic acids, such as oxalic, citric, malic, tartaric; (d) salts of organic acids; (e) sugars, such as grape sugar (glucose) and cane sugar (sucrose); (f) water-soluble proteins, alkaloids, and certain pigments, such as anthocyanin. Generally, the cell sap is slightly acid. The concentration of cell sap varies from cell to cell, and it may vary in the same cell during the course of the cell's life.

The most common pigments of the vacuolar sap are the **anthocyanins.** These pigments are responsible for the red, purple, or blue of the petals of many flowers or of other parts of plants. The yellow coloration of poppy flowers is also due to yellow vacuolar anthocyanin-like pigments called anthoxanthins. The red color of roots and leaves of garden beets is due to another vacuolar pigment called betacyanin.

Inorganic Crystals. Cells with crystals can be found in almost all plants and in many different plant tissues. Crystals vary in chemical composition and in form (Fig. 4.20). The most common crystals are of calcium oxalate; it is generally held that they are an excretory product of the protoplast formed by the union of calcium and oxalic acid. This acid, a byproduct of certain activities of the protoplast, is soluble in cell sap and is toxic to the protoplasm if it attains a high concentration in the cell. By its union with calcium, the soluble oxalic acid is converted into the highly insoluble calcium oxalate, which will not injure the protoplasm. In addition to calcium oxalate, crystals of calcium sulfate or of protein sometimes occur.

The Cell Wall. The protoplast is surrounded by a plasmalemma or cytoplasmic membrane. Outside this membrane, and surrounding the entire protoplast, is a rigid wall secreted by the protoplast that it encloses. When, as is usual, protoplasts are separated from each other by walls, the walls are cemented together by an intercellular substance, the **middle lamella** (Fig. 4.21), which is characterized by pectates and certain other substances. The first wall formed by the protoplast is the **primary wall** (Fig. 4.21) and it is composed mainly of **cellulose.** Aging of the cell may bring about the deposition by the protoplast of more wall material, which is laid down on the primary wall. Thus, a **secondary wall** (Fig. 4.21)

Figure 4.18 Above and top right: Freeze-etch preparation of an onion root tip cell, showing the organelles just discussed as they appear when the cells are examined after freezing, rather than after chemical fixation. Bacteria and other cells, after a similar cold treatment, are viable. The similarity between the chemical fixation image and the freezing (living) image is striking. *A*, freeze-etch preparation, $\times 10,000$; *B*, diagram of *A*, with labels (ER represents endoplasmic reticulum), $\times 7400$. (*A*, courtesy of D. Branton.)

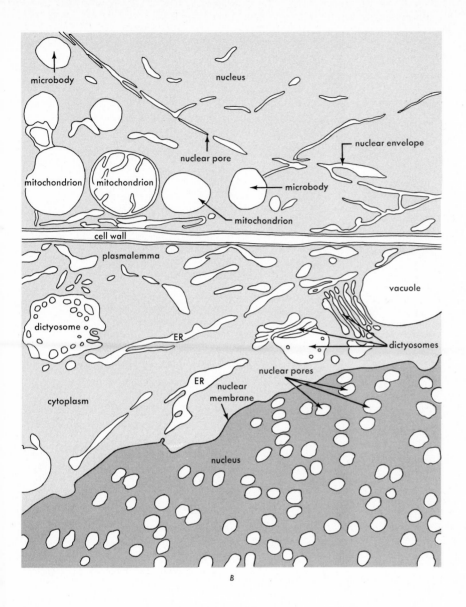

microbody

nucleus

mitochondrion

mitochondrion

nuclear pore

nuclear envelope

microbody

mitochondrion

cell wall

plasmalemma

vacuole

dictyosome

ER

dictyosomes

cytoplasm

ER

nuclear pores

nuclear membrane

nucleus

B

is formed, and the complete mature cell wall may finally come to have a thickness many times as great as the primary wall. In some tissues, the secondary wall is stratified and composed of several layers. In others, the cells do not lay down secondary wall material, in which event the common wall between two adjacent protoplasts is composed of the middle lamella, with primary wall material on each side. The secondary wall may be of cellulose or of cellulose impregnated with other substances. Some of these substances, notably **lignin,** lend resistance to decay to wood; others, like **suberin** and **cutin** (Figs. 4.21, 8.14, 10.9), are waxy and protect leaves and stems against water loss. In addition, certain other materials may enter into the composition of the cell wall—gums, tannins, minerals, pigments, proteins, fats, and oils. It should be emphasized that in mature hard tissues, such as wood, lignin may be deposited not only in the secondary wall but also in the primary wall and middle lamella.

Although the walls of cells vary considerably in composition in different species, and from one part to another in the same individual plant, **cellulose** constitutes the greatest percentage of the material of which most cell walls are made. It is synthesized by the protoplast.

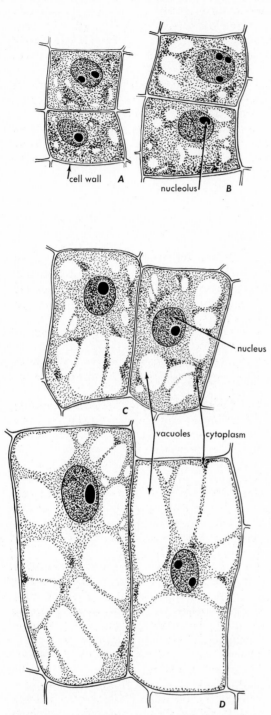

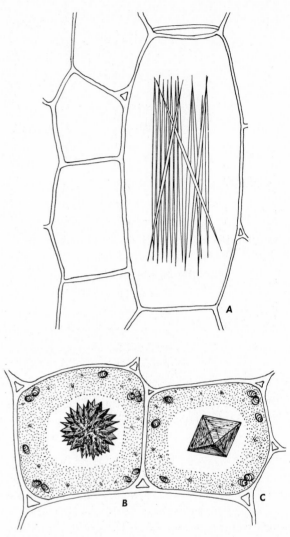

Figure 4.20 Types of inorganic crystals found in the vacuoles of living cells and in the cell walls of older nonliving cells. *A*, raphides; *B*, a cluster of crystals; *C*, a single crystal, ×2000.

Figure 4.19 Stages in the growth of a cell. Progressively older cells shown from *A* through *D*, ×2000.

Other than cellulose, the commonest and most widely found substance in the walls of plant cells is **lignin,** a material that adds to their hardness.

Suberin is a waxy substance associated with cellulose in the walls, particularly of cork cells. Since neither water nor gases can pass through suberin, cork tissue is an excellent protection against excessive loss of water.

Cutin is a fatty substance usually found as an external coating on the outer cellulose wall of the epidermal cells of leaves and stems.

Other substances, besides lignin, suberin, and cutin, that may be associated with cellulose in the walls of cells will be discussed at other points in this text.

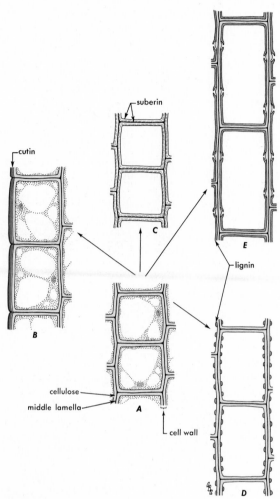

Figure 4.21 Development of cell walls during primary growth while stem is elongating. *A,* cell with cellulose walls. *B,* cutin may be laid down on the outside of an epidermal cell. *C,* in another location, suberin may be deposited within the cellulose primary cell wall to form a cork cell. Lignin is deposited within the primary wall and between the primary cell wall and the protoplast. *D,* during rapid elongation, it is laid down as rings or as a spiral band. *E,* when elongation stops (except for pits), the lignin forms a complete box—the secondary wall—about the cell. Each specialized cell is derived from its own antecedent meristematic cell and never gives rise to another specialized cell.

Summary

1. The cell may be considered the unit of life because it is the smallest unit known to be able to replicate itself, and to develop in an orderly fashion from a simple beginning to a very specialized end.

2. Cells may be divided into protoplast and cell wall.

3. Protoplasm is divided into many compartments by an array of particles and membranes or organelles.

4. The plasmalemma is the membrane separating the protoplast from the cell wall, and the tonoplast is the membrane separating the vacuole from the cytoplasm.

5. Ribosomes are small particulates having a high RNA content.

6. The endoplasmic reticulum consists of an array of extensive flattened vesicles. It may or may not have ribosomes associated with the outer surface of the vesicles. It appears to be involved in protein synthesis.

7. Mitochondria are small organelles about $0.5\,\mu$ in width and from 1 to $3\,\mu$ in length. They are bounded by a double envelope. The inner component invaginates to form cristae which are surrounded by a homogeneous matrix. Respiration is localized in the mitochondria.

8. Chloroplasts are approximately $5 \times 10\,\mu$ in size. They are bounded by a double envelope. The internal chloroplast lamellae aggregate to form cylindrical grana which are connected by intergranal lamellae. The photochemical reactions of photosynthesis are membrane-bound; the enzymatic reactions of photosynthesis are located in the stroma.

9. In addition to chloroplasts, cells may contain leucoplasts, amyloplasts, proplastids, and chromoplasts.

10. Chloroplasts contain fibrils of DNA and give evidence of having an inheritance pattern that is independent of nuclear control.

11. Dictyosomes are stacks of from 3 to 10 flattened cisternae. Each cisterna is surrounded by a peripheral net. Dictyosomes appear to be involved in the synthesis of various cellular products.

12. Microtubules are of indefinite length; they are about 28 nm in diameter, and have an internal core about 8 nm in diameter. Many functions are ascribed to them.

13. Microbodies are all bounded by a single membrane. They are variable in size and in morphology. They are closely associated with various types of intracellular enzyme activities.

14. Crystals, probably largely of protein, occur within the cytoplasm and cytoplasmic organelles. Their function is not clear.

15. The nucleus is bounded by a double envelope provided with pores. The nucleoplasm appears to be characterized by a closely packed array of unit fibers about 22.5 nm in diameter and of indefinite length. While cells may live and even differentiate for a short time without a nucleus, a nucleus is required for the continued life of a cell and for cell division.

16. Vacuoles are aqueous solutions within the protoplast, separated from the cytoplasm by the tonoplast. Inorganic crystals are located in vacuoles.

17. The cell wall, which bounds the protoplast, is formed of cellulose fibrils embedded in an amorphous matrix. Other compounds (suberin, pectin, cutin, and lignin) may become a part of it.

18. The type of cell just summarized is highly compartmentalized. It is known as a eukaryotic cell. More primitive cells are not compartmentalized; DNA photosynthetic processes, and respiratory activity all share a common cytoplasm. These are prokaryotic cells.

5

Chemistry and Physiology of the Cell

Chemistry of the Cytoplasm

Plants receive radiant energy from the sun. They convert it into chemical energy in the form of food, from which it is released to be used in the multiple activities of the living cell. These cellular activities—growth, duplication, and differentiation—all occur in the protoplasm of the cell. The astonishing properties of the protoplasm in which all this occurs are inherent in its structure. The structure of protoplasm may be studied morphologically under microscopes or chemically in test tubes. Significantly, since the advent of the electron microscope, the distinction between chemistry and morphology is becoming less and less clear. In other words, the chemical elements (Table 5.1) group together in a little-understood manner to give the visible structure of protoplasm and the intricate seriation of energy transformations that occur within it. Therefore, in discussing the activities of protoplasm, we must first take into consideration the arrangements of atoms and of molecules that are at the bases of its chemistry and morphology.

We have within the living cell a wide variety of inorganic molecules, as well as carbon-containing molecules, making up the cell wall, protoplasm, and vacuolar materials. Although a discussion of all these compounds would involve a whole course in plant chemistry, we should have a general understanding of the diversity of these chemical compounds in the cell and of the major roles that the more important compounds play in the life of the cell. In general, we may divide the chemicals in a plant into two groups: (a) inorganic (substances taken into the cell from the outside) and (b) organic (carbon-containing compounds synthesized by the living cell from inorganic raw materials).

Inorganic Ions and Molecules in the Cell

The most abundant inorganic compound in the cell is water, which permeates the cell wall, makes up more than 90% of the fresh weight of protoplasm, and is the chief constituent of the cell vacuole. Many inorganic mineral salts are essential for the growth of plant cells (Chap. 11) and are obtained by the plant from its external environment, particularly from the soil or water in which it is growing. These inorganic salts are found in the cell generally in low amounts. They serve as raw materials which, together with carbon dioxide and water, are used to synthesize the many complex organic materials in the cell. Some inorganic materials in the cell, such as potassium, may accumulate to relatively high levels as inorganic **ions** (charged atoms or groups of atoms).

Organic Compounds in the Cell

Most of the dry weight of a plant cell is organic matter synthesized from inorganic raw materials by a cell itself or by other cells in the plant and moved to that cell. A wide variety of organic compounds from relatively simple sugars to extremely complex proteins occurs in the cell. Some of these molecules are made up of hundreds of atoms. Organic compounds make up the food for the cell, the structural components of the wall and the protoplasm, and many of the more special compounds such as hormones, pigments, and enzymes. We will look at some of the principal types of organic compounds found in the cell.

Carbohydrates. Carbohydrates are organic compounds containing carbon, hydrogen, and oxygen atoms with the general formula $(CH_2O)_n$, where n may be any number. They make up the major part of the cell wall and are, in addition, one of the three general types of food. The major structural carbohydrate in the wall is cellulose, whereas the principal carbohydrate foods are sugars and starches. A **food** is any organic material that serves the plant as a source of energy and as a source of carbon from which new compounds may be made. A food also supplies animals with the same source of energy and carbon.

Sugars. Let us consider briefly the structure of a sugar molecule. In one group of simple sugars, the hexoses, each molecule contains six carbon atoms and has the general formula $C_6H_{12}O_6$. For example, glucose has the straight-chain structure shown in Fig. 5.1. The –CH_2OH end is an **alcohol** group; the –CHO end is an **aldehyde** group. The sugar molecule may occur in two forms; when not in solution the carbon atoms form a straight chain (Fig. 5.1). When the sugar is in solution, four or five of the carbon atoms (depending on the kind of sugar) and an oxygen atom form a closed ring (Fig. 5.2). In the straight-chain form, note the difference in the end carbon atom of glucose and fructose. As noted, the –CHO of glucose is an aldehyde. The –C=O of fructose characterizes a **ketone**. Aldehydes are of considerable importance in plant metabolism, and we shall meet with them in other topics; i.e., nucleic acid and lignin.

It is obvious that the –OH and –H groups of the sugar molecule may be arranged in different positions in the ring without changing the relative numbers of carbon, hydrogen, and oxygen in the formula. In fact, a shifting of these atoms, as in the examples of glucose and fructose, results in sugars of different properties. Actually, sixteen different hexoses are possible and all sixteen are known, although only a few occur naturally in plants. Molecules such as

Table 5.1   Some of the Common Chemical Elements Found in Plants

Element	Symbol	Approximate atomic weight	Common valence numbers	Common ions in which elements are found
Elements Essential for Plant Growth				
Carbon	C	12	$-4, +4$	CO_3^{2-}, HCO_3^{2-}
Hydrogen	H	1	$+1$	H^+, H_3O^+, OH^-
Oxygen	O	16	-2	OH^-, NO_3^-
Phosphorus	P	31	-5	$H_2PO_4^-$, HPO_4^{2-}, PO_4^{3-}
Potassium	K	39.1	$+1$	K^+
Nitrogen	N	14	$-3, +5$	NH_4^+, NO_3^-
Sulfur	S	32.1	$+6, -2$	SO_4^{2-}
Calcium	Ca	40.1	$+2$	Ca^{2+}
Iron	Fe	55.9	$+2, +3$	Fe^{2+}, Fe^{3+}
Magnesium	Mg	24.3	$+2$	Mg^{2+}
Boron	B	10.8	$+3$	BO_3^{3-}
Zinc	Zn	65.4	$+2$	Zn^{2+}
Manganese	Mn	54.9	$+2, +3$	Mn^{2+}
Molybdenum	Mo	96	$+3, +5$	Mo^{3+}, MoO_2^+
Chlorine	Cl	35.5	-1	Cl^-
Copper	Cu	63.6	$+1, +2$	Cu^+, Cu^{2+}
Some Other Elements				
Sodium	Na	23	$+1$	Na^+
Silicon	Si	28.1	$+4$	SiO_3^{2-}
Aluminum	Al	27	$+3$	Al^{3+}

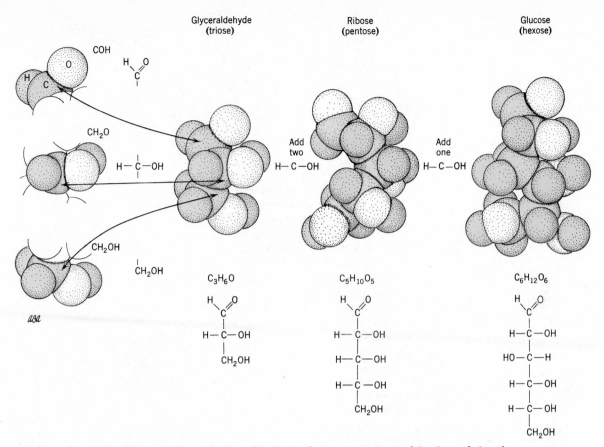

Glyceraldehyde
(triose)

Ribose
(pentose)

Glucose
(hexose)

COH

$$H-C=O$$

$$CH_2O$$

$$H-C-OH$$

Add
two

$$H-C-OH$$

Add
one

$$H-C-OH$$

$$CH_2OH$$

$$CH_2OH$$

C_3H_6O

$C_5H_{10}O_5$

$C_6H_{12}O_6$

Glyceraldehyde:
$$H-C=O$$
$$H-C-OH$$
$$CH_2OH$$

Ribose:
$$H-C=O$$
$$H-C-OH$$
$$H-C-OH$$
$$H-C-OH$$
$$CH_2OH$$

Glucose:
$$H-C=O$$
$$H-C-OH$$
$$HO-C-H$$
$$H-C-OH$$
$$H-C-OH$$
$$CH_2OH$$

Figure 5.1   Structural formulas and three-dimensional representations of 3-, 5-, and 6-carbon sugars.

these sugars with the same chemical composition but with different chemical properties are called **isomers.** The most common hexoses in plants are glucose and fructose. Sugars with three carbon atoms are called trioses; with four carbon atoms, tetroses; with five and seven carbon atoms, pentoses and heptuloses. **Ribose,** a constituent of nucleic acids and other complex protoplasmic molecules, is a pentose sugar (Fig. 5.1).

Two molecules of simple sugars, **monosaccharides,** may combine with each other. Thus, if two molecules of glucose are united, with the loss of a water molecule, a **disaccharide** is formed. One such disaccharide is known as **maltose.** The union of the monosaccharides, fructose and glucose, results in the commonest of all sugars, **sucrose** (Fig. 5.3). Sucrose can easily be split into fructose and glucose. One water molecule is required to split a molecule of sucrose and the process is called **hydrolysis.** While many different kinds of disaccharides are possible, sucrose, our common table sugar, is by far the most abundant sugar in plants, where it serves as a soluble stored or movable food. Maltose occurs in a free state only to a limited extent, but may be readily obtained by the hydrolysis of starch.

Polysaccharides.   Three or more monosaccharide molecules may join to form tri-, tetra-, or **polysaccharides.** The latter are composed of the union of many simple sugar molecules with the loss of a water molecule for each pair of simple sugar molecules united. Polysaccharides are not generally soluble in water, nor are they sweet. **Starch** and **cellulose** are the two most abundant polysaccharides in the plant. Each is composed of a long chain of many glucose molecules. In starch, the chain may be coiled because of the way the glucose units are linked together and some chains are branched, while in cellulose the chains are unbranched and more or less straight. Cellulose is a major structural material in the wall, while starch is a reserve water-insoluble food that is stored in many cells. Some cells, instead of storing starch, store a polysaccharide called **inulin** that is composed of fructose units instead of glucose units.

The union of relatively simple molecules, like sugar, into long-chain gigantic molecules composed of the repetition of simple units is a common chemical process known as **polymerization.** We shall meet it again in our discussion of proteins and nucleic acids.

Chemistry of the Cytoplasm   **55**

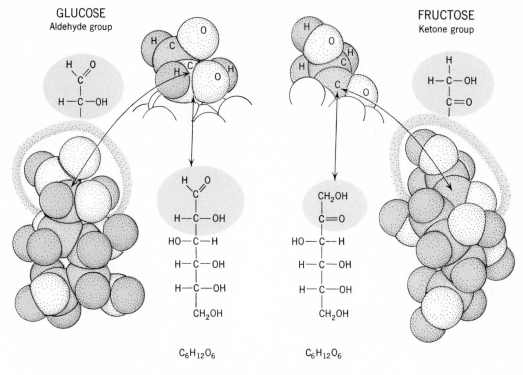

A   Straight-Chain Structures

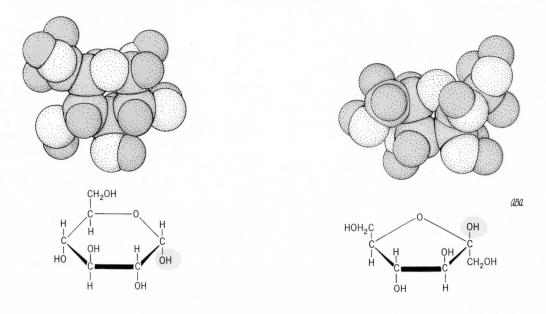

B   Ring Structures

Figure 5.2 Structural formulas and three-dimensional representations of glucose and fructose.

Glucose + Fructose ⟶ Sucrose + Water

$$C_6H_{12}O_6 + C_6H_{12}O_6 \longrightarrow C_{12}H_{22}O_{11} + H_2O$$

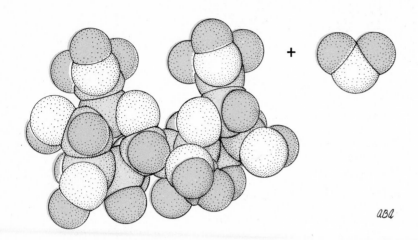

+

Figure 5.3 Formation, structural formula, and three-dimensional representation of sucrose.

The formula of cellulose is given in Fig. 5.4.

Lignin. A cotton fiber is almost pure cellulose. The secondary walls of various wood cells, however, are permeated by a second substance: **lignin** (Fig. 4.21). Lignin reduces infection, rot, and decay. It also is responsible for some of the strength of wood. Between 20 and 30% of the cell walls in wood is lignin. The unit structure in lignin is not a sugar but a complex alcohol. It does not polymerize in straight chains as glucose does in cellulose, but, instead forms a firm net. Lignin is thus a complex polymer with many slight variations in its molecular structure. Some plant taxa may contain only a specific type or variant of lignin. For example, oxidation of lignin from monocotyledonous wood yields, among other things, three different types of aldehydes; lignin from dicotyledons yields two of these types; and lignin from gymnosperm wood yields but one of these. The lignin in these three plant groups must therefore have different structures. Primitive vascular plants contain relatively little lignin, and oxidative

treatment suggests that it has a simple chemical structure. Lignin is among the most chemically inert of plant substances and it remains in fossils of woody stems (see Chap. 30).

Lipids. All living plant cells contain **lipids** (fats and fat-like substances) in their cytoplasm, and in some cells, fatty compounds are present in the cell walls. These fats serve as food. Phospholipids (lipids containing phosphorus) and other complex lipids are

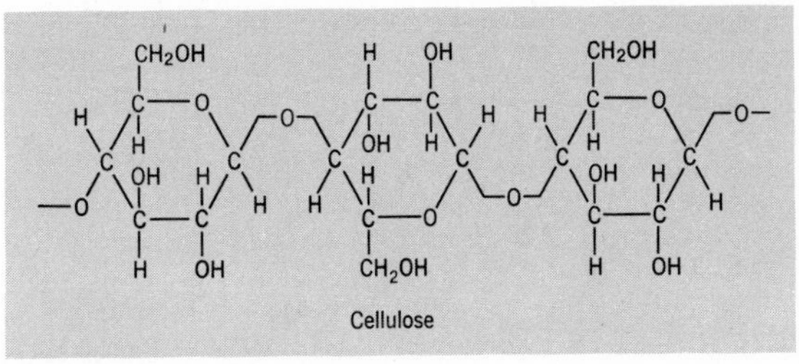

Cellulose

Figure 5.4 Structural formula of glucose units in cellulose.

Chemistry of the Cytoplasm   57

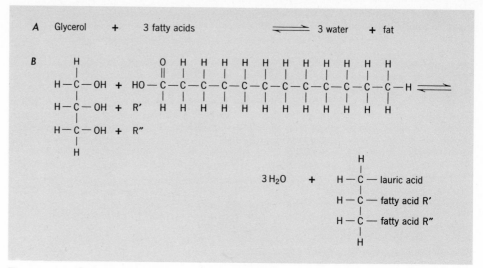

Figure 5.5 Structural formula for a fatty acid (lauric acid), glycerol, and a fat. *A*, in words; *B*, structural formula.

found in protoplasmic membranes. **Cutin** and **suberin** are lipid materials that function in protecting the plant against excessive water loss. One common property of lipids is their relatively poor solubility in water.

Fats, the most abundant and simplest of plant lipids, are composed of fatty acids united with the three-carbon-atom alcohol, **glycerol.** Fatty acids have very little oxygen and are generally relatively long chains of carbon with only hydrogen atoms as side attachments. In lauric acid, the fatty acid shown in Figure 5.5, the carbon chain is composed of twelve carbon atoms.

Glycerol is a rather simple three-carbon-atom compound. Its formula is $C_3H_8O_3$ and its structural configuration is shown in Fig. 5.5. When glycerol unites with three fatty acids, three molecules of water are lost and a fat is produced. Notice that in the fats there are only six oxygen atoms, while there may be many hydrogen atoms. This arrangement leaves many spots where oxygen may be introduced when fats are broken down in the cell. Fats are easily oxidized and as foods are very good sources of energy.

Proteins. One of the most important classes of compounds found in protoplasm is the **proteins.** Proteins are complex nitrogen-containing compounds of high molecular weight. Egg albumin and gelatin are good examples of proteins. While there are many kinds of plant proteins, they do not occur in such large masses in plants as they do in animals; hence, plant proteins are not so commonly known.

When proteins are taken apart, it is found that they are all formed of 20 different units, the **amino acids.**

Each amino acid has at least two linked carbon atoms: —C—C—. To the terminal carbon atom there are always attached (*a*) an oxygen atom, $=O$, and (*b*) an hydroxyl group, —OH. This terminal group

of atoms, $-C\overset{O}{\underset{OH}{\diagdown}}$, is known as a carboxyl group.

Its charge is negative, or acidic. To the second carbon there is always attached, at least, one hydrogen atom, —H, and an amino group, $-NH_2$. The charge on the amino group is positive, or basic. The terminal structure of an amino ($-NH_2$) acid (COOH) thus is

There remains one unsatisfied charge on the amino carbon. This is satisfied by a specific side chain which we may designate as R. There are 20 different side chains that occur in common amino acids; thus each amino acid is characterized by its own side chain. The simplest of these amino acids is glycine,

AMINO ACIDS

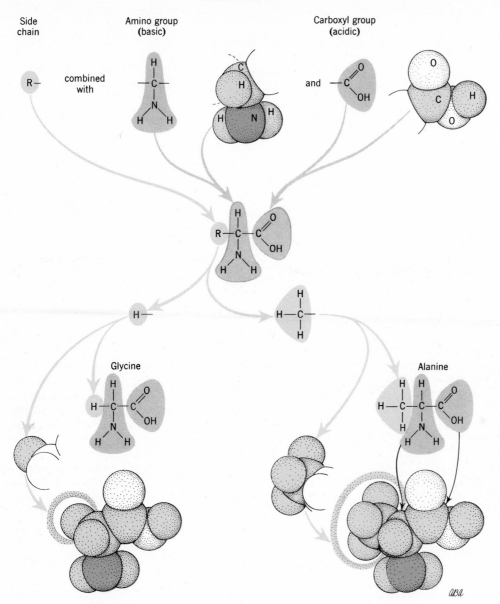

Figure 5.6 Structural formulas and three-dimensional representations of several amino acids.

whose side chain is a single hydrogen atom (Fig. 5.6). The structural formulas of three other amino acids, tryptophane, arginine, and cystine, are also shown in Fig. 5.7.

Polymerization of these amino acids takes place in the formation of protein, and a long chain is formed in which the side chains of the amino acids extend outwardly from the backbone of the molecule (Fig. 5.8).

Consider that each of the amino acids is a symbol so that we can form an alphabet with them of more than 20 characters. From the 26 letters of our alphabet, all of the countless words of all of the known languages can be formed. Furthermore, the great majority of these words have no more than five letters and very few of them have more than 20 letters. Even the simplest proteins have 1000 or more amino acids, and the common number of amino acids per protein is of the order of 100,000. The possible number of different kinds of proteins

Chemistry of the Cytoplasm

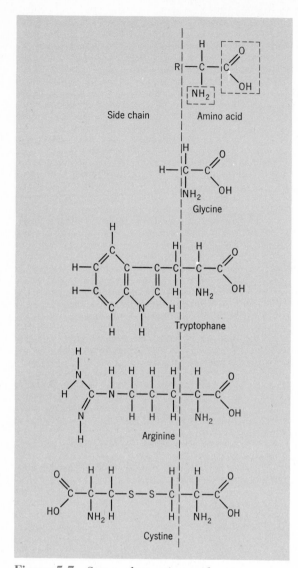

Figure 5.7 Some other amino acids.

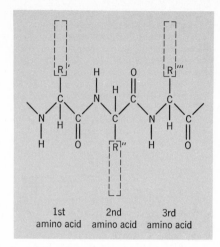

Figure 5.8 A polypeptide chain is a part of a protein molecule. The backbone of the protein molecule is formed by many amino acids joined by the union of the amino group (NH_2) of one amino acid to the acid group (COOH) of another amino acid by the removal of a water molecule. The R groups represent side chains of the different amino acids.

formed from these 20 different amino acids, taken in any order, is enormous. One of the most exciting fields of modern biology is in the determination of the exact sequence of amino acids in protein molecules.

Proteins are the basic building block of protoplasm. They account for its great activity and are the reason for our previous statement that the protoplasms of similar cells are always somewhat different.

Long protein molecules can fold, bend, and coil in many ways (Fig. 5.9). Particularly important is a folded or helical structure that many chains assume. In the cytoplasm of the cell, the long protein molecules tend to react with one another to form weak cross-linkages, and this increases the viscosity of the solution. The streaming of protoplasm in the cell must be viewed as involving the breaking and re-forming of many weak cross-linking bonds among the protein molecules. The primary valence bonds in the side chains or backbones of protein molecules remain intact under these conditions of streaming.

Enzymes. While there are many ways in which proteins play a part in the economy of the cell, let us consider for the moment only the important part they take in controlling the chemical reactions of the cell. As an example, we may look at one step in the energy flow in the cells of plants and animals, in which the energy stored in foods is released to perform useful work. Burning gasoline in a flask releases uncontrolled and therefore wasted energy. Burning gasoline in a car also releases energy, but the design of the motor controls the release so that useful work is performed. Still, much of the energy is lost as heat, as the temperature of the motor illustrates. Sugar is burned in the cells of your body and in the cells of a plant, again with the release of energy and the performance of useful work, although some energy is lost as heat. A primary function of certain specific proteins is to control and to direct chemical reactions in the cell, limiting them to very specific pathways and bringing them about at relatively low temperatures.

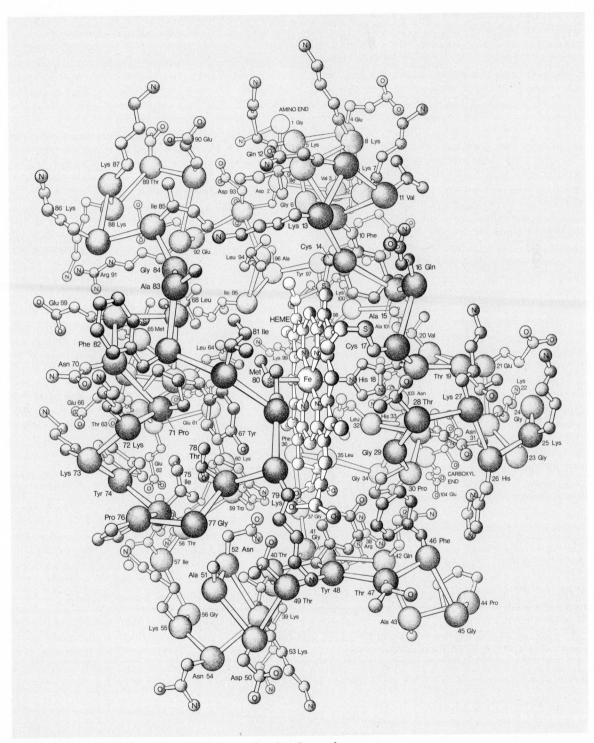

Figure 5.9 Proposed configuration of a molecule of cytochrome c, a relatively simple plant protein. (From *Scientific American*, April 1972. © Copyright 1972 by Richard E. Dickerson and Irving Geis.)

Substances that increase the rate of chemical change without themselves appearing to change in the reaction are **catalysts.** There are both *inorganic catalysts* and *organic catalysts.* Organic catalysts produced by living organisms are called **enzymes.** Enzymes are proteins. They are frequently very specific in their activity, catalyzing only one, or a few, chemical changes. Shape of the enzyme molecule plays an important role in both enzyme specificity and activity (Fig. 5.9). Shape specifically limits the type of combining material to only those molecules having a shape that complements that of the enzyme molecule. Thus, only highly selected molecules are brought into close proximity by combining with the enzyme molecule. Once they are closely associated, the combining molecules may react with each other, or with other molecules. At the completion of the reaction, the newly formed compound or compounds are released from the enzyme molecule. The enzyme molecule itself, being little if any changed during the reaction, is then ready for further activity. Thus, an enzyme greatly modifies the rate of a reaction without being much changed itself.

Enzymes have certain optimum conditions for their greatest activity. They may be destroyed by heat and by some metals and certain other substances. Many of the numerous steps involved in photosynthesis, respiration, and digestion are catalyzed by enzymes. Enzymes are active in practically all cellular processes.

Protoplasm may be thought of as a very complex colloidal system in which molecules of one sort are being constantly changed into molecules of another sort. Green plant cells store energy in one process and release it in another; both processes may go on within any green cell at the same time. Enzymes help to control the precision and the rate of the changes taking place in the protoplasm.

Coenzyme Activators. Some enzymes depend only on their protein structure for their activity, while others are unable to function in the absence of certain nonprotein substances called **cofactors.** In some instances, metal ions may serve as cofactors and either bind the substrate to the enzyme or actually serve as the catalytic group itself. One group of enzymes that have metal-containing active sites are the cytochromes (Fig. 5.9). We will see that cytochromes are very important in the oxidation-reduction reactions of photosynthesis and respiration. In some instances the nonprotein cofactors may be organic molecules called **coenzymes.** Coenzymes sometimes are bound tightly to specific enzymes, but in other cases the coenzyme may be very loosely bound to the enzyme. If the coenzyme is only loosely associated with the enzyme, it may actually diffuse away and take part in other enzyme reactions; that is, the same coenzyme may act in two different enzyme reactions. Thus, if the coenzyme is capable of accepting and then giving off hydrogen atoms (or electrons), it might serve as a carrier of hydrogen from one reaction to another.

Many **vitamins** serve as active groups of coenzymes. Thus, members of the vitamin B complex, riboflavine, thiamine, nicotinamide, and pantothenic acid all are found as active groups of certain coenzymes and in this way serve essential roles in the life of plants and animals. Four coenzymes that will be particularly important in our study of plant metabolism are listed in Table 5.2. You will note that the first three of these, NAD, NADP, and FMN, serve as carriers of hydrogen (electrons) between two different reactions in the cell. The fourth, CoA, acts as a carrier of an organic group containing two carbon atoms. It receives the group in one enzymatic reaction and may act as a coenzyme in another enzyme reaction where the two-carbon-atom group is used.

Pigments in the Cell. Pigments are colored compounds, and since they are colored they must somehow absorb certain wavelengths of light (see Fig.

Table 5.2 **Some Coenzymes Important in Metabolism and Having Vitamins as Reactive Groups**

Coenzyme	Abbreviation	Vitamin associated with activity	Reaction in which coenzyme functions
Nicotinamide adenine dinucleotide	NAD	Nicotinamide	Transfer of hydrogen (electrons)
Nicotinamide adenine dinucleotide phosphate	NADP	Nicotinamide	Transfer of hydrogen (electrons)
Flavine monocucleotide	FMN	Riboflavine	Transfer of hydrogen (electrons)
Coenzyme A	CoA	Pantothenic acid	Transfer of two-carbon-atom groups (acetyl groups)

13.8). Plant pigments may be divided for convenience into water-soluble pigments that are usually found in the vacuoles of the protoplasts, and lipid-soluble pigments that occur in the plastids within the cytoplasm of the protoplast. Vacuolar pigments (anthocyanins and anthoxanthins) have already been discussed on p. 47. Several kinds of other pigments occur in the plastids. The principal ones in the chloroplasts of higher plants are the green chlorophylls, chlorophyll a (Fig. 5.10) and chlorophyll b, and the yellow and red carotenoids, the carotenes and the xanthophylls.

We have noted that carotene apparently serves to trap the energy of certain wavelengths of light and to pass it on to the chlorophyll molecule. It also is an effective agent against photoxidation in the cell and a precursor of vitamin A. The carotenes and xanthophylls are responsible for much of the fall leaf coloring and the color of ripe fruits. This fact suggests that they have some functions in senescence.

There is an interesting group of water-soluble pigments important in the trapping of light in the blue-green and red algae. These are known as the phycobilins. They are able to trap light of low intensity and pass the trapped energy on to chlorophyll for the normal steps of photosynthesis. The phycobilins are responsible for some of the color of the blue-green and red algae (see Color Plate 12).

Chlorophyll has a molecular structure of moderate complexity. It consists of a head and a tail region. At the center of the head is a single atom of magnesium which is surrounded by four complex rings of carbon and nitrogen. The fatty tail is a long carbon chain flanked by atoms of hydrogen. The structural formula of chlorophyll a is shown in Fig. 5.10.

Other very important pigments occur in plant cells but are inconspicuous. One such pigment is phytochrome. It is found in extremely small amounts in leaves and is involved in a number of physiological processes including the flowering of plants. Other pigments that occur in minute amounts in cells are riboflavine, which is important in the bending response of plants to light, and the cytochromes (Figs. 5.5, 5.9), yellow pigments that are found in the cytoplasm and are concerned with certain steps in the processes of respiration and photosynthesis (p. 243).

Hormones. It is a common observation that if the tip of a plant is pinched or cut off, new shoots develop below the injured area. Why do the buds below the tips grow only after the tip is removed? Are they stimulated in some way by the injury? Does the presence of the tip prevent growth? It can be shown that a chemical produced in very small amounts in the growing shoot tip moves down the

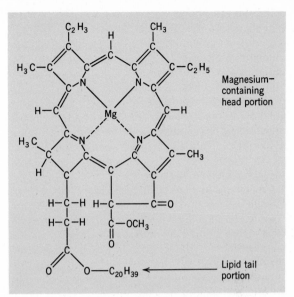

Figure 5.10 **Chlorophyll a molecule.**

stem and prevents the growth of the lateral buds. When the tip is removed, the inhibitory chemical decreases in concentration in the stem and the lateral buds begin to grow.

Thus, it can be demonstrated that plants, like animals, produce chemical messengers that move in very low concentrations from one part of the plant to another and bring about many important changes in growth and development. Such naturally produced growth regulators are called **hormones.** The hormone that is involved in the dominance of the apical bud over the lateral buds, as described above, belongs to a class of plant growth regulators called **auxins.** Auxins were the first plant hormones to be studied, isolated, and chemically identified. The most common natural auxin is a complex organic compound known as indole acetic acid or **IAA.** Indole acetic acid in very low concentrations has a wide range of effects on plant growth and development (Chap. 20) and is very important because it influences cell elongation.

The IAA molecule (Fig. 20.6A) is made up of two rings and an acid side chain. Although one of the rings contains nitrogen, other compounds that have a structure similar, in some respects, to the IAA molecule, but which lack nitrogen, are also effective growth regulators. Chemists are able to synthesize artificial growth regulators, such as naphthalene acetic acid, which in many ways act like IAA in affecting plant growth and development.

Soon after auxins were discovered, it became apparent that both plant growth and development were regulated not by one hormone but by several, and that a shift in the balance of hormones in a cell could profoundly affect the way in which the cell would develop and grow. Two other classes of hormones that have been studied extensively are **gibberellins** and **cytokinins.**

Gibberellins, like auxins, affect cell elongation (Chap. 20). However, gibberellins differ from auxins not only in their chemical nature but also in many of the other effects that they have on plant growth. A number of gibberellins have been identified in various plants and all have complex ring molecular structures (see Fig. 20.6B).

Cytokinins are particularly important in their influence on nuclear division; interestingly, they are chemically related to nucleic acid. The first cytokinins to be studied and identified was not a natural plant hormone, but a chemical that, almost by chance, was found to be effective in regulating the growth and development of cells in a tissue culture. When small amounts of this chemical were added to a tobacco tissue culture, cell division was stimulated. The chemical was called **kinetin.** A number of naturally occurring related compounds have been shown to be active in promoting cell division. Cytokinins have been shown to affect other growth phenomena such as cell elongation and development of buds.

In addition to the compounds that we have discussed, many other important substances are found in the cytoplasm, nucleus, and vacuoles of the cell. The chemistry of the nucleus in particular will be considered in more detail in Chapter 6.

The Physiology of the Cell

We have studied the structure and some of the chemistry of the individual plant cell and we have seen that most plants are composed of countless numbers of these cells. A knowledge of the structure of a cell, a tissue, or an organ is essential to an understanding of their activities or functions. An automobile is composed of several thousand parts, each of which has a particular composition, shape, and location upon which depends its special role in the proper operation of the machine. The human eye is a complex organ in which the many individual parts are "put together" in such a way that the function of sight is possible. Likewise, the cell is a most complex structure—a *living structure.* It is the structural and functional unit of the plant. It has work to do; that is, it has functions or activities. **Physiology** is the study of the functions or activities of cells or organisms.

Our study of the physiology of cells will be simplified if we consider a chlorophyll-containing plant such as a unicellular alga (Fig. 5.11, Color Plate 3)

floating freely in water. This one-celled plant carries on all the functions that are necessary to maintain its life. Within this microscopic bit of living substance a number of physical and chemical processes occur simultaneously. Each cell is (a) absorbing materials such as water, mineral salts, and gases from outside its own body, and simultaneously losing materials to the external environment; (b) building foods from the materials and light energy that it absorbs; (c) digesting foods (changing complex foods to a simple form); (d) respiring, thus releasing energy for various activities; (e) building protoplasm from the foods; (f) growing; and (g) producing new cells.

We shall briefly discuss each of these physiological processes. However, it is not possible to elaborate on several of them without a knowledge of the structure of the plant as a whole. Therefore, the following discussion merely defines some of these processes, reserving a more complete discussion for later chapters.

Since the life activities of all cells are carried out in a water environment and often involve the movement of solutes throughout the individual cell, between cells, and between the cell and its external environment, it is appropriate that we first consider how a cell obtains water and solutes. Why does water move into some cells and out of others? Can the cell regulate the flow of water through it? Does the cell select the solutes that it absorbs? How does the structure of the cell influence water or salt absorption? Is energy necessary for either process? The answers to these and similar questions lie in the ultimate relation between cellular structure and function.

Permeability of Plant Membranes

Diffusion. All molecules are in motion and may move from place to place. In liquids and gases, molecules move freely; they will move from a region of higher concentration to a region of lower concentration. This is **diffusion.** It is a very important basis for physiological activities in plants.

Permeable Membranes. Let us now consider the green plant floating in water. Bathing the cell on all sides is a dilute but exceedingly complex solution of molecules and ions that are constantly bombarding the outer surface of the cell, the cell wall. This wall, though apparently continuous when viewed through a microscope, is in reality quite porous, with **microcapillary spaces** existing between the interwoven cellulose **microfibrils** (Fig. 5.12). Many of these spaces are filled with substances such as pectates or lignin, but many are filled instead with liq-

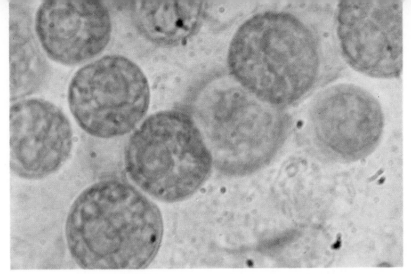

Figure 5.11   A free-floating algal cell.

Figure 5.15   Plasmolysis in the aquatic plant *Elodea*. *A*, living untreated cells; note the distribution of chloroplasts around sides, top, and bottom of the cell, with a vacuole occupying the rest of the cell. *B*, cells of *Elodea* are able to concentrate substances, such as the dye neutral red, in the vacuole. *C*, cells containing neutral red plasmolyzed with 10% $Ca(NO_3)_2$; reduction in size of vacuole indicates outward diffusion of water, but the dye is retained in the vacuole. *D*, in cells recently killed with formaldehyde, the protoplast may withdraw slightly from the cell wall, chloroplasts may form irregular clumps, and the nucleus frequently becomes visible. *E*, neutral red dye does not accumulate in dead cells, nor (*F*) do dead cells become plasmolyzed when placed in $Ca(NO_3)_2$.

A   *B*   *C*

D   *E*   *F*

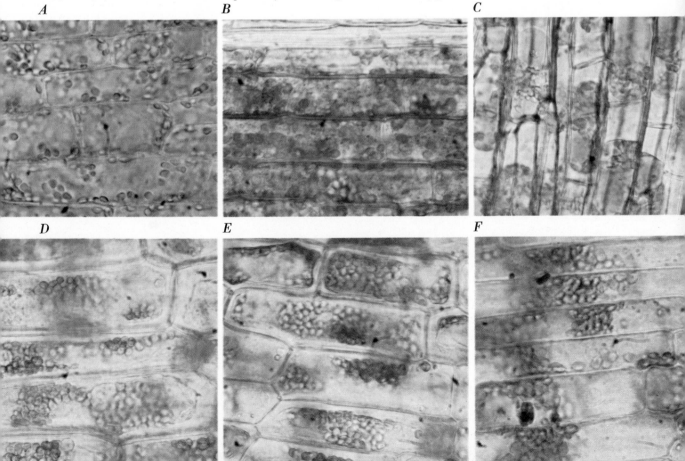

A

B

Figure 5.16 Osmosis in living and dead beet strips in water. *A*, in living cells, the membrane retains natural pigments within the cells. *B*, the pigments rapidly diffused from slices of beets placed in boiling water for 30 seconds, killing the cells.

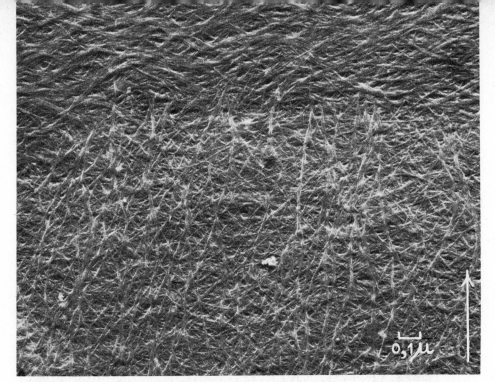

Figure 5.12 Electron micrograph of the cell wall of a cotton fiber, ×60,000. (Courtesy of P. A. Roelofsen.)

uid. There is a strong attraction between the insoluble cell wall materials and water. Consequently, water is found throughout the wall, forming a continuous pathway through which solute particles can freely diffuse. The process in which water is attracted to **(adsorbed on)** the surfaces of the cellulose microfibrils, causing them to move apart and the wall to swell, is called **imbibition.** Cellulose imbibes water and the cell wall thus allows the relatively free passage of water and dissolved materials. The cell wall is **permeable.** Cell walls found in the outer surface of higher plants may be impregnated with fatty materials such as cutin and suberin. These walls do not imbibe water, and are **impermeable** to water and dissolved substances, though they may be permeable to substances soluble in fat.

Differentially Permeable Membranes. Lying in intimate contact with the wall of a living cell is the outer cytoplasmic or plasma membrane, the **plasmalemma,** which is the first important barrier to the free passage of the molecules and ions that bombard the cell from the outside (Figs. 4.3, 4.5). This limiting surface of the protoplast is characterized by a different physical and chemical composition from the rest of the protoplasm. It is not a rigid, unchanging surface, but an ever-changing dynamic "guardian of life," for if it is destroyed, the cell dies. Although the exact nature of this membrane is not definitely established, it is thought to be a unit membrane (Figs. 4.5, 4.6), through which water can readily pass, but which restricts or prevents the passage of many

dissolved materials. It is **differentially permeable,** allowing some substances to pass freely, others to pass slowly, and still others to pass hardly at all. Although the cell cannot select those substances which are beneficial to it and prevent from entering those which are harmful or toxic, the presence of the cytoplasmic membrane does prevent the *free* diffusion of materials into and out of the cell.

The inner cytoplasmic membrane, the **tonoplast,** surrounding the vacuole, is also differentially permeable and is a unit membrane. It is more fatty in nature than the plasmalemma and consequently, certain substances diffuse through the cell wall and the plasmalemma into the cytoplasm but do not readily find their way into the vacuole. As we have already seen, the vacuolar sap frequently contains high concentrations of plant products (such as anthocyanins) that are not found in the cytoplasm. Destruction of the cytoplasmic membranes—by immersing the cell in alcohol, for instance—renders the membranes permeable, allowing the outward diffusion of vacuolar materials, and results in the death of the cell (Fig. 5.15, Color Plate 3).

Within the cell itself, solutes are not free to diffuse at random throughout the protoplast. In addition to the tonoplast which restricts the movement of materials between the cytoplasm and the vacuole, there are the nuclear and plastid membranes, which restrict the free diffusion of materials into and out of these bodies. Moreover, the colloidal nature of the cytoplasm itself presents a delicate but ever-changing architecture within which the foods, vita-

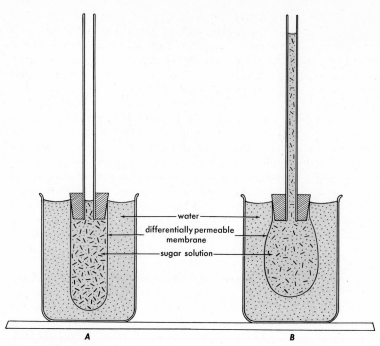

water

differentially permeable
membrane

sugar solution

A   B

Figure 5.13 Diagram of
osmometer and diffusion through
a differentially permeable
membrane. *A*, before osmosis has
occurred; *B*, after osmosis.

mins, hormones, enzymes, and inorganic ions are so distributed that the intricate yet precise processes of life proceed.

Diffusion through Differentially Permeable Membranes

Nonliving Membranes. The differentially permeable membranes of the cell play very important roles in the maintenance of life activities. Certain features of their behavior can best be explained by reference to the simple apparatus called an **osmometer** (Fig. 5.13). The cylindrical sac of an osmometer is made of parchment paper, plastic, or other material that is differentially permeable. This sac is filled with a concentrated sugar solution and stoppered with a tight-fitting rubber stopper through which a glass tube is fitted; the sac is then immersed in distilled water (pure solvent). In following this discussion we should keep in mind that the sap of a living algal cell is usually more concentrated in total solutes (salts, sugars, and the like) than is the pond water in which it may be floating. Movement of water into the plant cell obeys the same laws as those governing movement of water into an osmometer.

Turgor Pressure. Outside the osmometer are water molecules, and inside are water molecules and sugar molecules. The membrane, being differentially permeable, permits the free movement of water molecules inward or outward and retards or prevents the free movement of sugar molecules outward. After a

short time the liquid rises in the tube, and the sac becomes distended, or **turgid.** The pressure of the solution against the sac, the **turgor pressure,** is a measurable pressure opposed by an equal and opposite pressure of the wall of the sac against its contents (the **wall pressure**). Evidently, water molecules are moving inward more rapidly than outward. Under these conditions the **activity (chemical potential)** or "free energy" of water molecules outside the sac is greater than the activity or "free energy" of water molecules inside the sac; thus, water is driven into the sac. As water continues to diffuse into the sac, turgor pressure is built up inside it. This pressure increases the activity of water molecules in the sac, and the tendency for water to diffuse back out of the sac gradually increases.

This outward diffusion of water finally equals the inward diffusion. At this point, there is no further net movement of water into the sac, and the system is at equilibrium. Water molecules are passing through the membrane in both directions at the same rate. The activities of water on both sides of the sac are now equal. An important conclusion to be drawn from this demonstration is that (a) the activity of water is reduced by the presence of solute particles that reduce the concentration of the water, and (b) the activity of water is increased by the development of a physical turgor pressure in the solution. These two factors, solutes and turgor pressure, thus have opposite effects on the activity of water. In our example, equilibrium was attained when these two factors balanced each other and water within the sac had the same activity as the pure water outside.

The absolute value for the activity or chemical potential of water in a cell or in a solution is not easily measured. *Differences* in activity, however, can be determined, and pure water is the standard reference by which these differences can be conveniently measured. This difference between the activity of water molecules in pure water at atmospheric pressure (i.e., in an open container) and the activity of water molecules in any other system (i.e., water in a cell, in a solution, or in the sac) may be called the **water potential** of the cell or solution. Pure water has a water potential of zero under standard conditions of pressure and temperature. Since pure water is used as a standard and the activity of water in a cell or a solution is usually less than that of pure water, the water potential of a cell or a solution is usually a negative number.

The activity of water molecules, and thus the water potential in a system, is affected by several factors. The activity is decreased by the addition of solutes; it is increased by the physical pressure of cell walls on the aqueous cell contents and also by an increase in temperature. Water will diffuse from regions of high water potential to regions of lower water potential (i.e., from a dilute solution toward a more concentrated solution, when solute concentration is the only factor affecting the water potential). The water potential may be regarded as the driving force that causes a net movement of water in any system.

Osmosis. The diffusion of water (a solvent) through a differentially permeable membrane is called **osmosis.** Osmosis is not mysterious, but is simply a special example of diffusion in which the differentially permeable membrane of a sac, or of a cell, separates the internal water that has a given water potential from the water of the surrounding environment that has a different water potential. Osmosis occurs (i.e., water diffuses across the membrane) along the water potential gradient from a region of higher to a region of lower potential of water. The cell gains or loses water until the potentials of water on both sides of the membrane are equal.

Water movement into living cells is usually controlled by two major factors that affect the water potential and hence the direction of diffusion of water. These are (a) the presence of solute particles in the water and (b) the existence of a turgor pressure. Other factors, such as a change in temperature or the presence of water-attracting materials like some colloids, also affect the potential of water in a cell. As we have seen, the presence of solute particles (i.e., molecules or ions) lowers the activity of

the water in which they are dissolved. This effect on the water potential is proportional to the total number of dissolved particles (molecules and ions) in a given volume of water. Ten grams of sugar dissolved in 50 ml of water will lower the water potential approximately twice as much as will 5 grams of sugar dissolved in 50 ml of water.

This reduction of the activity of water caused by the presence of dissolved particles in it can be measured. If this solution is placed in an osmometer with pure water outside, osmosis will occur and turgor pressure will develop as long as there is any difference in the potential of water across the membrane. The theoretical maximum turgor pressure that could develop in this system equals the amount that the water potential in the solution is reduced by the solute particles in it.

Osmotic Pressure. The term **osmotic pressure,** sometimes called osmotic concentration or osmotic potential, is used to express these two properties of solutions, namely (a) the amount of reduction of the water potential in the solution caused by the presence of dissolved particles, and (b) the potential maximum turgor pressure that could develop as a result of osmosis. The osmotic pressure of a solution is usually expressed in units of atmospheric pressure. One atmosphere of pressure is approximately 15 lb per sq in. (Fig. 5.14). The osmotic pressure of a solution is not an actual physical pressure that exists in the solution. We might say that it is a power rating of the solution, just as an automobile engine has a

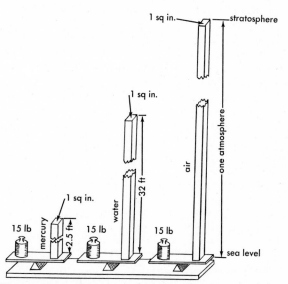

Figure 5.14 Diagram showing the relationship of one atmosphere pressure to the height of mercury and water columns that it will balance.

rating of 150 horsepower. We do not mean that when the motor is not operating it is doing the work of 150 horses. It is only when the proper conditions exist (i.e., when the motor is operating), that the ability to do work—the power—actually results in the work being done. A solution having an osmotic pressure of 5 atmospheres and standing in a beaker has the capacity to develop a turgor pressure of 5 atmospheres only when placed in an ideal osmometer with pure water outside. This turgor pressure will develop because the water potential of pure water is 5 atmospheres higher than the water potential in the solution. If, however, the solution were placed in an osmometer and immersed in another solution having an osmotic pressure of 3 atmospheres, osmosis would occur only until a turgor pressure of 2 atmospheres had developed in the osmometer, because in this case the water potential in the solution outside the sac would only be 2 atmospheres higher than the water potential inside the sac.

Living Membranes. Water moves along a gradient from the external environment into the plant cell—whether it is the single cell of an alga, a root cell of a flowering plant, or a seed—along a gradient of water potential. Usually, the concentration of the osmotically active solutes (the osmotic pressure of the cell sap) and the turgor pressure in the cell largely determine the water potential in the cell. In certain instances, such as air-dried seeds, there is present a high percentage of colloidal material such as starch, protein, and cellulose, all of which have a great affinity for water and hence a large capacity for imbibing water. When these seeds are planted in moist soil, the water in the seeds has a very low potential compared with the water in the soil. Consequently, water diffuses into these seeds, causing them to swell. In this case, the colloidal materials largely contribute to the low activity of water in the seed.

The activity of water in a pond is primarily determined by the concentration of dissolved material, but in the case of water in the soil an additional factor is involved, namely, the presence of colloidal soil particles that imbibe water and thus lower its activity. Nevertheless, the direction of water movement will be from a region where the water potential is high (near zero) to a region where the water potential is lower (more negative).

We have seen that turgor pressures result because of differences in solute concentrations. Although the pressure outward is often considerable, the protoplast is prevented from bursting by the elastic cell wall, which resists stretching and exerts a pressure inward. The cell wall serves the same purpose that

the casing of an automobile tire did in preventing the bursting of the inner tube.

It is obvious that in a cell the *turgor pressure outward* against the wall is equal to the *wall pressure inward* against the cell contents. A cell may have varying degrees of turgidity. The maximum turgor pressure that can be developed in a cell is equal to the osmotic pressure of the cell contents. That is, when a cell has absorbed all the water it can, osmotic pressure, turgor pressure, and wall pressure are equal. Normally, most living cells are in a condition in which turgor pressure is somewhat less than the maximum pressure possible. This means they are capable of taking in more water by osmosis. However, if they lose water, cells may become *flaccid*, and have a very low turgor pressure.

The crispness of the leaves and the rigidity of the young parts of plants are due to the turgid condition of the individual cells. A young bean seedling stands erect chiefly because of the turgor pressure in all the cells of the stem. In such a young plant, strengthening tissue is not plentiful. If the cells of the seedling lose water rapidly, the whole plant becomes flaccid and droops, but it may recover if water is resupplied.

Plasmolysis. If a cell or group of cells is immersed in a solution that has a higher solute concentration than that of the cell sap, water diffuses outward, and the turgor pressure in the cell is reduced. The volume of the cell decreases somewhat, but more striking is the withdrawal of the protoplast from the cell wall and the decrease in the size of the vacuole. This phenomenon is called **plasmolysis.** As shown in Fig. 5.15, Color Plate 3, the space between the cytoplasm and the cell wall in a plasmolyzed cell is filled with the plasmolyzing solution in which the cells are immersed. If plasmolyzed cells are immersed in water or in a solution whose concentration is less than that of the cell sap, the cells regain their turgor; water molecules diffuse inward. If cells remain long in a state of pronounced plasmolysis, however, death ensues. A normal, healthy, and functioning cell is one in a turgid condition.

It is instructive to plasmolyze living cells that contain a pigment dissolved in the vacuolar sap. For this purpose, cells of *Elodea* with neutral red in the vacuole may be used. When they are immersed in a strong solution, such as one of table salt, plasmolysis soon follows. It will be seen, however, that the red pigment is retained in the vacuole (Fig. 5.16, Color Plate 3). Obviously, the tonoplast is impermeable to the outward movement of this dye. If the cells are heated or treated with various chemicals, such as chloroform, alcohol, or ether, the red pigment readily diffuses from the protoplast. Cytoplasmic membranes manifest differential permeability *only when the cell is alive;* when the cell is killed, cytoplasmic membranes lose their differential permeability. As we shall see later on, however, many nonliving, as well as living, membranes have differentially permeable properties.

A cell may die if plasmolysis is pronounced and prolonged. For example, if heavy application of ordinary salt is placed on the soil where weeds are growing, so that the root cells are surrounded by a solution of high concentration, water diffuses from the cells, and they become severely plasmolyzed. If this state is prolonged, the roots die. The salt killed, not because it was toxic to root cells, but rather because severe plasmolysis was brought about by the high concentration of the soil solution. In parts of the western United States where rainfall is low and the evaporation rate is high, salts of the soil may accumulate on the surface and form what are known as "alkali flats." In such soils, the concentration of the soil solution may be so high that ordinary crop plants cannot grow; only species that are especially adapted to tolerate high salt concentration are able to survive. Plants of this type are known as **halophytes** (see Fig. 18.10).

Absorption of Dissolved Substances. In our discussion of absorption thus far, we have explained how water enters the living cell. In addition to water, the cell absorbs various inorganic ions such as potassium, calcium, nitrates, phosphates, and sulfates; and the gases, oxygen and carbon dioxide, both of which are soluble in water.

Water and the different solute particles move into the cell *independently.* That is, if water molecules are diffusing through the cell wall and the cytoplasmic membranes and are entering the vacuole at a certain rate, it does not follow that any particular solute particle is entering the vacuole at the same rate. Moreover, in the event that the solution surrounding the cell contains $NaNO_3$, the entering of NO_3^- ions into the cell does not necessarily mean that Na^+ ions are moving in at the same rate. Although loss of water in vapor form from a plant does affect the proportions of water and solutes absorbed, it appears that the kind and quantity of ions absorbed are chiefly determined not by the volume of water absorbed but by certain chemical and physical properties of root cells. Different solute particles diffuse independently of each other, and the direction of diffusion of any solute is independent of the direction of the diffusion of any other solute that may be a part of the same solution.

It has been found experimentally that the concentration of a solute particle (ion or molecule) may be greater in the vacuole than in the solution outside the cell (Table 5.3). How, then, do we harmonize *simple diffusion* with that in which solute particles move from a place where they are in low concentration to a place where they are in high concentration? It has been demonstrated that the **accumulation** of solute particles by plant cells is usually attended by high respiration rates. It may be assumed that the energy released by respiration is utilized by the cell to perform the labor of *forcing* the solute particles to move *against a concentration gradient,* and this same energy maintains the concentration difference. Thus, the living cell performs work, the energy for which is derived from respiration. Ions that are accumulated by root cells are apparently "pumped" into the vacuole of the cell from the soil solution.

Summary of Absorption

1. All substances that enter the cell, gases included, are in solution in water.
2. The cell wall is normally permeable to all substances in true solution.

TABLE 5.3 The Concentration of Various Ions in the Vacuolar Sap of *Nitella clavata* and in the Pond Water in Which It Was Growing[a]

Ion	Cell sap concentration, millequivalents per liter	Pond water concentration, millequivalents per liter	Accumulation ratio, sap concentration/ external concentration
Ca^{2+}	13.0	1.3	10.0
Mg^{2+}	10.3	3.0	3.6
Na^+	49.9	1.2	41.6
K^+	49.3	0.51	96.7
Cl^-	101.1	1.0	101.0
SO_4^{2-}	13.0	0.67	19.0
$H_2PO_4^{3-}$	1.7	0.008	212.5

[a]Data from Hoagland and Davis (1929), *Protoplasma.*

3. The cytoplasmic membranes are differentially permeable.
4. Cell colloids imbibe water and swell.
5. Osmosis is the diffusion of water through a differentially permeable membrane. Water diffuses along a gradient of water potential.
6. Any particular solute may accumulate in the cell and thus maintain in the vacuole a higher concentration of that substance than exists outside the cell. To accomplish and maintain this higher concentration, the cell must perform work. The energy for this work is furnished by respiration.
7. The diffusion of water and different solutes is independent.
8. Solute particles decrease the activity of water molecules; the solution then has a low water potential.
9. When water diffuses into a cell, the internal turgor pressure increases.
10. Turgor pressure increases the potential of water.
11. Water enters the cell when the water potential in the cell sap is less than that of the water in the solution outside the cell. Water leaves the cell when the water potential inside the cell exceeds that in the solution outside the cell.
12. Cells immersed in a solution of high concentration become plasmolyzed. Water diffuses out of the protoplast, and the protoplast withdraws from the wall.
13. Respiratory energy may be utilized by a cell to accumulate ions.

Metabolism

The sum total of all the chemical reactions that go on in the plant body is called **metabolism.** Food manufacture, formation of cell walls, and synthesis of protein all involve many chemical reactions and result in the production of new materials. Such synthetic metabolic reactions, **anabolic reactions,** contrast with the breakdown reactions such as are involved in respiration or the oxidation of foods. The phase of metabolism that involves breakdown or destructive reactions is known as **catabolism.** Both aspects of metabolism are essential for normal cell activities, since anabolic reactions are necessary for the continued production of new materials, and catabolic reactions release, in an available form, energy stored in foods.

Photosynthesis

Every day the combined oil wells of the world deliver some 2 billion barrels of oil for running the cars, boats, trains, and factories of our highly industrialized civilization. The energy this oil so conveniently stores came originally from the sun. It was "captured" by plants, and because of certain geological conditions was trapped and stored in the earth's crust for many millions of years. The process by which plants in the past captured, and today continue to capture, the radiant energy of the sun is known as **photosynthesis,** a term that literally means putting together (*synthesis*) by means of light (*photo*).

The principal features of this physiological process will be mentioned here, but a more complete discussion will be given in Chapter 13. Photosynthesis is a process that goes on only in cells that have **chloroplasts** and only when these cells are **illuminated.** Certain blue-green algae and purple bacteria are exceptional in carrying on photosynthesis in the absence of chloroplasts, but chlorophyll bound to membranes is present.

In photosynthesis, the simple compounds **water** and **carbon dioxide** are united to form **sugars** and **oxygen.** Glucose is one of the principal sugars produced during photosynthesis.

The oxygen may go into solution in the cell sap or diffuse out of the cell or be used in another cellular process, respiration. Carbon dioxide is dissolved in the water in which the plant is floating. As carbon dioxide is taken out of solution and used in photosynthesis, its concentration in the cell sap becomes less than that in the water outside the cell. As a result, diffusion of that gas into the cell goes on as long as active photosynthesis continues. In land plants, green cells obtain carbon dioxide from the atmosphere, but this gas must go into solution in the imbibed water of the cell wall and cytoplasm before it can diffuse to the chloroplasts.

Photosynthesis is an energy-storing process. Energy is required to bring about the synthesis of glucose from carbon dioxide and water. Light is the source of energy for photosynthesis, and the glucose molecule contains transformed light energy holding the atoms together. If the glucose molecule is burned, oxygen is consumed and carbon dioxide and water are formed. The heat energy evolved in burning represents the release of the glucose molecule's chemical potential energy, which is equal to the light energy transformed and stored as chemical energy in the photosynthetic process.

Photosynthesis may be simply represented as follows:

$$6\,CO_2 + 6\,H_2O + 686 \rightarrow C_6H_{12}O_6 + 6\,O_2$$

carbon + water + kcal → sugar + oxygen
dioxide

Glucose may be utilized in a number of different ways: (*a*) broken down in the process of respiration, yielding energy; (*b*) converted into some closely related carbohydrate, such as sucrose (cane sugar,

$C_{12}H_{22}O_{11}$), or cellulose (cell-wall building material), or starch (a reserve food supply); (c) converted into fatty substances; (d) united with the nitrogen, sulfur, and phosphorus derived from various inorganic salts absorbed from the water of the soil or the lake or stream in which the plant lives, thus forming proteins; and (e) employed as the chemical basis for a number of other substances that may be found in the cell, such as chlorophyll or alkaloids such as nicotine.

The conversion of glucose into other carbohydrates may be simply represented as follows:

$$n\,C_6H_{12}O_6 \rightarrow n(C_6H_{10}O_5) + n\,H_2O$$

glucose → carbohydrate + water

Digestion

Complex carbohydrates are formed by union of sugar molecules, fats are composed of fatty acids and glycerine, and proteins are polymers of amino acids. These materials are not only synthesized in cells but may be broken apart within cells and outside them. Enzymes are involved in these processes of building and degradation. The breaking apart of complex carbohydrates, fats, and proteins into sugars, fatty acids, glycerine, and amino acids occurs easily in the presence of specific enzymes and without the release of large amounts of energy. This process whereby complex organic compounds are broken down into simpler compounds is known as **digestion.** Since water molecules are generally used in the digestion process, the process is chemically one of hydrolysis.

Plants as well as animals **digest** foods; that is, they break down complex foods into simple compounds. Plants have no special organs for digestion; it is carried on in any cell that may store food, even temporarily. In most green cells, photosynthetic activity may proceed at such a rate during the day that the food (glucose) accumulates faster than it is used in those cells. When this occurs, the glucose may be changed temporarily to starch, which appears as granules in the chloroplasts. Glucose is soluble in the water of cell sap; starch is insoluble. When the time comes for this temporary starch reserve to be used by the cell or transported out of the cell, it must be changed back into sugar. This chemical transformation of the insoluble starch into soluble glucose is one example of digestion. The equation for starch digestion is as follows:

$$(C_6H_{10}O_5)_n + n\,H_2O \rightarrow n\,C_6H_{12}O_6$$

starch + water → glucose

Any other foods that are insoluble in the cell sap, such as proteins and fats, must be digested (rendered soluble) before they can diffuse and nourish the cell. Enzymes *facilitate digestion.*

Respiration

Many processes taking place in plants require energy. These include: the absorption of mineral salts against an osmotic gradient, the synthesis of complex compounds such as proteins, the maintenance of the protoplasm in a living state, cell division and cell growth, movement, photosynthesis, and translocation. Energy for photosynthesis is supplied by sunlight; energy for the translocation of water comes from the sun, which evaporates water from leaf surfaces. All other processes obtain energy from respiration.

Respiration may be defined as the **oxidation** of *organic substances, with the release of energy, within cells.* In its simplest form, oxidation involves chemical union of some substance with oxygen. However, in its broadest sense, oxidation covers the energy-releasing reactions in which molecular oxygen itself may not be involved. In the burning of a match, oxygen of the air unites with wood, heat energy is released, and the wood is broken down into carbon dioxide and water. We have seen that photosynthesis is the reverse: carbon dioxide and water unite, oxygen is released, and energy is stored. Photosynthesis represents a type of reaction termed **reduction.** Actually, the two reaction types are much more complicated than expressed here and will be considered in more detail in later chapters.

Sugar is the usual organic substance oxidized in all plant cells. The energy stored in the sugar is, as we have learned, derived from sunlight. Sugar is a **food,** and all foods are energy-rich compounds whose molecules are so constructed that the energy they store may be released with comparative ease.

Respiration is a cellular process. All living cells respire. Although respiration is a complicated chemical process, the overall reaction occurring in cells may be described as follows: 1 gram-molecule of glucose sugar (180 grams) combines with 6 gram-molecules of oxygen (192 grams) to form 6 gram-molecules of water (108 grams) and 6 gram-molecules of carbon dioxide (264 grams), releasing 686 kilocalories of heat. A **kilocalorie** (kcal) is the amount of heat required to raise the temperature of 1 liter of water (slightly more than a quart) 1°C.

This statement may be simplified as follows:

$$C_6H_{12}O_6 + 6\,O_2 \rightarrow 6\,CO_2 + 6\,H_2O + 686$$

sugar + oxygen → carbon + water + kcal
dioxide

Energy Flow. When sugar is burned, carbon bonds

are broken and heat is generated. The heat may be used to do work. If it were possible to use the re-arrangement of the atoms of the sugar molecules directly to do work, little heat would be generated, and our machine would be highly efficient. This is exactly what happens in living cells during the process of respiration; reactions that release energy are **coupled** with reactions that absorb energy. Molecules are rearranged, new materials are formed, and work is done. These **coupled reactions** are key reactions in biosynthetic processes.

When carbon bonds are broken in the cell during the oxidative reactions of respiration, small amounts of energy may be made available for work. Some phosphate bonds, on the other hand, are able to store large amounts of energy, and when broken release it easily to drive other processes along in **coupled reactions.** Two most important compounds associated with energy transfer in cells are **adenosine diphosphate (ADP)** and **adenosine triphosphate (ATP)**. There is also an **adenosine monophosphate (AMP)** with only one phosphate group on the molecule. When a second phosphate is added to the single phosphate of AMP in the formation of ADP, a bond is formed that concentrates much more energy than is present in the phosphate bond of AMP. When a third phosphate is added to form ATP, an additional large amount of energy is concentrated in this bond. These two bonds that hold the two terminal phosphate groups of the ATP molecule are known as **high-energy bonds** and are designated with the symbol ~P. These bonds may be easily broken by hydrolysis, and each then releases about 8000 cal of energy per mole of ATP hydrolyzed. When the carbon–oxygen bonds in a disaccharide or in a fat molecule are hydrolyzed, only about 1000–2000 cal are released.

Adenosine triphosphate is the principal compound that appears to be universally formed during respiration in the cells of microorganisms, plants, and animals, and in which the energy from the oxidized food is temporarily stored. ATP has been called the energy currency of the cell. It may be formed from ADP and inorganic phosphate in the mitochondria during respiration, or it may be formed in the chloroplasts of green cells exposed to light during photosynthesis. From its site of synthesis, ATP may move to other locations where it gives up its energy in the performance of cellular work and it is converted to ADP and inorganic phosphate. ADP may now migrate back to the mitochondria where, in cooperation with respiration, it is again combined with inorganic phosphate to form ATP, and the process is repeated.

Let us see how all these topics are related. Our problem is to follow the radiant energy of the sun as it is captured by the cells of green plants, stored in the carbohydrate molecules, and finally used to synthesize animal and plant protoplasm which in turn utilize the energy of the carbohydrate molecule in various ways, ranging from simple leaf movements to the intellectual activities of man. Light may be thought of as small packages of energy, **photons** (p. 217). The absorption of photons (light energy) by chlorophyll is thought to start a flow of electrons (electrochemical energy) through a series of compounds that are found in the chloroplast. This flow of electrons can be compared to the flow of electrons in a wire as an electric current passes through it. Just as work may be done by the passage of the current in the wire, so also work can be done as electrons are transported from one substance to another in the chloroplast under the influence of light energy. In the chloroplast, ATP is formed during this electron flow, but the energy that it temporarily holds is rapidly used during the synthesis of sugar. ATP is an excellent means of transferring cellular energy, but it is not a particularly good means of storing energy, since only a small amount of ATP is formed in the cell at any one time. The storage of chemical energy is a function of the sugar and other foods. Stored energy is unused energy; to be utilized it must be released in the cell, and this is done, as we have seen, during the process of respiration where there is again a flow of electrons from the food, this time to oxygen. During this electron flow, some of the energy is again trapped as ATP is formed. The ATP may be transferred throughout the cell and its energy used to perform work such as the synthesis of more protoplasm and many other cellular constituents.

Assimilation

We have seen how the green plant manufactures glucose from carbon dioxide and water. This sugar may be transformed into related carbohydrates, such as sucrose or starch; or it may be changed to fats; or, with the addition of chemical elements (particularly nitrogen, sulfur, and phosphorus) from the soil, it may be used to form proteins. Carbohydrates, fats, proteins, and simpler nitrogenous substances are the **foods** of plants, as well as of animals. They are *organic compounds*. The foods are manufactured from *raw materials,* such as water, carbon dioxide, nitrates, phosphates, and sulfates; these raw materials are *inorganic compounds*. The cell takes the foods (nonliving substances) and from them builds protoplasm (living substance). *The conversion of foods into the living material (protoplasm) of the cell is*

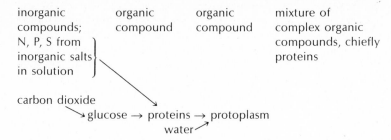

inorganic compounds; N, P, S from inorganic salts in solution

organic compound

organic compound

mixture of complex organic compounds, chiefly proteins

carbon dioxide → glucose → proteins → protoplasm

water

called **assimilation.** The chemical changes involved in this conversion are not understood, but energy is probably used in the process. Except in a very general way, the chemical nature of protoplasm itself is unknown.

Animal cells and nongreen plant cells can also convert foods into protoplasm. Also, any plant cell, with or without chlorophyll, can synthesize fats and proteins and can change glucose into starch and other carbohydrates. But only chlorophyll-bearing cells can manufacture glucose from carbon dioxide and water.

The steps leading up to the building of protoplasm in the free-floating water plant, or in any green cell, may be summarized as shown in the diagram above.

Control of Metabolic Reactions

Although protoplasm is complex, the wide variety of chemical reactions which occur in it simultaneously during the life of the cell are not haphazard. There are intricate spatial and time interrelationships among them. The regulated sequential nature of cellular reactions is one of the attributes of life. How are these reactions so precisely regulated and controlled in such a minute space as the volume of a cell? From the point of view of the size of an enzyme molecule, this volume is enormous; but our efforts toward miniaturization of electrical circuits in sophisticated electronic equipment can hardly be compared to the ultimate in miniaturization, the living cell.

Part of the regulation and order of metabolic reactions is dependent on cell structure. We have already seen (p. 37) that there are many organelles and membranes within a cell and that certain enzymes are confined, during the life of the cell, to very specific locations. This compartmentalization of enzyme activities can be readily demonstrated by separating the cell into its various parts (p. 43). Then it is found that specific metabolic reactions occur only in certain parts of the cell. For example, a sugar molecule synthesized in the chloroplast by photosynthesis may move into the cytoplasm, where it is changed and partially oxidized by the action of enzymes in the cytoplasm during a series of reactions called **glycolysis.** The oxidized product of glycolysis then moves into a mitochondrion where it is further oxidized. In this example, the sugar molecule has passed through three compartments (chloroplast, cytoplasm, mitochondrion) and has been treated uniquely in each. This beautifully precise compartmentalization in the cell is an important part of the mechanisms that control metabolic reactions.

Another aspect of metabolic control depends on genetic information in the nucleus that ultimately regulates the synthesis of specific enzymes in the cell.

A further form of metabolic control is dependent on the sensitivity of enzyme reactions to certain ions and molecules produced or used during metabolism. For example, if a plant is growing on an inorganic medium containing ammonium ions as a source of nitrogen, the plant may be unable to utilize nitrate ions rapidly when they are placed in the solution. However, in a short time, a rapid increase in the plant's ability to use nitrate is often observed. It can be demonstrated that the absorption of nitrate from the culture medium has stimulated the synthesis of an enzyme, **nitrate reductase.** This enzyme is said to be "induced" in the cell by the **substrate** (nitrate).

Often, the end product of a series of metabolic reactions can inhibit an enzyme catalyzing one of the early steps in the reaction sequence. This regulation of the course of a reaction through the action of one of the products of the reaction is known as **feedback.** It is a basic type of control in electronics. In cells, the concentrations of certain key metabolites play similar regulating roles in many phases of metabolism.

As an example of feedback control, we can look at some effects of ATP in the cell. During respiration, ATP is formed. The energy in ATP may be used to do useful work in the cell. The more work that is done, the more rapidly ATP is used. When little work is done, ATP tends to accumulate in the cell. It is known that a high level of ATP in the cell decreases the rate of at least one enzyme reaction in respiration. Consequently, an increase in ATP in the cell causes a decrease in the rate of respiration. Conversely, when a cell does more work, the level of ATP decreases, its inhibitory effect on the enzyme reaction involved in respiration decreases, and respiration increases. More ATP is generated. Thus, the rate of respiration is in part controlled by the level of cellular ATP, which in turn is influenced by the rate at which work is being done.

Summary of Cellular Physiology

1. Metabolism is the sum total of all catabolic or breakdown reactions and all anabolic or synthetic reactions in the cell.
2. Cells absorb water and mineral salts from the external environment.
3. The numerous chemical reactions that go on in the cell are controlled by organic catalysts called enzymes.
4. Green plant cells are able to synthesize carbohydrates from water, carbon dioxide, and light energy, through the process of photosynthesis.
5. Carbohydrate foods can be transformed into fats and, with the addition of certain chemicals, into proteins.
6. Complex foods are rendered simple through the process of digestion.
7. Respiration is the oxidation of organic substances within living plant cells.
8. Protoplasm is built up through the process of assimilation of foods.
9. Energy flows from sunlight through chlorophyll into sugar molecules where it is stored as chemical energy during the process of photosynthesis.
10. Some chemical energy in sugar is transferred to ATP during respiration.
11. ATP is used by the cell to drive the energy-consuming reactions of metabolism.
12. The control of metabolism may depend in part on compartmentalization, enzyme synthesis, enzyme sensitivity to ions and molecules, and may involve a feedback mechanism.

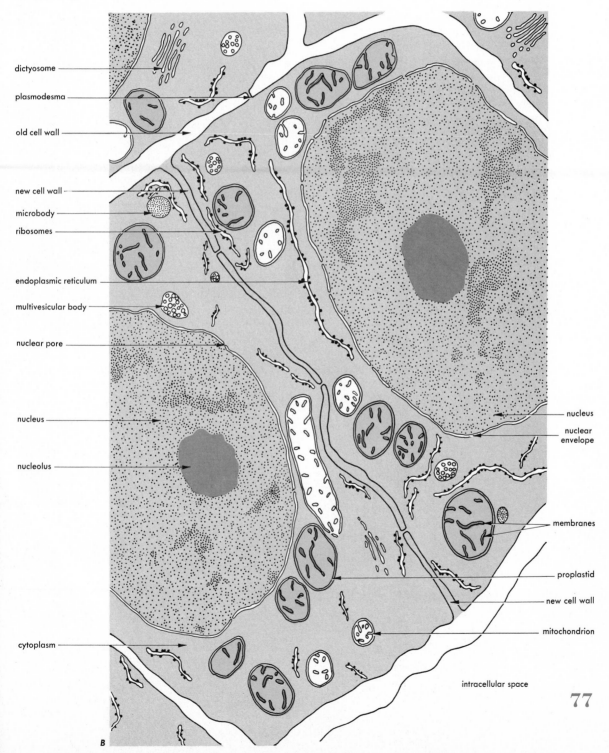

Figure 6.2 Below left and below: Meristematic cells in the shoot apex of *Chenopodium album*. *A* is an electron micrograph. *B* is a drawing of *A*. Note large nucleus, mitochondria, dictyosomes, proplastids, microbodies, and absence of vacuoles. Compare with Fig. 4.3. Magnification ×10,000. (Courtesy of E. M. Gifford and K. Stewart.)

dictyosome

plasmodesma

old cell wall

new cell wall

microbody

ribosomes

endoplasmic reticulum

multivesicular body

nuclear pore

nucleus

nucleolus

cytoplasm

nucleus

nuclear envelope

membranes

proplastid

new cell wall

mitochondrion

intracellular space

B

Cell division is the first step in growth and differentiation. The formation of new tissue cannot take place without cell division. During the period of division, both nucleus and cytoplasm divide. It is customary to designate the period of nuclear division as **mitosis** and to divide it into four phases: (*a*) **prophase,** (*b*) **metaphase,** (*c*) **anaphase,** and (*d*) **telophase.** The division of the cytoplasm is known as **cytokinesis.** Mitosis gives rise to daughter nuclei having identical gene complements. Cytokinesis gives rise to two new parcels of cytoplasm which are probably never identical (Fig. 6.1). The period of preparation for division is known as **interphase** and we shall see that the sequence of events taking place during interphase is equal in importance, in the complete process of cell division, to that occurring during mitosis and cytokinesis. We may summarize the complete process of cell division as follows:

Cell Division

Preparation for division, interphase
Period of division
    Mitosis → identical daughter nuclei
        prophase
        metaphase
        anaphase
        telophase
    Cytokinesis → different parcels of daughter cytoplasm

The Interphase Cell

An interphase cell in meristematic tissue has characteristic differences from an interphase cell of mature tissue. Portions of two interphase cells taken from the shoot apex of *Chenopodium album* are shown in Figs. 6.2*A* and *B*. They may be compared with the young leaf cell of Fig. 4.3. Meristematic cells of apical regions are almost spherical, and their shapes may be compared with a compressed mass of bubbles or lead BB shot. There is no large vacuole, and the nucleus occupies a major proportion of the cell volume. Mitochondria, dictyosomes, endoplasmic reticulum, and ribosomes are similar to those found in the leaf cell. The shoot apex of *Chenopodium* was killed in potassium permanganate, so the ribosomes do not show. However, microbodies (e.g., sphaerosomes and multivesicular bodies) are present. The plastids, which lack chlorophyll in the apical cells, differ from those found in the leaf cell, which do contain chlorophyll. In meristematic cells, the young plastids, or proplastids, are slightly larger than mitochondria and possess only a few paired membranes (Fig. 6.2).

The microtubules may be present and appear, at least just before the onset of mitosis, to form a narrow band encircling the nucleus in the plane of the forthcoming equatorial plate (Fig. 6.3).

Thin sections of interphase nuclei frequently show the nucleoplasm to be evenly and finely granular (Fig. 6.2). A coarser, more irregular clumping of granules is also seen (Figs. 4.3, 4.16*A*). In addition, denser aggregations of granular or fibrilar material may be observed (Figs. 4.3*A*, 6.2*A*). DNA may be localized in quantity in these areas and they may represent sections of chromosomes. Higher magnifications and special techniques reveal short, coiled fibrils in the nucleoplasm. Some fibrils are visible in the nucleoplasm of the nucleus in Fig. 4.16*A*. Measurements of well-preserved fibrils in the nuclei of grasshopper sperm show them to be about 10 nm in diameter.

Considerable detail is shown in the nucleolus from a root tip of the water plant *Hydrocharis* (Fig. 4.16*C*). Ribosome-like particles in spiral aggregations appear

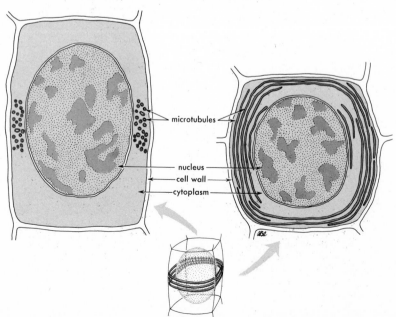

microtubules

nucleus
cell wall
cytoplasm

Figure 6.3 Meristematic cells in a young leaf of tobacco (*Nicotiana tabacum*). *A*, section at right angles to the plane of the future cell plate shows cross-section of microtubules, ×10,000. *B*, section in plane of future cell plate shows microtubules encircling nucleus, ×10,000. *C*, three-dimensional interpretation of *A* and *B*. (Redrawn after K. Esau and J. Cronshaw, *Protoplasma* **65,** 1. Copyright Springer-Verlag.)

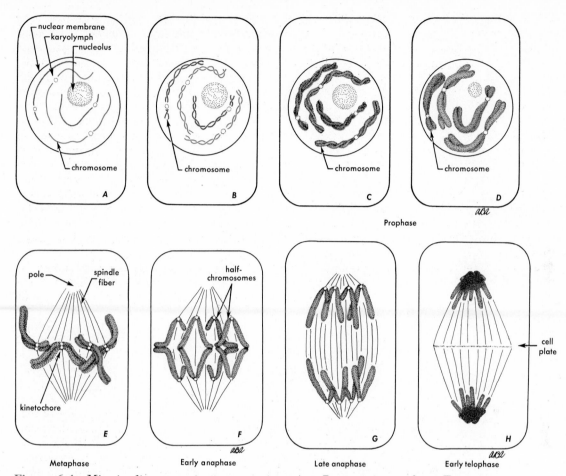

Figure 6.4 Mitosis, diagrammatic representation; *A* to *D*, stages in prophase. Two pairs of chromosomes are represented; the green pair have a median kinetochore, while in the gray pair, the kinetochore is close to one end. The chromosomes shorten and thicken, and chromatids become apparent. *E*, metaphase; *F*, early anaphase; *G*, late anaphase; *H*, telophase.

to form the basic nucleolar structure. Note the absence of a barrier between the nucleolus and nucleoplasm.

Nuclei isolated from young barley root tips and spread on distilled water will burst. The resulting film may be picked up, carefully dried, and observed on the electron microscope. The nuclei appear (Fig. 4.16*B*) to consist of loosely coiled or folded fibers about 17.5 nm in diameter. Both thin-sectioned and isolated nuclei indicate that the basic structure in interphase nuclei is an array of fibers called **unit fibers.** The DNA of these fibers is thought to complex with a simple protein, probably a histone.

Mitosis

Prophase

The unit fibers in the interphase nucleus are long, slender, and seem tangled. We do not know their exact disposition in the interphase nucleus nor their precise relationship with chromosomes. But the onset of mitosis, as seen in the light microscope, is heralded by the presence of definite chromatin threads (Figs. 6.4*A*, 6.5*A*). These threads gradually shorten and thicken and become easier to see (Figs. 6.5*B, C*). They stain more heavily with certain dyes. For this reason they are called colored (*chromo*) bodies (*soma*), or **chromosomes.** Each chromosome is derived from an individual thread. The nucleolus slowly decreases in size and finally disappears during this stage (Figs. 6.4*A–D*).

It becomes apparent that at late prophase each chromosome is composed not of one but of two threads coiled about each other (Figs. 6.4*B, C, D*). The nuclear membrane disappears toward the end of the prophase, but there is reason to believe that the nucleoplasm may not mix with the general cytoplasm of the cell and that the chromosomes may remain embedded within it.

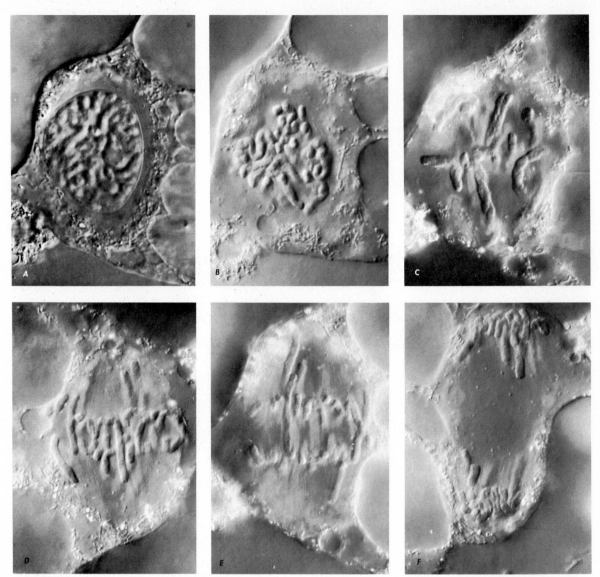

Figure 6.5 Mitosis as followed in living unflattened cells of the endosperm of seeds of the blood lily (*Haemanthus katerinae*). Photographs in Nomarski system optics. *A*, early prophase; the nuclear membrane is still present. Note the clear zone of cytoplasm surrounding the nucleus. *B*, the nuclear envelope has disappeared, and the clear zone still surrounds the chromosomes. A few short spindle fibers are already present. *C*, full metaphase; the coiled chromatids are distinct. The two chromatids moving to the upper pole have separated, and a few spindle fibers may be seen. *D*, early anaphase; the chromatids are moving to opposite poles of the cell. *E*, midanaphase; spindle fibers are in evidence between the chromosomes and the poles of the cell. *F*, telophase; the chromatids have aggregated at the opposite poles of the cell. A nuclear envelope has not yet formed. Time intervals after *A*, *B*, 14 min; *C*, 64 min; *D*, 74 min; *E*, 93 min; *F*, 107 min, ×2000. (Courtesy of A. Bajer, *Chromosoma* **25**, 249. Copyright Springer-Verlag.)

Metaphase

Forces active within the cell now arrange the chromosomes, or at least a specialized portion of each chromosome (the **kinetochore**), in the equatorial plane of the cell (Figs. 6.4*E*, 6.5*D*). The nucleoplasm appears to elongate slightly, and **spindle fibers** appear attached to the kinetochores. As the nucleoplasm elongates, the chromosomes, or at least the kinetochore, is moved to the equatorial plane of the cell (Figs. 6.4*E*, 6.5*D*, 6.6). Spindle fibers extend from the chromosomes to the opposite poles of the cells.

Other fibers apparently reach to the poles but are not attached to the chromosomes. The electron microscope demonstrates that these spindle fibers have the characteristics of microtubules (Fig. 6.6). This structure—chromosomes, microtubules, or spindle fibers—plus any adherent nucleoplasm is called the **mitotic spindle.**

Electron micrographs of isolated metaphase chromosomes shows them to be constructed as was the interphase nuclei of unit fibers. But here the fibers are more tightly folded than in the interphase nuclei (Fig. 6.7).

The chromosomes are now distinct bodies of two closely associated halves, each half being known as a **chromatid** (Fig. 6.4E). In plants, such as corn, which have been intensively studied, each chromosome can be recognized and numbered. There are 20 chromosomes in corn, but only 10 different types that can be distinguished by their size and form. There are thus 2 chromosomes of each type. The 20 chromosomes of corn may be arranged in 10 pairs. Maps have been prepared of corn chromosomes showing the relative positions of the genes along them. Since the chromosomes split longitudinally, each gene is replicated and each half-chromosome contains a full set of genes.

Anaphase

The chromosomes do not remain long in the equatorial plane. The chromatids soon separate from each other and move to opposite poles of the cell. This period of separation of half-chromosomes is **anaphase** (Figs. 6.4F, G, 6.5D, E).

Telophase

When the divided chromosomes have reached the opposite poles of the cell, they group together and spin out into long, thin threads of the reticulum. The nuclear membrane and the nucleolus again become apparent. This period of transition from chromosomes to interphase nucleus is known as the **telophase.** The half-chromosomes that have gone into the constitution of the new daughter nuclei grow, forming full-sized chromosomes once again (Figs. 6.4H, 6.5F).

Figure 6.6 Electron micrograph of thin section of metaphase chromosomes in the endosperm of *Haemanthus katherinae*, showing the aligned chromosomes with microtubules (spindle fibers) attached to the kinetochore. Magnification ×40,000. (Courtesy of A. Bajer.)

Interphase

Even in the most rapidly dividing cells, the telophase nucleus gradually returns to the interphase state. Either a new division must be prepared for, or the genetic information must be transmitted to the cytoplasm for the synthesis of new proteins, wall materials, and structure that accomplish differentiation in the new cell.

Polarity of Cell Division

If the spindles in a meristematic tissue were oriented at random, an irregular mass of tissue would result. This does occur when a single cell or small group of cells from a carrot root in tissue culture produces an irregular mass of cells called a **callus** (Fig. 6.8A). For orderly growth there must be a precise orientation of the mitotic spindles. A polarity is established so that, in general, the axes of spindles are parallel with each other and with the axis of the shoot or root (Fig. 6.8B). When the spindle axes are oriented parallel to the root–shoot axis, the divisions are called **anticlinal.** Less commonly, when the axes are perpendicular to the root–shoot axis, increasing the girth, the divisions are called **periclinal** divisions (Fig. 6.8B).

What is responsible for polarity? We do not know. Polarity seems to be inherent in cells and in the tissue of which it is a part. Changes in polarity may be induced by hormones. Occasionally, the orientation to be assumed by the spindle can be detected in plants at the cellular level before metaphase. In some instances, the cellular organelles will pass

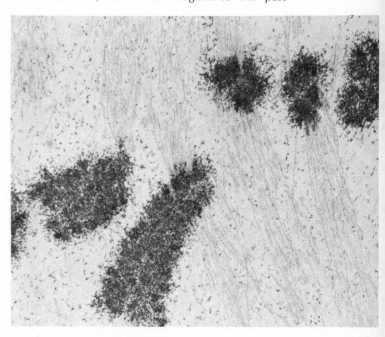

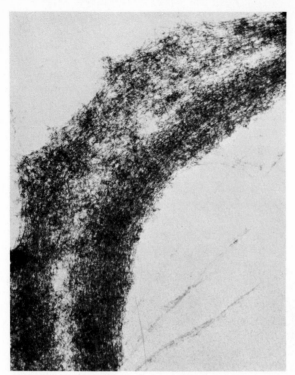

Figure 6.7 Isolated metaphase chromosome (*Vicia faba*) consists of unit fibers similar to those seen in the interphase nucleus, ×30,000. (Courtesy of S. Wolfe and R. G. Martin, *J. Exptl. Res.* **50,** 140. Copyright Academic Press.)

largely to one end of the interphase cell. Some observations report that microtubules form a circular band at the equatorial plane of the cell (Fig. 6.3) during interphase, just before the onset of mitosis.

In animal cells, polarity is generally indicated before the onset of mitosis by the presence and activity of the centrioles.

Cytokinesis

In the great majority of cases, the division of the nucleus is followed by the division of the cytoplasm. Light microscopy demonstrated that a cell plate forms at the equatorial plane of the cell at right angles to the spindle fibers (Figs. 6.9*A*, *B*). Electron microscopy has shown that the spindle fibers are microtubules. At the equatorial region, the spindle fibers or microtubules are gathered together in bundles and each bundle is surrounded by an amorphous material (Fig. 6.9*C*). Vesicles appear in this region and eventually fuse to form the first membrane dividing the daughter protoplasts (Fig. 6.9*D*). This membrane, formed of pectin, is known as the **middle lamella.** The new cell wall is formed by the deposition of cellulose by each protoplast on its side of the middle lamella. The resulting structure is the **primary cell wall.**

The formation of the primary wall, and subsequently, secondarily deposited materials poses an interesting problem of genetic control. The cellulose is laid down outside the protoplast, apparently re-

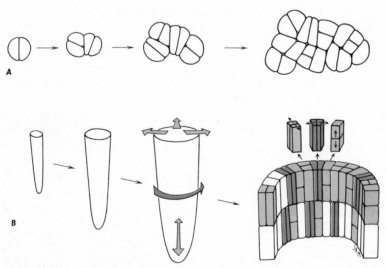

Figure 6.8 Influence of polarity of cell divisions on development of form. *A,* nonpolarized cell divisions result in an irregular mass of cells; *B,* cell division parallel with the axis of the root (gray arrow and cells) results in growth in length; cell division parallel with the circumference of the root (dark green arrow and cells) increases the circumference of the root; cell division at right angles to the circumference of the root (light green arrows and cells) increases the diameter of the root.

sulting from the polymerization of many sugar molecules to form long, unbranched molecules of cellulose. We shall see in Chapter 7 that cell walls have definite characteristics. In the case of the pollen grain, the wall has very intricate and species-specific designs (Fig. 15.28). How can genes determine a pattern laid down by inert cellulose outside the protoplast?

The molecules of cellulose, which may be several microns long, are held together in a crystalline array by hydrogen bonds to form long, unbranched fibrils. These fibrils are of indefinite length, somewhat flattened, from 10 to 30 nm in breadth and about 7.5 nm in thickness. A cross-section of a fibril having an area of 1 sq nm would show, on the average, the cross-section of three molecules of cellulose. Between 20 and 30% of the primary cell wall consists of fibrils of cellulose. The fibrils, embedded in an amorphous matrix, give support to the cell wall as do the steel reinforcement rods in a concrete wall.

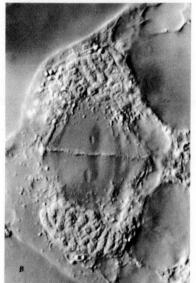

Figure 6.9 The formation of the cell plate in the endosperm of *Haemanthus katherinae. A* and *B*, living unflattened cells; photography by Nomarski system optics. *A*, telophase; the chromosomes are tightly clustered at opposite poles of the cell, and the cell plate has started to form; compare with *C*, ×4000. *B*, late telophase; the cell plate is almost continuous, and fibrils may be seen extending poleward a short distance from the plate; compare with *D*, ×4000. *C* and *D*, comparable stages of cell plate formation as seen with the electron microscope. *C*, early stage in plate formation; the microtubules are clustered into groups which are associated with a denser material, apparently in the plane of the future cell plate, ×26,000. *D*, late stage; the microtubules are fewer in number and are not clustered; vesicles are present and apparently fusing to form a continuous separation phase between the sister cells, ×26,000. (*A* and *B*, courtesy of A. Bajer, *Chromosoma* **25,** 249. Copyright Springer-Verlag. *C* and *D*, courtesy of P. K. Hepler and W. T. Jackson, *J. Cell Biol.* **38,** 437.)

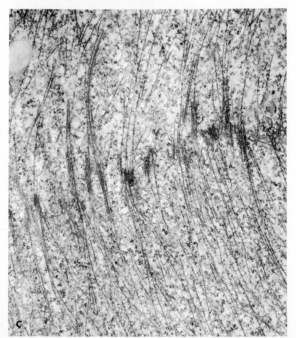

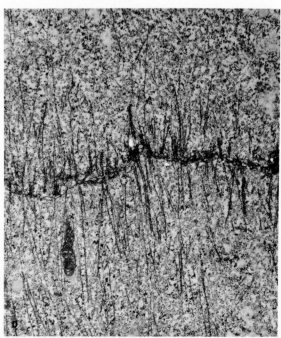

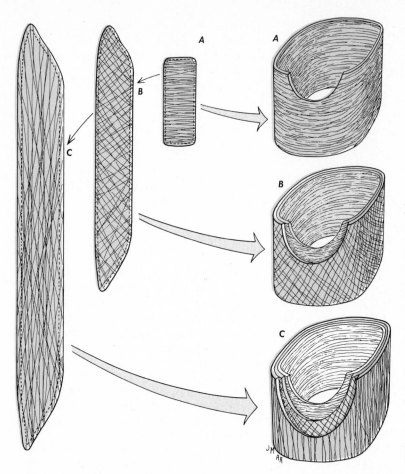

Figure 6.10 Diagram to show deposition and change in orientation of the cellulose fibrils in an elongating primary cell wall. *A*, *B*, and *C*, represent increasing age and length of the same cell. The cellulose fibrils are first deposited parallel with the circumference of the protoplast. They are then pulled out of this orientation as the cell wall elongates. (In the actual wall, the different stages of fibril orientation would grade into each other.) Green represents the earliest fibrils deposited, and light gray represents those most recently deposited.

After removal of the amorphous materials from the wall, the array of fibrils may be studied with the electron microscope. When first deposited by the protoplast, in cells destined to lengthen, the fibrils are in parallel array and form a band around the protoplast (Figs. 6.10*A*, 6.11). The mechanism bringing about this precise arrangement is not understood. However, the microtubules in the cytoplasm are also present in a similar parallel array. It is logical to postulate that the cytoplasmic microtubules may be involved in the orientation of the cellulose fibrils in the developing cell wall.

As the cell lengthens, the deposition of fibrils by the protoplast continues, with no change in the orientation of the newly deposited fibrils. However, elongation of the wall must change the orientation of older fibrils. They first separate to form a net more or less parallel with their original position (Fig. 6.10). As elongation continues, the fibrils, now further removed from the protoplast, assume a position more in line with the axis of the elongating cell. At a terminal position, the oldest cellulose fibers are all parallel with the elongating axis of the cell (Figs.

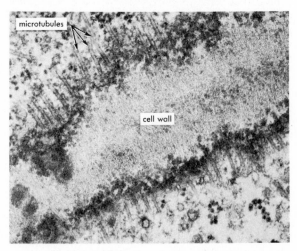

Figure 6.11 Microtubules in the root tip of *Arabidopsis thaliana*. The microtubules are oriented parallel with the circumference of the cell wall and thus parallel with the cellulose fibrils, ×50,000. (Courtesy of M. C. Ledbetter.)

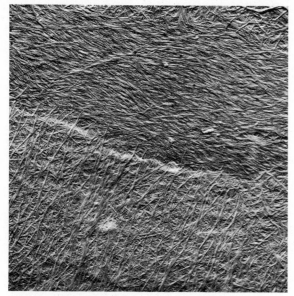

Figure 6.12 The wall of a hair cell from the base of a sedge (*Juncus*) leaf. Note that the innermost fibrils are parallel with the girth of the cell and that the outermost cellulose fibrils are parallel with the long axis of the cell, ×53,000. (Courtesy of A. L. Houwink and P. A. Roelofsen, *Act. Bot. Neerland.* **3**, 385.)

6.10, 6.12). This type of wall growth is known as **multi-net growth.** A section through a primary cell wall after completion of elongation shows the oldest cellulose fibrils to be parallel to the long axis of the cell; the most recently deposited cellulose fibrils encircle the cell at right angles to the long axis (Fig. 6.12). Intermediate fibrils have intermerdiate positions.

Summary of Cell Division

Cell division is now completed and its general aspects may be summarized.

1. Cell division increases the numbers of cells.
2. All cells have the potentiality to divide but are normally blocked from doing so.
3. Preparation for division takes place during interphase.
4. Nuclear division is called mitosis, cytoplasmic division is called cytokinesis.
5. The daughter nuclei are genetically identical.
6. The daughter parcels of cytoplasm may be unlike each other.
7. In general, mitotic spindles are oriented parallel with each other and with the axis of the shoot. A definite predetermined polarity brings this about.
8. The stages of mitosis are prophase, metaphase, anaphase, and telophase.
9. The spindle, which is the mechanism for chromosome movement, is thought to be constructed from many microtubules.
10. The middle lamella is the first membrane to divide the two daughter protoplasts.
11. Vesicles collect around the microtubules in the equatorial plane of the cell. They fuse to form the middle lamella.
12. Cellulose formed by the two daughter protoplasts and deposited on the middle lamella forms the primary wall.

DNA and Its Replication

Deoxyribose nucleic acid, DNA, is the relatively simple molecule in which, it appears, is stored all the information needed for the development of the several hundred thousand species, past and present. DNA is a prime biological universal. As far as we can tell, it has existed as long as life has existed. Although there may be variations in the type and distribution of the nitrogen bases in DNA from different species, there is no reason to suppose that there has ever been any fundamentally different kind of carrier of genetic information than the DNA known today.

As early as 1887, Weismann suggested that "qualities" responsible for inheritance had a linear arrangement in the chromosomes, which were just becoming known at that time. Between 1900 and 1904, the combined results of a number of investigations demonstrated the relationship between chromosomes, meiosis, and Mendel's newly rediscovered laws of inheritance. These observations ushered in a very exciting quarter century of research on inheritance and chromosomes.

Miescher, when he was but a 22-year-old Swiss physician, in 1868, was able to separate nuclear material from constituents of pus cells obtained from discarded bandages of soldiers wounded in the Franco-Prussian War. The characterization of the acidic component of this nuclear material as nucleic acid led to many studies aimed at determining the occurrence, nature, and chemistry of the nucleic acids. These nucleic acids have three parts: a phosphate group, a five-carbon-atom (pentose) sugar, and organic nitrogen-containing bases known as **pyrimidines** and **purines** (p. 87). It was first believed that there were two different nucleic acids, one derived from animals and the other from plants. Later work showed that animal and plant cells each contain the same nucleic acids.

The search for color reactions that would demonstrate nucleic acids led Feulgen, in 1924, to the discovery of the reaction that now bears his name. It had been known for some time that aldehydes such as the sugar aldehyde ribose, will reconstitute the color of a red dye, basic fuchsin, which has been decolorized by sulfurous acid. Aldehyde changes the basic fuchsin from colorless to violet. Now, both kinds of nucleic acids have pentose sugars (deoxyribose in DNA, ribose in RNA) that are aldehydes, but these potentially reactive groups are blocked, in the nucleic acids, by their association with the purine and pyrimidine bases. Feulgen wondered if the aldehyde group of the pentose sugar in a nucleic acid could be made available to bring about the color reaction with the decolored basic fuchsin. The answer was yes, and more. A mild hydrolysis with dilute hydrochloric acid broke the bond between the purines and the deoxypentose in DNA but did not readily break a similar bond between the nitrogen bases and the pentose of RNA. So the Feulgen technique of subjecting tissue or cells to mild acid hydrolysis followed by treatment with colorless fuchsin became, under properly controlled conditions, a specific and quantitative method of measuring the deoxypentose nucleic acid, DNA. The histochemical demonstration of DNA in nuclei was now a reality.

What is the relationship between DNA and the fibrils we have seen in the interphase nuclei and metaphase chromosomes? How is DNA replicated? What is the mechanism for the equal distribution of DNA in the dividing chromosomes? How does DNA code for the synthesis of proteins?

In 1953, Watson and Crick in a short, highly significant paper in *Nature,* presented the evidence for the molecular structure of DNA. The structure shown in Fig. 6.13 shows the now famous Watson–Crick model for the double helix of DNA. While a complete DNA molecule might be several million nanometers long, it is only 2.5 nm in diameter.

The fibrils found in isolated interphase nuclei and metaphase chromosomes are 12.5 nm in diameter. We do not know their length. The diameter and length of chromosomes varies with the stage of cell division, so a direct comparison of their diameters with those of the unit fiber is not profitable. However, a number of unit fibers could fit easily into an extended chromosome, and several unit fibers have been seen in chromosomes of *Vicia faba* (Fig. 6.7). It is not known whether these strands represent one unit fiber folded back on itself or several different strands. A similar situation exists for the state of the DNA molecule within the unit fiber. This hierarchy of structure complicates every level of chromosome study. How, for instance, are the many 12.5 nm unit

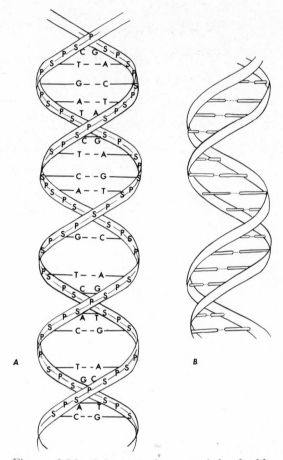

Figure 6.13 Schematic diagram of the double helix in a portion of a DNA molecule. Alternating sugar, S (deoxyribose); and phosphate, P, groups make up the backbones of the strands. Attached to each sugar unit is one of the purine or pyrimidine bases, adenine, A; thymine, T; guanine, G; cytosine, C. The strands of the helix are held together through hydrogen bonds between the base pairs, adenine to thymine and guanine to cytosine. *A,* one strand is displaced along the axis of the helix for convenience of diagramming; *B,* diagram of the parallel strands in the helix.

fibers, with their enclosed DNA double helix, divided equally between the two chromatids? Is a chromosome multistranded or not?

The DNA Molecule

The molecular structure of the DNA double helix is shown in Fig. 6.13. Each strand forming part of the backbone of the helix is made up of alternating sugar, S (deoxyribose), and phosphate, P, groups. These backbone strands are held together by hydrogen bonds between the nitrogen-containing organic

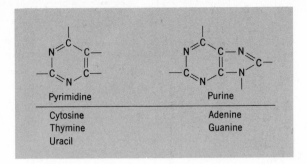

Pyrimidine

Purine

Cytosine
Thymine
Uracil

Adenine
Guanine

Figure 6.14 General formulas for purine and pyrimidine bases. Each particular base has characteristic groups attached at the unfilled valence bonds in the diagrams. These groups are shown in the complete formulas of Figs. 6.15 and 6.16.

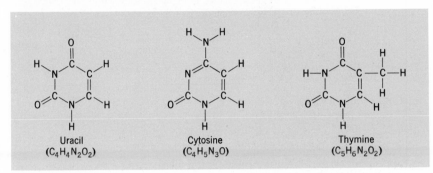

Uracil
$(C_4H_4N_2O_2)$

Cytosine
$(C_4H_5N_3O)$

Thymine
$(C_5H_6N_2O_2)$

Figure 6.15 Pyrimidines found in nucleic acids.

Adenine
$(C_5H_5N_5)$

Guanine
$(C_5H_5N_5O)$

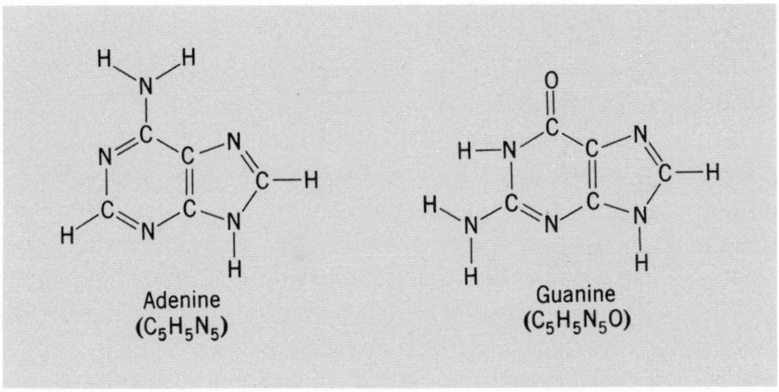

Figure 6.16 Purines found in nucleic acids.

bases (**purine** and **pyrimidine**). The bases of one strand are bonded to complementary bases of the other strands (Figs. 6.14, 6.15). When a purine or a pyrimidine base is joined with a ribose sugar and a phosphate, the resulting compound is a **nucleotide.** Although hundreds or even thousands of nucleotide units may polymerize to form the gigantic nucleic acid molecule (Fig. 6.13), only four different nucleotides are found in a nucleic acid molecule. The two pyrimidine bases occurring in DNA are **cytosine** and **thymine;** the two purine bases are **adenine** and **guanine** (Fig. 6.16). Notice in Fig. 6.13 that the bases have a linear arrangement along the background strands of the helix and the double helix is stabilized by hydrogen bonding between complementary base pairs. The sequence of purines or pyrimidines in one of the strands of the spiral determines the sequence of the bases in the other strand, since adenine of one strand always pairs with thymine of the other strand, and guanine of one strand always pairs with cytosine of the other. In order to illustrate this arrangement and the great variety within the nucleic acids that may arise from this simple set of four compounds, let us use the letters A, G, T, and C. We may first write them down in a line using each as many times as we wish and in any order we wish. Now, below these write again A, G, T, and C, but be careful that A always matches T and G always matches C.

Thus, if the first strand is represented by

A G C A A C A T C

the second strand would be

T C G T T G T A G

The two complementary bases (AT or GC) joined by hydrogen bonds between the complementary strands in the DNA molecule are frequently called **base pairs.**

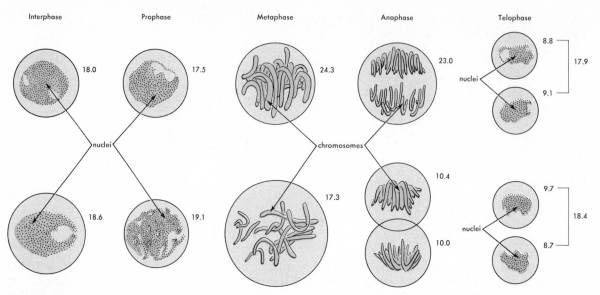

Figure 6.17 DNA content of nuclei at different stages of mitosis. Nuclei are shown for each of five stages. All pictures were made with the same magnification, so the circles show the relative sizes of the nuclei and chromosomes during mitosis. The numbers to the right of each circle indicate the amount of DNA present. (Redrawn after Patau and Swift, *Chromosoma* **6,** 149. Copyright Springer-Verlag.)

Enzymes regulate metabolic reactions in the cell. Each different kind of enzyme is a different protein and regulates a different reaction in the metabolism, growth, and development of the cell. Since each kind of protein has its own exact sequence of amino acids, the way in which the cell regulates the sequence of amino acid in proteins is very important.

The exact sequence of the amino acids in a newly formed protein molecule is specifically determined by the genetic information contained in the cell's nucleic acid molecules. We now know that the basis of this genetic information is the sequence of bases which each part of the DNA molecule contains. One of the most exciting accomplishments of modern biology has been the cracking of this genetic code.

Each amino acid is coded for by a short nucleotide sequence. In fact the four bases, in combinations of three, specify for each of the 20 amino acids found in proteins. The genetic code is made up of three "letters." It is a triplet code.

Replication of DNA

We shall now consider the manner in which DNA itself is replicated. From each double helix must come two new double helices without, at the same time, disturbing the precise ordering of the base pairs. The dividing double helix may be several million nanometers long, and since a single base pair occupies the space of only a few nanometers, there are many base pairs along a double helix.

DNA is localized in the chromosomes during mitosis. It ranges in amount from 2×10^{-9} to 10.36×10^{-9} mg per nucleus per cell. It is constant in amount for a given tissue and is definitely related to the number of chromosomes a tissue contains. Measurements of the DNA content in relation to stages of cell division are shown in Fig. 6.17. It should be noted that the late interphase and prophase nuclei have a value for DNA constant within the limits of the experiment. When sets of separating anaphase and telophase chromosomes are measured, each set now has half the amount of DNA found in the interphase, prophase, and metaphase nuclei. It follows that DNA must be synthesized during interphase. There is excellent evidence, not only that this is actually so, but that mitosis will not normally start until the DNA of the nucleus has been doubled.

Precise information telling us the exact time and location of the synthesis of new DNA may be obtained by using radioactive precursors of nucleic acid. A commonly used substance of this type is radioactive thymidine. It seems to enter cells and to be directly incorporated into nuclei of dividing cells of both plants and animals. Root tips of *Tradescantia* grown in a culture solution containing radioactive thymidine show that the interphase may be divided into three periods. During the first stages, the nucleus increases in volume, but there seems to be no incorporation of thymidine-^{3}H. Later, radioactive material is incorporated into newly synthesized DNA. A period follows in which the nucleus maintains a constant size without any further incorporation of labeled thymidine. One investigator calculated that it took about 20 hr for a complete cell division in *Tradescantia* root tips. The time intervals for each stage are as follows: the presynthetic interphase (a period during which the nucleus is increasing in size but no DNA is being incorporated), 4 hr; the period of DNA synthesis, 10.8 hr; the postsynthetic interphase (a period during which the nuclear volume is constant and there is no DNA synthesis), 2.7 hr; prophase, 1.6 hr; metaphase, 0.3 hr; and anaphase–telophase, 0.6 hr.

But these observations do not tell us the manner of replication of the double-stranded DNA helix. It is possible that a whole new double-stranded helix might be replicated, thus, totally conserving the old double helix intact. Or the coiled double strands could somehow separate and a new strand could be formed by each old strand. This is semiconservative; half of the new helix contains old DNA. Or, the replication could take place at random throughout the DNA molecule.

The following observations make it possible to set up an experiment to definitely demonstrate that the second, or semiconservative, method of replication actually occurs:

1. The DNA helix is double-stranded.
2. The time of replication is known: It occurs during the second 4 hr period of interphase and is followed by 6–8 hr of growth without replication of DNA.
3. Thymidine, even radioactive thymidine (thymidine-^{3}H) is incorporated into the replicating DNA unit.
4. Any radioactive material forms, as does light, a latent image on a photographic plate. Upon development, this latent image is changed to visible silver grains.

The following experiment was carried out by Taylor, working at Columbia University, with seedlings of *Crepis capillaris*. Roots were placed in a nutrient solution containing radioactive thymidine for 6 hr and were then transferred to a similar nutrient solution with unlabeled thymidine for another 6 hr. The root tips were squashed on a slide and then carefully pressed out onto a very thin photographic film, where they were left for about a month. The film was then developed while the root tips remained in place upon it. The chromosomes in the root tips were stained. The result was a slide with stained chromosomes lying above a thin photographic film

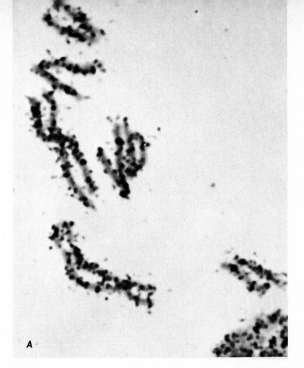

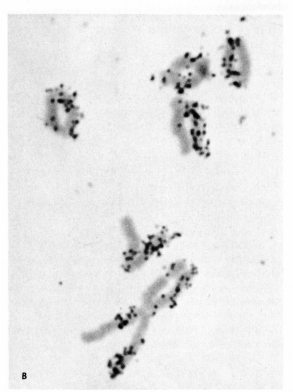

Figure 6.18 Autoradiograph showing presence of DNA synthesized with radioactive thymidine. *A*, first mitotic division after incorporation of thymidine-^{3}H; *B*, second division after incorporation of thymidine-^{3}H. (Courtesy of J. Herbert Taylor.)

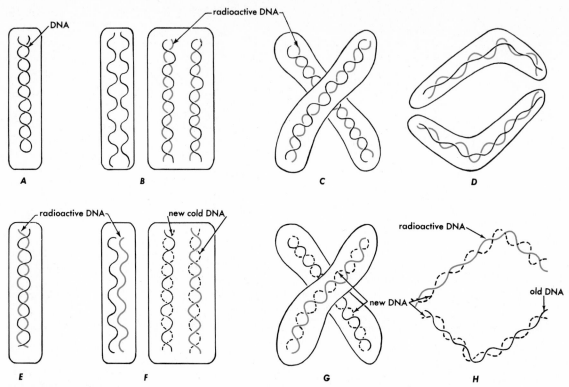

Figure 6.19 Diagram of postulated steps in DNA synthesis. Roots in thymidine-^{3}H 6 hr, in unlabeled nutrient solution during second mitotic division. Double helix represents DNA molecule; solid line indicates original molecule; green line indicates DNA formed in presence of thymidine-^{3}H; and dotted line shows DNA formed after incorporation of thymidine-^{3}H. *A* to *D*, first mitosis: *A*, chromosome in early interphase before replication; *B*, midinterphase replication of DNA; *C* and *D*, late prophase and anaphase, respectively, of first mitosis after thymidine-^{3}H incorporation; *E* to *H*, second mitosis: *E*, early interphase before incorporation of unlabeled thymidine; *F*, midinterphase after incorporation of unlabeled thymidine; *G* and *H*, late prophase and anaphase, respectively, of second division.

in which the silver grains present indicate the location of the radioactive thymidine taken up during the replication of the DNA molecule.

Examination under the microscope showed an equal concentration of silver grains under both chromatids of the late prophase (Fig. 6.18). Since each chromatid contains a complex double-stranded helix of DNA, this tells us that there must be newly replicated strands of DNA in each daughter chromatid. Each DNA helix in the daughter chromatids must contain one new and one old strand. In other words, replication has been semiconservative. If this is so, a second division in "cold" (nonradioactive) DNA should result in late prophase chromosomes with one labeled and one unlabeled chromatid (Fig. 6.19). This is what happened when roots were examined after a second division (Fig. 6.18*B*). But an unexpected thing occurred in some of the chromosomes. The radioactive thymidine was not confined

to one chromatid, but was present, at different levels, on both chromatids. Somehow during the replication, new and old strands had exchanged partners so that each strand consisted of alternating and complementary lengths of new and old strands. This happening is analogous to the phenomenon of crossing over, which will be discussed in Chapter 17.

The next problem is to explain the manner in which a single double-stranded helix can become two double-stranded helices, and how each helix can contain one strand of the old helix plus a new and complementary strand.

Assume that a double helix is present in early interphase at the onset of replication. This helix may untwist, separating the two strands (Fig. 6.20). Each strand now serves a template for the formation of a new strand. Thus, following replication with radioactive thymidine available, each chromatid will contain a strand bearing the radioactive material. In a

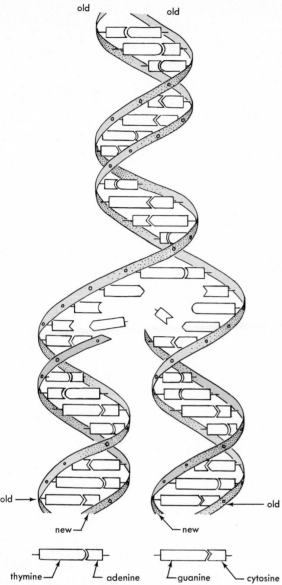

old   old

old →

old →

new

new

thymine — — adenine   guanine — — cytosine

Figure 6.20 Schematic diagram showing how one "old" double-stranded molecule of DNA could be replicated into two new double-stranded molecules of DNA. Each new molecule has one old and one newly synthesized strand.

subsequent division in the absence of radioactive thymidine, the radioactive strand will serve a template for a new nonradioactive strand as well as the old nonradioactive strand.

The one important aspect of the replication is the astonishing fact that a thymine molecule is always linked opposite an adenine base or vice versa. Also, guanine is always linked opposite a cytosine, or the reverse. The size relationships of the bases are such that adenine plus thymine is equal to guanine plus

cytosine. Thus, this combination of bases results in parallel strands of phosphate and ribose. Any other combination of bases would deform the helix.

The events taking place in cell divisions through DNA replication may be summarized as follows:

1. Cell division always precedes differentiation and growth.
2. Cell division appears to be required because only relatively small amounts of cytoplasm can interact with a single nucleus.
3. The presence in chromosomes of linearly aligned "qualities" controlling inheritance was first postulated in 1887. The importance of chromosomes as the physical basis of inheritance was established between 1900 and 1904.
4. Nucleic acid was first isolated in 1868; its location in the chromosomes was shown by Feulgen in 1924, and the molecular structure of the DNA molecule was described in 1953.
5. Replication of DNA involves the uncoiling of the DNA double helix, the separation of the complementary polynucleotide chains, and the formation of new polynucleotide chains adjacent to the old chains.
6. Both nucleus and chromosomes are characterized by fibers of undetermined length and about 17.5 nm in diameter. The diameter of the double helix of DNA is 2.5 nm.
7. The DNA molecule consists of two parallel helices formed by alternating pentose sugars (deoxyribose) and phosphorus. The helices are joined together through hydrogen bonding between pyrimidine and purine bases. The unit, deoxyribose, phosphate, and pyrimidine or purine, is a nucleotide.
8. The two associated pyrimidine and purine bases constitute a base pair. A pyrimidine base always joins with a purine base.
9. In DNA, the purine bases are adenine and guanine, and the pyrimidine bases are cytosine and thymine.

Protein Synthesis

RNA

So far we have followed the steps in the division of the nucleus through the steps of mitosis, and we have considered the structure and replication of the DNA molecule, and the manner in which it is enabled to serve as a storehouse of genetic information. Now we shall consider the manner in which this information is transmitted to the sites of protein synthesis in the cytoplasm. RNA is involved in this transfer of information. There are three major types

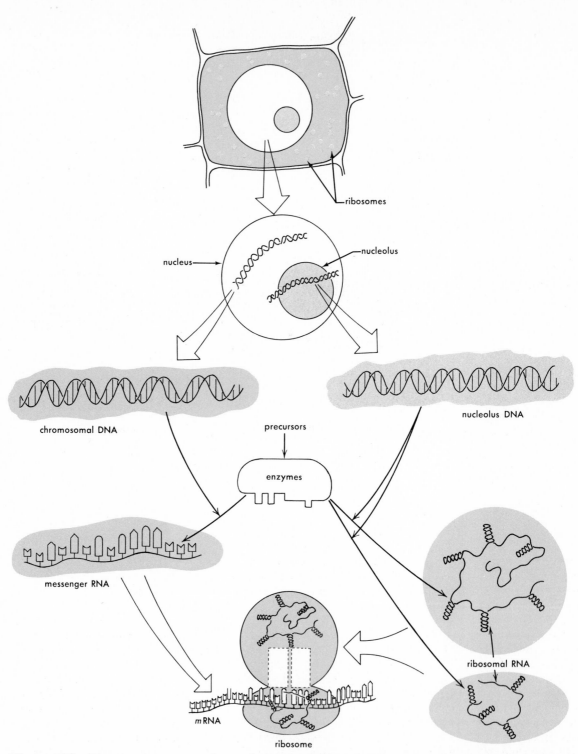

Figure 6.21 Diagrams representing the synthesis of *m*RNA and *r*RNA and their relationship. *m*RNA and *r*RNA are synthesized from precursors in reactions catalyzed by enzymes. Chromosomal DNA codes the *m*RNA. DNA in the nucleolus codes the *r*RNA. *m*RNA is single-stranded. *r*RNA is also single-stranded, but it doubles back upon itself to form regions of a double helix. *r*RNA strands are of two sizes and they become associated with proteins to form ribosomes which appear to have two components of different sizes. The *m*RNA becomes associated with the smaller component. The larger component has sites that function in the formation of the polypeptide chain (see Fig. 5.8).

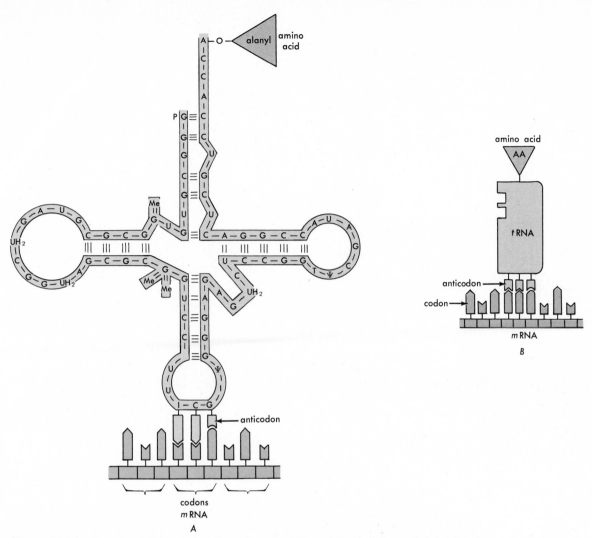

Figure 6.22 Transfer RNA. *A*, two-dimensional representation of the molecule showing a single strand of RNA folded back upon itself with specific cross-linkages (≡) between some nucleotides. One end of the molecule is keyed to bind with a specific amino acid. The opposite end (an anticodon) is coded to match a codon on *m*RNA. *B*, conventional shape frequently used to denote *t*RNA.

of RNA: messenger RNA (*m*RNA), transfer RNA (*t*RNA), and ribosomal RNA (*r*RNA) (Fig. 6.21). They are all synthesized in the nucleus under the direct influence of the DNA helix. In the cases of *m*RNA and *t*RNA, the DNA helix is located in the chromosomes. On the other hand, a portion of the DNA helix residing in the nucleolus is implicated in the synthesis of *r*RNA.

The molecule structure of RNA is similar to that of DNA in some ways and very different in others. Both have a similar backbone chain of alternating phosphate and sugar groups. Both have sequences of two pyrimidine and two purine bases linked to

the sugar so that in the resulting nucleotide the bases have position lateral to the main length of the chain. However, in RNA, the pentose sugar is ribose rather than deoxyribose. And the pyrimidine base uracil replaces the thymine that occurs in DNA. Perhaps the most striking difference between the two molecules is that a double-stranded helix is characteristic of DNA, while the RNA molecule exists as a single long strand, sections of which fold back to form a modified double strand.

The synthesis of RNA is directed by the DNA molecule. In the process the coded information carried in the sequential arrangement of the nucleotides in the DNA molecule is transcribed in new RNA molecules. RNA synthesis is initiated and terminated by an enzyme called RNA polymerase.

The synthesis of RNA molecules resembles the replication of DNA (a) in that the initial purine bases are joined with complementary pyrimide bases or vice versa, and (b) in the linking of new sequences of bases by ribose and phosphate groups to form a single continuous strand of RNA. This results, as in DNA synthesis, in a new strand bearing the complementary bases of the old strand in exactly the same sequence. A significant difference in the process of replication of DNA and synthesis of RNA lies in the disjoining of the newly formed sequences of RNA nucleotides from the DNA template to give a molecule of single-stranded RNA, portions of which fold back to form regions of a double helix.

Codons

Proteins are composed of amino acids linked together by peptide linkages (Fig. 5.8) to form complex

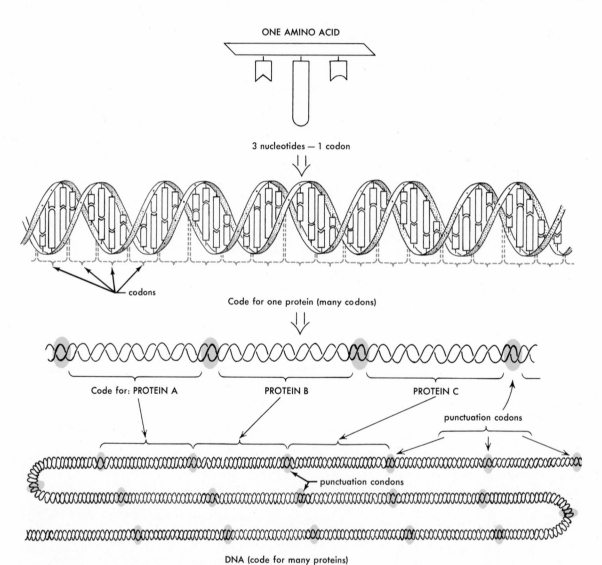

Figure 6.23 A sequence of three nucleotides is a codon. One codon codes for one amino acid: many codons code for a single protein. A single DNA molecule carries the code for many proteins. The protein codes are separated by special "punctuation" codons. The codon may be likened to a word, the protein to a sentence, and the DNA molecule to a paragraph.

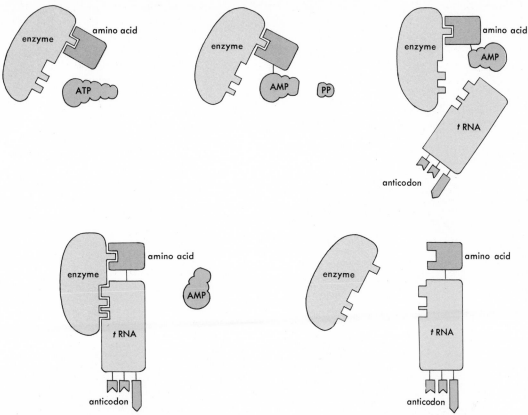

Figure 6.24 Diagram showing one manner in which *t*RNA could select a specific amino acid from an amino acid pool. Note that ATP and an enzyme are involved. The binding sites on the enzyme are specific for a certain amino acid and for the molecule of *t*RNA bearing the anticodon for that amino acid.

molecules similar to that illustrated in Fig. 5.9. The problem before us is to understand how a series of nucleotide bases in the DNA molecule can be responsible for the specific and accurate sequence of amino acids in a protein. A sequence of three nucleotides on a DNA molecule, known as a **codon,** codes for one amino acid. Sixty-four different codons may be formed from different combinations of four nucleotides. Three codons are punctuation codons marking the starting and end points of a given protein. All remaining codons code for the 20 or so amino acids, so several different codons may code for the same amino acid. The new molecules of RNA contain fewer nucleotides and are much shorter than the DNA molecules that serve as templates. Indeed, we may think of the DNA molecule as a paragraph. A sequence of codons coding for a protein would be a sentence, and a long one. Each codon itself represents a word (Fig. 6.23).

Figure 6.25 Electron micrograph of corn root tip showing ribosomes, some of which may be associated as polysomes.

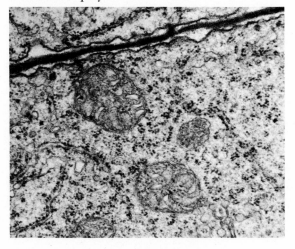

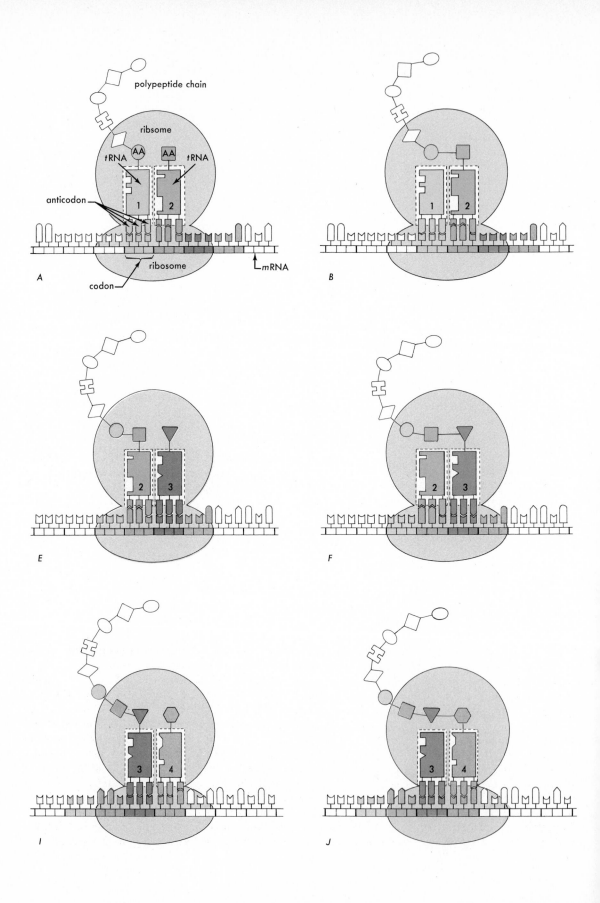

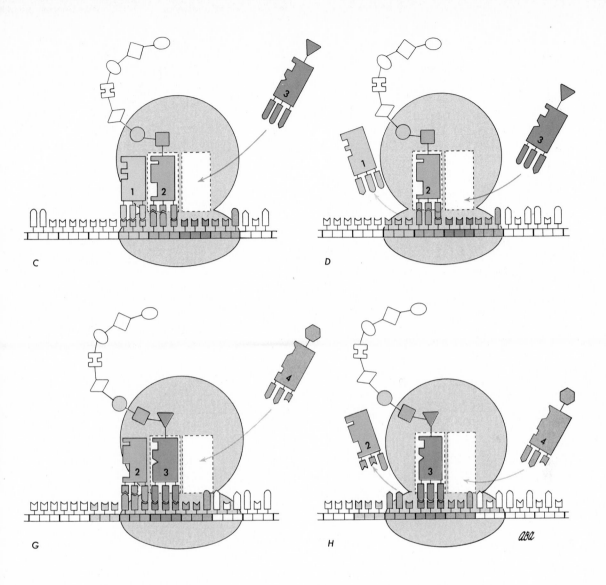

Figure 6.26   Protein synthesis. *A*, ribosome with *m*RNA strand in place in lower component, *t*RNA molecules in place in active sites, short polypeptide chain. The amino acid is bound to *t*RNA molecule 1. The polypeptide linkage with the amino acid on *t*RNA molecule 2 has not been formed. Note anticodon of the *t*RNA molecules and matching codons of the *m*RNA molecule. *B*, peptide linkage has formed between amino acids in the active sites, and bond linking amino acid with *t*RNA molecule 1 has broken. *C*, *t*RNA molecule 1 has left the reactive site, the ribosome has moved along the *m*RNA strand a codon to the right. *t*RNA molecule 2 is now in the left-hand reactive site and the right-hand reactive site is empty. *t*RNA molecule 3 is approaching the ribosome. *D*, *t*RNA molecule 1 leaves the ribosome and *t*RNA molecule 3 prepares to enter it. *E*, *t*RNA molecules 2 and 3 now occupy the reactive sites. Bond between *t*RNA molecule 2 and associated amino acid is intact. *F*, peptide linkage forms and bond between *t*RNA molecule 2 and its amino acid is broken. *G, H, I,* and *J*, ribosome reads *m*RNA and fourth amino acid is added to the growing polypeptide chain. It is important to understand that only *t*RNA molecules with anticodons matching the codons of *m*RNA can enter the active site. To form a polypeptide chain of 4000 amino acids takes about 1 second.

tRNA

There are three different types of RNA cooperating to synthesize a protein. *m*RNA carries the code for a specific codon sequence from the DNA to the site of protein synthesis. *t*RNA transfers amino acids from the cytoplasmic amino acid pool to the site of synthesis in the ribosome. *r*RNA somehow performs in fashioning the peptide linkage between two adjacent amino acids.

*t*RNA (Fig. 6.22) is easily isolated and is frequently also called soluble RNA. There are about 60 different *t*RNAs, each composed of about 80 nucleotides. A specific enzyme and ATP are involved in loading each *t*RNA molecule with its specific amino acid. We may diagram these steps by assuming that the molecules involved are shaped, or keyed, to match each other as are the pieces of a picture puzzle (Fig. 6.24). The first step, catalyzed by an enzyme, is the coupling of ATP with the amino acid (Fig. 6.24). The combination enzyme–amino acid is now keyed to a matching molecule of *t*RNA. The *t*RNA thus brought into contact with the amino acid becomes attached to it and the enzyme is released to catalyze other amino acid–*t*RNA combinations.

Three nucleotides of the *t*RNA molecule are free to combine with their complementary nucleotides on the *m*RNA molecule. Recall that in DNA and RNA synthesis purine combines with pyrimidine or vice versa. A similar relationship holds here. The nucleotides on the *t*RNA molecule are complementary to those on the *m*RNA molecule. Since the latter are known as a codon, the complementary nucleotides on the *t*RNA molecule are known as an anticodon (Fig. 6.22).

A conjectural structure of a *t*RNA molecule is shown in Fig. 6.22. Note that it is composed of a single strand, which doubles back upon itself forming definite regions of double helix. The anticodon site is indicated as well as the attachment point for the amino acid.

*t*RNA carries the amino acid to the ribosome. *m*RNA becomes associated with a ribosome. Subsequent steps in protein synthesis between *t*RNA and *m*RNA occur when both are associated with a ribosome.

mRNA

Molecules of *m*RNA vary greatly in size depending largely upon the length of the code message they carry. They may code for one, or several, proteins. The smallest may have a molecular weight of around 500,000 and carry about 900 nucleotides; others may have 12,000 nucleotides with a molecular weight of

4,000,000. There are many types of *m*RNAs; they are difficult to isolate, and only a few are known. The best examples come from bacteria with a virus infection. Some other well-known *m*RNAs have been isolated from red blood and from muscle cells. About 90% of the RNA of the cell is *m*RNA. This is understandable, because molecules of *t*RNA and *r*RNA may be used repeatedly, while many transcriptions are needed to carry the information for protein synthesis. *m*RNA may be thought of as a long backbone helix of phosphate–ribose linkages to which are attached the nucleotides, each three forming a codon (Figs. 6.21, 6.26).

rRNA

*r*RNA, as the name suggests, is associated with ribosomes (Fig. 6.26). It is formed on a DNA template just as are the two other types of RNA. There are two subunits in the *r*RNA molecule, both of which are required for the activity of *r*RNA. The nucleolus appears to play some part in bringing the two subunits together before they leave the nucleus (Fig. 6.21). The relative size of the *r*RNA subunits seems to depend upon their species origin. In the *r*RNA of *Escherichia coli*, which is the best-known, one unit has a molecular weight of about 600,000 and the other, of about 1,300,000. They both complex with proteins, about 10 different proteins being associated with the smaller subunit and 20 with the larger. The molecule of *r*RNA is composed of sections of both double and single helices (Fig. 6.21). x-Ray pictures of the molecule appear distorted, and a complete molecular model for *r*RNA has not been proposed. Nor has the precise function of *r*RNA been established. It seems to act mainly in a structural manner by bringing *m*RNA, with its loaded *t*RNA, into position for establishing the peptide linkage between two amino acids. This step occurs in the ribosome.

Peptide Linkage

The peptide linkage (Fig. 5.8) forms when the strand of *m*RNA with its loaded *t*RNA associates with a specific synthesizing site in the ribosome. This association may start at one end of the *m*RNA strand or at specific sticky points along the strand (Fig. 6.25). Only short portions of the *m*RNA strand become associated with a ribosome, so several ribosomes may be attached to the same strand of *m*RNA. Such an association may be referred to as a polysome (Fig. 6.25). When two adjacent *t*RNA molecules come to occupy the specific binding site, the peptide linkage is formed; the linkage between the amino acid and *t*RNA disappears and the hydrogen bonds holding *t*RNA anticodon to the *m*RNA codon are broken. The *t*RNA thus released diffuses to the cytoplasmic amino acid pool and is ready to transfer another amino acid to the ribosome binding site. Another amino acid has been added to the growing protein molecule. The process continues and as the protein molecule lengthens it becomes free from the ribosome (Fig. 6.26).

Under optimal conditions it takes about 10 sec for a polypeptide with a molecular weight of 40,000 to be synthesized in the bacterium *Escherichia coli*.

Summary of Protein Synthesis

1. RNA is a single-stranded molecule, composed of a backbone chain of phosphate and ribose units to which are attached the nucleotides uracil, guanine, pyrimidine, and cytosine.
2. There are fewer nucleotides in a molecule of RNA than there are in a molecule of DNA.
3. There are three types of RNA, all coded by the nucleotides on the DNA molecule.
4. The three types of RNA are transfer RNA (*t*RNA), messenger RNA (*m*RNA), and ribosomal RNA (*r*RNA).
5. There are 60 types of *t*RNA, each carrying about 80 nucleotides.
6. *t*RNA transfers an amino acid from a cytoplasmic amino acid pool to a specific codon on a molecule of *m*RNA associated with a ribosome.
7. There are many types of *m*RNA, each molecule being coded for a specific sequence of amino acids as determined by the nucleotide sequence of the DNA strand involved in the synthesis of *m*RNA.
8. The molecule of *r*RNA is composed of two subunits, both complexing with proteins. The molecular structure is not known.
9. A strand of *m*RNA and the amino acid-loaded molecules of *t*RNA are associated with a ribosome. At a specific combining site the codon and anticodon unit and the peptide linkage form. The protein molecule grows and the molecule of *t*RNA returns to the cytoplasmic amino acid pool.
10. *In vitro* studies on protein synthesis indicate that it takes about 10 sec to form a polypeptide with a molecular weight of 40,000.

7

Stems— Primary Growth

We have seen in Chapter 6 that mechanical strength of cellular structure was obviously a prerequisite for growth on land. Only when stems were supplied with supporting tissue could they rise into the atmosphere and display leaves, facilitating the trapping of radiant energy of sunlight. This increase in plant body size required anchorage in the soil and a conducting system for the transportation of water, food, and nutrients throughout the plant body (Fig. 3.2). Also required was a means of protecting meristematic and reproductive structures from desiccation. Furthermore, some mechanism must have evolved providing for survival over annually recurring periods unfavorable for growth. With the successful development of specialized structures to meet these requirements, there evolved the dominant present-day land flora.

Functions of Stems

Stems provide mechanical support and raise leaves into the air, thus facilitating photosynthesis. Flowers and fruits are also produced in positions facilitating pollination and seed dispersal. **Stems support.**

Stems provide a pathway for movement of water and mineral nutrients from roots to leaves, and for transfer of foods, hormones, and other metabolites from one part of the stem to another. **Stems conduct.**

The normal life-span of plant cells is 1–3 years. Water and mineral salts in dilute solution move in dead cells, but this movement depends upon the activity of living cells in leaves and roots generally less than 3 years old. Stems in herbaceous perennials (Fig. 7.1B) and in 2000-year-old redwoods, or bristle-cone pines (Fig. 7.1A), annually provide new living tissue for normal metabolism of the plant. **Stems produce new living tissue.**

Stems thus have three major functions: (a) support; (b) conduction; and (c) the production of new living tissue.

Growth in Stems

All stems grow in length; new living tissue is added at shoot tips. Some stems grow in girth; new living tissue is added to the circumference of the stem. All stems that increase in girth, first grow in length; hence, growth in length is **primary growth.** Growth in girth, which follows primary growth, is **secondary growth.** Secondary growth not only produces additional mechanical support but also, in both roots and stems, produces anew each year young and active cells for conduction. Secondary growth makes possi-

101

Figure 7.1 *A*, bristle-cone pine (*Pinus aristata*); gymnosperms and woody angiosperms achieve longevity through secondary growth, which annually results in a cylinder of new tissue around the trunk. The older tissues soon die. *B*, *Iris*; most other seed plants attain longevity by the continued growth of new primary tissue at the apex of the shoot. Older parts of the stem soon die. New leaves form at the apex, roots arise in nodes behind the apex, and older portions of rhizome die.

ble the attainment of great age by individual plants. The connection of new primary growth at the widely separated points of root and shoot apices (Fig. 3.1) by new secondary growth provides an efficient mechanism enabling even ancient trees to carry out all life processes in cells 1–5 years old (Fig. 7.1*A*).

In general, secondary growth is most prominent in woody angiosperms and gymnosperms. With few exceptions, the stems of all other land plants (ferns, horsetails, club mosses, herbaceous angiosperms, and monocotyledons) do not increase in girth year after year.

There are many annuals or biennials among the dicotyledons and monocotyledons. These plants, alive for 1 or 2 years, live out their life-span with tissues produced only by primary growth. They produce flowers and seeds and die. Longevity, with only primary growth, is attained in two general ways. First, the stem may continue to grow indefinitely in length. This generally results in a horizontal stem, either underground or creeping along the surface of the ground. Roots are formed at nodes close to the growing tip of the stem (Fig. 7.1*B*). Upright portions, leaves or shoots, may be produced only at the grow-

ing tip or also at special regions behind the tip, the nodes (Fig. 3.2). Old portions of these horizontal stems die and rot away, leaving separate plants. The example in Fig. 7.1*B* is an iris rhizome. A second method of continuing growth with only primary tissues is regular production of new short stems, from which grow upright flowering stems. Each year's crop of stems may die back during the winter. This occurs in bulbs, such as daffodil (Fig. 7.27*D*), or corms such as gladiolus (Fig. 7.27*B*). Here, the older bulb or corm is used up to produce the current flowering stalk, and a new bulb, or corm, is formed by primary growth in anticipation of the next season. In long-lived monocotyledons, such as the palms, Joshua trees, *Dracena*, and *Pandanus* (Fig. 7.29), other mechanisms are invoked (p. 125).

The principles of primary growth are generally similar throughout the vascular plants. We will first discuss primary growth as it occurs in a woody angiosperm. This will be followed by a discussion of primary growth in a monocotyledon and then by a fuller account of the ways longevity may be attained without secondary growth. Secondary growth is considered in Chapter 8.

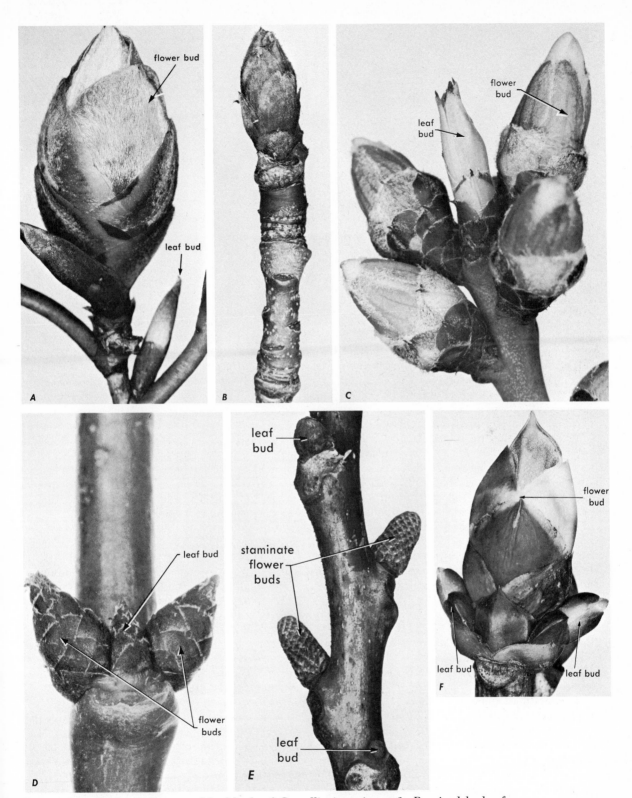

Figure 7.2 Buds. *A*, flower and leaf buds of *Camellia japonica*, ×1; *B*, mixed buds of pear (*Pyrus*) on short fruit spur, ×2; *C*, flower and leaf buds of almond (*Prunus amygdalus*), ×1; *D*, a single leaf bud of apricot (*Prunus armeniaca*) with two accessory flower buds at a node, ×2; *E*, the naked staminate bud of walnut (*Juglans regia*), ×2; *F*, a flower bud of lilac (*Syringa*) with two opposite leaf buds, ×2.

Figure 7.3 *A*, terminal mixed bud of buckeye (*Aesculus californica*), $\times 1\frac{1}{2}$; *B*, with bud scales opening, $\times\frac{1}{2}$; *C* and *D*, stages in expansion over a 24 hr period, $\times\frac{1}{2}$.

Buds

The basic structure of vascular plants is an axis with potentially immortal apical meristems at opposite ends of the axis. Specialized protection for the root apex involves only a root cap (Fig. 9.5), which serves to protect the root as it grows through the soil. More elaborate protection is needed for stem apical meristems, particularly in perennial stems which overwinter. In such cases, the shoot meristem is dormant during the unfavorable season. It may be protected from desiccation by small modified leaves known as **bud scales.** Such a dormant shoot apex with its protective scales is a bud (Figs. 7.2, 7.3). In a few instances, protective scales are lacking and the unprotected dormant apex is a **naked bud.** Actively growing shoot apices are also supplied with a protective covering of young leaves. Such active shoot apices are not generally called buds, and they occur most generally in the annual or biennial species.

Kinds of Buds

We shall see (Chap. 15) that growth terminates in shoot apices that produce flowers. Yet the production of flowers early in the season is frequently advantageous for woody plants; therefore, at the time of resumption of growth in spring, new primary growth must provide for the production of new shoots and flowers. Consequently, there are buds that produce only stems and leaves **(leaf buds),** buds that produce only flowers **(flower buds),** and buds that produce both leaves and flowers **(mixed buds).** *Camellia japonica* always produces, frequently at the same node, both flower and leaf buds (Fig. 7.2*A*). Apples and pears produce mixed buds on short spurs (Fig. 7.2*B*). These spurs grow less than 1 inch a year, but a severe pruning will induce the growth of a long

vegetative shoot rather than the usual production of a short shoot with fruit. Almonds produce both flower and leaf buds on short spurs (Fig. 7.2*C*). Apricots often exhibit three buds at a node; the middle bud is a leaf bud and the two **accessory** buds are flower buds (Fig. 7.2*D*). In walnut, there are both leaf buds and staminate catkin flower buds (Fig. 7.2*E*). The large flower bud of lilac is flanked by two small leaf buds (Fig. 7.2*F*).

A **terminal** bud of buckeye is shown in Fig. 7.3*A*. Separating bud scales reveals the largest rudimentary leaves (Fig. 7.3*B*). When all rudimentary leaves are removed, there remains a small floral apex and leaf bud primordium in an axil of a rudimentary leaf (Figs. 7.4*A*, *B*). This bud is a mixed bud. Figure 7.3 shows several stages, over a 24 hr period, in the expansion of a terminal bud. When expanding leaves are removed, only several small unexpanded rudimentary leaves remain (Fig. 7.4*A*). This bud is a leaf bud.

What determines whether a shoot tip will produce leaves alone, flowers alone, or both leaves and flowers? The answer is still very elusive. But it is known that, in some cases, induction of a floral apex is related to day length, to a hormone, and to two specific wavelengths of red light. The stimulus is transmitted from leaves to shoot apex.

A median, longitudinal section of a terminal bud of flax is shown in Fig. 7.8. The bud is well-protected by bud scales. Rudimentary leaves are attached at nodes (Fig. 3.2) to the conical shoot apex, which is protected, here, by several of the smallest leaves. Midribs and veins are already apparent in rudimentary leaves. Internodes (Figs. 3.2, 7.4) are very short in the bud. Elongation of these internodal regions will account for the growth of the shoot in length. Bud primordia in axils of rudimentary leaves are not commonly seen in leaf buds. However, if shoot growth is to occur in a mixed bud, provision must be made for vegetative growth. In the buckeye mixed bud with a terminal flower primordium, bud primordia do occur in axils of rudimentary leaves which in Fig. 7.4B have been removed in dissection.

Arrangement of Buds

A 3-year-old twig of walnut in winter condition is shown in Fig. 7.5. The tip of the twig generally bears a large terminal leaf bud (Fig. 7.5). At regular intervals along the stem, other buds may be seen; they are called lateral buds. Note that below the base of each lateral bud there is a scar that was made when a leaf fell from the twig; this is a **leaf scar.** Vascular bundle scars (Fig. 7.5) may be seen within each leaf scar; strands of food-and water-conducting tissues passing from the stem into the leaf stalk were broken when the leaf fell, leaving these scars. Buds and leaves are usually borne in this relationship to each other; buds form in the angle made by the stem and the leaf stalk. This angle is termed the **leaf axil,** and consequently these buds may also be called axillary buds.

The passage of strands of conducting tissue into leaves and buds must mean that the internal structure of the stem is so arranged as to make possible a continuous connection of leaves and buds with all other parts of the stem. We have thus a specialized stem region—a region to which are attached buds and leaves and within which conducting strands unite with other conducting strands of the stem. This region is known as **node** (Fig. 7.5). The region between any two adjacent nodes is an **internode.**

Protecting the young immature cells within the bud is a series of overlapping scales; **bud scales.** They are usually shed when the bud develops into a new shoot, and they also leave scars, bud-scale scars. The part of a stem or twig between sets of terminal-bud-scale scars is generally formed during one growing season. For instance, growth made by the twig this year is set off from growth made last year by means of a ring or girdle or terminal-bud-scale scars (Fig. 7.5). When scales of a terminal bud fall off in spring, they leave a number of closely crowded scars that form a distinct ring. Examination of several-year-old twigs shows that growth in length may vary from year to year. This is shown by the different spacings between the terminal-bud-scale scars. It is significant that there is no increase or decrease in the length of any portion of a stem after that portion is 1 year old.

The slightly raised areas on the bark are **lenticels.** They are composed of cells that fit loosely together, with air spaces between, which permit passage of gases inward and outward.

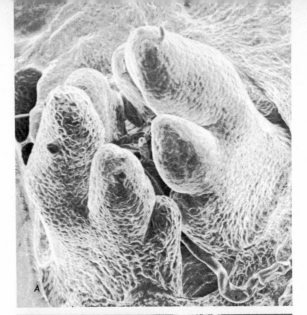

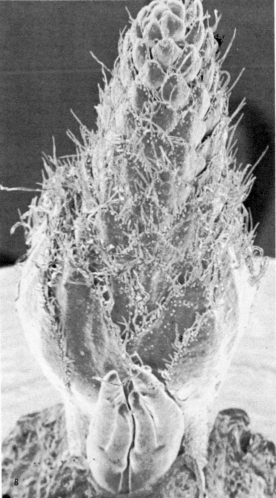

Figure 7.4 Stereoscan micrographs of shoot apices of buckeye (*Aesculus californica*). *A*, shoot apex, enveloped by opposing rudimentary leaves, ×250; *B*, floral apex, with adjacent small shoot apex, ×40. (Courtesy of D. Hess)

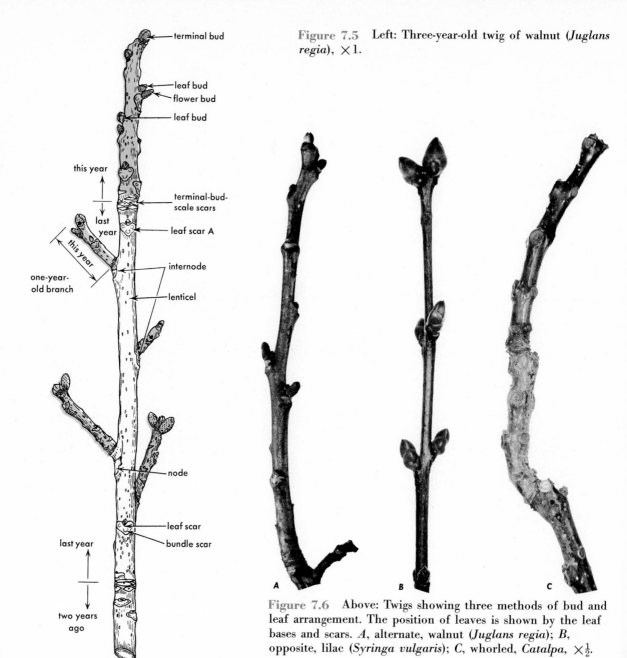

Figure 7.5   Left: Three-year-old twig of walnut (*Juglans regia*), ×1.

Figure 7.6   Above: Twigs showing three methods of bud and leaf arrangement. The position of leaves is shown by the leaf bases and scars. *A,* alternate, walnut (*Juglans regia*); *B,* opposite, lilac (*Syringa vulgaris*); *C,* whorled, *Catalpa,* ×½.

Position of Buds on a Woody Twig

In the walnut twig there are just *one leaf bud* and *one leaf at each node.* This arrangement of buds and leaves on the stem is spoken of as **alternate** (Figs. 7.6*A*, 7.7*A*, *C*, *E*). It is the most common type of bud and leaf arrangement. Ash, maple, lilac, and many other plants have two leaves opposite each other at each node, and a bud in the axil of each leaf (Fig. 7.6*B*, 7.7*D*, *F*). This arrangement of leaves and buds is spoken of as **opposite.** When three or more leaves and buds occur at each node, as in *Catalpa,* leaf arrangement and bud arrangement are said to be **whorled** (Fig. 7.6*C*).

Some plants have several buds in or near the leaf axil. For example, apricot often has a group of three buds in the leaf axil: a central bud, which develops into a side branch, and two lateral ones, which are flower buds (Fig. 7.2*D*). All but the central one are called **accessory buds.** Walnut may also have more than one bud in the leaf axil.

Not infrequently, buds may arise on the plant at

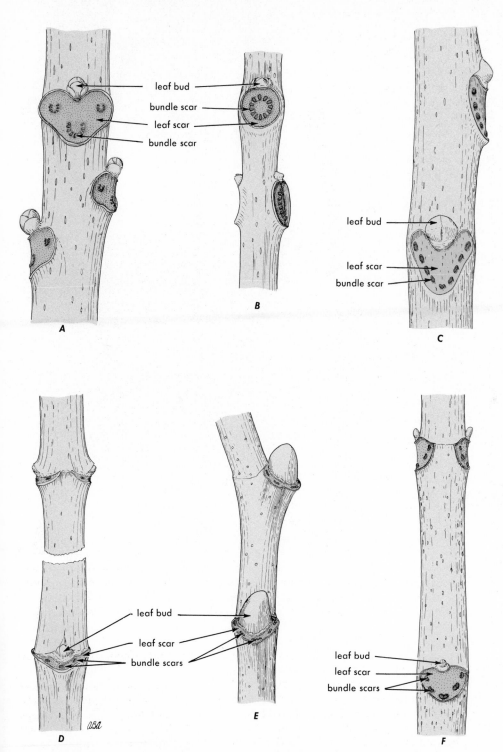

Figure 7.7 Leaf scars and bud arrangement of different species of woody plants. *A*, walnut (*Juglans regia*), ×2; *B*, catalpa (*Catalpa bignonioides*), ×2½; *C*, tree of heaven (*Ailanthus altissima*), ×2; *D*, box-elder (*Acer negundo*), ×2; *E*, European plane (*Platnus acerifolia*), ×3; *F*, buckeye (*Aesculus californica*), ×2.

places other than leaf axils. They may appear on stems, roots, or even leaves, and give rise to new shoots. Such buds are called **adventitious buds.** Their formation may be stimulated by injury, such as occurs in pruning.

Dormant or **latent buds** arise in a regular fashion in the leaf axil, but their development is usually inhibited by the dominance of the terminal bud.

From the above discussion it is seen that buds may be classified by their **arrangement on the stem,** which may be (a) **alternate,** (b) **opposite,** or (c) **whorled;** by their **position on the stem,** which may be (a) **terminal,** (b) **lateral (axillary),** (c) **accessory,** or (d) **adventitious;** and by the **nature of the organs into which they develop,** which may be (a) **leaf,** (b) **flower,** or (c) **mixed.**

As a rule, the terminal bud of a stem is the most active and grows more vigorously than any of the axillary buds. Usually, the lowest lateral buds on a year's growth of the shoot remain dormant and do not develop into branches. If the terminal bud is removed, however, as may be done in pruning, lateral buds, otherwise dormant, may become active.

Internal Characteristics of Buds

A longitudinal section of single leaf bud is shown in Fig. 7.8A, B. The bud is essentially a miniature shoot, with nodes and very short internodes and with leaf primordia at each node. Primordial buds may occur in the axils of the more advanced leaf primordia.

Note (Fig. 7.8A) that the cells of the shoot tip are not all alike. Groups of cells having a definite function to perform are known as **tissues.** For instance, some cells are constructed and grouped to protect, others to support or to conduct. If we have one cell type performing but a single function, we have a **simple tissue.** Frequently, however, several cell types will be organized into a tissue performing more than one function, such as support and conduction. In this case, we may speak of a **complex tissue.**

Let us examine the cells and tissues of the bud. At the apex of all shoots, whether terminal or lateral, there are dome-shaped masses of cells called **apical meristems.** Cells of this apical meristem have large nuclei, a compact cytoplasm, and small vacuoles (Fig. 4.19). While the bud is dormant these cells are relatively quiescent, but when conditions are favorable some cells divide rapidly and the daughter cells increase in size, becoming in some instances larger than the cell from which they originated. This increase in number and size of cells results in elongation of the young shoot. But in conjunction with

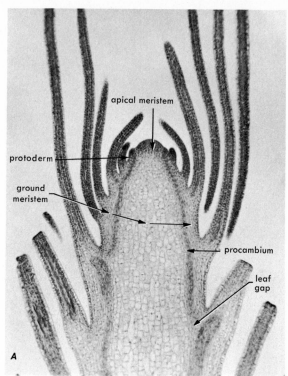

Figure 7.8 Above and right: Sections of shoot of flax (*Linum*). *A*, photomicrograph of longitudinal section of apex, ×300. *B*, diagram of longitudinal section from apex to completion of primary growth, length of stem shortened. Apical meristem, dark green; procambium, light green; primary xylem and primary phloem, light gray. *C* through *F*, diagrams of cross-sections of shoot taken from longitudinal section at positions indicated. (*A*, courtesy of E. M. Gifford.)

increase in number and size of cells, differentiation occurs; that is, cells change morphologically and physiologically from the meristematic cells from which they arose. Thus, a short distance (usually a few millimeters) below apical meristem we recognize three fairly distinct **primary meristematic tissues,** namely **protoderm, ground meristem,** and **procambium** (Fig. 7.8A–D). These three tissues are derived directly from apical meristem.

The term **shoot tip** is applied to the youngest part of the shoot and includes apical meristematic cells associated with leaf primordia. Some of these meristematic cells are elongating and others are in early stages of differentiation. Rapid elongation of stems results from the formation and growth of thousands of new cells in the shoot tip. It is in the shoot tip that we see the origin of leaf primordia, bud primordia, and primary meristematic tissues.

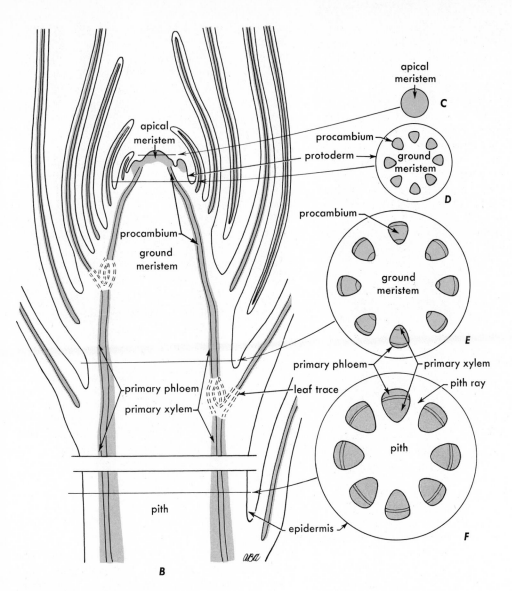

Development of Tissues of the Primary Plant Body of a Woody Stem

The Primary Meristems

The process of stem development is gradual. The youngest cells are those of the apical meristem, and they are similar morphologically. But in the process of differentiation, different cell types and tissues are formed. Thus, as seen in Fig. 7.8, and as mentioned previously, we recognize three different tissues, protoderm, ground meristem, and procambium. These three primary meristematic tissues differentiate into **primary tissues.**

Protoderm (Fig. 7.8A, D) is the outermost layer of cells. It develops into **epidermis**—the special primary tissue that covers and protects all underlying primary tissues. The epidermis prevents excessive water loss and yet allows for exchange of gases necessary for respiration and photosynthesis.

Ground meristem (Fig. 7.8A, B, D) comprises the greater portion of meristematic tissue of the shoot tip. Ground meristem cells are relatively large, thin-walled, and isodiametric. Primary tissues forming from the ground meristem are (a) **pith,** in the very center of the stem, and (b) **cortex,** in a cylinder just beneath the epidermis and surrounding the vascular

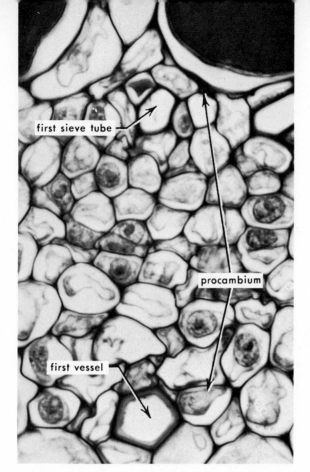

first sieve tube

procambium

first vessel

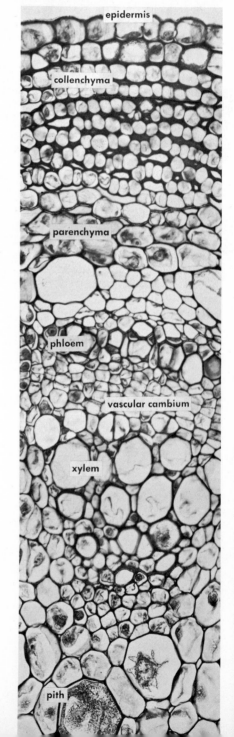

epidermis

collenchyma

parenchyma

phloem

vascular cambium

xylem

pith

tissues. Sometimes pith and cortex are connected by (c) **pith rays,** also formed from ground meristem (Fig. 7.8F).

Procambium cells usually appear first as strands among ground meristem cells (Fig. 7.8A). In cross-section (Fig. 7.8D), strands appear as isolated groups of cells arranged in a circle. Sometimes a continuous **procambium cylinder** is formed. As seen in a transverse section of a single procambium strand, procambium cells are smaller than those of the surrounding ground meristem; in longitudinal section, they are much longer, and some of them may be pointed at the ends. Procambium cells give rise to **primary vascular tissues** (Figs. 7.8, 7.9). These primary tissues carry out several functions and are divided rather rigidly into three groups according to these functions. Food is conducted in the outer group of primary vascular cells, which is **primary phloem.** Water and mineral salts are conducted in the inner group of primary vascular cells, which, together with strengthening cells, constitute **primary xylem.** In many stems, a meristematic region remains between primary xylem and primary phloem to become **vascular cambium** (Figs. 7.10, 8.1).

It is well to repeat here that **primary tissues** of the stem are those differentiated from the three primary meristematic tissues—protoderm, ground meristem, procambium—and that these three are derived from apical meristem of the shoot tip. In woody plants, we must look for primary tissues of the stem a very short distance behind the stem tip. Even before the end of the first season's growth, differentiation of these primary tissues from primary meristematic tissues is completed, and **secondary tissues** may be formed in abundance. Whereas primary tissues are derived from primary meristematic tissues of the shoot, secondary tissues are the result of production of new cells by **vascular cambium** and by **cork cambium**. The origins and nature of these two types of cambiums are discussed later.

A Summary of Primary Development

apical
meristem
{
protoderm   epidermis

ground {cortex
meristem {pith and pith rays

procambium {phloem
{vascular
     cambium
{xylem

The term **stele** is applied to the part of the stem that includes primary vascular tissues, pith, and pith rays. The **primary plant body** is composed of the above primary tissues.

The main functions of these primary tissues may be summarized as shown below.

Epidermis: Protects underlying tissues
Vascular tissues
  Phloem: Conducts food
  Vascular cambium: Produces secondary phloem and secondary xylem
  Xylem: Conducts water and mineral salts, and gives strength to stem
Cortex: Stores food and, in young stems, manufactures food, strengthens, and protects
Pith: Stores food
Pith rays: Store food, and conduct water, mineral salts, and food radially

Primary Tissues

The primary structural organization in the stems of most gymnosperms and dicotyledonous angiosperms consists of several tissues, possessing different cell types. The different types of cells described in the following few pages may be more or less isolated or they may be grouped to form a tissue.

The Epidermis. The epidermis is usually a single superficial layer of cells covering all other primary

tissues and protecting them from drying out and, to some extent, from mechanical injury. It is the limiting layer of cells between the plant and its environment. In surface view (Fig. 7.11), epidermal cells are elongated in the direction of the stem's length; in transverse section, they are usually isodiametric. Protoplasm forms a thin layer lining cell cavities and normally retains its living properties for a long period.

The outer tangential wall of cells exposed to air is usually thicker than the other walls, and its surface

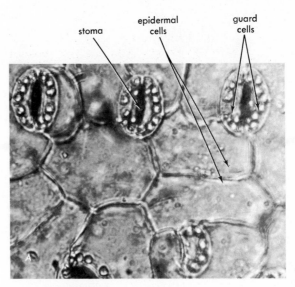

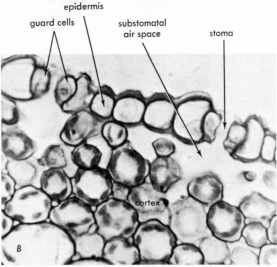

Figure 7.11 Epidermis. *A*, epidermis from bean stem showing epidermal cells, guard cells, and stoma; *B*, cross-section of alfalfa (*Medicago sativa*) stem showing epidermal cells, guard cells, stoma, and cortex with substomatal chamber, ×400.

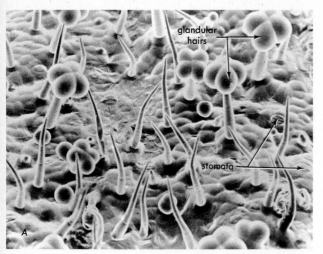

Figure 7.12 Stereoscan views of epidermal hairs. *A*, of tomato stem (*Lycopersicon esculentum*), ×130; *B*, lower epidermis of leaf of gambel oak (*Quercus gambelii*), ×230. (*A*, courtesy of D. Hess; *B*, courtesy of J. Tucker.)

layer is usually coated with a waxy substance called **cutin.** This superficial layer of cutin is termed **cuticle.** By means of certain stains and chemical tests, cutin can be readily distinguished from the cellulose of the rest of the wall. The cuticle is quite impermeable to water and gases. The inner walls, parallel to the stem surface, are thinnest, and radial walls, at right angles to the surface, often taper in thickness toward the inner wall.

Young stems usually possess specialized epidermal cells called **guard cells** (Fig. 7.11). Between each pair of guard cells is a small opening, the **stoma** (stomate) through which gases enter and leave the underlying stem tissues. Guard cells differ from ordinary epidermal cells in their crescent shape, as seen in surface view. Guard cells are usually thought of as structures peculiar to foliage leaves. While they are much more common in epidermis of leaves, they may occur also in epidermis of young stems, floral structures, and fruits.

Epidermal appendages, such as hairs, may occur on young stems (Fig. 7.12).

The Cortex. This complex tissue, derived from ground meristem, forms a cylindrical zone beneath the epidermis extending inward to the primary phloem (Figs. 7.8*F*, 7.10). The following simple tissues or cell types may be found within it: parenchyma, collenchyma, sclerenchyma, and secretory tissue.

Parenchyma. The principal tissue of the cortex is parenchyma (Figs. 7.10, 7.13*G*). It usually consists of isodiametric cells with thin walls, mostly of cellulose, and with protoplasts that remain alive for a long time. We may speak either of a parenchyma tissue

or of a parenchyma cell, one of the units composing this tissue.

Parenchyma tissue is characterized by the presence of intercellular air spaces which vary greatly in size; in some parenchyma tissues they are difficult to find, while in others they are very apparent (Fig. 7.13*G*). Because parenchyma cells retain active protoplasts, they function in the storage of water and food, or in photosynthesis, and sometimes in secretion. They also retain for a very long time the potential for differentiation. The green color of many stems is due to the presence of chloroplasts in parenchyma cells, **chlorenchyma,** of the cortex.

Parenchyma tissue is not confined to the cortex of the stem, but occurs in practically all tissues of the plant: flowers, fruits, seeds, leaves, and roots.

Collenchyma. The outermost cells of the cortex of young stems, lying just beneath the epidermis, often constitute a tissue known as **collenchyma.** This tissue may form a complete cylinder or it may occur in separate strands. The cells are elongated rather than isodiametric as are those of parenchyma, and have

Figure 7.13 Right: Development of tissues of the primary plant body of a woody stem. Cell types and tissues. *A* and *B*, fibers in lengthwise view; *C*, fibers in cross-section; *D*, collenchyma in cross-section; *E*, collenchyma in lengthwise view; *F.* sclereid; *G*, parenchyma; *H*, stone cells; *I*, woody parenchyma. (*A* and *B*, redrawn from Carl C. Forsaith, *The Technology of New York State Timbers*, New York State College of Forestry Publications 18; *F*, after A. S. Foster.)

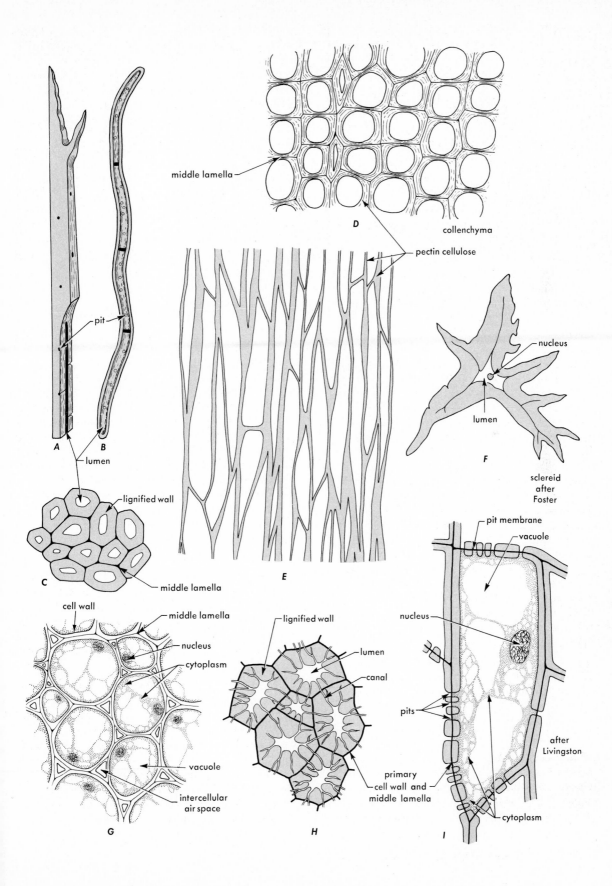

middle lamella

collenchyma

pit

lumen

pectin cellulose

nucleus

lumen

F

sclereid
after
Foster

A B

lignified wall

middle lamella

C

E

pit membrane

vacuole

nucleus

pits

after
Livingston

cell wall

middle lamella

nucleus

cytoplasm

vacuole

intercellular
air space

G

lignified wall

lumen

canal

primary
cell wall and
middle lamella

H

cytoplasm

I

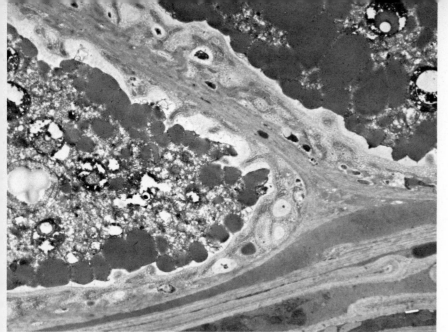

Figure 7.14 Transfer cells. *A*, portions of two cells from the aleuron layer of the grain of red foxtail grass (*Setaria lutescens*). *B*, xylem parenchyma transfer cells from the cotyledonary node of a seedling of *Lactuca*, ×275. (*A*, courtesy of T. L. Rost; *B*, courtesy of V. C. Pate and B. Gunning, *Ann. Rev. Plant Phsiol.* **23**, 173, 1972.)

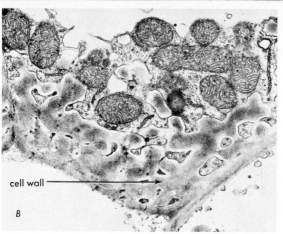

cell wall ⟶

B

pointed, blunt, or oblique ends. In the most common type of collenchyma, the cell walls are thickened at the corners (Figs. 7.10, 7.13*D*, *E*). Because of these thickenings, collenchyma serves as a strengthening tissue. The walls are composed of pectin and cellulose. As in parenchyma, cells of collenchyma have long-lived protoplasts. Chloroplasts are often present.

Collenchyma cells may also occur in other plant parts comprised of primary tissues. For instance, collenchyma is frequently associated with veins of leaves.

Sclerenchyma. The main functions of sclerenchyma cells are support and, in many cases, protection. Their shape and the thickness and toughness of their walls contribute to the ability of these cells to support and protect the young stem. Thickness and toughness of walls are increased by deposition, within the original cellulose wall, of a substance known as **lignin.** The process of deposition is called lignification. Lignin is elaborated by the protoplast, and when deposition ceases, the protoplast usually dies. There are two types of sclerenchyma cells: (*a*) **sclereids** and (*b*) **fibers** (Figs. 7.13*A*, *B*, *C*, *F*).

Of various types of sclereids, the most common are stone cells (Fig. 7.13*H*), which are more or less isodiametric. Other types of sclereids are much branched, resembling very irregular stars (Fig. 7.13*F*). Some sclereids are derived from parenchyma cells by pronounced thickening of cell walls; others arise from separate meristematic cells. Minute canals, **pit canals,** extend outward through thickened walls, being separated only by primary walls from ends of similar canals in walls of adjoining cells (Fig. 7.13*H*).

Sclereids occur not only in cortex of stems but also in hard shells of fruits, seed coats, and bark, in pith of stems, and in certain leaves.

Fibers. **Fibers** are elongated, thick-walled cells, usually pointed at the ends (Figs. 7.13*A*,–*C*). They give strength to the tissue in which they occur. Each fiber is **one cell.** Its walls may or may not be lignified. When lignified, the walls may be so thick that the cavity, **lumen,** of the cell almost disappears. Fibers are usually very elastic and can be stretched to a great degree without losing their ability to return to their original length. Various types of fibers occur in vascular tissue of stems as well as in the cortex. Protoplasts of fibers often disappear as they attain maturity.

Secretory Cells. Secretory cells are parenchyma-like, with dense protoplasmic contents. They secrete various substances, such as resinous materials (see Fig. 8.13) and nectar. Many epidermal hairs are secretory cells (Fig. 7.12*A*).

Transfer Cells. Frequently cells located in positions of active solute transfer will show irregular extension of the cell wall into the protoplast. Since the plasmalemma follows the contour of the wall, its surface is greatly extended. Mitochondria appear to aggregate adjacent to these areas of increased membrane surface. The morphological picture thus presented (Fig. 7.14) suggests an adaptation facilitating transport from one cell to another, or from the interior to the exterior of the plant. These cells have been called **transfer cells.**

Summary of the Cortex. Thus, we may find in the cortex of stems several kinds of tissues and cell types. Usually it is not possible to find all kinds in the cortex of any one species of plant. Parenchyma occurs in most species and usually predominates in vascular plants. Collenchyma, cortical fibers, transfer cells, and secretory cells are rarer.

The Primary Vascular Tissues

The term "vascular" pertains to tissues that conduct various substances in liquid form. For example, veins and arteries that carry blood in higher animals are vascular tissues. In vascular plants, water and different water-soluble inorganic salts from soil, as well as food substances, are conducted throughout the plant in well-defined vascular tissues.

In a young dicotyledonous stem, very near the bud (Figs. 7.8*A*, *B*, *D*), vascular tissues occur as separate bundles, **primary vascular bundles.** Each primary vascular bundle is differentiated from a procambium strand. Each mature primary vascular bundle of all vascular plants consists of primary xylem and primary phloem. If secondary growth is to occur, a thin band of meristematic tissue destined to become vascular cambium remains between primary xylem and primary phloem (Figs. 7.22, 8.1, 8.2). Between the vascular bundles are parenchyma cells forming **pith rays.** In the center of the stem **pith** is also composed of parenchyma (Figs. 7.8*F*, 7.22, 7.10).

The Primary Phloem

Phloem in angiosperms may possess several types of cells: **sieve-tube members, companion cells, fibers, sclereids,** and **parenchyma.** In gymnosperms, companion cells are lacking; there are **sieve cells** rather than sieve-tube members, and in primary phloem of many species there may be no fibers. An important role of phloem is the conduction of organic solutes. Sieve tubes, rather than other types of phloem cells, serve as the passageways.

A **sieve tube** is a vertical row of elongated cells; each cell is known as a **sieve-tube member** (Fig. 7.15). Among angiosperms, a sieve-tube member and a companion cell are sister cells; that is, they originate by division from the same procambial cell. Young sieve elements have the usual complement of organelles: nucleus, plastids, mitochondria, and dictyosomes. As the element matures, its protoplast becomes greatly modified. Its nucleus is thought to disintegrate. Plastids lose most of their internal membranes, but usually retain a carbohydrate granule. The mitochondria become small. The cytoplasm, much reduced in amount, becomes reduced to a thin peripheral layer. The central part of the cell is occupied by a large mesh or tangle of fibers or tubules (Fig. 7.15). This mass may be seen with the light microscope, and has been called **slime.** Since it is now known to be a protein, it is referred to as P-protein (Fig. 7.16). At maturity the sister cell (companion cell) has a normal protoplast with a full complement of organelles (Fig. 7.17).

A characteristic structural feature of mature sieve tubes is the **sieve plate** (whence the term sieve tube). It may occur in end or side walls (Figs. 7.15, 7.16, 7.18). In end walls, sieve plates are seen to the best advantage. The end wall between two adjacent sieve-tube members is thickened, and strands of cytoplasm pass through pores in it. Hence, protoplasts of adjoining sieve-tube members are connected.

The physiology of mature sieve-tube members, is still in doubt, even though they must move the main food stream of the plant body. They are living, functioning cells, but absence of nuclei and other modifications point to profound changes in metabolism. With the exception of a few trees such as palms, sieve-tube members live and function for about 3 years. New ones are formed annually. In many studies on structure of mature sieve-tube members, a carbohydrate known as **callose** is seen around margins of pores in the sieve plate (Fig. 7.18). In some instances, protein may collect at the sieve plate. Obviously, such a development would block the pores of the sieve plate and obstruct movement of food materials. It has been demonstrated in other tissues that callose forms very rapidly in response to wounding. Phloem cannot be studied without cutting and wounding cells. Furthermore, considering the relatively large amount of material moved through sieve tubes, it is quite possible that slime could pile up against a sieve plate when an osmotic equilibrium is upset by cutting a sieve tube. Because of the experimenter's own interference with the process of this experiment, it becomes practically impossible to know which appearance of fixed and stained images corresponds to living, uninjured sieve-tube members. The best guess is that in the

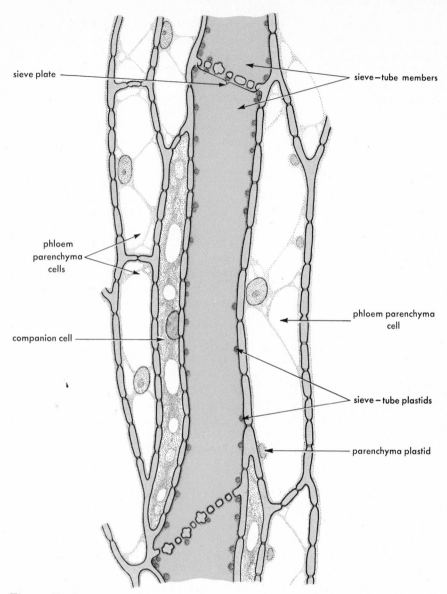

sieve plate

sieve—tube members

phloem
parenchyma
cells

phloem parenchyma
cell

companion cell

sieve—tube plastids

parenchyma plastid

Figure 7.15 Phloem tissue from the stem of tobacco (*Nicotiana*), ×400. (Courtesy of A. S. Crafts.)

living, functioning condition there is little or no callose on the margins of pores to obstruct movement of foods through sieve plates. P-protein, or slime, is present in the form of a fine reticulum or threads. It extends through the pores, and in some sieve-tube members, P-protein may be present as small P-protein bodies (Fig. 7.16).

Fibers of phloem have the same general characteristics as those described as occurring in the cortex. In most plants, development of fibers is greater in the phloem than in the cortex (see Fig. 8.14).

Xylem and phloem elements that have differentiated from procambium cells cease to be meristem-

atic. Certain cells derived from procambium, however, do not lose their meristematic character. They form a narrow band of tissue between xylem and phloem known as **vascular cambium** (Fig. 7.10). Procambium and cambium may be regarded as two developmental stages of vascular meristem.

The Primary Xylem

Conducting cells that occur in primary xylem of vascular plants are **tracheids** and **vessel elements.** These cells conduct water and mineral salts. Associated with them may be **fibers** (xylem fibers) and **parenchyma** (xylem parenchyma).

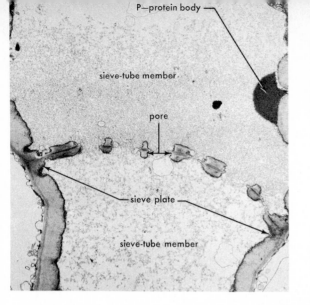

Figure 7.16 Left: Electron micrograph of sieve plate between two sieve-tube members of *Cucurbita maxima*, ×4600. (Courtesy of J. Cronshaw and K. Esau, *J. Cell Biol.* **38,** 292.)

Figure 7.17 Below: Longitudinal section through portions of three sieve-tube members and one companion cell. A nucleus and all cytoplasmic organelles may be identified within the companion cell. Most of the organelles are absent from one sieve-tube member, but may be recognized in the other two sieve-tube members, *Cucurbita maxima*, ×8300. (Courtesy of J. Cronshaw and K. Esau, *J. Cell Biol.* **38,** 25.)

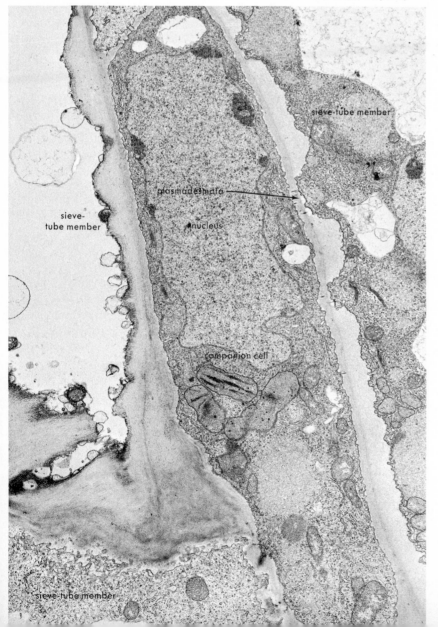

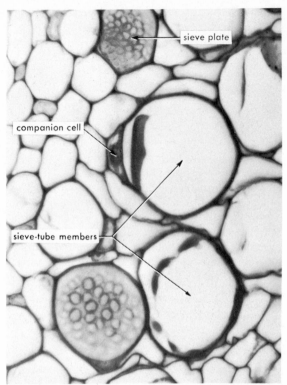

sieve plate

companion cell

sieve-tube members

Figure 7.18 Cross-section of phloem tissue of squash (*Cucurbita*). Sieve plates are visible in two sieve-tube members. The dark circles around the pores in the sieve plates are callose. Sieve plates do not appear in two other sieve-tube members because of sectioning. ×200. (Slide courtesy of Triarch Products.)

A **tracheid** is a single elongated cell more or less pointed at its ends (see Fig. 8.6). Functioning tracheids are not alive. The tracheid wall may not be the same thickness throughout. All the wall may be thickened except for numerous small, circular or oval areas called **pits** (Fig. 7.19). There are two types of pits in xylem cells, **simple pits** and **bordered pits.** A **simple pit** is shown in Fig. 7.13*l*. Here it will be observed that, where two cells lie side by side, the pit (depression or recess) in the wall of one cell is opposite the pit in the wall of the adjacent one. The term **pit-pair** is used to designate this condition. A pit-pair is *not* a hole in the wall; the two cells are separated by a middle lamella and the primary walls of each of the two cells. Plasmodesmata pierce this separating barrier. Pits of this type commonly occur in walls of parenchyma cells and vessel elements. The type of pit known as **bordered pit** (Fig. 7.19) occurs in tracheids, vessel elements, and xylem fibers.

A **vessel element** is a *single cell* with oblique,

pointed, or transverse ends. A **vessel** is a *series of vessel elements* differentiating end to end, with end walls perforated or dissolved. Thus, a vessel is a long tube (Fig. 7.20). Vessels are often several centimeters long and in some vines and trees they may be many meters in length. A row of procambium cells becomes transformed into a vessel. Before the protoplasts disappear, vessel walls become thickened, forming a secondary wall (Fig. 4.21); the thickening material is laid down on primary walls in various patterns so that, in places, secondary walls are thick and in others thin. The material deposited is cellulose; later, the layers of cellulose become lignified. The dissolution of end walls of vessel elements also takes place before the protoplasts disappear. Thus, the deposition of thickening material forming the secondary walls and the dissolution of end walls are functions of living cells. Following these processes, the protoplast dies. Therefore, functional vessels have no living contents.

The secondary walls of vessels in angiosperm stems are deposited in several different patterns (Fig. 7.21*A, B*). **Annular vessels** have lignin deposited as separate rings. In **spiral vessels,** lignin is laid down in the form of spiral bands. **Scalariform vessels** have the thickenings in the form of transverse, interconnecting bars. In **reticulate vessels,** the lignin principally forms a network on the wall. And in **pitted vessels,** the walls are pitted. The ends of the vessel elements (see Fig. 8.6) are generally on a slant, and although open, they may have bars of wall material across them (Fig. 7.21*B*). Vessels do not occur in small veins in leaves, and are lacking in most gymnosperms and lower vascular plants. In these forms, tracheids occur in elongating regions and they then may have annular and spiral types of secondary walls.

Xylem parenchyma cells outlive vessels, tracheids, and most xylem fibers. They function in the storage of water and foods, which, as we have learned, is one of the principal functions of parenchyma wherever it occurs in the plant. Parenchyma may also conduct materials for short distances.

Xylem fibers are similar to the fibers described elsewhere.

Pith and Pith Rays

Pith is composed of large-celled parenchyma with numerous intercellular spaces (Figs. 7.10, 7.13*G*, 7.22). Storage of food is its principal function. In some stems, primary vascular bundles are separated by wide strips of parenchyma that extend from the pith to the cortex. Such strips of parenchyma are called pith rays and may be considered radial extensions

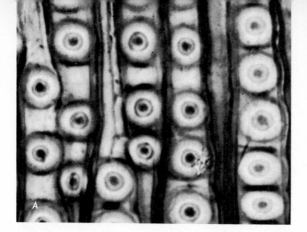

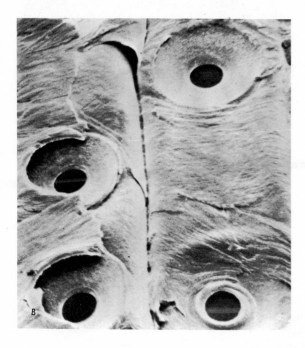

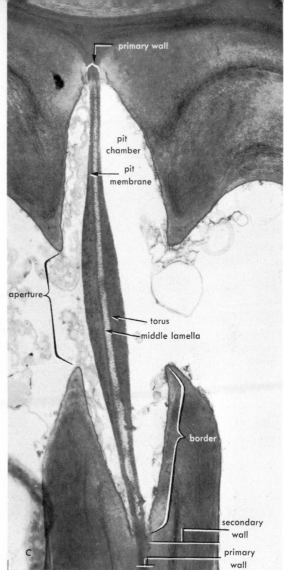

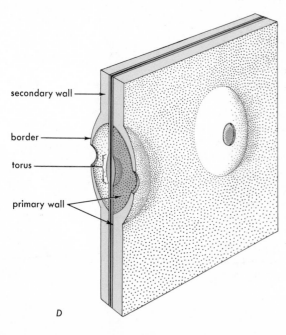

Figure 7.19 Bordered pits. *A* and *B*, in tracheids of pine wood. *A*, light micrograph, ×1800; *B*, stereoscan micrograph, ×2400; *C*, electron micrograph of bordered pit in ground hemlock (*Taxus canadensis*), ×4300; *D*, diagram. (*A*, courtesy of Artschwager; *B*, courtesy of S. Cook and D. Hess; *C*, courtesy of M. Ledbetter).

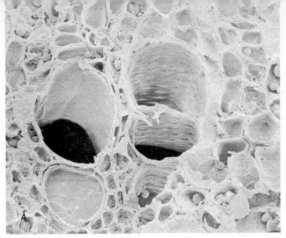

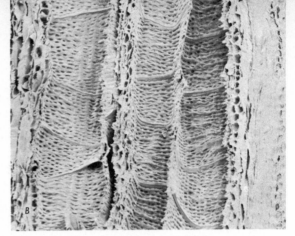

Figure 7.20 Stereoscan view of vessels. *A*, cross-section; *B*, longitudinal section (note the rims of old cross-walls now no longer present). (Courtesy of D. Hess.)

of the pith (Figs. 7.8*F*, 7.22). Like pith, one of their functions is food storage; they also conduct materials short distances radially. They merge with parenchyma of the cortex so that no line of separation is visible. In many woody species primary vascular bundles are separated by very narrow rays of perhaps two to four cells in width (Figs. 7.23, 8.2).

The Herbaceous Dicotyledonous Stem

Many species of angiosperms lack secondary growth and therefore do not become woody. They are gen-erally referred to as **herbs.** They are usually annuals or biennials, and in the temperate zones they are generally small plants. Their external appearance is similar to that of the young woody twig, except that, in keeping with their short life, buds do not form, and they do not achieve a winter condition similar to that of the woody twig. They do have apical growth, nodes, and internodes. Branches, or modi-fications thereof, may be present in the leaf axils.

Tissues of both herbaceous and woody stems originate in the same fashion. Primary tissues are arranged in bundles in alfalfa as they are in *Sambucus.* Slightly different arrangements occur in other species.

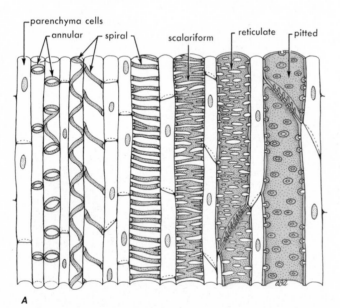

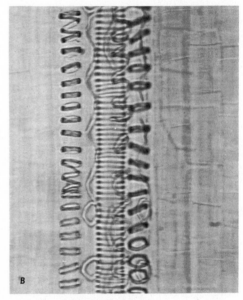

Figure 7.21 Types of vessels that occur in primary xylem in an elongating branch. *A*, diagram. Annular vessels were formed first, and are therefore the oldest and most stretched. The pitted vessel formed last is the youngest. Elongation has stopped, so the pitted vessels will not be stretched. *B*, photomicrograph of vessels during primary growth in cleared node of *Alternanthera*, ×30. (*B*, courtesy of L. M. Srivastava.)

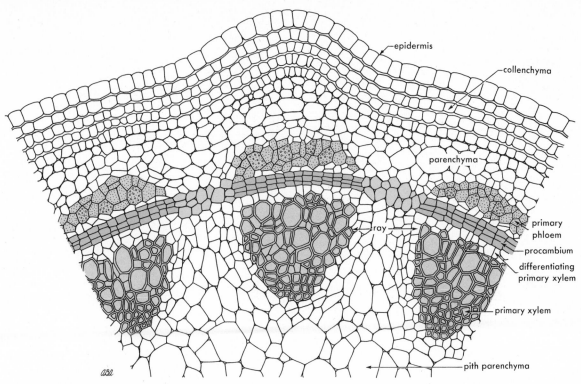

Figure 7.22 Diagram of a woody stem at the close of primary growth. Some procambium is still present, and the last primary xylem and primary phloem cells are still undergoing differentiation. Ray parenchyma separates vascular bundles. Cortex consists of collenchyma and parenchyma. Epidermis is intact.

Figure 7.23 Cross-section of elderberry (*Sambucus*) stem at stage of completion of primary growth, ×100. (Slide courtesy of Triarch Products.)

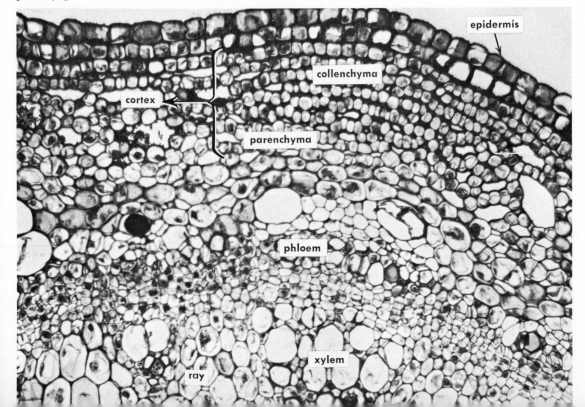

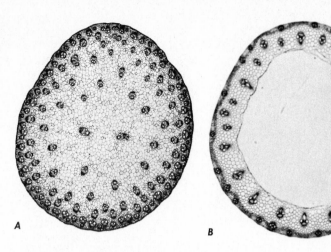

Figure 7.24 Cross-sections of monocotyledonous stems. *A*, corn (*Zea mays*); *B*, wheat (*Triticum*), ×30. (Courtesy of L. Feldman.)

These primary tissues are, from the epidermis inward: epidermis; cortex, composed of collenchyma, groups of fibers, and parenchyma; phloem with sieve tubes, companion cells, fibers, and parenchyma; xylem with vessels, tracheids, fibers, and parenchyma; pith and pith rays composed of parenchyma cells. The outermost cells of the primary phloem in alfalfa develop into fibers.

The Monocotyledonous Stem, as Exemplified by the Grasses

With few exceptions, monocotyledons do not increase in girth after their initial period of rapid growth. They do not have secondary growth. Some, such as palms and *Pandanus,* that do achieve considerable size, do not produce a woody stem suitable for lumber. There are many perennial monocotyledons, so mechanisms have been devised to circumvent the generally short life-span of primary tissues. Because of this, the external appearances and cellular anatomy of the monocotyledons show considerable variation. At this point, we will consider only monocotyledonous stems as they appear in the grasses, particularly corn (*Zea mays*). Corn is an annual. In external appearance its stem is not unlike that of an herbaceous dicotyledon. It has terminal growth, nodes, and internodes. Leaves, branches, flowers (tassel and ear), and fruits (ear) appear in leaf axils.

In stems of dicotyledonous plants, both herbaceous and woody types, and in those of gymnosperms, procambium strands, and hence vascular bundles, are usually arranged *in the form of a single ring*. With but few exceptions, procambium strands, and hence vascular bundles, in stems of monocotyledonous plants are *scattered* through out the ground meristem or at least through its outer region

(Fig. 7.24). However, in some grasses like wheat and barley, the stems are hollow, and vascular bundles are in a definite ring. In most monocotyledonous stems, *all* procambium cells differentiate into primary xylem and primary phloem elements; there is no vascular cambium and, as a consequence, no production of secondary tissues. Bundles of this sort which are "closed" to further growth are called **closed bundles** (Fig. 7.25). As has been observed, bundles of other seed-bearing plants that possess a vascular cambium are "open" to further growth; they are known as **open bundles.**

In transverse sections of internodes of most grasses (barley, rye, wheat), the following tissues are evident (Fig. 7.24*B*): (*a*) a single layer of epidermal cells, (*b*) strengthening tissue variously arranged beneath or near the epidermis, (*c*) ground parenchyma, and (*d*) vascular bundles. Stomata occur in the epidermis. Strengthening tissue consists of elongated fibers with thick lignified walls. In some grasses, vascular bundles are arranged in two rings near the periphery of the stem. In many grass stems, there is usually a continuous ring of fibers some distance from the epidermis, with the small bundles of the outer row embedded in it. On the outer sides of the small bundles of the outer ring occur strands of fibers that reach to the epidermis. Bands of chlorophyll-bearing ground parenchyma are enclosed between these strands. Ground parenchyma may extend to the center of the stem, as in corn (Fig. 7.24*A*), sorghum, and sugar cane. Or its central portion may become destroyed during the growth of the stem, leaving a hollow pith cavity, as usually happens in wheat (Fig. 7.24*B*), oats, barley, and rye.

The vascular bundle of corn stems is an example of a common type of bundle occurring in monocotyledonous stems (Fig. 7.25). Large pitted vessels of the xylem are prominent features. There are usually two of these, and one or two smaller annular

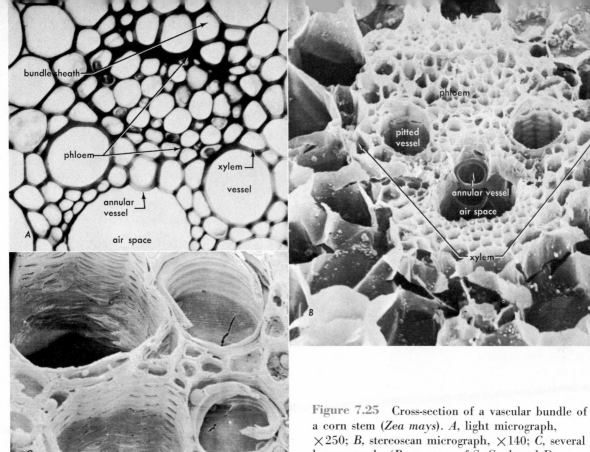

Figure 7.25 Cross-section of a vascular bundle of a corn stem (*Zea mays*). *A*, light micrograph, ×250; *B*, stereoscan micrograph, ×140; *C*, several larger vessels. (*B*, courtesy of S. Cook and D. Hess.)

or spiral vessels may also be present. In addition, older bundles invariably contain a large air space or intercellular passage. Close examination of this space is likely to reveal within it a lignified ring or portion of a ring. This is evidence that the air space was originally an annular vessel. Elongation of the young stem, which has stretched and broken the vessel, gives rise to the relatively large air space. Vessels, together with the intervening fibers, tracheids, and parenchyma cells, comprise the xylem tissue, which is always located on the side of the bundle toward the center of the stem. Phloem forms a regular pattern of thin-walled cells exterior to xylem. Sieve plates may usually be seen. Companion cells are small, generally square, or rectangular in cross-section, and, because of their cell contents, stain more heavily than the sieve tubes. The bundle is generally surrounded by lignified fibers forming a tissue called the **bundle sheath.**

In grasses, a tissue at the base of each internode usually remains meristematic long after the tissues in the rest of the internode are fully differentiated. Thus, such plants contain a meristem at the apex of the shoot and also a meristem at the base of each internode. Each internode has its own growing zone. These internodal meristems are called **intercalary**

meristems, and growth of the cells derived from them is termed **intercalary growth.** The flowering stems of such cereals as wheat, oats, barley, and rye shoot up very quickly. This rapid elongation results not only from the growth of cells derived from the apical meristem, but also from the growth of cells derived from intercalary meristems at the base of each internode.

Perennial Stems with only Primary Growth

Primary growth occurs most commonly at apices. Longevity of a primary plant body is then attained by continued elongation of the stem and root axis, resulting in a separation of root and shoot apices in space and time. Since herbaceous plants do not have secondary growth, there is no provision for connecting primary plant bodies at opposite ends of an elongating axis as the intervening primary tissues live out their life-spans and die. Instead, adventitious roots may arise at nodes a short distance behind the shoot apex. Nodes of perennial plants, lacking secondary growth, bear the usual leaves and buds, and also roots. Such stems are likely to have differen-

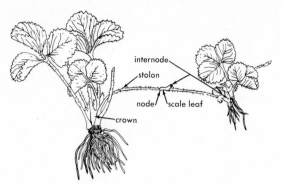

Figure 7.26 Runner of strawberry (*Fragaria*), $\times\frac{1}{8}$; roots and shoots are produced at every other node.

tiated upper and lower surfaces, with leaves and buds occurring on upper surfaces and roots on lower surfaces. Since these stems are not strengthened or enlarged by secondary growth, they are generally prostrate, frequently underground, and often small. There are numerous modifications among these stems. Comparable modifications of a primary plant body, which permit the attainment of longevity, also occur in such widely diverse groups as algae, horsetails, club mosses, ferns, and dicotyledons. A long-lived gymnosperm having only primary growth is unknown. Because of the wide diversity of these groups (see Chaps. 2, 28), it is thought that similar stem modifications must have evolved separately from several different starting points in different plant groups during different epochs, and therefore several times during the course of evolution. This is known as parallel evolution.

Stolon (or Runner)

Bermuda grass has above-ground horizontal stems called **stolons.** These stems creep along the ground, and at each node, shoots and roots arise. In strawberry (Fig. 7.26), roots and leaves arise at every other node.

Rhizome

In most iris species, leaves and flowering stalks are produced at the growing **rhizome** tip (Fig. 7.1*B*). The leaves die a relatively short distance back from the growing tip, so that many iris plants bear senescent leaves. Roots are also formed at nodes and they may remain for the life of the rhizome. Other rhizomes, like canna (Fig. 7.27*A*), produce upright leafy stems with terminal flowers at every third node. The intervening nodes are marked by only small sheath leaves.

Corms

A shortened vertical, thickened underground stem is a **corm** (Fig. 7.27*B*). In gladiolus, it consists of a short stem with much stored food. Nodes are, as usual, indicated by leaves; bases of some are shown in Fig. 7.27*B*. Small buds occur in axils of some of these leaves. A median section of a corm (Fig. 7.27*C*) distinguishes between stored food and the central portion containing a single bud that will produce a single leafy, flowering shoot. Food stored in a dormant corm is used in the production of the leafy shoot. New corms will develop from axillary buds. In addition, short underground stems may form, each giving rise, at its tip, to a single small corm.

Bulbs

A **bulb** differs from a corm in that in a bulb, food is stored in leafy scales. The stem portion is small and has at least one central terminal bud that will produce a single upright leafy stem. In addition, there is at least one axillary bud that will produce a bulb for the subsequent year. In the longitudinal section of the sprouting daffodil bulb shown in Fig. 7.27*D*, the stem is producing three leafy stalks, one of which is forming a new bulb.

Food stored in the leafy scales of a bulb is used up by the initial growth of a leafy shoot. Food to be stored for a new bulb is supplied from a leafy shoot. The table onion is a good example of a commercially valuable bulb.

Tubers

Tubers are enlarged terminal portions of slender rhizomes (Figs. 7.27*E*, 7.28). The potato, a dicotyledon (*Solanum tuberosum*), is a good example. The potato plant possesses three types of stems: (*a*) ordinary **aerial stems,** (*b*) slender rhizomes underground, which become enlarged at the tips and form (*c*) **tubers.** In the mature potato, the scar left where the tuber was broken from the rhizome is clearly visible. On the potato tuber there are nodes and internodes, lateral buds, and a terminal bud. Buds develop into stems (Fig. 7.27*E*). The "eyes" of the tuber are groups of buds; each group along the sides represents a lateral branch with undeveloped internodes. At the unattached "seed end" of the tuber, the "eye" is in reality a terminal branch on which only one bud is strictly terminal. In an elongated potato, it is possible to make out the spiral arrangement of the eyes, for there is only one eye at a node. Beginning at one end of a tuber and proceeding toward the other end, while at the same time turning the tuber, usually makes it possible to follow the spiral arrangement (Fig. 7.27*E*).

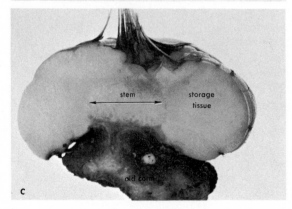

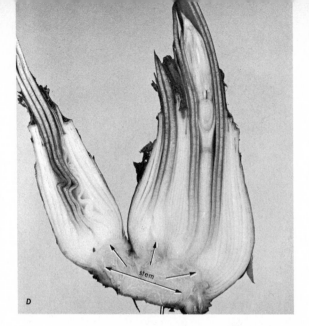

Figure 7.27 Plants illustrating continued development with only primary growth. *A*, roots and shoots are produced at every third node by the rhizome of *Canna*, $\times\frac{1}{2}$; *B*, corm of *Gladiolus*, $\times 1$; *C*, corm sectioned to show short stem with storage tissue of current corm and disintegration of corm of preceding year, $\times 1$; *D*, longitudinal section of young daffodil (*Narcissus*) plant, showing two bulbs, one with two shoots, united by the short stem; $\times 1$; *E*, potato tuber (*Solanum*), note spiral arrangement of the eyes and the short sprout, $\times 1$.

The Arborescent Monocotyledons

Palmaceae

Arborescent monocotyledons, in achieving longevity with only primary growth, pose a perplexing problem. Palms (Fig. 7.29*A*) have only primary growth. They continue to elongate, reaching a height of as much as 100 ft and living upwards of 200 years. Bases of their trunks are slightly swollen because the first internodes are shorter and broader than those produced later. There is also some continued enlargement of parenchyma cells in the trunk bases, with deposition of woody material in cell walls. There may also be an elaboration of a corky tissue (periderm). Otherwise, the diameter of a tall palm trunk

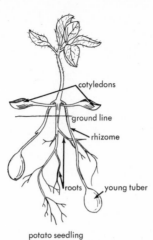

cotyledons

ground line

rhizome

roots   young tuber

potato seedling

Figure 7.28 A potato seedling showing development of young tubers at the end of slender rhizomes.

is constant. Palms have a massive apical cone from which new leaves and stem structure differentiate. Primary growth is presumably normal, with procambium strands branching out into differentiating leaves as well as growing upward in pace with the advancing apical meristem. New leaves are thus connected with the root system by way of living and functioning primary vascular tissue in palms over 200 years old. Even parenchyma cells in the oldest parts of the stem are still able to enlarge and elaborate new wall material. Presumably, sieve elements living and functioning during the 200 year life-span of palm trees lack nuclei. Though their ultrastructure is not known, the light microscope shows them to be similar to 1–2-year-old sieve elements of herbaceous plants. The only other living, functioning cell lacking a nucleus is the mammalian red blood cell, and its life-span is measured in weeks. The century life-span of the cells of a palm tree is without parallel.

Pandanaceae

Although many species of *Pandanus* develop a massive conical apex similar to that in palms, their continued arborescent growth is not anticipated (as it is in palms) by the initial production of an enlarged buttressed base (Fig. 7.29*B*). Instead, root primordia are produced in the axils of the leaves of the first five–ten nodes. At the summit of the stem, leaves are produced and die as in the palm. Buds, as usual, are present in leaf axils. While most of these buds remain dormant, a specific proportion grow, producing a regularly branched tree. Since branches also taper, many *Pandanus* trees resemble dicotyledonous trees. Each new branch has root primordia

at the first formed nodes. Additional support and probably increased capacity for conduction for the enlarging tree is obtained by development of roots from root primordia at lower nodes. Since secondary growth has not been observed in any of the Pandanaceae, they too must have long-lived primary vascular tissue in at least portions of their older stems. It is, however, not known whether, after formation of prop roots, primary tissue in the initial stem remains functional.

Agavaceae

In *Dracena, Yucca* (Fig. 7.29*C*), and other members of this family, a cylinder of cambium is present around the stem close to its surface. This cambium, however, produces secondary xylem and phloem and parenchyma cells mostly on the inner side of the cambium. At precisely separated points, small aggregations of parenchyma cells develop into a meristematic tissue resembling a procambium strand. From this, xylem and phloem differentiate in a manner similar to the differentiation of primary vascular tissue in a procambium strand. Thus, secondary growth occurs in the Agavaceae. Stems increase in girth, and primary vascular tissues at shoot and root tips are connected by young vascular tissues in the stem. However, in these cases a strand resembling a procambium strand is secondarily produced and the elaboration of xylem and phloem follows the pattern of primary development.

Other Arborescent Monocotyledons

There are still other small assemblages of arborescent monocotyledons, such as bamboo (*Bambusa*), *Philodendron,* and others that occur as isolated groups in large herbaceous families. Methods of production of vascular tissue in most of these forms is too unclear for consideration here.

Summary of Primary Growth in Stems

1. Buds are characteristic of woody stems. Bud scales enclose and protect rudimentary leaves surrounding an apex. The apex may be either a vegetative shoot or a floral apex. A floral apex terminates growth. Woody plants each year require new vegetative growth which is normally accompanied by flowers. Various devices have evolved that ensure the formation each year of buds enclosing floral and/or vegetative shoot apices.

Figure 7.29 Arborescent monocotyledons. *A*, California fan palm (*Washington filifera*) in a native stand; *B*, palms, and presumably *Pandanus*, have long-lived phloem; *C*, the Joshua tree (*Yucca brevifolia*) has secondary growth but produces typical closed vascular bundles from meristematic tissue resembling procambium strands. (*A*, courtesy of Richard Parker–National Audubon Society; *B*, courtesy of Diamon T. Smithers–National Audubon Society.)

2. Primary growth brings about elongation of stem and lays down the basic pattern of cells and primary tissues characteristic of the particular stem and upon which the functioning and future growth of the stem depend.

3. Primary meristematic tissues are: apical meristem, protoderm, ground meristem, and procambium.

4. The primary plant body is composed of the following different kinds of cells and tissues: (a) epidermis, with possibly three cell types (epidermal, guard cells, and epidermal hairs); (b) cortex, composed of collenchyma, sclerenchyma (fibers and sclereids), and parenchyma; (c) vascular tissue, composed of xylem (fibers, tracheids, vessel elements, and parenchyma) and phloem (fibers, sieve-tube members, companion cells, and parenchyma); (d) pith, composed largely of parenchyma with sclereids; (e) pith rays, composed of parenchyma cells.

5. The functions of these cells and tissues are as follows. Parenchyma is used for storage of water and food and conduction of materials for short distances. Collenchyma, sclereids, fibers, and tracheids are strengthening or mechanical tissue elements. Tracheids and vessels (series of vessel elements) conduct water and mineral salts. Sieve tubes (series of sieve-tube members) conduct foods. Vascular cambium is a meristematic tissue, the cells of which are capable of division.

6. Collenchyma and sclerenchyma may occur in patches or completely surround the stem just underneath the epidermis. Sclerenchyma is frequently associated with vascular bundles.

7. In woody and herbaceous stems, vascular tissues are arranged in bundles, generally forming a definite circle. In monocotyledonous plants, bundles are irregularly distributed throughout ground parenchyma.

8. In woody dicotyledonous plants, some procambium, together with parenchyma cells of pith rays, unite to form a continuous cylinder of vascular cambium which proceeds to lay down the secondary plant body.

9. With the exception of a few genera such as palms, the life-span of plant cells does not exceed 5 years. Longevity of individual woody species is attained by secondary growth. In species with only primary growth, longevity is attained by continued primary growth at the shoot apex and the formation of roots at nodes. Older portions of the stem die.

8

Stems— Secondary Growth

Primary growth is responsible for increases in stem length and laying down the basic tissue pattern. Secondary growth is responsible for development of new secondary vascular tissue, increasing the girth of the stem, and providing a continuous connection of living cells between newly formed primary tissues of the developing shoots and roots. This device makes possible attainment of great age by individual plants, although living tissue may be only 3 years old or less.

Secondary Growth

Formation of Vascular Cambium

This discussion is concerned with perennial woody stems like those of common shrubs and trees, including pines and their allies; but even some herbs, such as alfalfa, may develop some secondary tissues before the end of their relatively short lives.

Indeed, formation of a complete circle of cambium in alfalfa makes this plant a convenient one for an explanation of the manner in which cambium develops (Fig. 8.1). This occurs in the following manner: Not all procambium between primary xylem and primary phloem is used up; some remains and proceeds to divide to form phloem toward the exterior of the stem and xylem toward the interior. It becomes a vascular cambium and forms **secondary xylem** and **secondary phloem.** This causes the stem to increase in girth (Figs. 8.1*A*, 8.2). Since this first cambium forms within vascular bundles, it is called a **fascicular cambium.** What happens between the vascular bundles? In stems such as alfalfa, in which vascular bundles are separated by strips of parenchyma (pith rays), certain of these cells may become meristematic (Fig. 8.1*A*). Thus, cambium within bundles may be joined by cambium between bundles, forming a complete cambium ring (Figs. 8.1*B*, C). Cambium formed between bundles is called **interfascicular cambium.** The circle of vascular cambium thus formed in alfalfa and other herbaceous stems is not active for long. In woody stems it remains active as long as the tree lives, over 3000 years in redwoods. Interfascicular cambium thus formed joins fascicular cambium to form a complete ring of vascular cambium (Fig. 8.1*C*). It should be noted that fascicular cambium cells are derived from procambium cells that have never ceased to behave as a meristem. Interfascicular cambium originates from cells that have differentiated from ground meristem into a parenchyma tissue.

This development results in a circular sheet of

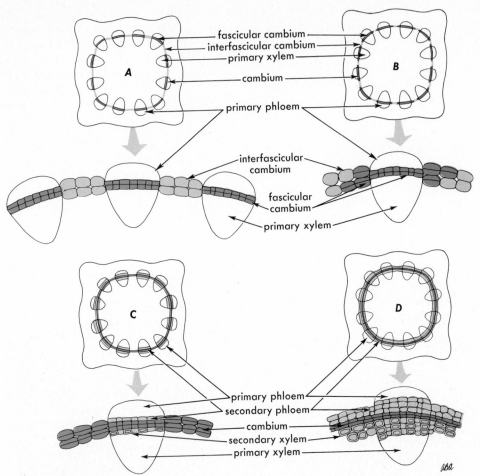

Figure 8.1 Formation of a complete cylinder of vascular cambium. *A*, at completion of primary growth, some meristematic cells remain between primary xylem and primary phloem (fascicular cambium, shown in dark green). Parenchyma cells appear in pith rays between vascular bundles (light green). *B*, some of the parenchyma cells of the pith ray become meristematic (interfascicular cambium; dark green). *C*, parenchyma cells of pith between the meristematic cells of the bundles have returned to a meristematic state, forming a cylinder of cambium (dark green). *D*, complete cylinder of secondary xylem and secondary phloem (light gray) is formed by the vascular cambium. Parenchyma cells are shown in light green, vascular cambium in dark green, and secondary tissues in light gray.

vascular cambium completely surrounding xylem and surrounded on the outside by phloem (Fig. 8.3*A*). In cross-section, these cambial cells are uniformly brick-like with thin walls (Fig. 8.3*C*). However, all cells of vascular cambium thus formed are not alike, as reference to the longitudinal view (Fig. 8.3*B*) will show. Small cells packed in regular groups will develop into vascular rays; they are **ray initials.** Elongated cambial cells, **fusiform initials,** will develop into fibers, tracheids, vessel elements, companion cells, or sieve-tube members; these cambium cells are called fusiform initials.

In most woody species, interfascicular cambium

differentiates typical secondary xylem internally and typical secondary phloem externally. A solid cylinder of xylem and phloem is thus formed (Fig. 8.1*D*). In many vine types of stems, the interfascicular cambium continues to produce parenchyma cells only; thus, these stems are composed of a number of well-formed vascular bundles separated by distinct pith rays.

Sections of Woody Stems

A woody stem may be viewed in three different planes (Fig. 8.4). A cross-wise view is logically termed a cross-section. There are two types of longitudinal sections. A section cut in a plane passing through both the center of the stem and the circumference is cut along a radius and is termed a radial section. The other longitudinal section is cut at right angles to a radius, some distance from the center, and is therefore called a tangential section.

Differentiation of Phloem and Xylem from Vascular Cambium

Fig. 8.5 shows stages in differentiation of vascular cambial cells to form secondary phloem and secondary xylem. Cambial cells divide most frequently by a tangential wall, forming two daughter cells. As a rule, an inner daughter cell, next to the xylem, develops into a secondary xylem cell. An outer daughter cell, next to the phloem, remains meristematic and again divides. When new phloem elements are produced, an inner daughter cell retains the power of division, and an outer daughter cell gives rise to one or more secondary phloem cells. Generally, more secondary xylem than secondary phloem is produced during the season. Cambium cells continue thus to divide throughout a growing season, adding secondary xylem on the *outside* of old xylem, and secondary phloem on the *inside* of old phloem. Thus, secondary xylem is superimposed upon primary xylem, and secondary phloem tends to exert pressure on primary phloem and cortex and to push them outward. Although cambial cells, by repeated divisions, are differentiating into xylem and

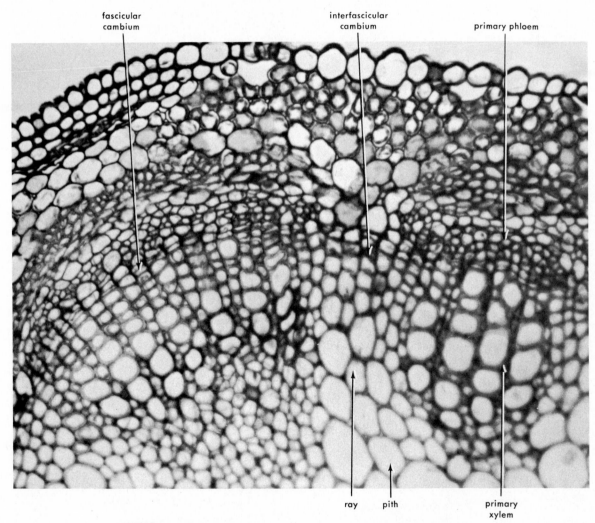

Figure 8.2 Cross-section of a stem of alfalfa (*Medicago sativa*), showing the fascicular cambium within the bundle and interfascicular cambium between the bundles. Some secondary vascular tissue has been produced, ×300.

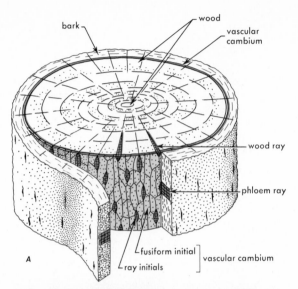

Figure 8.3 Above and left: Vascular cambium. *A*, diagram showing relationship of cambium and cambial initial to stem; *B*, tangential section; *C*, cross-section of quiescent cambium of locust (*Robinia*), $\times 3–3\frac{3}{4}$.

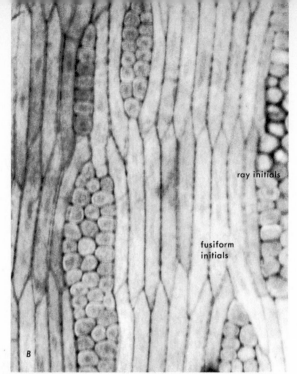

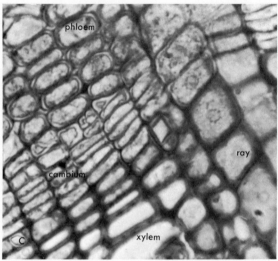

Figure 8.4 Below: Portion of stem of oak (*Quercus*), showing cross, radial, and tangential sections and their gross characteristics, $\times\frac{1}{2}$.

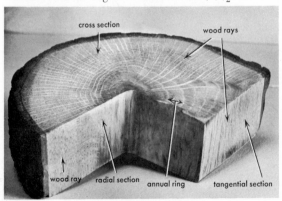

phloem elements, some daughter cells remain meristematic, and so a cambium is always between xylem and phloem. In addition, as the stem increases in circumference, the cambium keeps step with this increase by radial divisions of cambial cells, thus increasing the number of cells and enlarging the circumference of the cambium cylinder. During the winter season, cambium is inactive, only to begin divisions again in spring.

The resulting cellular composition of secondary xylem (Figs. 8.6, 8.9, 8.11, 8.12) and phloem is, in general, quite similar to that of the primary vascular tissues. It differs in that annular and spiral vessels are usually lacking, and secondary phloem contains more fibers.

The Tissues of a Young Woody Stem. In a section of a young woody twig, such as that of basswood shown in Fig. 8.7, the relationships and arrangements of various secondary and primary elements may be seen distinctly. Note layers of cork cells on the exterior of the stem. The collenchyma of the cortex (primary tissue) is intact, but there appears to be some crushing of cortical parenchyma. Primary phloem is the outermost tissue of the phloem region; it is not distinguishable in Fig. 8.7. There are two types of **phloem rays,** and sieve tubes and companion cells are sandwiched between masses of fibers. Individual

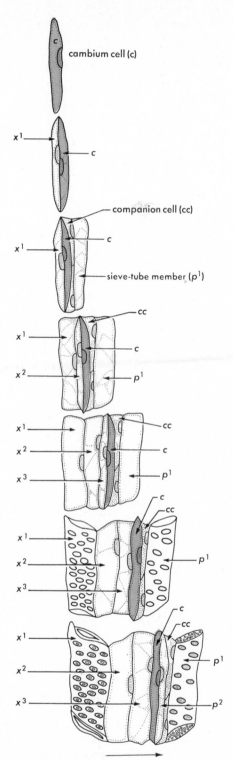

Figure 8.5 Diagram as seen in radial section, showing stages in differentiation of vascular cambium cells (c, cambium; cc, companion cell; p^1, p^2, phloem; x^1, x^2, x^3, xylem).

cambium cell (c)

x^1

c

companion cell (cc)

c

sieve-tube member (p^1)

cc

x^1

c

x^2

p^1

cc

x^1

c

x^2

x^3

p^1

c

cc

x^1

x^2

p^1

x^3

c

cc

x^1

x^2

p^1

x^3

p^2

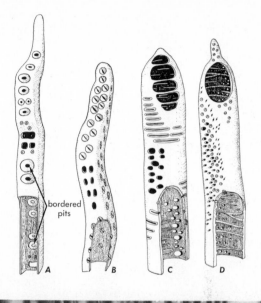

bordered pits

A

B

C

D

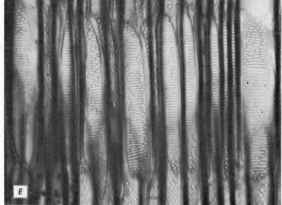

E

Figure 8.6 Tracheids and vessel elements from secondary wood. *A*, tracheid from spring wood of white pine (*Pinus*); *B*, tip of tracheid from wood of oak (*Quercus*); *C*, tip of vessel element from wood of *Magnolia*; *D*, tip of vessel element from wood of basswood (*Tilia*); *E*, vessel end walls in xylem of *Alseuosmia*. (*A* through *D* redrawn from Carl C. Forsaith, *The Technology of New York State Timbers*, New York State College of Forestry Publication 18; *E*, courtesy of L. M. Srivastava.)

cambium cells are difficult to distinguish, but a cambial zone is apparent. Secondary xylem is well-differentiated into **spring wood** and **summer wood,** which together form a year's growth, or an **annual ring.** Spring wood is frequently characterized by cells having larger lumens and thinner walls than those found in summer wood. Vessels are prominent, but in cross-section the distinction between tracheids and fibers is difficult to see. **Xylem rays,** composed

cork

cortex

sieve tubes

phloem rays

phloem fibers

phloem

cambial zone

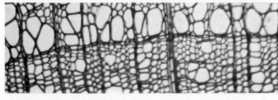

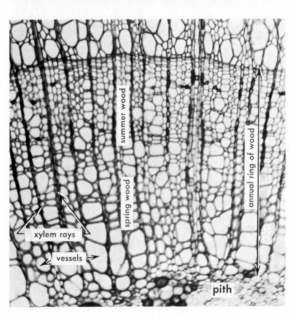

summer wood

spring wood

xylem rays

vessels

annual ring of wood

pith

of parenchyma cells, are conspicuous, although none occur which are as large as the larger phloem ray in the illustration. Points of primary xylem may be distinguished adjacent to the parenchyma cells of pith (primary tissue).

The Tissues of an Old Woody Stem. As a woody stem increases in girth, primary phloem and tissues external to it are completely lost, as will be described below. Primary xylem, on the other hand, together with the pith, remain in the center of the stem, but both are reduced to insignificance by great accumulation of secondary xylem. The possible variations in size, shape, and arrangement of vessels, tracheids, fibers, and parenchyma cells composing secondary xylem are so great that individual species may generally be identified by the structure of their wood. Secondary phloem, formed in lesser amounts and continuously sloughed off, constitutes a complex tissue.

Xylem. Cell types of the xylem have already been described in the discussion of primary xylem, and though detailed differences do occur between cells of primary and secondary xylem, we shall not consider them. Thus, in the following description of secondary xylem fibers, tracheids, vessels, and parenchyma cells, unless otherwise specified, will be considered as having the same characteristic as those of primary xylem.

A comparison of a cross-section of oak and a cross-section of poplar (Fig. 8.8) shows how differences in size and distribution of vessels may be responsible for important characteristics of wood. Note that summer wood and spring wood are present in both cases. In oak, the vessels are of two distinct sizes and less numerous than in poplar. Furthermore, the larger vessels of oak are grouped in spring growth. The character of rays and grouping of fibers are also different in each species. Because of the distribution of vessels in rings in oak, it is said to be **ring-porous,** while poplar is **diffuse-porous.**

Under higher magnification, cellular detail of oak wood becomes more apparent (Fig. 8.9A). Note the large vessels in spring wood and the compact summer wood. Rays are of two sizes: one but a single cell in width and the other massive. Radial views (Fig. 8.9C) show rays one cell in width lying perpendicular to a group of fibers. Note the intertwining fibers and the parallel rows of wood parenchyma interspersed among the fibers. A single large vessel is present in this section and, as it was not perfectly

Figure 8.7 Cross-section of 3-year-old stem of basswood (*Tilia*), ✕185. (Slide courtesy of Triarch Products.)

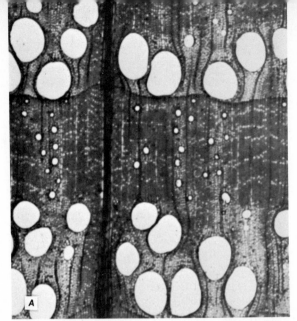

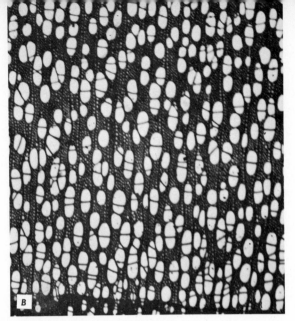

Figure 8.8   Cross-sections of *A*, oak (*Quercus borealis*); *B*, poplar (*Populus deltoides*), ×30. (Slides courtesy of Triarch Products.)

parallel with the section, the cut passes through it. Note the pits and overlying parenchyma cells. One complete vessel element and portions of two others are visible. Locations of original end walls are indicated by projections into the lumen of the vessel. The tangential section (Fig. 8.9*B*) shows again intertwining fibers and vertical rows of parenchyma cells.

Two types of rays are strikingly evident. Note that the vessel present in this section is smaller in diameter than that of the vessel seen in the radial section, but it is composed of elements of comparable length, as indicated by projections of the original end wall. The presence of large rays in radial section is a characteristic of quarter-sawed oak.

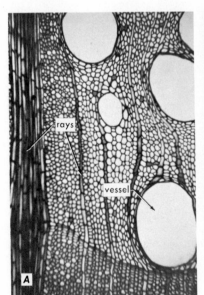

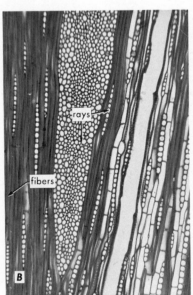

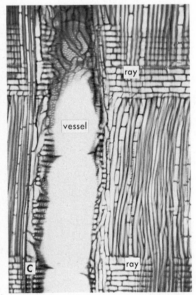

Figure 8.9   Sections of wood of oak (*Quercus borealis*). *A*, cross-section; *B*, tangential section; *C*, radial section, ×40. (Slides courtesy of Triarch Products.)

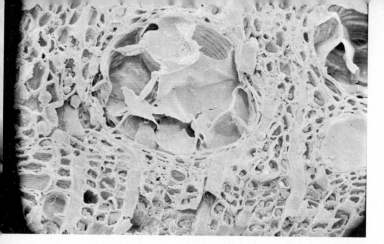

Figure 8.10   Stereoscan micrograph of tyloses plugging a vessel in secondary xylem of the locust (*Robinia*). (Courtesy of S. Cook and D. Hess.)

As a tree grows older, wood in the center of the stem changes in character; it becomes known as **heartwood,** and it no longer serves to conduct. Outer unchanged conducting wood is called **sapwood** (Fig. 8.19). Parenchyma cells adjacent to vessels may grow into the lumen of these vessels through pits and completely plug the vessels. Such cells form **tyloses** (Fig. 8.10).

Wood of gymnosperms, with few exceptions, has no vessels. A three-dimensional reconstruction is diagrammed in Fig. 8.12. A cross-section from a small piece of *Sequoia* is shown in Fig. 8.11A. It is extremely regular, almost like a wire netting. Tracheids

appear in cross-section as square hollow cells. Those formed in the spring are largest in diameter; as the season progresses, newly formed tracheids are smaller in diameter with thicker walls. The heavy-walled cell is called a **fiber-tracheid.** Note **xylem rays** or **wood rays** composed of parenchyma cells, and occasional small square cells with dark-staining contents. The latter are parenchyma cells. The radial view (Figs. 8.11, 8.12) shows mainly tracheids, with a small strand of fiber-tracheids. Pits are present in radial walls of tracheids. An elongated parenchyma cell is made apparent by its dark-staining contents. Wood parenchyma cells are represented by dark green in Fig. 8.12. They are shown in tangential section and cross-section. A horizontal ray in the radial section lies perpendicular to a tracheid. Pits occur between ray cells and tracheids. The ends of the rays are apparent in the tangential section (Fig. 8.11B). Here, fiber-tracheids have pits in their tangential walls. Pits in radial walls of tracheids are just visible.

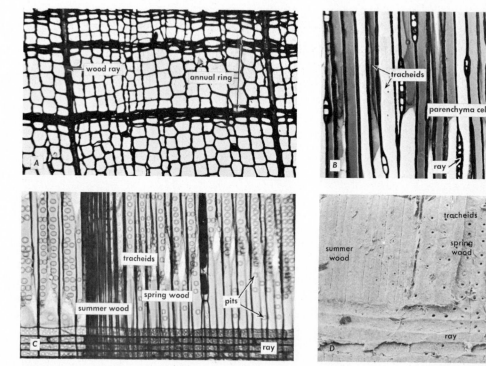

Figure 8.11   Sections of wood of redwood (*Sequoia sempervirens*). *A*, cross-section; *B*, tangential section; *C*, radial section; *D*, stereoscan micrograph of radial section, ×100. (Slides courtesy of D. Graham; stereoscan courtesy of D. Hess.)

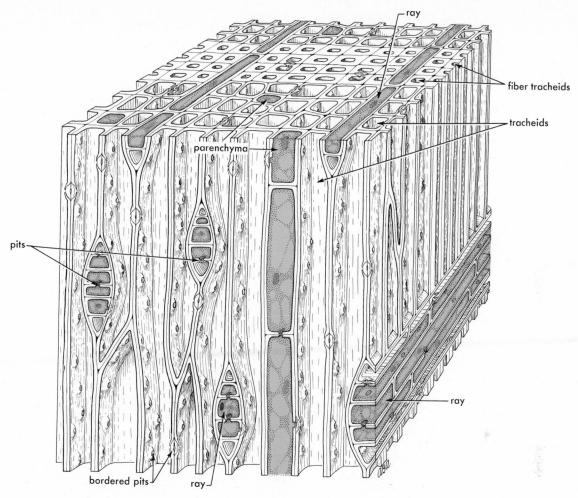

Figure 8.12 Block diagram of secondary xylem of redwood (*Sequoia sempervirens*).

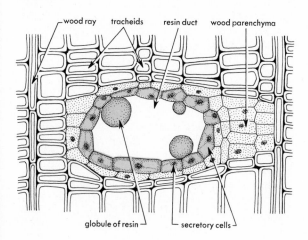

Figure 8.13 Resin duct in pine wood (*Pinus*) as seen in cross-section.

In this view, several parenchyma cells with heavily stained contents are present.

Sequoia does not produce resin as do many other gymnosperms. This substance occurs in specialized cavities which are surrounded by groups of parenchyma cells. Together they are called **resin ducts** and occur in many gymnosperms, where they may be continuous over considerable distances. One is shown in Fig. 8.13.

We have learned that water and mineral salts are conducted upward in xylem (in tracheids and vessels), and that foods are conducted upward or downward in phloem (in the sieve tubes).

Radial conduction occurs in *vascular rays*. These are composed of relatively long-lived parenchyma cells which are somewhat elongated radially.

Vascular rays in xylem are spoken of as wood or xylem rays, and those in phloem as phloem rays.

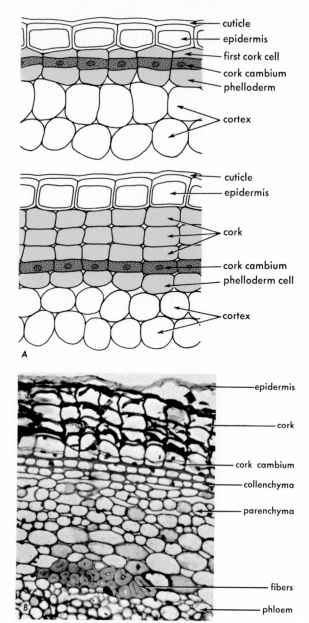

cuticle
epidermis
first cork cell
cork cambium
phelloderm

cortex

cuticle
epidermis

cork

cork cambium
phelloderm cell

cortex

A

epidermis

cork

cork cambium
collenchyma

parenchyma

fibers

phloem

B

Figure 8.14 *A*, diagram showing origin of the first cork cambium and the first layer of cork. New cork cambia and new layers of cork form in a similar manner each spring. *B*, light micrograph showing cork cambium and newly formed cork in an elderberry stem (*Sambucus*).

outer cortical cells. Cortex cells divide by tangential walls, producing daughter cells. The outer dughter cells generally differentiate into cork cells, thus forming a layer of **cork** beneath the epidermis. Should the outer daughter cells remain meristematic, inner daughter cells may give rise to a tissue known as **phelloderm,** which is composed of parenchyma-like cells. Cork cambium may originate also from epidermal or phloem cells.

Cork. This tissue (Figs. 8.7, 8.14) is composed of flattened, thin-walled cells with no, or small, intercellular spaces. A fatty substance called **suberin** is deposited in the walls, rendering the cells almost impermeable to water and gases. Hence, cork tissue provides protection for the stem against excessive loss of water and also against mechanical injury. The protoplasts of cork cells are short-lived. When cork tissue with its impermeable cell walls is formed, the epidermis and other tissues outside it are cut off from water and food supplies and soon die. In woody stems, cork replaces epidermis as a protective tissue.

In a young stem, the bark is made up of the following tissues, in order, from the *outside* to the *inside:* cork, cork cambium, phelloderm (if present), cortex, and phloem (Figs. 8.7, 8.15). Microscopic examination may reveal the presence of epidermal cells still clinging to the cork. In old stems, the epidermis, cortex, and primary phloem become separated from the adjacent inner tissues by successively deeper layers of cork formation. As this happens, tissues outside newly formed cork die for lack of water and nutrients. They dry up and eventually wither away.

Lenticels

An impervious layer of cork would effectively cut off the oxygen supply of the living tissues beneath, if it were not for the development, here and there, of groups of parenchyma cells called **lenticels** (Fig. 8.15). They frequently originate beneath the epidermal stomata (Fig. 7.11) of young stems. When a cork cambium is formed, it produces ordinary parenchyma cells outward, below these stomata. The resulting loose aggregation of parenchyma tissue bursts through the epidermis to form the lenticel. The air spaces between the parenchyma cells permit gaseous interchange.

Phloem rays are always contiguous with xylem rays. They both originate from ray initials of cambium.

Parenchyma cells in vascular rays may differ in detail from parenchyma cells of pith and cortex. Parenchyma cells of cortex, pith, xylem, and phloem form a living network extending throughout the stem, at least until the formation of heartwood.

Formation of Cork

Cork Cambium. Fig. 8.14 shows the origin of cork cambium and development of secondary tissues from it. Cork cambium in most plants arises from

Anatomy of the Stem at Nodes

Buds occur in axils of leaves, and this whole region is called the node. Now, since both buds and leaves are joined by vascular tissue to conducting tissue of the stem, anatomically the node must be characterized as a region where vascular bundles are connected to bring about a continuity of vascular tissue in leaves, buds, and stem. It is often a rather complex region. Bundles leading to leaves and buds are referred to as **leaf traces** and **bud traces,** respectively. Xylem of traces is continuous with xylem in leaf and stem, and phloem of the trace is continuous with phloem in leaf and stem.

Where vascular tissue leaves the vascular cylinder of the stem, a gap exists in the cylinder immediately above the point where the trace departs. This is called a **leaf gap** or a **bud gap.** These are shown diagrammatically in Fig. 8.16A. Branch gaps are already present in twigs in the winter condition and they may be seen by cutting a free-hand cross-section through a bud (the future branch) and its accom-panying node. Figure 8.16B shows a free-hand cross-section through a bud and node of a young branch of fig (*Ficus carica*). The section makes a nearly longitudinal cut through the bud (the future branch), and the continuity of vascular tissue of the bud with the vascular cylinder is apparent. The plane of the cross-section of Fig. 8.16B is roughly indicated by the line A—A on Fig. 8.16A. Gaps are also prominent features of herbaceous stems from which pith, phloem, and cortex have rotted away. The more resistant woody tissues remain, and the gaps are holes through the xylem cylinder. This is shown in Fig. 8.16C by the vascular cylinder of a tomato stem.

Now, as secondary development of branch and trunk proceed, both will increase in diameter; new vascular tissue of the stem must be laid down in such a manner as to bury the younger portions of the branch in the trunk (Fig. 8.16D). Furthermore, there must be a crowding of new tissue in the upper acute angle formed by the branch and the stem. This is shown diagrammatically in Fig. 8.17. This entire process results in a knot. Can you explain why it is that in some gymnosperm lumber knots are held firmly in place, while in others they fall out?

Primary and Secondary Growth

In spite of the relatively limited life-span of cells, some plants having only primary growth do reach considerable age. Primary growth, as usual, takes place at the shoot apex, but in such plants roots form at nodes a short distance behind the apex (Fig. 7.1B). Older portions of stems with associated roots die after several years. No provision is made to connect the most recently formed primary growth of roots and shoots that are separated by large distances of space or age. Secondary growth does just this; it connects active functional primary growth of shoots and roots separated by considerable distances and by many generations of living cells. This mechanism is illustrated by a longitudinal section of a 3-year-old shoot (Fig. 8.18). The center of the stem is occupied by pith. The strands of primary xylem form the innermost ring of vascular tissue for all 3 years. The cylinder of secondary xylem produced the first year extends for only a short distance up the stem (Fig. 8.18A1). In cross-section, it is represented only in the oldest section (Fig. 8.18D). At the level shown in Fig. 8.18C, only the most recently formed cylinder of secondary xylem is present. However, primary xylem extends upward from the base of the stem to the shoot apex (Fig. 8.18B). The significant observation is that the active functional primary xylem is always in contact with the most recently formed cylinder of secondary xylem. Furthermore, older portions of

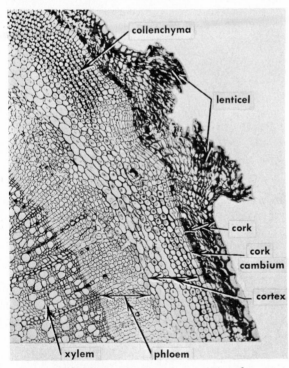

Figure 8.15 Cross-section of a portion of a young stem of elderberry (*Sambucus*). From the outside to the inside of the stem, the following tissues may be seen: cork, cork cambium, cortex, secondary phloem, and xylem. The vascular cambium is probably the two layers of thin-walled cells on the phloem side of the xylem. Note the lenticel in the cork, ×100.

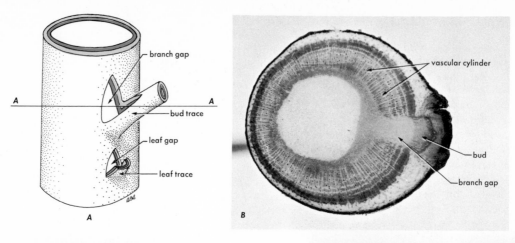

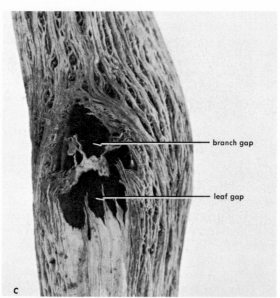

Figure 8.16 Above: Branch and leaf gaps. *A*, diagram of branch and leaf gaps. *B*, free-hand cross-section of a twig of fig (*Ficus carica*) through bud and accompanying node. Gap in vascular cylinder is apparent, ×15. *C*, gap in the vascular cylinder of a tomato (*Lycopersicon esculentum*) stem, ×3. *D*, branches form a continuous pathway from the center of the trunk to the exterior, thus constituting a gap in the xylem cylinder of the trunk (*Pinus jeffreyi*), ×1/10.

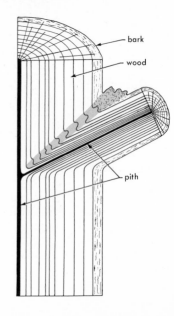

Figure 8.17 Right: Relationship of wood and bark in main stem to wood and bark in a branch. (Redrawn from A. J. Eames and L. MacDaniels, *Introduction to Plant Anatomy*. Copyright 1947 by McGraw-Hill Book Company. Used with permission of McGraw-Hill Book Company.)

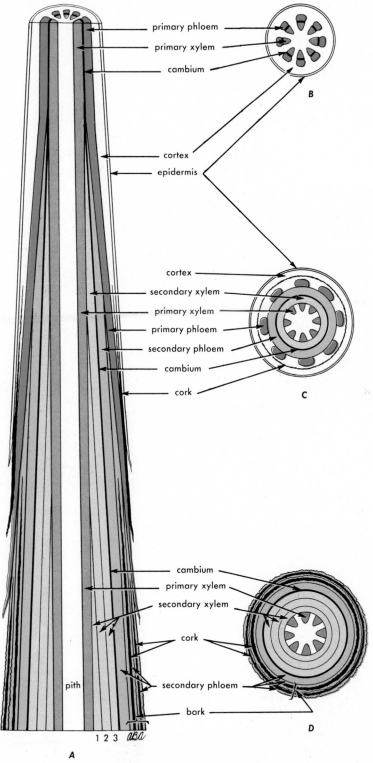

primary phloem

primary xylem

cambium

B

cortex

epidermis

cortex

secondary xylem

primary xylem

primary phloem

secondary phloem

cambium

cork

C

cambium

primary xylem

secondary xylem

cork

pith

secondary phloem

bark

D

1 2 3

A

Figure 8.18 Changes in a stem as it increases in age and in girth. *A*, longitudinal section through a 3-year-old stem; *B*, *C*, and *D*, cross-section at indicated levels.

primary xylem are also in contact with all other annually produced cylinders of secondary xylem. Primary phloem on the periphery of the expanding stem is continually cut off from inner portions of the stem by the stem's increase in girth and the successive formation of layers of cork (Fig. 8.18D). However, young functional primary phloem, too, will always be in contact with the most recently formed layers of secondary phloem (Fig. 8.18B, C). This arrangement provides a continuous pathway of living cells less than 5 years old from shoots to roots of trees which, as individuals, may be many hundreds of years old (Fig. 7.1A).

A Summary of Secondary Growth

1. Vascular cambium originates from procambium and adjacent parenchyma cells in rays between vascular bundles.
2. Vascular cambium lying between xylem and phloem produces new secondary xylem and secondary phloem each year; fusiform initials differentiate into elongated conducting and strengthening cells, and ray initials differentiate into ray cells.
3. Primary xylem (next to pith) becomes more and more widely separated from primary phloem (next to cortex) as secondary xylem and phloem are produced.
4. Vascular cambium persists as a meristem between the last-formed xylem and the last-formed phloem.
5. Cork cambium is short-lived; new cork cambiums may arise each year, producing new layers of cork.
6. Cork cambium may originate in successively deeper tissues from epidermis, cortex, and phloem.
7. Secondary xylem provides a continuous pathway of young functioning cells between active primary growth of shoot and root. This mechanism makes possible attainment of both great age and great size by individual plants, in spite of a normal 3–5-year life-span of plant cells.

Macroscopic Appearance of an Old Woody Stem

If we look at the cross-section of an older stem or at the end of a log, two prominent regions are distinguished—the bark and the wood (Fig. 8.19). Pith may be evident in certain species or in young stems. In old stems, it may be impossible to find even a trace of pith, and it may be also very difficult to find primary xylem, which is at the very center of the stem just outside the pith. The vascular cambium can be seen only by microscopic examination.

Figure 8.19   Cross-section of a branch of mulberry (*Morus*), $\times \frac{1}{2}$.

Bark. Bark (Fig. 8.19) includes all tissues outside the vascular cambium, and its exact cellular composition will depend upon the age and species of the twig or tree trunk being examined. In this discussion, we

Figure 8.20   Bark of incense cedar (*Libocedrus decurrens*). *A*, section of bark; *B*, three-dimensional view, $\times \frac{1}{2}$.

Figure 8.21   Bark of three species of trees. *A*, western yellow pine (*Pinus ponderosa*); *B*, cork oak (*Quercus suber*); *C*, plane tree (*Platanus orientalis*).

shall consider only bark as it occurs on an old woody tree trunk in which all epidermal and cortical tissues have disappeared and activity of cork cambium has resulted in formation of considerable cork tissue. Bark may easily be separated from the woody cylinder or trunk, especially in spring or early summer, because walls of cells in the "cambial zone" are thin and easily ruptured. When bark is peeled from wood, some cells of the cambial zone adhere to the bark, and some adhere to the wood. Examination of a piece of bark of incense cedar (Fig. 8.20) shows that it is made up of a series of layers of two different-appearing tissues. Recalling that the vascular cylinder of a tree is increasing in diameter, let us examine the manner of formation of this bark.

At the close of primary growth, primary vascular tissues, in strands, formed a ring around the peripheral area of the stem (Figs. 7.22, 8.18*B*). The central portion of the stem was occupied by pith. Cortex and epidermis surrounded the ring of vascular strands. Fascicular cambium joined with interfascicular cambium to produce a ring of secondary vascular tissue (Fig. 8.1). This increases the girth of the stem (Fig. 8.18). As this takes place, what will happen to the cortex and epidermis?

Cells composing the epidermis are fully differentiated, and except in rare cases (apple), do not divide. As the stem increases in diameter through the activity of vascular cambium, the epidermis must be stretched and torn. When this happens, delicate underlying cells would be exposed to drying out and weathering were it not for development of a cork cambium and a layer of cork as previously described (Figs. 8.14, 8.18*A*, *C*).

As the stem continues to grow in girth, the cylinders of cork, in turn, will be ruptured. Exposure of underlying tissue is prevented by generation each spring, or sometimes more frequently, of a new cork cambium. Finally, even parenchyma cells in secondary phloem may produce a new cork cambium (Figs. 8.18*A*, *D*). This is why bark from some old tree trunks is layered (Fig. 8.20). This layering is brought about by formation of successive layers of cork, cutting off exterior phloem which is thus deprived of water by the impervious nature of cork cells. Now, with further increase in circumference, dead exterior sheets of phloem and cork cells are stretched and torn, resulting in bark patterns characteristic of many trees (Fig. 8.21). Weathering causes a continual wearing away of the surface of bark, and in regions where sand particles are carried by wind, trees have a beautiful polished appearance. Commercial cork is bark of an oak (*Quercus suber*, Fig. 8.21*B*). It is not layered, because it develops in a somewhat different fashion; here cork cambium never develops within phloem tissue. Thus, outer bark of the cork oak may be comprised almost entirely of cork cells.

Wood.   The commercial product, **wood,** is xylem. It occurs in all vascular plants, even the tenderest herbs, but only trees that have lived for a number of years and gradually built up a large trunk capable of being converted into lumber furnish us with commercial wood (secondary xylem).

Annually, vascular cambium forms a new layer of wood which is usually many cells thick. As seen on the end of a log, these layers appear as concentric rings (Fig. 8.19), the **annual rings.** It should be kept

of the ring that is formed in the early part of the season is known as **spring wood.** That of angiosperms has larger and more numerous vessels than does **summer wood,** produced later in the season (Figs. 8.7, 8.8). Also, tracheids and fibers of spring wood are less abundant and thinner-walled than those of summer wood. As viewed with a hand lens, or even with the naked eye, summer wood appears darker and denser than spring wood. Of course, in any annual ring, summer wood lies outside spring wood, and summer wood of any year is bounded on the outside by spring wood of the following year. Therefore, small, thick-walled elements of summer wood of one year stand out in contrast to adjoining, relatively large, thin-walled wood elements of spring growth of the following year (Fig. 8.7). Thus, a sharp line of demarcation lies between them.

Annual rings, or growth increments, are just as conspicuous in gymnosperms, which possess no vessels. In gymnosperms, tracheids of spring wood are radially larger and have thinner walls than those of summer wood (Fig. 8.11).

In moist tropics, where there are no marked seasonal changes, growth rings are almost completely absent or, at best, poorly defined. If they occur, they do not necessarily represent annual growth increments and hence could not be used in determining the age of a tree.

Examination of the cut end of a log shows that in an individual tree, annual rings vary greatly in thickness. In other words, the amount of wood produced each year is not the same. As a rule, the first few rings produced in the early life of the tree are thicker than those formed later. Also, the thickness of rings, or, to state it differently, the amount of wood added during any season, is influenced by many factors, such as the age of the tree, light, temperature, moisture, leaf area, and competition both above and below ground. As would be expected, naturally fast-growing tree species have thicker annual rings than slow-growing species living under the same environmental conditions.

The age of a woody stem can be determined by counting annual rings of wood. It is obvious that the true age of a tree can be ascertained only by counting rings at the very base of the main trunk. As shown in Fig. 8.22, the number of annual rings decreases from the base of the trunk to its apex.

Infrequently, two or more rings of growth may be formed during one season. For example, in irrigated apricot orchards, an application of water may stimulate cambium activity, resulting in the differentiation of wood elements resembling those of spring growth. Then, as the soil dries out, the wood elements formed may be smaller and thicker-walled. A second irrigation again stimulates vascular cambium

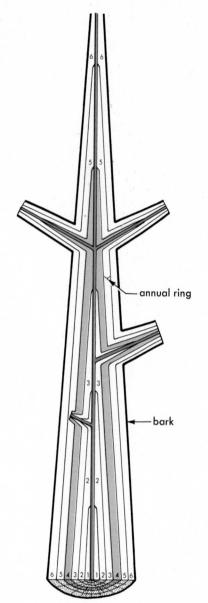

Figure 8.22   Diagram showing cross-section and longitudinal section of a 6-year-old tree.

in mind, however, that vascular cambium in a tree like pine, for example, has the form of a hollow cone; consequently, wood produced annually has the same form (Fig. 8.22). Wood produced in any one year is laid down over that of the preceding year.

Annual rings of wood are particularly conspicuous in woody plants of temperate climates, where a season of active growth alternates with a season of inactivity. In such plants, wood elements formed in spring and early summer are quite different from those produced in late summer and fall. The portion

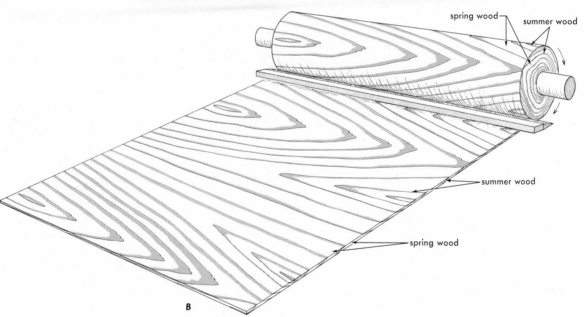

Figure 8.23 Manner of preparing plywood.
A, sheet of veneer being peeled from a slowly
revolving log; *B*, diagram showing relation of grain
in plywood to annual growth rings of the log.
(*A*, courtesy of American Forest Institute.)

commercial timber trees. It is of practical value to
know in terms of board feet the volume of wood
formed by a certain species, say over a 10 year period. Volume may be determined by counting the
annual rings on cross-cuts of tree trunks, or on
stumps of a considerable number of trees, or by
using an instrument known as the increment borer
to determine the age of a number of standing trees.
This borer has a hollow bit that bores out a core
of wood, on which rings may be counted.

Sapwood and Heartwood. The cross-sectional end
of a log usually shows the presence of two zones
of wood: (a) a light-colored outer zone, the sapwood, surrounding (b) a generally darker-colored
zone, the heartwood (Fig. 8.19). The sapwood functions in sap conduction and food storage, whereas

to greater activity, and thus a line of demarcation
is formed between two rings of growth, both made
in one season. Sometimes a drought or an insect
attack may destroy the foliage and thus interrupt the
activity of the cambium. Later, a new crop of leaves
may be produced, cambial activity renewed, and
another ring of wood produced. Under these conditions, two (or more) growth rings of wood will be
formed in one season.

The forester is primarily interested in the amount
and quality of wood laid down in main trunks of

the heartwood performs only the function of
mechanical support. Sapwood is gradually transformed into heartwood. In this transformation, various substances infiltrate into the cell walls and even
accumulate in cell cavities. The substances include
tannins, gums, resins, pigments of various sorts,
salts of organic acids, and other materials. These
depositions may impart a characteristic color to the
heartwood, and they also may increase its weight,
hardness, and durability and decrease its permeability. In some species, vessels, and even tracheids, may

A Summary of Secondary Growth 145

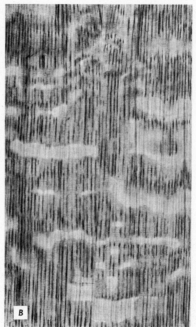

Figure 8.24 Characteristics of finished woods. *A*, plywood, a perfect tangential section; *B*, quarter-sawed oak, a radial section; *C*, knotty pine, $\times\frac{1}{4}$.

be plugged by **tyloses** (Fig. 8.10). The penetration of wood preservatives is rendered difficult by the presence of tyloses.

Industrial Products from Trees

The most obvious use of wood is as lumber. Trees, however, are grown in many countries for other purposes, some of which do not involve the destruction of the trees. For instance, turpentine, natural rubber, and chewing gum are derived, respectively, from the sap exudates of southern pine, particularly the long leaf pine (*Pinus palustris*) and the slash pine (*P. cubensis*); the rubber tree (*Hevea brasiliensis*); and the sapodilla tree (*Achras sapota*). In all three cases, gashes are cut in trunks of trees and sap is collected in buckets or on sticks, after which it is processed in factories into the well-known articles of commerce.

When wood itself is utilized, the trunk may be sawed into lumber or completely disintegrated into the cells of which it is composed. In manufacture of ordinary lumber, the log moves against the saw so that tangential cuts rip each board from the log. Large rays of oak wood give boards cut from these trees a characteristic pleasing appearance if they are cut in a radial plane (Fig. 8.24*B*).

Sheets for plywood manufacture are produced in a different manner. Logs are steamed and then placed on a large lathe and made to turn against a large and sharp knife which literally peels off thin

Figure 8.25 Manufacture of paper. The "wet" end of a Fourdrinier paper machine. A thin suspension of pulp flows onto a moving endless wire screen from the head box in the background. The water drains through the screen, leaving behind a mat of fibers in the form of a sheet of paper shown in the foreground. (Courtesy of Crown Zellerbach Corporation.)

sheets of wood from the slowly revolving log (Fig. 8.23*A*, *B*). Note (Fig. 8.24) the different appearances that these processing procedures give to wood. Can you account for the "grain" in Figs. 8.23*A* and *B*, in terms of spring and summer wood?

For other industrial uses, wood is not sawed into boards but placed into machines which tear it apart, breaking it down in some instances into fibers and tracheids. The paper you have in your hands is the end product of such a process (Fig. 8.25), and if examined under a microscope will reveal the presence of the cellular elements of the trees from which it was manufactured. Dacron shirts may have started as trees, although other cellulose products may have been the original raw material. In paper-making, lignin is undesirable and must be removed from fibers and tracheids. It is a waste product, and huge piles of it have accumulated outside paper mills. Finding a use for it has been a major research project of a number of large paper companies. It is now being marketed and used as, among other things, a soil conditioner, an aid in drilling oil wells, and an adhesive in manufacture of plywood and linoleum.

Summary of Stem Growth

1. Primary growth provides the basic requirements for the production of a complete plant body from seed to seed.
2. The predominantly short life-span of 3–5 years for cells, in general, limits the size and longevity of individual plants having only primary growth. The mechanism of secondary growth overcomes this limitation.
3. If all meristematic cells in a procambium strand become differentiated into primary vascular tissue, the vascular strand is closed to further growth.
4. If there remains an active meristematic region between primary xylem and primary phloem, continued growth is possible. These meristematic cells become the fascicular cambium.
5. An interfascicular cambium forms from ray parenchyma cells between vascular strands.
6. Union of fascicular and interfascicular cambium produces a uniform cylinder of vascular cambium.
7. Divisions in cambium are longitudinal, so the stem now increases only in girth.
8. The smooth structural transition from primary to secondary growth means that primary vascular tissue is always in physical and physiological contact with the most recently formed secondary tissues.
9. In a 3000-year-old redwood or bristle cone pine, primary tissues in young shoots and roots have been in continued physical and physiological contact with each and every layer of secondary vascular tissue as these layers have formed during the past 3000 years.
10. The production of secondary vascular tissue thus makes possible the attainment of great size and great age by individual trees, even though the functioning cells may be no older than those in a perennial herbaceous plant such as an *Iris*.

Stem Modifications

Stems may become adapted for functions other than support, conduction, and production of new growth. They may, for instance, become attachment organs for vines; they may carry on photosynthesis, store water in preparation for periods of water stress, and develop protecting devices.

Stem Tendrils

Tendrils are slender, coiling structures that are sensitive to contact stimuli and attach the plant to a support. Tendrils of plants, morphologically, are of two sorts: leaf and stem. In the trumpet flower, for example, several uppermost pairs of leaflets have no blades, but instead form very slender tendrils (see Fig. 10.15*E*). Obviously, these are leaf tendrils. In grapes and Virginia creepers, tendrils are modified stems (Fig. 8.26*B*). This is evidenced by their presence at nodes in leaf axils.

In the Virginia creeper, each tendril ends in a knob which flattens out when it comes in contact with a surface to which it adheres.

Cladophylls (Cladodes)

These are stems which are leaf-like in form, are green, and perform the functions of leaves. They may bear flowers, fruit, and temporary leaves. Examples of plants with cladophylls are *Ruscus* (Fig. 8.26*D*), asparagus (Fig. 8.26*E*), smilax, various species of cacti, and some orchids (e.g., *Epidendrum*).

Stem Spines and Thorns

Most spines and thorns of plants are modified stems or outgrowths of stems. Leaf spines, however, occur in certain plants, such as barberry and black locust, and in a few cases even roots become modified as spines. Good examples of stem thorns are those of *Pyracantha* (Fig. 8.26*F*) and honey locust (Fig. 8.26*G*). They are borne in the axils of leaves as ordinary branches are. Sometimes thorns bear leaves (Fig. 8.26*F*), which is further evidence that they are stems.

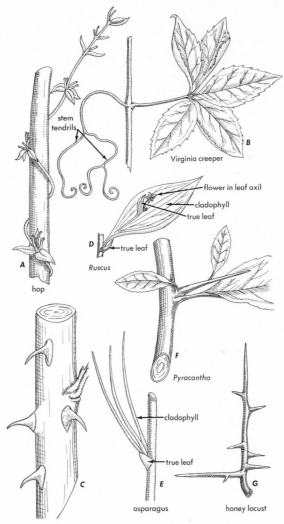

stem tendrils

Virginia creeper

B

flower in leaf axil

cladophyll

true leaf

D → true leaf

Ruscus

A

hop

F

Pyracantha

cladophyll

true leaf

C

E

asparagus

G

honey locust

Figure 8.26 Types of stem modifications. *A* and *B*, stem tendrils; *C*, *F*, and *G*, thorns; *D* and *E*, leaf-like stems.

In honey locust, stem thorns may be branched (Fig. 8.26*G*).

Storage of Water in Stems

Succulent plants are good examples of stems storing large quantities of water. In various cacti, for example, water-storage tissue constitutes a large proportion of the stem, and water stored therein may be as much as 90% of the weight of the entire plant (see Fig. 18.38). This water reserve may be used during drought.

Production of Other Materials by Stems

A great range of substances are produced in stems. A few of these may be mentioned.

Mucilaginous substances (complex carbohydrates very similar to gums) accumulate in stems of some plants, such as many ferns, cacti, and other succulents.

Tannins, substances that impart an astringent, bitter taste to tissues in which they are present, occur in bark and wood of many plants. They are derived commercially chiefly from bark of hemlock, tan oaks, mangrove, and wattle (certain *Acacia* species), and from wood of chestnut.

Latex is a milky secretion that occurs in many different families. It is a very complex mixture containing such materials as water, resin, oil, proteins, gums, tannins, sugars, alkaloids, and salts of calcium and magnesium. Latex may occur in special cells or in tubes (**latex tubes**), which are distinct from the vascular system. Common latex-producing plants are the fig, dandelion, spurge, and milkweeds. Of particular economic importance is latex of those plants from which crude natural rubber is obtained. The most important rubber-yielding species are as follows: (*a*) *Para rubber tree* (*Hevea brasiliensis*), a native of the tropical forests of the Amazon and Orinoco river valleys in South America, but since 1929 grown in large plantations, chiefly in Ceylon, Malaya, Java and Sumatra; (*b*) *Panama rubber tree* (*Castilloa elastica*), a native of Mexico and Central America; (*c*) *Manihot glaziovii*, a native tree of Brazil, yielding "ceara rubber;" (*d*) *guayule* (*Parthenium argentatum*), a native American shrub which, during World War II, was grown in large plantations in the western United States; and (*e*) *kok-saghyz*, or Russian dandelion (*Taraxacum koksaghyz*), a plant grown quite extensively in Russia, and to some extent, experimentally, in the United States during the period of rubber shortage.

Resins are complex substances, secreted by glands into special ducts known as **resin ducts** (Fig. 8.13). Crude resins contain, in addition to resinous materials, a considerable amount of essential oils. The principal product derived from resin is turpentine, obtained exclusively from coniferous trees.

9

Roots

The roots of a plant, considered collectively, form the **root system.** The two principal functions of roots are **anchorage** and **absorption.** In addition, roots of all plants usually **store** a certain amount of food, at least for a short time; roots of such plants as sugar beet, carrot, sweet potato, and others are specialized food storage organs. Besides anchorage, absorption, and storage, roots perform the function of **conduction.** Water and mineral salts absorbed from soil, and foods that may be stored in roots, are conducted by roots to stems, and thence to leaves and other organs above ground. Foods manufactured in leaves are conducted by stems to main roots, and then by the latter to branch roots, so that these foods are carried to the extremities (growing tissues) of all the smallest roots.

Balance between Shoot and Root Systems

In a normal, healthy plant there is a balance between shoot system and root system. Of particular importance is the relation of total leaf surface to total root surface. There is a balance between total surface exposed to the sun, from which energy is absorbed and used in the manufacture of carbohydrates, and total root surface in contact with the soil solution, from which the plant absorbs water and mineral nutrients. The root system must be able to supply the shoot with sufficient water and mineral nutrients, and the shoot system must manufacture enough food for maintenance of the root system.

Roots absorb water, stems conduct water, and leaves use water in photosynthesis and transpire to the air much of the water absorbed by roots. We shall see that **water absorption** is a principal function of young primary roots. Furthermore, the efficiency of the water-absorbing activity of a root system is of considerable importance in the successful competition of the plant. We have seen a great range in stem morphology. Root systems, too, show variation, which, in general, is more or less closely associated with stem morphology and the ecological niche the plants inhabit.

Kinds of Root Systems

Small garden plants or weeds easily pulled up by roots show a mass of soil-clad roots arising from the base of the stem. This constitutes a **fibrous root system** (Fig. 9.1A). Some fibrous root systems do not penetrate deeply into soil, but extend for a distance

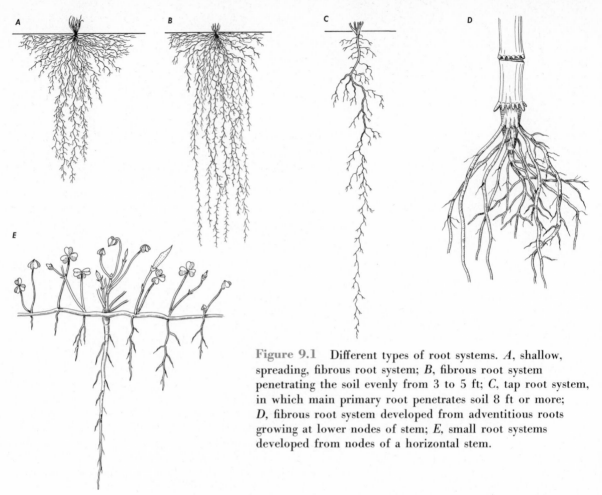

Figure 9.1 Different types of root systems. *A*, shallow, spreading, fibrous root system; *B*, fibrous root system penetrating the soil evenly from 3 to 5 ft; *C*, tap root system, in which main primary root penetrates soil 8 ft or more; *D*, fibrous root system developed from adventitious roots growing at lower nodes of stem; *E*, small root systems developed from nodes of a horizontal stem.

outward from the base of the plant (Fig. 9.1*A*). Other fibrous systems are more deeply rooted, the cluster of roots forming a rather uniform mass as they penetrate soil (Fig. 9.1*B*).

Main roots are approximately of the *same size*; each gives off numerous side roots, or roots of the second order; these bear branches of the third order, which, in turn, bear roots of the fourth order. All roots are slender and fiber-like; no one root is more prominent than others. Other young seedlings do not come up from the soil so easily. Stems of an oak or hackberry seedling 10 inches tall will break before the root system releases its hold on the soil. The oak root system must be dug up and it will then be seen to consist of a long slender root, with short lateral rootlets. It has grown directly downward deeply into the soil. This is a **tap root system** (Fig. 9.1*C*).

The fibrous root system is characteristic of many plants, including most of the monocotyledons. It is best adapted to shallow soils or to regions of light rainfall that may not penetrate the soil deeply. The tap root system is characteristic of most dicotyledons

and gymnosperms. It is well-adapted to deep soils and to those in which the water table is relatively low. Most tap roots are woody but some, like carrots and sugar beets, are fleshy storage organs.

Development of Root Systems

A seed contains a young plant, the **embryo.** Soaking a seed, such as a bean, in water to remove the seed coat more easily, makes it possible to observe the entire embryo. It will be seen that certain rudimentary organs are already in evidence. One of these organs is the **radicle** or **rudimentary root** (Fig. 3.1). When the seed germinates, the radicle is usually the first structure to appear; it becomes the *first* root and is called the **primary root.** It has branches and subbranches, all of which are **secondary roots.** In many plants, such as beet, radish, and carrot, the primary root remains the principal root throughout the life of the plant. A tap root system consists only of the primary root and its branches.

Development of a fibrous root system, like that of a cereal, is quite different from that of a tap root system. In the embryo of barley or wheat, for example, a root apex (see Fig. 16.24) is evident. When the seed germinates, this seedling root takes the lead. Soon it is followed by two pairs of thread-like roots, which, however, are not branches of the primary root; that is, they are not strictly secondary roots (see Fig. 16.32). The primordia of these two pairs of roots can be detected by microscopic examination of ungerminated seeds. Therefore, we may speak of the primary root and the two pairs of roots which follow it as **seminal (seed) roots.** The primordia of all these roots are evident in the embryos of the seeds. Thus, the seminal root system of barley or wheat usually consists of five roots (sometimes six). In most grasses, the seminal root system is usually temporary, although under some conditions it may function throughout the life of the plant.

Within a short time after the seminal root system is formed, permanent roots arise from the lower nodes of the young stem. The fibrous root system of a corn plant is formed from roots growing from the lowest nodes (Fig. 9.1D). These roots are not branches of the primary root. They are **adventitious roots.** It is customary to speak of all roots that arise from organs other than roots as adventititous roots.

Roots do not have nodes as do stems. Indeed, the lack of nodes is a distinguishing external trait differentiating roots from stems. But roots can, and frequently do, grow at nodes on stems (Figs. 7.1B, 7.27A). In fact, nodes of most monocotyledons, like that of the grain just discussed and the corn shown in Fig. 9.1D, are provided with root primordia which may grow or remain dormant. Bermuda grass, a monocotyledon, and *Oxalis,* a dicotyledon (Fig. 9.1E), both regularly produce roots at nodes. This device enables them to be active and invasive competitors in lawns and gardens.

Root systems are generally thought to occur only at the lower end of the plant axis (Fig. 3.1). Such a restriction is true of all gymnosperms and of many dicotyledons, but only of some monocotyledons, and probably not of any lower vascular plants. Roots, forming fibrous root systems, regularly occur at nodes along the stem (Fig. 9.1E). In many plants, like corn and wheat, there is also a fibrous root system at the lower end of the main axis of the plant, complementing the root system at designated nodes. Formation of roots at nodes is very frequently associated with some type of horizontal stem, rhizomes, or stolons (Figs. 7.1B, 7.26). Formation of roots at nodes of such stems, whether monocotyledon or dicotyledon, makes possible the continued development of primary stem and root systems without resorting to secondary growth. Roots forming at nodes a short distance behind the growing shoot apex assure a continued supply of water and nutrients to the developing apex. Secondary tissues in either stem or root are thus not needed.

Adventitious roots may form on stems at places other than nodes. Canes of roses and blackberries readily initiate roots along their length where they contact moist soil. Roots will also form on the ends of stem cuttings, and their formation may be stimulated by auxin.

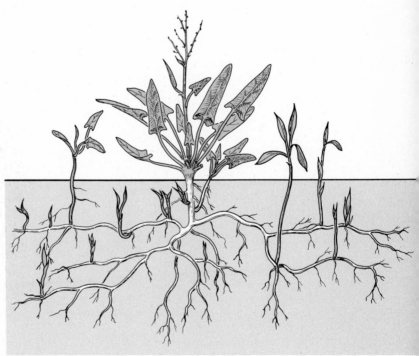

Figure 9.2 Although roots do not ordinarily produce buds, adventitious stems do form on some roots, as in *Rumex acetosella.*

While roots are not divided into nodes and internodes, shoot buds can form on roots. This adaptation seems to be species-specific, regularly occurring in some species and not at all in others. The common plant, sheep sorrel or dock (*Rumex acetosella*), produces long, somewhat horizontal, primary roots. These produce, at intervals, adventitious buds giving rise to shoots which grow upward to form the typical above-ground sheep sorrel shoot. A fibrous root system also grows downward from the primary horizontal roots (Figs. 9.1E, 9.2).

Root Systems and Plant Competition

Where rainfall and temperatures permit, plants of the same and different species grow close together and compete for available soil water, mineral nutrients, and light energy. As will be discussed in Chapter 18, plants reduce competition by utilizing different parts of the environment: Some utilize full sunlight, while others grow normally in shade; some are ephemeral and complete their life cycle only during periods of abundant moisture, while others are perennial and grow slowly even during periods of water shortage. In addition, as documented by the American ecologist J. E. Weaver, root systems of different species occupy and utilize different parts of the soil profile.

Weaver and his students carefully excavated root systems of hundreds of prairie plants and concluded that there were three general categories of rooting depths. As shown in Fig. 9.1, some grassland species are like blue grama (*Bouteloua gracilis*), possessing a very shallow root system. Most of the roots are within the top $\frac{1}{2}$ ft of soil. Another group of species is like buffalo grass (*Buchloe dactyloides*), having an even distribution of roots to a soil depth of about 5 ft. A third group is like loco weed (*Oxytropis lamberti*), having an even distribution of roots to soil depths much below 5 ft. Loco weed happens to have a tap root system, but there are some grasses with fibrous root systems that also fall into the last category. These three categories of plants, then, reduce competition for moisture when growing together by utilizing different regions of the soil profile.

Competition for moisture may be the reason one species is replacing another in the northern intermountain region of the United States. About 125 years ago, the dominant plant of this grassland area was bluebunch wheatgrass (*Agropyron spicatum*), a perennial. About the middle of the nineteenth century, an annual grass called cheatgrass (*Bromus tectorum*) was accidentally introduced to the area from Europe. From that time to the present, graziers have noticed an enormous increase in abundance of cheatgrass and an equally impressive decrease in the abundance of bluebunch wheatgrass. What caused the shift?

Both species have a similar life cycle. They germinate (or break dormancy if perennial) in fall, grow slowly during winter, grow rapidly in spring, form flowers in early summer, and die in June (or begin dormancy in mid-July if perennial). G. Harris of Washington State University studied growth and survival of these two species during a year, starting from seed for both. He found that the presence of cheatgrass greatly reduced growth and survival of bluebunch wheatgrass. In one field trial, bluebunch wheatgrass and cheatgrass were sown together in two plots; in one plot the amount of cheatgrass sown was much higher than in the other. After a normal growth period (October–June), the number and weight of wheatgrass plants in each plot were tabulated. Table 9.1 shows that survival and plant weight was much lower when cheatgrass was more abundant than when it was scarce.

Harris then planted seeds of each species in long glass tubes filled with soil. The tubes were inserted in the field, level with surrounding soil. Every month the tubes were lifted and depth of rooting was measured. The graph in Fig. 9.3 shows that root growth during winter was much greater for cheatgrass than for wheatgrass. At the start of rapid spring growth in April, cheatgrass roots had penetrated 90 cm, in contrast to 20 cm for wheatgrass. This difference in depth allowed cheatgrass to absorb water from a much greater part of the soil profile than wheatgrass could. Consequently, when the upper soil became dry in early summer (with both species drawing water from it), only wheatgrass suffered, for water still remained available at greater depths (where only cheatgrass roots penetrated). Harris measured soil water availability in the two soil regions at this time and found water potential was below −15 atmospheres in the upper region (near the wilting point of many species), but was only −1

TABLE 9.1 Survival and Plant Weight of Bluebunch Wheatgrass after a Season's Growth in Two Levels of Cheatgrass Competition

Relative cheatgrass competition	Survival of wheatgrass, %	Average dry weight of wheatgrass, g
Low	86	10.4
High	39	1.8

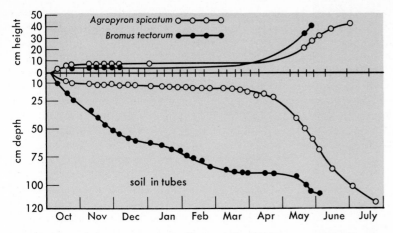

Figure 9.3 Seasonal leaf and root growth of *Agropyron spicatum* and *Bromus tectorum* in glass tubes in the field. (Redrawn from Harris, *Ecological Monographs* **37**, 102.)

atmosphere in the lower region (essentially still saturated).

The recent success of cheatgrass, then, seems due to its winter root growth, which gives it a spring and early summer advantage over wheatgrass that started from seed at the same time. Cheatgrass each year increases in density, and each year this results in greater competition for moisture in the upper soil and greater stress on wheatgrass.

Summary of Root Systems

1. The primary root develops from the radicle in the embryo. It generally penetrates the soil to some depth and may either form a tap root system or develop an extensive series of lateral roots to form a fibrous root system.
2. More frequently, the fibrous root system is formed by adventitious roots arising from nodes of the stem. This is the general situation in monocotyledons and occurs in many dicotyledons; it is most generally associated with horizontal stems.
3. Adventitious roots may arise in the internodal regions of the stem.
4. Shoots may arise from adventitious buds developing on roots.

Structure of Roots

The root, like the stem, grows, conducts water and nutrients, and stores food. Unlike the stem, it absorbs nutrients and water from the soil and anchors the plant. Accompanying similarities and dissimilarities in function are corresponding variations in structure. We shall find, in the root, meristematic and vascular tissues similar to those found in the stem. Both primary and secondary growth occur. There are three tissues in the root (root cap, endodermis, and pericycle) that are not present in the stem, and the arrangement of vascular and meristematic tissues is slightly different. A cuticle does not develop on epidermal cells of young roots. Instead, some root epidermal cells become hair cells and facilitate absorption (Fig. 9.9).

Primary Plant Body

External Features. The general structural features of roots can best be studied by means of seedlings that have been grown either in sand or on filter paper in a moist atmosphere. Fig. 9.4 shows such a seedling. Observe here the portion of the root that is clothed with root hairs; this region is known as the **root-hair zone.** Note that root hairs do not extend to the very tip of the root. Root-hair zone and hairless tip constitute a region of especial interest. It is in this part of the root that (a) growth in length, (b) most absorption of materials from soil, and (c) development of primary tissues take place. Below the root hairs there occur, in descending order, the **region of elongation,** a **meristematic zone,** and the **root cap. Differentiation** starts in the upper cells of the meristematic zone and extends upward through the region of elongation into the lower zone of root hairs.

In stems, the apical meristem is either naked or protected by rudimentary leaves and bud scales. In contrast, the apical meristem of roots is protected by a thimble-shaped mass of cells, the **root cap.** Root cap cells are constantly being sloughed off at the very tip, but at the same time new cells are being added to it by apical meristem cells. In the **apical meristem,** cells are actively dividing, adding new cells to the root cap and others to the region of elongation. Rapid growth in root length, however, is largely the result of elongation of cells behind the apical meristem. This is the **region of elongation** (Figs. 9.4B, C, 9.5A). Thus, it is seen that a very short

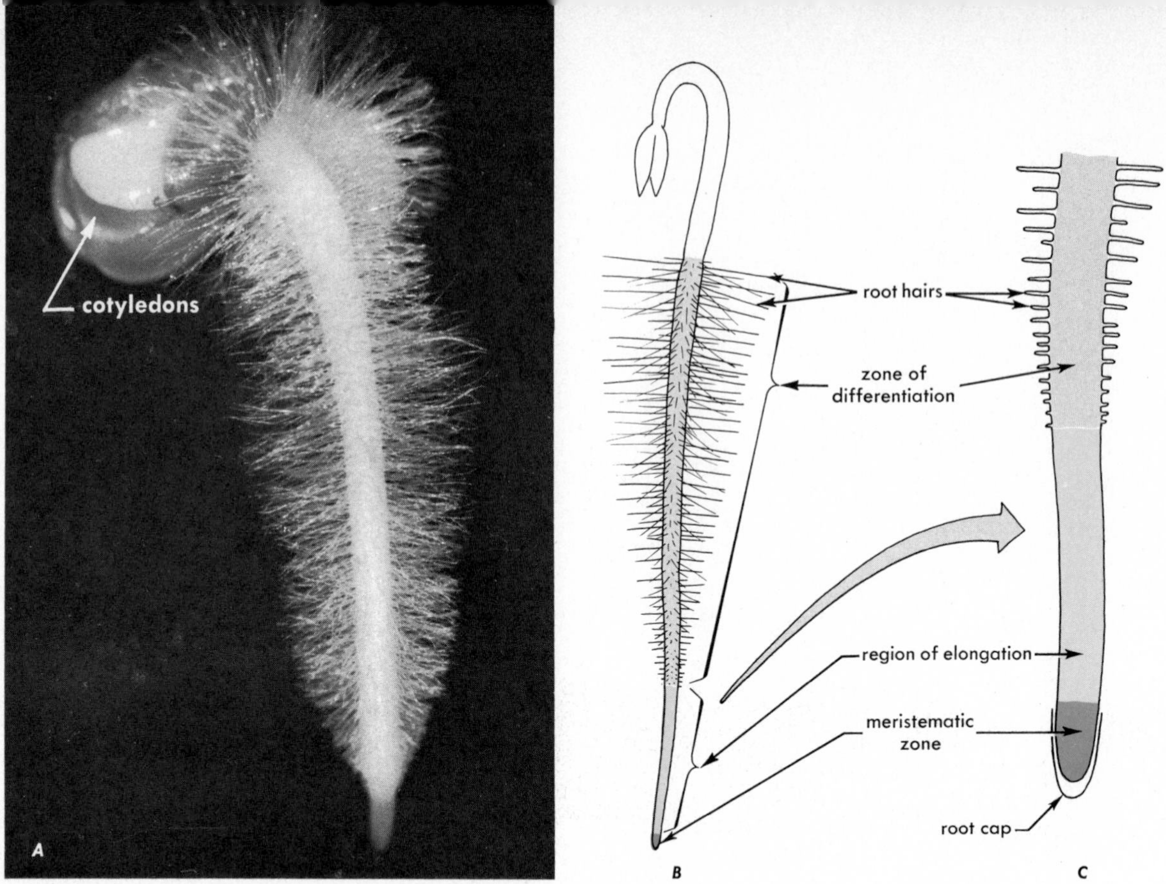

Figure 9.4 Radish seedling (*Raphanus sativus*).
A, photograph showing root hair growth, ×2; *B* and *C*,
diagrams showing growth regions.

portion of the root at the tip, usually 2–5 mm long, is constantly being forced through the soil. The method of growth in length of the root just described explains the protective function of the root cap; the delicate cells of the apical meristem are protected from mechanical injury as they are pushed through the soil. The region of elongation grades into the region of differentiation of primary vascular tissue internally and the formation of root hairs in the epidermis. Thus, the mechanism for absorption (root hairs) differentiates as conduction tissue is forming from procambium (Figs. 9.6, 9.7).

It is significant that elongation of cells does not occur in the root-hair zone. If elongation did occur in this zone, root hairs that wrap around and adhere to soil particles would be torn loose.

Internal Features. Meristematic cells of root apices have precise arrangements that differ from those of the shoot apex (compare Figs. 7.8 and 9.5). In the monocotyledons, as illustrated by apices of onion and corn (Fig. 9-6), five cell groups are discernible. The lowest of these groups forms (a) the **root cap.**

Meristematic cells which will give rise to mature tissues of the root form long rows of cells. Rows of cells destined to become (b) epidermis **(protoderm),** (c) cortex **(ground meristem),** and (d) the central vascular cylinder **(procambium cylinder)** may be recognized. These rows of cells diverge from a smaller fifth zone of less regularly arranged cells. There is good evidence that cell divisions rarely occur in this

Figure 9.5 Right: Longitudinal sections of root tips. *A*, section of an onion (*Allium cepa*) root tip showing growth regions. Processing emphasizes nuclei, ×85. *B*, region of apical meristems of corn (*Zea mays*) root tip, showing the very early differentiation of tiers of cells that eventually become specialized in the mature region of the root, ×300. *C*, diagram of apical region showing tiers of cells. The relative number of cell divisions in a given region is indicated as the incorporation of triturated thymidine is shown here by the relative depth of color. (*A* and *B*, courtesy of Lewis Feldman and Elizabeth Cutter.)

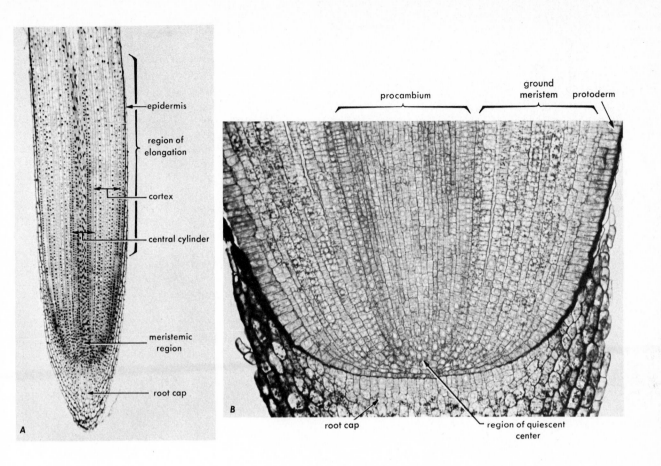

A

epidermis

region of
elongation

cortex

central cylinder

meristemic
region

root cap

B

procambium

ground
meristem

protoderm

root cap

region of quiescent
center

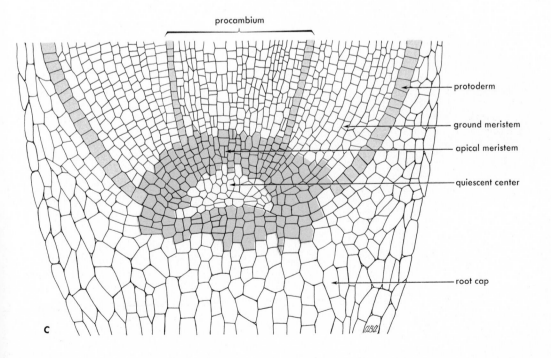

procambium

protoderm

ground meristem

apical meristem

quiescent center

root cap

C

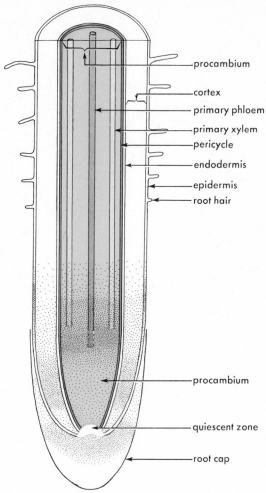

procambium

cortex

primary phloem

primary xylem

pericycle

endodermis

epidermis

root hair

procambium

quiescent zone

root cap

Figure 9.6 Diagrammatic representation of development of a root tip from the root cap to the initiation of the first cells of the primary phloem.

central cell mass or (e) **quiescent zone** (Figs. 9.5B, 9.6). Numerous divisions take place in the cells surrounding the quiescent zone. The function of the quiescent zone is not known: It may be of significance in determining the pattern of root development.

The root cap is composed of short-lived parenchyma-like cells. Since new cells are continually being added to it, the root cap persists throughout the growing life of the root. It is present in all common land plants, but is lacking in many aquatic plants and certain other specilized types.

The apical meristem is composed of thin-walled cells that are very much alike and are practically without intercellular spaces. Sections in this region usually show many cells in which the nuclei are in some stage of mitosis (Fig. 9.5A).

In the region of elongation, there is less uniformity in the appearance of cells than there is in the apical meristem. Some differentiation has taken place. As in the corresponding region of the stem, three primary meristematic tissues are evident, although the exact sequence of their development may vary in different plants: (a) protoderm, (b) procambium, and (c) ground meristem. These three meristematic tissues differentiate into primary tissues of roots (Figs. 9.5B, 9.6, 9.7).

Differentiation of Primary Tissues. Differentiation of meristematic cells into primary body proceeds in roots as it does in stems, with slight modifications. The protoderm gives rise to an epidermis devoid of a cuticle and guard cells, but well-provided with root hair cells (Fig. 9.4A). A cortical tissue of parenchyma cells arises from ground meristem. While strengthening cells, such as collenchyma, are not formed, the innermost layer of cortical cells does

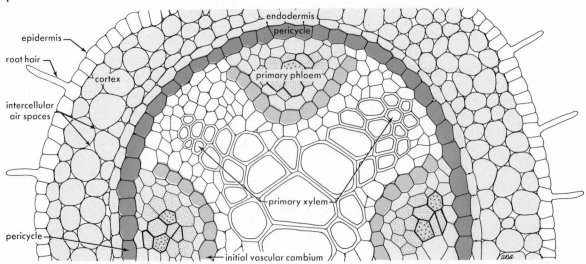

epidermis

root hair

cortex

intercellular air spaces

pericycle

endodermis

pericycle

primary phloem

primary xylem

initial vascular cambium

Figure 9.7 Cross-section of the primary plant body of a root at the close of primary development.

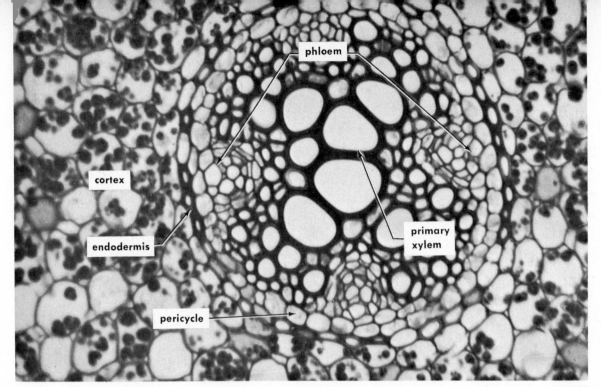

Figure 9.8   A photograph of the vascular cylinder, including pericycle and endodermis, of young root of *Ranunculus* at close of primary growth, ×200.

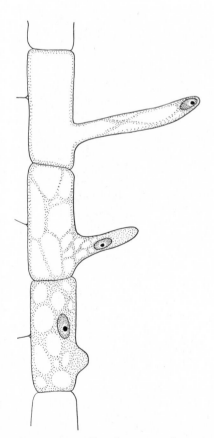

Figure 9.9   Development of a root hair.

become specialized to perform an important role in the regulation of movement of water and nutrients. This is the **endodermis** (Figs. 9.6, 9.7, 9.11). Note that it starts to form a very short distance behind the apex (Fig. 9.6). The procambium cylinder gives rise to a cylinder of vascular tissue (Fig. 9.6). There is no pith. The outermost layer of procambium cells differentiates early into a specialized cylinder of cells known as the **pericycle** (Fig. 9.6). The first cells of vascular tissue to form are two–four phloem cells (sieve elements) which arise singly at the outer periphery of the procambium cylinder (Fig. 9.6). The differentiation of sieve elements is soon followed by an equal number of vessel elements that arise at the periphery of the procambium cylinder in pockets between sieve elements (Figs. 9.6, 9.7, 9.8). Further differentiation of vascular tissue occurs toward the center of the root. When the primary plant is complete, it appears in transection as in Fig. 9.7. A cross-section through the primary vascular tissue of a young root at the completion of primary growth is shown in Fig. 9.8; the following primary tissues can be seen:

Epidermis.   This region, with its root hairs, is a single layer of cells derived from protoderm. Fig. 9.9 shows stages in the development of a root hair. Essentially, it is a lateral outgrowth of an epidermal cell. When grown in moist air, each root hair has the form of a slender tube, but in soil, it may be greatly contorted in its growth between and around soil particles. The root hair and the epidermal cell from which it grows constitute a single cell. The walls are thin,

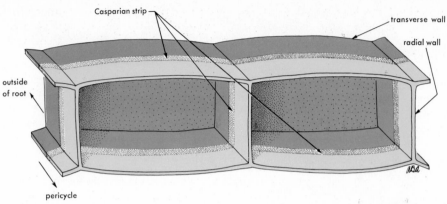

Figure 9.10   Three-dimensional view of two endodermal cells. The suberized Casparian strip is shown in gray. Water from solution outside the root wets the cellulose walls until it reaches the Casparian strip. Since water cannot wet and cross the Casparian strip, the cell walls inside the Casparian strip are wet by water that has passed through the protoplast of the endodermal cell.

composed principally of cellulose and pectic substances. The protoplast has a high water content. The nucleus is usually near the end of the hair, and there is a large central vacuole.

In most plants, the life of any one root hair is short; it functions only for a few days or weeks. New hairs are constantly forming at the anterior end of the root-hair zone, while those at the posterior end are dying. Thus, as the root advances through the soil, fresh, actively growing root hairs are constantly coming into contact with new soil particles. In the rye plant, it is estimated that new root hairs develop at an average rate in excess of 100 million per day. Root hairs of such plants as redbud, honey locust, and a number of others may persist for several years. *Root hairs do not become roots.*

Although nearly all ordinary land plants possess root hairs, a few, such as the firs, redwoods, and Scotch pine, are devoid of them. Also, many aquatic plants have no root hairs. Moreover, land plants (corn, for example) that normally develop root hairs when the root system grows in the soil or in moist air, develop no root hairs when the roots grow in water. In plants devoid of root hairs, absorption is accomplished entirely through typical thin-walled epidermal cells. Root-hair development is often inhibited by a concentrated soil solution and by high or low soil temperatures. Root hairs develop in light and dark about equally well if moisture and oxygen are adequate.

Cortex.   This region, relatively thicker than that in stems, is derived from ground meristem. The cortex is composed chiefly of storage parenchyma with large intercellular spaces (Fig. 9.7). In many species, secretory cells and resin ducts are present. The innermost layer of the cortex is a single row of cells, the **endodermis,** which is usually a conspicuous feature of roots. As a rule, in the primary state, endodermal cell walls are thin except for a band-like thickening

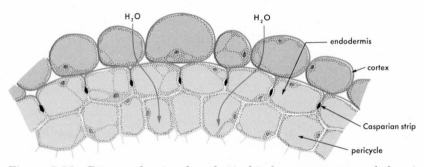

Figure 9.11   Diagram showing the relationship between cortex, endodermis, and pericycle. Water present in the walls of cortical cells cannot diffuse across the suberized Casparian strip. Water inside the endodermal layer has been filtered by the protoplasts of the endodermal cells.

running around the cell on radial and transverse walls. This thickened strip, the **Casparian strip,** is suberized (Figs. 9.10, 9.11).

Cellulose walls are permeable to water. Cell walls and intercellular spaces of the cortex may freely imbibe water and dissolved nutrients from the surrounding soil solution. The suberized strip that forms a complete ring around the endodermal cell in radial and transverse walls prevents water from reaching the pericycle except by passing through the protoplast of the endodermal cell. This situation is shown diagrammatically in Figs. 9.10 and 9.11. The Casparian strip in Fig. 9.10 is represented by a gray band completely encircling the cell. The dark green portion of the cell represents the side toward the soil solution, and consequently these walls have imbibed water directly from the soil solution. In Fig. 9.11, the Casparian strip is represented by a black bulge in the center of the radial walls of the endodermal cells. The dark green walls of the cortical cells and the outer portion of the endodermal cells freely imbibe water from the soil solution (dark gray). But water is not imbibed by the suberized Casparian strip. Consequently, any water entering pericyclic cells must pass through protoplasts of endodermal cells. Thus, the protoplasts of the endodermis may exercise a degree of control over water and dissolved substances entering and leaving the root. Cell walls and intercellular spaces (light green) of pericyclic tissue have imbibed water that has passed through protoplasts of endodermal cells into pericyclic cells (light gray). Frequently, in older root regions, inner, radial, and transverse walls of endodermal cells are thickened.

Pericycle. At an early stage of development, a special layer of parenchyma cells is differentiated from the outer region of the procambium cylinder. This is the pericycle (Figs. 9.6, 9.7, 9.8). It persists as a rather unspecialized type of meristematic tissue until secondary development is initiated. Then it may give rise to lateral roots; certain of its cells may develop into portions of vascular cambium, and others may give rise to cork cambium. In roots without secondary growth it may eventually form a sclerenchyma tissue.

Vascular Cylinder. (Figs. 9.6, 9.7, 9.8). This originates from the remaining internal portion of procambium. Since there is generally no internal ground meristem as occurs in tips of shoots, pith does not usually develop in dicotyledonous roots, although it may occur in roots of monocotyledons.

In some discussions of root structure, pericycle and vascular cylinder, which both originate from procambium, are considered a single general region called the **stele.**

It will be recalled that in stems, xylem and phloem of a vascular bundle occur in such a manner that a radial line cuts through both tissues (see Fig. 8.1). In roots, in contrast, primary xylem and phloem are arranged in such a manner that a radius passing through a xylem arms does not pass through phloem. Primary xylem usually consists of a central mass or "core" of xylem elements, with several radiating arms, between which are groups of phloem elements. Between these two tissues are one or more layers of procambial cells, represented by the light green line on Fig. 9.12A. In roots with secondary growth, these cells give rise to vascular cambium (which is shown in Fig. 9.12B as a light green line), while in roots without secondary growth, they often mature into sclerenchyma. Many roots have no pith, but in most monocotyledons and some herbaceous dicotyledons, the central core of the stele is parenchyma that resembles pith of stems. Whereas pith of stems is derived from gound meristem, that of roots is derived from procambium.

Cells that occur in primary xylem of stems may also be found in primary xylem of roots, although spiral and annular vessels are relatively rare in roots. Primary phloem of roots does not differ essentially from that of stems. It consists of sieve-tube members, companion cells, and parenchyma.

Summary of Primary Tissues

root cap	apical meristem	protoderm	epidermis	
		ground meristem	cortex, including endodermis	
		pericycle	lateral roots	
			cork cambium	
			vascular cambium	
		procambium cylinder	primary phloem	
			vascular cambium	
			primary xylem	
			pith (if present)	

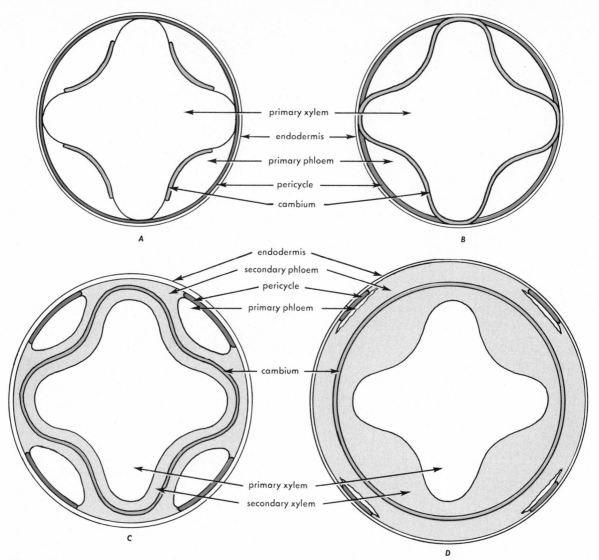

Figure 9.12 Diagrammatic representation of the development of the secondary plant body in a root. *A*, at the completion of primary growth a row of procambium cells remains (light green) and a complete circle of pericycle is present (dark green). *B*, the procambium (light green) joins with the pericyclic cells outside the xylem arms to form a continuous line of vascular cambium (light green). *C*, the vascular cambium forms secondary xylem internally and secondary phloem externally (light gray). The primary tissues are white. The primary phloem is being pushed outward and a small amount of pericycle (dark green) is still associated with it. *D*, a smooth circle of vascular cambium forms, producing secondary xylem and phloem. The primary xylem remains in the center of the stem, the primary phloem has been crushed, and only a small amount of the pericycle remains.

Origin of Lateral Roots. It will be recalled (p. 108) that lateral or branch stems originate from cell layers at or near the shoot tip. In contrast, lateral or branch roots in gymnosperms and angiosperms originate from cells of the pericycle. Fig. 9.13 shows the man- ner of origin of branch roots. Often, the point of origin of a branch root is opposite a primary xylem strand. For example, in the beet root, there are two primary xylem strands and two vertical rows of branch roots. It is seen from Fig. 9.13 that the tip

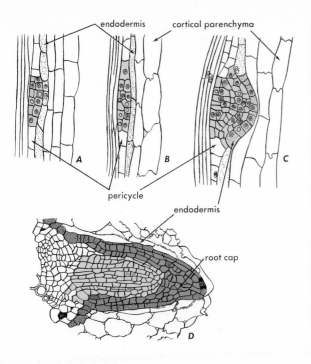

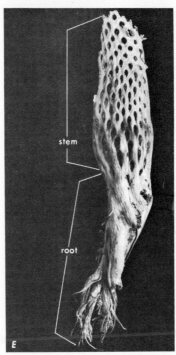

Figure 9.13   Branch roots. *A*, initiation of branch carrot (*Daucus carota*) root through formation of meristematic cells in pericycle; *B* and *C*, enlargement of meristematic region; *D*, young root pushing through cortex, ×50; *E*, roots forming in this manner do not leave a gap in the vascular cylinder, as is shown by the xylem framework of a rotted cabbage (*Brassica oleraceae*) at junction of stem and root, ×½. (*A* through *D*, after K. Esau.)

of the branch root must penetrate the cortex and the epidermis in order to reach the surface. These tissues are stretched and finally ruptured as a result of mechanical pressure of the growing root. It has also been suggested that cells of the branch root secrete substances which digest the tissues ahead of it. This method of root formation leaves the xylem portion of the vascular cylinder intact, in contrast with the stem where leaf and bud gaps result in openings in the xylem cylinder. This difference is shown in Fig. 9.13*E*, which illustrates the portion of cabbage xylem where root and shoot merge.

The Secondary Plant Body

If we examine old roots of woody plants, some of which may attain a diameter of many inches, we observe annual rings of growth resembling those in stems (Fig. 9.14). At completion of primary growth, procambium remains between radiating arms of primary xylem separating it from primary phloem. This remaining procambium is represented by light green cells of Fig. 9.7 and light green arcs of Fig. 9.12*A*. No procambium remains at the points of radiating primary xylem arms, so a continuous band of pro-

cambium is not present. In this early stage of development, no vascular cambium is present, but it is soon formed in the following manner: (a) Vascular cambium forms from procambium remaining between primary xylem and primary phloem, and (b) cells of the pericycle outside radiating arms of primary xylem become meristematic. This situation is represented in Fig. 9.12*B* by a continuous dark line. Note that dark green, representing pericylcle (Fig.

Figure 9.14   Cross-section of an old root of cherry (*Prunus avium*), ×$\frac{7}{8}$.

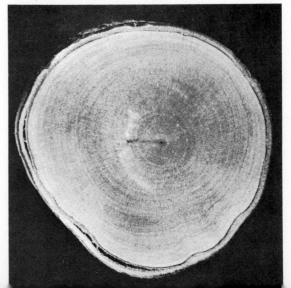

9.12*A*), becomes light green just outside the xylem arms (Fig. 9.12*B*). This change indicates its return to a meristematic condition.

Thus, a continuous layer of vascular cambium is formed from procambial cells that have not lost their meristematic character, and from pericyclic cells that have returned to a meristematic condition.

As seen in cross-section (Fig. 9.12C), this layer has the form of a wavy band; it passes inside each primary phloem group and outside each arm of the star-shaped primary xylem. This vascular cambium gives rise internally to secondary xylem and externally to secondary phloem, with initial production of secondary xylem being more rapid between the arms of primary xylem. This development soon results in a continuous smooth circle of vascular cambium (Fig. 9.12*D*). With continued secondary growth, epidermis and cortex are usually ruptured and finally sloughed off.

Cambium thus formed gives rise to secondary xylem and phloem after the manner of vascular cambium of stems. Annual rings are generally formed, and, macroscopically, wood of an old root cannot be distinguished from that of a stem. However, microscopic examination shows that, in general, secondary xylem of the root contains fewer fibers, more storage parenchyma, and larger, more numerous thin-walled vessels than does that of the stem. Secondary phloem of the root as compared with that of the stem has more storage parenchyma and less mechanical tissue.

Formation of cork cells accompanies development of secondary xylem and phloem. They are derived from a cork cambium which in most cases originates first in pericyclic cells. As the root increases in diameter, this first-formed cork is stretched and torn as in stems. Cork cambium forms anew from deeper-lying cells, generally in phloem. This results, as in stems, in development of a thick protective layer (Fig. 8.18).

Summary of Secondary Growth in Roots

A vascular cambium originates from procambium cells between primary xylem and phloem and from pericyclic cells exterior to the radiating points of primary xylem. This vascular cambium forms secondary xylem internally and secondary phloem externally. The resulting increase in diameter stretches and tears the endodermis, cortex, and epidermis. A cork cambium develops from the pericycle and forms cork. Woody roots consequently are very similar in structure to woody stems (Fig. 9.14).

Root Functions

The functions of root systems are **absorption, anchorage, conduction,** and **storage.** From the soil, roots absorb water, mineral salts, and oxygen. They anchor the plant firmly in place. They conduct water, mineral salts, and sometimes stored foods, *to* stems and leaves above ground. They conduct foods *from* stems to all parts of the system underground. Most roots usually store foods at least for a short period and in small quantities. Special storage roots accumulate rather large amounts of foods.

Absorption by Roots

Absorption of Water. In land plants, large amounts of water are absorbed by roots and lost from leaves. Most water that is absorbed moves upward to leaves and passes out as *water vapor* through stomata in epidermal layers of leaves.

We have seen that, while essentially similar patterns occur in both roots and shoots, there are differences related to their different functions. The response to water is different in the root and in the shoot, and some of the difference in the pattern of root and shoot appear to be related to the water economy of the plant. About 95% of the plant body is water. Water enters into all its metabolic processes. A part of the water absorbed by most land plants enters through the root hairs (Fig. 9.9); some enters through the thin walls of ordinary epidermal cells. Water entering through root hairs is conducted throughout the plant by the vascular system. Much water passes from leaves of the plant in the process of transpiration. The epidermis of the shoot is provided with a cuticle and guard cells to conserve water while allowing for exchange of gases. The epidermis of the root lacks a cuticle and is provided with root hairs for the absorption of water. In the shoot, primary xylem is internal to primary phloem. In the root, primary phloem alternates with arms of primary xylem. Only a pericycle and protoplasts of the endodermis intervene between the outermost vessel and the imbibed water of the cortex. Water need not pass through phloem tissue to reach xylem. We have learned that the area of contact of the root system with the soil may be enormous; that numerous rootlets are constantly elongating and exploring new soil areas; that root hairs are being formed anew just behind the growing tip of the root; and that these young, thin-walled root hairs are flattening out and surrounding soil particles, and thus coming into very close contact with the film of water that surrounds the soil particles.

Separating the protoplast of a root hair from the

soil solution is a cell wall chiefly made up of cellulose and pectic substances. There is usually a pectic coating on the outside of the wall. Due to the gummy nature of this coating, root hairs adhere closely to soil particles.

As seen from Fig. 9.7, the pathway of water from the root hair is through the walls, spaces, or cytoplasm of the cortex, then through the cytoplasm of the endodermis (Figs. 9.10, 9.11) and one or more layers of cells of the pericycle, to vessels in the xylem. Once in xylem vessels, water moves from very small root branches to larger roots, then to the stem, and on upward to leaves and other organs of the plant. From leaves and young stems, water passes off to the atmosphere as water vapor. Thus, there is a continuous stream of water through the plant that has been in existence since the germination of the seed.

Absorption of Solutes. Land plants derive their inorganic salts from soil. Usually, every kind of chemical element found in the soil can also be found in plants. Certain mineral elements, such as potassium, sulfur, phosphorus, calcium, and magnesium, constitute a relatively large portion of the inorganic components of plants. Other elements, such as iron, manganese, chlorine, molybdenum, boron, zinc, copper, iodine, and selenium, occur in small quantities or as mere traces.

At present, there is no reason to believe that every element found by chemical analysis in plants has an essential role in the plant's life. Although a number of elements are known to be essential, others are apparently carried into the plant but may be of no particular use. Certain elements, however, such as boron, zinc, copper, chlorine, molybdenum, and manganese, formerly regarded as nonessential, now are known to be indispensable. Some elements are absorbed in considerable excess of the plant's actual needs. It is very likely that most salts of the soil enter chiefly in the form of ions.

The process of absorption of solutes by roots will be discussed in Chapter 12.

Anchorage by Roots. We have seen (Fig. 9.1) the extent of the root systems of ordinary plants and how completely they may occupy the soil. Thus, the roots function as very effective anchorage organs.

Storage by Roots. All roots, even slender ones whose primary function is absorption, may have a small amount of food stored temporarily in them. For example, when sugar moves into roots more rapidly than it can be utilized by growing cells, it may be converted to starch and as such be stored for a time, particularly in cortical cells. During the

dormant season, rather large quantities of starch are stored in woody roots of orchard trees. This food constitutes a reserve that is called upon when active growth is resumed in spring. The roots of the wild morning-glory (bindweed) and other perennials store large quantities of food. This stored supply enables the plant to send up new shoots when the "tops" of the plants are destroyed. The seasonal trend in storage of foods in roots of wild morning-glory is seen in Fig. 9.15. It will be noted that, in undisturbed plants, readily available carbohydrates rapidly build up during summer, whereas, in plants cultivated at 2 week intervals, food storage is greatly hindered.

Food storage in roots may occur in cortex, phloem, and xylem. Among our native plants, the most striking examples of fleshy-rooted plants occur in arid regions. Such roots generally contain a large quantity of stored water that can be used by leafy shoots during periods of drought.

Reproduction by Means of Roots. Ornamental crab apple, cherry, plum, quince, hawthorn, and a number of other plants are usually propagated by root cut-

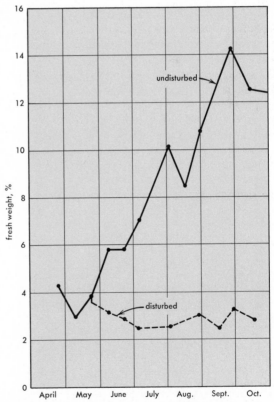

Figure 9.15 Seasonal trend in the storage of reserve carbohydrates in the roots of wild morning-glory (*Convolvulus arvensis*). (From Barr.)

tings. Shoots that arise from cuttings are from adventitious buds. Injury to roots may induce development of such buds. The tap root of a dandelion may be cut into many small pieces, and each section be capable of producing new shoots.

Summary

1. The principal functions of roots are absorption of water and nutrients, conduction of absorbed materials and food, and anchorage of the plant in the soil.
2. There are two general types of root systems, a fibrous root system and a tap root system.
3. Roots differ from all stems in lacking nodes and internodes. In some species, however, leafy shoots and associated root systems do form from adventitious buds.
4. In dicotyledons and gymnosperms, roots generally form at the lower axis of the stem; in many monocotyledons, roots or root primordia occur at nodes.
5. The root tip is divided into four zones of specialization: (a) root cap, which protects (b) meristematic region as it moves through soil; (c) a region of elongation; and (d) a region of differentiation characterized externally by root hairs and internally by formation of primary vascular tissues.
6. Soil water and nutrients move easily through epidermal and cortical tissues.
7. The endodermis forms the innermost cell layer of the cortex. A suberized band, the Casparian strip, in radial and transverse walls, completely encircles endodermal cells.
8. Water cannot move across the Casparian strip. Therefore, all water with dissolved nutrients inside the endodermis has passed through protoplasts of endodermal cells.
9. The pericycle, a row of cells internal to the endodermis, represents the first row of cells of vascular cylinder; they have differentiated from procambium. Cells of pericycle may eventually form part of vascular cambium or give rise to branch roots.
10. In cross–section, primary xylem is star-shaped and generally has from three to five arms. Primary phloem arises between arms of primary xylem. There is no pith in roots of dicotyledons and gymnosperms.
11. Water and inorganic nutrients enter primary xylem from parenchyma cells without passing through cells of phloem.
12. Water and inorganic nutrients enter root-hair cells, pass through the cortex, are filtered by protoplasts of endodermal cells, cross the pericycle, and move directly into the vessels or tracheids of primary xylem.
13. In roots having secondary growth, a vascular cambium is formed by pericyclic cells over xylem arms and procambial cells remaining between primary xylem and phloem. This vascular cambium at first forms a wavy line in cross-section. It eventually becomes circular, and there is little difference in the appearance of wood from root or stem.

10

Leaves

Green plants are the only producers that use sunlight as an energy source; all other inhabitants of the earth consume the production of green plants. In seed plants, leaves are the principal organs of production. Chloroplasts within leaf cells are the sites in which light energy is trapped and converted into chemical energy. Chlorophyll, membrane-bound within chloroplasts, performs the first step in trapping light energy. Leaves are green because wavelengths of red and blue light are absorbed by chlorophyll and wavelengths of green light are transmitted or reflected. The absorbed light gives up energy in exciting an electron in the chlorophyll molecule to a higher energy level (Chap. 13). This is the first step in **photosynthesis,** a primary function of leaves.

The water economy of plants calls for absorption of much more water than can be metabolized. This excess water is returned to the atmosphere by leaves. A second function of leaves is **transpiration.**

Form and anatomy of leaves are such that leaves are peculiarly adapted to carry on the two primary functions of photosynthesis and transpiration.

Types of Leaves

Leaves are of many shapes and sizes (Figs. 10.2, 10.3, 10.4, 10.5). Leaves of ferns are generally much dissected (Fig. 27.27). Those of conifers are needles or scale-like (Figs. 28.11, 28.16). Many monocotyledonous plants have strap-shaped leaves like those of corn and other grasses. While leaves of dicotyledons show great variation in shape, they almost all have a flattened blade and a stem-like petiole to attach them to the plant axis. Practically all leaves have veins for support and conduction and a chlorenchyma tissue containing chloroplasts. Shape and size are constant traits for many categories of plants. The shape of maple leaves (Fig. 10.3C) easily distinguishes them from oak leaves (Fig. 10.2C). The arrangement and shapes of leaves of different members of the pea family (Fig. 10.2F) are so constant that one may frequently place an unknown plant in that family simply if it has this type of leaf. Also, different species within a genus may each have a characteristic leaf shape as in *Populus* (Fig. 10.1). Leaf shape, however, is not always a good diagnostic trait for use in plant identification, because it may vary within a species (Fig. 10.1), and different species may have similar leaf shapes. The blades provide large surfaces for absorption of light energy and carbon dioxide, both of which are required for photosynthesis. Leaf blades are thin and hence no cells lie far from the surface. This form of the leaf facilitates absorption of light energy and

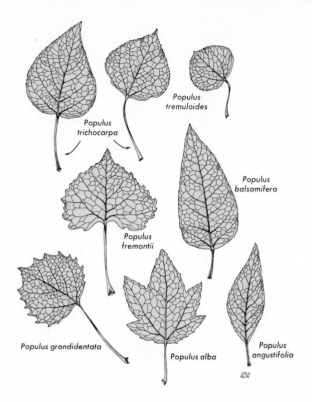

Populus trichocarpa

Populus tremuloides

Populus balsamifera

Populus fremontii

Populus grandidentata

Populus alba

Populus angustifolia

exchange of carbon dioxide, oxygen, and water vapor between the intercellular spaces of the leaf and the atmosphere. The water leaves utilize and transpire comes from roots. The food they manufacture must pass to all other parts of the plant. Veins support the thin parenchyma tissue. Xylem and phloem within veins perform their usual functions of support and conduction.

Dicotyledonous Leaves

The leaves of dicotyledons are generally distinctly different from those of monocotyledons. A typical foliage leaf of a plant belonging to the dicotyledons is composed of two principal parts: (*a*) **blade** or lamina and (*b*) **petiole** or stalk (Fig. 10.2*A*). The blade is thin and expanded, the petiole slender. The thin blade is supported by a very distinct network of

Figure 10.2 Below: Different kinds of leaves. *A*, poplar (*Populus deltoides*); *B*, castor bean (*Ricinus communis*); *C*, oak (*Quercus lobata*); *D*, rose (*Rosa odorata*); *E*, Virginia creeper (*Parthenocissus quinquefolia*); *F*, faba bean (*Vicia faba*), $\times\frac{1}{2}$.

Figure 10.5 Gymnosperm leaves. *A*, dawn redwood (*Metasequoia glyptostroboides*); *B*, *Ginkgo biloba*; *C*, pine (*Pinus*); *D*, *Podocarpus*, about ×¾.

they are tubular (Fig. 16.32), and in palms they are large and fan-like (Fig. 7.29A). The grasses have a very characteristic type of leaf. Crabgrass and corn (Fig. 10.4A, B) may be taken as examples. The grass leaf is divided into two parts, **sheath** and **blade.** The blade is the typical thin, expanded portion. The sheath is green, perhaps nearly as large as the blade, but it is not a flat expanded structure. Instead, it completely sheaths the stem, extending, in many species such as corn (Fig. 10.4B), over at least one complete internode. In crabgrass (Fig. 10.4A), the sheath covers only about one-half of an internode. If the region of union between the blade and the sheath is examined carefully, a small flap of delicate tissue extending upward from the sheath may be seen, closely investing the stem. This is called the **ligule** (Fig. 10.4B). It may, in some cases, serve to keep water and dirt from sifting down between stem and sheath. In many species, of which barley (Fig. 10.4C) is a good example, the base of the blade, at

its union with the sheath, is carried around the stem in two ear-like points. These points are called **auricles.** Ligule and auricles may both be present, or one or the other may be absent.

Gymnosperm Leaves

The leaves of all gymnosperms native to the United States and Europe are needle-like (Fig. 10.5C), or scale-like (Fig. 28.16B). They are discussed in some detail in Chapter 28, on the conifers. The needles of the dawn redwood (*Metasequoia glyptostroboides*) and the fan-shaped leaves of *Ginkgo biloba* are shown in Figs. 10.5A, B. Both of these trees are native to China and represent very primitive seed plants. *Ginkgo* is known only in cultivation, and the dawn redwood, like the California redwood, has a very restricted distribution. The leaves of many gymnosperms of the southern hemisphere are linear (Fig. 10.5D) or expanded.

Anatomy of the Foliage Leaf

The anatomy of a leaf blade is best shown in section (Figs. 10.6, 10.7A, B). In these figures, we observe three principal tissues: (a) **epidermis,** (b) **mesophyll** (middle of leaf), and (c) **veins** or **vascular bundles.** The epidermis usually consists of a single layer of cells that covers the entire leaf surface. It protects the tissues within from drying out and from mechanical injury. The mesophyll is composed of pa-

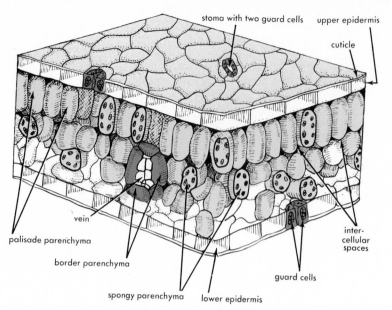

Figure 10.6 Three-dimensional diagram of a section of a foliage leaf, ×20.

renchyma cells, most or all of which contain chlorophyll and thus are able to carry on photosynthesis. Veins possess xylem and phloem elements and hence conduct water, inorganic salts, and foods. Fibers and collenchyma may be associated with conducting elements of midrib and larger lateral veins.

The petiole has its own specialized structures (Fig. 10.12), enabling it to support the leaf blade, conduct food, water, and inorganic salts, and disconnect itself from the stem at the close of the growing season without exposing living stem tissue to drying out or to infection.

Let us now consider each of these regions in detail.

Epidermis

The epidermis covers the entire leaf surface and

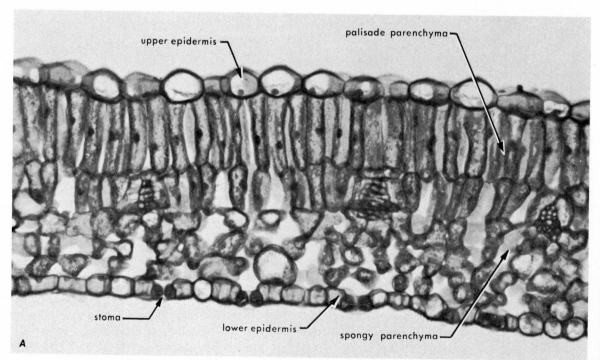

Figure 10.7   Above and right: Photomicrographs of sections of a lilac (*Syringa*) leaf, *A*, cross-section of leaf, ×75; *B*, section cut parallel to surface of leaf, ×75. (Slides courtesy of Triarch Products.)

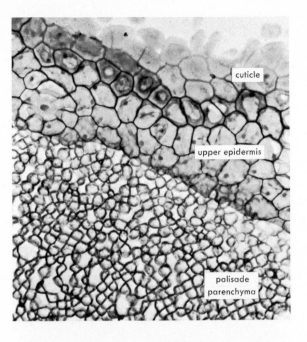

cuticle

upper epidermis

palisade parenchyma

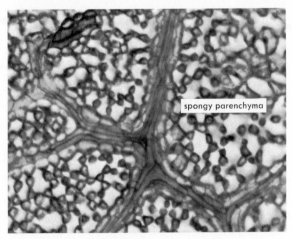

spongy parenchyma

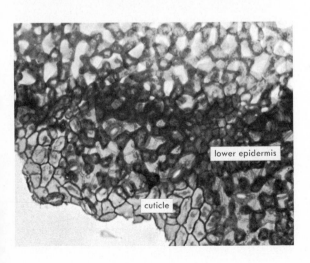

lower epidermis

cuticle

is continuous with the surface of the stem to which the leaf is attached. In most leaves, the epidermis is a single layer of cells. It may consist of several kinds of cells: (*a*) **ordinary epidermal cells,** (*b*) **guard cells,** (*c*) **hair cells,** (*d*) **glandular cells** (Fig. 10.8). Ordinary epidermal cells show a variety of shapes, depending upon the species. One form is shown in Fig. 10.8C. They are similar in shape to an irregular pavement block, whose depth is usually less than its breadth or length. Epidermal cells are generally covered on their outer surfaces by a waxy cuticle secreted by their protoplasts. The deposition of the cuticle frequently forms a pattern, quite specific for the leaves on which it is found. The cuticles on the upper and lower epidermis of leaves of *Cnidoscolus* (an herb from Mexico with stinging hairs, appropriately called mala mujer) are shown in (Fig. 10.9).

Guard Cells

Guard cells occur in pairs, and each is crescent-shaped or semicircular in form, as seen in surface view (Fig. 10.8A). Chloroplasts occur in guard cells but are lacking in ordinary epidermal cells. Both kinds of epidermal cells have long-lived protoplasts. The outer wall of epidermal cells, including guard cells, has a cuticle (Figs. 10.6, 10.7B) like that of the epidermis of stems. It is effective in limiting the movement, either inward or outward, of water vapor and other gases. The cuticle is usually thicker on the upperside of the leaf than on the underside.

A guard cell is a special type of epidermal cell. Guard cells occur in pairs, and between them is an opening or pore (Fig. 10.8C).

A cross-sectional view of the guard cells shows that the walls are typically unevenly thickened, with the thicker, less elastic walls, adjacent to the stoma. The stomata are the only openings in the leaf epidermis, and it is chiefly through them that gases pass into or out of the leaf. Although the cuticle is nearly impermeable to gases, small amounts of gases pass directly through the outer wall and the cuticle of epidermal cells.

Stomata vary considerably in size in different species of plants, and even in any one plant. Some representative measurements in microns are as follows (length × breadth): bean, 7 × 3; geranium, 19 × 12; corn, 19 × 5; oat, 38 × 8; sunflower, 22 × 8. The number of stomata per unit area varies widely, depending upon the species of plant and the environmental conditions under which it is growing. Usually more stomata are on the lower surface than on the upper. Table 10.1 gives the average number of stomata per square centimeter on the upper and lower surfaces of some common plants.

Anatomy of the Foliage Leaf   **171**

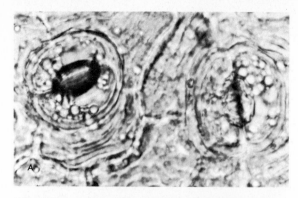

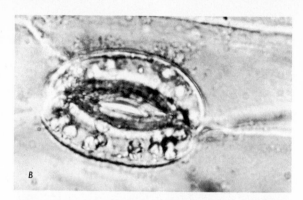

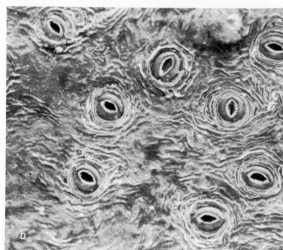

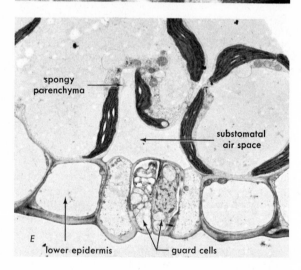

Figure 10.8 Stomata in leaves. *A* and *B*, light micrographs, ×710; *C* and *D*, stereoscan micrographs, ×400; *A* and *C*, ivy (*Hedera helix*); *B* and *D*, sugar beet (*Beta vulgaris*); *E* and *F*, electron micrographs; *E*, *Opuntia*, ×4000; *F*, rice (*Oryza sativa*), ×900; *G*, diagram. (*C*, *D*, and *F*, courtesy of D. Hess; *E*, courtesy of W. W. Thomson; *G*, redrawn from R. M. Holman and W. W. Robbins, *A Textbook of General Botany*, John Wiley & Sons, New York.)

TABLE 10.1   Average Number of Stomata per Square Centimeter

Plant	Upper epidermis	Lower epidermis
Alfalfa	16,900	13,800
Apple	0	29,400
Bean	4,000	28,100
Cabbage	14,100	22,600
Corn	5,200	6,800
English oak	0	45,000
Nasturtium	0	13,000
Oat	2,500	2,300
Potato	5,100	16,100
Tomato	1,200	13,000

Stomata are of widespread occurrence in the plant kingdom. With the exception of a few aquatic plants, stomata are present in all angiosperms and gymnosperms, and functional stomata have been found in liverworts, mosses, horsetails, ferns, and cycads. In the angiosperms, they can occur on stems, petals, stamens, and pistils as well as on leaves. The two major groups of plants that lack stomata are the

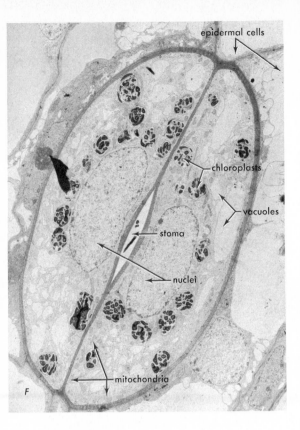

epidermal cells

chloroplasts

vacuoles

stoma

nuclei

mitochondria

F

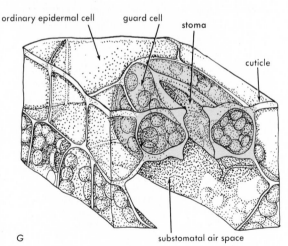

ordinary epidermal cell   guard cell   stoma

cuticle

G   substomatal air space

Figure 10.8   Continued.

algae and fungi. Stomata are basically similar in the many different groups in which they are found.

Opening and Closing of Stomata

The physical opening and closing of stomata is a result of changes in guard cell turgor, particularly with reference to turgor in adjacent epidermal cells. Water movement from these adjacent epidermal cells into guard cells results in an increased turgor

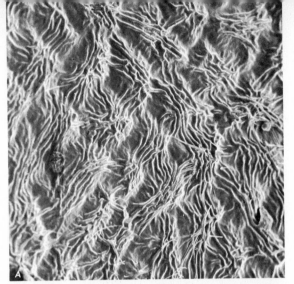

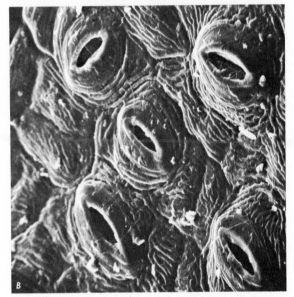

Figure 10.9 Above: Cuticle on the epidermis of mala mujer (*Cnidoscolus*) as seen with the stereoscan electron microscope. *A*, upper surface, showing cuticular ridges, ×650. *B*, lower surface at higher magnification, showing cuticular ridges between stomata, ×1100. (Courtesy of G. Breckon.)

in the guard cells and an elastic stretching of the guard cell walls. In two associated guard cells, thin walls adjacent to epidermal cells stretch more than the thicker inelastic wall bordering the pore. This results in the formation of an elliptical aperture. The change in shape of guard cells can easily be visualized by imagining that you are blowing up a long balloon with one side slightly thinner than the other. As air pressure increases in the balloon, the thin side will stretch more rapidly than the thicker side and the balloon will assume a kidney shape. The central regions of two similar balloons, attached only at their ends and with their thicker walls parallel and touching, would separate when blown up. Thus, if the structural features of guard cells are understood, it is easy to see how changes in turgor can cause either an opening or closing of stomata. But how are changes in turgor brought about? If one observes the influence of the external environment on stomatal opening, it soon becomes apparent that stomatal opening and closing can be triggered by a number of factors: Light generally causes opening and darkness, closing; wilting, or a water deficit in leaves, or an increase of CO_2 around leaves may induce closure; increase in temperature frequently results in stomatal opening.

Since such widely different environmental changes may result in opening or closing of stomata, it is evident that the osmotic regulatory mechanism of this stomatal action is complex—in fact it is still not clearly understood. One of the earliest suggestions made to explain changes in turgor associated with light-induced stomatal opening was that light caused photosynthesis in chloroplasts in guard cells. Sugar, an osmotically active solute, would be produced by photosynthesis and would then cause a decrease in water potential in the guard cells. Thus, water would diffuse into the cells, increasing turgor and causing stomatal opening. It was soon discovered, however, that the effects of light were very rapid and, in some instances, light intensity that induced opening was too low to produce significant photosynthesis.

An extension of this photosynthesis theory was based on observations that in light the sap of guard cells became less acid and the total amount of starch decreased. It was thought that decrease in acidity resulted from the use of CO_2 during photosynthesis (CO_2 combines with water to form an acid). In the presence of phosphate in the cell, decrease in acidity induced an enzymatic breakdown of starch to a sugar phosphate which in turn formed sugar and liberated the phosphate. Water would diffuse in and stomata would open. Stomatal closure was envisioned as a reverse process in which respirationally produced CO_2 would result in a more acid environment in guard cells. Starch synthesis from sugar glucose would be induced, the osmotic pressure would decrease, water potential would increase, and water would diffuse out of the guard cells. The resulting decrease in turgor would cause closure.

Although this theory is based on a number of valid observations, it cannot be used to explain all the reactions of the stomatal apparatus to environmental changes. It has been observed that both the opening and closing reactions may require cellular energy which may be supplied by the generation of ATP in the normal processes of photosynthesis and respiration (p. 233). Thus, stomatal opening in the light has been found to be correlated with an active accumulation of potassium ions in the guard cells. Energy for this ion accumulation comes from ATP formed during photosynthesis.

Epidermal Hairs

Several different types of hairs grow out from the epidermis of leaves and resemble those from the epidermis of stems. They may be unicellular or multicellular, simple or branched, scale-like or glandular. The unicellular hair is the simplest kind. It may be an extension of an epidermal cell, or it may be provided with a basal cross-wall. Unicellular hairs are sometimes branched. Multicellular hairs may consist of a single row of cells, or they may be branched. Glandular hairs bear, at the upper end, a single large cell or a group of cells; some such cells excrete ethereal oils, which often impart a stickiness to leaves (Fig. 7.12).

Mesophyll

Mesophyll is the photosynthetic tissue between upper and lower epidermis. It is parenchyma tissue, traversed by veins. Chloroplasts are present in the mesophyll cells, which may be divided in two distinct layers: **palisade parenchyma** and **spongy parenchyma** (Figs. 10.6, 10.7A). The palisade parenchyma is just below the upper epidermis and usually consists of from one to several layers of narrow cells with their long axes at right angles to the leaf surface. The spongy parenchyma extends from the palisade parenchyma to the lower epidermis. Cells of the spongy parenchyma are irregular in shape and loosely arranged. Large air spaces, or stomatal chambers, are generally present above each stoma (Figs. 10.8E, G).

A section cut obliquely through the leaf shows, essentially, cross-sections of these leaf tissues. Such a view is given in Fig. 10.7B. The cuticle occupies the upper portion of the figure, with sections of cells of the upper epidermis just beneath it. Below them, palisade parenchyma cells have a circular outline and fit together very loosely. One obtains the impression of a greater amount of air space in the palisade tissue than is apparent in the cross-section of leaf. Anastomosing veins, with their occasional endings, are very apparent.

Palisade parenchyma is, in general, found above veins, with spongy parenchyma below them. Note also that in this plane, cells of spongy parenchyma fit together to form an open network. The lower

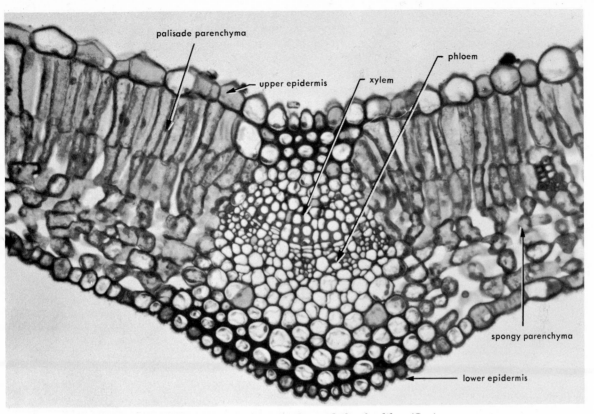

palisade parenchyma

upper epidermis

xylem

phloem

spongy parenchyma

lower epidermis

Figure 10.10 Photomicrograph of a cross-section of a midrib of a lilac (*Syringa vulgaris*) leaf, ×150.

epidermis is at the bottom of the section (Fig. 10.7*B*) and guard cells may be seen within it.

Intercellular spaces are prominently developed in the mesophyll but are much larger in spongy parenchyma than in palisade parenchyma. The air spaces between mesophyll cells are interconnecting, and many cells are in contact with an intercellular space. Thus, most food-making cells have free access to carbon dioxide and oxygen.

Veins of the Leaf

The veins or vascular bundles form a network extending throughout the leaf. The conducting elements are xylem and phloem, and fibers and collenchyma are sometimes associated with these. Hence, veins conduct water, mineral salts, and foods, and also mechanically support the mesophyll tissue. In addition to the midrib and larger lateral veins that are visible to the naked eye, innumerable minute branch veins occur that can be seen only with the aid of a microscope. The large veins contain vessels, tracheids, sieve tubes, and companion cells and also some mechanical tissue (Fig. 10.10). Such veins, in some leaves, may have both primary and secondary

vascular elements. Smaller veins have few vascular elements and few or no mechanical elements. The very end of a vein is usually a single tracheid of the spiral type (Fig. 10.11). Free ends of veinlets are usually surrounded by one or more layers of parenchyma cells, **border parenchyma,** which may or may not possess chloroplasts. Through these cells, water and solutes must pass from conducting elements of veinlet to cells with chloroplasts. The smallest veinlet has an unbroken connection with vascular elements of the midrib, petiole, and stem to which the leaf is connected.

In larger veins that have both xylem and phloem elements, the xylem is toward the upper surface of the leaf, the phloem is toward the lower surface (Fig. 10.10).

The Petiole or Leaf Stalk

In vascular tissues of the petiole, phloem and xylem maintain their relative positions, as in the stem (Fig. 10.12*B*); thus, phloem is on the underside of the petiole (and leaf blade) and xylem is on the upperside. One or more vascular bundles are embedded in parenchyma. Fibers may be associated

Anatomy of the Foliage Leaf **175**

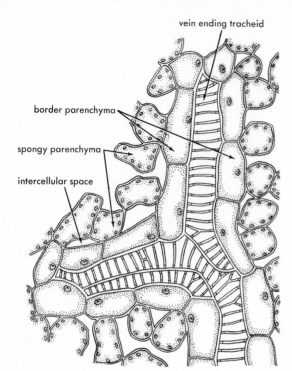

vein ending tracheid

border parenchyma

spongy parenchyma

intercellular space

Figure 10.11 Diagram of vein ending in a leaf. Note the parenchyma tissue bordering the vein and the presence of only annular tracheids in the vein.

with vascular tissues of bundles, and, not infrequently, groups of collenchyma cells occur beneath the epidermis (Fig. 10.12A).

Leaf Shape

Genetic Control

Leaf shape is under direct genetic control. Several single gene mutations of tomato (*Lycopersicon esculentum*) result in striking changes in leaf shape (Fig. 10.13). Cell cytoplasm, however, may also influence the final form assumed by the leaf. In hybridizing experiments with the evening primrose (*Oenothera*), the leaf shape of the progeny always resembles that of the maternal parent. The leaves of *Oenothera berteriana* are broader and longer than those of *O. odorata*. The pollen parent in breeding experiments between these two species has no influence on the shape of leaves borne by the progeny. When *O. berteriana* is the maternal parent, all leaves of the progeny resemble *O. berteriana* leaves. Conversely, when *O. odorata* is the maternal parent, leaves of the progeny resemble *O. odorata* leaves. The genes contributed by the paternal chromosomes

do not influence the leaf form of the progeny. This type of **maternal inheritance** demonstrates the dominant influence of the egg cytoplasm in determining leaf form in these species.

Environmental Factors

Leaf shape is also influenced by environmental factors such as light, moisture, and temperature. The

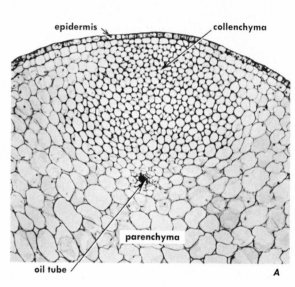

epidermis

collenchyma

parenchyma

oil tube

A

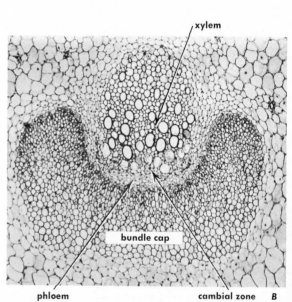

xylem

bundle cap

phloem

cambial zone *B*

Figure 10.12 Photomicrographs of celery (*Apium graveolens*) petiole. *A*, outer tissues; *B*, vascular bundle (vein), ×50. (Courtesy of K. Esau.)

176   Leaves

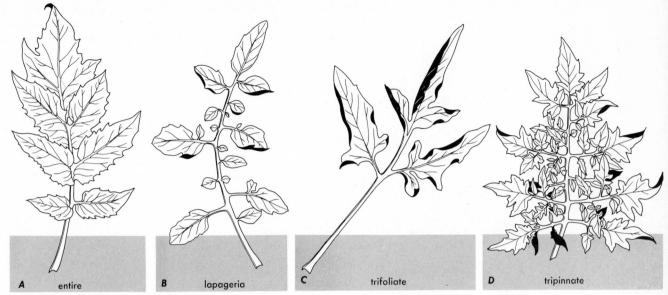

Figure 10.13 Different leaf shapes induced in leaves of tomato (*Lycopersicon esculentum*) by single gene mutations. (Courtesy of Tomato Genetics Cooperative.)

total intensity, wavelength, and daily duration of light have separate but interrelated effects. Plants growing in intense sunlight usually have thick leaves with a thick palisade tissue and a dense, spongy parenchyma. Their intercellular spaces are small, and the epidermis is heavily cutinized and generally glossy, with stomata confined to the lower epidermis. These plants may also have woolly epidermal hairs. Leaves of the same species growing in shade have contrasting traits. They are thin, with a single palisade layer and a spongy parenchyma with many intercellular spaces. The epidermis has a thin cuticle which is usually dull, and stomata may be present on both upper and lower epidermal surfaces.

Light is required for the normal development of leaves including the differentiation of chloroplasts to a state capable of carrying on photosynthesis. The leaves of a typical dicot remain small and pale yellow in darkness while the stems grow long and slender. This condition is called **etiolation.** Regular daily illumination of adequate intensity to support photosynthesis is required for the continued healthy existence of leaves. Excessive shading usually results in death of a leaf.

The duration of light may also influence the shape of leaves. When *Kalanchoë* was grown on a regime of 8 hr of light and 16 hr of darkness, the leaves were small, succulent, and sessile, with entire margins. Under the reverse situation, 16 hr of light and 8 hr of darkness, the leaves were large, thin, and had petioles. Their margins were notched.

Submerged leaves of semi-aquatic plants may be vastly different in shape from aerial leaves of the same plants. Similar changes may be induced by reducing the CO_2 content and lowering the temperature of the aerial portion of the shoot.

Age of Plant

Leaf shape may vary considerably with the physiological age of the plants. Thus, some plants have distinctive juvenile leaves for the first few years of growth. Subsequent adult leaves of a different form are characteristic of the older or adult plant (Fig. 10.14).

Leaf Modifications

Leaf shape may be considerably modified for functions other than photosynthesis (Fig. 10.15).

Bud scales are short, thick, sessile, often covered with dense hairs on the outer surface, and sometimes waxy or resinous. When present, they protect

Figure 10.14   Adult (*A*), ×$\frac{1}{4}$, and juvenile (*B*) leaves of *Eucalyptus*, ×$\frac{1}{2}$.

the delicate meristematic tissue of the shoot tip and the rudimentary leaves from drying out.

The **spines** of various species of cacti and those of *Fouquiera* represent entire transformed leaves. In the black locust (Fig. 10.15*A*), the stipules are spines, and in *Parkinsonia,* the stipules and midrib are spines (Fig. 10.15*B*).

In some species of *Lathyrus,* the **tendrils** are transformed leaflets; in others, the whole leaf is transformed into a single tendril, and leaf-like stipules perform the normal functions of the leaves. In *Bignonia carpreolata,* the third leaflet is transformed into a tendril (Fig. 10.15*E*). Both leaf tendrils or stem tendrils serve to attach the plant to a support. In certain species of *Acacia,* the petiole is leaf-like (Fig. 10.15*D*).

Leaves are sometimes modified as food or water storage organs. The thick, fleshy bases of leaves that make up much of the daffodil bulb (Fig. 7.27*D*) acccumulate large quantities of food. Succulents of deserts and of saline soils have thick, fleshy leaves with special water-storage tissue. This tissue consists of large parenchyma cells that usually lack chloroplasts. During the short period when water is available, it accumulates in the special storage tissue, and the plant draws upon it during periods of drought. Examples of plants bearing water-storing leaves are stonecrop (*Sedum*), ice plant, (*Mesembryanthemum*). Russian thistle (*Salsola*), *Sempervivum,* and other similar plants.

A striking adaptation of leaves to a special function occurs in insectivorous plants. In these plants, the leaves have taken on forms and various structural features that enable them to capture insects and obtain food from their bodies. Well-known insectivorous plants are sundew (*Drosera,* Fig. 10.16),

pitcher plants (*Darlingtonia,* Fig. 10.15*C*), and Venus' fly trap (*Dionae muscipula,* Fig. 10.16*B*).

Leaves generally do not function effectively in vegetative reproduction. However, African violets are regularly reproduced vegetatively by the placing of leaves in moist sand. In certain species of *Bryophyllum* (Fig. 10.15*F*), patches of tissue in notches along the leaf margins remain meristematic. In time, this tissue will develop small new plants while the parent leaf is still active. The little plants eventually drop from the leaf to the ground, where under favorable conditions they may develop into new individuals.

Leaf Development

Morphology of the Shoot Apex

Leaf development is intimately associated with the differentiation of the young shoot apex. The following section briefly reviews some of the problems of leaf and shoot development. The series of events is much more complex than in the root and is directly related to stem development. In spite of much work, many aspects of shoot development remain unclear.

The leaf is initiated in promeristem of the shoot, close to the apex (Fig. 10.17). It first appears on the flanks of the apex, as a slight bulge resulting from enlargement of several cells in the outer layer of apical promeristem (Fig. 10.17). Promeristem cells, including the leaf initials, are true meristematic cells, small in size, having large nuclei, and with a densely staining cytoplasm. Ground meristem rapidly differentiates from promeristem and may be recognized by the more elongated, more vacuolated cells, below the promeristem (Fig. 10.17). Directly below the leaf initials, however, cells remain truly meristematic; they are small, have large nuclei, and have a dense, heavily staining cytoplasm, with few, if any, vacuoles. This is the procambium, and it forms a continuous connection from leaf initials to primary vascular tissues below (Fig. 7.8). Differentiation of procambium strands and the leaf initials is very closely correlated.

Initiation of a Leaf

What causes the initiation of a leaf? Why, in lilac, do two leaf initials arise 180° apart and at the same time? The next two leaf initials will also be 180° apart, but oriented at 90° from the leaf initials below them. Leaves and axillary buds in lilac are opposite (Fig. 7.6B). Leaves in walnut are alternate (Fig. 7.6A). Taking bud A on a walnut twig (Fig. 7.5) as a starting point, it will be found that a line drawn through successively lower buds on the stem traces a spiral around it. The fifth bud in the series will be directly below bud A, and the spiral trace passes twice around the stem before arriving at the fifth bud. This relationship—the number of buds between two aligned buds and the number of turns around the

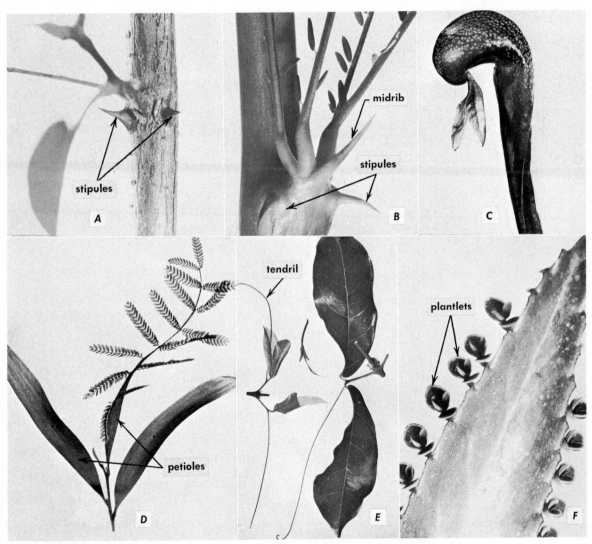

Figure 10.15   Leaf modifications. *A*, stipular spines of the black locust (*Robinia pseudoacacia*); *B*, spines from stipules and midrib in *Parkinsonia*; *C*, insect-capturing leaf of *Darlingtonia*; *D*, modified petioles of *Acacia*; *E*, tendrils from leaflets of *Bignonia*; *F*, plantlets on leaf of *Bryophyllum*, $\times \frac{1}{4}$ to $\times 1$.

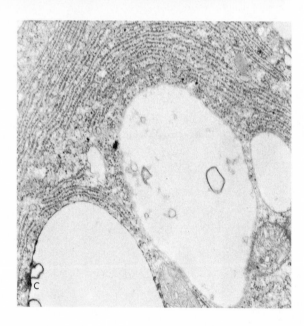

becomes visible to the eye. If that precise point of future leaf initiation is injured, that leaf primordium will not develop. It is currently thought that new leaf initials arise in the next available space. When, in the growth of the shoot apex, such a space becomes available, new leaf initials will appear.

The degree of autonomy of the shoot apex is graphically demonstrated by the ability of cultured apical tissue of a fern, consisting only of pro-meristem, to continue to differentiate and to finally produce a whole plant. In the angiosperms, excised promeristem alone will not differentiate when placed in a complex growth medium. However, if the first leaf initials are included with the excised apex, differentiation follows, with the eventual formation of a whole plant.

Development of a Young Leaf

The young leaf develops first as a slender pencil of tissue (Fig. 10.18). Becoming gradually flattened against opposing leaves, a marginal meristem appears (Fig. 10.18). The lamina of the leaf is produced by this marginal meristem, which is also apparently responsible for the shape of the leaf. Uniform activity of the marginal meristem results in an entire leaf like the poplar (Fig. 10.2*A*). If the activity is more closely related to the veins, lobed lamina arise (Fig.

stem required to arrive at an aligned bud—may be expressed as a fraction, which for walnut is $\frac{2}{5}$. The relationship in lilac is given by the fraction $\frac{1}{2}$. Other common bud arrangements are $\frac{1}{3}$ and $\frac{3}{8}$. The relationship is constant for a given species, and the arrangement makes it possible to precisely locate the point of leaf initiation on a shoot apex before it

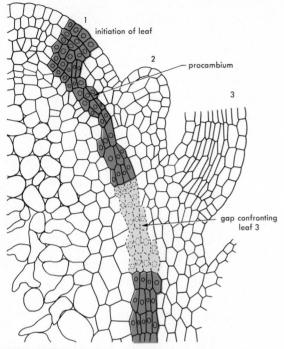

Figure 10.17 Diagram of a shoot apex showing the direct relationship between the initiation of a leaf and procambium strand. The second leaf shows a definite primordium and the third leaf is well-formed. The procambium strands associated with these leaves are not shown.

10.2C). The cellular organization of the leaf is completed while it is very small. In horse chestnut, cell divisions are even completed in the fall while the leaves are still enclosed within the bud. Expansion of the leaves as the bud opens in the spring (Fig. 7.3) is due entirely to cell enlargement.

Leaf Abscission

The separation of plant parts from a parent plant is a normal, continually occurring phenomenon. Leaves fall, fruits drop, flower parts wither and fall away, and even branch tips or whole branches may be separated normally from the parent plant. The autumn fall of leaves from woody dicotyledons is the most common example of this. In practically all cases, separation or **abscission** is the result of lack of differentiation in a specialized region known as the **abscission zone** at, or close to, the base of the petiole (Figs. 10.19A, C). Parenchyma cells comprising the abscission layer may be smaller and lack lignin, which may occur in significant amounts in the cells of adjacent tissues. Even vascular elements may be shorter,

and fibers may be absent from the bundle in the abscission zone. These anatomical features definitely make this zone an area of weakness.

Previous to leaf fall, changes may normally occur in the zone. Cell divisions, though apparently not necessary, frequently take place. When they do occur, a layer of brick-shaped cells is formed across the petiole. Actual separation of the leaf may be brought about in a number of ways. In some species, the middle lamella is dissolved away and cells simply fall apart. In other plants, walls and cells are dissolved, and thus the separation is effected. In a third small group of plants, a layer of cork forms across the petiole so that the leaf simply withers in place and is blown away. In all cases, a protective, corky layer of cells develops across the leaf scar. It is continous with stem cork. Furthermore, the vessels are likely to become plugged with tyloses or gums. Thus, the fall of a leaf does not leave an open wound or point of entrance for organisms that might cause disease.

The function of the abscission zone is twofold: (a) to bring about the fall of the leaf or other plant part, and (b) to protect the region of the stem from which the leaf has fallen against insect damage or rot caused by bacteria or fungi.

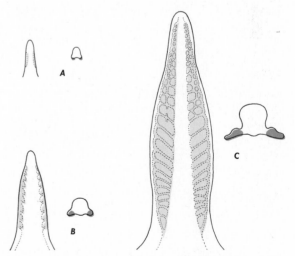

Figure 10.18 Growth of a young leaf. *A*, pencil stage, formation of marginal meristems; *B*, activity of marginal meristems initiates lateral expansion of a young leaf; *C*, continued lateral growth through activity of marginal meristems. (Redrawn after G. S. Avery, *Amer. J. Bot.* **20**, 565.)

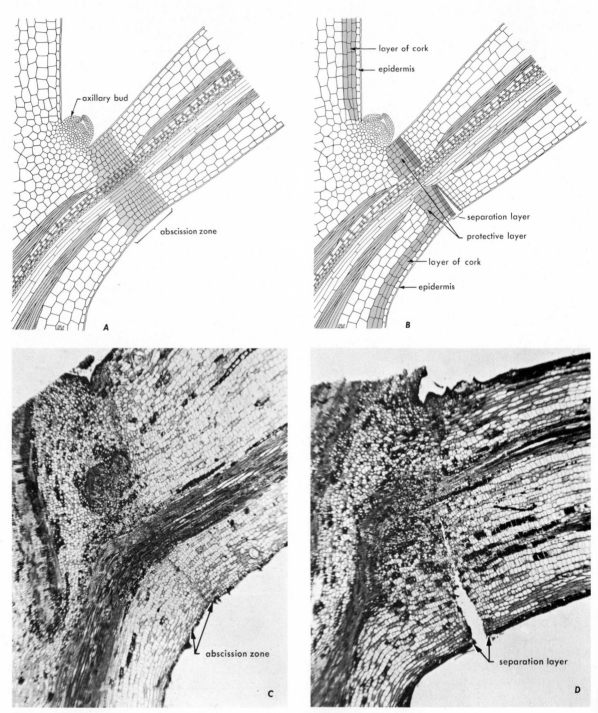

Figure 10.19 Leaf abscission. *A* and *C*, formation of the abscission zone; *B* and *D*, separation within the abscission zone and initiation of periderm. (*A* and *B*, courtesy of F. T. and A. B. Addicott; *C* and *D*, courtesy of Hall.)

Hormones and Leaf Abscission

It has been known since 1933 that auxin, particularly indole acetic acid (IAA), can inhibit abscission. It is orchard practice to spray with the synthetic auxin NAA to prevent fruit drop. This is particularly important in apples, where a large drop of young fruit is a common occurrence. In the course of studying the influence of auxin on cotton, a substance was isolated from cotton buds that accelerated abscission of cotton leaves and was antagonistic to auxin. This substance was chemically characterized in 1963. It is now called **abscisic acid** or **ABA,** and it promotes abscission in many other plants.

At the same time, other workers were isolating a material which was known to induce dormancy in buds. It was given the name dormin. It is now known that dormin and abscisic acid are the same substance.

Abscisic acid is widespread in the plant kingdom. In small amounts, it stimulates abscission of both leaves and fruits, induces dormancy of apical buds, prolongs dormancy of potato buds, inhibits seed germination, and plays some part in flowering.

The gibberellins will also inhibit leaf abscission, but are only about one-half as active as abscisic acid. The action of the two hormones is somewhat different and may account for the variation in the morphological changes during abscission that were noted above. When abscission is induced by abscisic acid, cell division is restricted and no definite separation layer is formed. The cell walls of the parenchyma cells of the abscission zone break down. Separation occurs about 24 hr after application of ABA and may be initiated at any place in the abscission zone. In contrast, gibberellin application causes a great increase of cell division and the creation of a definite separation layer. Actual cell separation occurs by a breakdown of the middle lamella. The leaf falls about 48 hr after the application of the gibberellin.

In general, gibberellins, auxins, and cytokinins inhibit stimulate growth. Abscisic acid and ethylene promote abscission and retard growth. It is tempting to speculate upon the possible interaction of ABA and ethylene with the auxins, gibberellins, and phytokinins as a partial explanation of the yearly rhythm of plant growth and senescence. During the fall, an increase of abscisic acid could account for leaf and fruit abscission and dormancy of buds, seeds, and other vegetative parts. The dominance of other hormones in the spring could induce the germination of seeds, break the dormancy of buds, and in general account for the flush of spring growth. But little is yet known about the detailed interactions of these hormones. (see also chapter 20).

Summary

1. The functions of leaves are photosynthesis and transpiration.
2. Angiosperm leaves are generally supplied with flat, thin blades attached to the stem by petioles or sheaths. Veins strengthen the blades and transport food and water.
3. Gymnosperm leaves of the northern hemisphere are needle- or scale-like.
4. A cross-section through the blade of a leaf shows the following tissues: upper epidermis, palisade parenchyma, spongy parenchyma, and lower epidermis. Generally, a waxy cuticle coats the epidermis. Guard cells of the epidermis form stomata which control gas exchange.
5. Increased turgor inside guard cells enlarges the stomatal opening; decrease turgor decreases the size of the stomatal opening. Increased turgor may result from increased intensity of sunlight and increased temperature around the leaves. Wilting, or an increase of CO_2 around leaves, brings about a reduction of turgor.
6. Mesophyll tissue is comprised of palisade and spongy parenchyma tissues and veins. Chloroplasts are present in both parenchyma tissues. Mesophyll is adapted for photosynthesis.
7. Leaf shape is under both nuclear and cytoplasmic control, and is also influenced by light, moisture, and the physiological age of the plant.
8. Leaves may become modified, serving as bud scales, food or water storage organs, and insect traps.
9. Leaf primordia develop in the next available space in a definite spatial sequence on the flanks of the shoot apex.
10. The excised promeristem of a fern apex will develop a new shoot, but leaf initials of an angiosperm shoot must be included with the excised apical meristems before shoot differentiation will occur.
11. The young leaf lamina is produced by marginal meristems which are also responsible for the shape of the leaf.

12. Cell divisions are completed while the leaves are enclosed in the bud; leaf expansion results mainly from cell enlargement.
13. The formation of a definite abscission zone across a petiole or fruit stem is responsible for leaf fall or fruit drop.
14. Abscisic acid induces the formation of the abscission zone; it is also responsible for dormancy of apical buds. It is antagonistic to the growth hormones.

11

Soil and Mineral Nutrition

We have been studying the morphology of seed plants—the structure of their organs, tissues, and cells. With this knowledge to help us, we are now ready to consider in more detail the physiology of the plant—the ways in which the various parts function together in the processes of absorption, conduction, transpiration, photosynthesis, and respiration.

Soil—The Environment of Roots

Soil may be defined as the weathered superficial layer of the earth's crust that typically is made up of decomposed and partly decomposed parent rock material with associated organic matter in various stages of decomposition.

Soil is the natural medium in which the roots of most plants grow. From soil the plant absorbs water and solutes necessary for its continued well-being. If a soil is fertile, it contains in a readily available form all the chemical elements essential for plant growth. It is through soil that the agriculturist can effectively alter the environment of roots and thus control plant growth, at least partially. The time and method of fertilizer used, as well as cultivation and irrigation practices, are all directed toward increasing production of plant products through effects that these practices have on soil and root relationships and ultimately on the growth and development of the plant. There are many kinds of soils and many different soil conditions. The character of natural plant covering and behavior of crops depend upon soil conditions as well as upon climatic conditions. Environmental factors that operate through soil are called **edaphic factors;** those that act upon the plant through the atmosphere are called **climatic factors.**

Soil Formation

Each environment creates a soil type unique to it. These soil types have their own history of development, morphology, and chemical attributes. Soil is a complex system which includes mineral (inorganic) matter, organic matter, water, air, and organisms.

Most soils consist largely of mineral particles formed by a slow continual process of weathering of the parent rock. Mineral matter is derived from fragmented rock. The kind of parent rock (granite, lava, sandstone, limestone, shale, etc.) and the degree of weathering determines the nature of the mineral or inorganic components of the soil.

Weathering may involve simply **mechanical break-**

185

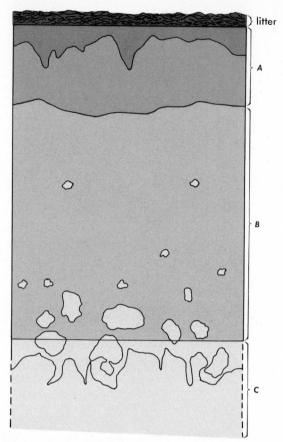

}litter

A

B

C

Figure 11-1 Profile of a typical soil of a wet temperate region. The *A* horizon, except for a thin region just beneath the litter, is leached of organic matter, clay, and salts. These leached substances accumulate in the *B* horizon. The *C* horizon, removed from weathering processes, represents parent material from which the *A* and *B* horizons have been formed.

ing of the parent rock. For instance, the action of strong winds or wave action may hurl sand against rock outcroppings and wear them down, or water collecting in pores of the rocks may expand at freezing temperatures to create fissures which ultimately fragment the rock. **Chemical weathering** results in more profound changes in the mineral matter itself. Atmospheric gases, such as CO_2 or SO_2, become dissolved in rainwater and produce acidic solutions which dissolve the parent rock material. Plant roots secrete weak acids. Certain algae, bacteria, and lichens hasten the decomposition of some of the least resistant of the mineral matter.

Weathering of the parent material under the influence of climate and organisms proceeds along a definite series of stages from a young soil in which the processes of soil development are continuing to a mature soil that has approached a state of equilibrium.

Soil formation may proceed relatively rapidly if the type of parent material and type of climate are favorable. Conditions that hasten the rate of soil development are a warm, humid climate, forest vegetation, flat topography, and a parent material which is easily broken down. Fort Kamenetz, built of limestone slabs in the Ukrainian part of Russia, was abandoned in 1699. Today, a mature soil 4–16 inches thick has developed from the limestone. Conditions which retard soil development are a cold, dry climate, grass vegetation, steeply sloping topography, and parent rock material that is not easily broken down. It has taken from 1000 to 10,000 years for mature soils to develop in northern areas scoured by the Wisconsin glaciation.

In a wet temperate region, soil development is principally caused by percolating rainwater that continuously dissolves nutrient salts in the upper portion of the soil and carries them, along with particles of finely divided, chemically weathered mineral matter and bits of organic matter, downward in a process called **leaching.** In time, the soil consists of a series of superimposed layers or **horizons** which differ in color, texture, and chemical attributes (Fig. 11.1).

The upper, or *A*, horizon is rich in organic matter only in the few uppermost inches, where freshly deposited organic litter is decomposing. The lower portion of the *A* horizon is sandy and light-colored, and may be deficient in nutrients, organic matter, and clay. Gradually, 1–2 ft below the surface, the *B* horizon begins. The *B* horizon is often reddish-brown from oxidized iron which has accumulated in it; it is relatively rich in mineral nutrients and clay and usually has a more neutral pH than the *A* horizon. The *B* horizon, then, is a layer of accumulation. Finally, 3–4 ft below the surface, the *C* horizon, made up of slightly weathered parent material, appears. The relative thickness of the *A* and *B* horizons depends upon the climate and amount of plant cover.

Mineral Matter of the Soil

The mineral fractions in most soils are primarily responsible for the texture of the soils.

In referring to the size of the mineral fractions of a soil, it is customary to speak of coarse sand, fine sand, silt, and clay. The size of soil particles decreases in the order given (Table 11.1). Clay is not only the smallest mineral soil particle, but it is also the most changed from the parent rock by chemical

TABLE 11.1 Classification of Soil Mineral Matter According to Size of Particles

Type of particles	Range in diameter of soil particles, mm
Coarse sand	2.0–0.2
Fine sand	0.2–0.02
Silt	0.02–0.002
Clay	0.002 and smaller

weathering. The clay fraction, together with some of the organic material, is of colloidal dimensions and, as we will see, gives to soils some of their most important properties, such as the ability to hold water and nutrients.

Organic Matter of the Soil

The organic matter of the soil is derived from plants and animals. Throughout the centuries of soil formation, the plants and animals that lived in and on the soil have left their residues. Annual herbs die each year; the whole plant, including the roots and tops, contributes to the organic matter of the soil. Also, trees and shrubs shed their leaves, twigs, bark, and fruits, and their roots die. All this plant material decomposes, and thus, with unimportant exceptions, soils are a mixture of mineral matter and organic matter, the latter in various stages of decomposition.

We might be led to believe that during the centuries the organic material in the soil would gradually increase. This material, however, is being continually decomposed by bacteria and fungi. Cultivation hastens the loss of organic matter from a soil by increasing aeration.

Generally, for good crop growth, most soils should have considerable organic matter. Soils naturally low in organic residues may be improved by adding barnyard manure or other fertilizers, or by plowing under green manure crops. Organic matter improves the physical condition of most soils, especially those with much clay. It makes them easier to cultivate, and it may increase slightly their ability to hold

TABLE 11.2 Composition of Three Soils According to Size of Mineral Particles

Soil type	Coarse sand, %	Fine sand, %	Silt, %	Clay, %
Sandy loam	67	18	6	9
Loam	27	32	21	20
Clay	1	9	22	68

water, but added organic matter may not increase the amount of water the plant is able to extract from the soil. The gradual decomposition of organic matter in the soil continually liberates mineral elements that are essential for plant growth. Sulfur, phosphorus, and nitrogen are among the most important elements thus liberated in forms available for plant growth.

The amorphous, dark-colored, partially decomposed organic matter in soils is called **humus.** The chemical nature of humus is very complex; in part, humus is a polymer of rings and straight chains of C, H, O, and N.

Mineral and Organic Soils

Although most soils contain both mineral and organic matter, in the majority of soils, organic matter constitutes only 2–5% of the soil by weight. These predominantly **mineral soils** may be classified according to the relative amounts of the sand, silt, and clay they contain. Soils which contain roughly equal amounts of sand, silt, and clay are called **loams.** Depending upon the relative amounts of each component in the soil, we speak of clay, clay loam, silt loam, loam, sandy loam, loamy sand, and sand (Table 11.2). Soil texture and the suitability of a soil for plant growth are greatly influenced by its sand, silt, and clay content.

Organic soils are defined as soils having a layer, 12 or more inches thick, that contains more than 30% organic matter. Such soils occur in wet, cool depressions that become acidic and anaerobic as the rate of litter accumulation becomes greater than the rate of decomposition.

Organic soils are divided into two categories, **peat** and **muck,** depending on the degree of decomposition. In peat, most of the original plant material can be identified as to species, while in muck the original plant material has been decomposed beyond recognition. Peat accumulates to such an extent that, in treeless areas, it is cut out in blocks and used for fuel.

Soil Water and Its Dissolved Substances

An important difference among various soils is their ability to hold water. It is greatest in clay and organic soils, and least in coarse sand.

If water is applied to a soil in the field, the spaces between the soil particles (**pore spaces**) become filled only for a short time to the depth wetted. With drainage, the water begins to move downward under the influence of gravity. After a while, this movement

Cation Exchange

$$NH_4^+ \quad K^+ \quad \left(\!\begin{array}{c} - \text{Soil} - \\ - \text{colloid} - \\ - \quad - \end{array}\!\right) \; Ca^{2+} \; + \; 7\,H^+ \;\rightleftharpoons\; \left(\!\begin{array}{c} H^+ \quad H^+ \quad H^+ \\ - \text{Soil} - \; H^+ \\ - \text{colloid} - \\ H^+ \quad - \quad - \; H^+ \\ H^+ \quad H^+ \end{array}\!\right) \; + \;\begin{array}{l} 2\,K^+ \\ \; + \; Ca^{2+} \\ Mg^{2+} \; + \; NH_4^+ \end{array}$$

Clay or organic matter with various ions as cations $+$ Hydrogen ions in soil solution $\rightleftharpoons$ Clay or organic matter with hydrogen as cations $+$ Various cations in solution

downward stops; the soil particles hold a certain amount of water against the pull of gravity. The amount of water held by the soil after drainage is called the **field capacity** of that soil. When a soil is at field capacity, or even at a moisture content well below this amount, a film of water completely surrounds each soil particle, and water also exists in the form of wedges between soil particles. Clay particles and organic matter in the soil constitute **colloidal systems,** and such particles hold water by **imbibition.** Water imbibed by soil particles is much more difficult to remove from the soil than water that exists as a film or as wedges.

The soil, then, is a reservoir of water, but in addition it is the reservoir of mineral nutrients required for plant growth. The water holds in solution various inorganic salts (nitrates, phosphates, sulfates, etc.) and other water-soluble substances. The composition and concentration of this solution is ever-changing, depending upon the higher plants and microorganisms growing on and in the soil. All substances that enter the plant must be in solution. This makes possible their passage through the cell wall and cytoplasmic membranes of root hairs. In most agricultural soils, the soil solution has a very low concentration—usually lower than that of the cell sap. The water potential in the soil is usually higher than that in the plant. As long as this relationship is maintained, water will diffuse inward. Absorption of ions by roots from the very dilute soil solution would soon exhaust all the available ions were it not for the continual release of nutrients from the solid phase of the soil. In the soil, nutrient elements may exist in a relatively unavailable form. For example, most organic nitrogen compounds are not directly available to plants; that is, nitrogen in the form of either plant or animal proteins cannot be absorbed by green plants. However, these organic compounds are acted upon by soil organisms, and some of the products of decomposition do become directly available. Organic sulfur compounds are also broken down by living organisms to inorganic compounds. In general, it can be said of soils that at all times chemical changes are taking place which set free inorganic nutrient substances that will dissolve in soil water

and thus assume a form that can be taken up by the roots.

Soil particles themselves, particularly finely divided clay and organic matter, act as giant negatively charged ions which attract a cloud of positively charged ions such as Ca^{2+} and K^+. Thus, colloidal soil particles, while not in solution themselves, serve as reservoirs to which many of the various ions essential for plant growth are loosely attached and from which these ions may be released into solution by a process known as **cation exchange** (see above).

Roots in the soil are covered by a thin film of the dilute soil solution. As respiration occurs in the living root cells, CO_2 is produced. Carbon dioxide combines with water, releasing hydrogen ions (H^+), bicarbonate ions (HCO_3^-), and carbonate ions (CO_3^{2-}):

$$CO_2 + H_2O \rightleftharpoons H_2CO_3 \rightleftharpoons$$
$$H^+ + HCO_3^- \rightleftharpoons H^+ + H^+ + CO_3^{2-}$$

These ions may pass out of the root cell into the surrounding soil solution. Here, the hydrogen ions may exchange with the cations associated with the negatively charged surfaces of soil colloids.

The cations now in solution may diffuse to the root surface and be absorbed. Actually, the cell wall itself is negatively charged and may bind cations in the same way that the soil colloids do. By the process of cation exchange, the soil continually supplies cations to the root in exchange for hydrogen ions. The soil itself becomes more acid in the process.

Mineral elements essential for plant growth cannot move rapidly in the soil for any appreciable distance, but plant roots have evolved the capacity to develop a large surface to contact the soil. Root hairs are particularly important, or some roots have developed a symbiotic association with fungi which may serve this need for an extensive soil surface contact. In addition, roots have the capacity to grow in and penetrate soil. In this way, the root system of an individual plant can tap the water and mineral reserves in a large mass of soil.

Negatively charged ions such as nitrate, NO_3^- and sulfate, SO_4^{2-}, are not held by the soil particles, but are in solution. Phosphate, PO_4^{3-}, presents a different

Relative acidity	pH range	Plants that grow well in soils of this acidity
High acidity	3.2–4.6	*Sphagnum* moss in peat bogs
Moderate acidity	3.5–5.0	Blueberry
Slight to moderate acidity	4.5–6.0	Azalea, camellia
Nearly neutral	6.0–7.5	Many forest and field plants
Alkaline	7.5–9.1	Greasewood, rabbit brush, stonewort, various plants growing around alkali flats

picture; these ions enter into the composition of certain clays or are easily precipitated and do not usually exist in high concentrations in soil solution.

Acidity of the Soil

The acidity of a soil influences the physical properties of the soil, the availability of certain minerals to plants, and the biological activity in the soil. It consequently strongly influences plant growth. Certain plants such as azaleas, camellias, and cranberries grow best in acid soils; most plants do best in soils near neutrality, while a few plants grow satisfactorily in slightly alkaline soils (Table 11.3). Soil acidity is a measure of the concentration of hydrogen ions in solution and is usually expressed in pH units. The pH scale is a logarithmic scale in which each unit represents a hydrogen ion concentration ten times more, or less, than the next unit. Thus, a solution having a pH of 7, which is the neutral pH, has ten times the hydrogen ion concentration of a solution of pH 8, a solution that is slightly alkaline. The same solution of pH 7 has one-tenth of the hydrogen ion concentration of a solution of pH 6, a slightly acid solution.

Acid soils, such as those beneath northern coniferous forest with acidic foliage, range from pH 3.5 to 5.0; agricultural soils in the humid areas range from pH 5.0 to 7.0; and soils of arid or saline regions may range as high as pH 11.0, but 8.0–9.0 is more common. Acidity *per se*—that is, H^+ or OH^- ions—is not directly responsible for limiting plant growth. Rather, soil pH affects the availability of plant nutrients in two ways. In an acid soil, hydrogen ions (H^+) replace other **cations** (like K^+, Ca^{2+}, and Mg^{2+}) on

the negatively charged clay particles, allowing the nutrient ions to be easily leached from the soil. Soil pH also affects the solubility of plant nutrients. As shown in Fig. 11.2, some nutrients, like iron, manganese, and aluminum, increase in solubility as pH decreases. Aluminum may reach toxic levels in some acid soils. Other nutrients, like calcium and magnesium, increase in solubility as pH increases. Soil pH may also affect plant growth indirectly by suppressing bacterial growth at pH extremes.

Acid soils may be improved by the application of lime (calcium hydroxide or calcium carbonate), which tends to neutralize the acid. Alkaline soils may be improved by the addition of sulfur or ammonium sulfate. If elemental sulfur is applied, it may become oxidized to sulfate and make the soil more acid. The ammonium of ammonium sulfate is either converted into nitrate by bacterial action or is absorbed more rapidly than sulfate by the plants, resulting in a net increase in sulfate and acidity in the soil.

Air of the Soil

We have spoken of the pore space in soils. The air of the soil contains the same gases as the atmosphere above the soil. However, the air of the soil is considerably richer in carbon dioxide and poorer in oxygen than that of the atmosphere. In the latter, the average percentage by volume of carbon dioxide is 0.03%; in soil air, the percentage may go as high as 5%. This high percentage of carbon dioxide in soil air is due to the respiration of soil microbes and,

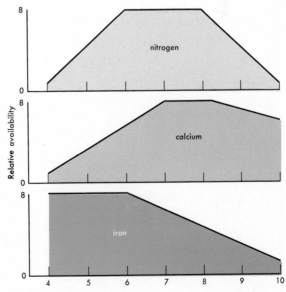

Figure 11.2 Relationship of relative nutrient availability to soil pH.

Soil—The Environment of Roots **189**

to a lesser degree, the respiration of roots them-selves. The soil air usually has a relative humidity near 100%, a condition very favorable to the growth of soil organisms.

The living cells of roots must have oxygen to sup-port respiration. Normally, the oxygen necessary for respiration reaches the roots by absorption through the root hairs and other epidermal cells. This inward diffusion of oxygen from the soil air ordinarily goes on readily because of the thinness of the walls of these cells and the absence of a cuticle. Roots at all levels in the soil secure oxygen directly from the air in the pore spaces and, to a limited extent, from the oxygen dissolved in water; in ordinary land plants, there is no system of air-conducting tubes that conveys oxygen from the atmosphere through the plants to the extremities of the roots.

The amount of air in the soil depends not only upon the pore space but also upon the water con-tent of the soil. If water occupies the pore space, air is forced out. A water-soaked (saturated) soil contains practically no air save that which is dis-solved in water. If the soil about the roots is contin-uously water-soaked, plants die because of insuffi-cient oxygen and possibly as a result of the accumulation of carbon dioxide. Most species of land plants will grow normally with their roots in a water solution, if it is well-aerated by bubbling air through it (Fig. 11.3). Evidence indicates that inade-quate soil aeration results in a reduction in the rate of water and mineral absorption.

Whereas most land plants, including agricultural plants, will not long survive with the root system submerged in unaerated water or surrounded by a soil that is water-soaked, some plants flourish under such conditions. Among them may be cited rice, various swamp and marsh plants, and the bald cy-press (*Taxodium distichum,* Fig. 11.4). Almost all such plants contain in the stem and roots large communi-cating air spaces. Thus, air absorbed into the leaves and stems may reach the living cells of the root in sufficient quantity. Bald cypress and certain species in tropical mangrove swamps develop special root branches that grow upward until their ends are above the water level. These special root branches have a central core of loose tissue through which air moves downward to the submerged organs.

Organisms of the Soil

Ordinary soils teem with living organisms, both plants and animals. These organisms are an impor-tant part of the environment of roots. The soil flora includes bacteria, fungi, and algae. Living roots of higher plants also are an important part of the soil. When the roots die, they leave a considerable quan-tity of organic matter. The soil fauna includes proto-zoa, nematodes, earthworms, various insects, and burrowing animals. Most of these are beneficial to man in some way or another; others may be harmful. Certain bacteria are absolutely essential in the main-tenance of soil fertility. This point will be pursued further in Chapter 18, but it should be indicated here that certain bacteria and fungi that live in the soil are responsible for the decomposition of organic matter (plant residues, manure, and other organic fertilizers), and for nitrogen fixation (see Fig. 18.39). As a result of their activity, they keep up the supply of soil nitrogen, one of the chief factors in soil fertil-ity. Earthworms affect soil structure by moving and

Figure 11.3 Effect of aeration on root growth in tomato (*Lycopersicon esculentum*). Left, plants growing in complete nutrient solution through which air was bubbled; right, plants growing in same solution without aeration. (Courtesy of D. R. Hoagland.)

Figure 11.4 Aerial "stump roots" of bald cypress (*Taxodium distichum*). (Courtesy of U.S. Forest Service.)

mixing the soil; they pass large quantities of soil through their bodies. Charles Darwin, in 1885, published *The Formation of Vegetable Mold,* in which the importance of earthworms in soils is stressed. He points out that earthworms in an acre of soil may pass through their bodies annually as much as 15 tons of dry earth.

Temperature of the Soil

The rate of absorption of both water and solutes by roots may be reduced by extremes of temperature. Plants native to cool climates absorb these substances more freely at low temperatures than do plants of warm climates. A plant may wilt in a soil containing ample water if the soil temperature sinks below or rises above, a certain degree. In cold, dry climates, winter-killing may be the result of a cold soil that slows up absorption, accompanied by a high transpiration rate. It is believed that in winter-killing, the plant is as frequently killed by direct drying as by actual freezing.

All chemical and biological activities of a soil are influenced by soil temperature.

The soil temperature is not always the same as the temperature of the air above it. It may be lower or higher than the air temperature (see Chap. 18).

Mineral Elements Essential for Plant Growth

For over 100 years physiologists have been studying the mineral nutrition of plants. Extensive investigations have been carried out to determine which elements are essential, how the plant absorbs them, how they are utilized, why they are essential, and what effects are produced in the plant when a particular essential element is lacking (see Fig. 11.5, Color Plate 4).

Since 1860, it has been known that plants required at least seven elements: nitrogen, sulfur, phosphorus, potassium, calcium, magnesium, and iron. Since then, nine more required elements have been discovered. These elements are required in such minute amounts that their importance remained long undetected because the air about the plants, the containers holding the growth medium, or impurities in the chemicals used supplied enough of them to support plant growth. There are other elements that are required by some plants in very small amounts but which most plants do not require. Sodium, for example, is required in very small amounts by the desert shrub *Atriplex,* but is not required by most other species. It is not considered an essential element.

Plants utilize elements in four basic ways: (*a*) The elements may form part of structural units, such as carbon in cellulose or nitrogen in protein. (*b*) Elements may be incorporated into organic molecules important in metabolism, like magnesium in chlorophyll or phosphorous in ATP. (*c*) Elements may function as enzyme activators, necessary as catalysts in certain enzymatic reactions. (*d*) Elements in ions help to maintain the osmotic balance, e.g., potassium in guard cells (p. 193). Magnesium is used as an enzyme activator in several of the enzymatic steps of glucose degradation in the process of respiration.

What methods have been employed to determine the chemical elements that are essential to plant growth? Of course, accumulated experiences of agriculturists have shown that applications of various nutrients to soil result in healthier plants, increased yields, and better quality. Plant physiologists have used more critical methods—careful weighings and measurements of quantities of elements used and results obtained. The solution-culture method has long been employed to determine what elements are indispensable in plant growth. Most plants that normally grow in soil can be grown to maturity (fruit and seed production) in water to which soluble essential nutrient salts have been added (Fig. 11.5; see Color Plate 4). By growing plants in a nutrient solution containing all elements believed to be essential, except the particular element being investigated, the response of plants when this element is absent can be ascertained. In this case, all plants grown in deficient solutions were first grown in a complete nutrient solution for 2 weeks before being transferred into the deficient solution.

From such experiments a list of elements that are known to be essential for the growth of plants has been compiled. Long recognized as essential to the continued growth and development of green plants are the following ten elements, with their appropriate chemical symbols: carbon (C), hydrogen (H), oxygen (O), phosphorus (P), potassium (K), nitrogen (N), sulfur (S), calcium (Ca), iron (Fe), magnesium (Mg). If the symbols are arranged in a line, they can be used to remember the elements thus:

C HOPK'NS CaFe Mg

"See Hopk'ns Cafe, mighty good." Of these ten elements, the last seven are in the nutrient medium; that is, their various salts are dissolved in the soil water. Relatively large quantities of nitrogen, sulfur, phosphorus, potassium, magnesium, and calcium are required; very small quantities of iron meet all plant needs for this element. These elements, with the exception of iron, have been called the **macronutrient elements,** because they are needed in relatively large amounts. More recently, it has been discovered that at least six additional elements are needed by higher plants but in much smaller amounts. The **micronutrient elements** (trace elements or minor elements) include, in addition to iron (Fe), the elements boron (B), manganese (Mn), copper (Cu), zinc (Zn), chlorine (C), and molybdenum (Mo). Table 11.4 shows the relative amount of macro- and micronutrients removed in pounds per acre by wheat during a growing season.

A plant growing in soil or in solution cannot distinguish between elements that are essential to it and those that are not essential or that might be harmful. If the element or ion containing it is in solution, it will probably be absorbed by the plant. Thus, we find in the plant almost all the elements present in the soil. Gold has been isolated from plants growing in gold-bearing soils. Selenium, an element poisonous to livestock, when present in soils is absorbed by certain plants in sufficient amounts to be harmful to animals grazing on them.

Macronutrient Elements

Carbon, Hydrogen, and Oxygen. The absorption of water, carbon dioxide, and oxygen brings into the plant large quantities of the elements carbon, hydrogen, and oxygen, from which is formed the major part of each of the hundreds of organic compounds found in the plant. We have seen how in the process of photosynthesis these elements are combined into carbohydrates. Simple fats also are composed entirely of these three elements, whereas proteins, alone of the foods, contain appreciable amounts of other elements, particularly nitrogen and, to a lesser extent, sulfur and phosphorus.

Nitrogen. This element is a component of proteins, which form an essential part of protoplasm and also occur as stored foods in plant cells. Nitrogen is also a part of other organic compounds in plants, such as chlorophyll (chloroplyll a has the formula $C_{55}H_{72}O_5N_4Mg$), amino acids, alkaloids, and at least some plant hormones.

Ordinary green plants cannot utilize elemental nitrogen, which represents about 78% of the air. It has been estimated that above every acre of land surface there are about 145,000–150,000 tons of this gas. Chief sources of nitrogen for green plants are the nitrates, such as sodium nitrate (Chilean saltpeter), potassium nitrate, ammonium nitrate, and calcium nitrate. Green plants can also utilize ammonium salts, and evidence indicates that some plants can derive nitrogen from certain low-molecular-weight organic nitrogenous compounds.

Nitrogenous fertilizers, both natural and commercial, are usually the most important fertilizers applied to growing plants. The chief commercial nitrogenous fertilizers are: (*a*) those with nitrogen in the nitrate form; (*b*) those with nitrogen in ammonia or its compounds; (*c*) those with nitrogen in organic compounds, such as tankage and cottonseed meal; and (*d*) those with nitrogen in the amide form, such as urea and calcium cyanamide. Organic nitrogen in complex molecules cannot be used directly by plants

TABLE 11.4   Amounts (in pounds per acre) of Macro- and Micronutrients Removed from the Soil in One Growing Season by a Wheat Crop

Element	Macronutrients						Micronutrients				
	N	K	P	Ca	S	Mg	Fe	Mn	B	Zn	Cu
Amount	76	42	14	12	11	8	0.7	0.5	0.3	0.2	0.03

Color Plate 4

Mineral Nutrition

Figure 11.5 Tomato plants grown in nutrient culture solutions to show visual symptoms of mineral deficiencies. *A*, complete solution; *B*, deficiency symptoms in leaves; *C*, minus phosphorus; *D*, minus potassium; *E*, minus nitrogen; *F*, minus sulfur; *G*, minus calcium; *H*, minus iron; *I*, minus magnesium; *J*, minus all micronutrients. All plants subjected to deficient solutions were first grown in a complete nutrient solution for 2 weeks.

A

B

C

D

Figure 12.1 Moisture content, dry weight, and ash content of fresh grass leaves. *A*, 100 g fresh leaves yield *B*, 82 g water and *C*, 18 g dry leaves; *D*, the dry leaves yield, on burning, 1.5 g ash.

Color Plate 5

**Transpiration;
Photosynthesis**

A   B

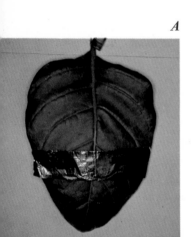

Figure 13.2 Light is necessary for photosynthesis. *A*, leaf from a bean plant which was kept first in darkness for 24 hr; a portion of the leaf was then covered by foil as shown, and the plant was kept in the light for 8 hr. *B*, the iodine test shows that starch was formed only in the portion of the leaf receiving 8 hr of light.

Figure 13.9 Chlorophyll is required for photosynthesis. Variegated leaf from *Coleus*. *A*, living leaf; *B*, same leaf after extraction of chlorophyll and iodine staining to show starch. Compare the blue areas indicating the presence of starch with the distribution of chlorophyll (green and brown parts of the living leaf).

A   B

but must first be converted into available forms through the action of organisms in the soil.

The rate of growth of plants is influenced to a large degree by the available nitrogen. In contrast to some other nutrients (e.g., calcium), nitrogen is very mobile in the plant and can be translocaled from mature to immature regions of a plant. An early symptom of nitrogen deficiency is a yellowing of leaves, particularly the older leaves; then follows a stunting in the growth of all parts of the plant (Fig. 11.5*E*). An excess of available nitrogen results in vigorous vegetative growth and a suppression of food storage and of fruit and seed development. The nitrogen cycle in nature is discussed more fully in Chapter 18.

Sulfur.   Sulfur forms a part of the molecules of proteins. Plant proteins containing sulfur may have from 0.5 to 1.5% of this element. The sulfhydryl, –SH, group is a very important group essential for the action of certain enzymes and coenzymes. In addition, sulfur is a constituent of ferredoxin and of some lipids found in chloroplasts.

Ordinary green plants cannot utilize elemental sulfur; it must first be oxidized to sulfates, in which form it is normally absorbed by the roots.

Certain agricultural soils may become deficient in sulfur. The most noticeable responses to applications of sulfur fertilizers are an increased root development and a deeper green color of the foliage. Sulfur deficiency symptoms in tomato are shown in Fig. 11.5*F*.

Phosphorus.   This element is also a component of some plant proteins, phospholipids, sugar phosphates, nucleic acids, ATP, and NADP. The highest percentages of phosphorus occur in parts of the plant that are growing rapidly, such as meristematic regions and maturing fruits and seeds. Some of the very important roles that phosphorus plays in plant metabolism in relation to energy transformation and biosynthetic reactions are discussed in Chapter 20.

Methods have been developed by which elements are "tagged." For example, radioactive phosphorus in the form of phosphate has been introduced into the culture solution in which tomatoes were growing. By means of the Geiger-Müller counter, or by radiograms, the path of radioactive phosphorus through the plant has been followed. After a time, the greatest accumulation of phosphorus was found in the young developing fruit; very little occurred in the ripe fruit. Within the fruit itself the greatest accumulation of phosphorus was in the developing seed.

Applications of phosphorus to soils deficient in this element promote root growth and hasten maturity, particularly of cereals. Phosphates are the principal source of phosphorus for plants. A tomato plant grown in culture solution without phosphorus is shown in Figs. 11.5*B, C.*

Potassium.   Potassium is the only monovalent cation that is essential for plant growth. This element, indispensable to plants, accumulates in tissues that are growing rapidly. It will migrate from older tissues to meristematic regions; for example, during the maturing of a fruit crop there is a movement of potassium from the leaves into the fruit. In plants, potassium occurs chiefly in inorganic form.

The primary role of potassium in the cell is as an enzyme activator. Over 40 different enzymes have been found to require monovalent cations for maximal activity. In plant cells, potassium serves this role. However, in contrast to most enzyme activators that are micronutrients, potassium is effective only at the relatively high concentrations that it occurs in plant cells. Processes in the plant that appear to require an adequate supply of potassium are: (*a*) normal cell division, (*b*) synthesis and translocation of carbohydrates, (*c*) synthesis of proteins in meristematic cells, (*d*) reduction of nitrates, (*e*) development of chlorophyll, and (*f*) stomatal opening and closing. In addition to affecting changes in sugar concentration in guard cells, illumination has been shown to stimulate the accumulation of potassium ions in guard cells. This light-stimulated ion accumulation is correlated with stomatal opening. When the leaf is darkened, the guard cells lose potassium and stomata close. This accumulation of potassium by illuminated guard cells can be dramatically observed by means of an electron probe analyzer.

Any water-soluble inorganic compound of potassium, such as potassium sulfate, potassium phosphate, or potassium nitrate, can be utilized by plants as a source of potassium. Potassium deficiency symptoms in tomato are shown in Fig. 11.5*D.*

Calcium.   All ordinary green plants require calcium. It is one of the constituents of the middle lamella of the cell wall, where it occurs in the form of calcium pectate. Calcium affects the permeability of cytoplasmic membranes and the hydration of colloids. Calcium may be found in combination with organic acids in the plant. Oxalic acid, for example, is a by-product of metabolism. It is a soluble substance and is toxic to the protoplasm if it reaches a high concentration in the cell. When united with calcium, however, the soluble oxalic acid is converted into the highly insoluble calcium oxalate, which does not injure the protoplasm. There is also evidence that calcium favors the translocation of carbohydrates and amino acids and encourages root development. Calcium deficiency is frequently characterized by a death of the growing points (Fig. 11.5G) because it is not readily translocated in the plant from mature to immature regions. General disorganization of cells and tissues also results from calcium deficiency. This effect is consistant with one of the key roles of calcium—maintaining the normal structure of cell membranes.

Magnesium. Magnesium is a constituent of chlorophyll, where it occupies a central position in the molecule (Fig. 5.10). Chlorophylls are the only major compounds of plants that contain magnesium as a stable component. Many enzyme reactions, particularly those involving a transfer of phosphate (energy metabolism), are activated by magnesium ions. Consequently, magnesium deficiency rapidly affects many aspects of metabolism.

A deficiency of magnesium results in the development of pale, sickly foliage, an unhealthy condition known as **chlorosis** (Fig. 11.5I). This disease is a common one of economic plants, and in many instances, applications of magnesium have effected a cure. However, chlorosis can be caused by deficiencies in other essential elements such as iron.

Micronutrient Elements

Until recently, it was not known that the elements boron, zinc, manganese, chlorine, molybdenum, and copper were indispensable for normal plant growth. The reasons were that they are needed only in very minute quantities (mere "traces") and that the chemcials used in culture solutions contained unsuspected traces of these elements sufficient to supply the needs of the plant's growth in the water culture. For example, a bottle of ferric sulfate labeled "chemically pure," taken from the laboratory shelf, may show on the label traces of chlorine, phosphoric acid, manganese, zinc, copper, sodium nitrate, and ferrous salts. If a culture solution containing this "chemically pure" ferric sulfate were used, most likely it would possess enough manganese, zinc, chlorine, and copper to meet the plant's requirements. In a study of micronutrient elements needed, or suspected of being needed, by plants, it has been necessary to employ highly refined techniques, employing redistilled water and highly purified salts. It may even be necessary to prevent atmospheric dust from falling into the culture, as the dust may contain traces of critical elements. A tomato plant grown in a nutrient medium lacking all the micronutrient elements except iron is shown in Fig. 11.5J.

Iron. A number of essential compounds in plants, such as cytochromes and ferredoxin, contain iron in a form that is bound rather firmly into the molecule. Iron plays a role in being the site on some electron carriers where electrons are absorbed and then given off during electron transport. The iron atom is alternately reduced and then oxidized. Several cytochromes, both in mitochondria and in chloroplasts, contain iron. Other iron-containing proteins include ferredoxin in the chloroplasts, and certain enzymes such as catalase and peroxidases. Thus, iron plays very important roles in energy conversion reactions both of photosynthesis and respiration.

Although iron is not a component of the chlorophyll molecule, it is essential for the synthesis of chlorophyll. Chlorosis may be caused by iron deficiency as well as by a deficiency of magnesium (Figs. 11.5B, H). The quantity of iron required is very small. As an example, chlorosis of pineapples in Hawaii, due to the unavailability of iron from the soil, is remedied by spraying the plants with iron salt solutions. Orchard trees suffereing from iron chlorosis may be cured by injecting iron compounds into the trunk, or by applying various kinds of iron salts to the soil.

Boron. That plants require boron is wellestablished. Symptoms of boron deficiency include inhibition of root and shoot elongation, inhibition of flowering, darkening of tissues, and various growth abnormalities and disturbances. Some of the physiological diseases of plants due to boron deficiency are internal cork of apples, top rot of tobacco, cracked stem of celery, browning of cauliflower, and heart rot of sugar beets. Very small applications (20–50 lb) of sodium tetraborate (borax) per acre may be sufficient to cure these diseases. In contrast, fertilizers containing the macronutrient elements are frequently applied to fields at rates of several hundred pounds per acre.

Although plants require boron for normal development, the quantity in the soil or culture solution must be very small or injury will result. In fact, in certain agricultural sections, severe injury to crops occurs because of excessive amounts of boron in the soil or in the irrigation waters. Plants vary in their tolerance of boron. Borates are used as weed-killers.

Although the exact function of boron in plant metabolism is still obscure, boron does play a regulatory role in carbohydrate breakdown. Boron deficiency tends to shift metabolism from glycolysis to the pentose phosphate pathway (pp. 237–243).

Zinc. Zinc is essential to the normal development of a variety of plants; probably it is required by all plants. As with other micronutrients, proof has come from careful water-culture experiments; also, certain diseases of plants have been cured by zinc appli-

Figure 11.6 Disease of peach (*Prunus persica*) known as "little leaf," caused by a deficiency of zinc. Branch at left untreated; branch at right cured by driving in zinc-coated nails. (Courtesy of E. L. Proebsting.)

cations. Large quantities of zinc are toxic to plants. Two well-known diseases caused by zinc deficiency are "little leaf" of deciduous fruit trees (Fig. 11.6) and "mottle leaf" of citrus trees. These abnormal conditions are corrected by spraying the trees with zinc salts, by injecting dilute solutions of zinc salts into the trunks, or by driving zinc brads into the trunks. These corrective measures show how minute are the quantities of zinc required by plants for normal development.

The very pronounced effect of zinc deficiency on growth, especially internode elongation, is a consequence of the importance of zinc in auxin synthesis. Indole acetic acid, auxin (see Chap. 20), is necessary for cell elongation. It is found to be in low concentration in zinc-deficient plants even before visible symptoms appear. If zinc is added to the culture solution, the auxin level increases and this is followed by a resumption of growth. Zinc is also an activator or a part of several enzymes.

Manganese. The most striking symptom of plants with a deficiency of manganese is chlorosis (Fig. 11.7), but it is a somewhat different type from that caused by iron deficiency. Whereas in iron chlorosis the young leaves may become yellow or white with prominent green veins, manganese chlorosis results in the leaf taking on a mottled appearance. Spraying or dusting crops suffering from manganese defi-

ciency with as little as 20 lb of manganese sulfate per acre frequently effects a cure.

Manganese's importance as an activator of several enzymes of the Krebs' cycle of aerobic respiration (Chap. 14) explains some of the disruptive effects of manganese deficiency on metabolism. A further function of manganese in green cells is concerned with the oxygen-liberating steps of photosynthesis (Chap. 13).

Copper. Abnormalities in the growth of many plants, especially those in marsh and peat soils, have been corrected by the application of copper compounds (Fig. 11.8). Copper is also a constituent of certain enzyme systems, such as ascorbic acid oxidase and cytochrome oxidase. In addition, copper is found in plastocyanin, part of the electron-transport chain in photosynthesis.

Molybdenum. Molybdenum is now recognized as an essential micronutrient element, at least to certain plants. For example, it was found that molybdenum is required for normal growth of tomato seedlings, that only 0.01 parts per million (ppm) of the element in the nutrient solution is needed, and that concentrations exceeding 10 ppm cause injury. For some Australian soils, only 2 oz of MoO_3 per acre, applied once every 10 years, produces optimum forage improvement.

Molybdenum is important in enzyme systems involved in nitrogen fixation and nitrate reduction.

Figure 11.7 Chlorosis of tomato (*Lycopersicon esculentum*) leaf caused by a deficiency of manganese in the nutrient solution. (Courtesy of D. R. Hoagland.)

Figure 11.8 Tomatoes (*Lycopersicon eculentum*) growing in solution with all essential chemical elements except copper. Leaves at left were sprayed with solution containing copper. (Courtesy of D. R. Hoagland.)

Plants suffereing molybdenum deficiency can absorb nitrate ions, but are unable to utilize this form of nitrogen. If nitrogen in the form of ammonium is given to these plants, many of the deficiency conditions are less severe.

Chlorine. Chlorine is present in the soil solution as the very soluble, negatively charged chloride ion, Cl^-. Because of the very small quantities required by plants and its almost universal occurrence in soils, chlorine is never added intentionally as a fertilizer, although it occurs in small amounts in all fertilizers. Rainfall may contain enough chlorine to supply plant requirements, and some plants may obtain enough from the air to prevent deficiency symptoms. It is very difficult to prepare a nutrient solution completely free of chlorine. Consequently, it is difficult to demonstrate chlorine deficiency symptoms in plants. Although the functions that chlorine plays in plant metabolism are obscure, it is known that chlorine takes part in the reactions of photosynthesis leading to oxygen evolution.

Summary

1. Soil is a complex system composed of rock particles and plant and animal remains in various stages of decay, and in which an abundance of soil microorganisms is growing. Throughout the soil structure there is an interconnecting system of large and small pores filled with air or with water and dissolved materials.
2. All the essential elements except carbon, hydrogen, and oxygen are absorbed through the roots, generally in the form of ions from the soil. Hydrogen and oxygen are absorbed by roots as water, and carbon enters the leaves as carbon dioxide.
3. Biologically produced hydrogen ions may replace cations associated with the negatively charged soil colloids: clay and organic matter.
4. The macronutrient essential elements are carbon, hydrogen, oxygen, nitrogen, sulfur, phosphorus, potassium, calcium, and magnesium.
5. The micronutrient essential elements are iron, zinc, copper, boron, manganese, chlorine, and molybdenum.

12

Transpiration, Conduction, and Absorption

The most abundant compound in an active cell is water. It plays a varied role in the life of a plant, for it serves as (a) a raw material in the synthesis of organic compounds, (b) the solvent in which vital reactions take place, (c) the medium through which solutes move from cell to cell, and (d) the source of turgor in plant cells. Plant tissues vary in water content; those actively growing have a water content equal to as much as 85–95% of their fresh weight, while a dormant structure such as a seed may have a water content as low as 5–10%. The entire shoot system of an herbaceous plant such as barley may be 82% water (Fig. 12.1; see Color Plate 5).

A plant from which the water has been removed is made up chiefly of organic material synthesized by the plant primarily from carbon dioxide, water, and inorganic nitrogen such as is found in nitrates or ammonium. Actually, the mineral matter absorbed from the soil is a small fraction of total plant weight but is, nevertheless, essential for the life of the plant. The small quantity of mineral matter contained in a plant can be readily demonstrated by burning the dry tissue. The carbon, hydrogen, and oxygen, as well as the nitrogen and some of the phosphorus and sulfur, pass off in the combustion process, while most of the minerals remain in the ash, combined with a small amount of oxygen (Fig. 12.1). These mineral elements may have occurred in the plant as inorganic ions dissolved in the cell or may have been a part of complex organic molecules. For instance, the magnesium in the ash may have come from the destruction of chlorophyll of which it is a part, while the carbon, hydrogen, oxygen, and nitrogen of the chlorophyll would have been lost during burning.

While minerals that are once absorbed by a plant are not generally lost again in large quantities, except during leaf fall, water is constantly being lost from the above-ground portions of the plant. Because this water loss profoundly influences the rates of water absorption and movement, it will be studied first.

Transpiration

By far the greater part of the water lost from the plant passes into the atmosphere as invisible water vapor. This loss of water in vapor form from a living plant is called **transpiration.** Although a small amount of water may be transpired directly through the cuticle, most of the water lost during the day diffuses out through the stomata. This is called **stomatal transpiration,** in contrast with **cuticular transpiration.** Leaves are the principal transpiring organs.

197

A knowledge of leaf structure is essential to an understanding of the mechanism of transpiration. It will be recalled that the mesophyll of most leaves is a very loose tissue with large intercellular air spaces. Even within the palisade parenchyma, a large portion of the walls of most cells is exposed to intercellular air spaces (Figs. 10.6, 10.7A). These are connected with similar air spaces throughout the leaf and, through the stomata, with the outside air. The extensive intercellular surface, loose internal structure, and numerous stomata of the leaf allow for rapid gaseous exchange between the internal leaf cells and the outside air. An appreciation of this structure is gained when we consider a large leaf such as a squash leaf that may be 70% internal air space, and may have 20 times as much total internal cell surface bordering this air space as leaf surface exposed to the outside air. In addition, a squash leaf may have as many as 60 million stomata connecting this internal air with the outside environment and may have 6000 vein endings per square centimeter of leaf surface.

Process of Transpiration

Water permeates the living turgid leaf cells, filling the vacuoles, making up a large part of the cytoplasm, and penetrating the walls. Thus, cell walls bordering on intercellular air spaces are moist, and because of this, air in these spaces is almost saturated with water vapor. The water potential or vapor pressure of water in these spaces is then generally higher than in the outside atmosphere. When the air around a leaf is not saturated but relatively dry, water vapor molecules diffuse from the saturated air in the leaf through the stomata into the less saturated outer air where the vapor pressure, and hence the water potential, is lower. This loss in water results in a slight drying of the air in the intercellular spaces (Fig. 10.6). Water molecules then evaporate from the wet walls and diffuse into this drier air in the leaf. As water leaves the walls, the walls in turn become drier and imbibe water from the enclosed cytoplasm and vacuole.

Thus, a diffusion gradient for water exists from the vacuole through the cytoplasm, the wall, and the intercellular air spaces to the drier outer air surrounding the leaf. Water diffuses along this gradient, in liquid form within the cell, and in vapor form once it evaporates from the cell walls. Each cell, however, is in contact with several other cells and is only a few cells distant from the water-filled xylem elements of a vein.

As water is lost from a cell by outward diffusion, the cell loses some of its turgor and the concen-tration of solutes in it increases; hence the water potential within the cell sap decreases. The cell is then able to gain water by osmosis from adjacent more-saturated cells. These in turn gain water from adjacent cells and eventually from the water-filled tracheid of a vein ending. A water potential gradient soon becomes established, along which water diffuses from the vein through one or several cells to a cell wall bordering an intercellular space, into and through the intercellular spaces, and through the stomata to the outside air.

Quantity of Water Transpired. The amount of water transpired by plants is very great. A single corn plant in Kansas, between May 5 and September 8, transpired 54 gal of water. An acre of such plants (6000 plants) would transpire, during that season, 324,000 gal of water, which is equivalent to a sheet of water 11 inches deep over the entire acre. It has been estimated that an acre of red maple trees, growing in a soil with ample moisture, may lose in a growing season an amount of water sufficient to cover each acre with 28.3 inches of water. A soil clothed with plants is depleted of its moisture at a much more rapid rate than one that is bare. Nearly all the water loss from a soil below the first 6–8 inches results from absorption and transpiration by plants.

Of the total quantity of water absorbed by the roots of plants, as much as 98% of it escapes from the plant by transpiration. The small quantity of the water that is retained by the plant includes the water of vacuoles and protoplasm, that in cell walls and in the conducting elements, and that entering into chemcial combination.

An alfalfa plant may transpire 900 units of water for each unit of dry matter produced and a millet plant may transpire 248 units of water for each unit of dry matter. Although these values vary with the environmental conditions under which the plants are grown, it is evident that plants differ in their water economy.

Evapotranspiration. In evaluating an area of land in terms of the amount of water needed to produce maximum plant growth, one must take into consideration both evaporation of water from the soil surface and transpiration of water. The total amount of water that a unit of land area loses to the atmosphere through evaporation and by transpiration from plants growing on the land is called **evapotranspiration.**

A number of rather complex and expensive instruments have been designed to determine evapotranspiration. One of these, the **lysimeter,** is shown in Fig. 12.2. The lysimeter, essentially, is a large metal tank containing a soil mass which may weigh (together with the soil water) as much as 100,000 lb. It is mounted in a pit so that the soil surface in the tank is level with the surface of the soil in the field. The tank is so arranged that a sensitive and automatic balance will continuously record any changes in weight of the container and its contents. The sensitivity of the balance may be to the nearest 2 lb, which is a sensitivity equal to 0.001 inch water applied to the soil surface. All water draining through the soil in the lysimeter is collected and measured. Hence, if plants are grown in the lysimeter, a continuous record of weight changes and of the volume of runoff or water percolating through the soil is recorded, and the evapotranspiration can be calculated. The gain in weight due to the growth of the plants must be taken into consideration, but this weight change is usually relatively small compared to the changes caused by moisture gain or loss.

Under conditions of adequate water supply to an area covered with vegetation, the plants will act as channels through which water passes from the soil to the atmosphere, and the amount of water lost by evaporation and transpiration will be primarily a function of the climate. Many attempts have been made to use weather data to calculate the maximum or potential evapotranspiration for a given area. These calculations have been only partly successful.

Factors Affecting Transpiration Rate. The factors influencing the rate of transpiration include two principal groups: **environmental factors** and **morphological factors.** The former include conditions external to the plant, whereas morphological factors include structural features and habits of growth of the plant, sometimes referred to as internal factors.

Environmental Factors. The important environmental factors that influence transpiration rate are (a) relative humidity of the atmosphere, (b) air movements, (c) air temperature, (d) light intensity, and (e) soil conditions. These factors affect transpiration through their effect on the vapor pressure of water in the intercellular space or of water in the air.

Atmospheric humidity. The humidity of the air surrounding the plant is an important external factor determining the rate of transpiration. The drier the air above the plant, the greater the transpiration rate. For example, air with a relative humidity of 20% is considered dry because it has the capacity for holding a great deal more water; air with a relative humidity of 80% is regarded as moist because its capacity for holding more water is not great.

The air in intercellular spaces of a turgid leaf is very moist—in fact, usually almost saturated. A

Figure 12.2   A lysimeter is a device for measuring the evaporation and transpiration from a given area of turf. *A* shows the appearance of the turf; the area being measured is delineated by the circle and the size of the area may be judged by the individual present; *B*, the lysimeter laboratory beneath the turf is crowded with electronic instruments. (Courtesy of W. O. Pruitt.)

diffusion of water vapor from intercellular spaces through stomata to the air surrounding the leaf takes place as long as the vapor pressure of water in the intercellular spaces exceeds that of the air outside. Diffusion becomes more rapid as the difference between these two pressures increases. When the difference is great, we say the **diffusion gradient** is steep. If the relative humidity of air surrounding the plant is low, the diffusion gradient between moist intercellular spaces and outside air is steep, and hence transpiration is rapid.

Air movements. As transpiration occurs, there is a tendency for a moist layer of air to form next to the leaf surface, particularly in still air. This will decrease the diffusion gradient between the leaf and the atmosphere, and transpiration will consequently decrease. On the other hand, air movement carries away this layer of humid air, replacing it with drier air, resulting in an increase in transpiration. The more rapid the air movement, the faster the moist air will be carried away and the faster the rate of transpiration. If the wind is quite strong, stomata may close, probably as a result of excessive water loss, and transpiration is then reduced.

It is a common observation that on a warm, bright, windy day, plants may lose water to the point of wilting. On a cool, cloudy day with little wind movement, however, seldom is the water loss sufficient to result in wilting. Even though plants may wilt during the hottest part of the day, this may be but a temporary condition, as is shown by the fact that the next morining the leaves are again fresh and turgid. Return to the turgid condition indicates that available water in the soil is not exhausted.

Air temperature. In direct sunlight, the temperature of a leaf is usually higher than that of the air about it; the difference may be as much as 10°C. About 80% of the radiant energy of sunlight that falls upon green leaves may be absorbed by them. Part of this absorbed energy is changed to heat and raises the temperature of the leaf; part is utilized in vaporizing water; and another small part is used in photosynthesis (see Fig. 13.21). When the leaf temperature rises, the vapor pressure in intercellular spaces becomes greater than that in the air surrounding the leaf; as a result, the diffusion gradient becomes steeper, and the rate of transpiration increases.

Light intensity. As light intensity increases, the internal temperature of the leaf is raised, with the result that water loss is accelerated. Another effect of illumination upon transpiration is that it stimulates, at least in many species of plants, the opening of stomata (p. 174). When the stomata are closed, transpiration virtually ceases. Thus light, through its effect upon leaf temperature and upon the opening of stomata, influences the rate of transpiration.

Soil conditions. Any soil condition that influences absorption of water by the roots affects the transpiration rate. When soil becomes very cold, absorption of water is retarded, even though soil water may be available. An increase in the concentration of the soil solution, as often occurs in alkali soils, reduces the rate of water intake. Poor aeration of the soil may result in diminished absorption. The availability of soil water also determines the rate of intake. The rate at which water is absorbed by the roots greatly influences the rate of transpiration. If water loss exceeds water absorption, wilting will occur. Under these conditions, the mesophyll cells do not give up water vapor as freely as when they are well-supplied with water. Consequently, the rate of transpiration is reduced.

Morphological Factors. On days when water loss is high, various kinds of plants growing side by side, exposed to the same environmental conditions, respond differently with respect to wilting. Some may be severely wilted; others may show no signs of distress. It is apparent that some plants have better provisions for transpiration regulation than do others. Thus, plants have certain structural features and habits of growth that influence the rate of transpiration. They may be referred to as morphological factors, that is, conditions in the plant itself.

Certain plants are able to survive in extremely dry habitats, whereas certain other plants succumb. Such surviving plants undoubtedly maintain a balance between water outgo and water intake, whereas in plants that fail to survive the dryness, the loss of water, at least for a period, exceeds the absorption of water. Plants of dry habitats often possess roots that penetrate deeply into the soil; or, like many desert annuals, they have an extensive surface root system that can rapidly absorb limited rainfall from spring or summer showers. Many desert plants possess special water storage tissue, as is well-shown in so-called succulents, that is, plants with fleshy stems or leaves (see Fig. 18.38).

Anatomical features that are advantageous from the point of view of prevention of water loss are (*a*) the cuticle of leaves, young stems, and fruits, (*b*) sunken stomata, (*c*) distribution of stomata, and (*d*) reduction of the transpiring surface. In addition, stomatal behavior is an important factor in controlling water loss.

Cuticle. Most of the water lost from a plant passes out through the stomata. Some water, however, is lost through the cuticle. Various modifications in leaf structure reduce cuticular transpiration; for example, thickening of the outer wall of the epidermal cells and the presence of a wax-like material, cutin, in this wall. Most plants of arid and semi-arid climates have a thicker cuticle than do those of humid climates.

Stomatal behavior. We have seen in Chapter 11 that a stoma may open and close as a result of changes in the turgor of guard cells. When guard cells become turgid, the stomatal aperture widens; when they become flaccid, the aperture narrows or completely closes. Often after a period of rapid transpiration, there is a deficit of water in all leaf cells, and stomata close, thus reducing the water loss due to transpiration. This is an advantageous behavior. But, from the standpoint of photosynthesis, the stomata must serve also as entrances for carbon dioxide. If they close during the daytime because of excessive transpiration, the diffusion inward of carbon dioxide, as well as the diffusion outward of water varpor, is restricted. Stomatal closure, in this event, works to the disadvantage of the plant by decreasing photosynthesis, although it is a favorable behavior in that it limits water loss. However, since the photosynthetic CO_2-trapping mechanisms in C-4 plants is much more efficient than that of the C-3 plants, partial stomatal closure does not reduce photosynthesis as much in a C-4 plant. As shown in Chapter 13, one characteristic of C-4 plants is their adaptation to arid conditions. This is an example of a biochemical rather than a morphological adaptation.

Although light is a factor that stimulates stomatal opeining, in many plant species stomata may remain closed throughout the day if the water deficit in leaf cells is pronounced. On the other hand, there are succulent species in which stomata are open all night, and in some instances stomata may remain open even though transpiration is proceeding at a high rate.

Sunken stomata. In some plants (Fig. 12.3), the stomata are below the general level of the leaf surface. When this condition occurs, the water vapor must diffuse through a relatively long and sometimes tortuous passageway, with the result that the diffusion rate is lessened.

Distribution of stomata. The leaves of many plants have stomata only on the undersurface, or, if on both surfaces, mostly on the lower. This distribution is shown in Table 10.1. Hence, loss of water from leaves that have stomata only on the lower surface, or fewer on the upper than the lower, is usually less than loss from leaves that have an equal number on both the upper and lower sides.

Reduction of transpiring surface. In most plants, the principal organs of transpiration are leaves. Therefore, any decrease in leaf surface will reduce transpiration and conserve water absorbed by roots. In corn and other grasses, and also in oleander, the leaves roll up during drought, thus exposing less surface to the air than do full expanded leaves. Cacti and some euphorbias have no foliage leaves; in these plants the transpiring surface is restricted to the stem surface (Fig. 18.38).

Epidermal hairs. The leaves of many plants are clothed with epidermal hairs (Fig. 7.12). In some plants, such as common mullein and dusty miller (*Centaurea cineraria*), the hairs may form a dense, cottony covering. There is no clear evidence that surface hairs reduce transpiration; in fact, experiments with some plants have shown that water loss from the leaves is greater when the hairs are present than when they are lacking.

Guttation. The loss of liquid, as contrasted with the loss of vapor, from leaves of *intact plants* is termed **guttation.** This will occur especially in herbaceous plants when conditions favor rapid absorption of water and low transpiration. For instance, when a well-watered, vigorously growing tomato plant is placed under a bell jar, transpiration ceases as the atmosphere in the jar becomes saturated. Continued water absorption then results in a slow exudation of water from the tips of the leaves (Fig. 12.4). Many plants have specialized openings called **hydathodes** at the tips of their leaves, through which the liquid passes outward. Experiments with barley plants show

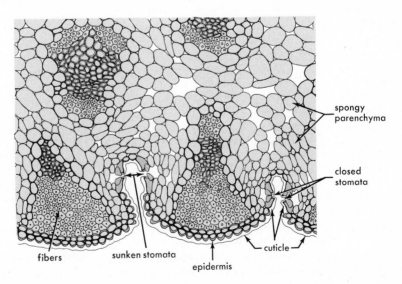

Figure 12.3 Sunken stomata in crypts of a yucca leaf. Note also the thick epidermal cuticle and the large bundles of fibers.

fibers · · · sunken stomata · epidermis · cuticle · spongy parenchyma · closed stomata

Figure 12.4   Guttation from tips of barley (*Hordeum vulgare*) leaves.

Figure 12.5   Deposit on blades of grass following the evaporation of guttation fluid. Deposit composed largely of glutamine with a small amount of potassium chloride and undetermined organic matter. (Courtesy of Curtis.)

that, when the roots are immersed in distilled water, guttation is very slight or ceases altogether. This is true even though the water is aerated amply. In dilute salt solution, but without aeration, guttation is also slight. On the other hand, if the roots are immersed in a dilute salt solution, with good aeration and favorable temperature, guttation is ample and continues for a long period in a humid atmosphere. Guttation is associated with salt absorption and salt movement into the xylem. The liquid of guttation is not pure water but a dilute salt solution (Fig. 12.5).

Summary of Transpiration

1. Transpiration is the loss of water vapor from plant surfaces, largely those of leaves.
2. Evaporation removes the water vapor from the external surface of the leaf.
3. The amount of water transpired during the life of a plant is large, from 200 to 1000 times its dry weight.
4. The rate of transpiration is affected by the relative humidity of the air, air movements, air temperature, light intensity, and soil conditions.
5. Morphological details such as the cuticle, epidermal hairs, and location and distribution of the stomata also influence the rate of transpiration.
6. Guttation is the escape of water as liquid from leaves; it may occur when the relative humidity is high. Guttation fluid contains solutes.

Processes Concerned in Ascent of Water

Water in the plant moves upward from the roots to the stems and thence into leaves, from which it escapes into the atmosphere. That the path of movement is through the xylem is shown by "ringing" or "girdling" the stem, that is, removing a complete ring of bark (thus all of the active phloem). This procedure has little immediate effect on the movement of water; the plant does not wilt. The path of movement of water may also be demonstrated by placing a leafy branch that has been cut underwater into an aqueous solution of a dye, such as eosin. If after a short period the stem is split, it can be observed that the walls of vessels and tracheids are stained by the dye, whereas other tissues are not stained.

Transpiration Pull

It is estimated that a force of 20 atmospheres (Fig. 5.14), almost 300 lb per square inch, is necessary to lift water to a height of 350 ft through the xylem of a tree. This estimate takes into consideration the resistance to water movement offered by the xylem, as well as the weight of the water column. How does water get to such a height in a tree? What are the forces involved? Although plant physiologists are not agreed on the relative importance of the various processes, the explanation that is now most generally accepted as most plausible will be described here. It is spoken of as the **transpiration-pull and water-cohesion theory.**

Evaporation

When evaporation of water occurs from the cell walls in a leaf during transpiration, forces are brought into play that result eventually in the movement of water through the plant. The diffusion of water molecules from the wet cell wall surfaces into the intercellular spaces of the leaf mesophyll results in partial drying of the walls.

Imbibition

The walls of a leaf cell, losing water as a result of evaporation, imbibe water from the enclosed cell contents or from adjacent cell walls with a higher water content. The forces of imbibition that hold the water in the walls are very great. Although the first water that is lost from a saturated mesophyll wall is held loosely, the forces holding the remaining water in the wall are stronger. Thus, during transpiration, the water potential in the wall gradually decreases because some of the imbibitional forces are not satisfied.

Osmotic Forces

This decrease in the water potential in the wall results in the movement of water from the protoplast into the wall. As water moves from the protoplast of a mesophyll cell into the walls, the turgor of the cell decreases. This causes a decrease in the water potential of the cell, and water will move into the cell from adjacent cells down a water potential gradient. Because of the relatively high osmotic concentration in mesophyll cells, these cells can have a lowered water potential and still retain considerable turgor pressure. The disturbance in water equilibrium is transmitted from cell to cell through both cellular contents and walls. Water moves into these cells from the tracheids in the veinlets of the leaves. Thus, a force equivalent to the imbibitional force created in the leaf cell walls, as a result of the loss of water from them during transpiration, is pulling on the water in the tracheids. Because of the relatively low osmotic concentration in the tracheids, as the water potential in the tracheids gradually decreases, the contents may be subjected to reduced pressures and to tensile pull.

Water Continuity and Cohesion

Water in the tracheids of leaf veinlets is continuous from the leaves to the roots. This continuity exists throughout the life of the plant. In a sense, the water columns "grow" with the plant. However, living cells of the root, xylem elements themselves, and living leaf cells offer resistance to the free flow of water. This resistance, and the gravitational pull on the water itself, are overcome by the forces developed in the leaf when water is moving up in the plant. The question now arises: Do these unbroken columns of water have sufficient tensile strength to prevent them from being broken by the forces acting on them? Experiments using both water and plant sap indicate that this requirement is met. As a matter of fact, plant sap in vessels and tracheids may withstand tensions greater than 300 atmospheres without being broken. Thus, it is evident that the attractive forces of the water molecules for each other, **cohesion,** and of cell walls for the water molecules, **adhesion,** are very great. It is because of these attractive forces that the liquid columns possess tensile strength.

The columns of water are raised by the force of evaporation of water from the leaf cell walls and the imbibitional forces that develop here.

Summary of the Transpiration-Pull Theory

1. The evaporation of water from the walls of mesophyll cells of leaves results in a decrease in the water potential in these walls.

2. This disturbance in the water balance of the cells causes a series of changes that set in motion the entire train of water through the plant.
3. The living leaf cells next to the tracheids in veinlets eventually lose water to neighboring cells.
4. Water moves from the veinlets into the adjacent cells. This results in the upward movement of water in continuous liquid columns in the xylem.
5. Continuity of water is maintained even under conditions of considerable strain because of the strong cohesive forces between water molecules.

Absorption

Absorption of Water

The large quantities of water moved through the plant body and lost by transpiration, then, must be replaced continually if the plant is to live.

In the absorbing region of the root (Figs. 9.4, 9.7), the phloem does not lie outside the xylem but occurs in groups of cells that alternate radially with xylem masses or with the radiating arms of a single central mass of xylem. Hence, it is possible for the water and mineral salts to be absorbed by epidermal cells or root hairs; pass through the cortex, endodermis, and pericycle into the primary xylem; and move upward without traversing the phloem (Fig. 12.6). This dilute solution of water and mineral salts that moves upward through the conducting elements of the xylem to the stems and leaves is called xylem sap.

Mechanisms of Absorption. The uptake of water by roots involves two mechanisms, which may or may not act simultaneously:
1. When transpiration is slow and water is available in the soil, absorption of water may exceed transpiration. This condition results in a pressure being developed in the xylem, and may lead to guttation. If the shoot of a plant is cut off a few inches above the soil level, absorption of water by the roots will often force the sap out of the cut ends of vessels and tracheids, a phenomenon known as **bleeding.** It appears that this type of absorption depends upon the *activity of living root cells.* The pressure developed in the xylem is called **root pressure.**

   This mechanism of absorption of water by roots is generally explained as resulting from the osmotic movement of water into the root. As a result of the activities of the living root cells, soil solutes are absorbed and accumulate in the cell in high concentrations (pp. 69, 163). The total concentration of the cell sap of the root hairs, as well as of the other root cells, is thus normally greater than that of the soil solution. The total amount of solutes present in the xylem elements of the root is also generally higher than that of the soil solution. The greater amount of solutes in the cell sap reduces the potential of water in the cell sap below that of the water in the soil solution. Under this condition, movement of water is inward. The rate of movement inward depends upon the difference in the water potential of the two solutions.

2. When transpiration exceeds absorption, no root pressure can be demonstrated. In the xylem elements of plants in this condition, the water is under tension, and the pressure on the water in the vessels is lower than atmospheric pressure. Under these conditions, a high rate of transpiration results in a water deficit in the plant tissues. Water then moves into the roots passively. In fact, under such conditions, water may even be absorbed by dead roots. Water absorption may then result from osmotic forces originating in roots or from forces that originate in the leaves —that is, forces set in motion by the loss of water in transpiration.

Factors Affecting Absorption. We have seen that roots grow in a very complex environment—the soil. The activities of roots are influenced to a marked degree by this environment. Many environmental factors determine the nature and extent of the root system, the rate of growth of roots, the rate at which they absorb water and mineral nutrients—in fact, all their activities.

Unavailable Soil Moisture. Plants cannot absorb all the water in a soil. There is always some water in the soil which is not available to the plant; it is held so tightly by the soil particles that the roots cannot absorb it rapidly enough to prevent wilting. If such a wilted plant is placed in a darkened moist chamber and will not recover, it is said to be **permanently wilted.** The only way of bringing about its recovery is to add water to the soil. At the time that the plant has permanently wilted, the percentage of water left in the soil may be determined. This percentage (based upon the dry weight of soil) is called the **permanent wilting percentage of the soil.** This water represents that which the plant cannot absorb readily from the soil. Only that part of the total soil moisture above the permanent wilting percentage is available for plant growth.

Different kinds of ordinary plants growing in the *same soil* reduce it to approximately the same permanent wilting percentage.

Influence of Soil Type.   *Different kinds of soils,* however, differ in their permanent wilting percentage. Let us perform an experiment to illustrate this point. Fill one container with a clay soil, and another with a sandy soil. Then wet each soil to field capacity. We known the clay soil will hold more water than the same volume of sandy soil. In each grow the *same kind of plant.* Allow each to reach permanent wilting, and then determine the permanent wilting percentage of the two types of soil. It will be found that the permanent wilting percentage of clay soil is much higher than that of sandy soil.

Moisture Content of Soil.   From the standpoint of plant growth, the readily available water of a soil is the important consideration. This is really of more significance than the water-holding capacity of a soil.

It has been determined experimentally for many crop plants, including orchard trees, that if a soil has

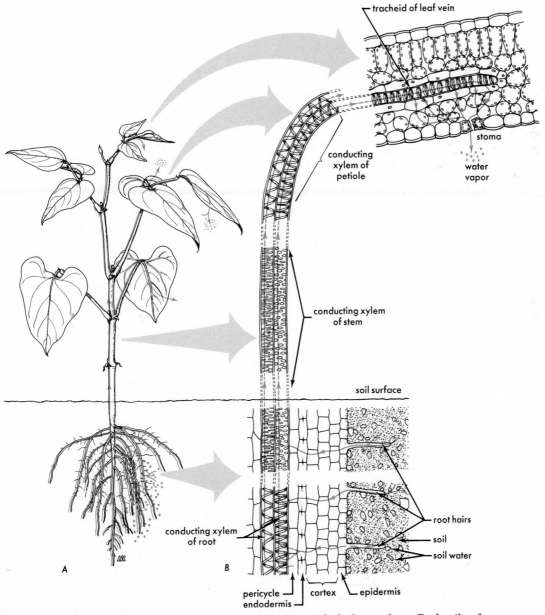

Figure 12.6   *A*, diagram showing water movement in a whole bean plant. *B*, details of the xylem pathway.

a moisture content *above* the permanent wilting percentage, there is no optimum moisture content, that is, a content at which the plants grow best. For example, if the percentage of water in a soil at field capacity is 30%, and its permanent wilting percentage is 15%, the plant growing therein will do equally well with the moisture of 16, 17, 18, or any other percentage up to 30%.

Under field conditions, soon after a rain or irrigation, followed by drainage, the soil to a certain depth (depending upon the amount of rain or irrigation water applied) is up to its field capacity of moisture. The roots of plants begin to absorb the water, and soon the soil is below its field capacity. The soil moisture constantly decreases in the zones occupied by roots. Moreover, the movement of water by capillarity, from moist soil to drier soil immediately surrounding the roots, is so slow that it may be regarded as negligible in influencing water available to the roots.

Root Growth and Soil Moisture. Since the capillary movement of water in soils is too slow to meet the requirements of growing plants, roots must continually penetrate into new soil if they are to utilize the water it holds. The total root elongation of a plant may be very great. Root extensions and the formation of new root hairs are constantly bringing about new contacts with moist soil.

The erroneous impression is sometimes held that roots "go in search of water." A more accurate statement is that they grow only in moist soil (soil above the permanent wilting percentage). In practice, this would mean that if the upper foot of soil is moist, and dry soil lies below it, the roots of plants will be confined to the upper, moist soil. In irrigation practice, this relationship of soil moisture and root growth is important. In irrigating a garden or lawn with a hose, one is easily misled into believing, because the soil surface is wet, that sufficient water has been applied. Examination may reveal that at depths of 5 inches or more the soil is very dry—too dry for the growth of roots. In growing plants, it is nearly always desirable to stimulate maximum root growth and a root system that attains its normal depth and spread. This may be accomplished in part by keeping the soil moist to the proper depth.

The loss of water from soils at levels below the first few inches is essentially due to transpiration from plants. Soils are dried out because of the presence of absorbing roots. One of the principal reasons for removing weeds from a growing crop is that they deplete the soil of water.

Solute Concentration of Soil Solution. As the concentration of the soil solution approaches that of the cell sap, the rate of water absorption declines. When the concentrations of the two solutions are the same, water intake ceases. And, if the soil solution becomes more highly concentrated than the cell sap, owing to excessive applications of fertilizers or of saline water, water will be withdrawn from the root cells. Moreover, root growth is inhibited at high salt concentrations, and the roots are not able to extend into new soil areas.

Classification of Plants According to Their Water Needs

In many agricultural sections of the country, and particularly in semi-arid and arid regions, water is the principal limiting factor in crop production. Water is a most important factor in determining both the distribution of plants over the earth's surface and the character of the individual plant. Probably no single factor is so largely responsible for the diversity of plants in various habitats as is the difference in supply of water. So important are the water relationships of plants that various attempts have been made to classify plants on the basis of these relationships. One such classification divides plants into (a) **xerophytes,** which are able to live in very dry places, (b) **hydrophytes,** which live in water or in very wet soil, and (c) **mesophytes,** which thrive best with a moderate water supply.

Plants with xerophytic characteristics, which limit transpiration or in other ways balance water outgo and water intake, occur in different climatic zones, but those of deserts are most typical. Xerophytic species do not necessarily have a lower transpiration rate than do mesophytes when water is ample, but they do possess one or more characteristics that enable them to survive periods of drought. The most effective of these are thick cuticle, stomatal closure, reduction of the transpiring surface, and water storage tissue.

It should be kept in mind that all plants that fall into any one of the above three classes cease active growth when the soil moisture reaches the permanent wilting percentage. Xerophytes and mesophytes alike, growing side by side, reduce the soil to approximately the same permanent wilting percentage. Mesophytes may not be able to long survive such low soil moisture, whereas plants with xerophytic characteristics will.

Absorption of Solutes

In the discussion of absorption by plant cells, it was pointed out that different solute particles diffuse independently of each other and of water. Solutes move by diffusion from the soil solution into the wet cell walls or root cells. Here, some of the cations may exchange for cations associated with the negatively charged sites on the wall. Although solutes may move relatively freely through the wall, ions meet a barrier to free diffusion, the plasmalemma, when they reach the protoplasmic surface. Cell membranes are not readily permeated by charged ions. Nevertheless, cells do absorb ions and actually accumulate them to concentrations many times higher than their concentration in the external medium. How do solutes enter cells? What is the method used by the cell to accumulate solutes?

Active Solute Absorption

It has been known for a long time that some ions may be taken up and stored in living cells so that their concentration is 1000 or more times that in the external medium. Such an accumulation against a concentration or diffusion gradient cannot be accomplished by simple diffusion. The ions are not precipitated or bound in the cell, but are still free ions. Diffusion would result in equal concentrations of a particular ion throughout the whole system. Accumulation of ions requires energy. It is an active process. What is the source of the energy used?

The major source of energy to nongreen cells such as root cells is food which is respired. Several experimental observations indicate that solute absorption is linked to respiration.

1. The absorption of many ions is markedly decreased under low oxygen conditions and essentially ceases when oxygen is completely removed from the cell's environment. This parallels the effect of oxygen on aerobic respiration.
2. Increasing the temperature around roots from 10 to 30°C increases the rate of respiration and the

rate of solute absorption similarly. Simple diffusion is relatively less sensitive to temperature change.
3. It is possible to use certain chemicals to inhibit respiration. Such studies show a close correlation between the production of ATP during respiration and ion absorption.

Many experimental observations such as these point to the fact that inorganic ions are accumulated in plant cells by the expenditure of energy released through aerobic respiration.

The Carrier Hypothesis

But how do ions pass the plasmalemma that represents a barrier to their movement? One hypothesis that has been advanced to explain this problem suggests that there are definite chemical compounds called "carriers" in the membrane. According to this idea, an ion that comes in contact with the plasmalemma may combine with a carrier that has a specific attraction for that ion. The carrier ion complex then moves across the membrane, releasing the ion into the cytoplasm. The carrier is then free to return to the outside surface of the plasmalemma and combine with another ion of the same kind (Fig. 12.7). The results of many experiments on the relation between the concentration of an ion and its absorption are in agreement with this carrier hypothesis.

Once an ion reaches the cytoplasm, it is free to diffuse or to be carried by protoplasmic streaming to other parts of the cytoplasm. It may diffuse from one cell to another through plasmodesmata. However, since cell organelles are separated from the rest of the cytoplasm by membranes, one might postulate that these membranes too may have specific carriers for specific solutes.

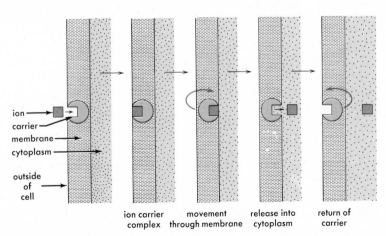

ion
carrier
membrane
cytoplasm
outside of cell

ion carrier complex   movement through membrane   release into cytoplasm   return of carrier

Figure 12.7 Diagrammatic illustration of the carrier hypothesis of ion transport across cell membranes.

Summary of Absorption

1. Absorption of water by roots when transpiration is low or absent occurs only in living root systems.
2. This absorption of water results when the solute concentration of the xylem sap is higher than that of the external solution.
3. Root pressure, guttation, and bleeding are expressions of active absorption.
4. Passive absorption of water may result from the transmission to the roots of forces that originate in the leaves; it may even take place if the roots have been killed.
5. A plant is said to be permanently wilted when it will not recover in a saturated atmosphere unless water is added to the soil.
6. The percentage of moisture remaining in soil in which a plant is permanently wilted is called the permanent wilting percentage of that soil.
7. Only that part of the total soil moisture which is above the permanent wilting percentage is available for plant growth.
8. The rate of water absorption is influenced by the available water in the soil, the air in the soil, the soil temperature, the concentration of the soil solution, and the rate of transpiration.
9. The absorption of ions by root cells may occur against a concentration gradient. This process is called accumulation and is dependent upon energy released by the living cells during respiration.
10. The carrier hypothesis states that solutes pass through cell membranes by combining with specific carriers in the membranes.

Conduction

Within a living plant, substances are constantly moving from one place to another. Although they are interrelated, we can detect four types of movement based on the rates at which movement occurs: (a) slow *diffusion* of molecules and ions; (b) moderate movement of materials carried by *cytoplasmic streaming* in living cells; (c) more rapid *flow* of material *in the sieve tubes;* and (d) very *rapid conduction* of water and mineral solutes *in the xylem.*

Diffusion and Protoplasmic Streaming

The movement of water or of solute molecules and ions into a cell across the cell wall is a relatively slow process. Diffusion occurs along activity gradients and may take place through plasmodesmatal connections between cells as well as directly through the permeable cellulose walls of most cells. Once a solute molecule or ion diffuses into a cell, its rate of movement may be increased manyfold as it is picked up by the protoplasmic stream. Cytoplasm may stream at rates of a few to several centimeters per hour. The highest rate of protoplasmic streaming that has been measured was observed in the protoplasm of a slime mold in which the cytoplasm was streaming at a rate of 48.6 cm per hr.

In terms of the dimensions of a single cell, cytoplasmic streaming is frequently very rapid, since a substance may be transported across the length of a cell in a matter of seconds. In terms of the whole plant, however, this is a very slow process. It would take days, for instance, for a molecule in the leaf to be carried upward to the growing shoot tip or downward to the roots. Fig. 12.8 diagrams the way in which one solute may be moving by diffusion and protoplasmic streaming in one direction through a series of cells and another solute may be moving in another direction. Diffusion and protoplasmic streaming are the chief methods by which solutes move through living plant tissues, except in the sieve tubes.

Conduction in the Phloem

The rapid translocation of foods throughout the plant takes place chiefly in the phloem and more specifically in sieve tubes. That the phloem is the major path of food transport is borne out by the results of ringing experiments followed by chemical analyses of tissues and cells above and below the ring. When a ring of bark is removed from a stem down to the cambium (thus removing the phloem), the carbohydrate content and especially the sugar concentration of the sap in the leaf, bark, and wood *above* the ring increases after a few hours. *Below* the ring, the total carbohydrate content and the sugar concentration in the sap of the bark and the wood decreases. These conditions would not prevail if foods were moving downward in the xylem tissue.

Both carbohydrates and nitrogenous substances move to tissues that are utilizing them. This means that foods may move upward or downward in the phloem. The direction of food movement is usually from a region of the shoot where excessive food is present to a region where a lesser amount is present. Food may move from the place of manufacture (leaves) to a storage place (roots, fruits, seeds) or to a region of growth (buds, cambium). It may also move from storage tissue to a region of growth.

Translocation of material in the phloem may reach a rate of 100 cm per hr. When this is compared to

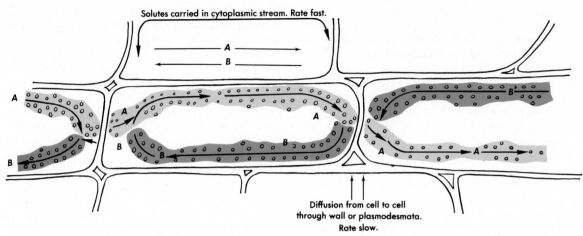

Solutes carried in cytoplasmic stream. Rate fast.

Diffusion from cell to cell
through wall or plasmodesmata.
Rate slow.

Figure 12.8 Diagram illustrating how two solutes may be moving in opposite directions by diffusion and protoplasmic streaming in the same tissue.

the much lower rates of protoplasmic streaming, it is evident that normal protoplasmic streaming alone is insufficient to cause this rapid transport. In fact, evidence indicates that the mature functional sieve-tube members have highly modified cellular contents (absence of nuclei and definite vacuoles, Chap. 7) and that protoplasmic streaming does not occur in them.

The contents of the sieve-tube members of one sieve tube make up a continuous liquid system. The pores in the sieve plates provide relatively large connections between adjacent sieve-tube members. It has been shown that at maturity the sieve tube is highly permeable in the longitudinal direction. The solution in the sieve tube is under positive pressure. This hydrostatic pressure is built up by the diffusion of water into sieve tubes in regions of the plant where food is produced or where insoluble food is solubilized. According to one of the most generally accepted theories, the mass flow theory, water and solutes move together (as one mass) in the sieve tubes in a longitudinal direction. This movement occurs from food production or solubilization regions to sites of food storage or utilization where solutes are removed from the sieve tubes. When solutes are removed, the water potential increases, and water diffuses out of the sieve tubes into adjacent cells and into the xylem. This results in a net movement of solution along the sieve tube.

Lateral Transport

There is lateral (radial) movement of foods in stems along the vascular rays, bringing the foods into close contact with conducting elements of both phloem and xylem. Thus, the rays carry foods from sieve tubes radially into the cambium and xylem.

In some plants, such as the sugar maple, the xylem parenchyma cells serve as places where large quantities of carbohydrates are stored. In the spring, the presence in the xylem of soluble carbohydrates results in a high osmotic concentration. This, in association with cold nights and warm days, causes sap to bleed from cuts made in the trunk in the spring of the year.

Conduction in the Xylem

If a stem is girdled by removing a layer of bark around the circumference of the stem down to the cambium, the mineral (inorganic) ions that are absorbed from the soil and conducted in the stems and roots move upward past this girdle. Girdling may or may not decrease this upward movement, depending upon the species and other factors. This fact would serve to indicate that normally the greater amount of ion translocation in the stem is in xylem elements. An especially convenient method of studying the path of movement of ions in stems is now being used. Certain ions, such as potassium, phosphate, and bromide, are rendered radioactive and added to the solution in which the plants are growing. By appropriate means, the path of movement of these radioactive **tracers** has been followed accurately, even for short periods of time, both in girdled and ungirdled stems. The results of these experiments seem to show that inorganic solutes move upward chiefly in xylem tissues. As the ions move upward, however, they may transfer radially from xylem through vascular rays into phloem tissues; and they may also accumulate in living cells along their path of movement. When radioactive phosphorus was supplied to the nutrient solution in which tomato plants 6 ft high were growing, it could be detected throughout the plants after 40 min. There is also evidence that certain minerals carried upward in the xylem to the leaves may be moved from the leaves and conducted downward via the phloem.

Frequently, not all of the xylem vessels or tracheids function in the conduction of water and mineral ions at any one time. In trees particularly, the heartwood ceases to play a role in conduction (pp. 136, 145). Sometimes, only the outermost layers of sapwood contain xylem elements filled with water.

The rate of movement of water and solutes in the xylem depends in large measure upon the rate of transpiration. If plants are growing in a very humid atmosphere, the movement of water through the xylem is very slow; if they are in a dry, warm atmosphere, which heightens the rate of transpiration, the movement may be as rapid as 131 ft per hr. In the tobacco plant, water conduction was found to go on at a rate of about 4.7 ft per hr. In the larch tree, the rate of water movement was found to be approximately 5.5 ft per hr at a period of the day when water loss was highest; but in the early morning, when the rate of water loss was at the minimum, the rate of water movement was but 0.2–0.4 inch per hr.

Absorption of inorganic ions by roots is *not proportional* to the absorption of water. Some increase in ion absorption accompanies an increase in water absorption, but not in the same ratio. Moreover, the different ions are not absorbed at the same rate.

Thus, the amount of water moving through the plant is not the only factor that determines the amount of ions or of any particular ion which is absorbed and conducted through the plants to the leaves. Other factors, such as the photosynthetic activity of leaves, respiration, and other chemical processes in the plant, determine to a degree the quantity and kind of ions absorbed and transported.

Summary of Conduction of Solutes

1. Normally, the greater amount of mineral salt (ion) translocation in the stem is in the vessels and tracheids of the xylem. The movement is usually upward.
2. Ions may also be conducted in the phloem.
3. The major path of food movement is through the phloem.
4. Food movement is from a region of high food content or place of manufacture or storage to a place of food utilization, or from a place of manufacture to storage tissues.
5. Rapid lateral movement of solutes takes place along the vascular rays.
6. The mass flow theory states that solutes and water move together in the sieve tubes as a result of the development of internal pressure in the sieve tubes.

13

Photosynthesis

The maintenance of life in a plant or animal cell requires the continual use of energy. This energy directly or indirectly comes from the sun. Two major processes that are carried out by terrestrial green plants and that use energy directly from the sun are transpiration and photosynthesis. Both of these processes use large amounts of light energy, but only in photosynthesis is a significant amount of energy from light actually stored for future use. Light influences other processes such as flowering, seed germination, certain growth curvatures, stomatal movements, and pigment production, but in these cases only very small amounts of light energy are involved.

During transpiration, energy from the sun evaporates water from cell walls. Although this results in a movement of water in the xylem, this energy is neither stored nor used to bring about the vital reactions involved in the synthesis of foods, in assimilation, growth, and reproduction. On the other hand, in photosynthesis, energy from the sun is transformed into chemical energy. In most photosynthesizing plants, photosynthesis results in the utilization of carbon dioxide and water and the production of complex sugar molecules and oxygen gas. During this process, light energy from the sun is changed into energy holding atoms together in the sugar molecule. Neither carbon dioxide nor water will burn, nor would either make a very good meal. Carbohydrates (sugars, starches, cellulose), on the other hand, burn easily and release considerable energy in the process. Furthermore, many carbohydrates are excellent foods. The great importance of photosynthesis lies in the transformation of low-energy compounds, carbon dioxide and water, into compounds (sugars) containing large amounts of energy. Energy stored as chemical energy in foods (carbohydrates, fats, and proteins) is continually released in living cells during the process of respiration and is lost as heat or converted into energy-rich bonds (for example, ATP). Many energy-requiring reactions carried out by living cells occur at the expense of energy temporarily stored in ATP. *Photosynthesis stores energy. Respiration releases it,* enabling cells to perform the work of living.

Perhaps some day man will use atomic energy to drive the energy-requiring steps in the production of food and fiber, but today photosynthesis is man's only source of food. Will the capacity of plants to carry out photosynthesis finally determine the number of people who can live on the planet Earth? There are now about 3.5 billion people living on earth; in 1900 there were about 2 billion, and in 1800 a little over 1 billion (Fig. 13.1). If the population

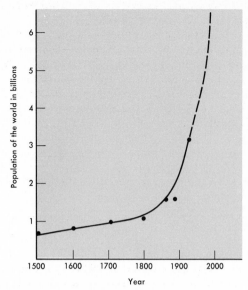

Figure 13.1 Graph showing the world population projected to the year 2000.

increases at its present rate, it will double by the year 2000, reaching between 6 and 7 billion.

All organisms, including man, increase in numbers according to well-established principles. While many factors may be involved, rate of growth is really a function of two factors, food supply and contamination of the environment. The familiar sigmoid curve of growth is shown in Fig. 21.19. It indicates a slow start when pioneers are paving their way, a rapid growth at a uniform rate when all conditions are favorable, and then a slowing down to a constant plateau when food and contamination become limiting factors. There is no evidence that twentieth century man's population increase will not follow this sigmoid growth curve.

If all the land surface of the earth could support plants, it would potentially be able to produce enough food for 1000 billion people. This figure is of course unrealistic, since only a small part of the land can be used for agricultural purposes. Land is needed for urban and recreational use, and some land is unsuitable for any of these uses. Even if only 7% of the earth's land surface is considered agriculturally productive, it should still be able to produce enough food to support 79 billion people.

Such calculations lead to two far-reaching conclusions: (a) the number of people who can live on the planet Earth ultimately will be limited not by the production of food but by the amount of space that a man needs to live in and to work in with reasonable comfort, and perhaps by his contamination of

his environment; and (b) starvation of thousands of people today is occurring in a world in which technical means of preventing this are known and available, given sufficient sources of energy.

The Discovery of Photosynthesis

Our present understanding of photosynthesis is closely linked with the development of modern scientific thought and modern chemistry. Prior to the early seventeenth century, the general belief was that plants derived the bulk of their substance from soil humus. This idea was overthrown by a simple experiment performed by Flemish physician and chemist Jan van Helmont, who planted a 5 lb willow branch in 200 lb of carefully dried soil. He supplied rain water to the plant as needed and in 5 years it grew to a weight of 169 lb. The soil had lost only 2 oz according to his measurements; consequently, van Helmont reasoned that the plant substance must have come from water. This was a logical deduction, though we now know it was not entirely correct.

Our knowledge of photosynthesis begins with the observations of a religious reformer, philosopher, and spare-time naturalist, Joseph Priestley. In 1772, Priestley reported that he had found almost accidentally that a sprig of mint could restore confined air that had been made impure by burning a candle. The plant changed the air so that a mouse was able to live in it. The experiment was not always successful, probably because Priestley, who did not know the role of light, did not always have adequate light. Seven years later, Jan Ingen-Housz noticed that air was revitalized only when green parts of plants were in the light.

It was 3 years later, in 1782, that a Geneva pastor, Jean Senebier, discovered another important part of the process—that "fixed air," carbon dioxide, was required. Thus, in the new chemical language of Antoine Lavoisier, it could be said at this time that green plants in the light use carbon dioxide and produce oxygen.

But what was the fate of the carbon in the carbon dioxide? Ingen-Housz answered this question in 1796 when he said that the carbon went into the nutrition of the plant. In 1804, 32 years after Priestley's early observations, the final part of the overall reaction of photosynthesis was explained by the Swiss botanist and physicist, Nicolas Th. deSaussure, when he observed that water was involved in the process. Now the experiment that van Helmont made almost 200 years earlier could be explained:

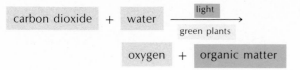

$$\text{carbon dioxide} + \text{water} \xrightarrow[\text{green plants}]{\text{light}}$$

$$\text{oxygen} + \text{organic matter}$$

Almost 50 years elapsed before the carbohydrate nature of the organic matter formed during photosynthesis was recognized.

The Products of the Reaction

Oxygen

The evolution of oxygen during photosynthesis in higher plants can be readily demonstrated. If an inverted glass funnel is placed in water over a mass of green water plants and the funnel is completely filled with water and then closed, gas bubbles given off by the plants may be collected in the funnel tube. If a glowing splinter is inserted into the tube, it will glow more brightly, showing that the gas contains oxygen.

The volume of oxygen liberated in photosynthesis is approximately equal to the volume of carbon dioxide absorbed. This ratio, $O_2/CO_2 = 1$, is known as the **photosynthetic quotient.** Under like conditions of temperature and pressure, equal volumes of different gases have the same number of molecules; hence for every molecule of carbon dioxide absorbed during photosynthesis, one molecule of oxygen is liberated. Thus, the equation

$$6\ CO_2 + 6\ H_2O \xrightarrow[\text{green plant cell}]{\text{light energy}}$$

$$C_6H_{12}O_6 + 6\ O_2$$

is supported by experimental data.

Sugar

The demonstrations that carbon dioxide and water are the raw materials and that oxygen is one of the end products of photosynthesis are easy to make and may be carried out in any botany laboratory. Proof that sugar is the other major product of photosynthesis was more difficult to obtain before $^{14}CO_2$ was available (p. 224).

It has been pointed out that the photosynthetic quotient, O_2/CO_2, is equal to 1; in other words, the amount of carbon dioxide taken up by the leaf is equal to the amount of oxygen given off. This condi-

tion is met by the elaboration of a carbohydrate as is seen from the overall equation.

Starch grains occur only in the chloroplasts of the higher plants, and if leaves containing starch are kept in darkness for some time, the starch grains will disappear (Fig. 13.2; see Color Plate 5); then if these leaves are exposed to light, starch reappears in the chloroplasts. Starch is the *first visible product* of photosynthesis.

Oxygen from Water

If we look at the overall equation for photosynthesis,

$$CO_2 + H_2O \xrightarrow{\text{light}} \underset{\text{carbohydrate}}{(CH_2O)} + O_2$$

a logical conclusion would be that the oxygen liberated is released from the carbon dioxide molecule. However, van Neil, working at Stanford University, made a comparative study of photosynthesis which occurred in a number of organisms belonging to different groups of plants. The green and purple sulfur bacteria (Figs. 21.15, 21.16) are able to use hydrogen sulfide instead of water to reduce carbon dioxide. Sulfur instead of oxygen is liberated.

$$CO_2 + 2\ H_2S \xrightarrow{\text{light}}$$

$$(CH_2O) + H_2O + 2\ S$$

Some other bacteria and some algae can use hydrogen instead of water to reduce carbon dioxide:

$$CO_2 + 2\ H_2 \xrightarrow{\text{light}} (CH_2O) + H_2O$$

By comparing similar reactions that occur in a variety of forms in different organisms, one is often able to clarify some of the puzzling parts of complex reactions. In the two examples above, CO_2 was still used up, but O_2 was not released. Van Neil concluded that a general equation for photosynthesis should be written as follows:

$$\underset{\substack{\text{carbon} \\ \text{dioxide}}}{CO_2} + \underset{\substack{\text{hydrogen} \\ \text{donator}}}{H_2A} \xrightarrow{\text{light}}$$

$$\underset{\text{carbohydrate}}{(CH_2O)} + \underset{\text{water}}{H_2O} + A$$

The hydrogen donator H_2A can be H_2O, H_2S, H_2, or any other substance capable of donating hydrogen to CO_2 in the process of photosynthesis. The donation of hydrogen is a reduction reaction and requires an input of energy. The product A is one of the results of oxidation. The validity of van Neil's hypothesis was demonstrated when Ruben, Hassid, and Kamen used the heavy isotope of oxygen (^{18}O) and determined that when water molecules contained ^{18}O, then the oxygen liberated by photosynthesis contained ^{18}O. In contrast, if carbon dioxide contained ^{18}O, then the oxygen liberated did not contain the heavy oxygen isotope. The oxygen liberated in photosynthesis comes from *water*. A more complete equation for photosynthesis that indicates this fact is:

$$6\ C^{16}O_2\ +\ 12\ H_2^{18}O\ \xrightarrow[\text{green plant cell}]{\text{light energy}}$$

$$C_6H_{12}^{16}O_6\ +\ 6\ ^{18}O_2\ +\ 6\ H_2^{16}O$$

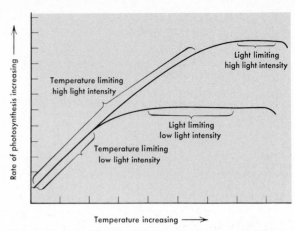

Figure 13.3 Graph showing interaction of light and temperature on the rate of photosynthesis. When light is limiting, the rate of photosynthesis is independent of temperature. When temperature is limiting, the rate of photosynthesis is independent of light intensity.

Light and Dark Reactions

Between 1883 and 1885 a German physiologist, Engelmann, in a remarkably simple experiment, demonstrated which *quality* (color) of light is effective in photosynthesis. He placed together on a microscope slide an algal filament and some bacteria that would migrate toward high concentrations of dissolved oxygen. He then placed the alga in different wavelengths of light. The bacteria were most active near the algal filament in red and in blue light. So red and blue light were trapped and supplied energy to drive photosynthesis and liberate oxygen.

At approximately the same time (1883), J. Reinke, another German scientist, was studying the effect of changing the *intensity* of light on photosynthesis. Reinke observed that the rate of photosynthesis increased proportionally to the increase in the intensity of light only at low-to-moderate light intensities. At higher light intensities, the rate of photosynthesis was unaffected by changing the light intensity; the reaction was then proceeding at a maximum rate and was unaffected by light—it had become light-saturated.

A further study and a more comprehensive interpretation of this phenomenon was carried out in 1905 by F. F. Blackman, a British plant physiologist. He reasoned that photosynthesis actually is made up of light-sensitive and light-insensitive reactions (Fig. 13.3). Thus, photosynthesis may be divided into two general parts: (a) photochemical reactions, and (b) "dark reactions," or enzyme reactions. These enzyme reactions are sensitive to temperature changes, but are independent of light and can occur either in light or dark.

Taking Photosynthesis Apart

In order to understand a complex series of reactions such as those that make up photosynthesis, it is necessary to take the process apart as much as possible and to study each reaction separately. This has been at least partially possible. As we have seen, Blackman's study on the effects of light and temperature on the rate of photosynthesis was an example of one method of demonstrating the involvement of two types of reactions. Later, the use of the heavy isotope of oxygen to trace the oxygen in the reaction added more knowledge to our understanding of the process. Another isotope, radioactive carbon, ^{14}C, was used by Melvin Calvin to work out the sequence of enzymatic reactions concerned with the carbon cycle of photosynthesis.

A further method of taking the process apart is to free chloroplasts or parts of chloroplasts from the cell and to study the role they may play in the complex process of photosynthesis. One of the earliest successful attempts to do this was made by Robin Hill in Cambridge, England. He demonstrated, in 1932, that chloroplasts isolated from the cell could still trap light energy and liberate oxygen. Later, Arnon, at the University of California, and others, demonstrated that isolated chloroplasts could convert light energy to chemical energy and use this energy to reduce CO_2.

With the advent of electron microscopy, it has been generally recognized that the photochemical reactions are bound to membranes found within the chloroplasts, while the enzyme reactions of the carbon cycle occur in the stroma.

Chloroplasts—The Site of Photosynthesis

A wide range of different plants are able to carry out photosynthesis: plants ranging from prokaryotes, single-celled purple and green bacteria, and blue-green algae, to the complex eukaryotes including such diverse forms as algae, mosses, ferns, herbs, shrubs, and trees.

In all photosynthetic organisms, the photosynthetic pigments, and consequently the light reactions of photosynthesis, are membrane-bound. In eukaryotes, these membranes are organized in chloroplasts. In the prokaryotes, the membranes are not organized into organelles but are located in the common protoplasm of the cell (Fig. 22.20).

Although chloroplasts in many plants have a characteristic globular shape, in some plants, especially the algae, they may have a wide variety of forms (see Figs. 22.4, 22.5). Chloroplasts seen with the light microscope in cells in a perfectly natural habit appear a homogeneous green. Such chloroplasts, in a filamentous stage in the life history of a moss, are shown in Fig. 13.5.

Chloroplasts, like most cell organelles, are very fragile and are extremely difficult to isolate from the cell in an intact and functional state. Special methods have been worked out for their isolation. When chloroplasts are isolated, two different kinds may generally be seen on a microscope slide; they may again be perfectly homogeneous or may show the presence of darker spots or grana. Electron micrographs of these plastids show some to be complete, with internal membranes embedded in stroma and surrounded by the outer envelope. The others have lost the envelope and stroma and consist only of a system of membranes (Fig. 13.4). Under certain conditions, these latter membranes will carry out part of the reactions of photosynthesis and will liberate oxygen as Hill demonstrated. Only the complete chloroplasts will both liberate oxygen and carry out the enzymatic steps that convert CO_2 into a carbohydrate. The internal membranes are called **thylakoids.**

A section of a plastid in a mesophyll cell of a corn leaf is shown in Fig. 13.6. The double nature of the envelope is apparent. The stroma is granular; some of the darker, larger granules are chloroplastic ribosomes. There appear to be two types of membranes, those forming the grana, and those interconnecting the grana (the fret membranes). The two end membranes of the grana are only about one-half as thick as the internal membranes (the partitions), and it is thought that the partitions are formed by the appression of two fret membranes. The location of the chlorophyll is not definitely known, except that it is membrane-bound.

We know that all biological membranes are high in lipids and proteins. In the case of the chloroplasts, it is believed that the chlorophylls and accessory carotenoid pigments are in close contact with lipid material and with some of the photosynthetic enzyme systems in the grana compartments in a definite protein, lipid, pigment pattern. Such a structural arrangement would facilitate the trapping of light energy and its transfer into chemical energy during photosynthesis. As we will discuss later, light energy,

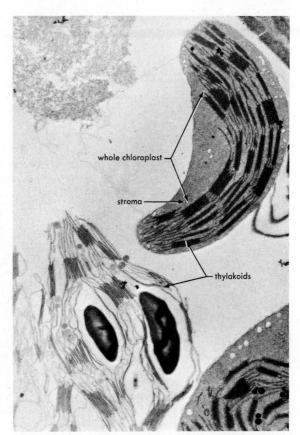

whole chloroplast

stroma

thylakoids

Figure 13.4 Electron micrograph of isolated chloroplasts of *Vicia faba*, showing one plastid with stroma and grana and one plastid that has lost its outer envelope and stroma, ×2500.

ing chloroplasts from living leaves in such a way that they are still able to carry out many of the reactions that they normally do in the leaf. When cells are cut or ground carefully in a dilute, buffered sugar or salt solution, many chloroplasts remain intact and can be separated from the rest of the cell contents by placing the ground material in a centrifuge and spinning it at high speed. If the correct speed is chosen, large and heavy particles, nuclei, starch grains, and cell wall fragments will collect at the bottom of the tube (Fig. 13.7). If the liquid containing the chloroplasts and other cellular components is now poured into another tube and centrifuged again at a slightly higher speed, the chloroplasts will separate out at the bottom of the tube.

These isolated chloroplasts (Fig. 13.4) may then be resuspended in a suitable solution, and in the presence of carbon dioxide and light, they will synthesize carbohydrate and liberate oxygen. Although carbon dioxide fixation by isolated chloroplasts occurs at a slow rate compared to the rate in an intact cell, it demonstrates that all the reactions of photosynthesis go on in the chloroplast. Chloroplasts isolated in this way have been used to study many of the reaction steps in the complex process of photosynthesis.

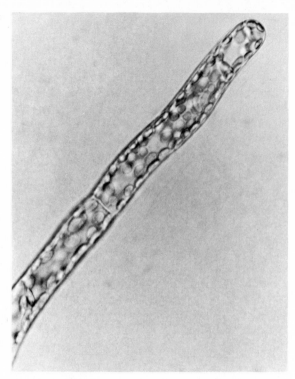

Figure 13.5   Protonema of a moss, showing numerous chloroplasts, ×30.

when trapped by chlorophyll, initiates a series of reactions in which electrons from chlorophyll may flow to other compounds **(electron acceptors)** found in the chloroplasts. It is known that such electron acceptors do occur in chloroplasts.

The chemical reactions going on in photosynthesis are directly associated with the structure of the chloroplast. This can be very clearly shown by isolat-

Light—The Energy Used in Photosynthesis

The Nature of Light

Before we can discuss the way in which light is used in the process of photosynthesis, we should consider the nature and properties of light.

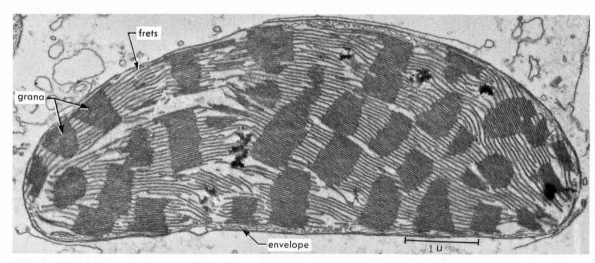

Figure 13.6   Electron micrograph of chloroplast from corn (*Zea mays*) mesophyll.

Figure 13.7 Diagram of the method of isolating chloroplasts.

White light, as it comes to us from the sun, is composed of waves of different lengths, ranging from relatively long waves of red light, through successively shorter waves to violet light. When passed through a glass prism, white light is resolved into these colors. The band of colors is the **visible spectrum.** The complete visible spectrum is composed of the following colors (starting with the longest rays): red, orange, yellow, green, blue, indigo, and violet. Wavelengths exist that we are unable to perceive with our eyes. Beyond the red are still *longer,* invisible rays, the infrared; and beyond the violet are *shorter,* invisible rays, the ultraviolet. Thus, the visible spectrum represents only a part of the radiant energy that comes to the earth from the sun. But only a part of the visible spectrum is effective in photosynthesis (Fig. 13.8).

Numerous experiments have shown that light not only acts as if it were traveling in waves but also resembles particles or packets of energy called **photons** or **light quanta.** The energy of photons determines the color of light.

Absorption of Light Energy

The fact that chlorophyll is green to the eye is evidence that some of the blue and red wavelengths of white light are absorbed, leaving proportionally more green to be transmitted or reflected and seen. It is the absorbed light that is used in photosynthesis. The part of the white light that is absorbed by chlorophyll can be determined if the light is passed through a solution of chlorophyll in alcohol and then caused to fall upon a glass prism so that it is broken up into its component colors. The spectrum thus formed is different from the spectrum of white light that has not passed through a chlorophyll solution. In the chlorophyll spectrum, the wavelengths that are absorbed appear as dark bands, the **absorption bands** (Fig. 13.8). The spectrum is known as an **absorption spectrum.** Thus, the positions of dark bands in the chlorophyll spectrum indicate which wavelengths are absorbed most strongly. They show that much of the red, blue, indigo, and violet are absorbed; these are the wavelengths that are utilized most in photosynthesis. Part of the red and most of the yellow, orange, and green are scarcely absorbed at all unless the chlorophyll solution is very concentrated.

Only absorbed light can be utilized in photosynthetic reactions. If a green leaf, instead of a chlorophyll solution, is placed between a light source and a prism, the absorption bands are quite similar to,

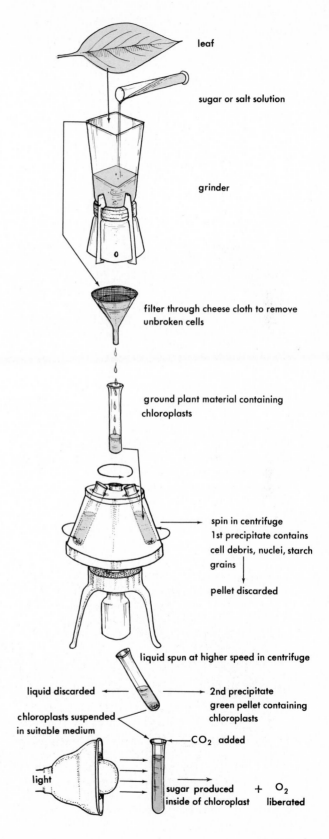

leaf

sugar or salt solution

grinder

filter through cheese cloth to remove unbroken cells

ground plant material containing chloroplasts

spin in centrifuge
1st precipitate contains cell debris, nuclei, starch grains

pellet discarded

liquid spun at higher speed in centrifuge

liquid discarded ←

→ 2nd precipitate green pellet containing chloroplasts

chloroplasts suspended in suitable medium

CO₂ added

light

sugar produced inside of chloroplast + O₂ liberated

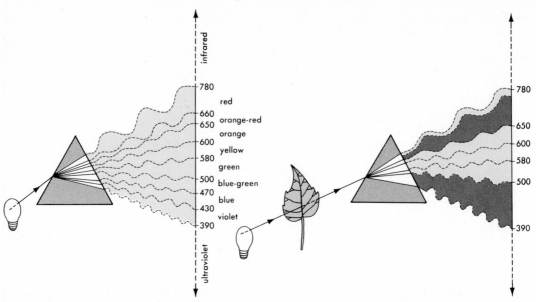

Figure 13.8 Left, visible white light passing through a prism is broken up into a range of wavelengths from the 390 nm of violet to the 780 nm of red; right, the sunlight passing through a leaf placed in front of the prism reveals dark bands in the spectrum between 390 and 500 nm and between 650 and 740 nm indicating that chlorophyll has absorbed light of these wavelengths.

though not identical with, those from the chlorophyll solution (Fig. 13.8). The leaf contains the yellow carotenoid pigments that absorb blue light; also, chlorophyll in the leaf is probably in association with protein and fat-like materials, while chlorophyll in the solution is not. These factors contribute to the difference between the absorption spectra of a green leaf and of a chlorophyll solution.

Some plants, such as certain varieties of *Geranium* and of *Coleus,* have variegated leaves, that is, leaves with white bands or blotches. Microscopic examination of mesophyll cells of the white areas shows them to be devoid of chloroplasts. These cells, however, are living and functioning. If a variegated leaf is given the starch test after several hours exposure to light, starch will be found only in the cells that contain chloroplasts (Fig. 13.9; see Color Plate 5), demonstrating that photosynthesis and starch storage have occurred in association with the chloroplasts.

What happens when light is absorbed by a chlorophyll molecule (Fig. 13.10)? How is it used in the process of photosynthesis? We know that molecules are composed of atoms having positively charged nuclei and one or more electrons spinning in certain definite positions, **orbitals,** around the nuclei. Energy is required to move an electron away from the positively charged nucleus. When a unit of light energy, a **quantum,** is absorbed by a molecule, the energy causes an electron of an atom in the molecule to be moved into an orbital farther from the atomic nucleus. The molecule now is said to be in an **excited state.** This is an unstable condition, and the electron tends to move rapidly (within a fraction of a second) back into its original orbital. As it does this, the trapped energy is released. It may be given off as heat, or as light, or as fluorescence. It is evident that in the process of photosynthesis, energy that has excited the chlorophyll molecule is in some way used to bring about a chemical reaction rather than being lost simply as light or heat.

The Conversion of Light Energy into Chemical Energy

Although we do not yet know exactly how light energy absorbed by chlorophyll in a chloroplast is trapped as chemical energy, a great deal has been learned in recent years about this process. Some of this energy we know is temporarily stored in two compounds, **ATP** (adenosine triphosphate) and **NADPH** (reduced nicotinamide adenine dinucleotide phosphate). These compounds are formed in photosynthesis and are used to bring about the reduction of carbon dioxide.

Photosynthesis is a complex part of metabolism in which the sequence of reactions is highly organized, and energy transfer proceeds in an orderly fashion from compound to compound in the sequence. Although photosynthesis occurs in plants that vary greatly in structural characteristics, there is a remarkable similarity in the chemical reactions of photosynthesis in all species. Photosynthesis is an oxidation–reduction process involving the transfer of large amounts of energy in relatively small energy steps. These energy transfers are brought about by the flow of electrons from reducing agents, electron donors, to oxidizing agents, electron acceptors. Unless outside energy is put into the system, the normal tendency is for electrons to move from a reduced form of an **electronegative** compound to an oxidized form of a more **electropositive** compound.

Compounds involved in the transfer of electrons can be compared on the basis of their tendency to accept or release electrons. The quantitative term for this comparison is the electrode potential, or oxidation–reduction potential. Since oxidation cannot proceed unless there is a corresponding reductant to accept electrons, we speak of an oxidation–reduction system composed of two pairs of compounds. The oxidized member of one pair accepts an electron and becomes reduced, while the reduced form of the other member donates the electron and becomes oxidized. A number of compounds in the chloroplasts have been identified as undergoing alternating reduction and oxidation as electrons pass through them during photosynthesis.

Although a knowledge of the minute details of the steps involved in the conversion of light energy to chemical energy during photosynthesis is not necessary for an appreciation of the significance of the process, a consideration of the major steps in the reaction as we now know them will help us to appreciate how the details of such a complex reaction are being unravelled. Eventually, when more is known about the process, man should be able to take this knowledge and to use light energy to produce food directly from the raw materials, carbon dioxide and water.

Under certain conditions, it is possible for a pigment molecule, excited by absorbed light energy, to transfer some of its energy of excitation to an adjacent molecule. In this way, a complex of molecules may act as a "funnel," transferring energy to a common molecule or "trap." This process of funnelling to a central reactive site all energy absorbed by many pigment molecules is a fundamental aspect of energy conversion in photosynthesis (Fig. 13.11). The absorption of light energy by chlorophyll starts a chain of events. An important step is the **excitation** of an electron in the photoreactive center (trap) of an assembly of pigment molecules. In one complex of molecules in the plastid—pigment system I— the photoreactive center is called P_{700} because light of 700 nm is strongly absorbed by this pigment. It is believed that the absorption of light energy by the pigment complex changes P_{700} from its ground state, with a standard reduction potential

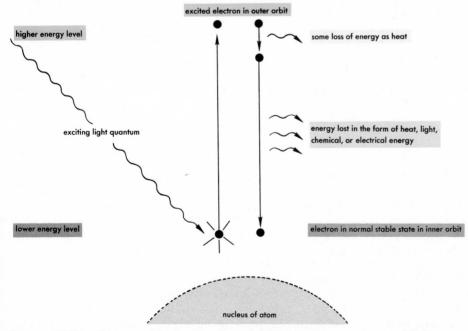

Figure 13.10 Diagram showing the fate of the energy in a quantum of light absorbed by a pigment molecule.

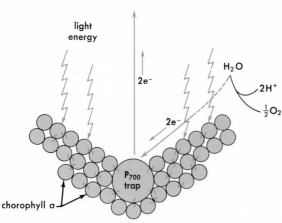

light
energy

$2e^-$

H_2O

$2H^+$

$\frac{1}{2}O_2$

$2e^-$

P_{700}
trap

chorophyll *a*

Figure 13.11 Diagram illustrating how a complex of pigment molecules may intercept light and funnel the absorbed energy to a common molecule, or trap. In this case, the trap is P_{700}.

of about $+0.4$ volt, to an excited state, with a potential of about -0.6 volt. In this excited state, the P_{700} molecule is strongly reduced and has a tendency to lose an electron to a suitable electron acceptor near it in the chloroplast. Once the electron acceptor has gained an electron, it is able to transfer the electron to an electron carrier chain. Note that the names of a number of compounds are given in this scheme (Fig. 13.12). These compounds are known to occur in chloroplasts and to undergo alternating reductions and oxidations as electrons flow through them. Their oxidation–reduction potentials are approximately shown in Fig. 13.12 by the scale on the left. The compounds with positive potentials are stronger oxidizing agents and tend, in their oxidized state, to accept electrons readily. The compounds listed opposite the more negative part of the scale are strong reducing agents in their reduced form and tend to give up electrons.

You will note also that there are two pigment systems and two light reactions shown in this scheme. Several lines of evidence indicate that there are two light reactions involved in photosynthesis in plants that evolve oxygen. First, these plants have two different chlorophylls, chlorophyll *a* and chlorophyll *b* or another form of chlorophyll. In addition, it was found by Emerson that the red light alone (light of 680 nm wavelength and longer) is very inefficient in photosynthesis, but the addition of light of shorter wavelength increased the efficiency of the long-wavelength light. Thus, it was postulated that there were two light reactions, one sensitive to longer-wavelength light, **photosystem I,** and the other to shorter-wavelength light, **photosystem II.** It is believed that photosystem II is associated with the evolution of oxygen from water.

It is also believed by most plant physiologists that these two photosystems are linked together in the chloroplast membrane by a series of electron carriers (Fig. 13.12). During the flow of electrons, some of the energy is trapped in the formation of ATP and some is trapped in NADPH.

In summary, light energy absorbed by photosystem I excites P_{700}, which loses an electron to an electron acceptor. Then the electron moves through an electron chain to NADP, which becomes reduced to NADPH. Some of the energy is stored in the NADPH. To replace the electron lost by P_{700}, an electron moves from an electron donor to P_{700} and electrons move through the electron carrier chain which in turn is supplied with an electron from the trap of photosystem II that has been excited by the absorption of light. In a way not yet clearly understood, an electron may move from water along another carrier system to replace the electron lost by photosystem II. The net result of this process is that light energy is used to transfer electrons from water to NADP. Oxygen is liberated.

Some of the energy is used to produce ATP. Here, energy is conserved in the phosphate bond formed by coupling of adenosine diphosphate (ADP) and inorganic phosphate (Pi). Water is eliminated in the process.

The two compounds NADPH and ATP are the chemicals in which the energy absorbed as light by chloroplasts is first stored as chemical energy. In final steps in the process of photosynthesis, the energy stored in these two chemicals drives the energy-consuming reactions that are involved in the reduction of CO_2 during the formation of sugar. ATP and NADPH are not accumulated in the cell in any appreciable amount, but are continually being formed and broken down as energy is transferred through them from chlorophyll to sugar molecules.

The production of ATP as a result of photochemical reactions like those just described is called **photophosphorylation.** Two types of photophosphorylation are recognized (Fig. 13.12). In **cyclic photophosphorylation,** the flow of electrons from light-excited chlorophyll molecules to electron acceptors proceeds in a cyclic fashion through the acceptors and back to chlorophyll. Only photosystem I is involved in this process. No oxygen is liberated and no NADP is reduced, since it does not receive electrons. ATP is formed during the electron flow, and light energy is converted into chemical energy in the ATP molecules (Fig. 13.13). However, since NADPH is not formed, cyclic photophosphorylation is not adequate to bring about CO_2 reduction and sugar formation. Cyclic photophosphorylation has been demonstrated, but its significance in photosyn-

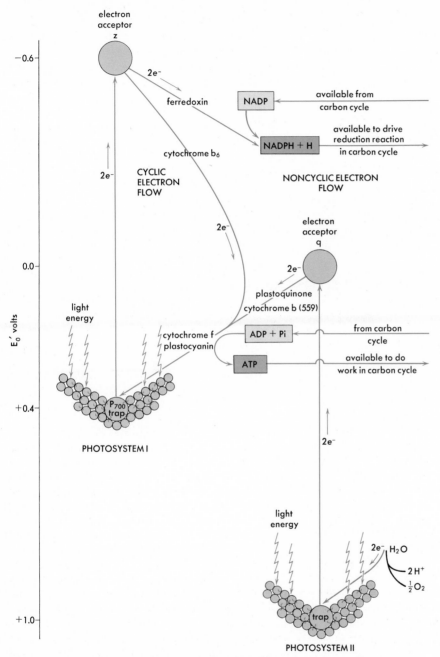

Figure 13.12 Diagram illustrating the probable arrangement of photosystems I and II. The two systems are connected by an electron-transport chain involving plastiquinone, cytochromes, and plastocyanin. Cyclic electron flow involves only photosystem I. ATP may be generated, but oxygen is not evolved nor is NADP reduced. Noncyclic electron flow involves both systems. Electrons from water move through the entire system of electron carriers. NADP is reduced, oxygen is liberated, and ATP may be formed.

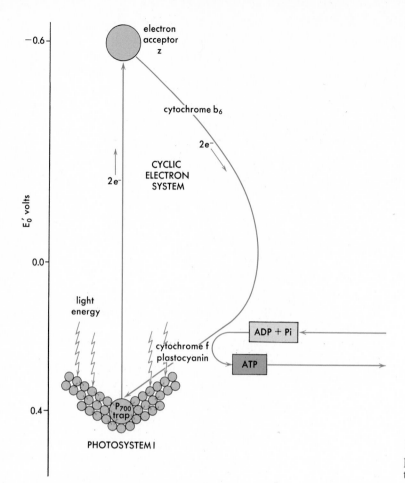

Figure 13.13 Cyclic
photophosphorylation, confined
to photosystem I.

thesis is limited to supplying energy in the form of ATP.

In the second type of photophosphorylation, both ATP and NADPH are produced. In this series of reactions, electrons from excited chlorophyll are trapped by NADP in the formation of NADPH and do not cycle back to chlorophyll. One postulated scheme for this **noncyclic photophosphorylation** is diagrammed in Fig. 13.14. Note that both photosystem I and photosystem II are involved in noncyclic photophosphorylation, and both ATP and NADPH are formed. These are the chemicals that are used to drive the CO_2 reduction reactions of photosynthesis.

Separating the Pigment Systems

In attempts to determine the relation of the photosystems to the ultrastructure of the thylakoid membranes in the chloroplasts in which they occur, investigators have used various techniques to break the membranes up into smaller particles and to test

these particles for the presence of one or both of the pigment systems. One of the most successful attempts has been accomplished by breaking up isolated chloroplasts with high-frequency sound and isolating the different-sized particles by density centrifugation. It was found that a preparation of small particles could be obtained that had the properties of photosystem I but not photosystem II, while another preparation of particles still contained both photosystems. Some research workers believe that the photosystem I particles were located in the stroma lamellae, while the other particles containing both photosystem I and photosystem II were located in the granal partitions (Fig. 13.6). This is one of many aspects of photosynthesis about which there is still much controversy, and active research is being conducted.

The Necessity of Enzymes for Photosynthesis

The complete process of photosynthesis will not take place in the dark, yet as we have seen, many

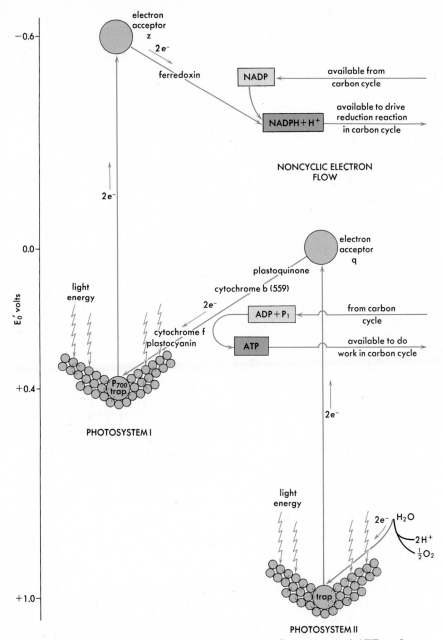

Figure 13.14 Diagram showing some steps in the process of ATP and NADPH production during noncyclic electron flow, involving photosystems I and II.

of the steps in the process are controlled by enzymes and are therefore not sensitive to light but are particularly sensitive to changes in temperature. Even in the production of NADPH, enzyme reactions as well as light reactions are involved. The process of carbon dioxide reduction to carbohydrates involves many enzyme reactions. All the enzymes directly concerned with photosynthesis occur in the chloro-

plast. Many of them, especially those linked to the carbon cycle, are water-soluble and are found in the stroma of the chloroplast.

One of the most extensively studied enzymes of photosynthesis (and the enzyme that probably occurs in higher concentration than any other enzyme in many leaf cells) is **carboxydismutase,** or ribulose diphosphate carboxylase. This is one of the most important enzymes in plants, since it catalyzes the first step in the entrance of CO_2 into organic compounds:

carbon dioxide + ribulose diphosphate

$$\xrightarrow[\text{carboxydismutase}]{} 2 \text{ phosphoglyceric acid}$$

In this reaction, CO_2 combines with the five-carbon atom sugar phosphate in the plastid to produce two molecules of the three-carbon atom acid, phosphoglyceric acid (PGA). This is a spontaneous reaction with little change in energy being involved.

Carbon Dioxide Reduction and the Formation of Sugar

Physiologists have long been interested in determining the sequence of carbon compounds that are formed in photosynthesis. Although early physiologists analyzed the carbohydrate content of photosynthesizing leaves and could easily recognize that starch as well as sugars usually accumulate during photosynthesis, it was not until two sensitive analytical techniques were developed that the actual sequence of reactions was understood. These new methods were the use of carbon dioxide in which the carbon was radioactive and the use of a special process, **chromatography,** with which the investigator could easily and accurately separate minute amounts of different organic compounds from one another. These techniques permitted the investigator to follow carbon in the series of reactions taking place in the plant.

The problem confronting physiologists was to determine the sequence of compounds (*A, B, C,* etc., Fig. 13.15) through which carbon atoms taken into the leaf pass during the synthesis of sugar. When the radioactive isotope of carbon (^{14}C) became available, scientists treated leaves with carbon dioxide containing radioactive carbon ($^{14}CO_2$) and

determined the carbon compounds formed. This may easily be done by the method outlined in Fig. 13.16. It is first necessary to prepare "maps" of the rates of movement of known compounds across the large sheets of filter paper. Once positions (on the filter paper) of known compounds, under controlled experiments, have been located, a drop of an unknown extract may be placed on the corner of the filter paper. After the usual washing and rotating, the presence of a spot at a definite position on the filter paper is taken to mean that a certain specific compound was present in the unknown sample. A group of scientists at the University of California under the direction of Melvin Calvin were particularly successful in using this technique for working out the early carbon pathways of photosynthesis. Calvin eventually received the Nobel Prize for this work.

The method finally adopted by this group as most successful was to grow algae in an inorganic nutrient medium, place such an algal culture in a glass flask in the light, expose the cells to radioactive $^{14}CO_2$, and kill the cells immediately in boiling alcohol. A similar technique of treating leaves with $^{14}CO_2$ could also be used (Fig. 13.17). The alcohol extract contained essentially all of the radioactivity incorporated into the cells after short time intervals. By varying the time to which the plant was exposed to $^{14}CO_2$, the first compounds synthesized during exposure to ^{14}C could be determined. Not only is photosynthesis very rapid, but also the first products are quickly changed into other substances. Thus, very short exposures to $^{14}CO_2$, in the range of 5–10 sec, were used. Fig. 13.18 shows an autoradiogram of the products formed when a tobacco leaf is treated with $^{14}CO_2$ for 10 sec.

If photosynthesis proceeds for an hour or so in an atmosphere containing radioactive carbon dioxide, most of the labeled carbon will be found in carbohydrates (sugar or starch). If photosynthesis is stopped after only a few seconds, most of the labeled carbon is found in a three-carbon-atom acid containing phosphorus, **phosphoglyceric acid** (PGA).

Phosphoglyceric acid, formed during photosynthesis, is thus an intermediate between carbon dioxide and sugar. By stopping photosynthesis at increasingly longer intervals (from a few seconds to several minutes), investigators have found that many carbon compounds become labeled with radioactive carbon.

It has been possible by means of such experiments to draw up an outline representing the various reactions that occur (Fig. 13.19). Here again it should be emphasized that a rote memorization of the sequence of all these reactions and of the intermediate compounds is not particularly useful, but an under-

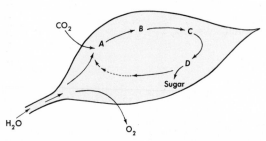

Figure 13.15 The carbon from carbon dioxide passes into several intermediate compounds (*A, B, C,* etc.) before sugar is formed in photosynthesis.

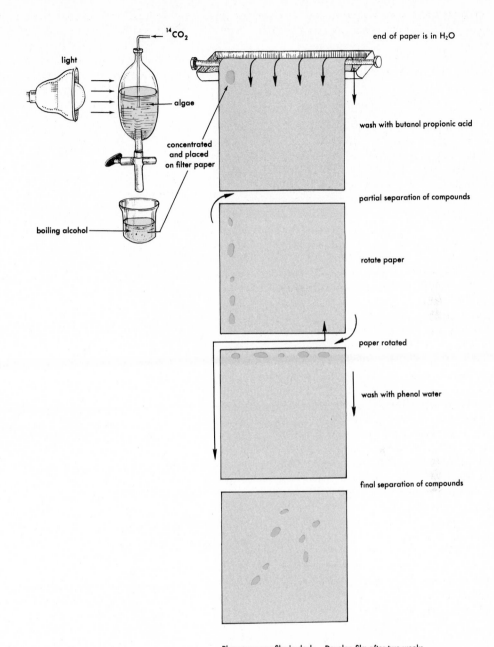

light

$^{14}CO_2$

algae

concentrated
and placed
on filter paper

boiling alcohol

end of paper is in H_2O

wash with butanol propionic acid

partial separation of compounds

rotate paper

paper rotated

wash with phenol water

final separation of compounds

Place on x-ray film in dark.   Develop film after two weeks.
Radioactive compounds on the paper are indicated by
darkened areas on the film.

final radiograph on the x-ray film

spots on film matched with paper
and compounds eluted and identified.

Figure 13.16   Diagram showing steps in
development of an autoradiogram as used in the
study of photosynthesis. Radioactive carbon is
supplied in the leaf, and metabolites are separated
by paper chromatography. ^{14}C produces a latent
image on x-ray film which, when developed, shows
the presence of the initial ^{14}C in the metabolites
formed by photosynthesis in the chloroplasts after
exposure to $^{14}CO_2$.

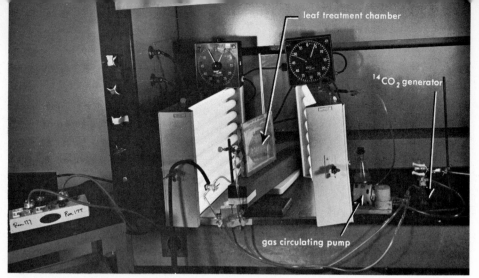

Figure 13.17 The apparatus used to treat a leaf with $^{14}CO_2$ in order to study the path of carbon in photosynthesis. The $^{14}CO_2$ is fed to the leaf in a closed system.

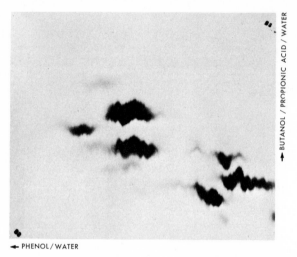

BUTANOL / PROPIONIC ACID / WATER

◄ PHENOL/WATER

BUTANOL / PROPIONIC ACID / WATER

◄ PHENOL/WATER

Figure 13.18 Autoradiogram of the products of photosynthesis after treating a tobacco leaf with $^{14}CO_2$ for 10 sec.

standing of the roles that this complex series of reactions play in the metabolism of the plant is important. The key points to note in this carbon cycle are the following:

1. Catalyzed by the enzyme carboxydismutase, CO_2 combines with **ribulose diphosphate,** a five-carbon-atom sugar phosphate continually being produced in the cell. Two molecules of phosphoglyceric acid (PGA), a three-carbon-atom compound, are produced.

2. The carbon cycle is associated at two points with high-energy products of photosynthesis, NADPH and ATP.

3. Two molecules of PGA are reduced to form two molecules of a three-carbon-atom sugar phosphate, phosphoglyceraldehyde. The energy that drives this reaction comes from NADPH and ATP, and energy from these molecules is stored in the newly formed triose sugar. ADP and NADP are regenerated.

4. Once the PGA is reduced, it may form another triose phosphate, dihydroxyacetone phosphate. The two triose phosphates may be combined to form a six-carbon-atom sugar phosphate, fructose diphosphate. The plant by this process has essentially added one CO_2 molecule to a five-carbon-atom sugar to produce one molecule of a six-carbon-atom sugar.

5. As this process continues, some of the fructose phosphate may be transformed through other reactions into other carbohydrates, including sucrose and starch.

6. Some of the fructose phosphate molecules are used to form more of the five-carbon-atom sugar phosphate, ribulose diphosphate. This in turn accepts more CO_2. Thus, a cycle of carbon com-

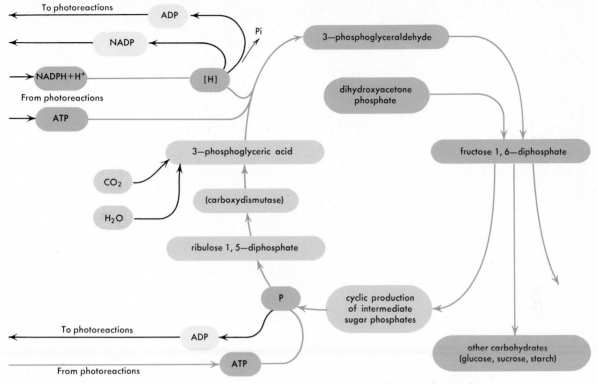

Figure 13.19 Diagram showing some steps in the carbon cycle of photosynthesis, the Calvin, or C_3, cycle.

pounds exists, with CO_2 from the air and hydrogen from water entering the cycle, and various sugars being produced.

This carbon cycle of photosynthesis is often called the Calvin cycle or the C_3 path of photosynthesis and appears to be of almost universal occurrence in plants that photosynthesize. However, it is reasonable to expect that during the evolution of plants in a wide variety of ecological situations, adaptations and modifications in metabolic processes would have occurred. Indeed, we do find that variations on the basic theme of photosynthesis have evolved. Two examples are found in succulent plants and plants of tropical origin.

Environmental Stress and Photosynthesis

Plants have adapted in a variety of ways to extreme environmental conditions existing in desert areas. In order to survive in these xeric conditions, plants must either endure recurrent drought or avoid drought by such means as carrying out the active part of their life cycle rapidly during the brief rainy periods. One group of plants, the succulents, have developed methods of storing and conserving water. The parenchyma tissue in succulent plants is highly developed, vacuoles are large, and intercellular spaces are reduced. When moisture is available, succulents absorb and store large quantities of water, and during periods of drought, they resist the loss of water to the environment.

In contrast to most mesophytic plants, many succulents have closed stomata during the day and open stomata at night. This adaptation is advantageous in reducing water loss during the day, when water stress is high, but it could be very disadvantageous for photosynthesis by reducing CO_2 uptake. However, these plants show a particular type of carbon metabolism. This is called **Crassulacean acid metabolism** (CAM), since plants belonging to the Crassulaceae were first seen to possess this process.

In the photosynthetic organs of at least some succulent plants, the total amount of organic acids in the vacuoles increases rapidly at night when stomata are open, and decreases during the day when stomata are closed. One of the major acids accumulated is malic acid, which is formed when CO_2 is absorbed in the dark. During the day, acid rapidly disappears, and carbon from it is incorporated into carbohydrate, probably through the addition of CO_2 to ribu-

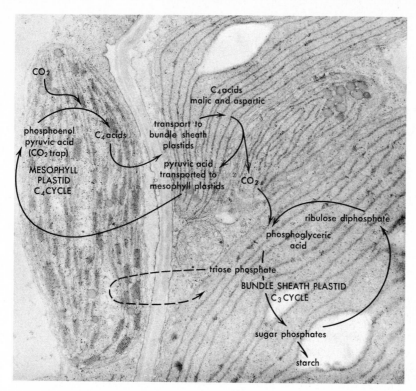

Figure 13.20 Distribution of C_4 and C_3 pathways in mesophyll and bundle sheath plastids of corn (*Zea mays*).

lose diphosphate to yield 3-phosphoglyceric acid in the usual C_3 cycle of photosynthesis. These plants thus have an effective mechanism of trapping CO_2 at night and releasing it and using it for photosynthesis inside the leaf in the day when the stomata are closed.

If we place a plant in the light in a closed chamber, we can determine how efficiently the plant can lower the CO_2 content of the air in the chamber. At some point, the CO_2 produced by respiration will just balance or compensate for the CO_2 absorbed during photosynthesis. The percentage of CO_2 remaining in the chamber under these conditions is known as the CO_2 compensation point. Thus, if a bean and a corn plant are placed together in a chamber in the light, the corn plant will successfully compete with the bean for the limited CO_2. Both will eventually die of starvation, but the bean plant will die before the corn plant does. Plants such as corn that have very low CO_2 compensation points also have other physiological as well as morphological features in common. In general, they seem to be adapted to grow in habitats with high light intensities and high temperatures, and they are more efficient in the use of water. They also have certain structural features in common such as the arrangement of large parenchyma cells in a sheath around the veins. Often, though not always, the plastids of the bundle sheath parenchyma cells lack or have

greatly reduced grana and frequently store starch. In contrast, the mesophyll cells between veins contain chloroplasts that have typical grana but little or no starch.

A special type of photosynthesis has been discovered in plants that have these specialized bundle sheath and mesophyll cells. Instead of CO_2 being taken up by the carboxydismutase reaction of the Calvin cycle, CO_2 is first fixed by another enzyme system, **phosphoenolpyruvate carboxylase.** This enzyme has a very strong affinity for carbon dioxide. In this system, CO_2 is added to the three-carbon-atom acid phosphoenolpyruvic acid (PEP) to form a four-carbon-atom acid, oxaloacetic acid. Subsequently, two other C_4 acids (malic and aspartic) may be formed from oxaloacetic acid. It is believed by many plant physiologists, though there is still strong difference of opinion, that the C_4 acids act as effective shuttles and move from mesophyll plastids to bundle sheath plastids where they release CO_2. They then return as a C_3 acid to the mesophyll, where another CO_2 can be picked up. The CO_2 released in the bundle sheath plastid can then enter the normal Calvin, or C_3, carbon cycle to form sugars (Fig. 13.20).

An interesting aspect of this idea is that there is a division of function between the two kinds of parenchyma cells. Mesophyll chloroplasts act as traps of CO_2 which is then (in the form of C_4 acids) shuttled to bundle sheath plastids, where it is reduced to sugars. This hypothesis requires a very extensive and rapid movement of solutes over considerable cellular distance and across a number of membrane barriers. This system was first extensively studied by Hatch and Slack, working in Australia, and it is known as the **Hatch–Slack,** or **C₄, pathway** of photosynthesis. It differs from the C₃, or Calvin, system in that it ensures a very efficient absorption of CO_2 and gives rise to a very low carbon dioxide compensation point.

Efficiency of Photosynthesis

Although the green leaf is the major organ in which light energy is used in the production of food used by man, it is not particularly efficient in utilizing the sun's energy. We speak of the efficiency of a machine, such as a diesel engine, a gasoline motor, or an electric motor. We calculate the energy value of fuel used and compare it with energy output of the machine. We know that much of the energy is lost. The ratio of energy outgo to energy intake represents efficiency. Of the total radiant energy that falls upon green leaves, about 80% is absorbed (Fig. 13.21). Of the remaining 20%, a part is reflected from the leaf surface and a part passes through the leaf. Part of the radiant energy absorbed is changed to heat and raises the temperature of the leaf; a large part of that absorbed is used up in transpiration; the remainder is utilized in photosynthesis and stored in carbohydrate molecules. Thus, only about 0.5–3.5% of all the light energy that falls on a leaf is used in the process of photosynthesis. By standards of machine efficiency, this is a very low percentage. But the supply of solar energy is continuous and abundant.

However, there is reason to believe that the leaves of higher plants, under the most favorable conditions, should be about ten times as efficient as this in converting solar energy into chemical energy by photosynthesis. This is a reasonably efficient chemical energy. Intensive research is needed to raise the low efficiency of photosynthesis of plants in the field to the efficiency that is theoretically possible.

The energy to form sugar from carbon dioxide and water can be measured. We know, for instance, how much energy is needed to separate the oxygen from the hydrogen in the water molecule. The oxygen-hydrogen bonds in water are among the strongest known. The bonds that hold together carbon and oxygen in carbon dioxide are also strong, but in sugar or starch they may be broken easily. Carbon dioxide and water do not burn, but sugar does. In other words, the heat of combustion of carbon dioxide and water is 0, but the heat of combustion of glucose is 686 kcal for each gram-molecule of glucose.

It is interesting to note that the plant is able to rearrange the elements carbon, hydrogen, and oxygen, which compose sugar, to form compounds of even higher energy content. For instance, sugar may be transformed into fats—compounds also containing carbon, hydrogen, and oxygen, but with relatively less oxygen and more energy per gram than is found in the carbohydrates.

Photosynthesis and Food Production

The large amount of research effort that has been devoted to solving the problems of photosynthesis has as its final goal the accumulation of adequate basic knowledge on the subject so that we will understand how nature uses energy to produce food from the abundant raw materials carbon dioxide and water. Once we understand this, a direct consequence will be our ability to artificially duplicate photosynthesis and hopefully to solve some of the problems of food production in the world.

Another approach to the solution of the problem of producing enough food for the world population is to develop, through plant breeding, varieties of plants that are very highly productive. This approach already has been so successful in some areas (e.g., the development of highly productive types of rice and wheat) that we speak of the "green revolution". The importance of this approach was recognized by the recent award of the Nobel Prize to Gustav Borlaug, who succeeded in breeding new strains of high-producing grains. Unfortunately, these strains require high levels of fertilizer application—an expensive and pollution-creating practice. There are also potential genetic problems associated with these grains (see Chap. 17).

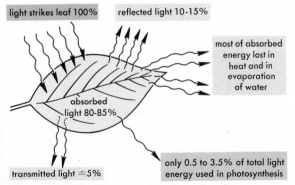

Figure 13.21 Diagram showing what happens to light that strikes a leaf.

A third approach to increasing the productivity of plants is to apply the principles of plant physiology to the cultural practices used in growing plants. This is the conscious or unconscious desire of everyone who grows plants whether he is a manager of an extensive farm or a home gardener.

Photorespiration

Scientists have wondered for a long time whether respiration proceeds in green cells at the same rate in the light as it does in the dark. Recently, it has been shown that light actually stimulates respiration in some plants. However, in these cases, respiration is different from the usual aerobic respiration carried out in these cells in the dark. Instead of mitochondria, specific microbodies called **peroxisomes** (Chap. 4) are involved in photorespiration.

Evidence indicates that glycolic acid (a two-carbon-atom acid) is an early product of photosynthesis in many plants. Glycolic acid production is stimulated under conditions of low O_2 and high CO_2 levels. Glycolic acid is able to diffuse from chloroplasts to peroxisomes, where it is rapidly oxidized. Since this oxidation is not coupled to the production of reduced nucleotide or to ATP synthesis, all the energy released is lost. Thus, it appears that photorespiration is a wasteful process. It has been shown that under some conditions over 30% of the carbon reduced during photosynthesis may be reoxidized to CO_2 during photorespiration. This is a very severe loss.

Some plants that have the C_4 acid cycle of photosynthesis show essentially no photorespiration under natural conditions. These plants, such as sugar cane and corn (Zea mays), are among the most efficient plants in trapping light energy.

Conditions Affecting the Rate of Photosynthesis

The rate of photosynthesis is of great importance in food and fiber production. The rate at which sugar or other carbohydrates is formed will affect the yield of any given crop.

Internal Factors

The rate of photosynthesis is influenced by a number of **internal factors,** that is, conditions inherent in the plant itself. Chief of these are (a) the structure of the leaf and its chlorophyll content; (b) accumulation within the chlorophyll-bearing cells of the products of photosynthesis; and (c) protoplasmic

influences, including enzymes. Considering the interplay of these various internal factors, it is evident that different species of plants vary considerably in their photosynthetic efficiency, even when growing under the same environmental conditions.

Leaf Structure. The structural features of the leaf influence the amount of carbon dioxide that reaches the chloroplasts. These features include size, position, and behavior of the stomata, and the amount of intercellular space. Also, the intensity and quality of the light that reaches the chloroplasts are influenced by thickness of the cuticle and epidermis, the presence of epidermal hairs, the arrangement of mesophyll cells, the position of chloroplasts in the cells, and so forth.

Products of Photosynthesis. With an increase in the concentration of the products of photosynthesis in mesophyll cells, there is a decrease in the photosynthetic rate. Starch may accumulate in chloroplasts during the day, when sugar is manufactured at a more rapid rate than it is transferred from the cell; a result of this accumulation may be a retardation in the rate of photosynthesis.

Protoplasm. The rate of photosynthesis is affected by conditions associated with the protoplasm itself. If the cells lack water and the protoplasm is dehydrated, photosynthesis slows down. Moreover, a disturbance of certain enzyme activities influences the photosynthetic rate.

External Factors

These factors as they influence the rate of photosynthesis are, to a degree, under the control of the plant grower. Crop yields can be influenced by modifying these factors. The principal external conditions that affect the rate of photosynthesis are (a) temperature; (b) light intensity, quality, and duration; (c) carbon dioxide content of the air; (d) water supply; and (e) mineral elements in the soil.

Temperature. Plants of cold climates carry on photosynthesis at much lower temperatures than do those of warm climates. The process is known to occur in certain evergreen species of cold regions, even at temperatures below 0°C. Algae in the water of hot springs may carry on photosynthesis at a temperature as high as 75°C. Most ordinary temperature-climate plants however, function best between temperatures of 10 and 35°C. If there is adequate light intensity and a normal supply of carbon dioxide, the rate of photosynthesis of most ordinary land plants increases with an increase in temperature up to about 25°C; above this range there is a continuous fall in the rate as the temperature is raised. At these higher temperatures, the time of exposure is of importance. At a given constant high temperature (for example, 40°C), the rate of photosynthesis decreases with time.

Under conditions of low light intensity, an increase in temperature beyond a certain minimum will not produce an increase in photosynthesis (Fig. 13.3). These conditions may occur in the winter in greenhouses. If the temperature is raised too high, the plants will suffer because the rate of photosynthesis has not been changed but respiration has been increased by the higher temperature.

Light. In discussing the effect of light upon the rate of photosynthesis, three elements must be considered: (a) intensity, (b) quality (wavelengths), and (c) duration. With proper temperature and sufficient carbon dioxide, carbohydrates produced by a given area of leaf surface increase with increasing light intensity up to a certain point (optimum light intensity), after which they decrease. It is not the intensity of light that falls upon the leaf surface that is of importance, as much as it is the intensity to which the chloroplasts are exposed. The light intensity diminishes from the leaf surface to the chloroplasts, owing to surface hairs, thick cuticle, thick epidermis, and other structural features.

Intense light appears to retard the rate of photosynthesis. Many plants that live in deserts and other places where the light is very bright often have structural adaptations which tend to diminish the intensity of light that reaches the chloroplasts. The usual light intensity in arid and semi-arid regions is well above the optimum for photosynthesis in many plants, especially introduced crop plants. In these regions on days when the sky is overcast, the light intensity is probably nearer the optimum for photosynthesis than on clear, sunny days. Leaves on the surface of plants receive light of greater intensity than those beneath that are shaded. Therefore, some of the leaves receive light of optimum intensity, whereas others may receive light either above or below the optimum.

Carbon Dioxide. The carbon dioxide utilized by land plants is absorbed by the leaves from the atmosphere. About 78% of the atmosphere is nitrogen and about 21% is oxygen, the remaining small percentage being composed of carbon dioxide, argon, and traces of hydrogen, neon, helium, and other gases. Surprisingly, only about 0.03% of the atmosphere is carbon dioxide.

Carbon dioxide in the air surrounding the leaves ultimately reaches the chloroplast. Its inward diffusion path is through the stomata to the intercellular spaces, through walls of palisade and spongy parenchyma cells, to the cytoplasm, and thence into the chloroplasts. The walls of palisade and spongy parenchyma cells contain water. Carbon dioxide is readily soluble in water; hence, carbon dioxide passes through the cell walls in aqueous solution.

In the process of photosynthesis, the cells remove carbon dioxide from solution in the cell sap. As a result, there is diffusion of that gas inward from the wet cell walls. This loss of carbon dioxide from the cell wall allows the water in the wall to dissolve more carbon dioxide from the air in the intercellular spaces. The carbon dioxide content of the intercellular spaces is lowered below that of the outside atmosphere. Diffusion of carbon dioxide inward through stomata tends to make up this deficiency. Thus, during active photosynthesis in the chloroplasts, a diffusion gradient for carbon dioxide is set up between the outside atmosphere and the chloroplasts. At the same time, oxygen liberated in photosynthesis is used in respiration, or it diffuses outward in aqueous solution through the cell walls to intercellular spaces and thence through stomata to the atmosphere surrounding the leaf (Fig. 10.6).

We may rightly consider the problem: If all the chlorophyll-bearing cells of the plants of the world are constantly taking carbon dioxide from the atmosphere during daylight, the quantities of this gas used must be enormous, and there must necessarily be processes in nature that are continually replenishing this supply. It is known that the amount of carbon dioxide in the air is low (0.03%) and that it remains fairly constant. It has been estimated that an acre of corn (10,000 plants) during a growing season of 100 days will accumulate 5585 lb of carbon; all this carbon is derived from the carbon dioxide of the atmosphere. It would require 20,480 lb of carbon dioxide to furnish this quantity of carbon. These estimates serve to emphasize the fact that enormous quantities of carbon dioxide are used in the photosynthetic process of green plants.

Obviously, the amount of carbon dioxide is limited. The present atmospheric supply would be used up in about 22 years were it not constantly being renewed. Several natural processes are continually releasing carbon dioxide to the atmosphere:

1. The living cells of all plants (both green and nongreen) and of all animals release carbon dioxide in the respiratory process (Chap. 14).
2. The dead bodies of plants and animals, and the excretions of animals, contain large quantities of carbon and other elements in the form of organic compounds; in the decay of these compounds, resulting from the activities of bacteria and fungi,

large quantities of carbon dioxide are released to the atmosphere.

3. Carbon dioxide is also added to the atmosphere when wood, coal, oil, gas, or any other carbon compound burns.
4. Carbon dioxide is released to the atmosphere from mineral springs and volcanoes.
5. The oceans are important reservoirs of carbon dioxide, and carbon dioxide probably escapes from the oceans whenever its concentration in the atmosphere decreases.

If light intensity and temperature are favorable, the carbon dioxide of the atmosphere frequently limits the rate of photosynthesis. This may be particularly true in greenhouses kept closed in the winter. Under these conditions, the carbon dioxide in the air may be reduced much below the 0.03% average.

It has been determined experimentally that at usual temperatures and light intensities an artificial increase of carbon dioxide up to a concentration of 0.5% may give an increased rate of photosynthesis, but only for a limited period. It appears that this high level of carbon dioxide is injurious to plants; after 10–15 days' exposure, the plants show injury.

Water. In land plants, the soil is the source of water. The water used by the green cells enters the roots from the soil and is conducted upward through the xylem of the roots and stems to the leaves. It moves through the xylem of the petiole and veins of the leaf blade to the finest veinlets. Then it passes from the tracheids of the veinlets to the mesophyll. Although the water content of an actively photosynthesizing cell is high and large amounts of water are lost from the cell by transpiration, only about 1% or less of the water absorbed by the roots is actually used in photosynthesis. However, the rate of this process may be changed by small differences in water content of the chlorophyll-bearing cells. In some instances, the rate of photosynthesis is increased by mild dehydration (15% water loss) and retarded by vigorous drying (45% water loss). Since stomata tend to close when the plant is deprived of water, conditions of drought tend to reduce the rate of photosynthesis. Thus, though water is one of the raw materials in the process, it rarely, if ever, is directly a limiting factor in photosynthesis.

Minerals. The chemical formula of chlorophyll a is $C_{55}H_{72}O_5N_4Mg$ and, of chlorophyll b is $C_{55}H_{70}O_6N_4Mg$. Thus, it is seen that the synthesis of chlorophyll depends upon a supply of nitrogen and magnesium, both derived from salts in the soil. Moreover, chlorophyll is not formed unless iron is available, although this element is not a component of the chlorophyll molecule. Leaves of plants deficient in nitrogen, magnesium, or iron are pale and yellow, a condition termed **chlorosis.** This abnormal condition may also be caused by other factors, but when it occurs, the rate of photosynthesis is lowered.

Summary of Photosynthesis

Some of the important points that have been mentioned concerning photosynthesis are summarized in the schematic diagrams shown (Figs. 13.12, 13.13, 13.14, 13.19, 13.20).

1. Photosynthesis is the major energy-storing process of life in which light energy is stored as chemical energy in organic compounds.
2. Carbon dioxide and water are the raw materials.
3. The products of photosynthesis are sugar and oxygen.
4. Light energy is absorbed by chlorophyll in the chloroplasts and drives the reaction of photosynthesis.
5. Two light reactions are involved.
6. Adenosine triphosphate (ATP) and reduced nicotinamide adenine dinucleotide phosphate (NADPH) are formed by photoreactions.
7. The hydrogen of the water is transferred to organic compounds and eventually reduces the carbon dioxide.
8. The oxygen of water is liberated as a gas.
9. Phosphorus is essential to the formation of sugars in plants and is found in some of the intermediate products, one of which is phosphoglyceric acid.
10. Many steps in the process of photosynthesis are controlled by enzymes.
11. Both external and internal factors affect the rate of photosynthesis.

14

Respiration

Every living cell in order to stay alive must break down complex organic molecules (foods) and obtain energy from them. This energy is supplied by cellular respiration. Differential permeability cannot be maintained without a supply of energy. Roots will not accumulate solutes, protoplasm will not move without energy from respiration. The synthesis of new cellular material such as amino acids, proteins, fats—growth itself—all require energy from respiration. The term respiration was first used to indicate the exchange of gases between an organism and its environment. Even today, respiration and breathing of animals are often popularly considered synonymous. Nevertheless, breathing is only an outward indication of the fundamental chemical reactions going on in the animal cells and characteristic of all life, i.e., the breakdown of food. This breakdown process, as we have seen in Chapter 5, is through oxidation reactions and often, though not always, yields carbon dioxide and water as end products. Today, the term respiration has come to have a wider and more fundamental meaning than mere gaseous exchange or breathing. In its broadest sense, **respiration** is defined as the oxidation of organic substances within cells and is accompanied by the release of energy. Some biochemists reserve the term respiration for the final sequence of reactions which terminate in the use of molecular oxygen. However, we will use respiration in its broader chemical sense as defined above.

The energy stored in food (largely sugars) is energy from sunlight that has been converted to chemical energy during photosynthesis. We shall see that some of the steps of respiration are the reverse of reactions occurring in photosynthesis. We will again meet sugar phosphates, three-carbon-atom compounds, ATP, and NADPH. But the energy pathways are very different. In photosynthesis, energy is stored during food synthesis. In respiration, energy is released during food breakdown.

In order for a cell to be able to use the energy stored in food, a mechanism must exist in the cell whereby some of the energy is captured in a form that can be delivered to the energy-consuming reactions of the cell. In other words, energy-yielding reactions are coupled or linked to energy-requiring reactions. Such mechanisms frequently result in the transfer of energy from food to the energy carriers ATP and NADH or NADPH (Fig. 14.1). To simplify Fig. 14.1, only ATP and NADPH have been shown. Whether NAD or NADP is reduced depends on the enzyme system.

Thus, cellular respiration is not a simple oxidation;

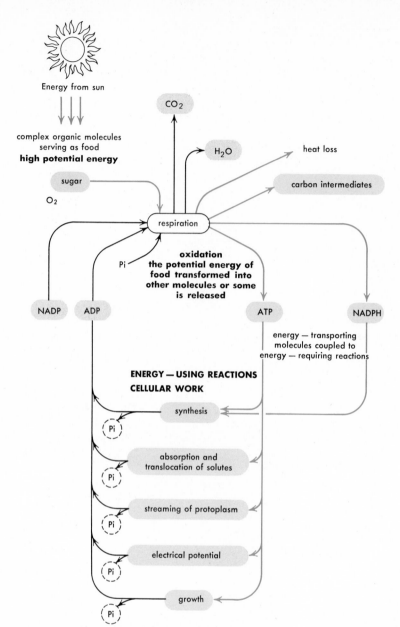

Figure 14.1 Energy from sunlight is trapped in food during photosynthesis. Some energy in the food is transferred during respiration to molecules (NADPH, ATP) that carry energy to sites of work in the cell.

it is a chain of chemical reactions marvelously linked to all cellular processes. To avoid the sudden release of large amounts of energy, the reaction is broken down into many small steps. The intermediate compounds formed along the way serve as starting points for other cellular processes. Specific enzymes catalyze each specific oxidation. ATP and NADPH cycle energy from energy-yielding sources to energy-requiring sites. The fact that specific enzymes are required for specific reactions and the fact that many steps in the whole sequence of events are bound to mitochondrial membranes ensures the orderliness characteristic of the process.

During respiration, some of the energy is liberated as heat which cannot be used by living cells and hence is lost. However, a significant amount of the energy in the food that is respired is conserved in energy carriers, e.g., ATP and NADPH or NADH. In these compounds, it can be transported to appropriate cellular sites where it can be utilized to perform the work of cellular metabolism.

When 1 gram molecular weight of glucose (180 grams) is burned to CO_2 and H_2O, 686 kcal of energy are released. Obviously, this amount of energy released at once would be destructive. Instead, energy transformations in cells occur slowly and in small steps through a sequence of many reactions. Each of these small reactions is catalyzed by its own specific enzyme. During several of these reactions, some of the energy in the food is used to combine a phosphate group with ADP, and ATP is formed. This phosphorylation reaction results in the storage of about 7 kcal of energy in the terminal phosphate bond in each mole of ATP formed. Since ATP is in solution in the cell, it may diffuse from one location to another, carrying its stored energy with it.

A second important link in the transport of energy between energy-yielding reactions of respiration and energy-requiring reactions of synthesis involves the reduction of NADP or NAD to NADPH or NADH and then their oxidation back to NADP or NAD. NADP, an electron-acceptor, can accept two electrons (and energy) given up by food during a few specific steps in respiration. The resulting NADPH molecule contains about 53 kcal (per mole) more energy than does NADP. This compound, like ATP, can diffuse from one point to another in the cell and carry energy from a source, oxidation of food, to a site where it is used to drive synthetic reactions. For example, the synthesis of hydrogen-rich compounds such as fatty acids uses energy obtained when NADPH is oxidized to NADP.

Thus, energy-yielding reactions are coupled to energy-requiring reactions by energy-transporting substances such as ATP and NADPH. Both ATP and NADPH yield energy that is used in the performance of cellular work. They become ADP and NADP. In these forms, they may move back to the site of energy-yielding reactions and capture energy again to become ATP and NADPH (Fig. 14.2).

Large amounts of ATP and NADPH are never formed in the cell at any one time, since the total amounts of ADP and NADP from which they are formed are small. The cell stores its energy reserves in the form of food.

If a cell is at rest and is not using energy rapidly, most of its ADP may be converted to ATP. Respiration may be very slow under these conditions. If such a cell is stimulated to do work (for example, to synthesize new materials, or to increase the rate of protoplasmic streaming, or the amount of salt uptake), then ATP is broken down. Some of the energy it contained is used to do work in the cell. ADP and inorganic phosphate are released. They become available to the respiratory machinery, where they may be synthesized again into ATP, trapping more energy. Under these conditions, the rate of respiration may be increased. This is one example of many control mechanisms that regulate the rates of various metabolic reactions in the cell. Fig. 14.3 shows some of these interrelationships.

When we consider the total energy cycle in the living cell, we see that stored chemical energy is moved from one part of the plant to another in the form of potential energy in the complex molecules that serve as food. At the cellular level, some of the potential energy in the food is trapped during respiration in high-energy, reactive molecules.

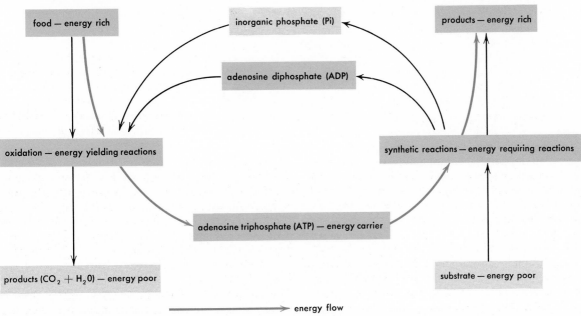

Figure 14.2   Diagram showing the coupling of energy-yielding reactions to energy-requiring reactions through ATP production and utilization.

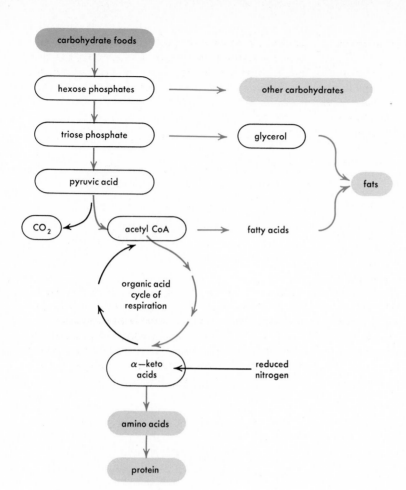

Figure 14.3 Some interrelationships between respiration and the synthesis of carbohydrates, fats, amino acids, and protein.

Respiration plays another very essential role in the life of the cell, for it is during the many steps in the respiratory breakdown of food that essential intermediate compounds are formed. Many of these intermediate compounds are used by living cells as sources of carbon building blocks from which new compounds such as proteins, fats, nucleic acids, and vitamins are synthesized. Respiration through the oxidation of food (a) supplies energy in an available form and (b) produces usable intermediate carbon compounds essential to the continued growth and metabolism of cells.

The Overall Process of Respiration

Let us now consider in more detail the mechanism by which the living cell breaks down food and traps energy in the form of the energy-transport molecules ATP and NADP. If one looks at the complete oxidation of a simple hexose (six-carbon-atom) sugar in the presence of molecular oxygen, only CO_2, H_2O, and energy (heat) are found as final products. This overall process may be written:

$$C_6H_{12}O_6 + 6\ O_2 \longrightarrow$$
sugar $+$ oxygen $\longrightarrow$

$$6CO_2 + 6\ H_2O + 686\ \text{kcal}$$
carbon dioxide $+$ water $+$ energy

The equation for respiration indicates the two reacting substances, sugar and oxygen, and the two usual end products, water and carbon dioxide; it tells nothing of the intermediate steps or of other possible end products.

In our consideration of this process, not all the detailed steps whereby sugar is completely oxidized will be discussed. Rather, the process of respiration will be considered as consisting of general phases, each of which represents a whole series of individual reactions. Furthermore, although these reactions have been studied in some organisms, it is not necessarily true that the course of respiration follows the same pathway in all plant cells.

Biochemists and physiologists have been able to take the process of respiration apart and thus have determined the individual steps in the reaction. They have characterized the enzymes and cofactors involved. However, it is more important for us to understand how the essential steps in the process fit together and result in the slow release and trapping of energy than it is to try to remember all the steps of the process.

Respiration may be separated into distinct phases, each of which involves many reactions. In the first phase, for every molecule of the sugar glucose which is oxidized, two molecules of **pyruvic acid** (a three-carbon-atom organic acid) are formed. No molecular oxygen is involved, and the energy change is small. This first phase is called **glycolysis.** The fate of the pyruvic acid depends upon the presence or absence of molecular oxygen. If oxygen is present, **aerobic respiration** occurs and the pyruvic acid formed by glycolysis is oxidized, stepwise, to CO_2 and H_2O, with the release of all the available energy that it contains. If molecular oxygen is not present, **anaerobic respiration** ensues in most plant cells, and CO_2 and alcohol are formed. Most of the energy remains in the alcohol and only a small amount is released.

Glycolysis

There are three distinct steps in the process of glycolysis (Fig. 14.4).
1. The preparation of the sugar for reaction by the addition of phosphorus—**phosphorylation.**
2. The splitting of the sugar into two fragments—**sugar cleavage.**
3. The oxidation of the fragments to form an intermediate product of respiration—**pyruvic acid formation.**

Preparation of Sugar for Reaction—Phosphorylation. Sugar itself does not combine readily with oxygen at temperatures at which life reactions proceed, nor is it easily broken down into intermediate products. You will recall that phosphorus compounds serve as intermediate steps in photosynthesis and that sugar

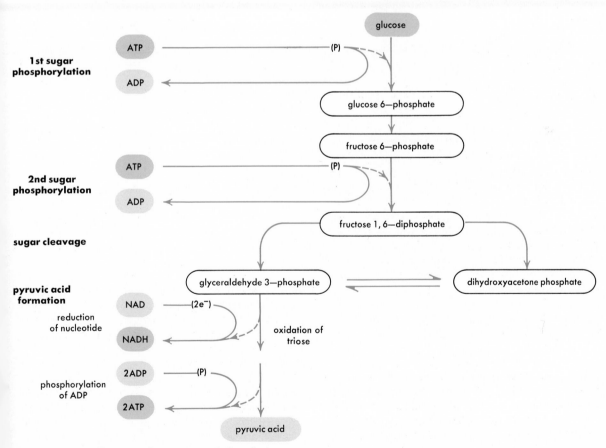

Figure 14.4 Steps in the process of glycolysis.

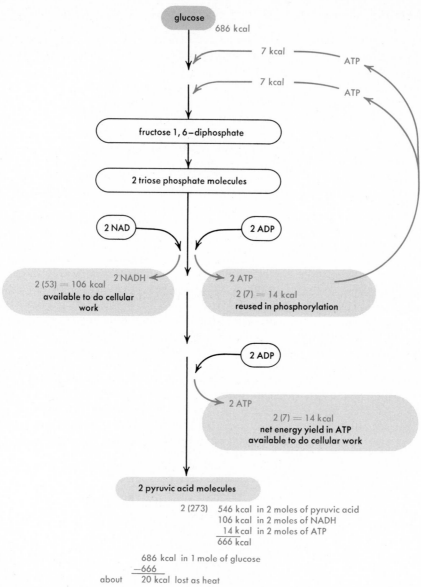

glucose
686 kcal

7 kcal — ATP

7 kcal — ATP

fructose 1, 6–diphosphate

2 triose phosphate molecules

2 NAD   2 ADP

2 NADH
2 (53) = 106 kcal
available to do cellular work

2 ATP
2 (7) = 14 kcal
reused in phosphorylation

2 ADP

2 ATP
2 (7) = 14 kcal
net energy yield in ATP
available to do cellular work

2 pyruvic acid molecules

2 (273)   546 kcal in 2 moles of pyruvic acid
          106 kcal in 2 moles of NADH
           14 kcal in 2 moles of ATP
          666 kcal

          686 kcal in 1 mole of glucose
         −666
about   20 kcal lost as heat

Figure 14.5 Energy summary of glycolysis of 1 mole of glucose. All energies are approximate kilocalories per mole.

phosphates are formed. So also in respiration, before sugar can be broken down and its stored energy utilized, it must be combined with phosphorus to form sugar phosphate. This preparation of sugar in the process of respiration is called **phosphorylation** and, like the other steps in respiration, is controlled by enzymes. Phosphorylation is an energy-requiring reaction, and during the process, one of the three phosphate groups of an ATP molecule is transferred to the sugar molecule that becomes phosphorylated. Actually, two phosphorylation steps are involved before a hexose sugar is ready for sugar cleavage in the normal glycolytic process. Thus, two ATP molecules are involved in providing energy for phosphorylation of the sugar molecule. Phosphorylation results in the formation of a sugar molecule linked to a phosphate group at each end (fructose 1,6-diphosphate).

Sugar Cleavage. An enzyme present in the cytoplasm of the cell is capable of catalyzing the splitting of fructose 1,6-diphosphate into two different three-carbon-atom sugars (trioses), dihydroxyacetone phosphate and D-glyceraldehyde 3-phosphate.

Pyruvic Acid Formation. These trioses are in equilibrium and may be converted one into the other. One of them, glyceraldehyde 3-phosphate, is oxidized to **pyruvic acid** and as it is broken down, the equilibrium shifts so that more glyceraldehyde 3-phosphate is formed. Thus, both triose phosphates are available for the reaction.

Several enzymatic steps are involved in pyruvic acid formation. Although molecular oxygen does not take part, the triose phosphate is oxidized by the transfer of two of its electrons and hydrogens to a hydrogen acceptor. In this case, NAD and not NADP is the usual hydrogen acceptor. Some of the energy originally in the triose is now in the newly formed NADH. Also, during the oxidation of each triose phosphate molecule to pyruvic acid, energy is used to form two molecules of ATP from two molecules of ADP and two phosphates. A total of four ATP and two NADH molecules are formed during the oxidation of two trioses.

The importance of glycolysis for the cell lies in the fact that during the stepwise phosphorylation, cleavage, and oxidation of sugar to pyruvic acid, usable intermediate compounds are formed, and although only a small amount (approximately 17%) of the energy has been trapped in energy-carrier molecules, only about 3% has been lost as heat (Fig. 14.5). The major part, almost 80%, of the energy still lies trapped in the final product, pyruvic acid.

The fate of the pyruvic acid depends on whether molecular oxygen is or is not available to the cell.

Anaerobic Respiration

Normally, higher plants are not able to live long in the absence of oxygen. Under this condition, the food is not sufficiently oxidized to yield enough energy for life processes, and certain products formed may be poisonous. In contrast, some fruits, notably apples, may be held for long periods in an atmosphere containing very small amounts of oxygen and continue to give off carbon dioxide. Yeast may live actively in an atmosphere with very small amounts of oxygen and produce relatively large amounts of carbon dioxide and alcohol.

But, yeasts have a limited tolerance for alcohol. When the alcohol concentration in the medium in which they are living reaches about 12%, the yeast cells are killed. This is why wines and other naturally fermented alcoholic beverages do not have an alcohol content above about 12%. Yeast grows much more luxuriantly under aerobic conditions than in the absence of oxygen. Indeed, there are only a relatively few organisms that grow only under strictly anaerobic conditions.

Some fungi, bacteria, and many animal cells may live and grow slowly under anaerobic conditions, where they can produce many products in addition to alcohol. When oxygen is not supplied rapidly enough to vigorously exercising muscles in your body, the muscle cells may produce, not alcohol, but lactic acid. It is the presence of lactic acid that makes you stiff after hard muscular work.

Under anaerobic conditions in higher plants, the pyruvic acid formed during glycolysis is usually converted into carbon dioxide and ethyl alcohol. This process, called **alcohol fermentation,** takes place in two steps. First, one carbon dioxide molecule is enzymatically split off from each pyruvic acid. This leaves a two-carbon-atom compound, acetaldehyde. Next, NADH, formed during glycolysis, is now used to reduce acetaldehyde to alcohol. NAD is released to take part in glycolysis again.

Thus, during anaerobic respiration, the energy trapped in NADH is used to form alcohol. It is significant in the life of the cell that for each glucose molecule fermented to alcohol a net of only two ATP molecules are available to do work in the cell. This is about 14 kcal, or less than 3% of the energy available in 1 mole of glucose. About 6% is lost as heat, while almost 84% is still locked in the alcohol and is unavailable to the plant (Fig. 14.6). In addition, the alcohol itself may be toxic to the plant cell. It is obvious that anaerobic respiration is not a very efficient way of utilizing energy in the food that is respired.

Aerobic Respiration

Since glycolysis results in the release of only about 20% of the energy available in a glucose molecule, 80% still is locked in the two pyruvic acid molecules formed. In contrast to anaerobic respiration, aerobic respiration in a plant cell yields adequate amounts of energy so that the cell can carry out the energy requirements of life. During aerobic oxidation of pyruvic acid to CO_2 and H_2O, between 40 and 50% of the energy is trapped in a potentially useful form in ATP or reduced nucleotide. How is this accomplished?

We may conveniently divide aerobic respiration into three main steps.

1. *Entrance of carbon into the organic cycle of respiration.* CO_2 is released from pyruvic acid, the remaining two-carbon-atom "fragments" enter into the organic acid cycle, and NADH is formed.
2. *The organic acid cycle of respiration.* The two-carbon-atom "fragment" is oxidized, CO_2 is released, and ATP, NADH, and NADPH are formed.
3. *Electron transport and terminal oxidation.* Re-

The Overall Process of Respiration   **239**

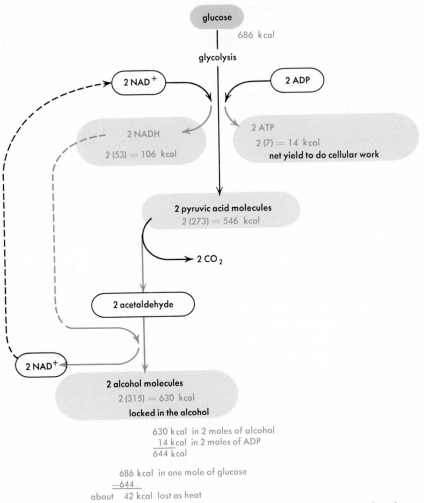

glucose

686 kcal

glycolysis

2 NAD$^+$

2 ADP

2 NADH
2 (53) = 106 kcal

2 ATP
2 (7) = 14 kcal
net yield to do cellular work

2 pyruvic acid molecules
2 (273) = 546 kcal

2 CO$_2$

2 acetaldehyde

2 NAD$^+$

2 alcohol molecules
2 (315) = 630 kcal
locked in the alcohol

630 kcal in 2 moles of alcohol
14 kcal in 2 moles of ADP
644 kcal

686 kcal in one mole of glucose
−644
about 42 kcal lost as heat

Figure 14.6 **Energy summary of the anaerobic oxidation of 1 mole of glucose.**

duced coenzymes are oxidized, ATP and water are formed, and molecular oxygen is used.

Entrance of Carbon into the Organic Cycle. One of the most complex series of reactions of respiration is involved in the oxidation of pyruvic acid. In these reactions, several vitamins, particularly those of the vitamin B complex (niacin, thiamine, pantothenic acid), serve as coenzymes or parts of coenzymes (p. 62). In the initial stages, for each molecule of pyruvic acid oxidized a molecule of CO_2 is formed, and two electrons, two hydrogen atoms, and about 53 kcal of energy are transferred to NAD, forming NADH. The remaining part of the pyruvic acid is now a two-carbon-atom "acetate fragment"

$$(H-\overset{\overset{\displaystyle H}{|}}{\underset{\underset{\displaystyle H}{|}}{C}}-\overset{\displaystyle O}{\overset{\displaystyle \|}{C}}-\text{ or acetyl group})$$. It becomes associated

with another coenzyme, coenzyme A, in the form of acetyl-CoA. In the presence of a specific enzyme, coenzyme A is capable of combining with (accepting) such acetyl groups and transferring (donating) them to other acceptor molecules.

In this way, the two-carbon-atom acetyl group (still containing about 20% of the energy present in the original glucose molecule) may be transferred from one series of reactions to another in the cell (Fig. 14.7). This is analogous to the transfer of electrons and hydrogen (energy) by the coenzymes NADH and NADPH.

The actual entry of the reduced carbon into the organic acid cycle, where its energy can be utilized to form ATP, is accomplished by the donation of the acetyl group from acetyl-CoA to an acceptor molecule, oxaloacetic acid, found in mitochondria. Oxaloacetic acid is a four-carbon-atom acid which combines with the acetyl group and forms the six-carbon-atom acid, citric acid. The acetyl group has now entered the organic acid cycle of respiration, where the mitochondrial machinery can act on it and release its potential energy.

Although pyruvic acid is the usual donor of acetyl groups to coenzyme A, it is not the only donor. The breakdown of fats and proteins (amino acids) results in the formation of acetyl groups that also may be donated to coenzyme A and then enter the acid cycle. Thus, fats and proteins may also serve as foods and supply energy for the energy-requiring processes of life.

The Organic Acid Cycle of Respiration. The new acid, citric acid, formed when the acetyl group of acetyl-CoA combines with the oxaloacetic acid, is gradually broken down. This process involves a cyclic series of reactions in which seven other organic acids including the acetyl acceptor, oxaloacetic acid, are formed (Fig. 14.8). During these reactions, pairs

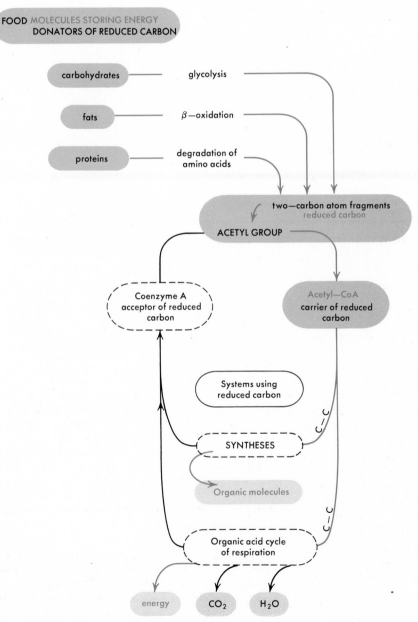

Figure 14.7   Diagram illustrating the role of coenzyme A as a carrier of acetyl groups in the process of respiration.

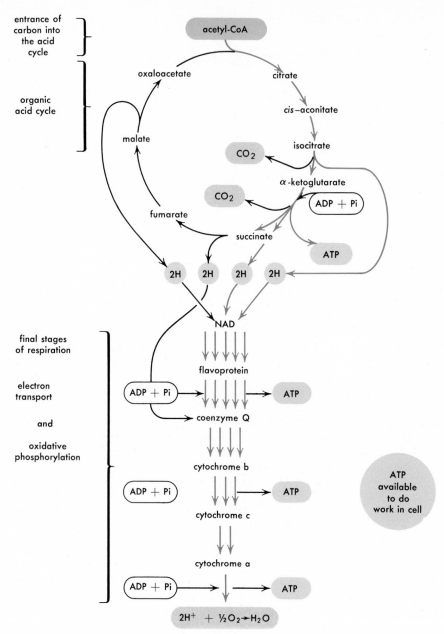

Figure 14.8 Diagram showing relation between organic acid cycle, electron transport, and oxidative phosphorylation.

of electrons (and hydrogen atoms) are transferred to electron carriers, the coenzymes NAD (NADP) and FAD (flavine adenine dinucleotide), see p. 62. Two molecules of CO_2 are produced, and a new oxaloacetic acid capable of accepting another acetyl group is formed.

Energetically, in one turn of the cycle, the potential energy in one acetyl group is released or trapped in other molecules. Part of it is lost as heat, but the rest is transferred to one molecule of ATP, three

molecules of NADH (NADPH), and one molecule of $FADH_2$. The ATP is free to carry its energy to other parts of the cell where work is being done, but the energy in NADH and $FADH_2$ generally is used in the synthesis of more ATP in the final stages of respiration. Of the total available energy in the original acetyl group, about 66% is trapped in ATP and reduced nucleotides that are produced when the acetyl group is oxidized.

Because this is a repeating process and because

organic acids are involved, it has been called the organic acid cycle of respiration. Sometimes it is called the citric acid cycle, the TCA cycle, or more frequently the Krebs cycle, after the physiologist whose research contributed a great deal to our knowledge of respiration. It is believed that in many plants, animals, and microorganisms, such a cycle occupies a central position in aerobic respiration. Although all the energy originally associated with the reduced carbon atoms in the sugar molecule (food) is released or transferred when the Krebs cycle is completed, only a relatively small amount, less than 10%, is directly trapped in the ATP formed during glycolysis and during the Krebs cycle. About 56% is first trapped in reduced nucleotides. How does the cell utilize the energy trapped in the reduced nucleotides?

The Final Stages of Oxidation—Electron Transport. The energy present in the reduced nucleotides usually is not utilized directly by the cell. In this case, the nucleotides are oxidized stepwise, and part of their energy is temporarily trapped in ATP. In this final series of reactions, pairs of electrons are transferred from the reduced coenzymes through a series of compounds that alternately are reduced and then oxidized. Several of these electron carriers are cytochromes and contain iron atoms that are alternately reduced to the ferrous (Fe^{2+}) form and then give up their electrons, becoming oxidized to the ferric (Fe^{3+}) form. The last step involves the transfer of a pair of electrons to an oxygen atom. Two hydrogen ions from the cellular environment then combine with the oxygen. Water is formed. If free oxygen is not present in the cell, this last transfer of electrons could not take place. The flow of electrons in the chain would cease and the entire aerobic respiratory sequence would stop.

In the electron transport process, three molecules of ATP may be synthesized for each NADH or NADPH that is oxidized, and two molecules of ATP may be synthesized for each $FADH_2$ oxidized. This ATP synthesis that is linked to oxidation reactions of respiration is known as **oxidative phosphorylation.**

Complete aerobic respiration of a glucose molecule thus involves a stepwise breakdown of the sugar molecules. Six CO_2 molecules and six H_2O molecules are formed. Six O_2 molecules are absorbed and about 40% of the potential energy in the glucose is transferred to ATP, which can be utilized to do work in the cell.

Alternate Pathways of Respiration

Although aerobic respiration is the most common method by which plant cells oxidize foods, alternate

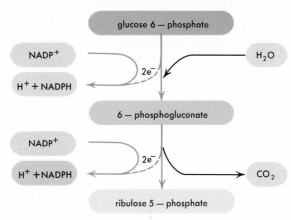

Figure 14.9 Diagram showing the first steps in the pentose phosphate pathway.

pathways do occur. One such pathway is known as the pentose phosphate pathway (PPP), or the hexose monophosphate shunt. An important result of this pathway is the production of NADPH and a five-carbon-atom sugar phosphate, ribose phosphate, in the cytoplasm of the cell. While the ribose phosphate may be broken down in later reactions, the NADPH is available for utilization in important synthetic reactions in the cell.

Biochemically, the PPP starts in the same way as glycolysis, i.e., with the phosphorylation of a glucose molecule to produce glucose 6-phosphate. In the PPP, however, in the presence of NADP and a specific enzyme, glucose 6-phosphate is oxidized (Fig. 14.9). Two molecules of NADPH are formed. One molecule of CO_2 is liberated, and a molecule of ribulose 5-phosphate is produced. The rest of the cycle involves a series of sugar transformations that is essentially identical to those found in the Calvin cycle of photosynthesis. Glucose may be regenerated in the cycle.

It is possible by studying the oxidation of glucose when the first carbon atom or the sixth carbon atom in the molecule is radioactive (^{14}C) to determine how much of the glucose is being oxidized by the PPP and how much is passing through the normal Krebs cycle. Young growing tissues appear to use the Krebs cycle as the predominant pathway of glucose oxidation, while aerial parts of the plant and older tissues seem to utilize the PPP as well as the Krebs cycle.

Respiration and the Synthesis of Organic Molecules

We have seen that all living cells respire and that respiration provides chemical energy and certain intermediate compounds, both needed for many cell functions. This intimate relationship between respiration and other cellular reactions can readily be seen when one considers the formation of carbohydrates, fats, amino acids, and proteins in the cell. All these compounds are ultimately derived from the sugars produced during photosynthesis.

During development of a fat-storing seed such as squash, the seed may store as much as 20% of its dry weight as fat (see Fig. 16.28). However, it is sugar and not fat that is translocated through the phloem into the seed. Obviously, a mechanism exists in the cells of the seed to convert sugar into fat. How is this done?

Fat molecules are made up of two parts: (a) a three-carbon-atom alcohol (glycerol), combined with (b) one or more organic (fatty) acids (Fig. 5.5). These two parts of a fat molecule are synthesized from respiration intermediates; glycerol may be synthesized from the cleavage products of glycolysis, and fatty acids are formed by a complicated series of enzymatic reactions acting on the acetyl-CoA formed when pyruvic acid loses CO_2 (Fig. 14.7). The production of fat is further tied to respiration, since the energy from ATP as well as from NADP is needed for fat synthesis.

Similarly, amino acid synthesis is linked to respiration. Some of the same organic acids that are formed as intermediates during the cycle of respiration are used in amino acid synthesis. Again, the ATP produced by respiration is used for the necessary activation of the amino acids used in protein synthesis.

Only a few of the many sugars and other carbohydrates found in a plant cell are directly products of photosynthesis. The interconversion of one sugar into another or the polymerization of a sugar in the synthesis of a high-molecular-weight carbohydrate such as starch or cellulose involves the activation of the sugar. Energy for these sugar interconversions is again derived from respiration.

Respiration and Cell Ultrastructure

We know that photosynthesis occurs within the chloroplasts. We may ask ourselves, "Where in the cell does respiration take place?" In a very active cell, such as a meristematic cell, the rate of respiration is very high. If the cell is in an inactive tissue, such as in a dried seed, the rate of respiration is very slow indeed. Just as investigators of photosynthesis have literally taken the cell apart in order to study its photosynthetic apparatus, the cell has been broken apart, and each part (soluble phase, nucleus, chloroplasts, dictyosomes, microbodies, and mitochondria)

has been studied to determine its role in cellular metabolism.

We have seen that the complete oxidation of a sugar molecule involves a whole series of reactions. Some of these, particularly those of glycolysis, and the PPP take place in the soluble phase of the cytoplasm. These reactions do not appear to be strongly associated with cell organelles or membranes. Others, particularly those involving the aerobic oxidation of pyruvic acid (the acid cycle, terminal oxidation, and oxidative phosphorylation steps of respiration), are carried out by mitochondria.

Mitochondria have been isolated from plant as well as animal cells and when properly prepared will break down pyruvic acid in the presence of oxygen. Carbon dioxide is produced, and ATP may be synthesized during the oxidation reactions. Thus, respiration can be studied outside the cell in the test tube, with preparations from living cells.

Mitochondria, though so small that little structural detail can be seen with the light microscope, nevertheless have a distinctive structure that can be recognized easily by use of the electron microscope. Fig. 14.10 shows a mitochondrion from a companion cell of squash (Cucurbita) and from the alga Carteria. Here, the characteristic double boundary membrane that separates the mitochondrion from the rest of the cytoplasm is clearly seen. The circular and elongated dark bodies inside the mitochondrion are sections through an internal membrane system that is made up of finger-like protuberances extending inward from the inner mitochondrial membrane. It is believed that many of the enzymes necessary for respiration are located on this membrane system in a definite relationship to each other. We see that the respiratory activity of mitochondria is closely associated with its structure in a manner similar to that in which the photosynthetic activity of chloroplasts is related to their structure. We see also that an integration of activities exists between the cytoplasm, where the early stages of food breakdown occur, and the mitochondria, where the final stages of aerobic respiration take place.

During the life of the cell, there is a continuous traffic of molecules across the mitochondrial membranes. Pyruvic acid, inorganic phosphate ions, and ADP molecules are three important substances moving into the mitochondria from the cytoplasm. Here, certain of the energy-releasing reactions during the final stages of respiration are coupled to the synthesis of ATP from the ADP and inorganic phosphate. The newly formed ATP, carrying with it some of this energy, may move from the mitochondria to some other place in the cell. As work is done there, the ATP is hydrolyzed to ADP and inorganic phosphate,

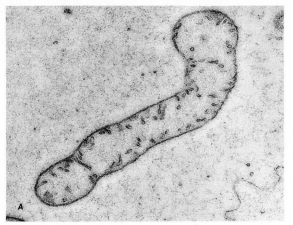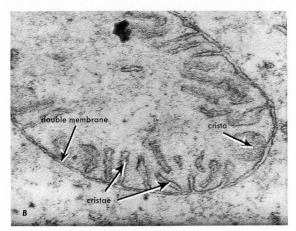

Figure 14.10 *A*, electronmicrograph of a thin section of a mitochondrion, the cellular site of aerobic respiration. Mitochondrion is from companion cell in phloem of squash (*Cucurbita*), ×8000. *B*, portion of mitochondrion from *Carteria* showing the double mitochondrial membrane and the inward extension of the inner component of the membrane to form the cristae, ×43,000. (*A*, courtesy of K. Esau; *B*, courtesy of Carole Lembi.)

which again may enter a mitochondrion. In this way, chemical energy in the substrate (pyruvic acid) molecules is continuously moving into the mitochondria, where it is transformed into a more readily usable form of energy (ATP). This in turn is constantly moving from the mitochondria to all parts of the cell.

The Environment and Respiration

Since respiration is the process by which some chemically stored energy in foods is transformed into a form immediately available to do work in the cell, any conditions either inside or outside the cell that affect the rate of respiration will influence cellular activities.

Cell Hydration

Many seeds remain viable for long periods of time when stored in relatively dry air. The water content of the cells at this time may be less than 10% of the seed weight. In contrast, the water content of active protoplasm may be as high as 90% or more. Under low moisture conditions, respiration and some other cellular activities slow to a very low rate; others stop completely. Respiration goes on at a very slow rate, so little oxygen is consumed, little carbon dioxide is given off, and only very minute amounts of heat are released. Energy is needed only to maintain a certain little-understood steady state in the quies-

cent protoplasm. Growth has ceased, mineral salts are not being absorbed, and cell divisions have stopped, as have many, if not all, reactions involving synthesis of new materials. Because of this dormant state, seeds may be kept in large storage bins such as grain elevators for long periods of time.

If even a small amount of water is added to the seeds, imbibition occurs, the seeds swell, respiration increases rapidly and, if the seeds are closely confined, as in a grain elevator, the temperature inside the mass of seeds that are insulated from the outside may increase to a point where the seeds will be killed. In this example, the rate of respiration is greatly influenced by the water content of the cell. A number of other factors, both external and internal, influence the rate at which a cell respires and consequently affect all the cell's activities that depend upon the use of energy released in respiration. In addition to the degree of hydration of the cell, factors such as temperature, oxygen supply, food availability, carbon dioxide concentration, the presence of certain chemicals, and the age of the cell commonly influence cell respiration.

Temperature

Temperature has a particularly marked effect upon most biological reactions, especially those that are controlled by enzymes. In the temperature range from near freezing (0°C) to about 30°C, an increase of 10°C approximately doubles the rate of respiration. Above 30°C the harmful effects of high temperature on the cell may become marked. At these higher temperatures, cellular enzymes progressively become inactivated and respiration decreases.

Nevertheless, over the long time of evolution, certain organisms have evolved characteristics that have enabled them to survive in otherwise hostile environments. In this way, certain species of algae and bacteria are adapted to respire and grow under temperature extremes that would kill unadapted species. Particularly impressive is the presence of some of these organisms in hot springs and streams where temperatures may exceed 60°C.

Food

Since respiration involves the utilization of food, the presence of easily available food is necessary if a cell is to continue to respire and stay alive. Photosynthesizing leaf cells, of course, are able to produce their own food in the light. Under normal conditions, they must also form enough food to supply the needs of all the nonphotosynthetic cells in the plant. If a plant is kept in the dark, continued food usage without food synthesis will rapidly deplete the available food. Stored food reserves, particularly starch and sugars, may be drawn upon by all living cells. If the plant is prevented from carrying out photosynthesis and if food is not supplied, the plant will eventually starve to death. In fact, any factor which limits photosynthesis and thus food availability must indirectly influence respiration.

Oxygen and Carbon Dioxide

Although most plant cells can continue for a time to oxidize foods even in the absence of gaseous oxygen (anaerobic respiration), molecular oxygen is normally necessary for the health of higher plant cells. Rarely does the concentration of oxygen in the atmosphere deviate enough from the normal 21% to appreciably affect the rate of respiration. However, underground stems, seeds, and roots may be in an oxygen-poor environment, since the microorganisms in the soil as well as the plant parts themselves may use the oxygen in the soil atmosphere faster than it is replaced from the air. Under these conditions, respiration in the roots may be decreased. The effect of aeration on root growth is shown in Fig. 11.3. Similar conditions of low oxygen level and high carbon dioxide concentration may occur in the internal cells in bulky plant organs such as large fleshy fruit.

A knowledge of the effects of the gas environment on respiration of stored fruits and vegetables has been of great practical importance to shippers and handlers of produce. Although the absence of oxygen is detrimental to the tissues of most plants, modified atmospheres made of low concentrations of oxygen, which inhibit both aerobic and anaerobic respiration, and higher levels of inert gases have been effective in prolonging storage life and in improving fruit and vegetable quality after shipping.

Summary of Respiration

1. Sugar is prepared for oxidation through the process of phosphorylation during which phosphorus is transferred from an organic phosphorus donor, ATP, to the sugar molecule. Sugar cleavage then occurs, resulting in the production

Comparison of Photosynthesis and Respiration

Photosynthesis	Respiration
1. CO_2 and H_2O are used	1. O_2 and food are used
2. Food (carbohydrate) and O_2 are produced	2. CO_2 and H_2O are produced
3. Energy from light is trapped in chlorophyll and in food	3. Energy in food may be temporarily stored in ATP or lost as heat
4. ATP is produced by use of light energy (photosynthetic phosphorylation)	4. ATP is produced by oxidation of food (oxidative phosphorylation)
5. Hydrogen is transferred from H_2O to NAD or NADP to form NADH or NADPH	5. Hydrogen is transferred from food to NAD or NADP to form NADH or NADPH
6. ATP and NADPH are used primarily to drive reactions involving sugar synthesis	6. ATP and NADH or NADPH are available to do many types of work in the cell
7. Only chlorophyll-containing cells carry out photosynthesis	7. Every living cell carries out respiration
8. Occurs only in light	8. Occurs both in light and in darkness
9. Occurs in chloroplasts in eukaryotic plants	9. Glycolysis occurs in cytoplasm, while the final steps of aerobic respiration occur in mitochondria
10. Total photosynthesis must exceed total respiration for growth to occur	

of two three-carbon-atom phosphate intermediates. These in turn are oxidized to the common intermediate, pyruvic acid.

2. If molecular oxygen is absent, pyruvic acid usually is changed to ethyl alcohol and carbon dioxide in the final stages of anaerobic respiration in plant cells. Only small amounts of energy are released in this process.

3. If molecular oxygen is present, the pyruvic acid becomes further oxidized to carbon dioxide and water through the organic acid cycle. In this process (aerobic respiration), large amounts of energy are stored in energy-rich compounds that are used when work is done in the cell.

4. The two major functions of respiration are (a) the transformation of potential energy stored in food to an energy form readily available to do work (ATP) and (b) the production of intermediate products that are used in the synthetic reactions of the cell.

5. Energy-yielding oxidative reactions may be coupled to energy-requiring reactions.

6. Respiration is carried out through an integration of reactions going on in the cytoplasm (glycolytic reactions) and the mitochondria (Krebs cycle and terminal oxidation reactions).

7. Alternate pathways of respiration, such as the pentose phosphate pathway, occur in some cells.

15

Flowers

The flower initiates the sexual reproductive cycle in all Anthophyta, and in so doing it terminates the growth of the shoot bearing the flower. Flowering poses some very interesting questions: (a) If flowering terminates the growth of a shoot, what provision is made for continued vegetative growth of a stem during and after flowering? (b) What determines the change from vegetative growth to flowering? (c) How are the flower parts adapted for sexual reproduction? (d) Are the flower parts modified vegetative leaves or is the flower a totally new structure unrelated to vegetative leaves? (e) What is the evolutionary history of the flower?

We have seen that in woody plants, provision is always made, through mixed buds or associated shoot buds, for continued vegetative growth at the same time that flowers are produced.

Flower induction of a vegetative apex is not yet understood. It seems as though the apical bud loses its ability to elongate so that the young floral apex has many crowded nodes with very short internodes (Fig. 15.26A). Furthermore, and more important, the potentiality for continued vegetative growth is lost. It is only regained after the fusion of sperm and egg cells (gametes), each with one-half the chromosome number of the parent plant. Meiosis, two cell divisions in which the chromosome number is halved, and fertilization, in which the normal vegetative chromosome number is restored, occur before vegetative growth is resumed. The function of the flower is to facilitate the formation and fusion of gametes.

The nature of the flower parts themselves, their relation to vegetative leaves, and their adaptation for sexual reproduction form the subject matter of this chapter and Chapter 16.

The evolutionary history of the flower is followed in some detail in Chapters 27–30, which deal with the comparative morphology of the vascular plants. It is possible to trace a satisfactory evolutionary development of the floral parts of the Anthophyta back to leaf-like structures in very primitive vascular plants.

Of all the characteristics of flowering plants, the flower is the least affected by changes in the environment. For instance, we have seen that leaf shape is influenced by age, light, water, and nutrition. The basic morphology and anatomy of flowers and fruits are not so affected. This constancy of floral structures makes flowers and fruits the most important plant parts for angiosperm classification.

In addition, flowers have a high esthetic value and the resulting fruits and seeds are of major importance in food production.

The essential steps of sexual reproduction, meiosis and fertilization, take place in the flower. The complete sexual cycle involves: (a) the production of special reproductive cells, accompanied by meiosis, (b) pollination, (c) fertilization, (d) fruit and seed development, (e) seed and fruit dissemination, and (f) seed germination. The seed completes the process of sexual reproduction in the angiosperms, and the embryo in the seed is the first stage in the life cycle of new individuals.

The Morphology of the Flower

Flowers are of many different forms. They also vary greatly in size, color, number of parts, and arrangement of parts. There are flowers so small that their organs are scarcely visible to the unaided eye; for example, the flowers of the duck-weeds, free-floating plants common in ponds throughout the world. Then there are the flowers of a plant (*Rafflesia*), growing on the floor of dark tropical forests of the Malay archipelago, that are 3–4 ft in diameter.

A typical flower is composed of four whorls of modified leaves: (a) **sepals,** (b) **petals,** (c) **stamens,** and (d) a **carpel** or **carpels,** all attached to the **receptacle,** the modified stem end that supports these structures (Fig. 15.1, see Color Plate 6; Fig. 15.2).

The **sepals** enclose the other flower parts in the bud. Generally, they are green. All the sepals taken collectively constitute the **calyx;** that is, the calyx of a flower is composed of more or less distinct parts, the sepals.

The **petals** are usually the conspicuous, colored, attractive flower parts. Taken together, the petals constitute the **corolla.**

The **stamens** form a whorl, lying inside the corolla. Each stamen has a slender stalk or **filament** at the top of which is an **anther,** the pollen-bearing organ. The whorl or grouping of stamens is called the **androecium.**

The **carpel** or **carpels** comprise the central whorl of modified floral leaves (Fig. 15.1, see Color Plate 6; Figs. 15.2, 15.3). Collectively, the carpels are spoken of as the **gynoecium.** Each *individual structure* in the gynoecium is commonly referred to as a **pistil.** A pistil may be composed of a single carpel, or of several united carpels in the center of the flower. Thus, a pistil may be formed from *one* or *more* carpels. The pistil may be dissected from a flower as a unit. When one carpel forms one pistil, a carpel may be physically separated from a flower as a whole unit (Figs. 15.2, 15.3, 15.5D, F). When several carpels form a single pistil, the pistil must be cut apart in order to separate out the carpels (Figs.15.5H, 15.6B). The name "pistil" is ancient and is derived from the resemblance of numerous commonly occurring pistils to the pestle of the mortar-and-pestle set always present in the medieval pharmacy, where it was frequently used to compound medicines from floral parts. The term is a convenient one, but modern concepts of flower structure have rendered its meaning somewhat inexact.

There are generally three distinct parts to each pistil: (a) an expanded basal portion, the **ovary,** in which are borne the **ovules;** and (b) the **style,** a slender stalk supporting (c) the **stigma** (Figs. 15.2, 15.3B). The pollen is deposited on the stigma.

The **receptacle** is the enlarged end of the flower stem or stalk to which the sepals, petals, stamens, and pistils are attached.

The term **perianth** is applied to the calyx and

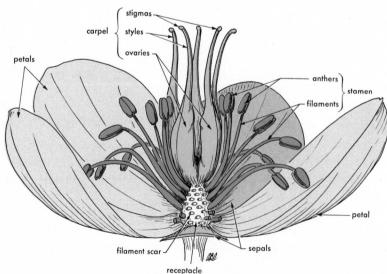

Figure 15.2 A diagram of a longitudinal section of a flower of Christmas rose (*Helleborus*). The perianth consists of two similar whorls; there are numerous stamens arranged in a spiral on a cone-shaped receptacle. Five separate carpels form the central whorl of floral parts.

A

B

Figure 15.1 Flowers showing three different arrangements of flower parts. *A*, *Lireodendron tulipifera*, three sepals, six petals, numerous stamens, many carpels compressed in inner whorl; *B*, *Sedum* sp., five sepals, five petals, ten stamens, five carpels; *C*, *Lilium tigrinum*, perianth consists of two whorls, each with three similar sepals and petals, six stamens, and three carpels coalesced in one pistil.

C

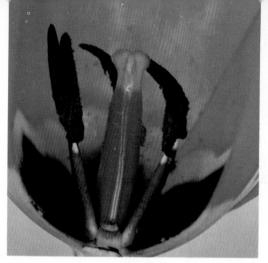

A

B

Figure 15.17 Elevation of floral parts. *A*, hypogyny in tulip flower (*Tulipa*); *B*, perigyny in almond flower (*Prunus amygdalus*); *C*, epigyny in daffodil flower (*Narcissus pseudonarcissus*). *A* and *C*, about $\times\frac{1}{3}$; *B*, $\times 1$.

Color
Plate 7

Arrangement of Floral Parts

B

C

C

A

Figure 15.29 Floral guides for insects collecting nectar or pollen. *A, Viola*, splashes of color and guide lines leading to throat of corolla; *B*, section of nasturtium flower showing tubular spur for nectar and guide hairs leading to entrance to tube; *C*, mimicry—the flower of the orchid *Ophrys* resembles a female wasp. (*C*, courtesy of R. Norris.)

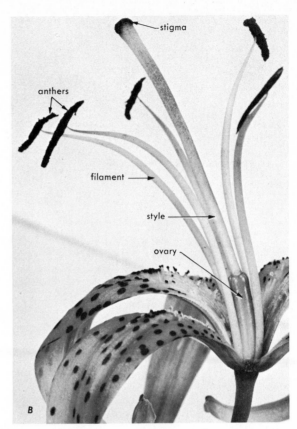

Figure 15.3 The essential floral organs. *A, Magnolia*, showing many separate stamens and separate carpels arranged in a spiral on the receptacle; *B*, regal lily (*Lilium regale*), showing six stamens with anthers and filaments, and three carpels in a single pistil with stigma, style, and ovary.

corolla collectively. It is frequently used to describe flowers, such as the tulip, in which the two outer whorls, though present, are morphologically indistinguishable. The individual parts of such a perianth may be referred to as tepals.

Sometimes individual flowers or compact clusters of flowers will have a whorl of small leaves or **bracts** standing close below them. Such a collection of bracts subtending flowers is called an **involucre.**

The parts of the flower in outline are as follows:

Receptacle

Calyx, consisting of sepals
Corolla, consisting of petals   Perianth

Androecium, consisting of stamens
Gynoecium, consisting of carpels

Essential Organs of the Flower

The essential floral organs are the androecium (stamens) and gynoecium (carpels, Figs. 15.2, 15.3*A*,

15.5). They are essential because they produce reproductive bodies. The perianth, composed of calyx and corolla, is a protective covering of the stamens and pistil and also serves to attract and guide the movements of pollinators.

The Androecium. Each stamen consists of an anther supported on a stalk or filament (Figs. 15.2, 15.3*B*, 15.5). The anther usually has two longitudinal lobes united by a band of tissue. Each lobe has two longitudinal pollen sacs, within which the pollen grains are produced. Before the distribution of pollen, the tissue separating the two pollen sacs in each lobe of the anther breaks down. In most species, a longitudinal slit then develops in the wall of each anther lobe. The edges of the slit separate and the pollen escapes through the opening thus formed.

The anther consists at first of a small mass of meristematic cells. Differentiated early in this mass are four separate groups of cells, called **pollen mother cells** (Fig. 15.4*A, B*). Each pollen mother cell undergoes meiosis by two successive divisions and

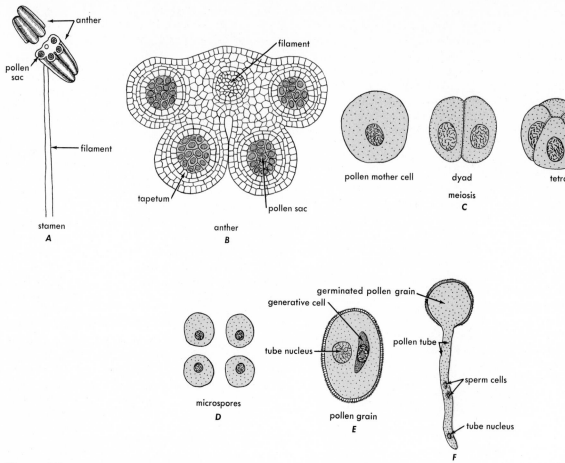

Figure 15.4 Development of pollen from a pollen mother cell to the pollen grain.
A, stamen; *B,* cross-section of anther; *C,* development of tetrad of cells from the pollen
mother cell by meiosis; *D,* four microspores; *E,* pollen grain; *F,* germination of pollen
grain into pollen tube.

forms four haploid cells, each with one set of chromosomes. These cells are called **spores** (Fig. 15.4*D*). Shortly after they are formed, the single nucleus divides by mitosis to form a two-cell haploid plant. The surrounding wall thickens and may become elaborately sculptured. This is the **pollen grain** (Figs. 15.4*E,* 15.28). The two cells are called the **generative** and **tube cells** (or sometimes nuclei), respectively (Fig. 15.4*E*). At this stage, the pollen is generally shed and in some manner transferred to the stigma of adjacent flowers. This process is **pollination.** On the stigma, the pollen grain germinates to form a **pollen tube.** Within the pollen tube the tube nucleus degenerates and the generative cell divides to form two **sperm** cells (Fig. 15.4*F*).

The Gynoecium. The structure of the gynoecium depends upon the number and arrangement of car-

pels comprising it (Fig. 15.5). In the pea flower (Figs. 15.5*A, B,* 15.6*A,* 16.2), there is a single carpel forming the gynoecium (Fig. 15.6*A*). It consists of three parts: (a) the **ovary,** an expanded basal portion, and (b) the **style,** a slender stalk that ends in (c) a hairy irregular portion, the **stigma.** In the flower of the Christmas rose (Figs. 15.5*C, D*) there are five separate and distinct carpels comprising the gynoecium, and each carpel has its own ovary, style, and stigma (Figs. 15.2, 15.5*D*). In the flower of *Cotyledon* (Fig. 15.5*F*) there are again five carpels which, while not actually fused together, grow so compactly as to appear on casual observation to form a single central pistil. In the tulip (Figs. 15.5*G, H*) there are three carpels so completely fused that only a single structure is present to represent the central whorl of floral leaves. A single pistil composed of one carpel is called a **simple pistil** (Fig. 15.2). When two or more carpels fuse to form a pistil, it is known as a **compound pistil** (Figs. 15.5H, 15.6B).

The ovary is a hollow structure having from one to several chambers, or **locules** (Fig. 15.6). The number of carpels in a compound pistil is generally related to the number of stigmas (Fig. 15.5G, H), the number of locules (Fig. 15.6), and, sometimes, the number of faces of the ovary (Fig. 15.6B). The pea has a single stigma and the ovary has one locule (Figs. 15.5A, B, 15.6A). A similar situation holds for each pistil in the gynoecia of Christmas rose (*Helleborus*) and *Cotyledon* (Figs. 15.2, 15.5D, F). There are three stigmas in the tulip gynoecium, the ovary is three-sided (Fig. 15.5H) and contains three locules (Fig. 15.6B); it is composed of three carpels.

Cross-sections of the gynoecia of the garden pea and of the tulip are shown in Fig. 15.6. Assuming the carpel to be a leaf, the structure of these two gynoecia may be interpreted as follows: The single modified leaf of the pea gynoecium has been folded along its midrib (Fig. 15.6A). The margins of the leaf are thus brought together. They infold slightly, and regions just back from the margins are fused together. There is a single vascular bundle for each of the fused margins. The ovules are attached to the margins. The mature pea pod will open by splitting both along the midrib and between the margins. These lines of splitting represent, respectively, the **dorsal** and **ventral sutures.** The region where the ovules are attached is the **placenta** (Figs. 15.6B, 15.8). In the tulip there are three infolded carpels (Fig. 15.6B); the margins of all three carpels fold inward into the locules and fuse to form a central placental region. Thus, here too, the ovules are borne on the margins of infolded modified leaves. The midrib of each carpel is marked by vascular tissue, as are the margins. When the tulip ovary is mature, it will split along the midrib or dorsal suture.

The table below shows the relationship between the parts of the gynoecia in the flowers shown in Fig. 15.5.

Plant	Carpels	Pistil
Pea	1 carpel	1 simple pistil
Christmas rose	5 single and separate carpels	5 simple pistils
Cotyledon	5 distinct but closely appressed carpels	5 simple pistils
Tulip	3 fused carpels	1 compound pistil

In all these cases the term gynoecium refers to the carpels taken collectively, without reference to their number or their manner of association with each other. The term "simple pistil" refers to a single carpel, while "compound pistil" denotes the union of several carpels to form a single structure.

The Carpel as a Modified Leaf. The question arises as to why a carpel is considered a modified leaf. There are two main reasons for so considering it. One of these reasons is its manner of development, which in early stages resembles that of a foliage leaf.

The second reason for considering the carpel a modified leaf concerns the evolutionary development of the flower. Fern leaves frequently bear spores or specialized reproductive cells resulting from meiosis (Chap. 27) on their margins or along veins (see Fig. 27.26). If a young pea pod, which is a single carpel, is opened carefully along its ventral suture, the margins may be folded back, showing the seeds, which have developed from ovules, along the margin of a leaf-like carpel (see Fig. 16.2). The situation is even more striking in *Sterculia platanifolia*, for in this plant there are five simple pistils united only by their stigmas (Figs. 15.7A, B), which, when mature, open to show five very leaf-like carpels bearing seeds on their margins (Fig. 15.7C). Carpels may thus be compared with the leaves of ferns, which are not only photosynthetic organs, but are reproductive structures as well. The evolutionary steps in the transition from leaves to the carpel of the flowering plants are taken up again in Chapter 30, Fig. 30.27.

Placentation. The tissues within the ovary to which the ovules are attached are called **placentae** (singular, **placenta,** Fig. 15.6). The manner in which the placentae are distributed in the ovary is termed **placentation,** and its determination is helpful in classification: When the placentae are on the ovary wall, as in the violet, currant, gooseberry, and bleeding heart (*Dicentra*, Fig. 15.8D), the placentation is **parietal.** When they arise on the axis of the ovary which has several locules, as in lilies and fuchsia, the placentation is **axile** (Fig. 15.8C). Less frequently, the ovules are on the axis of a one-loculed ovary, in which event the placentation is **central,** as in the primrose family (Fig. 15.8A, B).

Style and Stigma. The style is a slender stalk that terminates in the stigma. It is through stylar tissue that the pollen tube grows. In some flowers, the style is very short or entirely lacking; in others, it is long. In corn (*Zea mays*), the corn silks are the styles (Figs. 15.9, 15.21C, D). As a rule, the style withers after pollination, but in some plants (for example, *Clematis*) it persists and becomes a structure that aids in the dispersal of the fruit (Fig. 15.12B). The stigmatic surface often has short cellular outgrowths

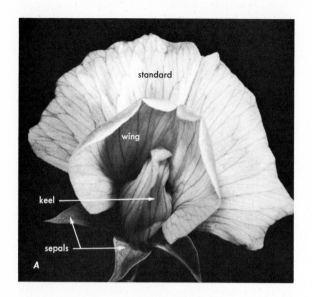

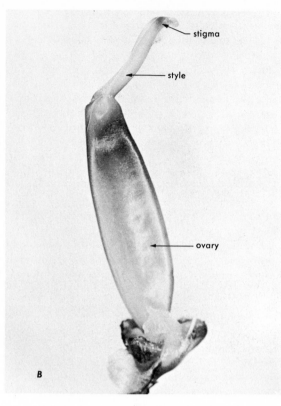

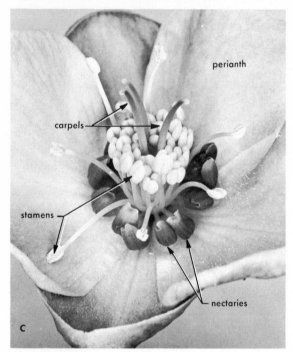

Figure 15.5 Flowers and essential organs.
A, flower of garden pea (*Pisum sativum*);
B, gynoecium of garden pea consisting of a single
carpel; *C*, flower of Christmas rose (*Helleborus*)
with stamens and carpels; *D*, maturing gynoecium
of Christmas rose consisting of several loosely
associated carpels; *E*, flower of *Cotyledon*;
F, opened flower of *Cotyledon*, showing stamens
adnate to corolla tube and five separate compact
carpels; *G*, flower of *Tulipa*; *H*, enlarged view of
stamens and compound pistil of *Tulipa*. *A, C, G,*
×$\frac{1}{2}$; *B,* ×2; *E, F, D, H,* ×1.

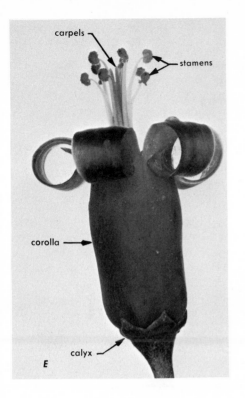

E

carpels

stamens

corolla

calyx

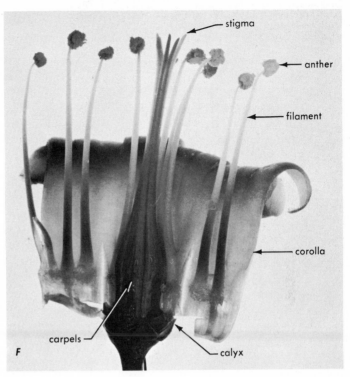

F

stigma

anther

filament

corolla

carpels

calyx

G

H

stigmas

anther with pollen

style

ovary

filament

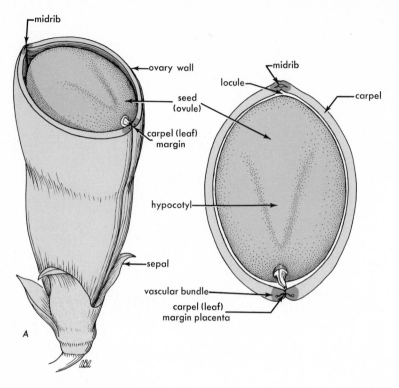

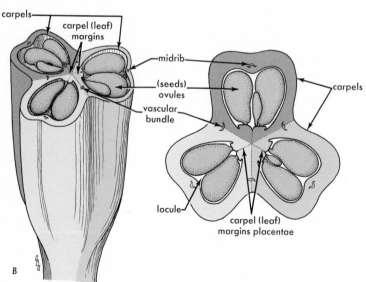

Figure 15.6 Diagram comparing gynoecia consisting of a single carpel and of three coalesced carpels. *A*, section of ovary of pea consisting of a single carpel; *B*, section of ovary of *Tulipa* showing three coalesced carpels.

that aid in holding the pollen grains; and sometimes it secretes a sugary and sticky solution, the **stigmatic fluid.** In many wind-pollinated plants, such as the grasses, the stigma is much branched, or plume-like.

The Ovule. The ovule, the structure which will eventually become the seed, arises as a dome-shaped mass of meristematic cells upon the surface of the placenta. Differentiation proceeds as outer

cells of the dome develop upward to form one or two protective layers, the inner and outer **integuments** (Fig. 15.10C). The integuments do not fuse at the apex of the ovule, thus they leave a small opening known as the **micropyle.** While this outer development is taking place, one of the internal meristematic cells, the megaspore mother cell, is enlarging. Preparations are going on within its nucleus for two successive divisions, the **meiotic** divisions. These

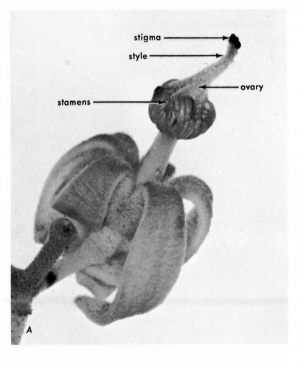

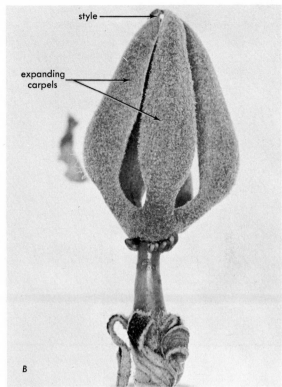

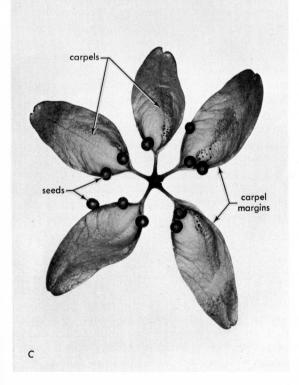

Figure 15.7 Carpels suggestive of foliage leaves. *Sterculia plantanifolia.* *A*, flower showing a single gynoecium, $\times 5$; *B*, after pollination, carpels separated except at expanding tips, $\times 3\frac{1}{4}$; *C*, dehiscence of matured ovary showing seed attached to the margins of the five leaf-like carpels, $\times \frac{5}{8}$.

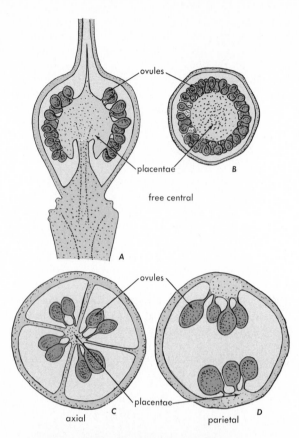

ovules

placentae

free central

A

B

ovules

placentae

axial   C   parietal   D

divisions will result in four cells, each having half the number of chromosomes of the megaspore mother cell. The tissue from which the megaspore mother cell has differentiated is known as the **nucellus.** The ovule then is comprised of one or two outer protecting integuments, with micropyle, nucellus, and megaspore mother cell (Fig. 15.10*E*).

Embryo Sac Development

Meiotic division of the megaspore mother cell may be considered as the first step in the development of a seed. As a result of these divisions, a row of four cells called the **megaspores** is produced in the nucellus. As a rule, three of the cells (the ones nearest the micropyle) disintegrate and disappear, whereas the one farthest from the micropyle enlarges greatly (Fig. 15.10*E*). This megaspore, with its one nucleus and cytoplasm, now develops into the mature embryo sac. The usual stages are as follows: (*a*) *a series of mitotic divisions;* the megaspore nucleus divides, forming a two-nucleate embryo sac, each of these two nuclei divides, forming a four-nucleate embryo sac, and then each of the four nuclei divides, forming an eight-nucleate embryo sac (Figs. 15.10*F, G, H*) (*b*) *migration of nuclei;* after the first division the daughter nuclei migrate to opposite poles of the embryo sac, and after the last division, one nucleus from the sets of four at the opposite poles of the embryo sac migrates toward the center (Fig. 15.10*I*); (*c*) *cell formation about nuclei;* the nuclei, with their surrounding cytoplasm, form cells. Six of these nuclei, each with its associated cyto-

Figure 15.9 Inflorescences of corn (*Zea mays*). Left, long silks (styles) arising from the kernels (ovaries), many of which make up the ear (inflorescence). Right, stamen-bearing flowers of the tassel, $\times \frac{1}{6}$.

plasm, become complete individual cells. The two polar nuclei form a single, binucleate cell. This results in a seven-celled embryo sac. These cells have names as follows: At the micropylar end of the embryo sac there is one **egg cell** associated with two **synergid cells.** Since it is frequently difficult to differentiate the egg cell from the other two, these three cells are sometimes referred to as the **egg apparatus.** The two nuclei that migrated centerward approach each other; they are called **polar nuclei** (Fig. 15.10*l*), and form a binucleate cell called the **endosperm mother cell.** The three nuclei remaining at the end of the embryo sac opposite the micropyle form the **antipodal cells.** The embryo sac, which may be considered to be a seven-celled haploid plant, is now mature and ready for fertilization (Fig. 15.10*l*).

While lily provides excellent class material for a study of embryo sac development, a similar sequence of stages occurs in only a relatively few species. In lily and other species having this type of development, all four megaspores, rather than a single megaspore, function in the subsequent development of the embryo sac. The sequence of stages results in polyploid endosperm tissue, a change which has genetic significance and means that the type of embryo sac development assumes practical importance for the improvement of foods derived from endosperm. The sequence of stages in embryo sac development as it occurs in lily is shown in Fig. 15.11.

Fertilization

Fertilization is a most critical stage of development in the life cycle of all sexually reproducing plants and animals (Fig. 15.10*J*). It involves the union of two cells called **gametes.** In flowering plants, gametes differ in size and form and are known as **sperms** and **eggs.** Fertilization is not completed until a sperm nucleus fuses with an egg nucleus to form a single nucleus. The cell resulting from the union of a sperm and egg is the **zygote** and its nucleus is the **zygote nucleus.** This nucleus must have two sets of chromosomes, one contributed by the sperm and another contributed by the egg. In higher plants, in addition to the union of a sperm and an egg, the nucleus of a second sperm cell present in the pollen tube unites with the two polar nuclei in the endosperm mother cell to form the **primary endosperm nucleus.** The cell

containing the *primary endosperm nucleus* is the **primary endosperm cell.** The zygote arising from union of egg and sperm will give rise to an embryo, which, having two sets of chromosomes, is *diploid*. The primary endosperm cell arising from the union of endosperm mother cell and sperm will develop into the **endosperm,** a food tissue of varying degrees of importance in different species. Note that the primary endosperm nucleus has three sets of chromosomes and is **triploid.** These relationships may be outlined as shown below.

Note that two sperm cells are discharged from the pollen tube and that the nucleus of one of them fuses with the egg nucleus and the nucleus of the other fuses with the other polar nuclei. This constitutes *double fertilization,* as both fusions are necessary for the development of fruit and seeds. Double fertilization occurs only in the Anthophyta.

After fertilization, synergids and antipodal cells usually degenerate, leaving the zygote and primary endosperm nucleus within the ovule (Figs. 15.10*J*, 16.21*A*). The other tissues constituting the ovule, the nucellus and the integuments, now enter into the further development of the ovule, eventually resulting in the seed (Chap. 16).

Variations in Floral Structure

General Description

Complete and Incomplete Flowers. The parts of a typical flower, sepals, petals, stamens, and carpels, are all attached to the receptacle. A flower with all four sets of floral leaves is said to be a **complete flower.** An **incomplete flower** is one in which one or more of the four sets are lacking. For example, there are (a) flowers that have no perianth, (b) flowers that have a calyx but no corolla, (c) flowers with carpels but no stamens, and (d) flowers with stamens but no carpels. The flowers of the calla lily are examples of flowers without a perianth (Fig. 15.12A). In the goosefoot family, which includes spinach, and beet, the flowers have greenish sepals but no petals; but in *Clematis* and a number of other plants, although there are no petals, the sepals are colored and petal-like (Fig. 15.12B). Unisexual flowers (with either carpels or stamens but not both) are fairly common in the plant kingdom. Examples are corn

| sperm | | egg | | zygote |
| one set of chromosomes (*n*) | + | one set of chromosomes (*n*) | → | two sets of chromosomes (2*n*) |

| sperm | | endosperm mother cell | | primary endosperm cell |
| one set of chromosomes (*n*) | + | two sets of chromosomes (2*n*) | → | three sets of chromosomes (3*n*) |

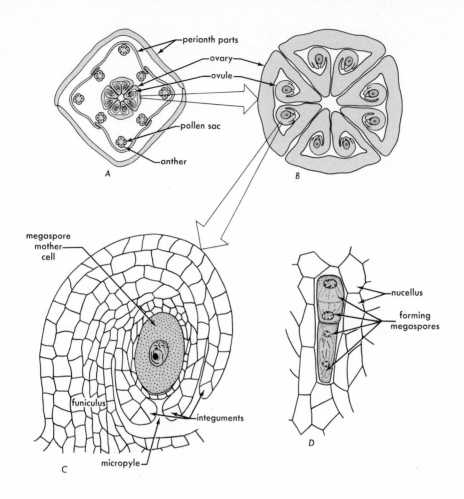

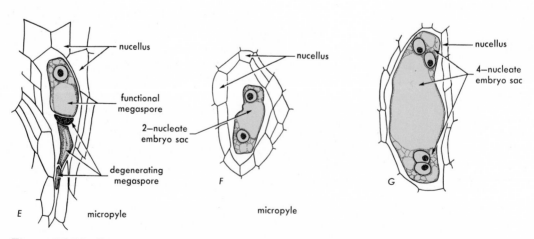

Figure 15.10 Diagram of embryo sac development. Above. *A*, cross-section of a flower bud showing the four whorls, four-carpelate ovary in center; *B*, enlarged view of ovary showing four carpels and ovules; *C*, an ovule, megaspore mother cell in prophase of meiosis; *D*, late telophase of second meiotic division; *E*, four megaspores, three degenerating; *F*, two-nucleate embryo sac; *G*, four-nucleate embryo sac; Right. *H*, eight-celled embryo sac; *I*, mature embryo sac; *J*, fertilization, synergids degenerating.

(*Zea mays,* Figs. 15.9, 15.21), oak, walnut (Fig. 15.13*A*, *B*), willow, poplar, squash (Fig. 15.13*C, D*), hemp, asparagus, hop, and date palm.

Perfect and Imperfect Flowers. Unisexual flowers are either **staminate** (stamen-bearing) or **pistillate** (pistil-bearing). Unisexual flowers are said to be **imperfect,** whereas bisexual flowers are **perfect** or **hermaphroditic.** When staminate and pistillate flowers occur *on the same individual plant,* as they do in corn, squash, walnut (Fig. 15.13*A, B*), and many other species, the species, or the plant, is said to be **monoecious.** In corn (Figs. 15.9, 15.21), for example, the tassel (borne at the top of the stalk) consists of a group of staminate flowers, and the young ear is a group of pistillate flowers. When staminate and pistillate flowers are borne *on separate individual plants,* as in asparagus, willow, and many other species, the species, or the plant, is said to be **dioecious.** For example, in a commercial asparagus field, approximately half the individual plants bear only staminate flowers, and half bear only pistillate flowers. In such circumstances, we speak of staminate

or "male" plants and of pistillate or "female" plants. Only pistillate (female) plants bear fruit (berries). Another example of dioecism is the date palm (Fig. 15.14). Some individual palms are staminate, others are pistillate. The edible fruit (date) is produced only by pistillate palms. In such instances, there will be no fruit production unless staminate plants and pistillate plants are growing near enough together for the pollen to be transferred. In commercial plantings of dates, most of the individual palms are pistillate, that is, fruit-bearing. Dates are propagated by offshoots, which arise chiefly near the base of the stem in the early years of the palm's life. Offshoots from a staminate palm grow into staminate palms, and offshoots from a pistillate palm into pistillate palms. In a commercial date garden, it is economically desirable to have as many pistillate or fruit-bearing individuals as possible, and this is secured by vegetative propagation. A relatively few staminate palms are scattered throughout the garden—only enough to supply pollen. Pollen may be carried naturally by wind or it may be collected and dusted by hand on the pistils (Fig. 15.14*B*).

Some species produce three kinds of flowers:

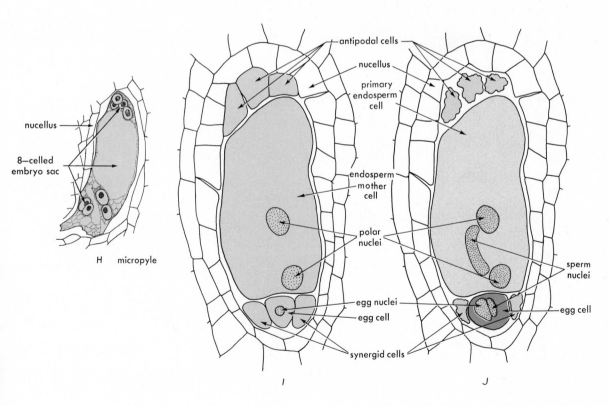

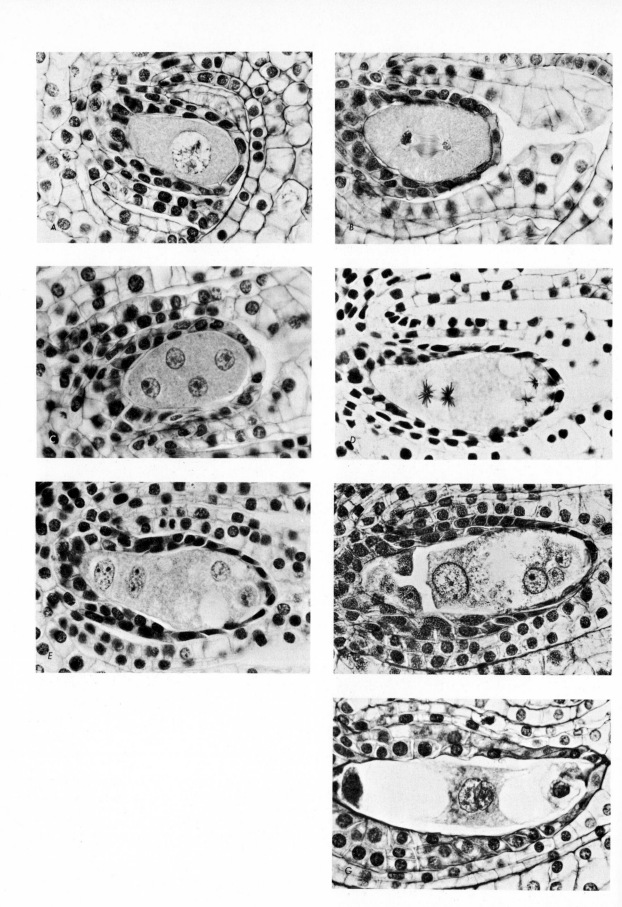

Figure 15.11 Embryo sac development is most generally illustrated by slides of lily ovaries. It shows to good advantage in lily, and good teaching material from other species is very difficult to obtain. *A*, cross-section of a lily ovule, megaspore mother cell in prophase of meiosis. *B*, telophase I. *C*, four megaspore nuclei. *D*, three megaspore nuclei move to antipodal end of cell, the fourth passes to the micropylar end. Mitosis ensues. The three antipodal nuclei share a common spindle. The separate sets of anaphase chromosomes are thus triploid. Compare the number of chromosomes in these triploid sets with the haploid number in the anaphase sets of the spindle at the micropylar end. *E*, second four-nucleate stage. The two large nuclei at the antipodal end each have three sets of chromosomes. Those at the micropylar end have one set of chromosomes each. *F*, a second normal mitosis ensues. There are four nuclei at each end of the embryo sac. The three antipodal nuclei have become disjoined from the main cell. The two large polar nuclei are evident as are the two nuclei that will form the synergids and the cell destined to become the egg. *G*, mature embryo sac. Because it is so large, it is almost impossible to get all eight nuclei of the mature embryo sac in a single negative. This figure illustrates the breakdown of the antipodal cells, it shows the two polar nuclei and what are probably an egg cell with one accompanying synergid.

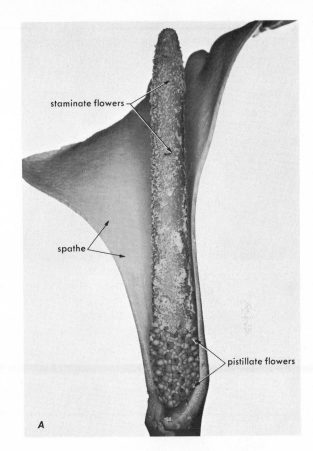

staminate, pistillate, and perfect (hermaphroditic). For example, in red maple, one may find both unisexual and bisexual flowers in the same flower cluster.

Often, staminate flowers have rudiments of a pistil, and pistillate flowers have rudiments of stamens.

Whorled and Spiral Arrangements. In most flowers, sepals, petals, stamens, and carpels are in **whorls** or circles on the axis of the flower. For example, in

Figure 15.12 Flowers with perianth parts lacking. *A*, spathe of calla lily (*Zantedeschia althiopica*), sepals and calyx absent; *B*, flower of *Clematis*, petals absent, sepals prominent.

Variations in Floral Structure 263

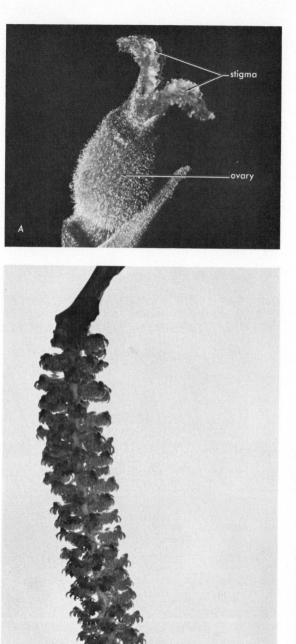

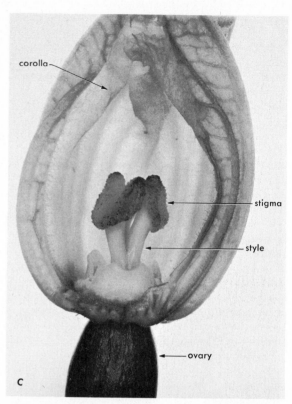

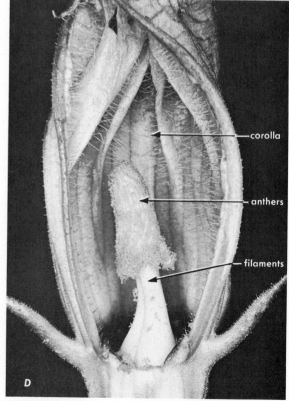

Figure 15.13 Imperfect flowers. *A*, pistillate flower, ×15; *B*, staminate flowers, in catkins in walnut (*Juglans regia*), ×½; *C*, pistillate and *D*, staminate flowers of squash (*Cucurbita*), ×½.

Figure 15.14 Date palms (*Phoenix dactylifera*). *A*, commerical grove. *B*, artificial pollination of date flowers: *a*, strands of male flowers being placed in the center of the female cluster; *b*, freshly opened flower cluster ready for pollination; *c*, flower cluster after pollination, with the strands tied to hold male flowers in place. (*A*, courtesy of California Date Administrative Committee; *B*, courtesy of Nixon.)

Sedum (Fig. 15.1*B*) and *Cotyledon* the sepals arise in a circle at the outer lowest level on the receptacle; slightly above this circle are the petals; above these is a circle of stamens attached to the petals; and in the center are two united carpels. In contrast with this **cyclic** or **whorled** arrangement of flower parts is the arrangement in which one or more sets of flower parts are in **spirals**. Examples of flowers with spiral arrangement of certain flower parts are Christmas rose (Figs. 15.2, 15.5*D*) and magnolia (Fig. 15.3*A*).

Floral Symmetry. In many flowers such as those of *Liriodendron* (Fig. 15.1*A*), lily (Fig. 15.3*B*), and *Sedum* (Fig. 15.1*B*), the corolla is made up of petals of similar shape that radiate from the center of the flower and are equidistant from each other. Other flower parts have a similar arrangement. Such flowers are said to be **regular**. In these cases, even though there may be an uneven number of parts in the perianth, any line (Figs. 15.1*B*, 15.15*A*) drawn through the center of the flower will divide the flower into two

Variations in Floral Structure   **265**

similar halves. They may be exact duplicates or mirror images of each other.

Irregular flowers, such as sweet pea (Fig. 15.5*A*) and mints (Fig. 15.15*B*), have whorls either (*a*) with dissimilar flower parts, (*b*) with parts that do not radiate from the center, or (*c*) not equidistant from one another. In most of these flowers, only one line will divide the flower in equal halves; the halves are usually mirror images of each other. Some flowers (bleeding heart, Dutchman's breeches), though irregular, may be bisected by any number of lines into similar mirror images.

The irregular bean or pea flower has a corolla

Figure 15.15 Floral symmetry. *A*, regular flower of columbine (*Aquilegia*); *B*, irregular flower of a mint (*Salvia*), $\times\frac{1}{2}$.

composed of the following (Fig. 15.5*A*): one broad conspicuous petal (the **banner** or **standard**); two narrower petals (**wings**), one on each side; and, opposite the banner, two smaller petals that are united along their edges to form the **keel.** Other examples of irregular flowers are mints, violets, orchids, and snapdragons.

Union of Flower Parts. In the flower of *Sedum*, illustrated in Fig. 15.1*B*, all parts of the flower are separate and distinct; that is, each sepal, petal, stamen, and carpel is attached at its base to the receptacle. In many flowers, however, members of one or more whorls are to some degree united with one another, or are attached to members of other whorls. Union with other members of a given whorl is termed **coalescence.** Partial or complete union of the sepals along their edges occurs in the flowers of mints (Fig. 15.15*B*), violets, evening primroses, peas (Fig. 15.5*A*), and many other plants. This condition is referred to as **synsepaly.** Petals may also be attached to one another (Figs. 15.5*F*, 15.13*C*, 15.16*B*, *C*, 15.17*C*; see Color Plate 7) and they are said to be **sympetalous** (the noun would be **sympetaly**). When the corolla is sympetalous, it may have the form of a bell (Canterbury bells, *Campanula*) or a tube (red hot poker, *Tritoma*) or a funnel (*Petunia*).

Stamens may coalesce, as is seen in the legume and cotton families (Figs. 2.4, 15.16*B*). This coalescence is **synandry.** In the cotton flower, the filaments of many stamens are united to form a sheath, whereas the anthers are separate. In the flower of orange, there are 20–60 stamens, united at their bases to form groups. And in the thistle family the anthers are united into a tube, whereas the filaments are distinct.

When the carpels of the gynoecium coalesce, the situation is called **syncarpy,** and one may say that the pistil is compound. This condition occurs in many families of plants and may be seen to advantage in such flowers as those of lily (Fig. 15.3*B*), tulip (Figs. 15.5*H*, 15.6*B*), and squash (Fig. 15.13*C*).

When parts of a flower have not joined, the prefix **apo-** (separate) may be used to describe the flower; for example, **apopetalous.** These situations are outlined below:

Whorl	No coalescence	Coalescence
Sepals	Aposepalous	Synsepalous, synsepaly
Petals	Apopetalous	Sympetalous, sympetaly
Stamens	Apoandrous	Synandrous, synandry
Carpels	Apocarpous	Syncarpous, syncarpy

We have just discussed the union of different members of the same whorl. Union of members of

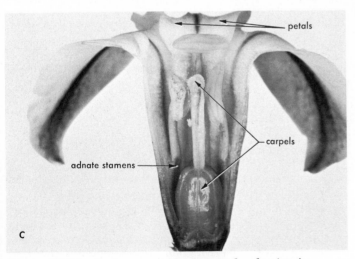

Figure 15.16 Coalescence, adnation, and reduction in number of floral parts. *A, Anemone,* many floral parts, no coalescence, no adnation; *B, Hibiscus,* reduction in number of perianth parts and coalescence of petals, stamens, and carpels; *C, Allium,* coalescence of perianth parts and of carpels; adnation of stamens to corolla tube, $\times \frac{1}{4}$.

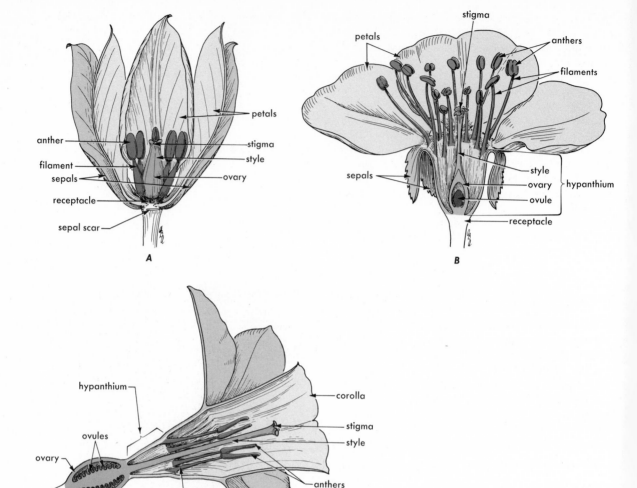

Figure 15.18 Diagram of elevation of floral parts. *A*, hypogyny (*Tulipa*); *B*, perigyny in cherry flower (*Prunus avium*); *C*, epigyny in daffodil flower (*Narcissus pseudonarcissus*).

two different whorls also occurs, **adnation.** For example, in such flowers as *Cotyledon* (Fig. 15.5*F*), almond (Fig. 15.17*B*), snapdragons, and honeysuckle, the stamens are attached to the corolla rather than to the receptacle. In these cases, the stamens are said to be adnate to the corolla.

Elevation of Flower Parts. In flowers of magnolia (Fig. 15.3*A*), lily (Fig. 15.3*B*), and tulip (Fig. 15.17*A*, see Color Plate 7; Fig. 15.18*A*), the receptacle is convex or conical and the different flower parts are arranged one above another. They occur in the following order (beginning with the lowest): sepals, petals, stamens, and carpels (Figs. 15.2, 15.18*A*). The gynoecium is thus situated on the receptacle above the points of origin of the perianth parts and androecium. An ovary in this position is said to be **superior.** In the daffodil (Fig. 15.18*C*) or sunflower (Fig. 15.25) the ovary appears to be below the apparent points of attachment of the perianth parts and the stamens. This is an **inferior** ovary. With an inferior ovary, the anatomy of the flower parts indicate that the lower portions of the three outer whorls, calyx, corrolla, and androecium have fused to form a tube or **hypanthium.** The ovary in the daffodil is completely adnate with the hypanthium.

In a flower with a superior ovary, sepals, petals, and stamens arise from the outer, lower portion of the concave receptacle, below the point of origin of the carpels (Fig. 15.2). The perianth and stamens are **hypogynous,** or with reference to the three outer whorls, a condition of **hypogyny** exists. In a flower with an inferior ovary, the perianth and stamens appear to arise from the top of the ovary, and they are **epigynous.**

In some flowers such as the plum and apricot, the hypanthium does not become adnate to the ovary. The perianth parts and stamens arising from the rim of cup-like hypanthium around the ovary are **perigynous.** Sometimes the hypanthium will be adnate to only the lower half of the ovary. This ovary is **half inferior.**

We have discussed in some detail the various aspects of flower structure as it occurs throughout the groups of flowering plants. We shall now proceed to describe the specialized flowers of two very important families, the Gramineae (grasses) and the Compositae (the sunflower family). Several members of the Gramineae are of great economic importance, supplying our single largest source of food directly in the form of flour and indirectly as feed for cattle, sheep, and hogs. The Compositae family, while not a major source of food products, is made up of a large number of species with a wide distribution and colorful flowers; zinnias, marigolds, dahlias (Fig. 15.22), dandelions, and many similar flower types belong to this family.

The Grass Flower

Grass flowers generally grow in a **head** or **spike** (Fig. 15.19), as in wheat or barley, or are loosely grouped in a **panicle** (Fig. 15.20), typical of oats. In all instances, individual small flowers or **florets** are associated in groups known as **spikelets** (Fig. 15.20). Each spikelet is separated from its neighbors by two small modified leaves or bracts known as **glumes.** They are found at the base of the spikelet and in some cases (wheat, oats, barley), they may fairly well enclose it. A separation of the glumes reveals the individual florets of the spikelet attached to a slender stalk, the **rachilla.**

Each floret is in turn protected by two additional bracts, the **lemma** and the **palea** (Fig. 15.20C). In wheat, barley, oats, and other grains, the lemma is large and may have a long slender **awn** attached to it. The palea is small and frequently enclosed by the lemma. The essential parts consist of the androecium, composed of three stamens, and the gynoecium, composed of two fused carpels (Fig. 15.20D). When flowers are mature, the palea and lemma separate slightly, exposing two feathery stigmas and allowing the anthers to hang free from the spikelet

on greatly lengthened filaments (Figs. 15.19, 15.21B). At the base of the floral parts, two small protuberances, the **lodicules,** may be distinguished. These are thought to be greatly modified perianth parts and are supposed to function in the separation of lemma and palea at pollination time.

In corn (Fig. 15.21), the flowers are imperfect, with staminate flowers grouped in the tassel, and pistillate flowers eventually developing into the ear. The peculiar and very important difference between corn flowers and those of other grasses is the absence of bracts of any sort around pistillate flowers.

The Composite Flower

As the name implies, the composite flower is a group of small flowers arranged to give the appearance of a single typical flower. Several types of com-

Figure 15.19 Head, or spike, of barley (*Hordeum vulgare*), ×1.

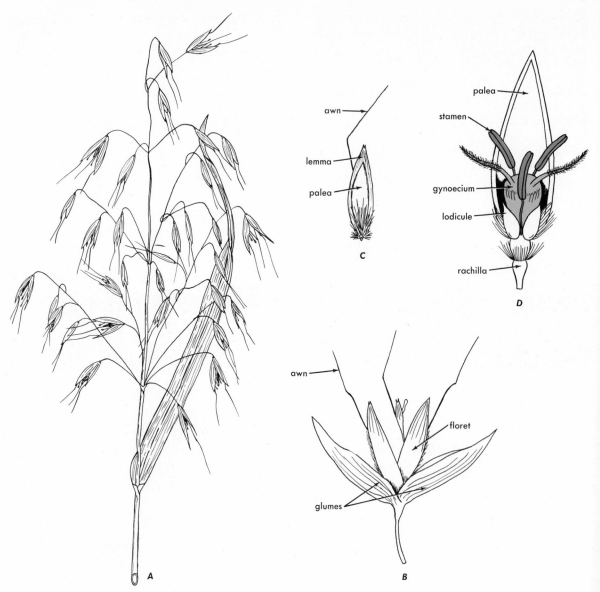

Figure 15.20 Diagram of inflorescence and flower of oats (*Avena sativa*).
A, inflorescence; *B*, spikelet; *C*, floret; *D*, opened flower. *A*, $\times \frac{1}{10}$; *B, C,* and *D*, about $\times 1$.

posite flowers are shown in Fig. 15.22. This characteristic grouping of flowers is called a **head.** In dahlia (Fig. 15.22*B*), sunflower (Figs. 15.23, 15.24), and many other composite flowers, there are two distinct types of individual flowers, **ray flowers** and **disk flowers.** Some species may, however, have all ray flowers (cactus dahlia, Fig. 15.22*A*), and others all disk flowers (globe thistle, Fig. 15.22*C*). Ray flowers are frequently sterile or pistillate.

The disk flower is generally complete. It is epigynous; the small tubular corolla arises from the top of the inferior ovary (Figs. 15.23, 15.24). There are two carpels, as evidenced by the forked stigma (Figs.

15.23, 15.24*C*, 15.25*B*), which, when receptive to pollen, protrudes from the corolla tube. The stamens, five in number, have separate filaments, but their anthers are coalesced to form a tube around the style. In sunflower, stamens usually mature before the stigma and protrude from the corolla tube. These details are shown in Figs. 15.23, 15.24*B*, 15.25*A*. Referring to the figures, note that flowers on the periphery of the flower head have been pollinated, and both stigmas and stamens have withered. Adjacent flowers show protruding stigmas dusted with pollen. Next come several rows of dark anthers capped by white masses of pollen. The flowers in the center

270 Flowers

Figure 15.21 Flowers of corn (*Zea mays*). *A*, the tassel, $\times\frac{1}{5}$; *B*, exserted anthers of staminate flowers, $\times 1$; *C*, pistillate flowers forming a very young ear, $\times\frac{1}{2}$; *D*, several pistillate flowers with attached styles, $\times 1$.

have not yet opened. A calyx is not present as such, but many composite flowers are surrounded by a conspicuous tuft of hairs or otherwise modified calyx parts. This is known as **pappus.** In sunflower, these are simply two small sepal-like structures (Fig. 15.25). In dandelion, the pappus is feathery and aids in the dispersal of the fruit.

Two kinds of **floral bracts** sometimes occur in composite flowers. The whole head may be sur-

rounded by green **involucral bracts.** Each individual flower may have its own **receptacular bract** arising from the receptacle near the base of the flower (Fig. 15.23C).

The Development of the Flower

As noted previously, the floral apex loses its ability to elongate, and all of it eventually differentiates

Figure 15.22 Three types of composite flowers. *A*, cactus dahlia (*Dahlia juarezii*), all ray flowers; *B*, dahlia (*Dahlia pinnata*), both ray and disk flowers; *C*, globe thistle (*Echinops exaltatus*), all disk flowers, $\times\frac{1}{2}$.

into floral leaves. Moreover, the floral leaves are very much crowded together, not being separated by long internodes (as for foliage leaves), and they lack axillary bud primordia. Compare Fig. 7.8*A*, which shows the shoot tip of flax, with Fig. 15.26*A*, which is a floral apex of *Ranunculus* (buttercup). As a rule the sepals are in a whorl; that is, they arise on the stem axis at approximately the same level. Petals and stamens also are usually in one or more whorls; and, if there are several carpels, they too may be in one or more whorls. In some of the more primitive types of flowers, such as *Ranunculus*. (Fig. 15.26*A*), the stamens and pistils are in a close spiral on the receptacle. In these, the distance between two successive individual stamens or individual pistils is very short; in other words, the internodes are very short.

The development of lettuce flowers (*Lactuca sativa*) illustrates, in a general way, some of the steps of floral development (Figs. 15.26*B, C, D, E*). The floral apex of lettuce, a composite flower, supports many flower primordia (Fig. 15.26*B*). Examination of these primordia shows that petals appear first and rise above the developing stamens and carpels (Fig. 15.26*C*). Lettuce is epigynous, and the two carpels form a depression in the center of the flower primordium (Fig. 15.26*C, D*). However, regardless of flower type, the carpels always originate as hollow flasks in the center of the flower primordia. As development continues, the petals and stamens elongate. The carpels also elongate, and their margins grow toward each other, finally fusing. The result is the hollow ovary which, at this early stage, may already contain the primordium of an ovule (Fig. 15.26*E*). Upward growth of the carpel margins continues and results in the style and stigma.

Inflorescences

In most flowering plants, flowers are borne in clusters or groups. Morphologically, an **inflorescence** is a flower-bearing branch or system of branches. In manuals of flowering plants, there are many different terms descriptive of various kinds of inflorescence in a flower-bearing branch or system of branches. Only the most common ones are discussed and illustrated here (Fig. 15.27).

Raceme Type of Inflorescence

A very simple type of inflorescence may be found in such plants as currant, hyacinth, wild mustard, and radish. It is called a **raceme.** In this type, the main axis has short branches, each of which terminates in a flower. Thus, each flower is on a short branch stem, the **pedicel.** The main axis of a raceme continues to grow in length more or less indefinitely; the apical meristem persists. Primordia of leaves arise in the usual manner along the margin of this apical meristem, and in the axil of each leaf a flower is borne. The oldest flowers are at the base of the inflorescence and the youngest are at the apex. In such an inflorescence (mustard, for example), mature fruit may be near the base, while at the upper end are minute buds containing very rudimentary flower parts. In plants like cabbage, mustard, and others with inflorescences of this type, seeds may be shattering from the fruits at the base, while flowers are still forming near the tip.

In a simple raceme, the flowers are on pedicels which are about equal in length. In a **spike** (Fig. 15.27), the main axis of the inflorescence is elon-

Figure 15.23   The composite inflorescence. *A*, head of golden glow (*Rudbeckia laciniata*), ×½; *B*, median section of head of sunflower (*Helianthus annuus*), ×1; *C*, enlarged view of flowers of sunflower head, ×2. In all figures, fertilized flowers are outermost, the stigmas reflexed; the next inner row shows exserted recurved stigmas; next lies a circle of flowers with stamens ready to shed pollen; and in the center are the young unopened flowers of the head.

Figure 15.24 Individual flowers of a sunflower head (*Helianthus annuus*). *A*, a ray flower; *B* through *E*, disk flowers. *B*, flower after fertilization, both stigmas and stamens have withered; *C*, flower with stigmas exserted and recurved; *D*, flower with staminal column exserted; *E*, unopened flower, ×2.

gated, but the flowers, each in the axil of a bract, are sessile (without a pedicel). The **catkin** is a spike that usually bears only pistillate or only staminate flowers. The inflorescence as a whole is shed later. Examples are willow, cottonwood, hazel, and walnut.

In all the foregoing kinds of inflorescences, the flowering axis is elongated; that is, internodes are quite long. If the internodes are very short, flowers appear to be arising umbrella-like from approximately the same level. An inflorescence of this kind, in which pedicels are of nearly equal length, is called an **umbel** (Fig. 15.27). The onion is a good example. The **head** is an inflorescence in which the flowers are sessile and crowded together on a very short axis. Members of the family Compositae, including thistle, sunflower, cosmos, dahlia, etc., have this type of inflorescence (Fig. 15.27).

Inflorescences of the raceme type may be compound, that is, branched. A branched raceme is called a **panicle** (Fig. 15.27), as in oats and rice. Compound spikes occur in wheat, rye, and certain other grasses. Umbels also may be compound, as in carrot, parsnip, and other members of the family Umbelliferae (which gets its name from its characteristic umbel inflorescence).

Cyme Type of Inflorescence

In contrast to the raceme types of inflorescenes, just described, is the **cyme** (Fig. 15.27). In the cyme, the apex of the main axis produces a flower which involves the entire apical meristem; hence, that particular axis ceases to elongate. Other flowers arise on lateral branches farther down the axis of the inflorescence, and thus usually the youngest of the flowers in any cluster occurs farthest from the tip of the main stalk. The flower cluster of chickweed is an example of the cyme.

The Pollen Grain

The function of the external floral parts is the formation and transfer of pollen from anthers to stigmas of the same species of flower. The pollen grain (Fig. 15.4E), at the time of its discharge from the anther, consists of a two-celled male haploid plant enclosed in a thickened and elaborately sculptured cell wall (Fig. 15.28). The role of the male haploid plant is to produce the two sperm nuclei required for double fertilization. The function of the pollen grain wall is to protect the two-celled haploid plant as it is being transferred from the stamen to the stigma.

The elaborate cell walls of pollen grains present many intriguing problems (Fig. 15.28). Their intricate details are species-specific and their chemical constitution is so resistant to decay that pollen may still be present in fossil and subfossil plant deposits when all other traces of biological material have been lost (Fig. 18.22). How do genes exercise control over the deposition of such inert materials outside the protoplast? The deposition of the wall begins during meiosis and passes through a number of steps, some of which occur after developing pollen grains lie free in the pollen sac in the anther. There is evidence of

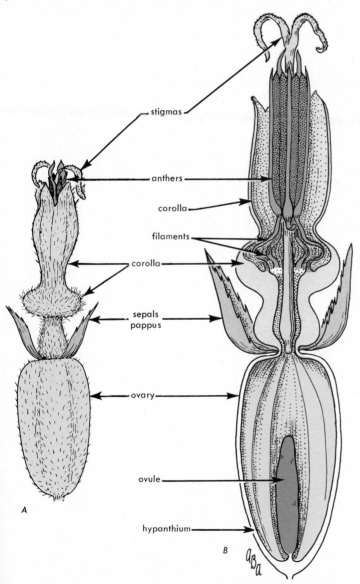

A

B

Figure 15.25 Disk flowers from a sunflower head. The pollen has been shed, and stigmas are raised and exserted. A, external view, $\times 2\frac{1}{2}$; B, enlarged longitudinal cross-section. $\times 4$.

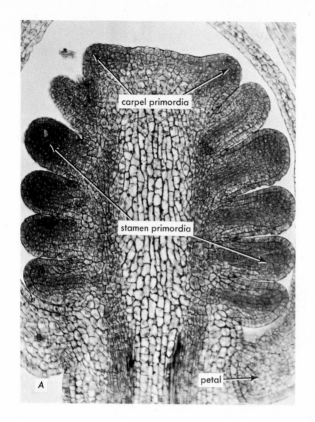

carpel primordia

stamen primordia

petal

Figure 15.26 Development of the floral apex. *A,* photomicrograph of floral apex of *Ranunculus* showing numerous stamen primordia and the earliest carpel primordia, ×200. *B* through *E,* diagram of the developing floral apex of lettuce (*Lactuca sativa*); *B,* young head showing several flower primordia; *C,* young flower showing carpel, stamen, and corolla primordia; *D,* slightly more advanced stage showing carpel just prior to fusion of the margins; *E,* carpel margins have fused, forming the hollow ovary in which occurs a primordium of an ovule; corolla tube, stamens, and stigma are in early stages of differentiation. (*A,* courtesy of S. Tepfer; *B* through *E* redrawn after H. A. Jones, *Hilgardia* **2,** 454.)

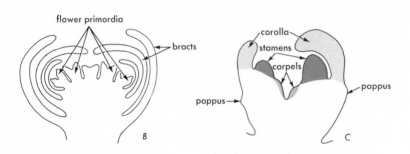

flower primordia

bracts

corolla

stamens

carpels

pappus

pappus

B

C

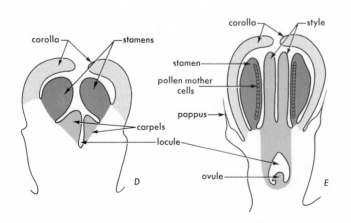

corolla

stamens

corolla

style

stamen

pollen mother cells

pappus

carpels

locule

ovule

D

E

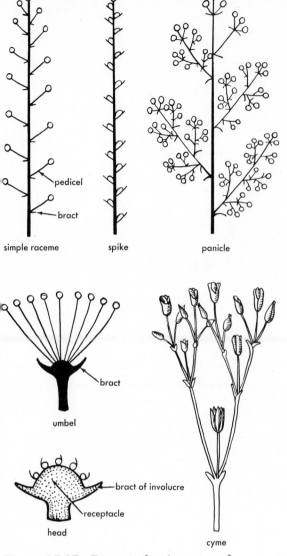

Figure 15.27 Diagram showing types of inflorescence.

some change in ribosomes in developing pollen and of ribosome association with the plasmalemma. There is also evidence that anther cells surrounding pollen sacs may be influential in wall pattern formation. Sculpturing of pollen walls is so precise and consistent in its major features that it forms the basis for a well-developed field of **pollen taxonomy.**

Pollination

Pollination may be brought about by either **wind** or **insects,** and occasionally by water or birds. Flower structure and pollen type is generally adapted to one

or the other of these pollinating vectors. In the case of insect pollination, adaptation may be very complex, indicating a long association between insect vector and plant.

Pollinating Vectors

Wind. Pollen is carried chiefly by wind and insects. Rarely, water and birds are agents of pollination. **Wind pollination** is the common type in plants with inconspicuous flowers, as in grasses (Figs. 15.20, 15.21), poplars, walnuts (Figs. 15.13A, B), alders, birches, oaks, ragweeds, and sage. Such plants usually produce pollen in enormous quantities. Flowers of grasses are inconspicuous, and most of them lack odor and nectar and hence are unattractive to insects. Furthermore, their pollen is light and dry and is easily blown. Their stigmas, in many cases, are feathery and expose a large surface to flying pollen. In cottonwoods, alders, birches, oaks, walnuts, and hickories, the flowers are in **catkins** (Fig. 15.13B). Staminate catkins are pendulous and move easily in wind. The light pollen is shaken from anthers and is readily carried away by breezes. In many catkin-bearing trees, flowers open before leaves unfold so that pollen movement is unhampered. In pines, pollen grains have two wing-like structures (see Fig. 28.23G), which assist in their dispersal by wind. Pine pollen is produced in such tremendous quantities that showers of pollen may be witnessed in a pine forest.

Living Vectors. Living vectors include a variety of insects, a few species of birds, and even some bats. Plants which are pollinated by living vectors usually share certain characteristics. They possess a colorful perianth which, together with other floral parts, may be arranged into a complex architecture (Fig. 15.29; see Color Plate 7). They also produce both a sweet-tasting fluid, called nectar, and volatile compounds having unique odors. Work of many investigators has shown that these characteristics are highly adapted for attraction of pollinators.

Bees, for example, are able to detect and distinguish between many odors and colors and degrees of sweetness. Bees can distinguish at least three colors in addition to ultraviolet, and these colors include the most common perianth shades (yellow and blue). Bees are directed to nectar-producing glands by splashes or lines of contrasting color (Fig. 15.29A). When the insect attempts to reach nectar or collect pollen, floral architecture ensures transfer by the insect of pollen to the stigma. In nasturtium (Fig. 15.29B), for example, nectar is held in a long narrow tube or spur. A foraging bee directs his proboscis down the tube and usually rubs it against the

Figure 15.28 Pollen grains as seen with the stereoscan electron microscope. *A, Iris*, ×210; *B*, day lily (*Hemerocallis fulva*), ×275; *C*, cucumber (*Cucumis sativus*, ×870; *D*, ragweed (*Ambrosia*), ×1000. (Courtesy of D. Hess.)

stigma along the way. In this way, pollen, which may have been picked up in earlier visits to other flowers, is deposited on the stigma.

Orchid flowers have many very specialized mechanisms to ensure pollination. The petal shape and color of species of *Ophrys* closely resemble the appearance of a female wasp or fly (Fig. 15.29C). Male

wasps or flies, emerging from the pupal stage before the females, mistake the *Ophrys* flower for females. The male lands on the flower and attempts copulation; repeated pseudocopulation results in pollination.

There is a very close interdependence between the life cycle of the commercial fig (*Ficus carica*) and that of its pollinator, the fig wasp (*Blastophaga psenes*). There are two types of fig trees, one which produces edible fig fruit and one which produces only small, hard, inedible fruit which is nothing more than an incubator of fig wasp larvae.

An immature, inedible fig is an inflorescence. Small, unisexual flowers line the interior of a hollow, flask-shaped structure which has a pore at the enlarged end (see Fig. 16.8A). Male flowers lie near the pore and female flowers line the rest of the cavity (Fig. 15.30A). The female flowers have pistils with short styles (Fig. 15.30B), and female wasps are able to enter the inflorescence through the pore and deposit eggs in each ovary. The eggs hatch into larvae and later the larvae change to wasps, eat their way out of the ovaries, and mate inside the immature fig. The males usually remain inside and die, but the females escape through the pore, becoming dusted with pollen from the male flowers (Fig. 15.30C) as they do so.

Searching for ovaries in which to deposit eggs, the female may enter the inflorescence of an edible fig. This inflorescence differs from that of an inedible fig by having only female flowers, and the pistils have much longer styles (Fig. 15.30D). The wasp is unable to oviposit because of the elongated styles, but she succeeds in pollinating flowers. Seeds and an edible fig fruit will later result. Eventually, she will find an inedible fig inflorescence, lay her eggs, and start the cycle over again. A single branch of an inedible fig can be grafted to an edible fig tree, and it will provide sufficient numbers of pollinators for the rest of the tree.

Types of Pollination

There are essentially two different kinds of pollination, determined by the genetic similarity of the plants involved. If anthers and stigmas, essential organs in pollination, have the same genetic constitution (that is, have chromosomes bearing identical genes, whether or not they are produced on the same or different individual plants), transfer of pollen from anther to stigma is **self-pollination.** If two parent plants with different genetic constitutions are involved, transfer of pollen from the anther of one to the stigma of the other is **cross-pollination.**

Forms of Self-Pollination

1. Transfer of pollen from an anther to a stigma of the same flower. This is the normal method in

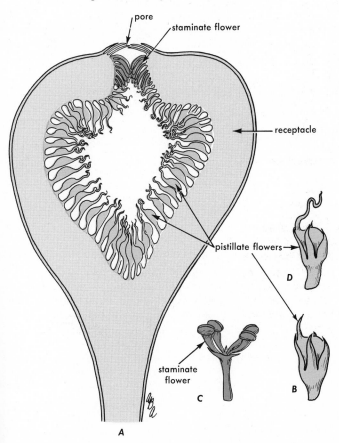

Figure 15.30 Diagram of a fig. *A*, *B*, and *C*, inedible caprifig; *A*, section of receptacle showing location of staminate and pistillate flowers, ×1; *B*, short-styled abortive pistillate flower, ×3; *C.* staminate flower, ×3; *D*, fertile, long-styled pistillate flower of the edible Calimyrna fig, ×3. See also Fig. 16.8.

cereals (except rye and corn), garden peas, and some other plants.

2. Transfer of pollen from anthers to stigmas of other flowers on the same plant. This occurs in many plants with perfect flowers and is usual in monoecious plants.

3. Transfer of pollen from anthers to stigmas of flowers on other plants.

   (a) For centuries many horticultural plants have been reproduced vegetatively. For example, potatoes are usually propagated by cutting tubers into a number of sections. Obviously, all plants growing from tubers or sections of tubers from one plant have the same genetic composition. Such a group of asexually reproducing plants is called a **clone.** Another example: All individual pear trees in an orchard may be parts of a single parent tree; that is, they may have been propagated by cuttings or buds from one parent tree. In this event, these individual pear trees have the same genetic composition. Pollination within clones is essentially self-pollination in that, although pollen goes from one plant to another plant, both plants

have the same genetic constitution because they are reproduced by vegetative means.

   (b) Two individual plants may belong to the same pure line. In wheat, for example, the stigma of a flower usually receives pollen from its own anthers. Following fertilization, the ovary of the self-pollinated flower develops into a grain. The plant that develops from this grain in turn produces many flowers. If these flowers are self-pollinated, the many individual plants that develop from the grains constitute a **pure line.** If pollen from one individual of a pure line is placed on the stigma of a flower of another individual of the same pure line, the result, genetically, is the same as when pollen is placed on the stigma of the flower that furnished the pollen.

Cross-Pollination. Cross-pollination always involves two plants with different genetic constitutions. The differences may involve one or many characters. The two parents may be of the same variety or of the same species; they may be of different species or even of different genera.

Germination of the Pollen Grain

Pollen grains adhere to stigmatic surfaces. In many flowers, this surface has short outgrowths to which grains adhere; and in some species, the stigma produces a sticky secretion, the **stigmatic fluid.** The presence of this fluid, which may give the stigmatic surface a shining or glistening appearance, is evidence that it is receptive to pollen. The hybridizer soon learns to recognize a receptive stigma and thus the proper time to apply pollen artificially. The pollen grain germinates on the surface of the stigma. The protoplast of the grain absorbs water and swells, breaking the outer wall and forming a protoplasm-lined tube. The pollen tube is a slender thread-like growth; it penetrates the tissue of the stigma, grows down through the style (Fig. 15.31), and enters the ovary. Styles of some flowers are hollow, but most of them are not, and so passage of the tube apparently involves the secretion of tissue-dissolving enzymes by the advancing tip of the tube.

In short-styled flowers, pollen tubes need to grow only a short distance. For example, in pea and tulip, the style has a length of only 2–3 mm (Figs. 15.5B, H). On the other hand, in common corn, the distance from the stigmatic surface at the end of the corn silk to the young corn grain (ovary) may be as

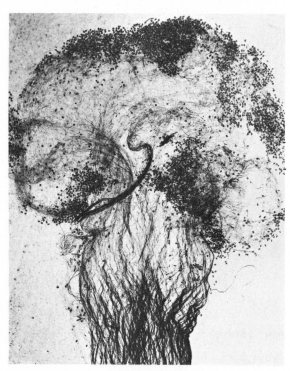

Figure 15.31 Crushed stigma and style stained to show pollen grains and innumerable pollen tubes in the style, ×100. (Courtesy of Triarch Products.)

much as 50 cm (Fig. 15.21C). In a long style, the pollen tube is not a continuous, unbroken tube extending from stigmatic surface to ovary. It consists only of a very short, advancing apical portion containing two sperm cells. The older tube, through which the sperms have passed, gradually disintegrates.

The rate of growth of pollen tubes varies widely. In wheat, for example, fertilization normally occurs 1–2 days after pollination. Under favorable conditions, pollen tubes of corn may grow 15 cm in 24 hr. In some oaks, the tube is almost a year in growing 2–3 mm.

As a rule, when the pollen grain is shed from the anther, it is a two-celled haploid plant, containing a generative cell and a tube cell (Fig. 15.4E). Before or during growth of the pollen tube, the generative cell undergoes division, forming two sperm cells (Fig. 15.4F). The two sperms, and in some species, the tube nucleus, move to the tip of the pollen tube, retaining that position as the tube grows. In other species, the tube nucleus degenerates either before the tube starts to grow or shortly thereafter. The two sperm nuclei bear the hereditary characters of the male parent.

The pollen of many plants will germinate in water. That of other plants requires a nutrient medium, such as a solution of cane sugar. For example, it was found that sugar beet pollen germinates abundantly on culture media containing 1.5% agar and 40% cane sugar. But the pollen of some species will not germinate in any artificial medium thus far tried.

Pollen varies considerably in the length of time it will remain viable, depending particularly upon moisture and temperature conditions surrounding the grains. Plant breeders may desire to keep pollen viable for a considerable length of time in order to make use of it in cross-pollination. It is sometimes shipped long distances for special hybridizing purposes. Pollen of date palms will retain its viability for several months, if kept dry. Experiments show that sugar beet pollen will germinate fairly well after 50 days if kept at low temperature and low humidity. Apple pollen has been successfully stored at $4\frac{1}{2}$ years at 50% relative humidity and at a temperature of 2–8°C. Pollen kept dry wiill withstand greater temperature extremes than pollen kept moist.

Fertilization

Reaching the ovary, the pollen tube grows toward one of the ovules, usually enters the micropyle, penetrates one or more layers of nucellar cells, and enters the embryo sac (Fig. 15.10I), approaching, in one manner or another, the egg cell and the endo-sperm mother cell. These stages are not easy to study, but it appears that the tip of the tube ruptures, one sperm cell enters the egg, and the other sperm enters the endosperm mother cell. Within the egg, sperm and egg nuclei fuse. Within the endosperm mother cell, the other sperm nucleus and two polar nuclei fuse (Fig. 15.10I). This double fusion of egg with sperm and polar nuclei with sperm is sometimes called **double fertilization.** The zygote and primary endosperm cell have now been formed, the antipodal cells and synergid cells will degenerate, and conditions are set for further development of seed and fruit.

Only one pollen tube functions in fertilization of each embryo sac. If an ovary contains many ovules, each of course with one embryo sac, one pollen tube normally enters each ovule. Sometimes, however, fertilization of the egg is not effected, in which event it does not develop into an embryo. The presence of 100 or more mature seeds in a watermelon fruit is evidence that at least as many individual pollen tubes grew down the style of the flower and discharged their contents into separate embryo sacs. As a matter of fact, however, usually many pollen tubes disintegrate at some point along their path of growth from the stigma to the embryo sac.

The immediate external evidence of fertilization is withering of stigma and style and, in many flowers, dropping of petals. If flowers are bagged and pollination is prevented, petals usually remain fresh for a much longer time than in pollinated flowers.

The fusions that occur in the embryo sac initiate a number of changes in the entire ovary, which are as follows:

1. Development of zygote to form the embryo plant.
2. Development of primary endosperm cell to form endosperm (reserve food supply of the seed).
3. Development of integuments to form seed coat.
4. Absorption or disintegration of nucellar tissue. In some plants, however, a portion of the nucellus may become storage tissue of the seed rather than the endosperm. Such tissue of nucellar origin is called **perisperm.**
5. Development of ovary tissue to which ovule or ovules are attached to form the fruit.
6. Possible stimulation of accessory flower parts, such as the receptacle, sepals, or petals, to increased growth and incorporation into the fruit.

Thus, the importance of fertilization is to stimulate growth of certain floral parts and to bring about a withering of others. The net result is the development of the ovary wall into the fruit and the ovule into the seed.

Apomixis

This term refers to a type of reproduction in which no sexual fusion occurs but in which structures usually concerned in sexual reproduction are involved. Normally, egg nuclei will not start on the series of changes that result in embryo plants unless fertilization has occurred. In other words, in most plants, the embryo is the result of a sexual process (the union of male and female cells, gametes). Each gamete contributes n chromosomes, and hence the zygote and all the cells derived from it have $2n$ chromosomes. Rarely, however, an embryo develops from an *unfertilized* egg. This occurrence is called **parthenogenesis.** Two types of parthenogenesis are as follows: that in which the embryo develops from an n, or haploid, egg cell (the individuals are usually sterile), and that in which the embryo develops from a $2n$, or diploid, egg cell. In the latter type, meiosis is omitted in the development of the embryo sac. Another form of apomixis is that in which the embryo plant arises from tissue surrounding the embryo sac. These "adventitious" embryos (diploid) occur in *Citrus, Rubus,* and other plants.

Apomixis has variations other than those given above, but all forms of this phenomenon involve the origin of new individuals without nuclear or cellular fusion.

In some plants, deposition of pollen on the stigma is a prerequisite to apomictic embryo development, even though a pollen tube does not grow down the style and nuclear fusion does not take place. In such instances, there is evidence that hormones formed in the stigma, or furnished by the pollen, are transferred to the unfertilized egg cell, initiating there the changes which result in embryo development.

Summary

1. The flower, the distinguishing structure of the Anthophyta, is formed of four whorls of leaves specialized to carry out sexual reproduction, including pollination, fertilization, and seed production.
2. A floral apex has lost its ability to elongate and its potentiality for vegetative growth. The ability for vegetative growth is not regained until the completion of the two nuclear phenomena, meiosis and fertilization.
3. The four whorls of floral leaves are (a) the calyx, composed of sepals; (b) the corolla, composed of petals; (c) the androecium, composed of stamens; and (d) the gynoecium, composed of carpels.
4. There are two parts to a stamen, the pollen-producing anther and the filament.
5. A single unit of the gynoecium is frequently called a pistil. If it consists of one carpel, it is a simple pistil; if it consists of two or more fused carpels, it is a compound pistil.
6. A pistil consists of three parts, the stigma, the style, and the ovary. The stigma is receptive to pollen, and the ovary encloses the ovules. At the time of fertilization, the ovules consist of integuments, nucellus, and embryo sac.
7. Meiosis takes place both in the anthers and in the ovule.
8. Pollen may be considered a two-celled haploid plant protected by a cell wall which is frequently elaborately sculptured. Two sperms are produced by one of the two cells.
9. The embryo sac may be considered a seven-celled haploid plant. Two of its cells, the egg cell and the primary endosperm cell, are stimulated to development by fusion with sperm cells.
10. Pollination is transfer of pollen from anthers to stigmas. A pollen tube growing down the style to the embryo sac serves as a pathway by which the two sperms reach the embryo sac.
11. Floral architecture facilitates pollination, which is most frequently carried out by wind or insect vectors.
12. Double fertilization, the union of one sperm with the egg cell and of a second sperm with the primary endosperm cell, occurs only in the Anthophyta.
13. The embryo plant develops from the zygote. The endosperm, developing from the fusion of sperm and primary endosperm nucleus, supplies some nourishment for embryo and sometimes for seed development.
14. Variation in floral architecture arises (a) from variation in number of parts in a given whorl; (b) from regularity in floral parts; (c) by coalescence of floral parts; (d) by adnation of floral parts; (e) by elevation of floral parts; and (f) by the presence or absence of certain whorls.
15. Flowers are either solitary on flower stalks or grouped in various inflorescences such as heads, spikes, catkins, umbels, panicles, racemes, or cymes.
16. The primordia of the floral leaves are crowded on the floral apex. The carpels, the central whorl, originate as depressions or cups; their margins fuse during floral development, enclosing the ovules.
17. The potentiality for vegetative growth has been

transferred from the root and shoot apex to the apices in the embryo.

18. There are essentially two types of pollination, cross-pollination, which results in much variation in the progeny, and self-pollination, which results in progeny having great similarities.

19. Occasionally, seeds may be formed without fertilization. This is a form of vegetative reproduction and is known as apomixis.

16

Fruit, Seed, and Seedling

A fruit is a ripened ovary of a flower. Sometimes, other floral parts may be closely associated with the mature ovary and thus be part of the fruit. In many cases, fruit development depends upon pollination. Sometimes, a large amount of inactive, or dead, pollen will stimulate fruit development, and indole-acetic acid, or other growth substances, may substitute for pollen. In the pineapple, banana, and navel orange, fruits normally develop without pollination. Without pollination there can be no fertilization, so these fruits must generally be seedless, and frequently they differ from normal seeded fruits. Fruits are auxiliary structures in the sexual life cycle; they occur only in the Anthophyta. Their development has unquestionably been an important factor in the successful evolution of land plants. Fruits protect seeds, aid in their dissemination, and may be a factor in timing their germination. Fruits are highly constant in structure and are thus very important plant parts in the classification of the Anthophyta. In everyday usage, the term "fruit" usually refers to a juicy and edible structure, such as an apple, plum, peach, cherry, orange, or grape. Such structures as string beans, eggplant, okra, squash, and cucumbers, which are commonly called "vegetables," and "grains" of corn, oats, wheat, and other cereals are not popularly thought of as fruits. However, all the above are **fruits** in a botanical sense.

Fig. 16.1 shows changes in the appearance of an almond as it matures from flower to fruit. Note that the petals fall, leaving the calyx tube with attached stamens intact (Fig. 16.1B). Soon the developing ovary bursts the calyx tube, after which the ovary rapidly enlarges and differentiates into three distinct layers. If this ovary is split longitudinally while still green (Fig. 16.1C), several regions become clearly visible. An unripened seed is present in the center. The enlarged ovary wall, which is now called the **pericarp,** is divided into a firm **endocarp,** a spongy **mesocarp,** and a skin or **exocarp.** As the maturing almond dries, mesocarp and exocarp split (Fig. 16.1D). These parts, commonly called the shucks, fall away, leaving the mature almond of commerce which is comprised of endocarp or shell and enclosed seed (Fig. 16.1D).

Development of the Fruit

Development of the fruit (pod) of pea and bean is here described as an example of the various changes that may occur during transformation of the ovary into fruit. Pistils of bean and pea flowers are each composed of one carpel (Figs. 15.5B, 16.2); that is,

Figure 16.1 A fruit is a ripened ovary. The ovary of an almond flower (*Prunus amygdalus*) develops into the almond. *A*, almond blossom, ×1; *B*, developing ovary bursting calyx tube, ×1; *C*, section of a maturing fruit showing divisions of pericarp and seed, ×1; *D*, mature almonds showing split outer portion of ovary wall (shucks) and intact inner portion (shell), ×$\frac{1}{2}$.

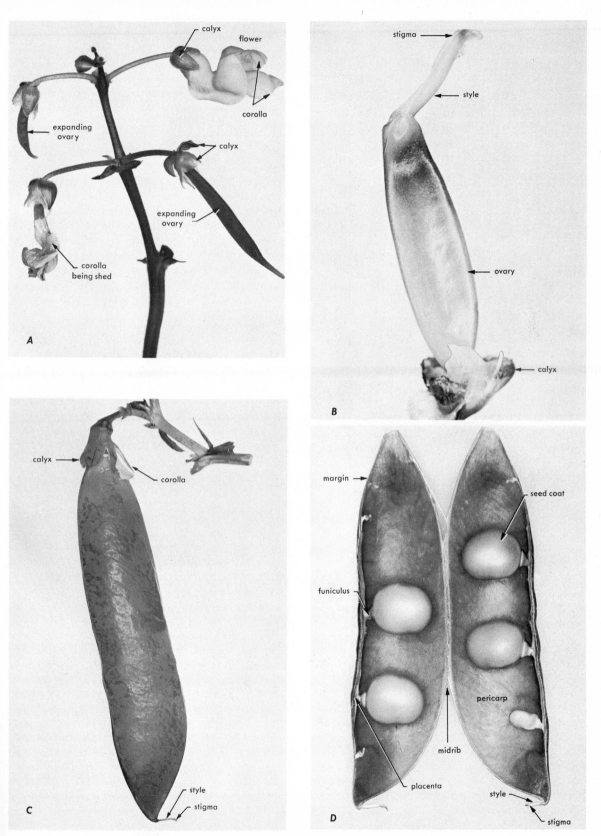

Figure 16.2 Development of legume fruits. *A*, bean (*Phaseolus vulgaris*) from flowers to young pods, ×1; *B*, pistil from a pea (*Pisum sativum*) flower, ×5; *C*, pea pod, unopened; *D*, opened pea pod showing developing seeds attached to carpel margins, ×1.

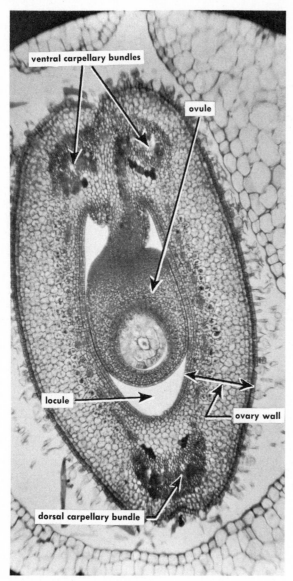

ventral carpellary bundles

ovule

locule

ovary wall

dorsal carpellary bundle

Figure 16.3 Cross-section of bean ovary after fertilization, ×20.

one ovule-bearing leaf. Recall that the ovary of bean is formed by fusion of margins of the carpel and that ovules are attached to these margins (Fig. 15.6A). At the time of fertilization, ovary and ovules within are small. A cross-section of the bean ovary (Fig. 16.3) shows ovary wall, carpellary bundles, ovule, and locule. The ovary wall may be divided into three distinct layers, (a) an outer epidermis, (b) an inner epidermis, and (c) a middle zone consisting of several layers of cells. There are three carpellary bundles, one for each margin and one for the midrib opposite them.

Fertilization initiates a series of changes in the embryo sac and other tissues of the ovule which lead to development of the seed. The stimulus of fertilization extends not only to ovules; other parts of the flower are also influenced. Tissues of the ovary wall undergo marked changes. Three layers of the ovary wall, however, are evident in the mature fruit. The fruit wall (developed from the ovary wall) is called the pericarp, and the three more or less distinct parts, in order beginning with the outermost, are exocarp, mesocarp, and endocarp. When the pod is mature, floral structures such as pedicel, calyx, withered stamens, and even remnants of the corolla may also be present. Quite generally, withered remains of style and stigma persist at the tip of the ovary (Fig. 16.2). Seeds have developed with the matured ovary.

In all kinds of mature fruits, various structures and tissues can be traced back to those of the ovary or ovaries, and to any other floral organs that may constitute a part of the fruit.

Mature pods usually show the presence of small, underdeveloped (abortive) ovules. It is probable that these ovules were not fertilized. If none within the bean ovary is fertilized, the ovary does not enlarge. In most plants, normal fruit development takes place only if pollination is followed by fertilization and if fertilization is followed by seed development. The fruits of tomato are sometimes distorted and irregular in form; this may be due to the lack of seeds in one or more locules and to the resultant poor development of such locules.

Parthenocarpy

The stimulus of fertilization usually exerts an influence on all parts of the flower, particularly on the ovary and its contained ovules. In Fig. 16.4, normal and seedless fruits of tomato are compared. Note the greatly increased development of the fruit that accompanies normal seed development. In some plants, *normal fruit development may take place without fertilization.* Such unfertilized fruits may or may not be seedless, depending upon the occurrence of apomixis and the subsequent development of embryos. On the other hand, fertilization may occur, but ovules may fail to develop into mature seeds even though fruit may form in a normal fashion. In a strict sense, only fruits that develop without fertilization are called **parthenocarpic fruits.** Practically, it is not always possible to know without extensive study, whether a given fruit is parthenocarpic. For instance, Thompson seedless grapes were thought to be parthenocarpic until it was shown that fertilization takes place, but ovules fail to mature into seeds. In many citrus fruits, seeds formed by

Figure 16.4 Tomato (*Lycopersicon esculentum*) fruits. *A* and *B*, normal development after fertilization; *B*, cross-section; *C* and *D*, seedless or parthenocarpic fruits; *D*, cross-section, $\times \frac{3}{4}$.

fertilization are accompanied by seeds with adventitious embryos. To complicate definitions further, sexually formed seeds are not as robust as adventitious seeds and frequently do not develop. This situation has led to a loose use of the word parthenocarpy to mean simply seedless fruits.

Parthenocarpic (seedless) fruits are quite regularly produced in such cultivated plants as English forcing cucumber, certain varieties of eggplant, navel orange, banana, pineapple, and some varieties of apple and pear.

In certain plants, seedless fruits may be induced by pollen that is incapable of fertilizing the ovules. For example, in some orchids, the placing of dead pollen or a water extract of pollen upon the stigma may start fruit development.

Recently, parthenocarpy has been induced in some plants by spraying the blossoms with dilute aqueous solutions of certain growth substances. Commercial application of this technique is being made.

Kinds of Fruits

The flower of garden beans has a single ovary that is composed of one carpel. After fertilization, seeds develop, the ovary enlarges, and the ovary wall becomes a dry, parchment-like hull. When the fruit is mature, it splits open along both the midrib of the carpel and the line of fusion of the margins. These splitting lines are **sutures;** that along the midrib is the dorsal suture, and the ventral suture designates the separation of the carpel margins. Fruits of peas and beans are called **legumes** or **pods** (Fig. 16.2). Thus, the characteristics of the legume or pod are as follows: (a) it is derived from a single ovary, (b) the ovary is composed of one carpel, (c) the pericarp is dry, and (d) the pericarp usually splits along both sutures at maturity.

The lily type of flower has a single ovary (Figs. 15.1C, 15.3B), but, unlike that of the bean, it is composed of three united carpels (Fig. 16.5). As in the bean, the ovary wall develops into a dry structure,

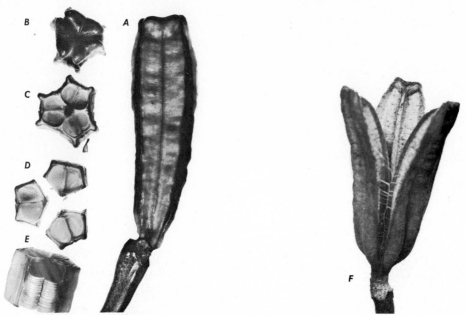

Figure 16.5 Capsule of lily (*Lilium regale*). *A*, nearly mature capsule; *B*, view of three stigmas; *C*, section; *D*, section separated into portions of three carpels *E*, longitudinal view showing ovules; *F*, dehisced capsule, ×1.

Figure 16.6 Aggregate fruit of the strawberry (*Fragaria*). *A*, flower; *B*, fruit, achenes with enlarged receptacle, ×2. (Courtesy of R. S. Bringhurst.)

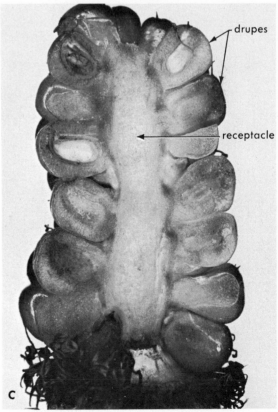

Figure 16.7   Aggregate fruit of the blackberry (*Rubus ursinus*). *A*, flower; *B*, flower shortly after petals have fallen; *C*, section of fruit, ×2.

each locule of which splits open at maturity, allowing the seeds to escape. This type of fruit is called a **capsule.**

These two types of fruit are discussed to point out some of the characters used in describing and classifying fruits. In classifying the different kinds of fruits, taken into account are: (a) structure of the flower from which the fruit develops, (b) number of ovaries involved in fruit formation, (c) number of carpels in each ovary, (d) nature of the mature pericarp (dry or fleshy), (e) whether or not the pericarp splits (**dehisces**) at maturity, (f) if dehiscent, manner of splitting, and (g) the part that sepals or receptacle may play in formation of the mature fruit.

Simple fruits are derived from a *single ovary*. They may be dry or fleshy; the ovary may be composed of one carpel or of two or more carpels; and the mature fruit may be dehiscent or indehiscent. On the other hand, aggregate fruits and multiple fruits are formed by *clusters* of simple fruits. The difference between these types of fruits depends upon the number of flowers involved in their formation. In strawberry (Fig. 16.6) and blackberry (Fig. 16.7), there are many simple fruits, each derived from the

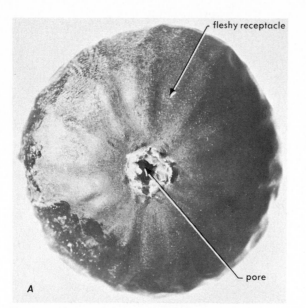

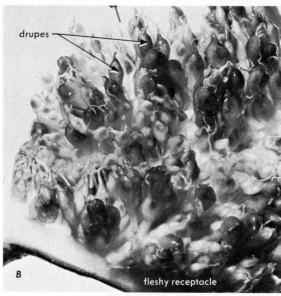

Figure 16.8   The fruit of the fig (*Ficus carica*); *A*, end of fleshy receptacle, showing pore, which is lined with staminate flowers; *B*, receptacle turned inside out, showing pistillate flowers that have matured into small drupes, ×1.

individual ovaries of a *single flower.* These matured ovaries or fruits are all attached to a common receptacle. Such groupings constitute *aggregate* fruits, and the simple fruits comprising them may be classified according to the scheme of classification of simple fruits given below. Mulberry, fig, and pineapple are *multiple* fruits. Multiple fruits consist of the enlarged ovaries of *several flowers,* more or less grown together into one mass. In some plants—mulberry, for example—associated floral structures form a part of the fruit, and in the fig the receptacle enlarges to become the sweet edible portion (Fig. 16.8).

Simple Fruits

Legume or Pod.   This type of fruit, characteristic of nearly all members of the pea family (Leguminosae), arises from a single carpel, which at maturity generally dehisces along both sutures (Figs. 16.2 and 16.9). In the bean or pea pod, the "shell" is pericarp, and the "beans" or "peas" are seeds. Pods may be spirally twisted or curved as in alfalfa. A number of legumes, however, such as honey locust, alfalfa, and bur clover have pods that do not dehisce.

Figure 16.9   Pod of Scotch broom (*Cytisus scoparius*). *A*, unopened; *B*, partially split; *C*, dehisced down both sutures.

Fruit, Seed, and Seedling

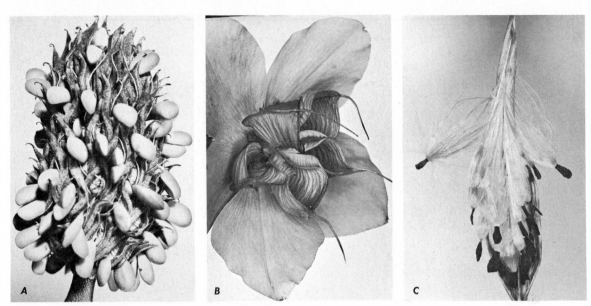

Figure 16.10   Follicles, *A, Magnolia*, each carpel shown in Fig. 15.3*A* has now developed into a follicle, $\times\frac{1}{2}$; *B*, Christmas rose (*Helleborus*), five follicles present in one flower, $\times\frac{3}{4}$; *C*, dehiscing follicle of a South American milkweed (*Oxypetalum caeruleum*).

Follicle. Examples of follicles are fruits of Christmas rose and magnolia (Figs. 16.10*A, B*). The follicle develops from a *single carpel,* and opens along *one suture,* thus differing from the pod, which opens along both sutures.

Capsule. Capsules are derived from compound ovaries, that is, an ovary composed of two or more *united carpels.* Each carpel produces a few to many seeds. Capsules dehisce in various ways: (*a*) *lengthwise* (*Amaryllis, Iris,* and *Datura,* Figs, 16.11*A, B, C*); (*b*) by *pores* toward the top of each carpel (poppy, Fig. 16.11*E*); (*c*) by a *transverse lid* (plantain, Fig. 16.11*H*).

Silique. This is the type of fruit (Fig. 16.12) characteristic of members of the mustard family (Cruciferae). One interpretation suggests that the silique is a dry fruit derived from a superior ovary consisting of two carpels. At maturity, the dry pericarp separates into three portions; the central persistent portion has the seeds attached to it.

Achene. Examples of this type of fruit are buckwheat, strawberry (Fig. 16.6), and members of the Compositae family, such as sunflower. These fruits are commonly called "seeds," but as in sunflower (Fig. 16.13), a carefully broken pericarp reveals the seed within, attached only by its stalk to the placenta. This pericarp may be separated easily from the seed coat, that is, from the layer of cells just beneath it. Achenes are indehiscent.

Grain or Caryopsis. This is the fruit of the grass family (Gramineae), which includes such important plants as wheat, barley, oats, rye, corn, and rice. Like the achene, the grain is a dry, one-seeded, indehiscent fruit (Fig. 16.14). It differs from the achene, however, in that pericarp and seed coat are firmly united all the way around and it is difficult to separate the two except by special milling processes.

Samara. This is a dry, indehiscent fruit, which may be one-seeded, as in the elm, ash, and "tree of heaven" (*Ailanthus*), or two-seeded, as in maple (Fig. 16.15) and box elder. These fruits are typified by an outgrowth of the ovary wall, which forms a wing-like structure.

Schizocarp. This is the fruit characteristic of the carrot family (Umbelliferae), which includes such common plants as carrot, parsnip, celery, and parsley. The schizocarp is a dry fruit that consists of two carpels which split, when mature, along the midline into two one-seeded indehiscent halves. The carrot "seed" of commerce, for example, is the one-seeded half of the schizocarp. The one true seed in each carpel completely fills the whole cavity and is usually grown fast to the pericarp.

Nut. The term nut is popularly applied to a number of hard-shelled fruits and seeds. A typical nut, botanically speaking, is a one-seeded, indehiscent dry fruit with a hard or stony pericarp (the shell). Examples are chestnut, walnut, hickory nut, acorn (Fig.

Kinds of Fruits   293

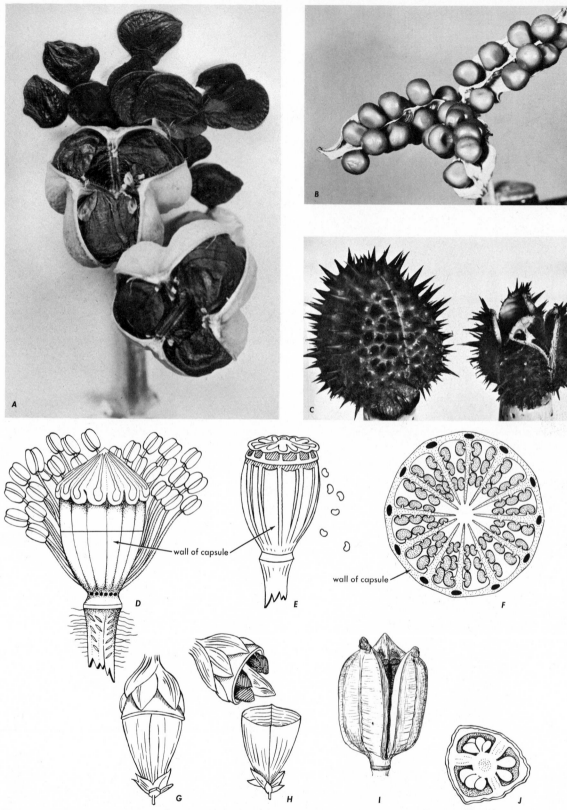

Figure 16.11   Capsules. *A, Amaryllis; B, Iris; C, Datura; D,* poppy capsule, side view before dehiscence; *E,* mature poppy capsule dehiscing by pores at the top; *F,* cross-section of poppy capsule, showing its many carpels; *G,* mature capsule of plantain; *H,* capsule of plantain opening by transverse lid; *I,* capsule of tulip dehiscing lengthwise; *J,* cross-section of tulip ovary. All approximately ×1. (*A, B,* and *C,* courtesy of W. Russell; *D, E,* and *F,* redrawn after Korsmo.)

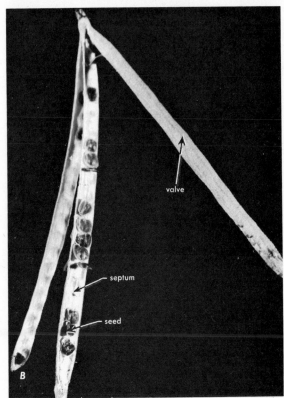

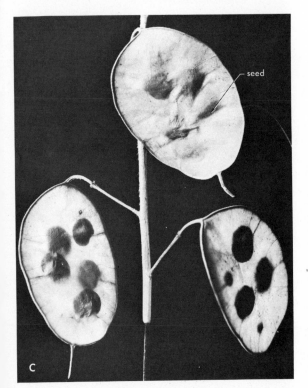

Figure 16.12 Siliques. *A* and *B*, of stock (*Matthiola incana*); *A*, unopened, ×¾; *B*, partially split to show orientation of valves and septum, ×¾; *C* and *D*, of moonwort (*Lunaria annua*); *C*, unopened, ×¾; *D*, opened to show valves and septum, ×1. (Courtesy of W. Russell.)

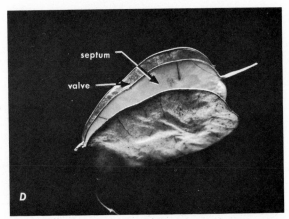

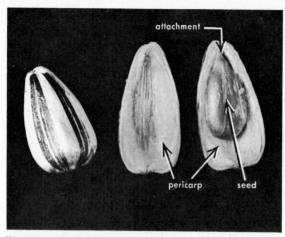

Figure 16.13   Achene of sunflower (*Helianthus annuus*), unopened, and opened to show attachment of seed, ×1.

16.16), hazelnut, and beechnut. The chestnut develops within a bur or prickly involucre. An acorn, the fruit of the oak, is partially enclosed by a hardened involucral cup. The husk or shuck of the walnut, which has been removed before the product reaches the market, is composed of involucral bracts, perianth, and the outer layer of the pericarp. The hard shell is the remainder of the pericarp. Unshelled almonds are drupes from which the hulls (exocarp and mesocarp) have been removed. Brazil nuts are seeds, not fruits. And the unshelled peanut is a pod, corresponding to the fruit of bean or pea; the edible portions are seeds within the pod.

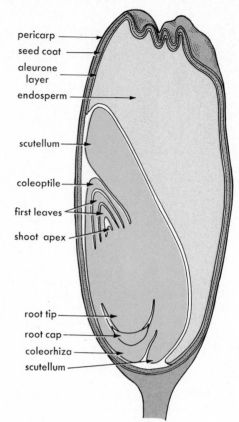

Figure 16.14   Median lengthwise section of mature corn (*Zea mays*) grain.

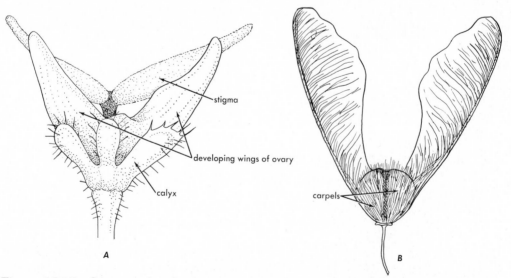

Figure 16.15   Samara or key fruit of maple (*Acer*). *A*, flower; *B*, mature fruit. (Redrawn from A. M. Johnson, *Taxonomy of Flowering Plants.* Copyright 1931, 1959 by Meredith Corporation. Used with permission of Appleton-Century-Crofts.)

Figure 16.16 Acorns of oak (*Quercus*). The fruits are a type of nut and the cups are fused involucrol bracts, $\times \frac{3}{4}$.

Drupe. Examples of this type of fleshy fruit are plum, cherry, almond, peach, and apricot, all of which are members of a genus in the Rosaceae (rose family). The olive fruit is also a drupe. The drupe is derived from a *single carpel,* and is usually one-seeded. However, if one examines young ovaries of flowers of almonds, cherries, plums, or other members of the group to which they belong, two ovules will be found, one of which usually aborts (fails to develop into a seed, Fig. 16.17). The drupe has a hard endocarp consisting of thick-walled stone cells. The exocarp is thin, forming the skin, and the mesocarp forms the edible flesh. The pit or stone of a cherry, for example, is one seed, with thin seed coats, surrounded by a stony endocarp. In other words, the pit is composed of the seed plus the stony inner layer (endocarp) of the ovary wall.

In almond fruits (Figs. 16.1, 16.17), the mesocarp is fleshy like a typical drupe when the fruit is young, but it becomes hard and dry as it develops, and forms the hull. The shell of the almond is endocarp. In coconut, also a drupe, the pericarp tissues become dry at maturity.

Berry. This fleshy type of fruit is derived from a *compound ovary.* Usually, many seeds are embedded in a flesh, which is both endocarp and mesocarp,

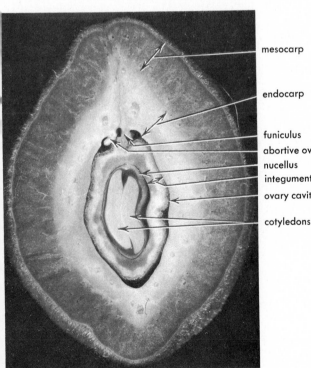

mesocarp

endocarp

funiculus
abortive ovule
nucellus
integuments
ovary cavity

cotyledons

Figure 16.17 Median transverse section of young almond fruit (*Prunus amygdalus*) (drupe). (From R. M. Holman and W. W. Robbins, *A Textbook of General Botany,* John Wiley & Sons, New York.)

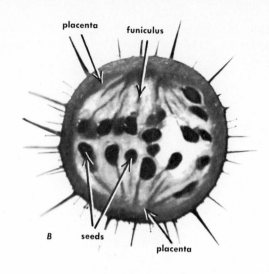

Figure 16.18 Gooseberries (*Ribes*). True berries, *A*, typical appearance, $\times\frac{3}{4}$; *B*, cross-section, showing parietal placentation, $\times 1\frac{1}{4}$.

although the line of demarcation may be difficult to discern (Fig. 16.18). The tomato is a common example of a berry. The wild form of the garden tomato has a two-celled fruit; cultivated forms have several locules in the fruit, and the placentae are fleshy (Fig. 16.4).

The date is a one-seeded berry, derived from a three-carpelled ovary. After pollination, one of the carpels rapidly enlarges and suppresses the other two, which soon dry up. The stone of the date is the seed, so it is not equivalent to the stony endocarp of peaches, plums, cherries, and the like.

The citrus fruit (lemon, orange, lime, and grapefruit) is a type of berry called a **hesperidium.** It has a thick, leathery rind (peel), with numerous oil glands, and a thick juicy portion composed of several wedge-shaped locules (Fig. 16.19). The peel of a citrus fruit is exocarp and mesocarp; the pulp segments are endocarp. The juice is in pulp sacs or vesicles; they are outgrowths from the endocarp walls (Fig. 16.19C), and each mature vesicle is composed of many living cells filled with juice. The common sweet orange usually has ten two-seeded locules, each representing a carpel. The navel orange is a seedless variety, an example of a parthenocarpic fruit.

Another berry-like fruit is that of members of the family Cucurbitaceae, which includes watermelon, squash, pumpkin, cantaloupe, cucumber, and others. It is called a **pepo** (see Fig. 29.11). The outer wall (rind) of the fruit consists of receptacle tissue that surrounds and is fused with the exocarp. The flesh of the fruit is principally mesocarp and endocarp.

Pome. This type of fruit is characteristic of a subfamily of Rosaceae, to which belong apple (Fig 16.20), pear, and quince. The pome is derived from an inferior ovary (Figs. 16.20A, B). The flesh is enlarged hypanthium, and the core has come from the ovary (Figs. 16.20C, D).

Aggregate Fruits

An aggregate fruit is one formed from numerous carpels of one individual flower. These fruits, considered individually, may be classified as types of simple fruits. Thus, fruits of strawberry are achenes (Fig. 16.6B) and fruits of raspberry and blackberry are small drupes (Fig. 16.7). The strawberry flower has numerous separate carpels on a single receptacle. The ovary of each carpel has one ovule, and the ovary develops into a one-seeded dry fruit (achene). The receptacle to which these fruits are attached becomes fleshy; the whole structure, which we call a strawberry, is an aggregate of simple fruits, each an achene. The receptacle is stem tissue and consists of a fleshy pith and cortex with vascular bundles between them. The hull of the strawberry fruit is composed of persistent calyx and withered stamens. The achenes are usually spoken of as seeds.

Flowers of dewberry, raspberry, blackberry (Fig. 16.7), and other species of *Rubus* have essentially the same structure as those of strawberry. In these flowers, many separate carpels attached to one receptacle develop into small drupes (Fig. 16.7C), instead of into achenes. An aggregate fruit is derived from a *single flower* having many simple pistils (carpels).

The various kinds of aggregate fruits differ in the structure of their individual fruitlets, which may be classified as simple fruits.

Multiple Fruits

A multiple fruit is one formed from individual ovaries of several flowers. These fruits, considered individually, may be classified as types of simple fruits. Fruits of mulberry, fig, and pineapple are examples of multiple fruits, and individual fruits composing them are nutlets in mulberry and fig, and parthenocarpic berries in pineapple.

Flowers of mulberry are of two kinds: staminate and pistillate. Pistillate flowers have a deeply four-lobed calyx surrounding a single pistil, the ovary of which is one-seeded. The ovary develops into a nutlet and is enclosed by persistent juicy calyx lobes. Separate fruits become very much crowded together, making up a collection of fruits, the mulberry.

The fig fruit we eat is an enlarged fleshy receptacle (Fig. 16.8). Its flowers are very small and are attached to the inner wall of this receptacle. Both staminate and pistillate flowers occur and may be borne in the same or in different receptacles. The pistillate flower has a single one-celled (and one-seeded) ovary, surrounded by a two- to six-parted calyx (the corolla is absent). Each ovary develops into a nutlet that is embedded in the wall of the receptacle. Thus, the fig is derived from many flowers, all attached to the same receptacle. Common edible figs (Mission and Kadota) are parthenocarpic. Smyrna figs and several other varieties do not mature their fruits unless fertilization and seed formation have taken place (Fig. 15.30).

Accessory Parts of Fruits

In the pineapple of commerce, the edible portion is largely greatly thickened pulpy central stem in which berries are embedded. Tissues, other than the ovary wall, which form part of a fruit, are referred to as **accessory.** Thus, much of the fleshy fruit of pineapple, apple, and strawberry is accessory.

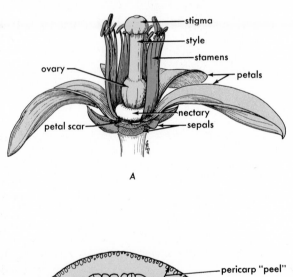

A

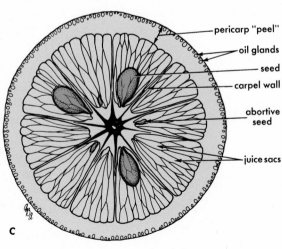

C

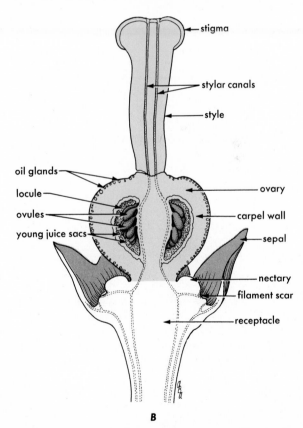

B

Figure 16.19 Flower and fruit (hesperidium) of orange (*Citrus sinensis*). *A*, flower, dissected to show pistil and stamens; *B*, lengthwise section of maturing ovary; *C*, cross-section of mature fruit.

Kinds of Fruits   299

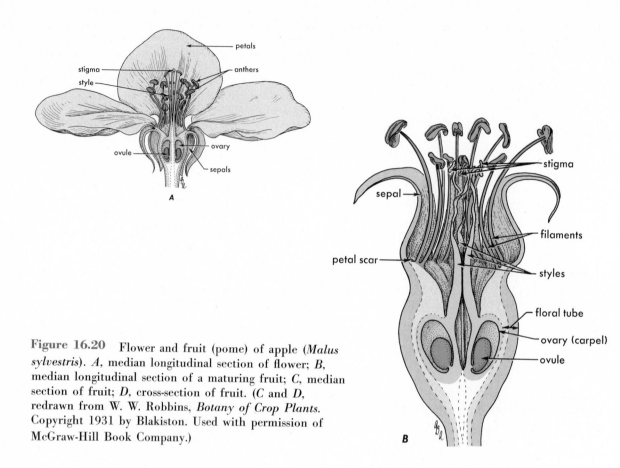

Figure 16.20 Flower and fruit (pome) of apple (*Malus sylvestris*). *A*, median longitudinal section of flower; *B*, median longitudinal section of a maturing fruit; *C*, median section of fruit; *D*, cross-section of fruit. (*C* and *D*, redrawn from W. W. Robbins, *Botany of Crop Plants.* Copyright 1931 by Blakiston. Used with permission of McGraw-Hill Book Company.)

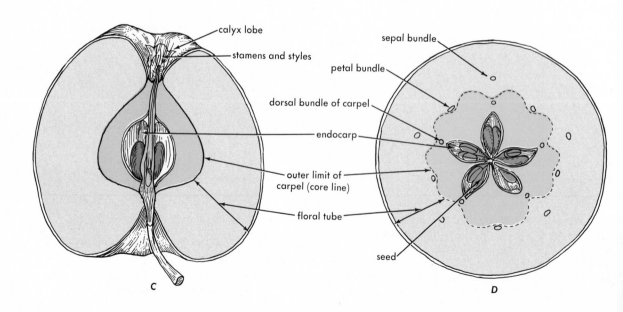

Summary of Fruits

1. A fruit is a ripened ovary, plus any other closely associated floral parts.
2. The initiation of fruit development generally depends upon pollination and fertilization.
3. Fruit development is accompanied by the development of the ovule into the seed.
4. There are *three different kinds of fruits,* classified on the basis of the number of ovaries and flowers involved in their formation:
   (a) *simple fruits,* derived from a single ovary;
   (b) *aggregate fruits,* derived from a number of ovaries belonging to a single flower and on a single receptacle;
   (c) *multiple fruits,* derived from a number of ovaries of several flowers more or less grown together into one mass.
5. Simple fruits may have a dry pericarp or a fleshy pericarp. If the pericarp is dry, it may be dehiscent (splitting at maturity, allowing seeds to escape) or indehiscent (not splitting).

The Seed

Seed Development

The seed completes the process of reproduction initiated in the flower, and it always consists of an embryo surrounded by seed coats. The embryo developed from the zygote, and the seed coats developed from the integuments of the ovule. Recall (p. 259) that at the time of fertilization, egg cell, primary endosperm cell, synergids, and antipodals constituted the embryo sac. The embryo sac is surrounded by nucellus, and all these cells and tissues are enveloped by one or two integuments. This complete structure is the ovule (Figs. 15.10*B,* 16.21*A*). Double fertilization involves (a) fusion of egg and sperm nuclei to form a zygote nucleus, and (b) fusion of polar nuclei with a second sperm nucleus to form a primary endosperm nucleus (Fig. 16.21*B*).

After fertilization, the zygote nucleus remains quiescent for a time. The primary endosperm nucleus, however, divides rapidly and soon builds up an **endosperm tissue** which, at first, may lack cell walls (Fig. 16.21*C*). Division of the zygote nucleus awaits development of the endosperm. The first cell divisions of the new generation result in a filament of from four to eight cells (Fig. 16.21*D*). Now the cell closest to the micropyle elongates, pushing the other cells of the filament into the endosperm. At about the same time, the cell furthest from the micropyle divides at right angles to the axis of the filament (Fig. 16.21*E*). This is the first in a series of divisions that will finally result in the embryo of the seed. Further divisions result in a spherical or globose stage (Fig. 16.21*F*). The major distinction between embryos of monocotyledons and of dicotyledons is the presence of one or of two cotyledons. There is always a root apex and a shoot apex. In dicotyledons, two cotyledons overshadow the shoot apex (Fig. 16.21*H*), while in monocotyledons, only a single cotyledon rises above the shoot apex (Fig. 16.21*G*). The course of events just described is very general, for there are many variations in details of embryo development.

While development of the embryo is taking place, nucellus, endosperm, and integuments are also undergoing changes which are characteristic of the group of plants to which the seeds belong. In the great majority of plants, nucellus and endosperm are required only for initial stages of development. This is particularly true of the nucellus, which is generally used up in early stages. It persists as a food storage tissue, the **perisperm,** in seeds of sugar beet, and a few other species. The persistence of endosperm as a food reserve occurs in many monocotyledons (Fig. 16.21*G*). It is highly developed in grasses, some of which are of major economic importance (Fig. 16.14). The endosperm persists as a food storage tissue in relatively few dicotyledonous seeds. Seeds of castor bean (*Ricinus communis,* Fig. 16.22) and coffee (*Coffea arabica*) are examples of a persistent endosperm in seeds of dicotyledons.

In most seeds, integuments of the ovule become hard and horny seed coats in the mature seed. The stereoscan microscope shows the seed coats to be variously and sometimes beautifully sculptured (Fig. 16.23).

Stimuli in Seed Development

The seed thus completes the process of sexual reproduction initiated in the flower. This development has been marked by a series of crises in which further development is blocked and cannot proceed without precise stimulation. Meiosis in the anthers results in meiospores, which go on to divide and differentiate into pollen grains (Fig. 15.4). Pollen is dormant and is stimulated to produce pollen tubes only when brought into contact with a specific stigmatic surface (Figs. 15.5*B, F, H*). Polar nuclei do not undergo further development until fusion with a sperm nucleus takes place. The egg nucleus does not divide until it fuses with a sperm nucleus, and even then, the first zygote division may still await the stimulus provided by an endosperm tissue (Fig. 16.21). The seed, too, blocked from further development, is a dormant structure, awaiting definite stim-

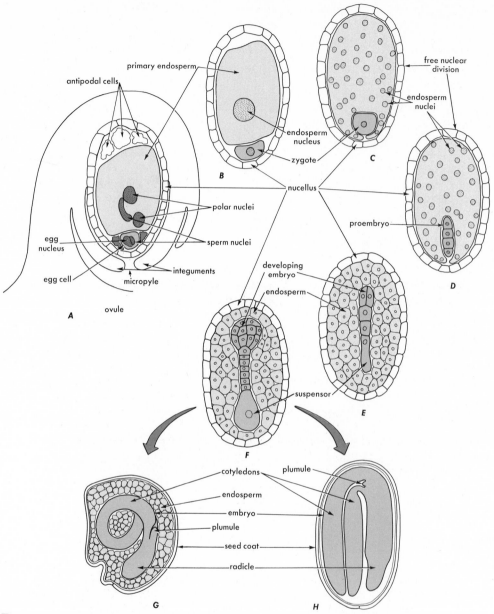

Figure 16.21 Diagram of the development of an angiosperm seed. *A*, ovule after fertilization; *B*, after fusion of gamete nuclei to form zygote and endosperm; *C*, free nuclear divisions in the endosperm; *D*, filamentous stage of the proembryo; *E*, elongation of suspensor cell and division of proembryo cell; *F*, globular stage of the embryo; *G*, embryo of a monocotyledon in seed; *H*, embryo of a dicotyledon in seed.

uli before (Fig. 16.21) it again becomes active in germination.

The nature of the zygote, coupled with the blocks to growth which occur at various stages of development, pose definite questions and have stimulated much experimentation. What keeps the zygote from dividing? What stimulates the zygote to divide? What induces the changes in cell polarity to produce

changes in form of developing embryos? What is responsible for dormancy in seeds? How long can a seed remain dormant? What breaks seed dormancy?

Kinds of Seeds

The seed provides a highly efficient mechanism to: (a) carry plants over periods unfavorable for growth;

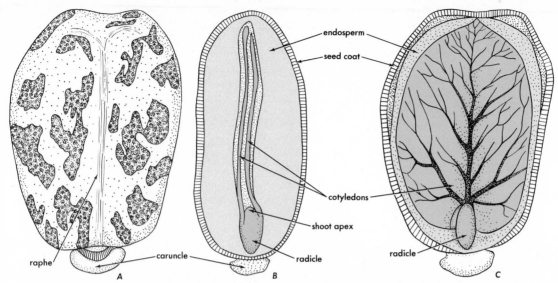

Figure 16.22   Castor bean (*Ricinus communis*) seed. *A*, external appearance; *B*, section showing edge view of embryo; *C*, section showing flat view of embryo.

(*b*) maintain a reserve supply of potential plants in case of successive unfavorable seasons; and (*c*) disseminate plants. Plants are unusually profligate in the production of reproductive cells. The tremendous production of seeds ensures the renewal or continuance of plant populations. Many animals, including man, cooperate in this population renewal and dissemination.

The large amount of food stored in fruits and seeds may be comprehended by considering the production of fruits and seeds (apples, grapes, oranges, grain for animal consumption, and grain for human consumption) in the United States.

Seeds of angiosperms differ in two ways: (*a*) they

have one or two cotyledons and (*b*) they store food either in the embryo, in the endosperm, or more rarely, in nucellar tissue, the perisperm, which lies between embryo and seed coat. In most dicotyledons, food is stored in cotyledons (Fig. 16.24). The endosperm is the principal food storage tissue in seeds of monocotyledons (Figs. 16.14, 16.26).

When food storage occurs within the embryo, the normal vascular tissues of the embryo convey the solubilized food to meristematic regions, where it is required for growth. Food stored in the endosperm, outside the embryo, is in many cases absorbed through normal protodermal or epidermal cells. In other cases, the cotyledon becomes special-

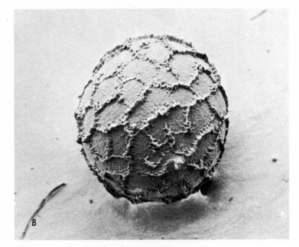

Figure 16.23   The coats of many seeds are characteristically sculptured. *A*, morning glory (*Convolvulus arvensis*), ×23; *B*, California poppy (*Eschscholtzia californica*), ×62. (Courtesy of D. Hess and H. Drever).

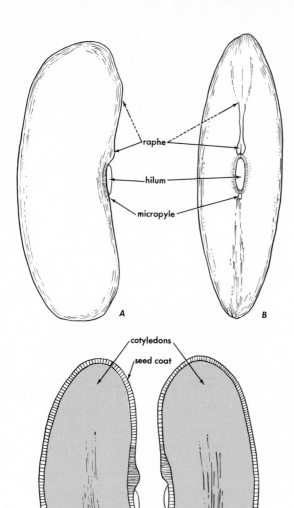

Common Bean. The fruit of the bean is a pod. Within it are seeds (beans). The external characters of the bean seed are more clearly seen after soaking it in water. Points of interest are the **hilum, micropyle,** and **raphe** (Fig. 16.24). The hilum is a large oval scar, near the middle of one edge, left where the seed broke from the stalk or **funiculus** when the beans were harvested. The micropyle is a small opening in the seed coat (integument) at one side of the hilum and was observed in the ovule, as the opening through which the pollen tube entered the ovule. The raphe is a ridge at the side of the hilum opposite the micropyle. It represents the base of the funiculus, which is fused with the integuments. Conducting tissue present in the funiculus will be continued in the raphe. At the end of the raphe, this conducting tissue may fan out over the ovule or seed and lose its identity as conducting tissue. This region is known as the **chalaza.** When present, it occurs at the end of the ovule or seed opposite the micropyle. An elevation or bulge of the seed coat adjacent to the micropyle marks the position of the radicle (embryo root) within the seed.

When the seed coat of a soaked bean is removed, the entire structure remaining is embryo; no endosperm is present. The following parts of the embryo

Figure 16.24 Bean (*Phaseolus vulgaris*) seed. *A*, external side view; *B*, external face or edge view; *C*, embryo opened.

ized to act as an absorbing organ. In lilies, onions, and others, the tip of the cotyledon is modified for food absorption. The seed coats of seedlings with an enclosed endosperm remain attached to the tip of the cotyledon for some time after it has emerged (Fig. 16.31). In most grasses, the entire cotyledon has become a highly specialized absorbing organ known as the **scutellum** (Fig. 16.25).

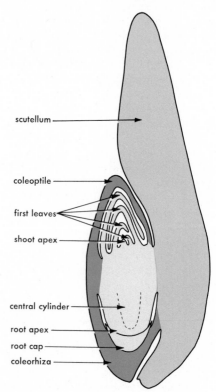

Figure 16.25 Longitudinal section of an embryo of corn (*Zea mays*).

can be observed: (a) a shoot consisting of two fleshy cotyledons, a short axis (the hypocotyl) below the cotyledons, and a short axis (the epicotyl) above the cotyledons, bearing several minute foliage leaves and terminating in a shoot tip; and (b) the root or radicle. In Fig. 16.24, the bean seed is opened to show the relationship of some of the parts just mentioned.

Castor Beans.   The external points of interest of this seed are: (a) **caruncle,** a spongy structure, an outgrowth of the outer seed coat; (b) hilum and (c) micropyle, which are covered by the caruncle; and (d) the raphe, which runs the full length of the seed (Fig. 16.22).

The chalaza region is marked, in the castor bean, by a protuberance at the end of the seed opposite the caruncle. In contrast to the bean seed, the castor bean seed has a massive endosperm in which the embryo is embedded. The shoot of the embryo consists of (a) two thin cotyledons with conspicuous veins, (b) a very short hypocotyl (shoot axis below the cotyledons), and (c) a minute epicotyl (shoot axis above the cotyledons). The root portion of the embryo axis consists of a small radicle.

Grasses.   Wheat is a member of the grass family (Gramineae). The so-called "seed" of corn, or of any other grass, is in reality a fruit (grain), as explained on p. 293. It is a one-seeded, dry, indehiscent fruit with the pericarp (ovary wall) firmly attached to the seed. Pericarp and seed coats are so firmly attached to each other, and to other tissues of the grain, that it is impossible to peel them away, as can be done with the seed coats of the common bean and castor bean. Therefore, in order to study the internal structure of the grain, it is necessary to employ sections.

A longitudinal section of a grain of corn is shown in Fig. 16.14. Note here that endosperm constitutes the bulk of the grain. The endosperm is composed of (a) an outermost layer (single row of cells) known as the **aleurone layer** and (b) a starchy endosperm. Cells of the aleurone layer contain proteins and fats but little or no starch. Starchy endosperm cells are filled with starch grains.

The embryo of grasses is quite unlike that of most other plant families. Like all embryos of seed plants, it has an axis with a **shoot apex** and a **root apex** (Fig. 16.25). The shoot apex of the corn embryo, together with several rudimentary leaves, are surrounded by a sheath, the **coleoptile.** The rudimentary root (radicle) is also surrounded by a sheath, the **coleorhiza.** At the juncture of shoot and root is a very short stem structure. A relatively large part of the grass embryo is a single cotyledon, which has for a long time been called the **scutellum.** It is a shield-shaped structure

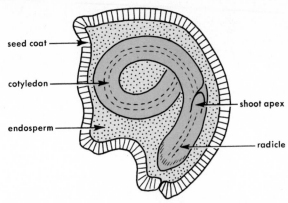

Figure 16.26   A longitudinal section through an onion seed (*Allium cepa*) showing the embryo coiled within the endosperm. (Redrawn after Wilhelm Troll.)

that lies in contact with the endosperm. The outer cells of the scutellum secrete enzymes that digest the stored foods of the endosperm. These digested foods move from endosperm cells through the scutellum to the growing parts of the embryo. Unlike the bean and castor bean seed, the cotyledon of grasses remains within the seed during germination and never develops into a green leaf-like structure above ground.

Lily or Onion.   The onion or lily seed consists of seed coats enclosing a small amount of endosperm. The embryo is a simple axis, the radicle and single cotyledon being quite prominent. The shoot apex is located at about the midpoint of the axis and appears as a simple notch. The embryo is coiled within the seed coats, the radicle usually pointing toward the micropyle in the seed coats (Fig. 16.26).

Dissemination of Seeds and Fruits

Agents in Seed and Fruit Dispersal.   The chief agents in seed and fruit dispersal are **wind, water,** and **animals,** including **man.**

Wind.   The structural modifications of seeds and fruits that aid in dissemination by wind are of several types. The most common of these are winged types (Figs. 16.15B, 16.27C) and plumed types (Figs. 16.10C, 16.27A, B).

Water.   Fruits with a membranous envelope containing air like those of sedges, or with a coarse, loose, fibrous outer coat, as in the coconut, are well adapted for dispersal by water. A great variety of fruits and seeds float in water, even though lacking

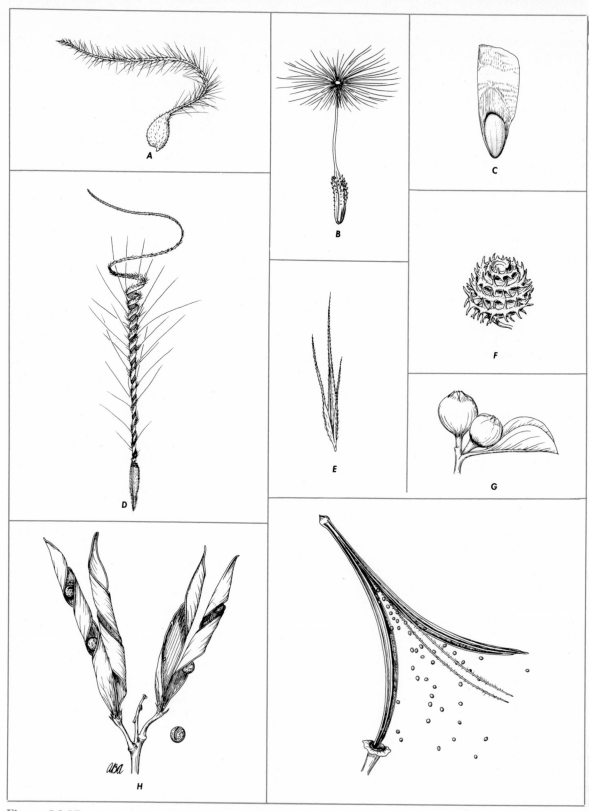

Figure 16.27 Fruits and seeds showing various devices aiding in dissemination. **Wind:** *A, Clematis,* ×4; *B,* dandelion (*Taraxacum vulgare*), ×10; *C,* seed of Coulter's big-cone pine (*Pinus coulteri*), ×2. **Attachment:** *D,* cranesbill (*Geranium*), ×3; *E,* foxtail (*Hordeum hispida*), ×3; *F,* bur clover (*Medicago denticulata*), ×5; *G,* fleshy **edible** fruit of *Cottoneaster,* ×4. Violent **dehiscence** of pericarp: *H,* vetch (*Vicia sativa*), ×2; *1,* California poppy (*Eschscholtzia californica*), ×2.

pecial adaptations to ensure buoyancy, and are readily transported long distances by moving water in the form of surface runoff, natural streams, irrigation and drainage channels, and floods. In irrigated districts, irrigation water is a very important means by which weed seeds are distributed.

Animals. Many seeds and fruits are carried by animals, both wild and domesticated. Seeds with beards, spines, hooks, or barbs adhere to hair of animals (Fig. 16.27D, E, F). Examples of seeds carried in this manner are wild barley, puncture vine, star thistle, sandbur, and cocklebur.

Seeds of many plants pass through the digestive tract of animals without having their viability impaired. Fleshy, edible fruits may be eaten by birds (Fig. 16.27G) and carried by them long distances, and then the seeds regurgitated or discharged with the excrement. Squirrels carry nuts, such as those of walnut and hickory, and the seeds of pines. Seeds of some aquatic and marsh plants and of mistletoe, which are covered with a sticky material, are carried on the feet of birds.

Summary of Seeds

1. Four types of seeds have been described in some detail; other variations occur. Seeds normally have one or two cotyledons, and they may store food within or outside the embryo, generally in an endosperm. In most dicotyledons such as cucumber and bean, food is stored in two cotyledons. Seeds of coffee have two cotyledons that serve as absorbing and, later, as photosynthesizing organs, and food is stored in an endosperm. In monocotyledonous seeds, food is stored in an endosperm.

2. In grasses similar to corn, a single cotyledon has become a highly specialized absorption organ which does not emerge from the seed. In seeds like those of lilies and onion, the cotyledon emerges from the seed coats and becomes green, but its tip continues to absorb food from the endosperm. In many monocotyledonous seeds, the food is absorbed through epidermal cells.

Seed Dormancy

If the seed is the culmination of the process of sexual reproduction initiated in the flower, it is also the start of a new generation! How long will seeds remain viable? In the fall of 1879, Beal, at the Michigan Agricultural College, buried 1000 freshly gathered seeds representing 20 different plants. Fifty seeds of each species were mixed with sand and placed in a pint bottle, so that the 1000 seeds were distributed equally among 20 bottles. These bottles, each containing 50 seeds of each kind, were buried at a depth of about 18 inches. Every 5 or 10 years until 1931, one bottle was dug up and the percentage of germination of each species was determined. Then the time interval was changed and bottles were dug up each 10 years. After 50 years, the seeds of 5 of the 20 original species were still viable, and after 60 years, seeds from three species were found to be living. In 1970, after 90 years of dormancy, seeds of only one species (*Verbascum blatteria*) were still viable—10 of the original 50 seeds germinated.

One of the most authentic cases of longevity of seeds under natural conditions is that reported for seeds of lotus (*Nelumbo nucifera*). Seeds of this plant are enclosed in a hard shell that is almost impervious to water. This hard seed coat must be ruptured before the seeds can germinate. Viable seeds of lotus buried for over 2000 years have been found in peat near Tokyo, but this is very unusual longevity.

What induces germination? Many seeds will germinate when provided with moisture, oxygen, and a favorable temperature. The water content of seeds is low, between 5 and 10%. The cytoplasm with contained organelles is scarcely recognizable crowded between the large amount of reserved food materials (Figs. 16.28A, B). Water is, at first, imbibed very rapidly, which results in the swelling of the protoplasm with a reappearance of organelles (Fig. 16.28C). Water uptake now slows down. Increase in water content makes possible an increased metabolic activity. From this time on the seed must have an adequate water supply for survival.

Many seeds will not germinate even when supplied with water, oxygen, and a favorable temperature. There are a number of different factors in different plants that are associated with the breaking of dormancy. For instance, light is necessary for the germination of some grass seed and for some strains of lettuce. Water and gases must pass through the seed coats. If seed coats restrict the movement of water and gases or are impermeable to one or the other or to both, the seed coats must decay, be broken, or be scratched, allowing water and oxygen to reach the embryo before germination can start. In some seeds, the embryo itself fails to germinate when provided with favorable conditions. In orchids, the seeds are disseminated while embryos are rudimentary and the embryos must develop further before germination.

Many kinds of seeds are known that will not germinate unless the seeds, while on a moist substrate, are subjected for a time to temperatures close to freezing. At the other extreme, some seeds will not

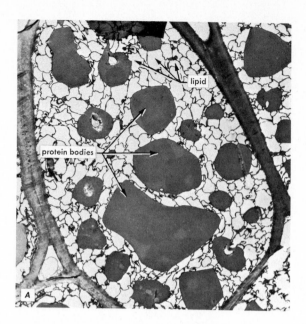

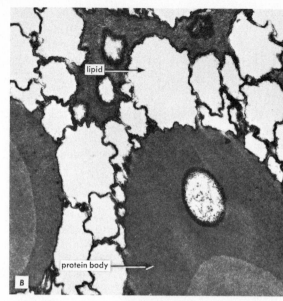

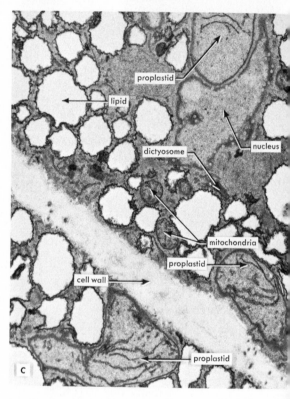

Figure 16.28 Electron micrographs of cotyledons of a squash (*Cucurbita maxima*) seed. *A,* food stored in the mature dry cotyledon crowds the dehydrated protoplasm into narrow strips between the lipid and protein granules, ×3000. *B,* after 24 hr hydration, the protoplasm has imbibed water; lipid and protein bodies still crowd the cytoplasm into narrow strands, and organelles are not apparent, ×41,000. *C,* after 72 hr hydration, organelles become apparent, ×41,000. (Courtesy of John Lott.)

germinate unless they have been subjected to the rather high heat of a flash fire. Another type of dormancy is brought about by the presence of natural chemical inhibitors. These occur in many fruits and serve to keep the seeds dormant while they are enclosed by the fruit. In other cases, inhibitors are present in the seeds themselves or may be produced by decaying leaves, or forest litter. The influence of the plant growth substances on dormancy and seed germination is still unclear.

Patterns controlling dormancy and germination are complex. They have evolved over a long period and confer certain competitive advantages that ensure survival.

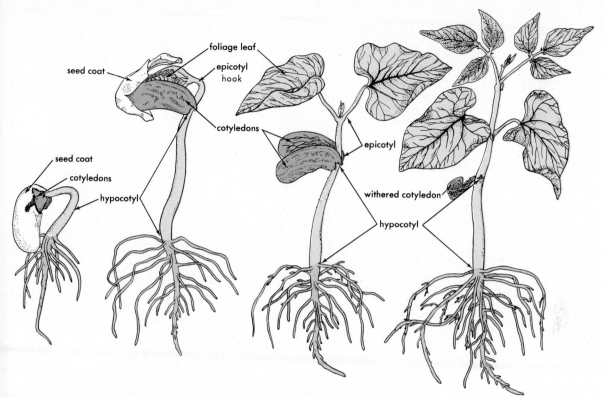

Figure 16.29 Stages in the germination of a bean (*Phaseolus vulgaris*) seed.

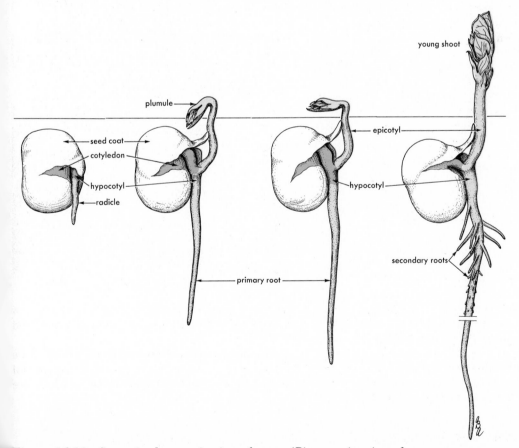

Figure 16.30 Stages in the germination of a pea (*Pisum sativum*) seed.

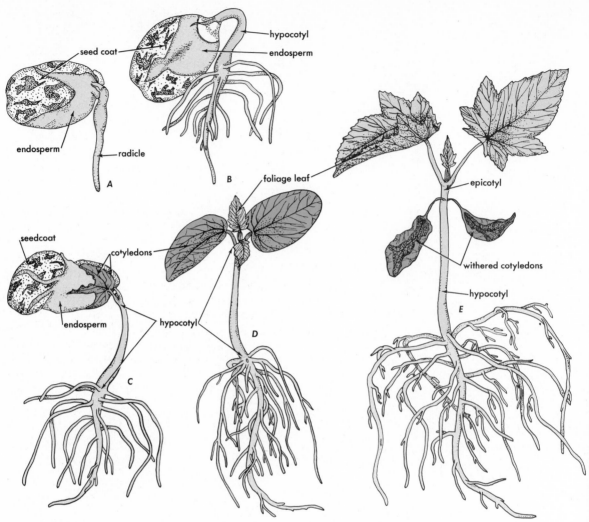

Figure 16.31   Stages in the germination of a castor bean (*Ricinus communis*) seed.

Seed Germination

That the processes of germination have started is generally first made evident by the swelling of the radicle. In all cases, the radicle imbibes water rapidly and, bursting the seed coats and other coverings that may be present, starts downward growth into soil. This helps to assure the young seedling an adequate supply of water and nutrients when the shoot breaks through the surface of the soil (Figs. 16.29, 16.30, 16.31, 16.32). While the succeeding steps of germination are essentially similar, there are variations. For instance, in germination of beans, peas, and onions, a structure with a sharp bend or **hook** is first forced through the soil (Figs. 16.29, 16.30, 16.32), but the structure forming the hook is different in each case. In bean (*Phaseolus vulgaris*), the hypocotyl (the part

of the shoot below the cotyledons) elongates (Fig. 16.29). In pea (*Pisum sativum*), the epicotyl (the shoot above the cotyledons) elongates (Fig. 16.30). In both cases, cotyledons and shoot apex remain below the ground for a short time. Above the ground, the hook now straightens. In the case of bean, a straightening of the hypocotyl raises cotyledons and shoot apex above ground (Fig. 16.29); a lengthening and straightening of the epicotyl will now pull the shoot apex away from the cotyledons (Fig. 16.29).

Upon straightening out of the pea epicotyl, however, the cotyledons will remain in the ground, and only the apex and first leaves will be raised upward (Fig. 16.30). In most cases, cotyledons, whether raised above ground or remaining below, do not carry on effective photosynthesis for a long time.

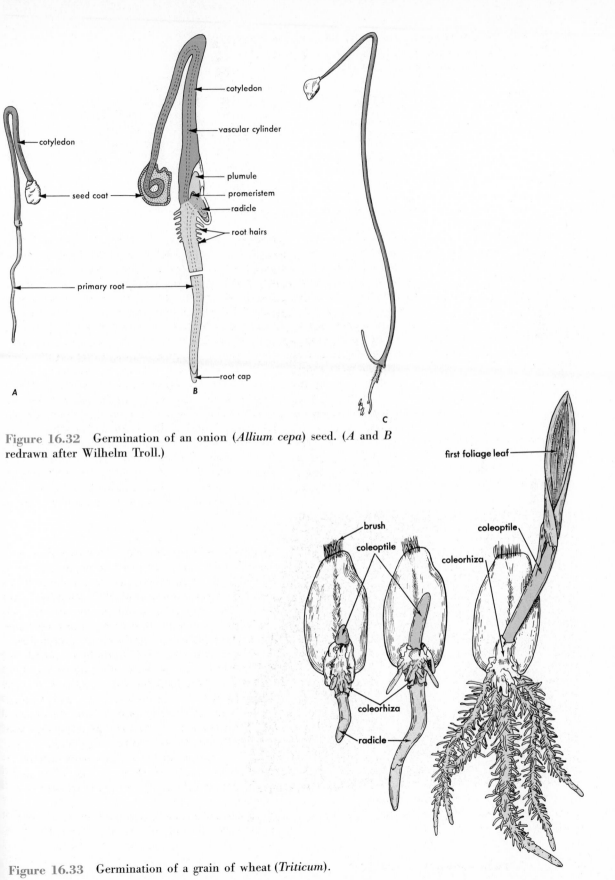

Figure 16.32   Germination of an onion (*Allium cepa*) seed. (*A* and *B* redrawn after Wilhelm Troll.)

Figure 16.33   Germination of a grain of wheat (*Triticum*).

In onion or lily, the primary root first penetrates the soil. Then a sharply bent cotyledon breaks the soil surface. The cotyledon slowly straightens out—in many instances, before all the nourishment has been absorbed from the endosperm. In such a case, both endosperm and surrounding seed coats may be lifted upward (Fig. 16.32A, B). The cotyledon of the onion is tubular and its base encloses the shoot apex (Fig. 16.32A, B). There is a small opening at the base of the cotyledon through which the first leaf finally emerges. In grasses, the situation is more complex. The shoot and root apices are enveloped by tubular sheaths known as the **coleoptile** and **coleorhiza,** respectively (Fig. 16.25, 16.33). The primary root rapidly pushes through the coleorhiza. The root continues to grow, but is eventually replaced by adventitious roots that arise from the lower nodes of the stem (Fig. 9.1D). The coleoptile elongates and emerges above ground, becoming 1-1$\frac{1}{2}$ inches long. At this time, the uppermost leaf pushes its way through the coleoptile and, growing rapidly, becomes part of the photosynthesizing shoot.

The life cycle of a flowering plant is now complete. We started with a seedling in Chapter 3 and we have now returned to a seedling.

Summary of Dormancy and Germination

1. Mature seeds are dormant and, depending upon the species and the immediate environment of the seeds, they may remain viable and dormant from a few months to many years.
2. Dormancy is usually broken by the provision of moisture, oxygen, and a favorable temperature. Other factors, such as light, removal of chemical inhibitors, or destruction of the seed coats, may be required in some instances.
3. In germination, the cotyledons may be elevated above the ground, sometimes to become photosynthetically active, or in the case of some monocotyledons, to continue absorption of the endosperm. In other cases, the cotyledons remain below the ground.
4. In the grasses, the single cotyledon functions as a specialized food-absorbing organ.

17

Inheritance

The individuals of a single-celled species living in a pond have been multiplying by dividing in two since life started, and they will continue to divide and grow as long as they, or other species to which they may give rise, continue to exist on this earth. The simplest type of life cycle is this division of a single-celled plant to form two new plants. Most individuals, plant or animal, are not so simply constructed; specialized cells are set aside for reproductive processes. For example, in flowering plants, flowers and seeds possess such specialized cells. Even in these plants, the special reproductive cells are in a sense immortal, for one can trace their history back to the beginning of life through a series constituted mainly of meristematic and reproductive cells. Every plant and animal living today must have a chain of ancestors stretching back to the beginning of life. During this long period, the characteristics of present-day plant and animal life have been derived. How has this happened? Are species of animals and plants constant? Do they change? How similar to each other are individuals of a given species? If individuals of a given species vary, what are the causes of variation? Can this variation be controlled or predicted? How do the specialized reproductive cells function? What is the importance of sex? These and many other questions arise when one considers the reproductive cycle of plants and animals. Partial answers may be given, but to no question stated above is there, as yet, a final and complete answer.

An Angiosperm Life Cycle

Briefly in the life cycle of a flowering plant—corn, for example—seed is planted and gives rise to the vegetative plant, which in time produces staminate flowers in the tassels and pistillate flowers in the ears. Pollen is formed by staminate flowers and carried to stigmas of pistillate flowers in young ears, either on the same individual plant or on adjacent individuals. Pollen tubes grow down long styles and eventually penetrate embryo sacs. Here, sperms and eggs fuse to form zygotes, which develop into embryos. A second sperm from the pollen tube fuses with the two polar nuclei of the embryo sac. The body resulting from this triple fusion develops into endosperm. Embryo and endosperm constitute the major portion of corn "seed," or, more properly, fruit (Fig. 16.14). Of very great importance in this cycle is the behavior of **chromosomes,** which are the carriers of **genes,** the units responsible for the morphological and physiological characteristics of every individual. During initial formation of pollen and

313

embryo sac, the chromosomes go through a characteristic division, the **meiotic division** or **meiosis.**

At fertilization, nuclei, with the gene-carrying chromosomes from two parents, are generally combined to form a zygote nucleus. These two phenomena, **meiosis** and **fertilization,** involving, as they do, the assortment and combination of chromosomes, form the basis of the sexual reproductive cycle. Indeed, the great diversity of plant and animal life is due largely to changes that occur in a chromosome set during meiosis and fertilization. In many plants (asparagus and date palm, for example), two separate parents are involved in sexual reproduction; in others, such as pea and wheat, the reproductive cells are produced on the same parent. Thus, chromosome behavior at meiosis and fertilization constitutes a mechanical basis for inheritance, and inheritance itself may be placed on a firm mathematical foundation.

Review of Mitosis

In prophase of mitosis, extended chromosomes divide and contract to form two chromatids per chromosome. The kinetochores then become aligned at the equator of the cell; during anaphase, sister chromatids move to opposite poles of the cell; during telophase, daughter nuclei are reconstituted; and the development of a cell wall completes the formation of two identical daughter cells. In this process, there has been no change in the chromosome number, and no reassortment of chromosomes; each daughter nucleus has the same complement of genes. It should be emphasized, however, that all cells containing $2n$ chromosomes have two sets of chromosomes. In corn, for instance, there are 20 chromosomes, two sets of 10 each. The chromosomes in corn have been numbered from 1 to 10. Each nucleus of the corn plant therefore contains the following chromosomes:

$$1 \quad 2 \quad 3 \quad 4 \quad 5 \quad 6 \quad 7 \quad 8 \quad 9 \quad 10$$
$$1 \quad 2 \quad 3 \quad 4 \quad 5 \quad 6 \quad 7 \quad 8 \quad 9 \quad 10$$

The two number 1 chromosomes are said to be **homologous** chromosomes, as are the two number 2 chromosomes, and so on. Each pair of chromosomes $(3 + 3, 4 + 4, 5 + 5,$ and so forth) constitutes a pair of homologous chromosomes. Each member of the homologous pair carries, at identical positions, or **loci,** a gene affecting the expression of the same trait. Such a pair of genes are known as **allelomorphs,** or **alleles.**

Characteristics of Meiosis

The division stages of chromosomes during meiosis are identical to the division stages of chromosomes undergoing mitosis. Meiosis begins with the chromosomes present as long, slender uncoiled threads (Figs. 17.1A, 17.2A). They proceed to shorten and thicken and split into chromatids as in mitosis (Fig. 17.2C). The chromatids separate at anaphase. Replication of DNA occurs in the interphase preceeding meiosis and again in the interphase after the completion of meiosis, when one chromatid is present in each chromosome. The single basic difference between mitosis and meiosis lies in the *pairing of homologous chromosomes* so that four chromatids, two derived from each of the pairing homologous chromosomes, are associated at midprophase (Figs. 17.2B, C). In mitosis, only the two sister chromatids, derived from a single chromosome, are associated at this point (Fig. 6.4D). This introduces a complication, for the chromatids of paired homologous chromosomes are able to exchange partners with each other (Fig. 17.2E). Such exchanges between chromatids derived from opposite members of a homologous pair of chromosomes (nonsister chromatids) (Fig. 17.2F) may be expressed as visible changes in the offspring. In order for this to take place, there appear to be two divisions in meiosis, which result in four cells from the initial mother cell. However, DNA replication does not take place during meiosis, so each of these four cells has only one complete set of chromosomes. Thus, during meiosis the two sets of homologous chromosomes found in the **diploid** parent are reduced to a single set of chromosomes, and the cell or plant bearing this single set is said to be **haploid.** The cells in which meiosis occurs are **meiocytes.** If they produce **spores**—cells able to produce new plants directly— as generally happens in plants, the meiocytes may be called **spore mother cells.** The spores resulting from meiosis are **meiospores.** Since meiospores give rise to a new generation of plants directly, it follows that plants arising from meiospores must be haploid. In order to regain the diploid state, two haploid cells must fuse. Cells capable of fusing are generally specialized cells and are called **gametes.** The diploid cell resulting from the gamete fusion is a **zygote** and will give rise by mitosis to a diploid, or $2n$, plant body.

Sometimes conditions arise that result in three, four, or more sets of homologous chromosomes per cell. This condition is known as **polyploidy** and, depending upon the number of homologous chromosomes, the plants are **triploid** ($3n$), **tetraploid** ($4n$), **hexaploid** ($6n$), etc.

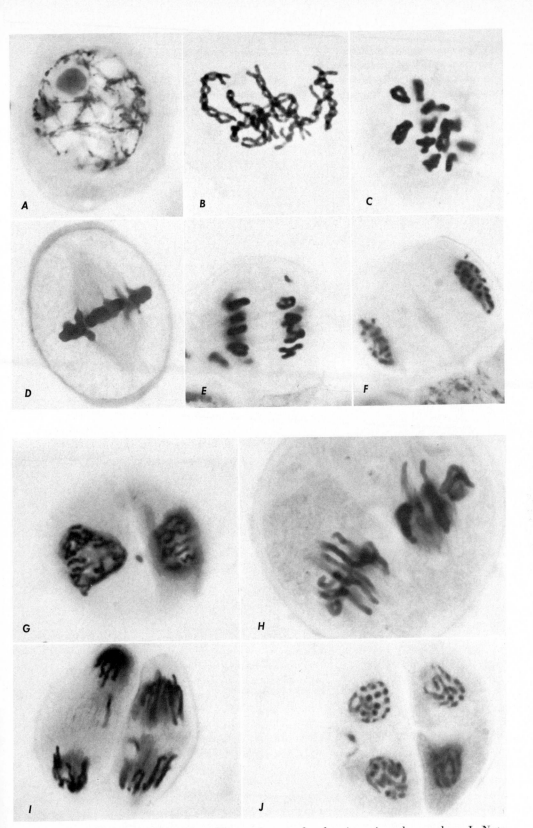

Figure 17.1 Meiosis in lily anther. Photomicrographs showing: *A*, early prophase I. Note paired threads. *B*, late prophase I. Each body represents two paired chromosomes; note chiasmata. *C*, late prophase I. Paired chromosomes; two chiasmata present. Compare with chromosomes of Fig. 17.2*E*. *D*, metaphase I. Note tractile fibers extending from chromosomes. *E*, anaphase I. *F*, telophase I. *G*, prophase II. *H*, metaphase II. *I*, anaphase II. *J*, telophase II. (*B*, courtesy of Gankin.)

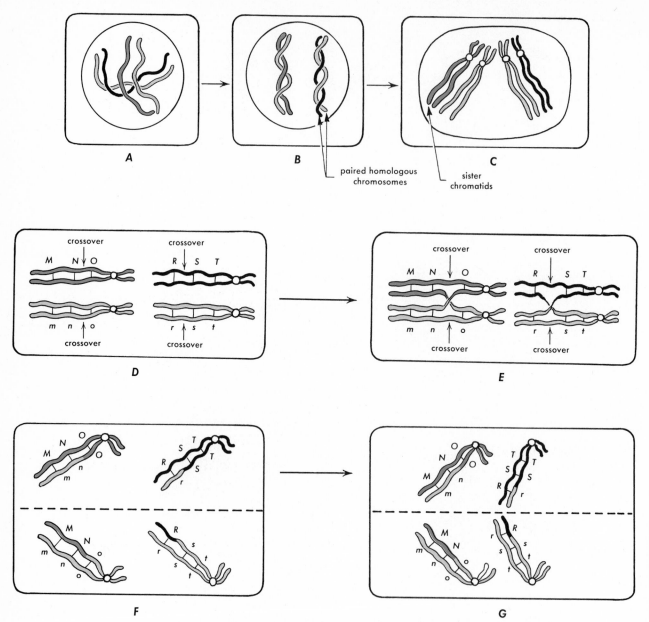

Figure 17.2 Diagram of meiosis, two pairs of homologous chromosomes are shown. *A*, early prophase with unpaired chromosomes. *B*, the homologous chromosomes have paired. *C*, the chromatids have formed. Those of a single chromosome are sister chromatids. Here and in the following diagrams they are represented as paired parallel lines (this is not the normal situation). *D*, markers on the chromosome, future location of crossover indicated by arrows. *E*, metaphase I. Nonsister chromatids of homologous chromosomes have broken and, at the crossover region, have exchanged pieces with each other. *F*, early anaphase I. Kinetochores separate; Sister chromatids are still associated at the kinetochore, but because of the crossover, nonsister chromatids are associated distally to the kinetochore. *G*, early telophase I. the chromosomes start to regroup into a nucleus; note that there is only one member of a homologous pair represented. *H*, interphase. Two sister chromatids attached at the kinetochores. *I*, early anaphase II. The chromosomes have moved to the metaphase II plate and the kinetochores have split. Here chromatids are moving to opposite poles of the cell. Note the disposition of the markers. *J*, early telophase II. *K*, four meiospores are formed. *L*, meiospores that would arise if the crossover had occurred distally to the third marker or in the short arm of the chromosome; the three markers in each chromosome would have remained linked in their original order.

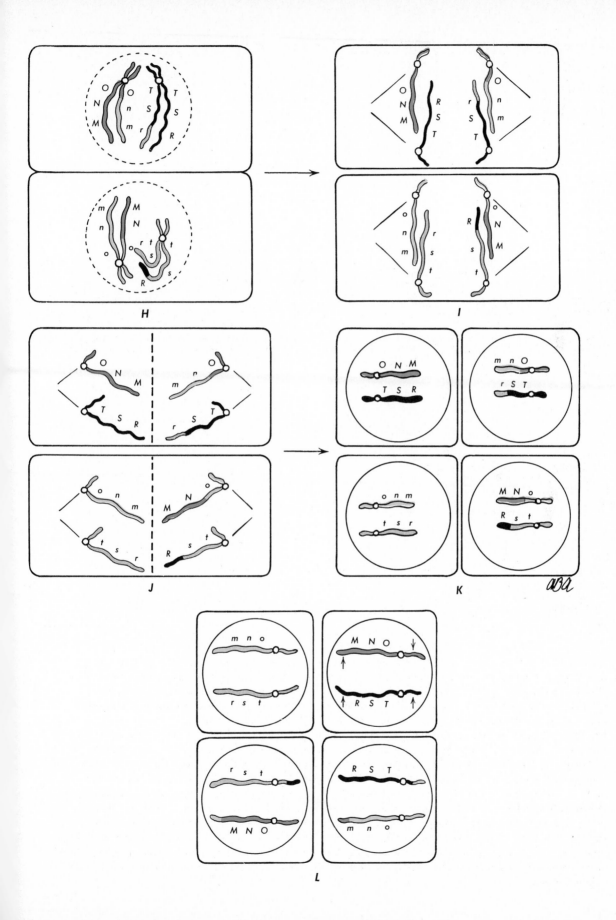

H

I

J

K

L

Since, apparently, two divisions are required for the completion of meiosis, it is common to designate the two divisions by the numerals I and II and the phases as prophase I, metaphase I, anaphase I, telophase I, prophase II, metaphase II, anaphase II, and telophase II.

Meiosis—First Division

Prophase I. During prophase I of meiosis (Figs. 17.1A, 17.2A, B, C), the nuclear threads of the vegetative nucleus, present in the diploid number, contract to form chromosomes, each with two chromatids; in this respect the process resembles mitosis. The process is complicated, however, because before the chromatids become apparent, the homologous chromosomes pair (the two number 1 chromosomes, for example). In so doing, they approach and coil about each other (Figs. 17.1B, 17.2B). The two chromatids of a chromosome appear only after the pairing of the homologous chromosomes is well-advanced (Fig. 17.2F). The resulting figure is composed of two paired homologous chromosomes and four chromatids (Figs. 17.1B, 17.2C). This stage of prophase I involves two phenomena: (a) the pairing of homologous chromosomes and (b) the formation of chromosomes with two chromatids each. Chromatids thus formed do not remain unchanged; they normally break at several points and rejoin in such a way that a given reconstituted chromatid may be composed of parts of four chromatids (Fig. 17.2C). This breaking and rejoining of the chromatids is called **crossing over,** and the cross formed by the chromatids involved in the interchange is known as a **chiasma (chiasmata,** plural) (Fig. 17.2E). Chiasmata may occur at any point along the paired chromosomes. Fig. 17.2D represents two pairs of homologous chromosomes, each with three pairs of genes to serve as markers; Mm, Nn, and Oo on the green homologs and Rr, Ss, and Tt on the gray homologs. If the chiasmata occurred between Oo and the kinetochore, or to the left of Mm, the relationship between these three pairs would not change: M, N, and O would remain **linked** together as would m, n, and o on the chromatids of the original chromosomes. However, in Fig. 17.2E, chiasmata are shown between Nn and Oo on the green chromosome and between Rr and Ss on the gray chromosome.

Thus, a third step in the prophase I of meiosis is (c) the formation of chiasmata due to the breaking and rejoining of chromatids from homologous chromosomes.

As in mitosis, the nucleolus disappears. The nuclear membrane also disappears during the later stages of prophase I (Figs. 17.1B, 17.2C).

Metaphase I. At metaphase I (Figs. 17.2 C, D), the kinetochores of the paired homologous chromosomes pass to the equator of the cell. Tractile fibers grow toward the poles (Fig. 17.1D).

Anaphase I. In anaphase I, the kinetochores of *whole chromosomes* separate from each other and, with their associated chromosomes, move to opposite poles of the cell. However, because of chiasmata formation, whole, complete chromosomes are not separated from each other, for on the side of the chiasma away from the kinetochore, sister chromatids will separate just as in mitosis (Fig. 17.2F). Anaphase I of meiosis is thus the separation of *two chromatids.* Since the chromatids have become variously modified, the separation involves both a separation of whole chromosomes (at the kinetochores) and a separation of sister chromatids as in mitosis (across the chiasma, Fig. 17.2F).

Telophase I. Following anaphase I, the chromatids group together at opposite poles of the cell (Figs. 17.1F, 17.2G) and immediately prepare for the second meiotic division. Each telophase chromosome consists, as usual, of two chromatids. In this sense, the n, or haploid number of chromosomes, is present. However, it must be remembered that at some levels this chromosome consists of two chromatids derived from the same parent (sister chromatids), and at other levels it consists of two chromatids derived from each of the two parents (nonsister chromatids). Are both these arrangements strictly haploid?

Meiosis—Second Division

A second meiotic division now follows, which separates these rearranged chromatids. During prophase II (Figs. 17.1G, 17.2H), chromosomes again form, each with two composite chromatids. The kinetochores approach the equatorial plate, forming metaphase II (Figs. 17.1H, 17.2I). As in mitosis, the kinetochores now split and separate. In anaphase II, single chromatids move to opposite poles of the cell (Figs. 17.1I, 17.2J) and are reconstituted in telophase II into nuclei (Fig. 17.2K). Each nucleus thus formed contains one of the four chromatids derived from the pairing of homologous chromosomes in prophase (Fig. 17.2L).

Walls develop about each new nucleus and associated cytoplasm, thus forming cells with the n or haploid number of chromosomes. Since the chromatids within each of these four cells have been variously modified by crossing over, each of the four cells may be genetically different. In just what way do they differ and of what importance is this variance in the life cycle of plants?

Distribution of Genes

The distribution of genes during meiosis depends upon the separation of the chromatids, and the location and amount of interchange taking place between the chromatids. If, for instance, as in Fig. 17.2D, the interchange took place to the left of Mm, then M, N, and O would move with the kinetochore at anaphase I, and the two telophase I nuclei would contain the markers M, N, O, and m, n, o. At metaphase II and anaphase II, the chromatids, bearing M, N, O and m, n, o, would separate and the three markers, M, N, O, m, n, o, would remain linked together so that only two types of meiospores M, N, O and m, n, o (Fig. 17.2L) would result. The same would happen if the exchange occurred to the left of Rr. T, S, and R would remain linked, as would t, s, and r. The number of different kinds of meiospores to be derived from one to several linked allelic pairs is two or four, depending upon whether crossing over has occurred between the alleles. Without crossing over, two of the meiospores would be similar; with crossing over, all four meiospores would be different. The proportion of meiospores having crossover chromosomes in a large number of meiotic divisions will depend upon the number and location of interchanges.

For example, if exchanges occur between Oo and Nn in four out of 100 meiocytes, eight of the resulting 400 meiospores would show interchanges. This is because one meiocyte gives rise to four meiospores, and because a single interchange results in two crossover and two noncrossover chromatids. In this case, 2% of the resulting meiospores would show interchanges. The allelomorphs Mm are further removed from Oo than are Nn (Fig. 17.2D) It turns out that the greater the distance separating genes, the greater the number of chromatid exchanges. We might expect that exchanges would occur between Mm and Oo in eight of 100 meiocytes. This is twice as many as occurred in the shorter distance between Oo and Nn. Sixteen of the 400 meiospores would show exchanges resulting from crossing over between Oo and Mm. The percentage of exchanges between genes linked together on the same chromosome is taken to indicate the relative distances separating them and is expressed as **crossover units.**

Now, consider the distribution of the genes, Oo and Tt, which are on nonhomologous chromosomes. Each of the four meiospores are different in the example shown; they are OT, Ot, oT and ot.

Combining segregation of chromatids during meiosis with chromatid interchanges, or crossing over, brings about variation of the genetic constitution of the resulting meiospores.

Meiospores give rise to haploid plants of varying genetic constitutions; they will all produce gametes that fuse to give rise, once again, to a new diploid generation. The gametes carry the genetic constitutions endowed them by the meiospores, and if meiospores differ, gametes must also differ. The fusion of gametes having different genetic constitutions further greatly enhances the possibility for variation.

Fertilization

While crossing over is probably always present, an explanation of the combined phenomena of meiosis and fertilization is greatly simplified by ignoring it for the present.

Let us assume that the Tt genes occur in a cross-pollinating plant. Recall that meiosis occurs in the formation of pollen and also in the development of the embryo sac. One-half the pollen grains and eggs formed will carry the gene T; and the other half will carry the gene t. Since large numbers of both pollen and egg cells are formed, and since the fusion of any two may occur, we should expect the following combination:

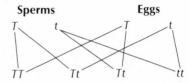

The many hundreds of sperm cells produced in hundreds of pollen tubes will carry either T or t genes in approximately equal numbers. These sperms will unite with eggs carrying either the T or t gene.

Notice that there is one chance for the formation of zygote TT, one chance for the formation of zygote tt, and two chances for the formation of zygote Tt. If the genes Tt governed height, we should expect to find plants of different heights occurring in certain very definite ratios in a breeding plot. That this situation actually does occur has been well-established by a long series of experiments.

Breeding Systems in Plants

The art of plant and animal improvement is very ancient. It has been practiced by many different races of men for thousands of years. Wheat, corn, and rice are examples of plants that have been improved by men of three different continents. The domestication of these grains was accomplished before the beginning of recorded history and ranks as an accomplishment of major importance. The task was so well done that the tribes that grew these grains were assured of a food supply and could turn their energies to activities other than getting food. Civilization resulted. Attempts to increase crop yields and the nutritive values of plants, or to introduce new and better food plants, have been continued down through the ages to the present time. The last 50 years has seen a great increase in plant and animal improvement and in a knowledge of the mechanism of inheritance in plants and animals.

Within the past decade some dwarf varieties of wheat, developed in Mexico under the direction of Victor Borlaug, have doubled the yields of wheat in many countries. Similar advances have been made with rice. High-yielding dwarf strains of this grain have been introduced into almost all the tropical countries of Asia. The introduction of these two crops has so enlarged the food production of the developing countries, that we are in the midst of a "Green Revolution."

Production of hybrid corn in the United States led the way to the development of these high-yielding crops. Only a few genes account for high yields, and the great majority of high-yielding strains carry the same genes. Thus, only a handful of varieties of some important crops are cultivated. For instance, in the United States, two types of peas and nine varieties of peanuts constituted 95% or more of their respective crops in 1969. This means a great genetic uniformity of certain particular genes. Most hybrid corn previous to 1970 carried the T gene which resulted in a male sterile plant. It turned out that this gene also rendered the plants very susceptible to the fungus causing corn blight. In 1970, about 15% of the corn crop was lost to this fungus. This occurrence points up the hazard of genetic uniformity in agricultural plants. Very little is currently known about the interactions between wheat and its parasites, and there definitely exists the probability that the new wheats, and rices too, are very vulnerable to damage by disease. Thus, while great increases in production have been achieved, genetic similarity has resulted, introducing the very great possibility of widespread crop damage with the danger of famine. It has occurred before.

Mutation

Most genes have been discovered by examining the individual plants in large uniform populations to find the rare individual that differs from its neighbors. It has been found that such differences occur in rather characteristic ratios. When these new plant types are bred to the standard wild type, they are generally recessive to the wild type and the new traits seem to be deleterious to the plants that bear them. These new genes are called **mutations.** They may be due to a variety of causes, but a common explanation is that an error has occurred in the duplication of the DNA helix. How might this occur?

In our discussion of the DNA molecule and its replication, it was pointed out that gene specificity resided in the arrangement of the base pairs, adenine–thymine and cytosine–guanine, in the DNA helix. It was further pointed out that the helix uncoiled in replication and, as the two strands separated, a new strand formed exactly like the disjoining member. The perfect replication of thymine opposite adenine and cytosine opposite guanine results in the new identical DNA helix. Sometimes there may arise a **copy error** caused perhaps by a modification of the sugar–phosphate bonds, perhaps by a small deletion of a portion of a chromosome. Such a copy error would give the replicating DNA helix a slightly different base sequence than that found in the original helix. Now, if our hypothesis is correct that the DNA molecule is involved in the production of specific cellular protein through mRNA, we would expect the change of sequence in the base pairs to be reflected in all events leading to the final form of the trait related to the gene in question. Such a change in the replication of the DNA molecule may be one cause of mutations.

Variation

Variation in a population of plants or animals may be due to a number of conditions. Variation is probably always related to genes. The number of genes involved in the expression of a trait, and the relationships between genes, play an important part in the expression of the trait and in the relative importance of the environment upon the type and amount of variation. Mendel's revolutionary approach to plant breeding lay in his plan to follow the numerical presence or absence of single traits through a number of generations of carefully controlled pollinations and seed collections. The differences between the progeny were large and easily separated the progeny into distinct categories. This is **discontinuous variation.** It may be contrasted with **continuous variation,** in which the variation is evenly and widely distributed about a mean (Fig. 17.7). Continuous variation is a very important aspect of modern genetics. However, the principles established through studies on discontinuous variation form the foundation for explaining the mechanism underlying continuous variation as well as other breeding systems.

Discontinuous Variation. Modern **plant breeding** and **genetics** date from 1900, when Gregor Mendel's

Figure 17.3 Gregor Mendel. (Courtesy of The Bettmann Archive, Inc.)

(Fig. 17.3) experiments on hybridizing garden peas were first appreciated. The importance of his experiments lies in the introduction to plant breeding of several procedures which, in 1900, were new and revolutionary. They are as follows:

1. The study of the inheritance of single or **unit** characters.
2. The keeping of accurate records of the number of times a given unit character appears in the offspring of selected parents.
3. The maintenance of the pollination of the experimental plants under the complete control of the investigator at all times.

While these procedures are accepted today as self-evident, they were so revolutionary in 1865 when Mendel's work was done that the standard scientific magazines of the time refused to publish Mendel's results. Instead, his experiments were presented before a small society of men in his home town. This society was not greatly unlike some of the present-day luncheon clubs and, like some of them, it published its proceedings. Thus, Mendel's work was placed on record, where it lay unnoticed for 35 years. In 1900, it was discovered by three prominent European plant breeders and it has become the cornerstone of all modern work in plant and animal genetics.

The Monohybrid Cross. Mendel selected garden peas for his experimental plants. He carefully took pollen from the anthers of a dwarf-growing variety and dusted it on the stigma of a tall-growing variety. The seeds resulting from this cross-pollination were collected and planted the following season. All the plants that grew from these seeds were tall. The same results were obtained when the tall variety supplied the pollen:

$$\text{tall plant} \times \text{dwarf plant (pollen parent)}$$
$$\downarrow$$
$$\text{all tall plants}$$

or

$$\text{dwarf plant} \times \text{tall plant (pollen parent)}$$
$$\downarrow$$
$$\text{all tall plants}$$

Then flowers of the tall progeny were self-pollinated. Peas are normally self-pollinated, and so in order to obtain this seed Mendel merely had to keep stray pollen from reaching the stigmas of his experimental plants. The seeds resulting from these self-pollinated flowers were collected and planted the following spring. Upon counting the plants that grew from these seeds, he found that 787 were tall plants and 277 were dwarf plants. In other words, about three-fourths of these pea plants resembled one of the original pair of parents and about one-fourth resembled the other parent (Fig. 17.4).

It has been found convenient to give designations to these generations of plants. The two original parents are designated by **P.** The first generation, comprising only tall pea plants, is the **first filial generation,** or the **F₁**; the second generation of three-quarters tall and one-quarter dwarf pea plants is the **second filial generation,** or the **F₂**. Subsequent generations of self-pollinated plants would be consecutively the F_3, F_4, F_5, and the like.

Note in the cross that, even though we are dealing with two different plants, we are concerned with only **one character**—height of growth. A cross dealing with a *single unit character* is a **monohybrid cross.** Mendel studied, in all, seven monohybrid crosses. The results were similar to those just described for height. They were as follows:

1. The F_1 always resembled one of the parents: tall × dwarf gave tall; wrinkled × round seeds gave round seeds; and the like. From this result, Mendel concluded that one expression of a given character was **dominant** (tall, or round seeds), whereas the other aspect of the same

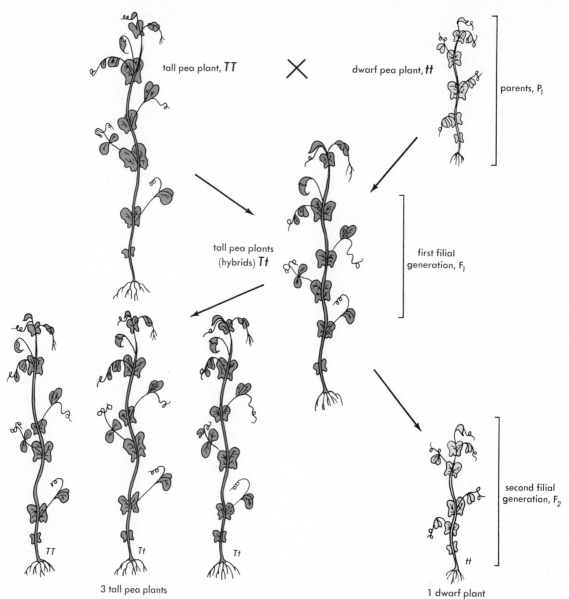

tall pea plant, *TT*

×

dwarf pea plant, *tt*

parents, P₁

tall pea plants (hybrids) *Tt*

first filial generation, F₁

second filial generation, F₂

TT   *Tt*   *Tt*

tt

3 tall pea plants

1 dwarf plant

Figure 17.4 Diagram of monohybrid cross between tall (*TT*) and dwarf (*tt*) pea plants.

character was **recessive** (dwarf, or wrinkled seeds).

2. Selfing of the F_1 resulted in an F_2 generation in which the original forms of a given character **segregated** to give three times as many dominant plants as recessive plants.

Mendel further postulated a possible mechanism for this genetic behavior. He knew that the F_1 plants must contain factors, now called **genes**, responsible for both tallness and dwarfness, even though all the plants were tall, because (a) tall and dwarf plants were crossed to produce the tall F_1, and (b) upon selfing, the tall F_1 gave rise to both tall and dwarf individuals. If we then represent the gene for tallness by T and the gene for dwarfness by t, we may diagram the genes determining height that are present in the tall F_1 by Tt:

$$Tt = \text{tall-growing } F_1 \text{ pea plant}$$

Since the F_1 hybrid contains two genes, T and t, determining height, each true breeding parent plant should contain two similar genes, thus:

$$TT = \text{tall-growing parent}$$
$$tt = \text{dwarf-growing parent}$$

Mendel assumed that the pollen would contain but one gene, and that the eggs similarly would contain but one gene. Thus, the tall parent, *TT*, would produce *T* pollen and *T* eggs, and the dwarf parent, *tt*, would produce *t* pollen and *t* eggs.

The parent plants whose nuclei contain the homologous chromosomes with the identical genes *TT* or *tt* are said to be **homozygous** for either the dominant or recessive genes. The F_1 plant with the homologous chromosomes carrying the genes *Tt* is said to be **heterozygous.**

Pollen produced by the heterozygous tall F_1 plant, *Tt*, will contain either gene, *T* or *t*. Likewise, eggs produced by this F_1 plant will contain either gene, *T* or *t*, and the sperm nuclei formed from the pollen will contain either gene, *T* or *t*. Many thousands of pollen grains, sperms, and eggs will be formed by the tall heterozygous F_1 plant. A sperm containing the gene *T* may unite with either a *T* or a *t* egg, as follows:

T sperm + *T* egg = *TT* zygote
T sperm + *t* egg = *Tt* zygote

A sperm containing the gene *t* may unite with a *T* egg or a *t* egg, as follows:

t sperm + *T* egg = *Tt* zygote
t sperm + *t* egg = *tt* zygote

These four possible combinations of genes will occur at fertilization and may be diagrammed as follows:

Possible sperms

	T	*t*
T	Zygote 1 *TT*	Zygote 2 *Tt*
t	Zygote 3 *Tt*	Zygote 4 *tt*

Possible eggs

The possible types of eggs are arranged along the left of the checkerboard, and the possible types of sperms at the top. The squares of the checkerboard represent the possible zygotes, and they are derived by combining the gametes opposite the squares. For instance, zygote 1 is derived from sperm *T* and egg *T*; zygote 2 is derived from sperm *t* and egg *T*, and so on. The genes, *Tt*, or *TT*, or *tt*, for example, present in the plant, constitute the **genotype** of the plant. Reference to the diagram will show that there are three different genotypes: *TT*, *Tt*, and *tt*. The appearance of the plant is the **phenotype.** For instance, the appearance or the phenotype of the plant *TT* is tall; of the plant *Tt* is tall; and of the plant *tt* is dwarf.

Note that there are three times as many tall plants represented on the checkerboard as there are dwarf plants. Furthermore, if Mendel's assumption regarding these factors or genes is correct, the homozygous tall plant *TT*, when self-pollinated, should give only tall offspring; the homozygous dwarf plant *tt*, when self-pollinated, should give only dwarf offspring; the tall heterozygous *Tt* plants, when self-pollinated, should give tall and dwarf plants in the ratio of 3 tall plants to 1 dwarf plant. This is what actually happens. A complete monohybrid cross is shown in Fig. 17.4.

This cross gives us additional information about these genes. It might be thought that they could interact with each other to bring about some modification in their genotypes, either in the zygote nucleus, or in the line of vegetative nuclei leading to the next generation of spore or pollen mother cells, or that the cytoplasm could act upon them as they are replicated at each somatic division in the interval between fertilization and meiosis. But none of these things happens. Genes *T* and *t* retain their identity and separate from each other inviolate at each succeeding meiosis, generation after generation. Another important fact about them is that they occupy equivalent spots or loci on homologous chromosomes. Such pairs of genes maintaining their integrity and having the same loci on homologous chromosomes are known as **alleles.**

The Dihybrid Cross. Let us consider a cross concerning two characters, height of the plant and form of the seeds. This is a **dihybrid cross,** that is, a cross involving *two unit characters* (Fig. 17.5). One parent plant is tall and has round seeds; the other parent plant is dwarf and has wrinkled seeds. We know that both these parent plants are homozygous for these particular characters because upon self-pollination they breed true. When cross-pollinated, all the resulting F_1 plants are **tall** and have **round** seeds. We know from this outcome that tallness is dominant over dwarfness and that roundness is dominant over wrinkledness. We may therefore write the genotype of the parent plants as follows:

The tall plant with round seeds,   *TTRR*
The dwarf plant with wrinkled seeds, *ttrr*

Since all plants possess the characters of height and seed form, factors determining these characters must always be present in every plant.

After meiosis, the pollen and eggs produced by the tall plant with round seeds will contain the genes *TR*. The pollen and eggs produced by the dwarf plant with wrinkled seeds will be *tr*. Union of sperms from a pollen grain *TR* with an egg *tr* (or a pollen grain

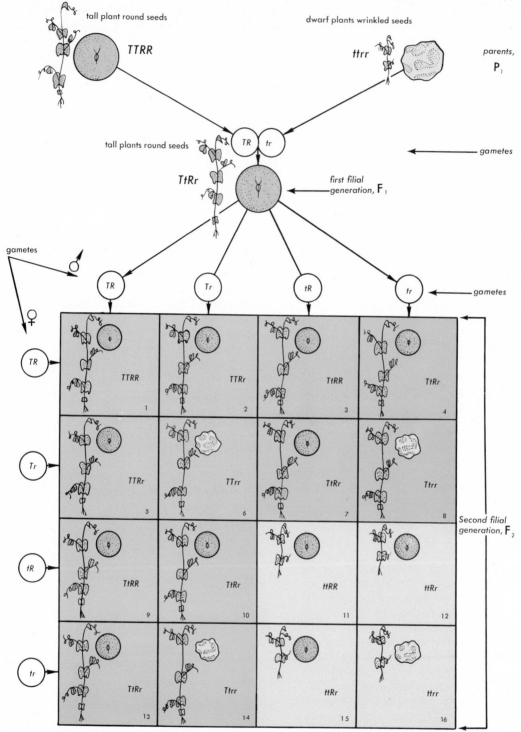

Figure 17.5 Diagram of a dihybrid cross between a tall pea plant with round seeds (*TTRR*) and a dwarf pea plant with wrinkled seeds (*ttrr*).

tr with an egg *TR*) will result in a zygote having the genotype *TtRr* (Fig. 17.5). This plant will be tall and have round seeds. It is heterozygous for both height and seed shape.

The four possible types of eggs and sperms that may be formed during meiosis from a plant with the genotype *TtRr* are shown in Fig. 17.6. The sperms and eggs that result are:

Sperms *TR, Tr, tR, tr*
Eggs *TR, Tr, tR, tr*

Any one of the four types of sperms may unite with any one of the four types of eggs. The number of possible combinations is best shown on the checkerboard of Fig. 17.5. The four types of sperms are listed along the top of the checkerboard; the eggs are placed at the left. Each square represents a possible zygote. Examination of these squares will show that four phenotypes are possible:

1. Nine tall plants with round seeds, squares numbered 1, 2, 3, 4, 5, 7, 9, 10, 13.
2. Three tall plants with wrinkled seeds, squares numbered 6, 8, 14.
3. Three dwarf plants with round seeds, squares numbered 11, 12, 15.
4. One dwarf plant with wrinkled seeds, square number 16.

A dihybrid cross carried through the F_2 generation may be simply diagrammed as follows:

Parent genotypes	*TTRR*	×	*ttrr*
Parent phenotypes	Tall plants with round seeds		Dwarf plants with wrinkled seeds
Gametes	*TR*		*tr*
F_1 genotype		*TtRr*	
F_1 phenotype		All plants heterozygous, tall with round seeds	
Self-polination of F_1	*TtRr*	×	*TtRr*
Gametes	*TR, Tr, tR, tr*		*TR, Tr, tR, tr*
F_2 phenotypes	9 tall plants with round seeds		
	3 tall plants with wrinkled seeds		
	3 dwarf plants with round seeds		
	1 dwarf plant with wrinkled seeds		

In this cross, it is important to note that the genes for height and seed shape separate or *segregate*

independently of each other during the formation of meiospores from the spore mother cell.

Soon after Mendel's work was discovered, it was noticed that his results could be explained by assuming that the genes were located in the chromosomes. It has now been well-demonstrated that the genes are carried in the chromosomes, and maps have been prepared showing their relative positions in the chromosomes of many plants.

The Backcross. In a backcross (also called a test-cross), offspring are crossed to one of the parents. Usually, a parent having homozygous recessive alleles is selected, because such a cross yields much information regarding the genotype of an offspring. For instance, there are four different phenotypes in a dihybrid cross and nine different genotypes. There are nine tall F_2 plants with round seeds, and they may have any one of four genotypes: *TTRR, TTRr, TtRR, TtRr* (Fig. 17.5). If the progeny of a backcross between one of these F_2 plants and a homozygous recessive parent (*ttrr*) contains equal numbers of tall plants with round seeds, tall plants with wrinkled seeds, short plants with round seeds, and short plants with wrinkled seeds, the genotype of the F_2 plant must have been *TtRr* (Table 17.1).

The backcross also provides information as to whether genes are linked. The crossover units separating them may be determined. For instance, there are genes in corn that are expressed as white sheaths on leaves, and liguleless blades. The backcross here consists of crossing a plant homozygous for the recessive white sheaths (*ws ws*) and liguleless blades (*lg lg*) with a plant heterozygous for these traits: *Lg lg* and *Ws ws*. If these alleles are on different chromosomes, the four types should appear in the progeny in approximately equal numbers. If the alleles are linked on the same chromosome, the numbers of wild-type plants should approximately equal the numbers of plants showing white sheaths and blades without ligules. A small number of plants with the dominant ligule leaves, but with the recessive white sheath, accompanied by an equal number of plants with the complementary traits, would indicate some crossing over between linked genes. It turns out that about 46% of the progeny of such a cross resemble the heterozygous parent and another 46% have white sheaths and ligule blades. About 8% were crossover types. Therefore, these genes, *lg* and *ws*, are linked and separated by eight crossover units.

In nature, backcrossing of hybrids with parental types results in introgressive hybridization: a continuous gradient of plant forms between the two parental types. In this case, the parental types are species and the intermediates are difficult to classify taxonomically (see Chap. 30).

TABLE 17.1 Progeny of Backcross: *TtRr* × *ttrr*

Gametes of *TtRr*	Gametes of *trtr*	Genotype in progeny of backcross	Phenotype of progeny of backcross
TR	tr	TtRr	Tall, round seeds
Tr	tr	Ttrr	Tall, wrinkled seeds
tR	tr	ttRr	Dwarf, round seeds
tr	tr	ttrr	Dwarf, wrinkled seeds

Other Breeding Systems

The breeding experiments so far discussed have involved one or two genes directly related to one or two traits. Height was controlled by one gene, leaf color or flower color, by another, and so on. These genes were independent of each other in activity, segregation at meiosis, and combining at fertilization. This situation probably occurs in only a relatively few cases. A variety of other more complicated breeding systems are known: (a) Numerous genes may play a part in determining the expression of a trait. This results in **continuous variation,** rather than in the **discontinuous variation** just discussed. (b) We have considered the presence of a single pair of alleles, at one chromosome locus, which showed expression in the phenotype as the simple presence or absence of a trait. Sometimes there are several

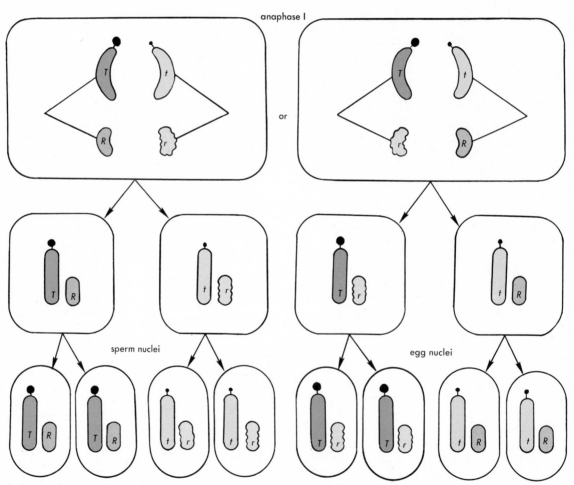

Figure 17.6 Types of sperms and eggs that may be formed from a plant having the heterozygous genotype *TtRr*.

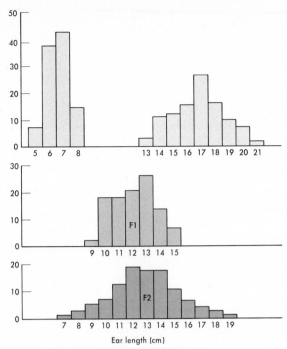

Figure 17.7 Continuous variation. Two varieties of corn are crossed. Black Mexican has long ears and Tom Thumb has short ears. The lengths of the ears of the F_1 show continuous variation between the ear lengths of the parents; there is continuous variation over a greater range in the F_2. (Information from E. M. East, graph adapted from S. L. Brewbaker, *Agricultural Genetics*, Prentice-Hall, Englewood Cliffs, N.J.)

5–8 cm long with a variety (Black Mexican) having ears 13–21 cm long, the ears of the F_1 generation were 9–15 cm long. The ears were 7–19 cm long in the F_2 generation. A classification of the ears of the F_1 and F_2 generations by length showed a normal distribution: the largest number of ears were close to the average length, with the number of ears gradually decreasing as extremes of ear length were reached (Fig. 17.7). Most traits of plants and animals involving growth rates, vigor, coloring, quality and amount of oil or starch, or quality of fiber or flour, show continuous variation. Traits showing continuous variation must be studied by statistical methods. It is necessary to count many individuals and to take many measurements, weights, and evaluations of color on a colorimeter. The collected figures are then processed by computers. These are **metrical traits.** They are influenced by numerous genes in different locations in the genotype. As the number of genes contributing to the expression of a trait increases, the importance of the contribution of an individual gene to the trait decreases. The many genes thus contributing to the expressing of a trait are **polygenes.** Polygenes are no different from other genes; some may show dominance, and they may interact with each other or mutate. Traits controlled by polygenes more easily respond to the environment than do traits controlled by a single or a few genes. These factors make continuous variation difficult to study, but in the age of the computer, investigations on continuous variation are providing significant information on the mechanism of inheritance.

Multiple Alleles and Pseudoalleles. On p. 323 we defined alleles as a pair of genes that retained their identity and separated inviolate at each succeeding meiosis. While it is true that genes do not appear to react with each other or to be greatly influened by their environment in their normal process of replication, we are no longer sure that a gene can be strictly defined as a unit of crossing over. This is because all alleles do not neatly determine the expression of a character.

Sometimes a given locus may not consist of a single allele, but may apparently contain an allelic series in which one allele is dominant over a series of recessive alleles. Such alleles are frequently called **multiple alleles.** It is important to remember, however, that only two alleles are present at a time, one on each homologous chromosome.

Alleles are paired genes and, by definition, at a given locus only two, one on each chromosome, can be present at a given time. Generally, one of these is dominant, the other recessive, and in an allelic series this same relationship holds. One gene is

alleles at one locus and their action may show a series of slight differences in the expression of the trait. Several alleles at one locus are known as **multiple alleles.** (c) The number of heterozygous alleles in a genotype seems to be directly related to the vigor of an individual or population. This is **heterosis.** (d) More than two sets of chromosomes occur in many plant groups. This is known as **polyploidy,** and it influences the vigor of a plant, as well as the genetic ratios to be expected in breeding experiments. Polyploidy rarely occurs in animals. (e) Genes interact with each other. (f) Portions of chromosomes may become **inverted** and, in the heterozygous condition, show interesting and unexpected expressions. (g) Parts of nonhomologous chromosomes may be exchanged, and such **translocated** portions of chromosomes greatly change the genetic expression of genes. We shall briefly consider selected aspects of continuous variation, multiple alleles, heterosis, and polyploidy.

Continuous Variation. When E. M. East (in 1905) crossed a variety of corn (Tom Thumb) having ears

generally dominant over all others, but the remainder show varying degrees of dominance over each other.

In corn, at least eight alleles forming an allelic series control the expression of pigment in the leaves, stem, aleurone, and pericarp. They are all located on a chromosome at a point known as the *A locus*. At least three different types of pigments are influenced by the activity at this locus; purple anthocyanins, yellow-brown anthocyanins, and a red-brown pericarp pigment. The allele A^{st} is dominant over all other alleles with respect to plant and aleurone pigmentation, giving purple pigmentation in the leaves and aleurone and red in the pericarp. However, several in the series tend to dilute these colors when present in a heterozygous condition with A^{st}.

A type of recombination may rarely occur between the alleles of such a series, or in situations resembling such a series. In this case, we may speak of **pseudoalleles.** In *Drosophila*, about once in 500,000 crosses, crossing over between multiple alleles occurs. This means that these multiple alleles are not strictly alleles, for they behave like separate genes in that crossing over occurs between them at rare intervals. They are pseudoalleles, and they prove that the classical gene may itself be subdivided into physical units. Indeed, in very rapidly multiplying organisms such as bacteria, it is possible to study chromosome interchange within the confines of a gene. In corn, pseudoalleles are known that are separated by 0.0015 of a map unit, a distance that matches the space occupied by about 1,500,000 of the nucleotide pairs of the DNA molecule. In bacteria, pseudoalleles separated by only 0.000001 of a map unit are known, and this figures out to be about 10 nucleotide pairs in the DNA molecule.

Heterosis.   Heterosis refers to the presence of a large number of heterozygous alleles in a genotype. The recessive alleles are generally less desirable. Therefore, an increase in the vigor of a plant should occur when one member of a pair of homozygous recessive alleles is replaced by a dominant allele. Thus, an increase in heterozygosity should increase vigor. An increase in heterozygosity is easily obtained. Inbreeding rapidly increases the homozygosity of a genotype. By crossing carefully selected inbred lines, it is possible to attain a very high degree of heterozygous alleles, and the progeny of such a cross exhibit great vigor, being larger and more robust than the initial parents of the pure lines. If the increased vigor were due simply to the dominant genes, genotypes having large numbers of homozygous dominant alleles should be as vigorous as plants with an

equal number of heterozygous alleles. This is not the case. Vigor seems to depend upon the heterozygous condition and not upon the presence of many dominant alleles. A number of hypotheses have been developed to account for the relationship between heterozygosity and vigor, but the reason for the relationship remains unclear.

How then may vigor be fixed in a species? Given lines of hybrid corn certainly produce year after year in a given predictable way. Asexual reproduction will, of course, fix the number of heterozygous alleles in a population. This does seem to be of importance because a great many plants do reproduce extensively by asexual means.

Strains of hybrid corn are obtained first by establishing pure lines of known constitution. This step requires about seven years of self-pollination, resulting in the development of many homozygous strains. Any two homozygous strains are then crossed to form F_1 hybrids, known to commercial corn breeders as *single-cross hybrids*. Since there are many homozygous strains differing in some respects, it is possible to obtain a large number of different single-cross hybrids, and, therefore, the selection of the best homozygous strains requires a thorough knowledge of the traits desired in the commercial strain. The seed produced in a field of such single-cross hybrids, $A \times B$ (Fig. 17.8), for instance, would give an F_2 generation with its resulting segregation of characters. This is impractical commercially and necessitates a further step. Two carefully selected single-cross hybrids, $A \times B$ and $C \times D$ (Fig. 17.8), are crossed, given what is known as *double-cross* seed. This is sold commercially and produces *double-cross* hybrid plants of remarkable uniformity, capable of yielding large amounts of grain. The steps in the production of double-cross hybrid corn seed are shown in Fig. 17.8 and are outlined in the diagram on p. 330.

Polyploidy.   Polyploidy is the increase in the number of chromosomes, or of chromosome sets, above two. Polyploidy may arise in two ways: by the doubling of an homologous set of chromosomes (**autopolyploidy**); or by combining two complete sets of chromosomes from genetically different parent plants (**allopolyploidy**).

Autopolyploidy.   The drug colchicine, dervied from a crocus growing in the Middle East, inhibits spindle formation when applied to cells during mitosis. The separation of chromatids occurs at anaphase, but they remain within the nuclear envelope. When the nucleus is reconstituted, it contains four sets of homologous chromosomes. This is a **tetraploid,** and since its four sets of chromosomes are homologous, it is an **autotetraploid.** Such a doubling of the chro-

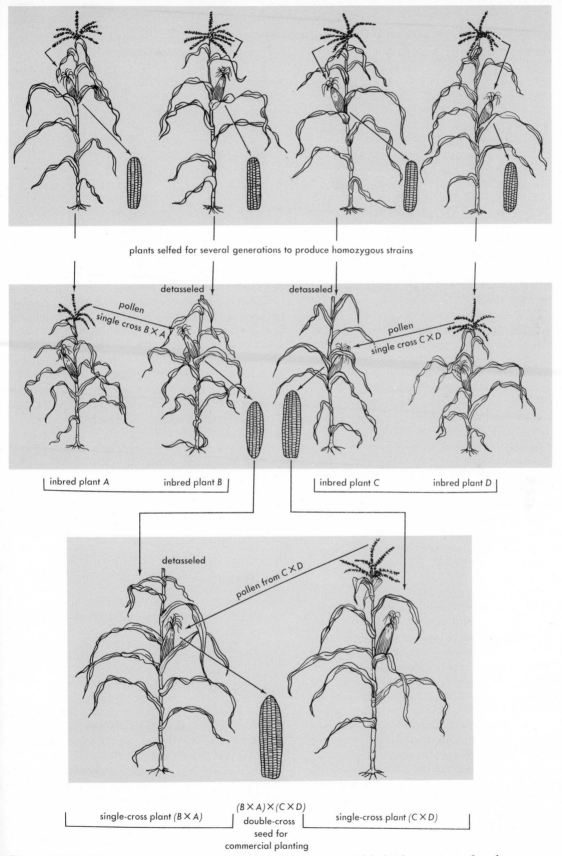

plants selfed for several generations to produce homozygous strains

detasseled

pollen
single cross B × A

detasseled

pollen
single cross C × D

| inbred plant A | inbred plant B | inbred plant C | inbred plant D |

detasseled

pollen from C × D

single-cross plant (B × A)

(B × A) × (C × D)
double-cross
seed for
commercial planting

single-cross plant (C × D)

Figure 17.8 Diagram showing the manner in which commercial hybrid corn is produced (Redrawn from *Farmers' Bulletin* 1744, U.S. Department of Agriculture, Bureau of Plant Industry, Soils and Agricultural Engineering.)

Steps in Production of Hybrid Corn Seed

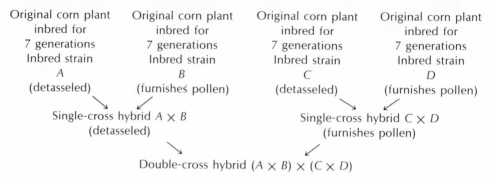

mosome number in a zygote would result in an autotetraploid plant, and such plants are known. This plant would have four chromosomes with the *A* gene. If it is homozygous, there would be four *A* genes or four *a* genes. There would be varying degrees of heterozygosity, *Aaaa, AAaa,* or *AAAa.* Pairing between four homologous chromosomes is irregular, and since gametic ratios depend upon the amount of pairing and crossing over, these ratios vary considerably. If a particular autotetraploid has the constitution *AAAa,* eight chromatids would arise during meiosis and they could show random segregation. Thus the gametic ratios of *AA, Aa,* and *aa* could range from the extremes of *3AA:3Aa:Oaa* to *15AA:12Aa:1aa.* In addition, one gamete might receive three *AAA* chromosomes and its counterpart one *a* chromosome. Such irregularity renders autotetraploids unstable genetically, and unless they reproduce asexually, they seldom become permanently established. Sugar cane is an autopolyploid that is reproduced asexually. Breeding experiments require the annual screening of some million seedlings derived from crosses between plants having different genotypes.

Allopolyploids. Hybrids between species are usually sterile. They do not have sets of homologous chromosomes; thus, pairing is very irregular, and viable seed is seldom set. However, should a doubling of chromosomes occur in a zygote having one complete set of chromosomes from each parent, the resulting tetraploid cell would have only two sets of homologous chromosomes. The plant arising from this kind of doubling of chromosome numbers is an **allotetraploid.** Pairing and crossing over are now normal; the parental chromosomes pair with each other as they would in the parent plants. This allotetraploid is a fertile plant. It will not successfully backcross with its parents and does not look like them. It is essentially a new species or even a new genus.

Aneuploids. It is not necessary that whole sets of chromosomes be added to those already present. One, two, three, or four chromosomes may be added to a plant with a diploid number. If one chromosome is added, a pair of homologous chromosomes will become a triplet of three homologous chromosomes. Instead of two *A* chromosomes (*AA*), there will be three *A* chromosomes (*AAA*). This plant is a **trisomic** plant. A diploid chromosome number of 30 indicates 15 pairs of homologous chromosomes. Since a single chromosome may be added to each of these homologous pairs, a plant having 15 pairs of chromosomes will have 15 possible different trisomic plant types. Numerous cases are known in which all the possible trisomics have been obtained. Two or more extra chromosomes may be present, and different arrangements are possible. Plants having extra chromosomes are **aneuploids.**

More information on polyploidy is given in Chapter 30.

Summary

Summarizing, the following important facts have been established regarding the mechanism of inheritance:

1. The factors responsible for inheritance are located in small, definite regions of chromosomes. In other words, inheritance is due to small regions of chromosomes. These factors or chromosome regions are called **genes.**
2. Most higher plants have the diploid number of chromosomes, the vegetative cells containing two sets of homologous chromosomes.
3. Two genes thus interact in the development of many characters in the diploid plant.
4. The gametes are haploid and consequently carry only one member of each pair of homologous chromosomes.

5. Members of a gene pair, *TT* and *tt*, while influencing a single character (height), determine different aspects of that character, such as tallness and dwarfness.

6. When different forms of a character are involved, such as tallness and dwarfness, one form is generally dominant, and the other recessive.

7. When a plant, heterozygous for a given character, such as *Tt*, is self-pollinated, the resulting progeny will number three times as many dominant plants as recessive plants.

8. When a plant, heterozygous for two given characters, such as *TtRr*, is self-pollinated, the resulting progeny will be of four different types, in the following ratio: nine plants will show only the dominant characters (tall and round), three will show one dominant and one recessive (tall and wrinkled), three will have the second dominant and second recessive (dwarf and round), and one plant will have both recessives (dwarf and wrinkled).

9. The chromosomes carrying the genes segregate independently of each other; thus, *T* may go with *R* or with *r*.

10. Genes are linked together on chromosomes. While such genes normally segregate as a unit, crossing over involving a recombination of characters occurs regularly between them.

11. Mutations may arise by an error in the replication of the nucleotide pairs in the DNA molecule.

12. The sum total of genes (particularly those under discussion) within a nucleus constitute the genotype, while the appearance of the plant due to these genes is called the phenotype.

13. The segregation of a progeny into sharply defined classes is discontinuous variation and is the result of the expression of one or two genes in the development of a precisely definable trait.

14. The expression of a trait may result from the interaction of many genes. The progeny in this case will show a continuous variation about a mean.

15. A series of modifications of a character occurring at a single locus is known as a multiple allele.

16. At rare intervals, recombination takes place between what appear to be genes in an allelic series; these alleles are known as pseudoalleles.

17. A high degree of heterosis or a large number of heterozygous alleles in a genotype greatly increases vigor.

18. The diploid number of chromosomes is the optimum condition for the normal steps of meiosis. Polyploid individuals of groups of plants have numbers of chromosomes other than the diploid numbers. An increase in the number of sets of homologous chromosomes is autopolyploidy. Allopolyploidy occurs when the chromosome number is changed by the addition of sets of nonhomologous chromosomes. If two full sets of chromosomes are not added, the plants are said to be aneuploid.

19. Changes in chromosome number by polyploidy may result in increased vigor and great disturbance in meiosis and fertilization.

18

Plant Ecology

Existing plant species have achieved—by evolution—a balance with their environment which has resulted in stability for both plants and environment. In each type of habitat, certain species group together. Fossil records indicate that some of these groups of species have lived together for millions of years. The species of these groups share incoming solar radiation, soil water, and nutrients to produce a constant biomass. They recycle nutrients from the soil to living tissue and back again, and they alternate with each other in time and space. Plant ecology attempts to discover what this balance between plants and their environment involves. What stresses does the environment put on plants, and how do plants respond to them? Plant ecology also attempts to determine how this natural balance may be artificially imitated or manipulated to achieve ends desired by man.

Plant ecology deals with levels of biological organization beyond the organism level: the population, the community, the ecosystem. A **population** is a group of closely related, interbreeding organisms. A **community** is composed of all the populations in a given habitat. An **ecosystem** consists of the living community above plus the nonliving factors of the environment. We will discuss ecology at each of these levels in this chapter. But first let us examine factors of the environment that affect populations, communities, and ecosystems.

Components of the Environment

The environment of an organism includes all the living and nonliving things about it. Some important components of the environment are moisture, temperature, light, soil, and living organisms. All components work separately and jointly to influence plant distribution and behavior. Species differ in their range of tolerance to these environmental factors. For example (Fig. 18.1), coast redwood (*Sequoia sempervirens*) is restricted to a narrow strip of California, some 5000 square miles, which experiences heavy summer fog. Coast redwood is not tolerant of variations in moisture or temperature. Red maple (*Acer rubrum*), however, occurs over some 1,500,000 square miles of eastern North America in wet, dry, cold, or warm environments. It is tolerant of wide variations in moisture and temperature.

There are two types of environment. The **macroenvironment** is the environment influenced by the general climate, elevation, and latitude of the region. Weather Bureau data on rainfall, wind speed, and temperature are measures of the macroenvironment.

333

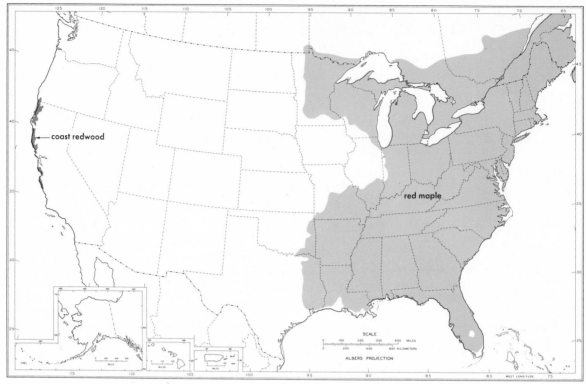

Figure 18.1 Distribution of coast redwood (*Sequoia sempervirens*) and red maple (*Acer rubrum*) in the United States.

Measurements are taken at a standard height of 4.5 ft above ground in a clear area away from buildings or trees.

The **microenvironment** is the environment close enough to the surface of an organism or object to be influenced by it. For example, bare soil tends to absorb heat, and consequently temperature just above or below the soil surface is much higher than nearly air temperature (Fig. 18.2). Light quality and

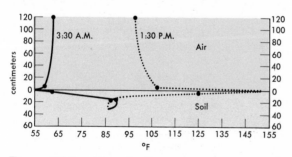

Figure 18.2 Air and soil temperatures at several positions just above and just below the soil surface. Temperatures were measured during the warmest and coolest parts of the day. (Redrawn from W. D. Billings, *Plants and the Ecosystem,* Wadsworth Publishing Company, Belmont, Ca.)

quantity is much different for herbs beneath a forest canopy than for leaves of the canopy (Fig. 18.3). Air as far as 10 mm from the surface of a leaf on a still day is less turbulent and higher in humidity than free air further away from the leaf. In marshy areas, where the water table is close to the surface, minor dips or rises in the topography greatly influence the root microenvironment. Plants growing in shallow depressions are often subject to more frequent frost than those growing on higher ground, because cold air will settle in the depressions. Ecologists are convinced that microenvironment is as important for plant growth as macroenvironment.

Moisture

The distribution of precipitation in time is as important to plant growth as total annual amount. For example, equatorial and tropical areas* may both receive over 70 inches of rainfall a year, but the equatorial area receives an equal share of the total each month while the tropical area to the north or

---

*For this chapter, equatorial means within the equatorial zone of latitude, and tropical means within the tropic zones of latitude. The "equatorial rain forest" of this chapter is often called the "tropical rain forest" in other books.

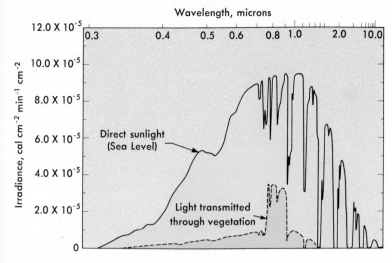

Wavelength, microns

Figure 18.3 Distribution of solar radiation energy by wavelength. Energy was measured in direct sunlight and beneath a forest canopy. (Redrawn from D. M. Gates, *Ecology* **46**, 1–13, 1965.)

south experiences pronounced dry and wet seasons (Fig. 18.4). The result is that plants of the two areas are very different; equatorial regions support tall evergreen trees and vines in profusion (Fig. 18.34; see Color Plate 9), but tropical areas support grasses, shrubs, or short trees which are often deciduous (Fig. 18.35; see Color Plate 9).

Plants adapt to dry climates in several ways. Cactus plants (Fig. 18.38) develop water storage tissue in the stem and reduce transpiration by the loss of leaves; all their photosynthetic activity is performed in stem tissue. Mesquite shrubs (*Prosopis*) and salt-cedar trees (*Tamarix*) possess long roots which tap ground water, sometimes at depths below 175 ft. **Epiphytes,** growing on tree trunks above the soil (Fig. 18.5), trap falling rain in leaf axils or in dead cells on the surface of roots. These epiphytes are not parasites; they

depend on trapped water and debris for all growth requirements. Annual herbs adapt to drought by completing a brief life cycle only during periods of sufficient soil moisture. *Boerhaavia repens* of the Sahara Desert is a small annual which can go from seed to seed in 10–14 days.

There is some evidence to show that plants can absorb dew through their leaves, but the amount of

Figure 18.4 Monthly rainfall of a rain forest near Georgetown, British Guiana and of a savannah near Bomako, Mali.

Figure 18.5 *Tillandsia*, an epiphyte native to Central America and Ecuador. (Courtesy of the New York Botanical Garden.)

Components of the Environment 335

water obtained this way is not great. Ponderosa pine (*Pinus ponderosa,* Fig. 2.16), a common tree of the western mountains, is one species which may take advantage of fog, mist, dew, or high humidity. This species sometimes grows far downslope in semiarid climates, where woil moisture can be low for several months. In one experiment, 1-year-old seedlings, growing in pots of soil, were exposed to fine mist every night. A control group was not exposed to mist. Both groups of pots were surrounded with waterproof material so that spray could only fall on the foliage and not on the soil. Seedlings lived and transpired until soil moisture was depleted and death resulted. The group exposed to mist lived, on the average, a month longer than the control group, indicating that the fomer used atmospheric moisture. The moisture could have been used directly by leaf absorption, or indirectly by slowing transpiration.

Plants may also utilize fog by condensing the mist into large drops which trickle down the stem or drip from the branches, increasing soil moisture considerably. The amount of water added to soil this way by trees on the San Francisco Peninsula was measured as 2–59 inches, depending on the type of exposure. Normal precipitation (rain) for the area is only 25 inches.

Temperature

Temperature influences moisture availability. At freezing temperatures, water changes to ice and is unavailable for plant growth; consequently, in cold climates one of the principal factors limiting growth may be physiological drought. At high temperatures, evaporation may remove much of the soil moisture before new growth of shallow roots can reach it. Transpiration will also be stimulated, resulting in further rapid depletion of soil water.

Brief extremes in temperature may be more important in determining plant distribution than are long-term moderate temperatures. Palms and many cactus palants seem excluded from areas which experience a certain frequency of frost. California lilac (*Ceanothus megacarpus*) seeds germinate poorly unless they have been exposed briefly to high temperatures. Many temperate zone plants are not frost-resistant until they have been exposed to moderately low temperatures for a short time and thus hardened.

Topography greatly influences soil and air temperature. Latitude and elevation combine, in mountains, to restrict species to certain belts (as shown for mountain hemlock in Fig. 18.6). Even at the same latitude and elevation, minor differences in the direction of slope (**aspect**) create different micro-

environments. A gorge running east–west through an Indiana forest was studied to illustrate the effect of aspect on plant growth. The gorge was 200 ft wide at the top and 150 ft deep; its sides supported scattered trees, shrubs, and herbs. Meteorological instruments were placed 6 inches above and below the soil surface midway down each side. During spring, the south-facing slope was found to exhibit a greater daily range of topsoil temperature, a higher mean air temperature, a higher rate of soil water evaporation, and lower relative air humidity than the north-facing slope. Of nine spring-flowering species present on both banks, the average flowering time was 6 days earlier on the south-facing bank. A similar difference in flowering time could be achieved on level ground only over a distance in latitude of 110 miles.

Temperature is only an estimate of the *heat energy* available from solar radiation. Solar radiation at the limits of our atmosphere is equivalent to about 2 cal per sq cm per min. Much of this radiation is absorbed, scattered, or reflected within the atmosphere, and only half may reach the ground and heat it. In turn, the warm earth reradiates energy back to space—*terrestrial radiation*. The difference between solar (incoming) radiation and terrestrial (outgoing) radiation is termed **net radiation.** During a clear day, solar radiation is greater than is terrestrial, and thus net radiation is positive; during the night, the reverse is true and net radiation is negative. Over a 24 hr period, a month, or a year, net radiation may be positive or negative, depending on the location. For example (Fig. 18.7), at Aswan, Egypt—a warm desert climate—net radiation is positive each month, with a peak in June. Net radiation is positive only 5 months of the year at Turukhansk, Siberia—a cold subarctic climate—and the annual net radiation is very close to zero.

Light

Fig. 18.3 shows how solar radiation is changed in quality and quantity by passage through green leaves. Light intensity on the forest floor may be only 1–5% that of full sunlight. Some herbs of deciduous forests complete their life cycle in early spring, before the trees above them completely leaf out and reduce light intensity. Leaves of plants which develop in shade exhibit different morphology, anatomy, and physiology than those which develop in sunlight, even if they are attached to the same plant. Shade leaves are larger, thinner, contain less chlorophyll per gram of tissue, have less well-defined palisade and spongy mesophyll layers, and achieve maximum rates of photosynthesis at much lower light intensities than do sun leaves (Fig. 18.8). Re-

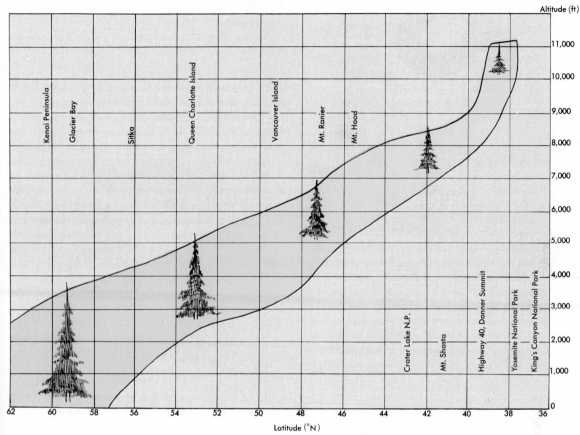

Figure 18.6 Distribution of mountain hemlock (*Tsuga mertensiana*), showing the relationship between altitude and latitude.

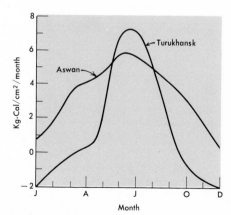

Figure 18.7 Monthly net radiation of Aswan, Egypt, and Turukhansk, Siberia. (Redrawn from D. M. Gates, *Energy Exchange in the Biosphere*, Harper and Row, New York.)

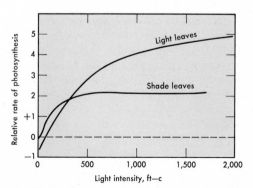

Figure 18.8 Photosynthetic response to different light intensities for sun and shade leaves of beech (*Fagus sylvatica*). (Redrawn from P. Boysen-Jensen and D. Muller, *Jahrb. Wiss. Bot.* **70**, 493–502, 1929.)

cent biochemical evidence has shown that the differences in photosynthetic rates is due to differences in the activity of the enzyme responsible for CO_2 fixation.

When a large tree dies in a mature forest, the canopy is opened and increased light intensity reaches the ground beneath. This increased light radiation induces seedlings there to begin rapid growth. Seedlings must be able to survive many years of slow growth in shade before such an opening "releases" them. Eastern hemlock (*Tsuga canadensis*) is remarkably shade-tolerant. Mature trees of this species live for 1000 years. They can retain the capacity for rapid growth when released from shade to an age of 400 years. Saplings only 6 ft tall and 1 inch in diameter may have a ring count indicating an age of 60 years.

Trees which are not shade-tolerant may become established and grow to maturity if some disturbance such as fire or logging first removes the shade. Seedlings of shade-tolerant species grow beneath them, however, and replace them as they die, while their own seedlings do not remain alive. For this reason, it is difficult to maintain pure stands of valuable shade-intolerant species such as teak, mahogany, Douglas fir, or the southeastern yellow pines. These species only invade disturbed sites and do not long maintain themselves.

Light is also reduced in intensity and quality by passage through water. Algae are not commonly found below a 200 ft depth except in very clear water. A record depth for algae may have been found in Lake Tahoe, Cal. A sample of water and bottom mud from a 450 ft depth contained *Chara*. Light intensity at that depth, despite the water clarity, is only 0.1% of full sunlight.

Brief exposure to light may be as important for some plants as long-term exposure. Seeds of some species (Chap. 20) can be stimulated to germinate in the dark if first exposed briefly to full sunlight by simply shaking them in soil in a glass container for a few seconds. This treatment could be duplicated in nature by a plow turning the soils and seeds in it, briefly exposing the seeds to sunlight, then reburying them. Plowing aerates the soil, makes it more permeable for root growth, and removes plants at the surface—all conditions of advantage to an emerging seedling.

Soil

Soil is the part of the earth's crust that has been changed by contact with the living and nonliving parts of the environment. Plants influence soil development in a number of ways. They move ions through the soil by absorbing them through their roots. These ions accumulate in the leaves and are returned to top soil after leaf abscission. Plant roots and shoots decay and add nitrogen to the soil. And plant canopies shelter the soil surface, reducing erosion and water loss. But soil development depends on its physical environment as well. From unconsolidated deposits such as glacial debris, volcanic ash, and wind-deposited loess, young soils can develop in less than 100 years, and mature ones in 1000 years. On the Kamenetz fortress in the Ukraine, a soil 4–16 inches thick has developed from limestone slabs since the fortress was abandoned in 1699. Soil development on harder rock, however, takes many thousands of years. Lava flows formed in the Pacific state approximately 1000 years ago are devoid of soil except for pockets where ash has accumulated. Factors which affect the rate of soil development are discussed in Chapter 11.

The soil profile discussed in Chapter 11 is by no means typical of the earth's surface. Temperature and moisture factors of the environment combine to mold many variations of that basic profile. Figure 18.9 summarizes the relationship of major **soil groups** to moisture and temperature. Let us examine three of the major soil groups: podzol, chernozem, and laterite.

Podzol soils sevelop in cold, wet areas which support dense coniferous forests (Fig. 18.32; see Color Plate 8). Acidic foliage accumulates as a thick layer of litter, and its slow decomposition helps create an acid soil of pH 3–5. Earthworm and bacterial activity is much less than in warmer soils. Below the litter is a sandy A horizon, the uppermost part of which

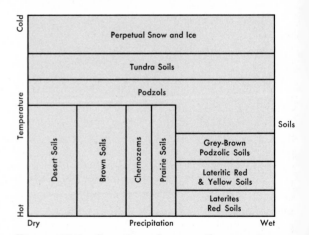

Figure 18.9 Location of great soil groups in relation to gradients of temperature and moisture. (Redrawn from H. J. Oosting, *Study of Plant Communities*, second edition, W. H. Freeman and Company, San Francisco, 1956.)

is stained gray from organic matter, but the rest of which is bleached white. Moisture is abundant enough to leach this horizon of all organic matter, clay, aluminum, and iron. A brown B horizon begins abruptly 1–2 ft below the surface and may exhibit accumulations of iron, aluminum, and organic matter in alternating bands. The C horizon, of partially weathered rock, gradually merges with the B horizon.

Chernozem soils develop beneath grasslands which experience moderate rainfall and high summer temperatures (Fig. 18.36; see Color Plate 9). Moisture is no longer sufficient to severely leach the soil, and the pH is 7–8. Fibrous root systems of grasses thoroughly permeate this soil and their decay evenly enriches it with organic matter. An A horizon, containing about 10% organic matter by weight, occupies the top 2 ft. It may be almost black in the upper part, but becomes dark brown below. Just beneath it is a layer of calcium carbonate accumulation, leached from above. This layer is sometimes called the A_c horizon. Below it is the parent material; there is no B horizon.

Laterite soils form in regions of high rainfall with continuously warm temperatures, such as those that support equatorial rain forests (Fig. 18.34; see Color Plate 9). Mineral cycling is rapid because of fast plant growth and fast litter decomposition. Soil is nearly neutral in pH. Removal of the forest canopy quickly leads to soil deterioration because litter decomposition no longer compensates for leaching. Excessive amounts of fertilizer must be added if the land is farmed. Silicates (sand) are leached, but clay and iron are not; the iron in clay is oxidized to reddish Fe_2O_3, giving topsoil a brilliant color.

Depending on the type of parent material, soils may be abnormally low or high in particular salts. For example, **serpentine** soils, among other characteristics, are very low in calcium, and many plants will not grow on them unless calcium is added. Some plants which do grow on serpentine are **endemic** to that soil (*restricted* to it and not normally found growing on other soils). Transplant tests, however, have shown that the serpentine endemics will grow perfectly well—in fact better—on normal soil *if* the surrounding plants are first removed. It appears, then, that the endemics are restricted to serpentine because they are poor competitors with other plants in nonserpentine areas, but they can tolerate serpentine soil and here find fewer competitors.

In arid regions, short and violent rainstorms create streams of excess water that run off and collect in low basins. In time, the ponds evaporate and leave the topsoil permeated with salts which had been in solution. The soil in the center of the basin (which lay under the most water) is highest in salt content; the soil at the edge is lowest. Plant distribution parallels salt distribution (Fig. 18.10): there may be no plants at all in the very saline center, but salt-tolerant plants such as salt grass (*Distichlis*) grow in a zone further out; less tolerant shrubs such as greasewood (*Sarcobatus*) grow still further out; and the usual shrubs of the area, such as sagebrush (*Artemisia*), surround the sink. Transplant experiments indicate that seeds of salt-tolerant plants will germinate and grow very well in nonsaline soil, and they appear to be restricted to saline soils only because they are poor competitors with other plants growing on nonsaline soil.

Fire

In the past 40 years, fire has been rediscovered as a major ecological factor. We say "rediscovered" because primitive man was well-aware of its importance and learned to use it to his own ends. It may well be that man's most important food crops, his domestic animals, his routes of migration, and some of his cultural attributes have been molded by natural and man-made fires. Certainly, many animals are today attracted to fires and exhibit behavioral patterns in relation to fire that imply thousands of years of evolution in response to fire.

Most natural fires are started by lightning, and there is evidence that in North America many vegetation types, including grassland, chaparral, and conifer forest, owe their distribution and community structure in large measure to lightning fires. On U.S. National Forest land, during the 22 year period 1945–1966, lightning fires accounted for 64% of all fires, an average of 5000 fires each year. In addition, early explorers and settlers invariably commented on the frequency of fires in forests and grasslands. In the words of E. V. Komerek, a recent investigator: "Lightning fires are an integral part of our environment and though they may vary in both time and space, they are rhythmically in tune with global weather patterns. Our environment can be called a fire environment."

One of the first vegetation types shown to be dependent on frequent fire for its maintenance in this country is the pine-grass savannah along the southeastern coastal plain (Fig. 18.11). Tall, scattered loblolly, slash, shortleaf, and longleaf pines dominate the area, and a thick growth of grasses (mainly *Andropogon* and *Aristida*), with some herbs, cover the ground beneath. During the nineteenth century, when this vegetation was protected from fire, it was noticed that the pines gave way to oak, and in time the result was a thick oak woodland with little grass. As the pines were valuable for lumber and turpen-

Figure 18.10 Zonation of plants around a saline basin, Grand Coulee, Washington. The saltiest soil in the center of the basin is devoid of plants. Surrounding this is a broad belt of salt grass (*Distichlis stricta*), then a narrow belt of dark shrubs (greasewood, *Sarcobatus vermiculatus*), and finally sagebrush (*Artemisia tridentata*). From R. F. Daubenmire, *A Textbook of Plant Autecology*, second edition, John Wiley & Sons, New York.)

tine, and the grass for forage, landowners were concerned about this trend. But the importance of fire was not conclusively demonstrated until 1930, with longleaf pine (*Pinus palustris*).

One of the most tolerant pines to fire, and dependent on it, is longleaf pine. The seeds germinate in the fall soon after they drop to the ground from the cones. During their first year of growth, the seedlings are very sensitive to even the lightest fire. However, during the next 2-4 years, longleaf pine seedlings are in the "grass" stage (Fig. 18.12). Most of the growing is done by the roots, and the stem apex remains close to the ground. The terminal bud is covered with a dense, protective pubescence sufficient to protect it from surface fires which may sweep the area during this time. Fire is even bene-

Figure 18.11 The pine savannah of the southeastern coastal plain. Pine shown is longleaf (*Pinus palustris*); area is in North Carolina.

Figure 18.12 Close-up of longleaf pine seedling in "grass" stage.

340 Plant Ecology

icial at this stage, because it kills a particular fungal disease that parasitizes the long needles.

By the time the seedling is 3–5 years old, a surface fire is desirable to remove grasses that may have covered the seedling and shaded it from the sun. The seedling will then make a spurt of stem growth so rapid, that by the age of 8–9 years the young tree's canopy will be high enough above the ground to be out of reach of surface fires. In addition, its thick bark can endure the heat of a fire without permitting damage to the cambium. The sparse branching habit of mature trees prevents crown fires—should one start—from spreading from tree to tree. However, if fire is kept from such an area for 15 years, grasses and young oak and pine saplings of other species become so dense that longleaf reproduction is suppressed. If fire continues to be kept out beyond that point, oak trees gradually replace the pines and form a closed forest.

Standard practice today is to burn the grass and pine debris every 4 years or so. The grass is quickly able to regenerate itself, and pine reproduction is favored, while oak seedlings are killed. If longer periods go by without fire, too much dry litter collects on the ground, increasing the hazard of setting a fire so hot, mature trees are damaged. Of course, no matter how "cool" the fire, great care is taken to control it. The cost per acre for control burning amounts to considerably less than $1 per acre.

West coast recognition of the importance of controlled burning lagged behind the east coast. In California, the absence of fire in some areas has produced potentially catastrophic conditions, especially in the mixed-conifer forests of the Sierra Nevada mountains. Before the arrival of the white man, natural fires swept these forests on the average of once every 8 years (as revealed from dating fire scars in tree trunks by ring age). However, those lightning fires did not create raging, rampaging, extremely hot fires, because the amount of litter on the ground was relatively little. The appearance of those fire-adapted forests was stately and open, according to early reports by travelers. John Muir had this to say:

> The inviting openness of the Sierra Woods is one of their most distinguishing characteristics. The trees of all the species stand more or less apart in groves, or in small irregular groups, enabling one to find a way nearly everywhere, along sunny colonnades and through openings that have a smooth, park-like surface, strewn with brown needles and burrs. . . . One would experience but little difficulty in riding on horseback through the successive belts all the way up to the storm-beaten fringes of the icy peaks.

As shown in Fig. 18.13, much of the mixed-conifer belt of the Sierras (about 5000 ft elevation) today no longer matches that description. We now have evidence that in the absence of fire, these stately forests are transformed into crowded fir and incense-cedar forests, and that the longer fire is excluded the more catastrophic is the eventual, in-

Figure 18.13 Thick understory of shade-tolerant trees that has developed in a Sierran mixed-conifer forest as a result of fire exclusion. (Courtesy of H. Biswell.)

TABLE 18.1 Appearance of Sequoia (*Sequoiadendron gigantea*) Seedlings on Burned Plots and on an Adjacent Unburned Plot; Seedlings Were Counted 1 Year after the Fire, and Are Tabulated on a per Acre Basis

Type of fire	Number of trees per acre	Number of seedlings
Light	2.9	7,514
Medium	4.6	18,350
Hot	9.3	40,130
Unburned	5.8	0

From B. Kilgore and H. Biswell, *California Agriculture* **25**(2), 8–10, 1971.

evitable fire. California experiences a high frequency of lightning strikes during the dry, late summers, and it is impossible to prevent fires from starting.

A considerable amount of litter—needles, twigs, and bark—is shed each year. So long as fires move through an area frequently, say every 5–8 years, the fire does not become hot enough to reach the upper canopy and materially damage the trees. The native trees are adapted to such frequent, light burns. The seedlings of many of them require contact with mineral soil in order to become established because litter does not retain moisture. Fire, of course, clears the ground of litter. In controlled experiments, seedlings of ponderosa pine, Jeffry pine, and sequoia all are much more abundant on burned plots than on unburned plots (Table 18.1). If the seedlings are protected from fire for 10–15 years, they become tall enough and develop a thick enough bark to with-

stand a light surface fire. However, should fire be excluded for much longer periods, so much brush and litter collects that even mature trees cannot withstand the flames which eventually come. Many magnificent stands of sequoia—the most massive tree in the world, and found nowhere else in the world—are now, for this reason, in great fire danger.

In the past 20 years, experimental burning of ponderosa pine in Arizona and California has been taking place, and the results are encouraging. In Arizona, 250,000 acres have been managed at a cost of only 5–13¢ per acre. In giant sequoia, the cost may be considerably higher (as high as $150 per acre for areas with considerable brush to be cleared and burned), but the preservation of this unique tree and the reestablishment of an impressive forest is essential at almost any cost (Fig. 18.14). In 1973, millions of acres were lost to fire, resulting in a financial loss much greater than the cost of controlled burning.

Biological Factors

Typically, several plant and animal species occupy a given habitat. Very rarely does a single species, to the exclusion of all others, occupy a habitat. If the species in a group are individually examined, it will be seen that they utilize different parts of the environment or alternate with each other in time. Some plants (green plants) are producers of carbohydrates, but others (parasitic and saprophytic fungi) are consumers or decomposers of it; some animals feed on plants, others feed on insects, others feed on small mammals; some plants are trees which utilize full sunlight, others are ferns which utilize weak light; some animals are nocturnal, others are diurnal; some plants are evergreen, others are deciduous; and so on. The portion of the environment utilized by each species is referred to as its **niche**. The niche has often been called the "occupational address" of a species.

In a particular area at a given time it is thought that one and only one species can occupy or fill each

Figure 18.14 A grove of giant sequoia (*Sequoiadendron gigantea*) in Yosemite National Park, Ca.

available niche. A laboratory experiment with two species of duckweed illustrates this point. Both species are small monocotyledons which float on the surface of quiet bodies of fresh water; their niches are identical. When *Spirodela polyrhiza* was grown in nutrient culture alone, its growth rate was spectacular (Fig. 18.15). But when it was grown with *Lemna gibba* it grew very poorly and eventually only *L. gibba* survived. The successful competitor in this case possessed a special type of parenchyma tissue which stored air, enabling it to stay on top of the

TABLE 18.2 Categories of Biological Interaction in Terms of the Effect the Relationship Has on Each Partner; "On" Indicates that the Two Organisms (#1 and #2) Are in Contact, while "Off" Indicates that the Two Are Not in Contact; No Effect = 0, Stimulation = +, Depression = −

| | Effect | | | |
| | On | | Off | |
Name of interaction	#1	#2	#1	#2
Neutralism	0	0	0	0
Competition	−	−	0	0
Mutualism	+	+	−	−
Protocooperation	+	+	0	0
Commensalism	+	0	−	0
Amensalism	0	−	0	0
Parasitism/predation	+	−	−	0

water and not be submerged by the crowding population. *Lemna gibba* was the better competitor, and it filled the available niche.

Competition is one form of biological interaction. It may be defined as decreased growth of two species or plants because of insufficient supply of some necessary factor(s). It is diagrammatically defined in Table 18.2. Sometimes only a single factor, such as a mineral element, is lacking, but usually two or more factors are limiting; it is difficult to separate them and determine which creates the greater competition stress. Competition is very important in determining plant distribution. It has already been mentioned that serpentine endemics and salt-tolerant species are restricted to their niches because they cannot compete with the numerous plants which populate more moderate niches. These poor competitors will die because the other plants have faster root and shoot growth rates, and remove water and sunlight. Sometimes, introduced plants become widespread pests and actually replace native species because they are *better* competitors than the native species. This seems to be the reason for the enormous increase of cheatgrass (*Bromus tectorum*) in the intermountain region of the United States (see Chapter 9).

Amensalism, another form of biological interaction, may be defined as the inhibition of one species by another. In the coastal hills of southern California, sage shrubs (*Salvia*) cover the slopes, and grasses carpet the valleys. Occasional pockets of shrubs occur in the grassland. Fig. 18.16 is an aerial view of these pockets of shrubs. The ground beneath and about the shrubs is bare of grass, and the grass is stunted as far as 30 ft from the shrubs. The zone

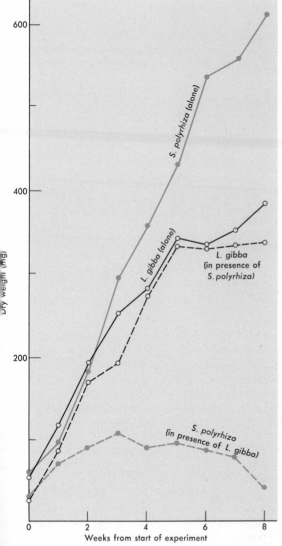

Figure 18.15 Growth of two species of duckweed in pure and mixed cultures. The species are *Spirodela polyrhiza* and *Lemna gibba*. (From John L. Harper, *Symposia of the Society for Experimental Biology* **15,** 1–39, 1961.)

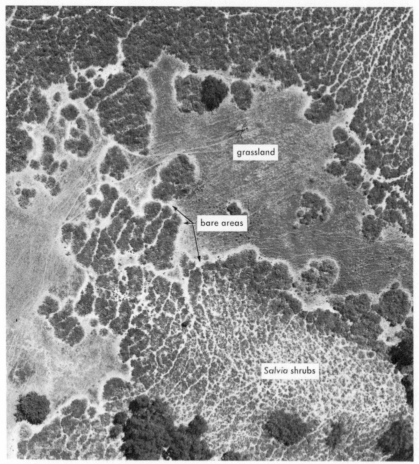

Figure 18.16 Aerial photo of *Salvia* shrubs adjoining grassland. The light bands between and around the shrubs are bare of grass. (Courtesy of C. H. Muller.)

of stunting is well beyond the limits of shrub root growth, so competition between shrubs and grass for soil moisture is not the cause of stunting. If clumps of grass are transplanted to the bare zone, their growth is severely retarded. Analysis of the soil beneath the shrubs, the bare zone, or the grasses, showed no major differences in physical or chemical attributes. Many volatile oils are emitted from sage leaves, and two of the oils in particular—cineole and comphor—are very toxic to grass seedlings. When fresh sage leaves were placed in a closed chamber with seeds of the grasses (Fig. 18.17), germination was often reduced to zero, and radicle growth was considerably reduced. These same toxins were iso-

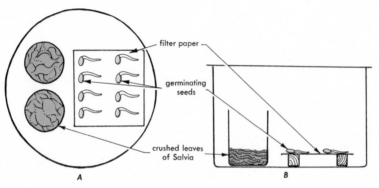

Figure 18.17 Chambers for the bioassay of the effect of volatile substances evolved from *Salvia* leaves on germination of various grass species. *A* shows a top view and *B* shows a side view. (Redrawn from C. H. Muller, *Bulletin of the Torrey Botanical Club* **93**, 332–351, 1966.)

lated from the air about sage shrubs, from the natural soil, and from seed coats of grasses in the soil. The conclusion was that the bare areas were caused by toxins emitted by *Salvia* which inhibited germination and growth of grasses. More recent studies indicate that a high intensity of foraging by birds and small mammals, which nest in the shrubs, also contribute to maintaining the bare zones.

Amensalism by chemical means may be very common in nature, for plants are leaky systems, passively contributing all sorts of substances to their environment. One investigator grew seedlings of 150 species in nutrient culture which included several isotopes as markers. The plants took up the isotopes through their roots and transported them to all parts of the plant, including the leaves. The plants were then exposed to a mist, and the water which condensed and ran off the leaves was collected for analysis. He found that 14 elements, 7 sugars, 23 amino acids, and 15 organic acids had been leached from the plants. In nature, these substances would have been leached by rain from the leaves and would have accumulated in the soil.

Predation, another form of interaction, plays an obvious part in plant distribution. A glance along pasture fences reveals that normally abundant but palatable plants are absent from the pasture, yet are common outside it; weed species (which are not palatable) are common in the pasture, yet are rare outside it. Grazing animals are responsible. Parasitism is another form of predation.

Animal pollinators form a **mutualistic** relationship with plants. The bee, moth, and beetle, which cross-pollinate flowers while feeding on nectar or pollen, are familiar examples of mutualism. A more complex mutualism, exhibited by the fig tree and its wasp pollinator, is described in Chapter 15.

Ecology and Plant Populations

The ecologist would like to use species as indicators of particular environments. He would like to known, for example, that a certain species only grows in a certain range of temperature or moisture. Then the temperature and moisture patterns of an area could be accurately predicted by just noting whether that plant species grows there. Unfortunately, most species are tolerant of a wide range of environments because they are not genetically uniform. Individuals of species *A* which grow in one environment are slightly different from others of species *A* which grow in another environment. Their genetic variation makes it possible for them to compete successfully in several different environments.

In the early twentieth century, the Swedish bota-

nist Göte Turesson collected seeds of species which had a wide range of habitats—from lowland, southern, and central Europe to northeastern Russia in the Ural Mountains. When he germinated the seed and grew the plants in a uniform garden in Åkarp, Sweden, he found that members of widespread species were not uniform. Plants which grew from seeds collected in warm lowland sites often were taller and flowered later in the year than those which grew from seeds collected in cold northern sites. Yet these variants had very similar flower morphology and were interfertile, traits which indicated that the variants were all part of one species. Turesson called these variants within species **ecotypes.** He hypothesized that they were ecologically and genetically adapted to particular environments.

Evidence accumulated since that time indicates that most wide-ranging species are composed of a continuum of ecotypes, each differing slightly in morphology and/or physiology. Alpine sorrel (*Oxyria digyna*) is a small perennial herb which grows in rocky places above timberline in mountains and north of timberline in the arctic. Both the alpine and arctic environments experience long periods of freezing weather, but alpine areas receive much higher light intensities during the growing season than do arctic areas. Physiological ecotypes have resulted. For example, the optimum light intensity for photosynthesis in alpine plants is much higher than that for arctic plants (Fig. 18.18). The two ecotypes also differ morphologically in leaf color, leaf shape, and the frequency of rhizome production.

Applications of the ecotype concept have been

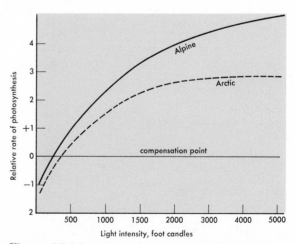

Figure 18.18 Photosynthetic response to different light intensities for alpine and arctic ecotypes of alpine sorrel (*Oxyria digyna*). (Redrawn from H. A. Mooney and W. D. Billings, *Ecological Monographs* **31**, 1–29, 1961.)

especially useful in forestry. The U.S. Forest Service conducts tests to determine the best ecotype of a species to be used for reforestation in a given area. Seed is collected throughout a species' range and is germinated in a uniform garden; seedling growth is carefully watched, and the ecotype best suited for the garden climate is selected for reforestation programs.

Ecology and Plant Communities

Some environments support only one species. Dense clusters of cat-tail (*Typha*) grow in water-filled ditches, and all other vascular species are excluded; widely spaced spinifex (*Triodia*) covers hundreds of square miles of arid Australia, and all other perennial species are absent (Fig. 18.19). Most environments, however, support groups of species called **communities.**

Provided that distances are not too great, a given community repeats itself wherever the environment is suitable. For example, high peaks in the Smokey Mountains of North Carolina all exhibit red spruce (*Picea rubens*), mountain ash (*Sorbus americana*), Fraser fir (*Abies fraseri*) and a ground carpeted with several moss and fern species (Fig. 18.20). The spruce, fir, moss, and fern species are important members (but not the only ones) of this high-elevation forest. High elevations at different latitudes or on different continents support other species, so there are other types of high-elevation forest communities. Along slopes of the California Coast Range at moderate elevation, a shrubby community of scrub oak

Figure 18.19 A community composed of essentially a single species: spinifex or porcupine grass (*Triodia basedowii*) in central Australia.

Figure 18.20 Subalpine forest in the mountains of North Carolina. The trees are red spruce (*Picea rubens*) and fraser fir (*Abies fraseri*); ferns carpet the floor of a local clearing caused by wind-throw of the trees. (Courtesy of U.S. Forest Service.)

(*Quercus dumosa*), scattered pine and cypress trees, and some annual herbs consistently repeats itself. Black spruce (*Picea mariana*), *Sphagnum* moss, and the shrub Labrador tea (*Ledum palustre*) are repeatedly associated together in virtually every bog across Canada.

Communities are named after their largest, most common species—the plants which receive the full force of the macroenvironment. The spruce and fir of the high-mountain community in the Smokey Mountains occur in greater numbers than does the mountain ash, and their leaf canopies influence the microenvironment in which the carpet of ferns and mosses grow. The spruce and fir are said to **dominate** the community, and the community name is **spruce-fir.** For similar reasons, we have **black spruce** communities and **scrub oak** communities.

Communities exhibit a unique architecture or structure, produced by the leaf canopies of their dominant and subordinate species. In the equatorial

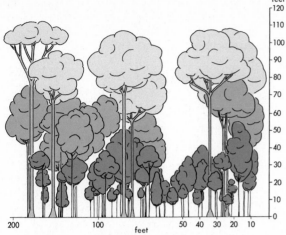

feet
120
110
100
90
80
70
60
50
40
30
20
10
0

200   100   50 40 30 20 10
                    feet

Figure 18.21 Profile diagram of the architecture of an equatorial rain forest in British Guiana. There are three tree strata, at approximately 105 ft, 60 ft, and 30 ft; shrub and herb species are uncommon. Small saplings are included in the 30 ft stratum. (Redrawn from P. W. Richards, *Tropical Rain Forest*, Cambridge University Press.)

rain forest of British Guiana, one group of species forms a canopy at 105 ft, a second group at 60 ft, and a third at 30 ft (Fig. 18.21). The ground floor is almost bare of shrubs or herbs. In deciduous forest communities of the United States, there is one tree canopy layer at about 60 ft, a layer of short trees (like dogwood) at 10 ft, and a rich shrub layer at 3 ft. Some desert communities exhibit only a single (shrub) layer.

Fossil impressions and preserved pollen in soil, rock, or bog deposits millions of years old sometimes reveal plant communities virtually the same as those which still exist today, but the deposits are usually geographically displaced from the living community. These fossil communities are so similar to present communities that we are able to reconstruct past climates from them. A fossil community near the town of Goshen in central Oregon, dated to the Eocene epoch (50 million years ago), included the following genera: *Drimys, Magnolia*, holly (*Ilex*), oak (*Quercus*), *Ocotea*, and fig (*Ficus*). The closest living approximation of this fossil community today occurs 1500 miles south of it in the temperate uplands of Central America. Apparently, the climate in North America is now much colder than it was in the Eocene. The reconstruction of past vegetation types and climates by use of fossil evidence is called **paleoecology.**

Another form of fossil evidence is pollen that has been trapped in lake sediments. As pollen is shed near a lake, some sinks to the bottom and is incorporated with silt and organic matter that forms sediment. A chronological sequence of pollen is thus preserved; the deeper the pollen occurs in the sediment, the older the pollen. Pollen of many species is resistant to decay under such anaerobic conditions, and may remain intact for thousands of years. Cores of sediments can be examined under the microscope and the pollen identified as to genus or species (Fig. 18.22). Within limits, paleoecologists assume that the abundance of pollen in the core is proportional to the abundance of that species in the past.

Fig. 18.23 summarizes the pollen profile of sediment from a lake in Nova Scotia. The present vegetation consists of a deciduous forest on hilltops with sugar maple (*Acer saccharum*), beech (*Fagus grandifolia*), and yellow birch (*Betula lutea*); and a coniferous forest in the valleys, with balsam fir (*Abies balsamea*) and white spruce (*Picea glauca*). But Fig. 18.23 shows that this pattern of vegetation is relatively recent. Sediments at a depth of 5-6 meters, corresponding to an age of about 9000 years, reveal pollen mostly from herbs and shrubs, which are

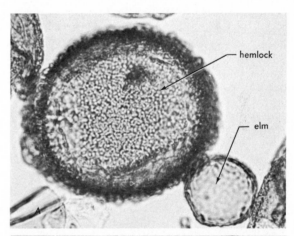

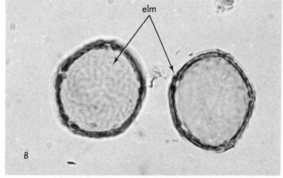

Figure 18.22 Fossil (*A*) and modern (*B*) pollen of the same genus.

Ecology and Plant Communities 347

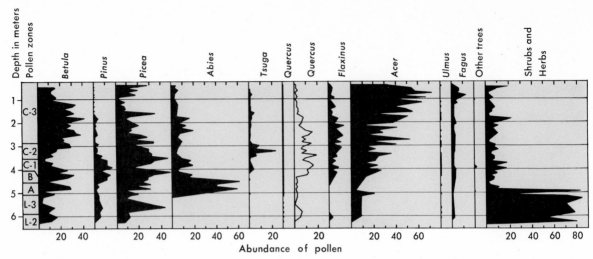

Figure 18.23   Pollen profile of sediment from a lake in Nova Scotia. Numbers along horizontal axis refer to relative abundance of each type of pollen. (From D. A. livingstone, *Ecological Monographs* **38,** 87–125, 1968.)

species characteristic of meadow vegetation near permanent ice. The nearest such vegetation today occurs several hundred miles to the north of the lake. Shallower depths, corresponding to an age of about 6000 years, show peaks in spruce and fir, indicating dominance of typical northern coniferous forest. In yet shallower depths, deciduous species of birch and maple increase in abundance and coniferous species decline, indicating a continual warming trend in the climate.

Aerial photographs, using infrared-sensitive film, can reveal the species composition of communities. Different species stand out vividly because they reflect different amounts of infrared (Fig. 18.24; see Color Plate 8). Such photographs can be used to map the distribution of species or communities over large areas.

How May Communities Be Measured?

If communities are to be useful as tools with which to estimate the nature of environment, they must be described more completely than just by their name, because minor variations in species composition indicate different environments. For example, where spruce and fir are both dominant and very abundant, the environment is probably different from where they are still dominant but less abundant. How can the abundance of a species or the composition of a community be measured?

1. One way to characterize a community is to measure the amount of ground covered by each species. The amount of ground cover is impor-

tant because it affects the microenvironment of the community. To sample the ground cover, one could stretch a tape measure across the ground; the length of tape covered by each plant canopy is proportional to the total ground cover of each species. This method of determining ground cover is called a **line transect.** Numerical information from line transects often contradicts common sense. In desert shrub communities, for example, a person looking across the landscape would think that the shrubs covered much of the ground (Fig. 18.37; see Color Plate 9), but this estimate is in error because of the low perspective. If seen from above, when the canopy edges can be projected perpendicularly to the ground, it is apparent that actually only 10–20% of the ground is covered. Forest communities, on the other hand, exhibit more than 100% cover because of overlap in canopy layers at different levels.

2. Another method of estimating cover, especially useful when plants are low to the ground, is by use of quadrats. **Quadrats** are variously shaped frames which can be placed on the ground right over the plants. The amount of ground covered by each species can be estimated as a percentage of the area enclosed by the quadrat frame. A community can be sampled by placing quadrats at random or at regularly spaced locations throughout it. Usually, 10% or less of the total community area is included in the quadrat samples.

3. Other methods of sampling, based on mathe-

A

B

Figure 18.24 *A*, normal and *B*, infrared aerial photographs of the same area, showing mountain bog (red), sagebrush (green), and pine forest (dark brown) communities. Notice how different species show up in contrasting colors with infrared, and how rivulets of water in the bog become more obvious. (Courtesy of S. Rae.)

Figure 18.28 Succession of communities about a pond near Juneau, Alaska. Standing water in the pond supports leaves of cow lily (*Nuphar*); mats of *Sphagnum* moss and sedge (*Carex*) encroach on the pond along its edge; beyond is a shrub-dominated community (*Empetrum, Cassiope, Kalmia*) with some mountain hemlock (*Tsuga mertensiana*); finally, sitka spruce (*Picea sitchensis*), the tall, dark-green trees, dominate the rest of the area. (Courtesy of J. Major.)

Figure 18.31 Taiga, northern Canada. (Courtesy of W. D. Billings.)

Figure 18.33 Change of leaf color in the deciduous forest during autumn.

Figure 18.32 Coniferous forest, Olympic Peninsula, Washington. Notice the moss- and lichen-covered trunks and branches and the dense carpet of ground vegetation.

Figure 18.34 Equatorial rain forest, New Guinea. The dense lower vegetation is absent farther into the forest. Notice the hanging lianas. (Courtesy of G. Webster.)

Figure 18.35 African savannah in Kenya. The flat-topped trees are *Commiphora*, source of the biblical myrrh. (Courtesy of O.A. Leonard.)

Figure 18.36 The original steppe, or prairie vegetation of central California, dominated by the bunch grass *Stipa pulchra*. (Courtesy of J. Major.)

Figure 18.37 Desert scrub in southern New Mexico, dominated by creosote bush (*Larrea divaricata*).

Vegetation Map of North America

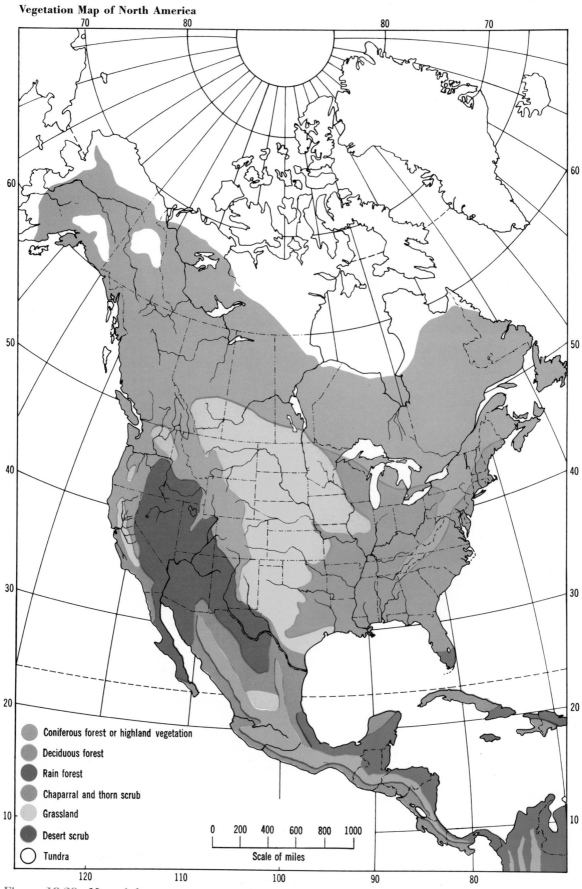

Coniferous forest or highland vegetation

Deciduous forest

Rain forest

Chaparral and thorn scrub

Grassland

Desert scrub

Tundra

0   200   400   600   800   1000

Scale of miles

Figure 18.29 Map of the vegetation types of North America.

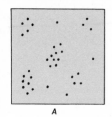
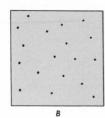
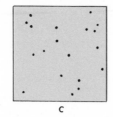

A

B

C

Figure 18.25 Diagrammatic representation of three types of plant distribution: *A*, clumped; *B*, regular; *C*, random.

matical assumptions, can be used to determine whether individuals are distributed at random, regularly (like trees in an orchard), or in clumps (Fig. 18.25). Nonrandom distribution (regular or clumped) may result from propagation by asexual means, the method of seed dispersal, competition, or some pattern (like terracing) in the microenvironment. Most species are distributed in clumps.

Succession

As we have already mentioned, the microenvironment beneath a plant community is very different from that in the open; temperature, humidity, soil moisture, and light are all affected by the canopy. A stable community consists of species whose seedlings can survive in its unique microenvironment; seedlings of other species do not survive. If the stable community is removed by some catastrophe, leaving the soil exposed to full sunlight, the first species which colonize the site are not those of the old community. They are seedlings of species adapted to growth in full light intensity. In parts of the southeastern United States where oak–hickory communities are stable, the first species to invade following such disturbance, are annual and perennial herbs growing best in high light intensity, such as horseweed (*Conyza canadensis*), *Aster*, and broomsedge (*Andropogon virginicus*). Pine seed blows in from neighboring areas during the first several years after the disturbance, and within 5 years there are many pine seedlings. As the seedlings grow, they compete for moisture with the herbs, and their expanding canopies begin to change the microenvironment. Within 30 years of the disturbance, a tall stand of pine results. Examination of the forest floor, however, shows many oak and hickory seedlings and very few pine seedlings. Pine seedlings grow poorly in shade and under conditions of root competition, but those of oak and hickory do well. Within 50 years of the disturbance, the hardwood seedlings form a well-defined understory beneath the pines (Fig. 18.26). Whenever a pine tree dies, it is replaced in the upper canopy by an oak or a hickory. Within 200 years of the disturbance, the forest again consists only of mature oak and hickory trees, and the forest

floor is covered with many seedlings of the same trees. The shrubs and herbs associated with an oak–hickory community are also present. Such a sequence of change in plant communities is called a **succession,** and the stable community which is capable of indefinitely maintaining itself is called a **climax community.** In this case, oak–hickory is the climax community.

Sometimes, climax communities are prevented from becoming established because of a recurring disturbance. There is some evidence that some of the prairie of the central United States which greeted pioneers could have been a forest were it not for recurring prairie fires. The fires destroyed slow-growing tree seedlings, but grasses quickly regener-

Figure 18.26 A stand of longleaf pine (*Pinus palustris*), about 50 years old, showing a well-developed understory of broad-leaved species of oak and hickory. In time, the pines will die and be replaced by maturing members of this understory. (Courtesy of U.S. Forest Service.)

Figure 18.27   Present-day shrub-dominated vegetation of the Mediterranean area. (Courtesy of H. A. Mooney.)

ated or germinated. When fire is controlled, prairie near the forest edge soon supports many tree seedlings. The Mediterranean area was once covered with a closed-canopy oak forest. The ground was densely shaded and soil moisture was conserved. Beginning several thousand years ago, the forests were cut for lumber and fuel, and the cut-over land was at first cultivated and then grazed by domestic animals. The exposed soil dried and eroded for many years until today it is so poor that the original forest species cannot grow on it, and will not return even if the area is left completely undisturbed. In place of the forest is a scrub community dominated by holly, heather, and gorse (Fig. 18.27). Here again, the potential climax community is prevented from establishing itself.

The first ecologists to discover successional pathways—men like Josef Paczoski in Europe and Frederick Clements in the United States—had no experimental evidence to go on. They could not record changes as time passed, for most successions take hundreds of years to complete. The determination of successional pathways was arrived at by a comparison of a climax community with adjacent small areas that had been disturbed at known times in the past. Examination of freshly disturbed or denuded areas (due to ice, landslides, fire, logging, wind-throw, etc.) revealed the likely **pioneer community,** or the first stage of succession. Sites which had been disturbed at some time earlier (verified from records of fires, grazing, or logging) exhibited later successional communities because the passage of time was greater. Complete sequences of succession were sometimes represented as bands of different communities about a gradually filling bog or

behind a retreating glacier (Fig. 18.28; see Color Plate 8). The community closest to the lake or glacier was the most recently established and represented the pioneer community. The next closest community occupied soil which had been exposed for a longer time; at some earlier time it had supported the pioneer community, but enough time had passed for the next successional stage to appear. Beyond the second community was a third, representing the third successional stage, and so on until the climax community was reached. Ring counts of trees closest to the pond or glacier, coupled with estimates of rate of silting or of glacial movement, gave clues as to the time necessary for each successional stage to be reached.

The conclusions of these early workers have yet to be experimentally or empirically verified, and only recently have some of the causes of succession (such as competition, amensalism, and growth requirements of seedlings) been experimentally dealt with.

Vegetation Types of the World

The term **vegetation** refers to the life form of plants dominating a community: trees, shrubs, or herbs; deciduous or evergreen; coniferous or broadleaved. A community dominated by evergreen angiosperms belongs to a different vegetation type than one dominated by shrubs or one dominated by conifers. On the other hand, many different communities dominated by different conifers belong to the same vegetation type. We will briefly examine six major vegetation types: tundra, taiga, deciduous forest, equatorial rain forest, savannah-prairie, and scrub.

The distribution of these vegetation types in North America is shown in Fig. 18.29 (see Color Plate 10).

Tundra

As one travels north or south toward the poles, trees gradually become stunted and less common, and finally disappear completely. At the same time, low shrubs, perennial herbs, and grasses become dominant. The resulting meadow-like vegetation is called **tundra** (Fig. 18.30). Annual plants are very rare, and the reproduction of perennials is chiefly by vegetative means such as rhizomes. Shrubs are dwarfed, gaining normal height only in the protection of the lee of boulders or small hills. Winter wind, carrying ice, acts like a sand blast and prunes back stems wherever they project beyond the boulder or hill.

The warmest month has an average daily maximum temperature of 50°F or less and the growing season is short. Only 2–4 months of the year have average daily temperatures above freezing. The top foot or so of soil thaws during the growing season and roots may freely penetrate it, but below this level soil water may be permanently frozen, and the soil is called **permafrost.** Annual precipitation is about 10 inches, very little falling as snow. The terrain is generally flat, but small depressions are common and water tends to stand in them during the growing season. Day length fluctuates from 24 hr in June (northern hemisphere) to 0 hr in December.

A similar tundra vegetation occurs in mountains at elevations above tree-line. The climate, however, is slightly different. Solar radiation is more intense, day length does not fluctuate so extremely, and frosts are more common during the growing season.

Many plants are tinged with red from excess anthocyanin, which may serve to protect the photosynthetic apparatus from damaging high light intensities by absorbing some of the light. The lower limit of these **alpine** areas depends upon latitude. At 60°N in Alaska, tree-line is at 3000 ft; at 45°N in the Rocky Mountains, it is at 10,000 ft; at 20°N in central Mexico, it is at 13,000 ft; at 5°S in the Andes of South America, it is back to 12,000 ft; and at 50°S in Chile, it is back to 3000 ft.

Taiga

Just south of the tundra is the **taiga,** a broad belt of communities dominated by conifers. A similar belt occurs in mountains, just below the alpine zone (Fig. 18.31; see Color Plate 8). Trees are slender, short (40–60 ft tall), and relatively short-lived (less than 300 years), but the forests are dense. The taiga is a patchwork of forests on upland sites alternating with bogs in depressions. Soils are acidic and relatively infertile, and in northernmost or uppermost parts of the taiga the soil is permafrost below a surface layer.

The growing season is as short as that of the tundra, but mean daily maximum temperature in the warmest month is much higher than 50°F (temperatures as high as 90°F are occasionally reached). Winter temperatures are very severe; the mean daily temperature is below freezing for 6 months of the year, and at Yakutsk, Siberia, the average January daily temperature is −46°F. Annual precipitation is less than 20 inches, and the air is exceptionally dry. The taiga is monotonous and silent; animal life is not abundant.

At lower elevations in some mountains, and along the coast (where temperature extremes are not so

Figure 18.30 Alpine tundra, Beartooth Plateau, Wyoming, 10,690 ft elevation. (Courtesy of J. Major.)

severe), coniferous forests are much more luxurious. Trees are taller, undergrowth denser. Many favorite recreation areas of the western United States are situated in these richer forests. The Olympic Peninsula of Washington receives over 80 inches of annual precipitation and supports the most spectacular coniferous forest in the world (Fig. 18.32; see Color Plate 8). Douglas fir, the most heavily cut lumber species of the United States, reaches its best growth in this area.

Deciduous Forest

In contrast to the taiga, the **deciduous forest** supports a diversity of plant life and is vibrant with animal activity. Deciduous angiosperm trees dominate communities of the deciduous forest. When day length becomes shorter and nights cooler in late summer, hormone concentrations in such trees change, and destruction of chlorophyll follows, allowing other pigments to show through in brilliant fall colors of red, yellow, and orange (Fig. 18.33; see Color Plate 9). In spring, before new leaves have fully expanded from buds, beautiful and distinctive herbaceous perennials and annuals come into flower. Shrubs are common, but not dense.

These forests occur in areas with a cold (though not severe) winter, and a warm, humid summer. Snowfall may be heavy, but most of the annual precipitation falls as summer rain. Annual precipitation ranges from 30 to 50 inches. The growing season may be as long as 200 days, so it is not surprising that much of the deciduous forest in Europe, Asia, and North America has been cut and replaced with cropland.

Equatorial Rain Forest

No other vegetation can quite equal the **rain forest** in sheer diversity of species and complexity of community structure (Figs. 18.21, 18.34; see Color Plate 9). The deciduous forest may exhibit 5–25 different tree species per acre, but the equatorial rain forest supports 20–50. The tree canopies typically form three layers, and beneath this dense overstory there are very few shrubs and herbs. Light intensity on the ground is less than 1% of full sun. Most of the herbaceous plants grow as epiphytes on trunks and branches of trees (Fig. 18.5). Other plants cling like vines to tree trunks and wind their way up to the canopies. Seeds of a few species, such as the well-named strangler fig, germinate in the canopy and grow down to the soil. Many of the tree trunks flare out at the base in perculiar buttresses.

The species are evergreen, that is, a given leaf may be shed or a given bud may break at any time of the year. The leaves tend to be very large, with long, tapering tips that may serve to drain the surface of excess moisture. Rainfall is high, often over 100 inches a year, but it is evenly distributed throughout the year (Fig. 18.4). Temperature is also uniform throughout the year, averaging 80°F. Warm temperatures and high humidity couple to produce an oppressive climate. Afternoon showers, which reduce the temperature, offer only short relief.

Farming peoples of the equatorial rain forest have for centuries practiced agriculture on a cut-burn-cultivate-abandon plan. The vegetation is removed in a small area by cutting and burning, cultivation follows for a few years, and the plot is abandoned. Abandonment follows because crop growth declines each year. The rain leaches nutrients from the exposed topsoil and the nutrients are not replaced by litter as they are beneath the normal forest canopy. Much of the equatorial rain forest of Asia and Africa has been disturbed in this way. More recent disturbance involves bulldozing and clear-cutting, especially in South America; rain forest may be incapable of reinvading these areas. Vegetation which invades abandoned plots is shrubby, dense, and often includes bamboo. It is this successional stage, which eventually will give way to rain forest, that serves as the "jungle" of television and films.

Savannah and Prairie

Forests give way to grassland as rainfall decreases and becomes more seasonally distributed. At first, trees remain close enough to retain a closed canopy, but gradually they become more widely spaced, and eventually they drop out altogether except along river banks and canyons. **Savannah** is the term applied to grassland with widely spaced trees (Fig. 18.35; see Color Plate 9); **steppe** or **prairie** is the term applied to grassland without trees (Fig. 18.36; see Color Plate 9).

Grasslands of the United States are heavily used for agriculture and grazing purposes, and little of the original vegetation or animal life remains today. Only from the records of explorers and early settlers are ecologists able to reconstruct in their imagination what these thousands of square miles once looked like.

The eastern part, with relatively high rainfall, supported a dense "sea" of grass some 3–6 ft tall. Here is a description from a journal dated 1837:

The view from this mound . . . beggars all description. An ocean of prairie surrounds the spectator whose vision is not limited to less than 30 or 40 miles. This great sea of verdure is interspersed with delightfully varying undulations, like the vast swells of ocean, and every here and there, sinking in the hollows or cresting the swells, appears spots of trees, as if planted by the hand of Art for the purpose of ornamenting this naturally splendid scene.

This is now the corn belt and is probably one of the most fertile areas of the world. Productivity is so high that it permits us to export grain to many countries around the world.

The drier, western part supported shorter and sparser stands of grass, but growth was still sufficient to allow graziers to cut and bale it as hay. Today, the western prairie supports such a poor growth of grass that ranchers often plant seed to ensure enough fodder, and cutting and baling native grass is unheard of. Reasons for such a dramatic change include too much grazing pressure by herds of cattle and a slight drying trend in climate of the area over the past 75 years.

Grasslands are the home of the largest herbivores and carnivores. Vast herds of bison which roamed the North American prairie are gone, but animal preserves in African savannahs still excite visitors by a spectacular animal cast that includes eland, wildabeest, giraffe, elephant, and lion. Umbrella-shaped trees are characteristic of African savannah (Fig. 18.35); see Color Plate 9).

Scrub

Scrub is vegetation dominated by shrubs, and it occurs in areas with hot, dry summers and moderate winters. In areas with high, but seasonal, precipitation, the shrubs are tall and thorny, but if rainfall is 15–20 inches a year, scrub can be a nearly impenetrable thicket of shrubs with evergreen, hard, spiny, small leaves (see Fig. 18.27). Scrub of this sort is found around the Mediterranean Sea, on coastal hills of California and Chile, in southwestern Australia, and at the tip of South Africa. Despite the fact that the species of plants in these widely separate areas are entirely different, the vegetation has come to look strikingly similar. This type of scrub is called **chaparral.**

Desert scrub (Fig. 18.37; see Color Plate 9) grows in areas of rainfall below 10 inches a year. Shrubs are widely spaced and have far-reaching root systems (much of which runs just beneath the soil surface). The shrubs may be associated with **succulents,** plants which store water in stems or leaves, like the common cactus of the southwestern United States (Fig. 18.38). Some shrubs have deep roots which tap permanent supplies of ground water, while others reduce transpiration by dropping their leaves every dry period and developing new ones every rainy period.

Figure 18.38 Desert scrub in southern Arizona with succulents (cactus species). (Courtesy of Ansel Adams.)

Seeds of desert annuals may remain viable for long periods in the soil until a favorable year when rainfall is sufficient to trigger germination and a colorful show of flowers a few weeks later. Although desert scrub seems to be made up mainly of perennial species, an actual count of all species appearing through the year reveals nearly as many annuals as perennials.

Ecology and the Ecosystem

Energy Flow

We have examined plant ecology at the population level, community level, and vegetation level. A final level of organization to consider is that of the ecosystem. The ecosystem of a particular habitat includes the plant and animal communities, the physical (nonliving) environment, and all the interactions between them.

The organisms of an ecosystem can be divided into three categories: producers, consumers, and decomposers. **Producers** are green plants and certain bacteria which perform photosynthesis. **Consumers** are parasitic, herbivorous, or carnivorous plants and animals which feed on other plants or animals. **Decom-**

posers are saprophytic bacteria, fungi, and certain animals which obtain nutrition by breaking down dead organic matter. The three groups form a complex web of interdependence: consumers feed on producers, and decomposers feed on organic remains of both and convert them to smaller molecules that are taken up again by plants. The path of nitrogen through the ecosystem illustrates this interdependence (Fig. 18.39). Similar diagrams could be drawn for other nutrients like calcium, phosphorus, or sulfur.

Caloric energy is also passed through the ecosystem, and in this transfer organisms are linked to each other by a **food chain.** Food chains are short in arctic tundra: meadow plants trap light energy and convert it to chemical energy (protein, carbohydrate, etc.); caribou transfer some of this energy to themselves, retaining some of it, excreting some, and respiring some; and men transfer some of the caribou caloric energy by eating a fraction of the total caribou. Aquatic ecosystems exhibit longer food chains: algae (plankton especially) produce the caloric energy; some is passed to small animals which feed on plankton, then to small fish which eat the plankton feeders, then to larger fish, and then to shore birds or land animals like bear or man.

There is tremendous variation in net productivity from one ecosystem to another. **Net productivity** equals calories produced in photosynthesis minus calories lost in respiration. Desert ecosystems, for

example, exhibit less than 0.5 gram dry matter produced per square meter per day; grasslands, 0.5–3.0; deciduous and coniferous forests, 3–10; agricultural areas utilized all year (as in sugar cane production), 10–25.

Energy transfer through the ecosystem seems very inefficient. Fig. 18.40 summarizes a very simple food chain consisting of a pasture on which steers are grazed. Sample areas within the pasture were clipped of vegetation at the start of the experiment and were clipped again after a year's growth. The clippings were dried and weight converted to caloric content using a conversion factor of 4.3 kcal per gram of tissue. Root growth was estimated and also converted to calories. Total root and shoot growth represented net productivity. Weight increase of the steers during the year was noted and also converted to calories. Loss of plant material to insects, rodents, and rabbits was estimated as minor and ignored.

As shown in Fig. 18.40, annual solar radiation was 1,600,000 kcal/m² (of ground area). Of this, about 44%, or 700,000 kcal/m², was both within the spectrum of wavelengths which plants can use for photosynthesis, and actually reached the ground. Annual plant growth (net productivity) was 1410 kcal/m², or only 0.2% of available light energy. Steer growth (net consumption) was 69 kcal/m², or only 5% of available plant energy (0.004% of annual solar radiation). From the energy standpoint, then, it is easy to see why large carnivores (like lions) require considerable ter-

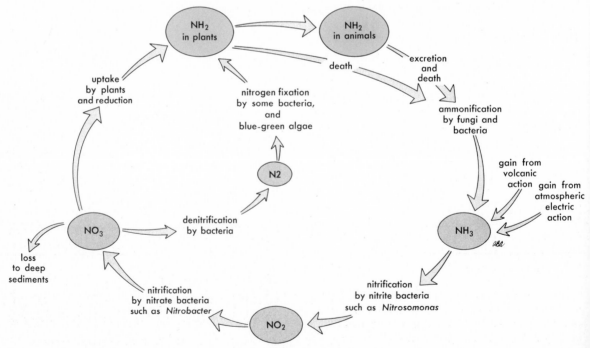

Figure 18.39 Diagrammatic path of nitrogen through an ecosystem.

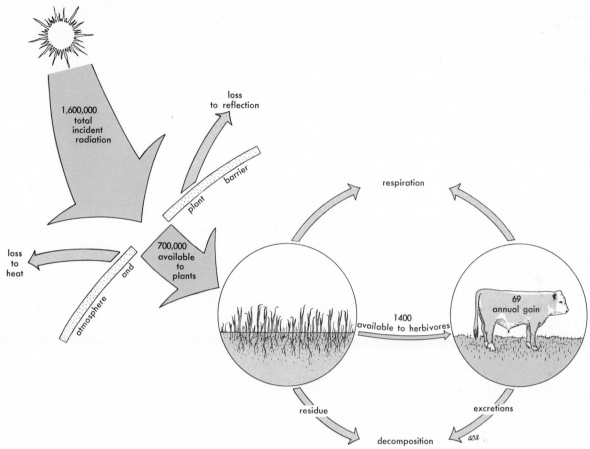

Figure 18.40   Transfer of caloric energy in a pasture–steer ecosystem. Numbers are calories per square meter. (Redrawn from W. A. Williams, *Journal of Range Management* **19,** 29–34, 1966.)

ritorial ranges; it takes a lot of savannah vegetation to support the herbivores, which in turn support large carnivores.

Pollution

Man is adding to the environment many gases, liquids, and solids whose effect on plants and animals is poorly known (Fig. 18.41). If organisms in food chains or ecosystems shared by man are damaged by these substances, then man may be indirectly, but nevertheless severely, damaged. The results of a spray program carried on at Clear Lake, Ca., illustrate this concept.

Clear Lake was an excellent recreation area except for intolerable summer gnats. The insecticide DDD (a close relative of DDT) was mixed into lake water in 1949, 1954, and 1957 to kill gnat larvae on the lake bottom. The amount of insecticide used was quite

Figure 18.41   Air pollution of a Tennessee foothill valley.

small—only enough to achieve a concentration of 0.02 part per million (ppm)—but larval kills were 99%. In 1954, a few months after the second application, over 100 dead grebes (birds which feed on fish) were found around the lake. There was no external evidence of disease. Another large number of dead grebes was found after the 1957 application and this time fatty tissue was removed from some of them and analyzed for DDD. The tissue was found to contain 1600 ppm DDD—a concentration some 80,000 times that of the lake water!

Components of the food chain leading to grebes were then analyzed: water = 0.02 ppm DDD; plankton = 5 ppm; plankton feeders = 15 ppm; fish = 1000 ppm; grebes = 1600 ppm. Each component of the chain had ingested large masses of the component below it, but had not metabolized or excreted the exotic new chemical, so a chain of accumulation resulted. In 1949, there were 1000 pairs of nesting grebes around the lake with many young; in 1961, there were 32 pairs and no young. Had man been the ultimate carnivore in this food chain, he undoubtedly would have found the high level in DDD in fish or birds toxic to himself. There is some recent evidence, however, to show that man is able to metabolize DDD instead of accumulating it.

Large amounts of home and industrial sewage which are added to lakes and streams also affect food chains and, ultimately, man. Lake Erie covers some 10,000 square miles and has a shore lined with high population areas that introduce organically rich waste to it. Sewage settles on the bottom and bacterial activity is greatly increased. As bacteria respire they use up oxygen. In 1928, the amount of dissolved oxygen in bottom water was 78% of saturation, but in 1960, it was only 25%. It is on the lake bottom that commercially valuable fish such as blue pike lay their eggs, and their eggs require a certain minimum oxygen concentration in the water to survive. Recently, that minimum concentration may have been reached and passed. In 1956, the blue pike catch was 6.8 million lb, worth $1.3 million; in 1963, the catch was 200 lb, worth $120! Other commercial fish (trout, whitefish, herring, sauger, walleye) have suffered a similar decline, resulting in an overall loss of $3 million a year. Fish species now prevalent include commercially undesirable carp, shad, alewives, smelt, and goldfish. Algal species have changed or increased in number to the detriment of fishing and boating. Bacterial counts occasionally reach levels which cause the closing of beaches. The time may not be distant when Lake Erie water is placed off-limits for drinking. Such an end came to the lower part of the Hudson River in 1950 as a result of industrial pollution from Albany. Twenty billion gallons a day flow out the mouth of the Hudson, all of it unusable by New York City.

It is hoped that ecologists will eventually show us how ecosystems can be rationally manipulated to simultaneously satisfy all our needs: industry, urbanization, agriculture, conservation, and recreation. Knowledge must be gained and compromises made between our needs and what the environment can support.

Summary

1. Plant ecology is the study of how the environment influences plant distribution and behavior. It is also the study of the structure and function of nature. Ecology deals with the population, community, and ecosystem levels of organization.

2. Components of the environment include moisture, temperature, light, soil, and living organisms. Each species utilizes a different part of the environment, its niche.

3. A taxonomic species includes genetically similar plants which are interfertile. Within any wide-ranging taxonomic species, however, are many ecotypes which have become adapted to different environments.

4. A plant community is comprised of a group of associated species or populations which occupy a particular type of environment. One or more of the species dominate the community, but all the species together combine to form a unique architecture of canopy layers. The structure and composition of communities can be objectively determined with the use of sampling methods such as line transects or quadrats.

5. Climax communities perpetuate themselves, but successional communities are transitory.

6. Vegetation types of the world include: tundra, taiga, deciduous forest, equatorial rain forest, grassland, and scrub.

7. An ecosystem consists of the plant and animal communities of a habitat, their abiotic environment, and all the interactions between these components. Organisms of an ecosystem are tied together by food chains. Damage done to one component of a food chain (as through the action of pollution) may cause indirect but serious damage to the entire ecosystem.

19

Principles of Plant Taxonomy

Plant taxonomy deals with the **identification** and **classification** of plants. The term identification implies that many plant groups exist, and each group possessing unique characteristics can be separated from all other groups. These characteristics, as we shall see, involve morphology, anatomy, physiology, cytology, biochemistry, and geographic distribution. The term classification implies that plant groups can be ranked in some hierarchical relationship, on the basis of similarities in their characteristics. This ranking attempts to reflect real genetic relationships.

As mentioned earlier (Chap. 2), taxonomy is building a library of the world's plant resources. Taxonomists are making a census of the unique characteristics of each of the half-million species that make up the world's flora. This census is far from complete for two reasons. First, many areas of the world have not been botanically explored and their floras can only be estimated; second, the process of evolution regularly produces new species even in well-known areas.

The important, exciting goal of the census is not identification, but rather classification. Taxonomists hope that, as the census becomes complete, patterns of similarity will become apparent within this maze of diversity; and that these patterns will enable them to reconstruct the evolutionary history of the present flora. They hope it will be possible to determine which characteristics—out of the hundreds available—are most reliable in showing pathways of evolution. Are they characteristics of flower morphology, wood anatomy, chemical composition, geographic distribution, breeding behavior, or chromosome number? Which traits in each category are **primitive** (imitating those of early, now-extinct plants)? Which traits are **advanced** (derived in time from primitive traits)? When the answers to such questions are known, it will be possible to construct a **phylogenetic classification**—one based on genetic, evolutionary relationships. Such a classification, in addition to summarizing the past, might prove useful in predicting future pathways of evolution.

Historical Aspects

Pre-Darwinian Period

The history of taxonomy can be very roughly divided into a pre-Darwinian (300 BC–1753 AD) period and a post-Darwinian (1753–present) period.

Before the time of written history, of course, man had learned to identify many plants at what we now consider the genus or species level. The plants

357

named were those particularly useful as spices, medicines, foods, drugs, or religious symbols. The written history of botany begins with the Greek Theophrastus (ca. 300 BC), a student of Aristotle. He described and classified almost 500 kinds of plants in the book *De Historia Plantarum*. He utilized morphological characters like habit (tree, shrub, herb), length of life (annual, biennial, perennial), corolla form (petals fused together or free), and ovary position (superior, inferior) to distinguish among his plants.

Many of the taxonomists who followed the lead of Theophrastus through the dark ages and Renaissance are typified by Otto Brunfels (ca. 1500 AD) of Germany. He too classified plants on the basis of gross morphology, but he contributed to the record by expanding the list of known plants and accompanying many of the descriptions with drawings. Hans Weiditz was Brunfels' illustrator, and he did a magnificent job of carving accurate and intricate woodcuts from which prints were made (Fig. 19.1). The descriptions and drawings of many species were compiled in books referred to as **herbals.** Herbaceous plants used for medicinal purposes were often emphasized in the herbals. It was common to assign medicinal properties to a plant on the basis of its form: If the plant or parts of it imitated the shape of an organ, then it was used for correcting ailments of that organ. For example, the thallus of liverworts,

which resembles the shape of a liver, was used to treat liver ailments.

Carolus Linnaeus (ca. 1750) represents the culmination of the pre-Darwinian period. Born Carl von Linné in southern Sweden, he attended the University of Uppsala and received an M.D. degree from the University of Haderwijk in Holland. During his early days at Uppsala, however, he had become interested in the classification of plants and had developed his own system, called the sexual system. It was based almost entirely on the morphology and number of stamens and pistils, and proved very efficient in classifying the many new plants then being brought to Europe from Africa, America, and Asia and propagated in botanic gardens. After working on the collections of many large botanic gardens, Linneaus published the book *Species Plantarum* in 1753. The work included about 6000 species and 1000 genera.

Each species was described in Latin by a sentence limited to twelve words which began with the genus name. Linneaus considered this sentence (a **polynomial**) to be the official, scientific species name, but he also coined a shortened form consisting of the genus name and one additional word from the polynomial. This shortened form (a **binomial**) served as his common name for the plant. He described spiderwort by this polynomial: *Tradescantia ephemerum phalangoides tripetalum non repens virginianum gramineum,* common name *Tradescantia virginiana.* Loosely translated, the polynomial means: The annual, upright Tradescantia from Virginia which has a grasslike habit, three petals, and stamens with hairs like spider legs, common name Tradescantia of Virginia. The genus name was in honor of John Tradescant, a gardener to King Charles I.

Not surprisingly, the binomial became favored over the polynomial by later taxonomists, and was accepted as the official, scientific name. Common names were left to individual choice and not restricted to Latin. The usual English common name for *Tradescantia virginiana* is spiderwort (spider plant), which comes from the jointed hairs on stamen filaments that resemble the jointed legs of a spider (Fig. 19.2).

Post-Darwinian Period

After Darwin published *On the Origin of Species* in 1859, emphasis in taxonomy shifted to (*a*) a search for characters which reflected genetic (evolutionary) relationships, and (*b*) the construction of a phylogenetic classification scheme. Construction of phylogenetic schemes demanded that decisions be made as to which traits are primitive and which are

CXCVI   Contrafayt

Storckenschnabel.

❡ Von dem Nammen.

Storckenschnabel/Gottes gnad/Kranch,

Figure 19.1 Part of one page from Brunfels' sixteenth century herbal. The plant illustrated is called stork's bill (probably *Erodium cicutarium*).

Figure 19.2 *Tradescantia virginiana* (spider wort). Note the jointed staminal hairs, from which the common name derives, ×2.

advanced. As there was little or no fossil evidence to support any of several hypotheses, early schemes were both numerous and often in conflict. For example, the American C. E. Bessey started with the assumptions listed in Table 19.1.

Although Bessey's assumptions are still widely accepted today, some 60 years after he proposed

TABLE 19.1 Assumptions for the Besseyan Phylogenetic Scheme

Primitive characters	Advanced characters
Plants woody	Plants herbaceous
Flowers bisexual	Flowers unisexual
Floral axis elongated	Floral axis short
Floral parts spirally arranged	Floral parts whorled
Floral parts numerous	Floral parts few
Sepals or petals free	Sepals or petals fused
Floral symmetry radial	Floral symmetry bilateral
Fruit single	Fruit aggregated

them, disagreements continue to arise over specific interpretations. For example, some angiosperms such as willow (Fig. 19.3A) have flowers which lack petals or sepals. Are these plants primitive, never having evolved a complete flower, or did they once possess flowers with petals and sepals but since lost them, hence really being advanced? Bessy thought they were advanced and consequently regarded plants like willow (*Salix*), birch (*Betula*), and maple (*Acer*)

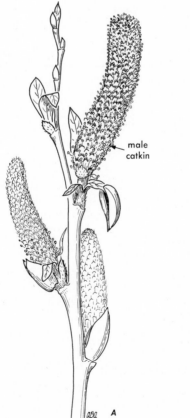

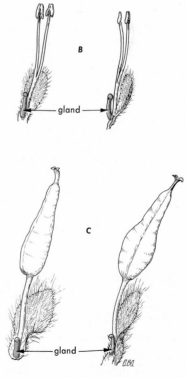

Figure 19.3 Willow (*Salix*) inflorescence and individual flowers; *A*, male catkin, ×1; *B*, single male flowers, ×15; *C*, single female flowers, ×15.

as closely related to advanced families. In contrast, equally competent taxonomists like Adolph Engler and Karl Prantl, living at about the same time, placed these plants among the most primitive angiosperm families.

Whether the absence of parts represents a primitive or derived condition can sometimes be determined by examination of the flower's vascular system. Vascular bundles which branch in the receptacle and extend to petals or sepals may still show the original branching even though the petals or sepals have been lost through time. Apparently, the vascular system is very **conservative** (does not readily change). In other cases, the missing parts have not been lost altogether and are represented by small projections of glandular or apparently functionless tissue. Evidence of this sort, gathered since the time of Bessey, has confirmed his hypothesis that plants like *Salix* are indeed advanced (Figs. 19.3*B, C*).

Another source of disagreement arises from answers to this question: Do species which now resemble each other represent progeny from a common ancestor, or could it be possible that they started from quite different sources and simply evolved to look alike? These two possibilities are diagrammed in Fig. 19.4. It is seen that species *A* and *B*, which are very similar today, actually arose from quite different stocks. Similar environments, however, have shaped them along convergent paths. This pattern is termed **convergent evolution.** A striking example of convergent evolution is the present similarity between cactus plants of southwestern United States and certain *Euphorbia* species of Africa. Both groups grow in hot, arid climates, and as seen in Fig. 19.5 both have a similar morphology: They have no leaves, their stems are ribbed and swollen and con-

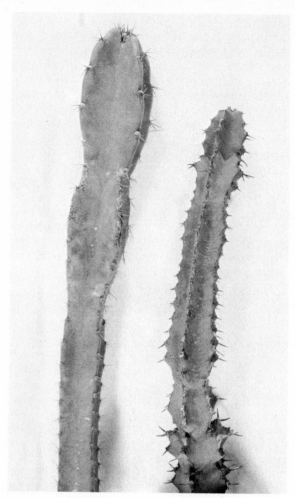

Figure 19.5 **Convergent evolution of members of the Cactaceae (left) of the southwestern United States and members of the Euphorbiaceae (right) of Africa,** $\times \frac{1}{10}$.

tain water-storage tissue, and they are armed with spines. Yet, other traits, like flower morphology and geographic distribution, are so different that it is clear they must have arisen from different stocks.

On the other hand, *B* and *C*, which are today less similar than *A* and *B*, actually did evolve from a common ancestor. This pattern is called **divergent evolution.** As time passes, *B* and *C* will become more and more different. The same pattern will be repeated by *C* and *D*, but at this point in time they are still quite similar.

If a certain group of plants arose from a single ancestral type, the group is referred to as **monophyletic** (Chap. 2). If a group arose from several different ancestral types, the group is referred to as **polyphyletic.** Bessey considered each separate group of seed plants (the Anthrophyta, Coniferophyta, Gink-

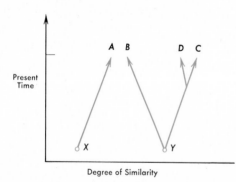

Figure 19.4 Hypothetical pathways of species evolution. *X* and *Y* represent ancestral forms from which *A, B, C,* and *D* have since evolved. The paths leading to *A* and *B* represent convergent evolution; the paths leading to *B* and *C* represent divergent evolution.

gophyta, etc.) to be monophyletic; some taxonomists considered all the seed plants together to be monophyletic; still others did not think that even one group, the Anthyophyta, was monophyletic. Unless fossil evidence is at hand, it is very difficult to separate the effects of convergent and divergent evolution. At present, most taxonomists agree that the largest group of plants which is *definitely* monophyletic is the family.

Considering the rich variety of hypotheses and plant characters upon which one could base a classification scheme, it is not surprising that the names of many plants have changed with time, being taken from one genus to another or being split into varieties or subspecies. *Tradescantia virginiana* L. is still an accepted name, more than 200 years after it was coined, but others that Linneaus selected have been changed. For example, in 1753 Linneaus described a plant from America as Canadian pine, *Pinus canadensis*. Later, the taxonomist Carriere realized it was a hemlock, not a pine, and transferred the species to the genus *Tsuga*. The new name was *Tsuga canadensis*, Canadian or eastern hemlock. How does this affect the old, valuable herbarium collections of the same plant which had been labeled *Pinus canadensis*? How would later taxonomists realize the connection between the two? There would be references, in older books, to something called *Pinus canadensis* which no longer is a name found in any taxonomic reference book. Who would know what plant was meant? There was a need, then, to allow for changes in name but at the same time preserve continuity and prevent a haphazard proliferation of new names.

The answer to the need was a set of international rules on botanical nomenclature, adopted by the International Botanical Congress of 1930, held in Cambridge, England. These rules were explicit guidelines for the determination and coining of valid plant names. Some rules dealt with the categories of classification available, others with the procedure of identifying new species, and others with the procedure for changing names.

One rule provided that, to be valid, a plant name must be published in Latin, accompanied by a short, descriptive paragraph which sets the new species off from related species, and that this description must be published in some recognized journal or flora.

PINUS :—

Boursieri, Carr. Conif. ed. I. 398 = contorta.
brachyphylla, Parl. in DC. Prod. xvi. II. 424 = Abies brachyphylla.
brachyptera, Engelm. in Wisliz. Tour N. Mexico, 89 = ponderosa.
bracteata, D. Don, in Trans. Linn. Soc. xvii. (1836) 443 = Abies bracteata.
Brunoniana, Wall. Pl. As. Rar. iii. 24. t. 247 = Tsuga Brunoniana.
bruttia, Tenore, Prod. Fl. Nap. p. lix = pyrenaica.
bullata, Roezl, Cat. Sem. Conif. Mexic. (1857) 16 = Montezumae.
Bungeana, *Zucc. in Endl. Syn. Conif.* 166.—China.
Buonapartea, Roezl, ex Gord. Pinet. 218 = P. Ayacahuite.
cairica, D. Don, ex Gord. l. c. 166 = halepensis.
calabrica, Hort. ex Gord. l. c. 168 = P. Laricio.
californica, Loisel. in Duham. Arb. ed. Nov. v. 243; Hartw. in Journ. Hort. Soc. ii. (1847) 189 = insignis.
Calocote, Roezl, ex Gord. Pinet. Suppl. 73 = P. Teocote.
canadensis, Bong. in Mém. Acad. Pétersb. Sér. VI. ii. (1833) 163 = Tsuga Mertensiana.
canadensis, Duroi, Obs. Bot. 38 = Picea alba.
canadensis, Linn. Sp. Pl. ed. II. 1421 = Tsuga canadensis.
canaliculata, *Miq. in Journ. Bot. Néerl.* i. (1861) 86. —China.
canariensis, *C. Sm. in Buch, Beschr. Canar. Ins.* 159. —Ins. Canar.

TSUBAKI, Adans. Fam. ii. 399 (1763) = **Camellia**, Linn. (Ternstr.).

TSUGA, Carr. Conif. ed. I. 185 (1855). *CONIFERAE*, Benth. & Hook. f. iii. 440.
    MICROPEUCE, Gord. Pinet. Suppl. 13 (1862).
Balfouriana, *McNab, in Journ. Linn. Soc.* xix. (1882) 211.—Am. bor. occ.
Brunoniana, *Carr. Conif.* ed. I. 188.—Reg. Himal.
canadensis, *Carr. l. c.* 189.—Am. bor.
caroliniana, *Engelm. in Coult. Bot. Gaz.* vi. (1881) 223.—Am. bor.
diversifolia, *Mast. in Journ. Linn. Soc.* xviii. (1881) 514.—Japon.
Douglasii, Carr. Conif. ed. I. 192 = Pseudotsuga Douglasii.
Hookeriana, *Carr. l. c.* ed. II. 252.—Calif.
Lindleyana, Roezl, Cat. Grain. Conif. Mex. 8 = Pseudotsuga Douglasii.
Mertensiana, *Carr. Conif.* ed. II. 250.—Am. bor. occ.
Pattoniana, *Engelm. in Gard. Chron.* (1879) II. 756.— Am. bor. occ.
Roezlii, *Carr. in Rev. Hortic.* (1870) 217.—Calif.
Sieboldii, *Carr. Conif.* ed. I. 186.—Japon.
Tsuja, A. Murr. in Proc. Hort. Soc. Lond. ii. (1862) 508 = Sieboldii.

TSUSIOPHYLLUM, Maxim. in Mém. Acad. Pétersb. Sér. VII. xvi. (1871) n. IX. 12. *ERICACEAE*, Benth. & Hook. f. ii. 602.
Tanakae, *Maxim. l. c.*—Japon.

Figure 19.6 Part of two pages of *Index Kewensis*. Note that *Pinus canadensis* L. is shown to be a defunct synonym (by the = sign) for *Tsuga canadensis* Carr. The abbreviations refer to the publications where the original species description occurs. For example, *Pinus canadensis* is described in *Species Plantarum*, p. 1421.

A book that summarizes all the plant species of a region is termed a **flora.**

Another rule provided that "each **taxon** (a general word for any plant group; pl. = taxa) . . . can bear only one valid name, the earliest that is in accordance with the Rules of Nomenclature." If two or more people independently proposed names for the same plant, then the earliest one proposed would be the valid one. Common sense indicated that there should be some limitation to this principle of priority, and for vascular plants the year 1753 (date of publication of *Species Plantarum*) was selected. Names coined before that year were not valid, and the principle of priority only applied to names coined after that year.

Other rules dealt with changes of name. One of these is quite important, for it helps solve the problem of *Pinus* (*Tsuga*) *canadensis*. It provides that when a genus or species is changed, the name of the original author be placed in parentheses and be followed by the name of the author who has now changed it. Thus, *Pinus canadensis* L. becomes *Tsuga canadenis* (L). *Carr.* It is now immediately apparent that *Tsuga canadensis* is not the original name for a plant and Linneaus had given it another name. The original name can be found in references like the Kew or Gray Herbarium Indexes which list all the previous names (sometimes there are several) for presently named species. Portions of pages of *Index Kewensis* are shown in Fig. 19.6.

Methods of Taxonomic Research

Traditional (Morphological)

About 20 years ago, the geneticist C. H. Waddington said it is "an empirical fact that living organisms do not vary continuously over the whole range, but they fall into more or less well-defined groups, which are commonly called species." This philosophy is quite appropriate for a traditional taxonomist. He realizes that man's love of classifying nature often puts boundary lines between things where boundaries simply do not exist; but when it comes to species, he is convinced from extensive field observations that discrete species do exist and that classification at this level is not arbitrary or artificial.

A traditional taxonomist primarily examines plant morphology, searching for a few traits which consistently enable him to separate plants into "well-defined groups." In pre-Darwin days, a variety of traits were utilized, but now the choice is limited to those which presumably are (a) genetically controlled (rather than environmentally controlled) and (b) conservative in the evolutionary sense. Flower and fruit morphology, for example, meet these criteria, but leaf size does not.

It should not be supposed that the traditional taxonomist limits himself at the start of a study to the examination of only a few traits. He examines many characters, measures many plants, and keeps records of all these. He eventually, however, subjectively selects only a few characters to serve in determining the number of taxa. The characters he selects are those which show discontinuities and are most helpful in separating related taxa.

Anatomical

We have already seen, with *Salix*, that the vascular system of a plant is very conservative and may be used to show genetic relationship between taxa. Anatomical studies of particular cell types within the vascular system reveal evolutionary patterns. Vessel elements, for example, show the changes summarized in Table 19.2 (see also Fig. 19.7).

Evidence for this set of hypotheses comes from examination of fossils of lower plants that have vessels (such as *Selaginella*), and of *correlations* between changes of vessel characters with changes in other characters such as flower morphology (that is, plants with primitive flower morphology also show primitive vessel characters). Similar hypotheses of evolution have been applied to other cell types, such as fibers.

Differences in anatomy are usually minor at the level of genus or species; nevertheless, these small differences can show evolutionary relationships. Creosote bush (*Larrea divaricata*) is a common desert plant in Argentina and the southwestern United States (Fig. 19.8). Although members of the species are very similar in morphology across its range from California to Texas, the species is not genetically homogeneous. Populations of creosote bush occur in three separate desert regions. The plants in each region have different numbers of chromosomes.

TABLE 19.2 **Characteristics of Primitive and Advanced Vessel Elements**

Primitive	Advanced
Similar in appearance to a tracheid	Not similar to a tracheid
Slender in cross-section ($100\,\mu$)	Broad in cross-section ($250\,\mu$)
Long about ($800\,\mu$)	Short (about $300\,\mu$)
Tapered ends	Truncated ends
Angular in cross-section	Round in cross-section
End wall multiperforated	End wall with single perforation

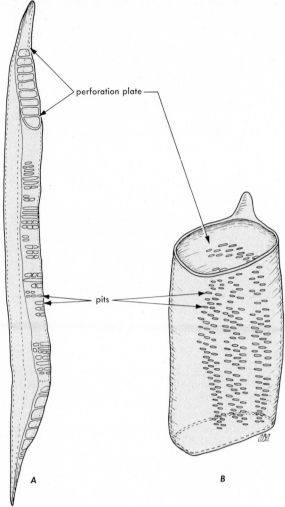

Figure 19.7 Advanced and primitive vessel elements: *A*, primitive type, from tulip tree (*Liriodendron tulipifera*); *B*, advanced type, from oak (*Quercus* sp.), ×100.

(Labels in figure: perforation plate; pits; A; B)

Figure 19.8 Creosote bush (*Larrea divaricata*), an abundant warm desert plant of North and South America. (Courtesy of Allan B. Lang–National Audubon Society.)

related of United States subspecies to Argentina stock. Fiber length shows the same pattern. Table 19.3 summarizes average fiber length of *Larrea* plants grown from seed in a uniform garden, all tested at the same age. The standard hypothesis for direction of fiber evolution is that advanced fibers are longer than primitive ones. Hence, the data in Table 19.3 indicate that western subspecies of *L. divaricata* have evolved from eastern ones.

Embryological

The pattern of embryo development is thought by some to be extremely conservative and to actually reflect in some measure the evolutionary history of the organism. For example, the embryo of man passes through early stages which are almost identical to those of lower groups of chordates (such as fish), and this is taken by some people as evidence of the evolutionary relationship between man and these other chordates.

Taxonomists have found that the pattern of embryo development in the Anthophyta is not uniform, and sometimes differences in the pattern correlate with differences in plant morphology. The position of the ovule within the ovary can be very distinctive and characteristic for entire families or orders; the ovule can be shaped so that the micropyle is directly opposite the point of ovule attachment to the placenta (**orthotropous**), or the micropyle can be facing the placenta (**anatropous**), or be somewhere between the two extremes (**campylotropous**) (Fig. 19.9). The number and position of cells in the embryo sac (prior to fertilization) can vary tremendously; there are eight cells in the embryo sac of lily (*Lilium*), four in evening primrose (*Oenothera*),

Each desert, then, has its own subspecies of *L. divaricata*. The eastern subspecies, in Texas and New Mexico, has the same number of chromosomes as the Argentina plants, indicating it is the most closely

TABLE 19.3 Length of fibers in Creosote Bush Seedlings

Area where seed collected (desert)	Average fiber length (μ)
California, Nevada (Mojave)	873
Arizona (Sonoran)	799
New Mexico, Texas (Chihuahuan)	628
Argentina (Argentinian)	533

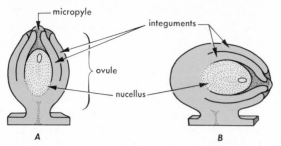

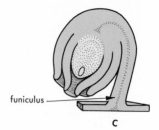

Figure 19.9   Three types of ovule position within the ovary; *A*, orthotropous; *B*, campylotropous; *C*, anatropous. (Redrawn from G. H. M. Lawrence, *Taxonomy of Vascular Plants*, The Macmillan Company, New York.)

and sixteen in *Peperomia* (Fig. 19.10). The number of cotyledons, of course, is one of the criteria for separating members of the Anthophyta into two classes, Monocotyledoneae and Dicotyledoneae. The degree of embryo differentiation at the time of seed dispersal may vary (members of the orchid family exhibit very rudimentary and incompletely formed embryos). The amount of endosperm may also be characteristic of large groups. For example, the endosperm of the grains is copious.

Biochemical

We have already seen (Chap. 2) that algal divisions are separated by biochemical characteristics, such as the type of chlorophyll or accessory pigments, the major component of the cell wall, or the principal form of food stored. Only within the past 20 years, however, has biochemistry been regularly applied as

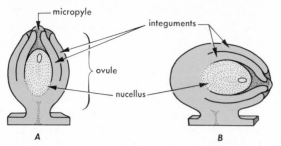

Figure 19.10   Number and position of cells within the mature embryo sac of three species: *A*, evening primrose (*Oenothera*); *B*, *Peperomia*; *C*, lily (*Lilium*). (Redrawn from A. S. Foster and E. M. Gifford, *Comparative Morphology of Vascular Plants*, W. H. Freeman and Company, San Francisco.)

a taxonomic tool for higher groups of plants. Members of closely related taxa are chemically treated to extract particular compounds and the amounts (or presence versus absence) of the compounds extracted are compared. The more similar the taxa are chemically, the closer their genetic relationship is presumed to be.

Almost every category of biochemical compound has been used at one time or another in these studies—sugars, amino acids, fats, oils, alkaloids, alcohols, terpenes, phenols—and they have been isolated by paper chromatography, electrophoresis, fractionation, or, more recently, gas chromatography. Many of these compounds are high in molecular weight and are structurally complex. It is reasoned by biochemical taxonomists that these complicated molecules probably evolved at only one time and in only one original group of plants; therefore, species or genera or families which possess the same complicated molecules are undoubtedly related to this original group of plants.

The existence of hybrids can sometimes be dramatically shown from biochemical evidence. Natural hybrids of two *Eucalyptus* species (Fig. 19.11) were suspected to occur in parts of Australia. The hybrids were hypothesized because unidentifiable *Eucalyptus* plants, somewhat intermediate in morphology, existed near the parent species. Leaf samples were collected from random members of each category and particular oils were extracted. Table 19.4 presents the results. The biochemical evidence of hybridization supported morphological evidence, because the suspected hybrids were intermediate in oil composition.

One of the most widely used methods for comparing taxonomic similarity of organisms at the molecular level, is **gel electrophoresis.** This method detects differences between enzymes (or other proteins) on the basis of differences in their charge and shape. Such differences are the result of changes in the amino acid sequence of the molecule (see Chap.

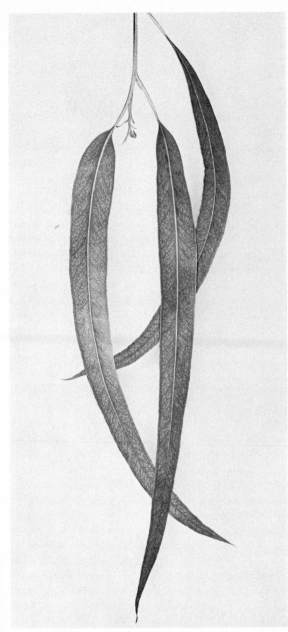

Figure 19.11 Branch and foliage of a typical *Eucalyptus* tree, $\times\frac{1}{2}$.

5). Since amino acids are coded for by small segments of the DNA molecule, amino acid differences are equivalent to single gene differences.

A small amount of liquid extract of crushed leaves (or any other plant organ desired) is placed in a notch cut into a gelatinous material (Fig. 19.12). This gel, often made from potato starch, either fills a tube or coats a glass plate. The extract is placed at one end of the gel, and an electric potential difference is set up across it. After several hours, the current is turned off, and the gel is "developed." The gel is immersed in a solution that contains a substrate upon which the enzymes under study act. The solution also contains a stain that couples with the product of the enzyme–substrate reaction. The result is that bands of color on the gel mark the locations of very specific enzymes. The distance that the enzymes have moved from their origin depend upon their charge and shape; the number of bands reveals the number of enzymes capable of reacting with the substrate (Fig. 19.13).

When members of a species are analyzed in this way, many of the bands are identical, revealing a high degree of genetic similarity, but there still may be considerable variation. The degree of genetic relationship between different species, genera, families, and even divisions, can be estimated in the same way: The closer the banding patterns compare, the closer the genetic similarity is presumed to be.

There are some limitations to the method. Basically, it underestimates the amount of genetic variation. First, not all types of enzymes are equally variable. If a researcher picks a "conservative" group of enzymes, he will not show as much variation as if he had picked another group. Second, not all changes in amino acid sequence of an enzyme will result in a change in the electrical charge or shape of the molecule. This means that two slightly different molecules will still migrate to the same point on the gel, and in those cases, they will be scored by the experimenter as identical, even though they actually are different.

Despite these drawbacks, gel electrophoresis allows taxonomists to estimate the degree of genetic similarity between organisms quickly and conve-

TABLE 19.4 Oil Content (in Relative Units) of Two *Eucalyptus* Species and Suspected Hybrids

Category	Geranyl acetate	Cineole	Eudesmol
E. macarthuri	4.0	0.0	3.0
E. cinerea	0.0	4.0	0.5
Suspected hybrids	0.8	1.6	1.4

From Pryor and Bruant, *Proc. Linn. Soc. N.S. Wales* **83**, 55, 1966.

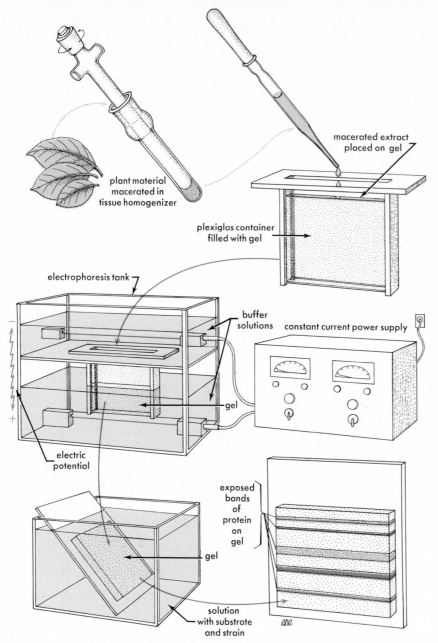

Figure 19.12 Diagram of a gel electrophoresis experiment.

niently. Such an estimate is much more accurate than one based on morphological similarities (like leaf shape and flower structure). Most morphological traits are controlled by many genes, not single genes, and can be strongly affected by the environment. But with amino acids, we are dealing with single genes, and environmental influences are largely eliminated.

Biological

Biological taxonomists are interested in determining what the **natural biotic units** are. By natural biotic units, they mean populations of plants which maintain their distinctiveness because of **biological barriers** that genetically isolate them from other populations. These isolating barriers may be due to breeding behavior (time of flowering, type of pollinator), habitat or geographic isolation, or inability to form fertile hybrids with closely related groups (sometimes because of differences in chromosome number or chromosome morphology).

Biological taxonomists often deal with taxa at, and below, the rank of species because they are vitally concerned with detecting divergent evolution at an

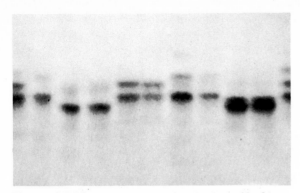

Figure 19.13 A "developed" gel of 11 *Clarkia* genotypes. The enzyme bands have migrated from the "top" of the gel shown, downward. (Courtesy of L. D. Gottlieb.)

early stage. They are interested in discovering how species become distinct and how they are able to maintain their distinctiveness even though living in close proximity or in similar habitats. Biological taxonomists are often referred to as **biosystematists,** and their field as **biosystematics.**

Conflicts continue to arise between biosystematics and traditional taxonomy because their conclusions are not always the same. The "natural biotic units" do not always correspond to "well-defined groups" of morphological distinctiveness. The conflicts have, in part, a historical cause. As a result of world explorations in the eighteenth and nineteenth centuries, it became clear that some species within a genus differed only slightly in morphology and habitat from region to region. One species might be in the mountains, another in the lowlands; one species might be along the coast of Europe, another along the coast of eastern North America. Were these really distinct species or should they be considered only as variants of a single species? The early answer was yes, they were distinct species. Much later, biosystematic research sometimes showed that fertile hybrids can result (or at least potentially can, once distances are overcome) from crosses between these species. The biosystematists then concluded that these "species" should be classed as subspecies of a single species because the morphological and geographic differences were less important indicators of relationship than breeding behavior.

On the other hand, biosystematists have sometimes discovered that breeding barriers exist between nearby populations of what traditional taxonomists had described as a single species (that is, the populations are morphologically identical). Yet, the genetic isolation undoubtedly is the first step of divergent evolution, and these populations represent distinct species at a very early stage. They are referred to as **sibling species.**

Many species which are widespread and occupy a variety of habitats may be composed of several genetically distinct subspecies. The subspecies are all interfertile, but hybrids are not common because hybrids are less suited to the habitats than are parental types. Clausen, Keck, and Hiesey of Stanford University have shown the existence of subspecies in plants such as five-finger (*Potentilla glandulosa*), which grows in California from the coast near Stanford to high elevations near timberline in the Sierra Nevada mountains. Clausen, Keck, and Hiesey dug up plants of five-finger that were growing all along this range and transplanted them together in a uniform garden at Stanford. In the uniform garden, any differences among the plants from different locations would be due to genetic control. They found morphological differences; timberline plants were much shorter than lowland plants. They also found physiological differences; the timberline plants flowered earlier in the summer than lowland plants. When five-finger plants from different locations were grown together in a garden near timberline, these differences proved critical to survival: lowland plants which did not become dormant were killed during the winter, and lowland plants which flowered late in summer were nipped by frost before seeds could be produced. Clausen, Keck, and Hiesey named the lowland group *P. glandulosa* ssp. *typica* and the timberline group *P. glandulosa* ssp. *nevadensis* (Fig. 19.14).

Biosystematists are also interested in comparing taxa on the basis of their survival strategies. How much metabolic energy do species invest in seed production, vegetative reproduction, accumulating a large bulk, and attracting pollinators or seed-dispersing animals? Why are some species genetically quite variable while others are very constant? What mechanisms do plants have to avoid or minimize competition and ensure survival? Experiments designed to answer these questions combine the talents of ecologists and taxonomists.

One example of different strategies can be seen by comparing an oak with a dandelion. Oaks and other large perennial plants devote a lot of energy to the accumulation of great bulk which can withstand many environmental stresses. The price paid is a long juvenile period of development when reproduction is nil, and even after that the amount of energy put into flower and seed production is low, compared to the total mass of the plant. Seeds (fruits) are usually large and are disseminated by gravity, water, or animals. Seed production and seedling establishment can be very low for many years in succession due to poor environmental factors, but the population density remains constant because of the long life span of individuals. This is called a **K** strategy by ecologists. Dandelions and most herbaceous annuals, on the other hand, invest a high proportion of their energy in seed production and have a short life cycle. Each seed is small and often wind-disseminated, which means there is very little food reserve to maintain the young seedling during establishment. The result is that most seedlings perish. But, because the reproductive capacity of the few survivors is so high, the population density is maintained. This is called an **r** strategy. Obviously, both K and r strategies can lead to success in an evolutionary sense. The question is, under what set of environmental conditions will each strategy be the most successful?

Numerical

All the approaches to taxonomy just discussed share one common feature: each subjectively selects one form of similarity on which to base conclusions, and each weights its own choice of evidence as being more important toward understanding relationships than any other form of evidence. Traditionalists weight morphological evidence, biosystematists weight breeding behavior, and so on.

Another approach to taxonomy is to consider all forms of evidence with equal emphasis; all evidence has equal weight. This is a statistical approach and is called **numerical taxonomy.** A basic tenet of numerical taxonomists is that a great deal of evidence is required to objectively separate taxa from each other. They utilize 50–300 characteristics for each study.

As an example, let us say that the number and relationship of species within one genus is under study. To begin, the numerical taxonomist arbitrarily divides the genus into a large number of "species;" this number may have no similarity to the number arrived at by traditional methods, and as we shall see, it has no effect on the outcome of the study. These arbitrary or operational species are starting points examined for each of the 50–300 traits and the data are recorded in relative terms (+ or − for present or absent and 1, 2, 3, or 4 for amount). Then all possible pairings between operational species are made and their similarity to each other is computed by a standard formula. Similarity is calculated as a percentage, 100% meaning complete similarity (identical) and 0% meaning no similarity (opposites).

The pairings are then ranked in order of increasing similarity and this ranking is diagrammed in a **den-**

| 98 ft at Stanford | 4592 ft at Mather | 10,000 ft at Timberline |

Figure 19.14 Growth of *Potentilla glandulosa* ssp. *typica* after one growing season at each transplant garden. This coastal subspecies, or ecotype, grew best in its normal environment at Stanford; it grew very poorly at timberline and was killed by the following winter temperatures. (Courtesy of W. M. Hiesey.)

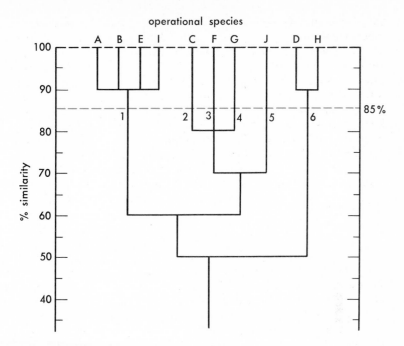

operational species

85%

Figure 19.15 Dendrogram, summarizing percentage similarity between pairs and groups of operational species. If members of the same species were defined as showing at least 85% similarity (dotted line), then the 10 operational species would be lumped into six species as indicated by the intersections of the lines, 1 to 6, with the dotted 85% line.

drogram (Fig. 19.15). The number of species can now be objectively determined by selecting an acceptable level of similarity. In Fig. 19.15, one taxonomist might select the range of 100–85% similarity as appropriate for members of the same species; if a line is drawn across the graph at the 85% level, we see that there would be six species. Another taxonomist might choose 65% similarity, and there would be three species. And so on.

The numerical taxonomist, then, does not admit that discrete species exist in nature. He insists that when one considers many traits instead of a few, one will see a continuum of variation. Boundaries between species can be erected by taxonomists, and these boundaries may be objectively selected; but the boundaries are arbitrary and artificial in the final analysis. The viewpoints of traditional and numerical taxonomists, then, are quite different. Interestingly, classification schemes arrived at by numerical means often show good agreement with those arrived at by traditional means. Can we conclude that experienced, traditional taxonomists are just as efficient as the computers used in numerical studies?

Summary

1. As a floral survey of the world becomes more complete, the goal of taxonomy becomes the construction of a phylogenetic classification scheme which summarizes the evolutionary history of our flora.

2. The history of taxonomy can be divided into a pre-Darwinian period and a post-Darwinian period. Linneaus culminated the pre-Darwinian period (ca. 1753) and organized plant nomenclature; nearly 200 years later, the rules of plant nomenclature were further elaborated.

3. Taxonomic research may take any of several directions, utilizing patterns of morphology, anatomy, embryology, biochemistry, and breeding behavior, any of which may be used to show genetic relationships between taxa. The numerical taxonomist utilizes information from all fields and does not give extra weight to any particular type of information. This is in contrast to other taxonomists who do choose to subjectively weight particular types of information.

20

Plant Growth and Development

All the genetic information required to build and operate a plant is carried in each individual cell encoded in the DNA of the nucleus, mitochondria and plastids (Chap. 5). During the life of a plant, this genetic information is used to direct increases in size (**growth**) and changes in form (**development**). As we have seen in looking at the architecture of cells, tissues, and organs in previous chapters, this growth and development requires a division of labor among cells, tissues, and organs. The formation of these differences is called **differentiation.**

The challenge of plant growth and development is to understand how particular bits of that genetic information are selected to chart a cell's growth and differentiation into a mature form that may differ in many ways from the undifferentiated meristematic cell from whence it ultimately derived. How is the reading of that genetic information accomplished so as to result in a red-colored cell in a rose petal, a green leaf cell or a tracheid? Why doesn't a buried seed, upon germination, form enlarged leaves underground? While we may at times emphasize the processes of differentiation at the cellular level, because cells are the basic functional unit, it should be understood that differentiation can be coordinated at the tissue and organ level as well.

Stored in this genetic information there must exist a library of programmed sequences of differentiation and development. One obvious example of such a sequence is the process of mitosis where, upon some signal, there is initiated a precisely orchestrated series of biosynthetic events, organelle movements, and membrane transformations. The choices of which program to follow, of when to do what, are made in response to appropriate cues. What constitutes an appropriate cue? Appropriate cues may be (a) signals from the environment, (b) hormonal signals from other parts of the plant or neighboring cells, (c) activities of neighboring cells or tissues, (d) position factors expressing the location within the plant body, and (e) nutritional factors, to name a few. A cue does not have to carry much information in itself in order to release a complex sequence of development. A period of low temperatures may trigger the change in the shoot apex that stops leaf initiation and allows the flower-making program to be read. The same low-temperature cue may be used to initiate the metabolic changes making a frost-sensitive bud which would be killed by $-5°C$ into a frost-hardy bud capable of surviving temperatures of $-30°C$.

The interval between perception of the signal and full development of the response can vary from seconds to months. Certain wavelengths of light can alter the movement of ions across membranes to a detectable degree in 10 sec. Application of auxin, a plant hormone (p. 375), to some tissues can result in increased growth within 1–2 min. The effects of low temperature on growth, however, may require weeks or months to complete. In the case of very rapid responses, the response may be so set up that there is no need to involve the direct intervention of RNA or protein synthesis. In other cases, new RNA molecules are doubtless formed, and corresponding new proteins appear (see Chap. 6). In others, cell divisions form new cells and even new organs are initiated. Thus, the extent of involvement of the metabolic machinery of the cell varies from one programmed sequence to another.

Variation in the numbers and enzyme complement of cell organelles constitutes an important aspect of differentiation. A leaf mesophyll cell, a root parenchyma cell, and a tomato fruit mesocarp cell all have the information in the nucleus and proplastids necessary to make a green chloroplast. Yet, the leaf mesophyll cell usually forms the complete photosynthetic apparatus (a mature fully green chloroplast) only if it is exposed to light; the root cells normally do not differentiate chloroplasts even when exposed to light; and the tomato fruit cell makes functioning green chloroplasts early in the life of the fruit. Later, during ripening of the fruit, a new course of differentiation besets these chloroplasts whereby the thylakoid membranes are rearranged, chlorophylls disappear, and red carotenoid pigments are synthesized in large amounts (Fig. 4.10D). In the fruit, this second differentiation of the chloroplasts into red chromoplasts is only one aspect of the whole complex differentiation process that is programmed to change a fruit from a seed-manufacturing organ into a seed-dispersing organ, attractive to the eye and taste of birds and animals.

Environmental Cues and the Selection of Programmed Patterns of Response

The development of a typical plant is much more the result of interaction with the environment than is the morphological development of one of our animal friends. The control or regulation of development may be considered at three levels: (a) within the individual cell, **intracellular,** (b) between cells, **intercellular,** and (c) from the outside of the organism, **environmental.** This is represented schematically in Fig. 20.1.

The next few paragraphs outline the kinds of problems and choices faced by a plant and how a variety of environmental cues elicit programmed patterns of response that maximize chances of survival of the species. In later sections, we will examine the internal mechanisms of control and their relationship to these environmental factors.

A wide array of blocks to seed germination are programmed to enhance the probability of germination under favorable conditions. These blocks can start in the fruit before seed maturity. Fleshy fruits frequently provide the seed with an inhibitor of germination such as abscisic acid (see later pages in this chapter) that prevents germination in the fruit prior to dispersal, fruit breakdown, and leaching away of the inhibitor. Occasionally, one can see a malfunctioning of blocks to germination in grapefruit and certain new tomato varieties where seeds do germinate within the fruit.

A requirement for exposure to a period of low temperature may favor germination in the spring. Some seeds will germinate only after sufficient rain, when the accompanying leaching action has washed out an inhibitor from the seed coat. This appears, for instance, to delay germination of some desert annuals until they have had a rainfall of adequate amount to support a few weeks' growth. Other seeds germinate only when buried (light inhibits germination), while some, especially small seeds with limited food reserves, may germinate only near or at the surface (light is required for germination). This light requirement is often the block removed when soil is disturbed and formerly buried seeds are stimulated to germinate, a characteristic of certain weeds familiar to gardeners.

The force of gravity acting upon organelles within cells gives directional information to the root enabling it to grow downward as soon as it emerges from the seed. In response to the same stimulus, the shoot is programmed to grow upward, the most likely location of light. Rhizomes may respond to the gravity vector by growing horizontally. Branches on trees often grow at a particular angle to the vertical. The direction of gravity is a very reliable cue on which to base these growth responses.

Why don't leaves unfurl and enlarge beneath the soil? The shoot system as it emerges from a germinating seed is a quite delicate structure that must push its way through soil of varying composition. It is a characteristic that dicotyledonous plants do not appreciably expand their leaves until they reach the soil surface (see Color Plate 11, Figs. 20.2, 20.3). The cluster of leaf primordia and apical meristem are usually inverted and *pulled* up through the soil behind the recurved portion of the hypocotyl or epi-

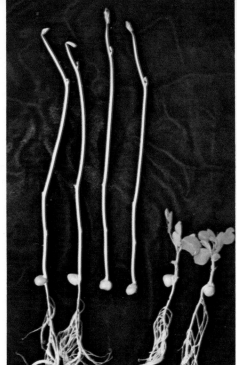

Figure 20.2 Above: Pea seedlings grown in light and in darkness. Note minimum stem elongation and maximum leaf development in light. The center two plants have had only 1 day in light, which caused straightening of the stem (hook-opening) and start of leaf expansion and greening, as compared to the fully etiolated seedlings on the left.

Figure 20.3 Above: Pea leaf development as a function of light. Fully developed green leaves from plants in continuous light; early stage of development after 1 day in light (center); etiolated leaf from dark-grown plant protected by plumular hook. All 8 days old.

Figure 20.4 Right: Corn seedlings grown in light or in darkness showing the long mesocotyl growth that pushes the coleoptile and its enclosed first leaf and shoot apex to the soil surface. In the light-grown seedlings the mesocotyls are only a few millimeters long.

Figure 20.5 Right: Corn (*Zea maize*) leaf development. A series showing the leaf enclosed in the coleoptile, emergence from the coleoptile in darkness where unrolling is prevented, an artificially unrolled leaf, and a similar leaf grown in light, fully unrolled with greening completed.

Figure 21.2 Left and below: *A*, Pierce's disease of grapes. The infective agent is a *Rickettsia*. *B*, yellowing and dwarfing of an orange tree infected with *Spiroplasma citri* (right). *C*, normal (large) and stubborn (small) Valencia oranges, the latter infected with *Spiroplasma citri*. *D*, fire blight of pear, the disease is caused by the bacterium *Erwinia amylovora*. (*A* and *D*, courtesy of Joe Ogawa, *B* and *C*, courtesy of E. C. Calavan.)

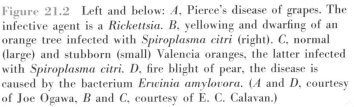

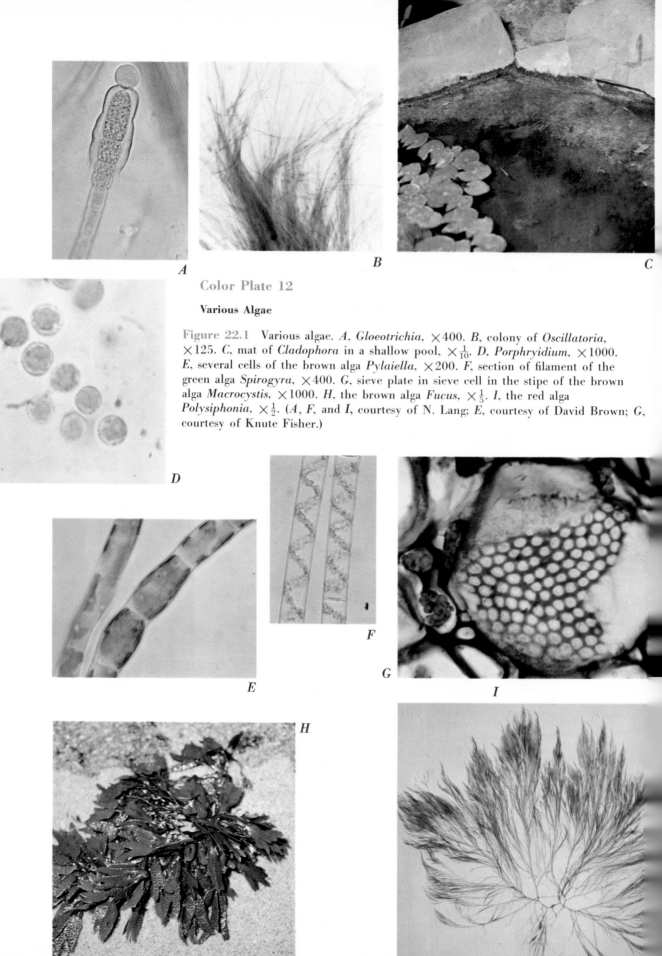

A

B

C

Color Plate 12

Various Algae

Figure 22.1 Various algae. *A, Gloeotrichia*, ×400. *B*, colony of *Oscillatoria*, ×125. *C*, mat of *Cladophora* in a shallow pool, ×$\frac{1}{10}$. *D, Porphryidium*, ×1000. *E*, several cells of the brown alga *Pylaiella*, ×200. *F*, section of filament of the green alga *Spirogyra*, ×400. *G*, sieve plate in sieve cell in the stipe of the brown alga *Macrocystis*, ×1000. *H*, the brown alga *Fucus*, ×$\frac{1}{5}$. *I*, the red alga *Polysiphonia*, ×$\frac{1}{2}$. (*A, F*, and *I*, courtesy of N. Lang; *E*, courtesy of David Brown; *G*, courtesy of Knute Fisher.)

D

E

F

G

I

H

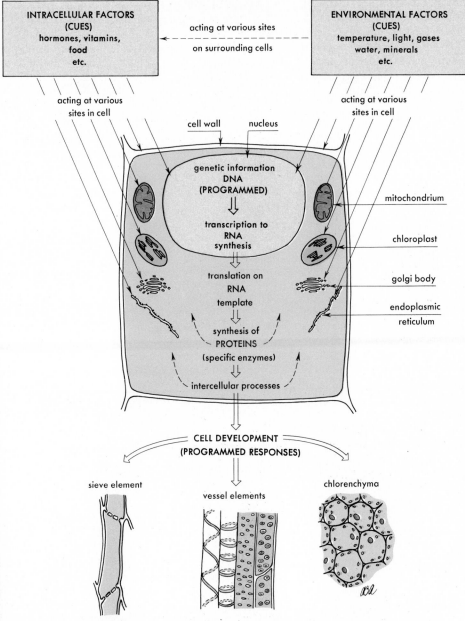

INTRACELLULAR FACTORS (CUES)
hormones, vitamins, food etc.

acting at various sites
on surrounding cells

ENVIRONMENTAL FACTORS (CUES)
temperature, light, gases water, minerals etc.

acting at various sites in cell

acting at various sites in cell

cell wall   nucleus

genetic information
DNA
(PROGRAMMED)

transcription to
RNA
synthesis

mitochondrium

chloroplast

translation on
RNA
template

golgi body

endoplasmic
reticulum

synthesis of
PROTEINS
(specific enzymes)

intercellular processes

**CELL DEVELOPMENT
(PROGRAMMED RESPONSES)**

sieve element

vessel elements

chlorenchyma

Figure 20.1 Diagram showing different levels of regulation of plant development, such as intracellular, intercellular, and environmental.

cotyl (the **hook**). The delicate leaves and apex are thus protected. In monocotyledonous plants, the young leaves are rolled up and completely en-sheathed in a protective tube, the **coleoptile.** When the coleoptile reaches the surface, the leaves push out of the splitting coleoptile and unroll. The cue which signals arrival at the soil surface is light received by the pigment, **phytochrome.** Phytochrome is a special light-absorbing molecule that serves as

a light-sensing system, making metabolic and devel-opmental processes responsive to the presence or absence of light. Growing a seedling in absolute darkness results in the kind of growth pattern adapted for movement through the soil (see Color Plate 11, Figs. 20.4, 20.5). The seedling is **etiolated.**

In addition to the leaf expansion or unrolling, a number of other changes are initiated by light. The stem usually elongates very rapidly in darkness (or in the soil) and light inhibits elongation. Supporting tissues of the vascular system are weakly developed in darkness when the soil would normally give all the support necessary. Light stimulates increased development of xylem. The initiation of growth can be traced to the appearance of enzymes that catalyze lignin synthesis and that make their appearance 2–3 hr after the start of illumination. The growth of the stem is programmed to develop the necessary strength to support itself and its leaves when growing above ground. While light slows the elongation of much of the stem, certain cells inside the hook respond by increased elongation, causing the stem to straighten out and the developing leaves to be oriented upward (Figs. 16.28, 20.2). These morphogenetic effects of light are all initiated by activation of phytochrome. A leaf cell is triggered into one pattern of behavior while a procambial cell takes off on another course and a cell in the hook on another programmed sequence of development, all in response to the same stimulus.

It may happen that the first light perceived by the emerging plant is coming from only one side. The usual response is for the stem or coleoptile to initiate a bending growth which orients the tip toward the light. This bending is initiated by absorption of more light on one side than on the other. The light-absorbing pigment is distinct from phytochrome.

The cells of the mesophyll of a leaf and the cortex of a young stem are programmed to delay differentiation of the photosynthetic apparatus until they are exposed to light (Figs. 20.3, 20.5). Then chlorophyll synthesis starts, membranes are organized into thylakoids, and the full battery of chloroplast enzymes is synthesized. Furthermore, the regulation is responsive to light intensity in many cases so that leaf morphology (sun and shade leaves) and enzyme content (ribulose 5-phosphate carboxylase) are adjusted to the amount of light available for photosynthesis (see Chaps. 10, 18).

Light filtered through the leaves of other plants is of lowered intensity and different spectral composition (Chap. 18). This appears to constitute a cue or cues to increase stem elongation, a response raising the chances of growing out of the shade of other plants and increasing light capture for photosynthesis.

On much of the land surface of the earth the suitability of conditions for plant growth varies seasonally with respect to temperature, water availability and total sunlight. Whatever the unfavorable season, it is typical that plants native to the area characteristically manage to time their activities in such a way as to complete the formation of seeds or resistant stages prior to the onset of the unfavor-

able season. The unfavorable season for one plant need not be the same for another species in the same area.

The key to much of this timing is the initiation of the formation of resistant stages or flowering early enough to complete the process prior to arrival of, for instance, the earliest winter frost in the locality. What are reliable cues? The seasonal change in day length is absolutely reliable and both plants and animals have evolved systems for measuring relative lengths of day and night. The ability to control developmental processes in relation to the lengths of day and night is called **photoperiodism**. The lengthening days of late spring are a certain indicator of an upcoming summer season just as the shorter days of late summer and early fall predict the oncoming winter. Thus, **long-day plants** and **short-day plants** usually bloom at different times of the year.

Frequently, photoperiodic signals are combined in various ways with temperature signals. A biennial plant may combine a requirement for a considerable period of low temperature (winter) with a requirement for a subsequent long day signal (the second spring) to initiate flowering. It thus grows vegetatively the first season and flowers only in the second. Photoperiodic and/or temperature signals can initiate flowering, the onset of dormancy, the development of frost resistance, leaf senescence, and the initiation of tubers and bulbs.

Plant Hormones as Agents of Control of Growth and Development

From many observations and some experiments, it was apparent to nineteenth century plant physiologists that the growth of one part of a plant was closely correlated with the growth or activities of another. They suggested that this communication must involve the movement within the plant of unknown chemical substances. Several direct lines of experiments can be traced down to recent years. Each line has yielded the identification of a particular compound which occurs in plants in very low concentrations. When these compounds have been purified carefully and applied to plants, they exert powerful controls on functions and plant development.

A set of five types of plant growth hormones seems to be nearly universal in seed plants. These are the most important agents involved in coordinating the growth of the plant as a whole. They frequently constitute the cues that determine the course of development and differentiation. Many of the effects of the external environment on development are mediated by changing the synthesis or distribution of these hormones within the plant. The five plant growth hormones are **auxins, cytokinins, gibberellins, abscisic acid,** and **ethylene,** gas. (Fig. 20.6). Depending upon the system, they may act essentially alone or in some sort of balance with one or another.

There are enough occurrences of hormones and hormone responses among more primitive plants to suggest that the evolutionary progenitors of seed plants had at least several of the important control systems. It is very likely that increasing importance and diversification of hormonal control mechanisms was an important component of the evolution of the seed plants.

Some hormones conform to the animal physiology definition of a hormone: a substance produced in one place in the organism which moves to another site, where it exerts its action. But that definition is unnecessarily restrictive for plant hormones. For example, one hormone, auxin, is generally produced in a shoot apex and then transported by a specific transport system to act on the growth of the stem and buds further down. However, there is a reasonable basis for saying that sometimes auxin must function in the same cells which synthesize it. Ethylene, another hormone, is a gas and is probably not transported in the plant in the dissolved state in any specific way.

Plant hormones sometimes initiate a sequence of developmental changes such that they must certainly be acting to alter the pattern of gene expression. In other cases, plant hormones appear to be controlling the rate of some enzymatic activity without directly involving the gene-directed synthesis of RNA and the RNA coding of protein synthesis (Chap. 6). In some cases, we can identify specific proteins synthesized as a result of application of the hormone to certain tissues. On the other hand, the response elicited in one plant system by a given hormone may be quite different from that produced by the same agent in another plant or in a different part of the same plant. So one hormone may initiate a number of different seemingly unrelated responses. One might take as a convenient working idea the proposition that the concentration of a given hormone is

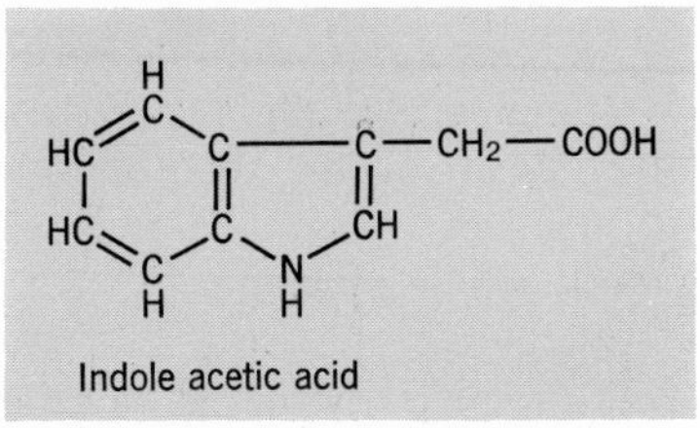

A: Molecule of the auxin, indole acetic acid (IAA), made up of two rings and a side chain

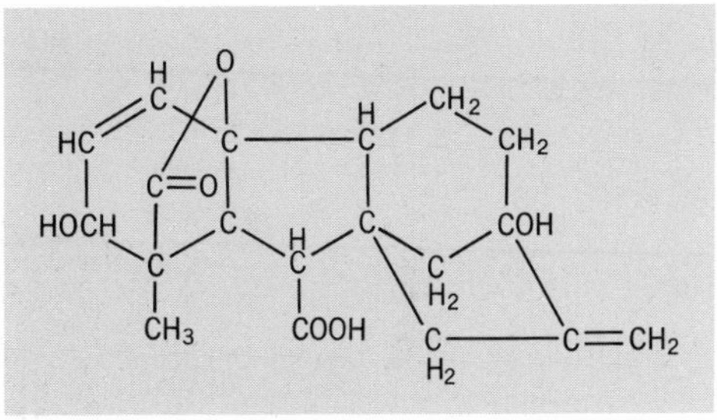

B: Molecule of gibberellic acid

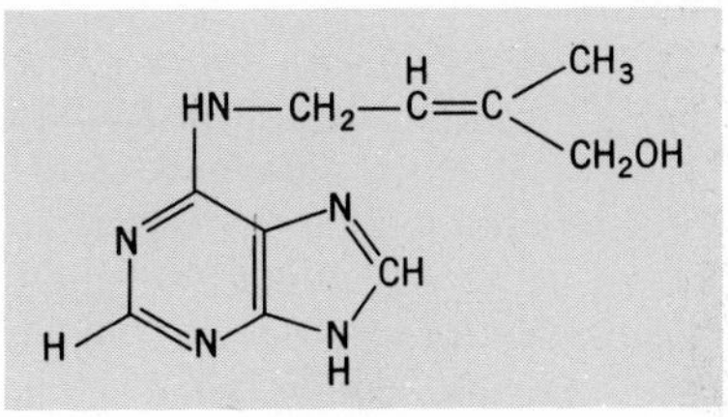

C: Zeatin

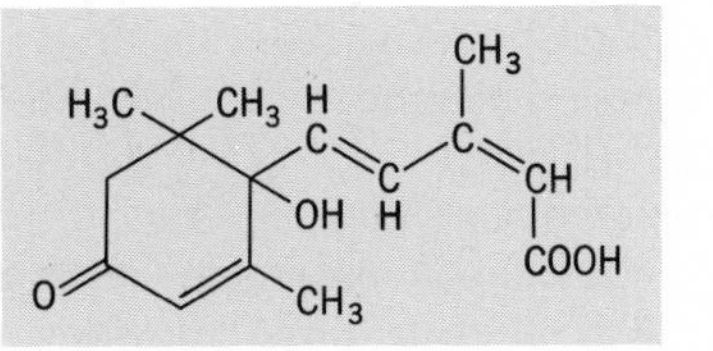

D: The structure of abscisic acid (also called dormin or abscisin II)

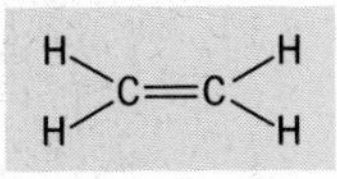

E: Ethylene

Figure 20.6 Plant Hormones. *A*, indole acetic acid, IAA, the most common naturally occurring auxin; *B*, gibberellic acid, GA_3, one of more than 40 very similar natural gibberellins; *C*, zeatin, one of several natural cytokinins, *D*, abscisic acid, ABA; *E*, ethylene, a gaseous growth regulator.

a meaningful cue about the internal environment of the plant and one that might initiate numerous programmed responses. This postpones the necessity of trying to consider that there is one mechanism of action of a given hormone before any one mechanism is known. It is easier to consider that these hormones are like radio tubes or transistors in that the kind of control function they exert depends on the kind of circuit they are plugged into.

It is most probable that these hormones, with the possible exception of ethylene, bind with protein receptor sites in the cell to carry out their action. Exhaustive studies of variations on the molecular architecture of the auxins, cytokinins, and the others to a lesser extent, indicate that the sites of activity are structurally very precise as are the sites on proteins which bind specific substrates. Other types of molecules do not generally have the structure-recognition capabilities that proteins have.

One actively pursued proposal is that a plant hormone may bind to a receptor protein. This receptor protein has a specific function in inhibiting or activating the transcription of a given gene (controlling the synthesis of mRNA on the nuclear DNA) or set(s) of genes. This hypothesis fits one notion of the regulation of development, but is not compatible with other knowledge about growth (Chap. 6).

Auxins*

Control of Cell Enlargement by Auxin

Cell enlargement is a process critical to the life of almost every cell in a plant, and the ability of the plant to control this process precisely is central to growth and morphogenesis. A plant cell may be thought of as an inflatable bag (the protoplast) surrounded by a supporting cover (the cell wall) which may be rigid or may be expanded under pressure, depending upon its makeup. By using respiratory energy and carrier systems, ions such as K^+, Na^+, and Cl^- are pumped into the vacuole, and water follows osmotically creating a turgor pressure. The turgor pressure in a typical cell might be maintained at 5 atmospheres or about 75 lb per square inch. Compare this to an automobile tire held at 28 lb per square inch pressure. Consider the surrounding wall. Is the cell wall going to expand while under this internal pressure? Obviously, some cells do expand under these conditions. Does the plant control the

rate of elongation by changing the turgor pressure against walls of unchanging properties? Or does the cell use a more or less constant force against a wall whose extensibility can be changed either to allow expansion or not? Plants predominantly use changes in wall extensibility to regulate growth, but they frequently use turgor changes to accomplish readily reversible swelling and shrinking of cells that underlie, for instance, the sleep movements of leaves, the opening and closing of the leaflets of the sensitive plant (*Mimosa pudica*) and, of course, the guard cell swelling and shrinking that controls stomatal opening. Auxin plays a role in changing wall extensibility.

The coleoptile of grasses (Fig. 16.33) has been considered since Charles and Francis Darwin's time as especially favorable material for studying this elongation process per se and the systems which control it. We now see that the coleoptile is used in these studies at a point in development where no more cell divisions will occur, there is little response to growth-promoting hormones other than auxin, and the cells are pretty well synchronized in a state of readiness to expand. Other plant materials tend to have cell divisions and enlargement proceeding at the same time and to be sensitive to the action of other hormones, simultaneously making them more complex and difficult to analyze. Although not properly a stem, the coleoptile at this stage grows in much the same way as the rapidly elongating portion of a stem.

Under normal turgor pressure, the cells of the coleoptile will expand rapidly if an optimal concentration of auxin is present, and slowly if it is too low or very high. Elongation is dependent upon an active metabolism. Inhibition of respiration causes nearly complete cessation of growth in a few minutes. In order to continue extension, the walls must constantly be worked upon by the protoplast through the plasmalemma. New materials are secreted through the membrane and deposited around the inside surface of the wall (cellulose microfibrils, p. 65) or well into the interior of the wall (other polysaccharide and protein components). There must be a breaking and reformation of cross-linking bonds between various of these macromolecules, making the framework of the wall (Fig. 6.10). How the cell controls these changes which occur outside its plasmalemma is one of the most intriguing problems in plant physiology. Auxin regulates some aspects of this wall metabolism to promote the loosening of the structure to allow the turgor force to stretch the wall.

Whether a cell expands in all directions equally to form an isodiametrical cell as in cortical parenchyma or whether the expansion is restricted to one

---

*Auxin is a general term which applies to a group of compounds which all affect plants the same way. It includes indole acetic acid (IAA), a naturally occurring compound, and many synthetic compounds. In this chapter, the term auxin will usually mean IAA.

axis to give an elongate, fusiform or tube-like cell depends upon the wall. If the supporting cellulose microfibrils are oriented randomly around the cell surface, expansion occurs uniformly, forming a larger sphere. If they are predominantly oriented more or less perpendicularly to one axis as are the hoops on a barrel, the wall will expand along one axis toward the ends, but the hoops prevent increase in diameter (Fig. 6.12). The result is formation of a tubular cell.

Another important point is that the growth process is very tightly controlled. Auxin presented to an oat coleoptile section deficient in auxin causes a growth rate increase in 1–2 minutes. At moderate auxin concentrations, the response is evident in 10 minutes in several tissues studied. Removal of the auxin supply results in a slowing of elongation in less than 30 minutes. Therefore there must be a continuous supply of auxin in the growing tissue to maintain growth. The rapidity of the response and other experiments are taken together to indicate that in this kind of effect on the cell wall, auxin does not act primarily through some direct effect on RNA or protein synthesis. There is, however, a requirement for protein synthesis if a cell is to continue growth over any period of time.

Controlling Auxin Concentration

The demonstration that elongation rates respond so quickly to removal or resupply of auxin makes it obvious how important the factors must be that control the concentration of auxin. The coleoptile and the growing stem generally separate the site of auxin synthesis from the zone of rapid elongation. The tip of the coleoptile and the stem tip with its cluster of young leaves synthesize indoleacetic acid (the major natural auxin) from the amino acid, tryptophan. The auxin is then moved down the stem through the elongating region (Fig. 20.7). The movement is called **polar transport** because it is a one-way energy-requiring active movement away from the tip. Typical rates of movement are 10–15 mm per hour, much slower than movement of materials in the xylem and phloem which may exceed 1 meter per hour. Further, polar transport in the youngest stem regions is counter to the direction of phloem and xylem movement and has been shown to occur even in stem tissue from which all vascular elements were removed.

Thus, auxin is swept from the tip through the growing zone, and the concentration in this stream sets the growth rate of the tissue. Remove the tip of a coleoptile, the auxin source, and the remaining auxin is rapidly drained away from the growing zone. The growth rate drops in a few minutes. Put back the tip or replace the auxin supply with a block of agar containing indoleacetic acid and growth is soon restored (Fig. 20.7). If, however, the auxin source is placed on only one side of the cut-off stump, the cells lined up below the auxin supply grow more than those on the other side. This leads to a bending of the coleoptile or stem.

Tropic Curvatures

Plants use this system of unequal supply of auxin on the two sides of a stem to orient the direction of growth in response to environmental cues. These curvature responses are called **tropic curvatures.** When the shoot and root emerge from the seed it is of obvious advantage to direct shoot growth upward and root growth down. The direction of gravitational force is a reliable cue. In growing regions in both shoots and roots, there are particles heavy enough to sink to the bottom of the cell. These sedimenting particles are called **statoliths,** and in higher plants much evidence indicates that amyloplasts perform this function. On reorientation of a cell from its original position (e.g., put a plant on its side), experiments have shown the amyloplasts to be deposited on the new bottom side in a few minutes. In some unknown manner, this reorientation of particles within the cell changes the direction of auxin transport progressively to the lower side so that by the time the auxin has traveled from the tip to the main growing zone there may be twice as much auxin moving through the lower half of the stem compared with the upper side (Fig. 20.7). The greater growth on the lower side pushes the tip until it is pointing up. The statoliths settle back to their original location, and auxin transport becomes uniform on both sides again. The original change in statolith position takes a few minutes, and by 15 minutes there is enough auxin concentration difference to measure the beginning of curvature. In some rapidly growing seedlings, the stem may turn upright within an hour from the time of putting it horizontal. This tropic curvature in response to gravity is called **geotropism.**

The geotropic curvature of roots downward has many aspects in common with the shoot response, except that there is considerable doubt that auxin is the hormone involved. It is clear that the root cap produces some controlling factor which moves away from the tip to control curvature. However, auxin is transported *toward* the root tip. The identity of the growth substance involved in root geotropism is being actively investigated.

Another stimulus that achieves lateral redistribution of auxin is light striking one side of the stem

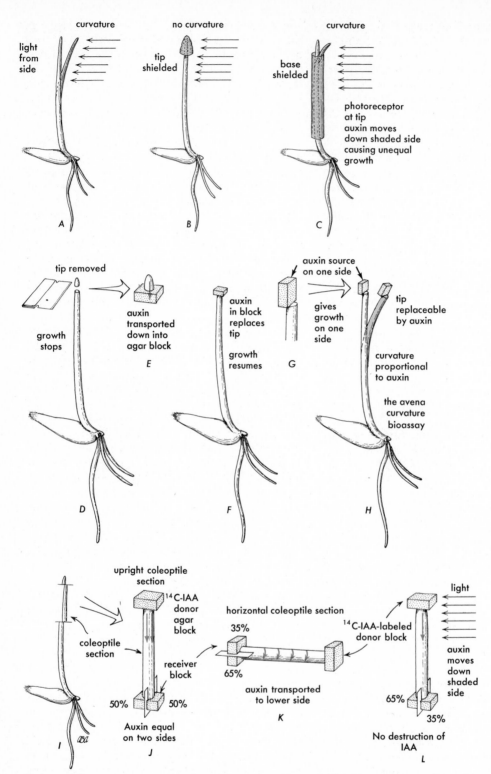

Figure 20.7 Diagram of experiments showing how regulation of auxin transport can result in phototropic curvature (*A, B, C, L*) or in geotropic curvature (*I, J, K*), and the *Avena* curvature bioassay used in estimating minute amounts of auxin (*D, E, F, G, H*). Experiments in which IAA labeled with radioactive carbon was used (*I, J, K, L*) show how geotropic and phototropic stimuli cause auxin to be transported laterally. Unequal amounts of IAA accumulate in the receiver blocks.

(unilateral illumination). The process of growing toward or away from unilateral light is called **phototropism** (Fig. 20.7). A pigment is relatively concentrated in cells just below the tip of coleoptiles and stems. Absorption of blue wavelengths of light somehow alters the pigment in a way that is linked to auxin transport. Dim light coming from one side of an axis results in more pigment activation on the lighted side than on the shaded side. Less auxin is moved down the lighted side, and a curvature of the tip toward the light results. We do not know the steps which link the phototropic pigment to the lateral redistribution of auxin. However, careful experiments have shown that indoleacetic acid (auxin) labeled with radioactive carbon and applied to the tips of coleoptiles moves symmetrically down the coleoptile in the dark. Light treatments which cause curvature result in as much as double the auxin moving down the shaded side as moves down the illuminated side, but the total amounts transported are the same as in the dark controls. Phototropism does not, then, normally involve auxin destruction.

One might suppose that continued production of auxin at the tip of a stem and its transport down ought to lead to accumulation somewhere. However, no region of accumulation has been detected; therefore, auxin must be destroyed or inactivated somewhere along the line. There are enzymes present in plants capable of destroying much more auxin than the plant could ever produce. The real question is how is auxin protected from destructive oxidation long enough to act. There are also enzymes that inactivate auxin by tying another molecule to it to form an inactive compound:

indoleacetic acid + aspartic acid →
$\qquad$ indoleacetylaspartic acid (inactive)

This latter system functions especially to prevent auxin concentration from rising too high. In fact, supplying high concentrations of auxin to tissues frequently results in the synthesis of the enzyme system responsible for inactivation of IAA.

Apical Dominance

In the typical pattern of growth, the growing tip of the stem with its young leaves exerts an inhibitory effect on the sprouting of lateral buds on the stem below the apex and on subsequent growth of lateral branches that do sprout. This phenomenon is called **apical dominance.** The relative effectiveness of this dominance of apical growth over lateral growth varies with the distance from the tip of the stem, the age of the plant, the genotype, and nutrition and other environmental factors. A plant with strong apical dominance has little or no branching, e.g., the sunflower. Weak apical dominance leads to a bushy appearance with numbers of side branches, as in tomato plants. In grasses, branching occurs at the very basal nodes of the stem. These branch shoots are called **tillers** and the process tillering. Relative rates of tillering are important characteristics of cereal grain varieties.

The existence of a reservoir of suppressed lateral buds provides preformed young shoot tips that are capable of rapid assumption of growth to replace a damaged stem tip. Experiments with pea seedlings have shown that by about 4 hr after removal of the growing stem tip there is the beginning of increased metabolic activity in the lateral bud next below the tip. One of the early events in the activation of lateral buds is the differentiation of the xylem to form a connection between the bud trace and the xylem in the bud itself. Completion of vascular connections to the stem is followed by growth of this new branch.

Removal of the apical bud or shoot tip removes the source of an inhibiting influence that passes down the stem. If the tip is replaced by a supply of auxin in a lanolin paste or agar block, the lateral buds remain inhibited. Thus, auxin can "replace the stem tip." If a cytokinin is applied directly to the lateral bud that is suppressed by the presence of an active shoot tip, outgrowth of the lateral bud is stimulated. After the branch has started to grow in response to cytokinin, its continued elongation is dependent on its own supply of auxin in the branch tip. Application of auxin to the recently sprouted lateral bud increases its growth. Thus, auxin functions in different ways before and after sprouting of the bud.

Auxin coming from the stem tip has been shown to inhibit the formation of vascular connections to the lateral bud. Cytokinins, on the other hand, promote the vascular differentiation. It is probable that the cytokinins originate in the roots and move up in the xylem sap. While we know some of the major aspects of the control system, it is not certain that differentiation of vascular connections is the first event that occurs in the bud as it is released from the influence of the apex.

There are many other "growth correlations" where the growth and development of one plant part is related to that of another (Fig. 20.8). Apical control may cause lateral organs such as branches, leaves, rhizomes, and stolons to grow horizontally or at some angle with respect to gravity that is different from the upright habit of the leader shoot. Removal of the dominant leader is often followed by bending of a nearby branch into an upright direction as it

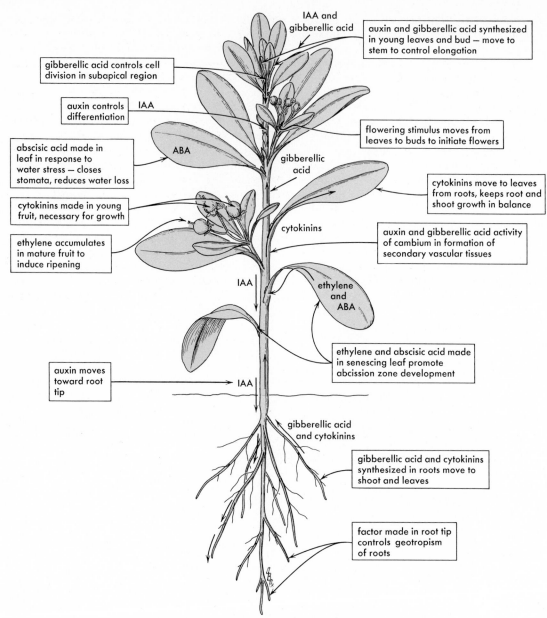

IAA and
gibberellic acid

auxin and gibberellic acid synthesized
in young leaves and bud — move to
stem to control elongation

gibberellic acid controls cell
division in subapical region

auxin controls
differentiation

IAA

flowering stimulus moves from
leaves to buds to initiate flowers

abscisic acid made in
leaf in response to
water stress — closes
stomata, reduces water loss

ABA

gibberellic
acid

cytokinins move to leaves
from roots, keeps root and
shoot growth in balance

cytokinins made in young
fruit, necessary for growth

cytokinins

auxin and gibberellic acid activity
of cambium in formation of
secondary vascular tissues

ethylene accumulates
in mature fruit to
induce ripening

IAA

ethylene
and
ABA

ethylene and abscisic acid made
in senescing leaf promote
abcission zone development

auxin moves
toward root
tip

IAA

gibberellic acid
and cytokinins

gibberellic acid and cytokinins
synthesized in roots move to
shoot and leaves

factor made in root tip
controls geotropism
of roots

Figure 20.8 A diagram illustrating some typical hormonal interrelationships among various portions of the plant.

takes over the function of dominant shoot. The direction of growth of some rhizomes and stolons has been shown to be dependent on hormones coming from the shoot tip.

Cell Differentiation in the Xylem

Differentiation of tracheary elements is one of the most extensively studied examples of cell differentiation in plants. The results of differentiation are easily seen because the changes in the cell are extreme. They result in extensive wall thickenings enabling the cells to withstand water stress, and they also result in loss of the protoplast to allow unimpeded water flow. One experimental approach has been to perform delicate operations on the shoot tip to study how young leaf primordia alongside the apex influence differentiation of procambial cells into tracheids or vessel elements. A leaf primordium promotes differentiation in the procambial strand leading to it (Fig. 10.17) and this promotory effect of the primordium can be replaced by auxin (Fig. 20.9).

In *Coleus* stems, a wound that severs a vascular bundle is followed by cell division and differentiation of tracheary elements from parenchyma cells in a path around the wound to reconnect the interrupted bundle. This regeneration process requires a supply of auxin that is normally transported polarly out of leaves above the wound and then down the stem. If the leaves are removed at the time of wounding, regeneration does not occur. The leaves can be "replaced" by auxin applied to the petiole stump.

Further down the stem where elongation has ceased, even an herbaceous plant like the sunflower may have some secondary growth from a vascular cambium. Here again, auxin from the young leaves of the shoot apex continues to function in promoting xylem differentiation. Auxin applied to the stump after removal of the apex can maintain divisions in the cambium and can also maintain the differentiation of the interfascicular cambium. In woody plants, cambial activation in the spring as growth is resumed involves the production of auxin by newly activated buds. A wave of cambial divisions moves down the stems stimulated by auxin moving down from the buds. Furthermore, continued cell division in the vascular cambium and normal differentiation of secondary xylem and phloem requires a supply of both gibberellin and auxin.

Further advances concerning the differentiation process have come with tissue cultures. It is now possible to regulate growth conditions, principally with auxin and cytokinins, to induce pea root cortex cells to differentiate into tracheids in tissue culture. As it occurs in these cells, the process includes an early synthesis of DNA, formation of polyploid nuclei, cell division, and finally rapid differentiation into tracheary elements. In these experiments, 16% of the new cells formed were tracheary elements. Doubling of the chromosome number is a part of differentiation of some other species and cell types, too, as is the case, for instance, in some root hair cells. The ability to induce specific cell differentiation at will in large quantities of cells provides a powerful means of studying differentiation and the role of hormones in the process.

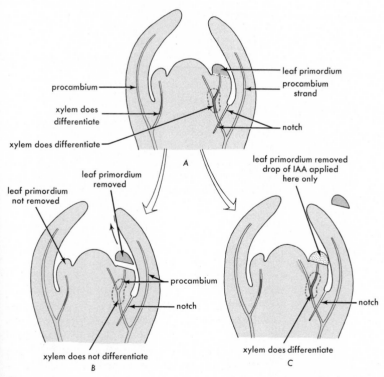

Figure 20.9 Diagram of an experiment showing how a leaf primordium provides a stimulus for xylem differentiation in the procambium. The notch severing the procambium isolates the tissue of interest above it. *A*, control with intact leaf primordium; *B*, leaf primordium removed; *C*, leaf primordium removed and a drop of auxin substituted. Auxin is effective replacement for stimulus from the primordium.

Gibberellins

Gibberellins and Stem Elongation

The shoot tip with its cluster of leaf primordia exerts control over growth activities below it in more ways than just supplying auxin. In sunflower, for instance, **gibberellins** (Fig. 20.6) can pretty well substitute for the shoot tip if it is removed. Removal of the shoot tip stops growth; gibberellins restore growth whereas auxin does not. Gibberellins are synthesized in the young petiolate leaves that surround the shoot tip. Sometimes they are made also in root tips, but it is not yet clear just what is the role of the gibberellins from the roots. Here we are dealing with the activities of a stem over a period of days. Gibberellins have been shown to promote cell division in the stem immediately below the apical meristem. It is in this region that enough divisions occur to provide a large fraction of the total cells involved in primary growth of a stem. There are synthetic chemicals, some of commercial agricultural importance, that act in the plant quite specifically to inhibit the synthesis of gibberellins. When these growth retardants (for example, CCC, AMO–1618) are applied to plants, cell division in the subapical region practically stops, whereas the leaf-initiating activity of the apical meristem continues. The result is a stem which essentially stops elongating but continues to make leaves separated by extremely short internodes. A plant of this form is called a **rosette,** whereas plants with elongate internodes are called **caulescent** (see both forms in Figs. 20.16, 20.18). The application of gibberellin to a retardant-treated plant causes the resumption of both cell division and elongation in the internodes and subapical region. Thus, gibberellins are an important factor in stem elongation.

Many plants naturally grow in a rosette form for at least part of their growth cycle. They typically respond to gibberellin application by rapid stem elongation, **bolting,** that is a part of flowering (Figs. 20.16, 20.18). As will be discussed later, the bolting response can be caused by an environmental factor such as cold treatment (winter) or long days, through triggering an increase in gibberellin.

As to the relation between auxins and gibberellins in stem elongation, it is apparent that the two frequently function simultaneously in the same stem. In some cases, gibberellins may stimulate cell division, thus providing more cells on which auxin can act. In the coleoptile, gibberellins act at an early stage of development, and the auxin-sensitive stage described above comes later in the life of the cells.

In Europe, commercial use is made of the growth retardant CCC or Cycocel (2-chloroethyltrimethyl ammonium chloride) to inhibit stem elongation in wheat plants. Shorter, stronger stems result and the plants are much more resistant to lodging, being knocked down by wind and rain. Lodging makes harvesting difficult. Short stems in grains are also achieved genetically in breeding programs and are important factors in Mexican wheats and Philippine IR8 rice, plants of the "green revolution," that have raised crop productivity in Asia and Central America.

"Dwarf" forms in a variety of plants are frequently due to a diminished gibberellin synthesis or to an enhanced production of natural antagonist of gibberellin action. Dwarfism is sometimes dependent on light for expression, as in peas where both dwarf and tall varieties are equally tall in darkness (Fig. 20.10). Dwarf forms of corn and peas are used in estimating the concentration of gibberellins in extracts of plants or plant parts. Fig. 20.11 shows such a bioassay wherein quantities of gibberellins are estimated by comparison to the ability of known concentrations of the hormone to stimulate stem elongation. Bioassays have been instrumental in the

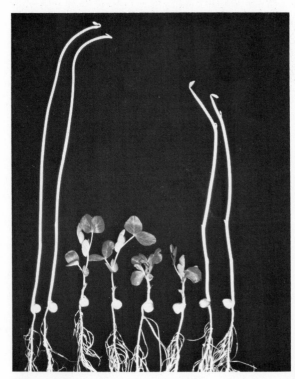

Figure 20.10 Dark- and light-grown pea seedlings of tall (left four plants) and dwarf (right four) varieties 8 days old. The etiolated plants were grown in the dark, the middle four were grown in the light.

0.0 0.0001 0.0005 0.001 0.0025 0.005 0.01 0.025 0.05 0.1

μg GA$_3$/ PLANT

Figure 20.11 Dwarf peas (*Pisum sativum*) showing the promotion of stem elongation by gibberellic acid applied 7 days prior to photograph. This response is used to bioassay the gibberellin contents of plant extracts. (Courtesy of L. Rappaport.)

purification and identification of all the plant growth hormones. The appropriate plant material may respond in hours or days to minute quantities of hormones that would be impossible to identify chemically.

Gibberellins and Enzyme Synthesis

As a grain such as barley or corn starts to germinate, the embryo begins to grow but has limited food reserves in itself. The main reserves are in the starchy endosperm, a collection of cells loaded with starch, reserve proteins, and some nucleic acids. A special layer of living cells, the aleurone layer, surrounds the main endosperm tissue and is instrumental in digesting the stored materials to soluble forms that can diffuse to the embryo. Early in germination, the embryo synthesizes gibberellin (GA) which diffuses to the aleurone layer where the GA acts as a signal to activate the synthesis and secretion of the enzymes necessary to digest stored materials in the endosperm. If the embryo is removed from the seed prior to germination, very little digestion of the remaining endosperm takes place. Adding gibberellin in minute quantities to the embryo-less endosperm induces the synthesis and secretion of enzymes just as if the embryo had provided the stimulus (Fig. 20.12).

The barley aleurone layer can be isolated and studied independently of the embryo and storage endosperm. It is a collection of cells with no growth activities and no division, but with a very active ability to synthesize a few proteins and to secrete them. As such it has been used to study the mechanism of action of gibberellin, and it has yielded much useful information, however, without as yet yielding the molecular basis for GA action. About 6 hr after adding gibberellin, the aleurone cells start to make α-amylase, an enzyme which breaks starch into soluble sugars, and other enzymes involved in breakdown of proteins and nucleic acids. The appearance of α-amylase is due to *de novo* synthesis (the enzyme is assembled from amino acids). Earlier, it was thought that GA might be specifically activating the gene coding for α-amylase. This is apparently not the case. Gibberellin application causes a number of dramatic changes in the aleurone cells prior to the appearance of amylase. New ribosomes and endoplasmic reticulum membranes appear, as well as enzymes involved in the synthesis of membrane lipids. The addition of GA to the aleurone unleashes the development of the endosperm digesting system, but we do not know just how it does this.

In the manufacture of beer, the starch stored in the grain endosperm must be hydrolyzed to a soluble form before its conversion into alcohol by the glycolytic enzymes of yeast. The natural production of amylase occurs in the malting barley. The addition of extra GA speeds up amylase synthesis by the aleurone enough to be used commercially.

Cytokinins

Cytokinins, Cell Division, Organ Initiation, and Senescence

It has long been recognized that cell division in many tissues is blocked or limited by insufficient concentrations of some hormonal factor. Coconut milk was shown to be a rich source of a cell division-inducing factor. A suitable bioassay was developed in which parenchyma cells in chunks of tobacco pith tissue responded to the factor by division and growth. First, a synthetic, and later, naturally occurring cell division-stimulating compounds of a class called cytokinins were isolated, identified, and studied. Zeatin (Fig. 20.6) and isopentenyladenine (IPA)

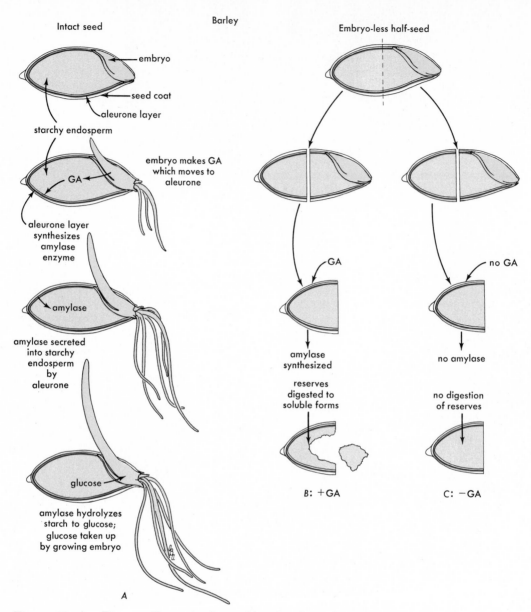

Figure 20.12 Diagram illustrating how gibberellin from the embryo induces the synthesis of the starch-degrading enzyme, α-amylase, in the aleurone layer.

are two naturally occurring cytokinins. Kinetin is a synthetic cytokinin. Actually, cytokinins exert control over many processes, and some do not involve cell division.

Cytokinins have been proved to occur in high concentration in rapidly dividing tissues, especially young fruits where they are apparently synthesized. Thus, corn grains at the milk stage are a rich source of zeatin, and young apples, plums, and other fruits contain high concentrations of cytokinins at the time when cell divisions are maximum.

The tobacco pith tissue culture has yielded much information on the hormonal control of morphogenesis. Freshly isolated cells will enlarge somewhat if supplied with nutrients and auxin, but will not divide unless minute amounts of a cytokinin are added. Thus, cytokinins can release the process of cell division. Moreover, by varying the balance between auxin and cytokinin, it is possible selectively to initiate the development of root or shoot primordia (Fig. 20.13). High auxin-to-cytokinin ratios cause root initials to differentiate; a suitably treated block

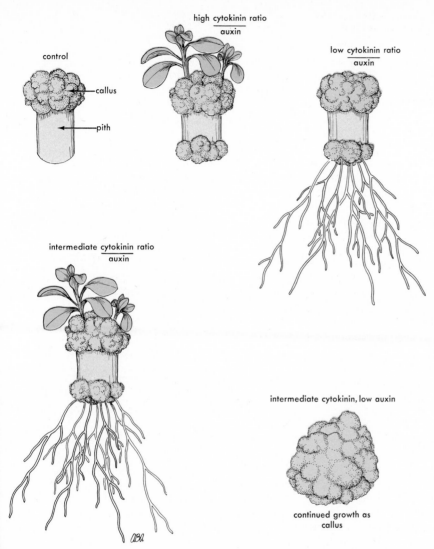

control

callus

pith

high cytokinin ratio
—————
auxin

low cytokinin ratio
—————
auxin

intermediate cytokinin ratio
—————
auxin

intermediate cytokinin, low auxin

continued growth as
callus

Figure 20.13 Diagram of the control of differentiation exerted by interaction of auxin and cytokinin. Pieces of tobacco pith tissues were aseptically grown on nutrient medium (tissue culture) supplemented with various levels of the two hormones.

of pith tissue can be covered with roots. A low auxin-to-cytokinin ratio causes differentiation of clumps of cells into apical meristems; the resulting shoots can cover the original block of pith. Intermediate concentration ratios stimulate a growth of relatively undifferentiated masses of cells, a callus tissue. Shoots can be removed, rooted, and grown to mature plants. Even single cells from the pith have been grown into mature plants, showing that a parenchyma cell contains all the genetic information necessary to form a plant: The original differentiation of the parenchyma cells did not involve a loss of genetic material.

Many observations point to a close balance between shoot and root growth. Certainly, a part of cytokinin physiology must involve the supply of cytokinins to the shoot through the xylem from the root tip where at least a portion of the plant's cytokinins are made. Cytokinins have been found in the xylem sap exuding from cut-off stumps of several species. If a leaf is removed from a stem, a sequence of changes is initiated. The changes can be called **senescence,** and they may include the breakdown of storage carbohydrates, cessation of protein synthesis, and an increase in the breakdown of proteins, nucleic acids, and chlorophyll. These changes re-

Cytokinins

semble somewhat those typically occurring during natural leaf senescence. If the leaf is put under conditions where it forms adventitious roots, the senescence is halted. The roots can thus substitute for attachment to the plant. Furthermore, cytokinin application to the leaf can substitute for the roots, too, in preventing or delaying much of the senescence. In practice, the delay or prevention of chlorophyll loss in isolated leaves or disks cut from leaves is a sensitive bioassay for cytokinins.

The maintenance of active RNA and protein synthesis appears to be an essential part of cytokinin action in delaying senescence. It is easy to visualize a strong correlation between root and top growth involving cytokinins. Treatments that slow root growth, such as water-logging or flooding the root zone, cut down the cytokinin concentration in the xylem sap.

A fascinating puzzle has developed concerning the mechanism of cytokinin action. Cytokinins occur free or simply combined with ribose or ribose phosphate, on the one hand, and they also occur as a functional part of tRNA molecules that carry certain of the amino acids to the ribosomes for assembly of proteins. It would be attractive to hypothesize that the dramatic effects one sees by adding cytokinins to the tobacco pith tissue or to isolated leaves are due to cytokinin action as part of tRNA involved in protein synthesis. But the available biochemistry says that one cannot add a preformed cytokinin and expect it to get into tRNA. The cytokinin in tRNA is made after the tRNA is assembled. One proposal is that while a cytokinin as part of certain tRNA's has a certain function to perform, the role of added cytokinins is to act in another way. This proposal suggests that free cytokinins are made in cells by breakdown of tRNA. This is just one example of the probability that successful application of the concepts of molecular biology to problems of plant development requires more skill than did the original development of the concepts with bacteria.

Ethylene

Ethylene as a Natural Growth Regulator

Ethylene is a simple gas, $CH_2{=}CH_2$ (Fig. 20.6), that is made in small quantities by many plant tissues in which it serves as a powerful natural regulator of growth and development. It is also a common product of combustion and a frequent air pollutant. It can create havoc in a florist's greenhouse, it is routinely used to bring about the uniform ripening of bananas, and in uncontrolled application it can lead to fruit spoilage. Since it is naturally involved in triggering the fruit-ripening process in so many fruits, it has been called the fruit-ripening hormone. More recent work has established many roles for ethylene in normal vegetative growth, too, so it has broad importance in growth regulation as do the other hormones we consider in this chapter.

The fact that ethylene is a gas requires a few explanations about how it normally functions and how it can be controlled or manipulated to modify its concentration in tissues for experimental or practical purposes. It is synthesized in the plant from the amino acid, methionine, a constituent of all cells. Since ethylene has only limited solubility in the aqueous phase of a cell, it volatilizes into the intercellular air spaces and diffuses away into the surrounding atmosphere as does carbon dioxide formed in respiration. Natural ethylene action in a tissue is dependent upon its continued synthesis. Its concentration in a tissue is determined by the relative rates at which it is synthesized and at which it diffuses away. If ethylene is introduced into the air surrounding a tissue, it diffuses in and raises the internal concentration to equal the external concentration. Enclosing plants or especially fruits in a container or space with restricted air circulation can cause the concentration of ethylene made by the plant itself to build up to the point that it has drastic effects on growth and development. Effective levels are frequently in the range of 0.1–1 part per million of air. Even the levels of ethylene present in urban air pollution are sometimes sufficient to cause measurable biochemical changes in plants. One successful method of achieving ethylene application under field conditions is to spray plants with a compound that spontaneously breaks down in the tissue to release ethylene. One such compound, chloroethylphosphonic acid, yields ethylene, chloride, and phosphate, which are all natural plant constituents.

An example of ethylene action in controlling vegetative growth is illustrated by a dicotyledonous seedling such as the pea during its early growth. As the plumule emerges from the germinating seed in the soil, the tip of the seedling shoot maintains a mode of growth that protects the delicate apical meristem and prevents young leaves from being damaged or torn off as the hook, the recurved upper portion of the stem, is pushed through the soil (Fig. 16-28). A high rate of ethylene synthesis in the young plumule in darkness maintains the hook and prevents leaves from enlarging appreciably (Figs. 20.2, 20.3). The transition from the underground mode of growth to the above-ground mode is triggered by absorption of light by phytochrome. One of the changes induced by the activation of phytochrome is the shutdown of ethylene synthesis to a much lower level. The leaves are thus relieved of ethylene inhibition of expansion. The hook "opens" or straightens out to present the expanding leaves to sunlight, which further stimulates leaf development (Fig. 20.3).

In many ways, growth in darkness is essentially equivalent to growth underground. However, growth underground may provide additional problems—environmental cues and plant responses, that are not illustrated by simply growth in darkness. If a restricting obstacle such as a soil crust or stone is encountered by the hook, pressures are generated in the tissue, and within a few hours a dramatic rise in ethylene synthesis can be measured. The result is an ethylene-induced swelling of the stem and an inhibition of elongation. Enlargement of the stem diameter enables it to exert a greater upward force against the obstacle. If the stimulus is prolonged, the stem geotropic response is modified and it starts to grow almost horizontal. These responses to extra-high ethylene increase the likelihood of penetration through, or growth around, a restriction and thus, successful emergence from the soil. They can easily be induced in dark-grown plants by gassing with ethylene (Fig. 20.14).

Ethylene may promote the germination of certain seeds.

In the growth of young trees, the relative height and diameter of the trunk is very much a function of the motion it undergoes as a result of wind. If a young tree stem is tied to a rigid stake, as is frequently the case in container-grown trees, or is supported by other plants in a crowded nursery, the usual response is a tall, slender, weak stem. If the stake is flexible enough to allow some movement in response to gentle winds, the plant has a sturdier form more like one grown without support. The plant that grows without support from a young stage typically has a shorter stem, larger in diameter, and with a tapered form coming from a wide base at the soil level. Routine exposure to high winds can greatly increase this response, as can be seen in trees on windy sea coasts or mountain ridges. Controlled gassing of tree trunks with ethylene stimulates enlargement, and it is thought that stress-induced ethylene synthesis in the trunk contributes to this growth response.

Ethylene and Auxin Activity

Ethylene and auxin physiology are often intertwined. It seems to be nearly universal that high concentrations of auxins induce the synthesis of ethylene in tissues where ethylene synthesis is low. It is not easy to tell how frequently the endogenous production of ethylene is controlled by endogenous auxin, but many of the responses to the auxin type of herbicides are in reality responses to ethylene. In turn, ethylene inhibits the transport of auxin in many tissues. The result is an elaborate control of the effective auxin level in a tissue.

Two examples of the relatedness of some (but certainly not all) auxin and ethylene responses are the control of flower morphogenesis in cucurbits (cucumbers, squash, melons), and in the initiation

Figure 20.14 **The response of dark-grown pea seedlings to various levels of ethylene during 4 days (ppm = parts per million ethylene in air). Internally generated ethylene under stress can induce similar responses. (Courtesy of J. Goeschl and H. Pratt.)**

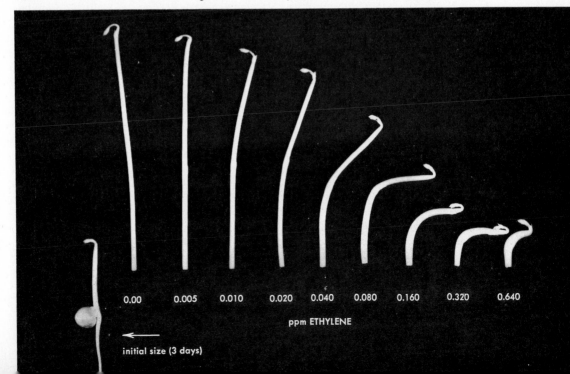

0.00   0.005   0.010   0.020   0.040   0.080   0.160   0.320   0.640

ppm ETHYLENE

initial size (3 days)

of flowers in bromeliads (e.g., pineapple). A typical cucurbit plant first forms male (staminate) flowers, and at later nodes it forms perfect flowers. There are some strains which have only female (pistillate) flowers. It has been shown that applied auxin or some other treatment to raise internal ethylene favors the initiation of female flower parts on flowers that would normally lack them, or it tends to make pistillate flowers from normally perfect flowers. Gibberellin, either naturally high concentrations or artificially applied, promotes maleness, the initiation of anthers. There appears to be an ethylene–gibberellin balance that determines, at the time of flower initiation, which organs the flower will have. Being able to control the sex of flowers chemically has simplified the commercial production of hybrid seed by restricting the parentage of seed to the desired cross.

Pineapple plants can be induced to flower and thus form fruit on a precise schedule by applying either auxin or ethylene or an ethylene-releasing substance to the plant. This is of great commercial importance in coordinating harvesting and processing or shipping activities. On a somewhat more humble scale is the trick of inducing ornamental bromeliad house-plants to flower by putting them in a transparent air-tight container with a ripe apple as an ethylene source. Bromeliads are seemingly unique in that ethylene is not generally an initiator of flowering.

Abscission of leaves and the role of abscisic acid in the process was discussed earlier, in Chapter 10. Ethylene treatment can also be very effective in causing abscission of leaves and fruits. Ethylene can interfere with the normal transport of auxin out of leaves. A regular auxin supply to the petiole is a necessity to prevent the formation or maturation of the abscission layer at the base of the petiole. When high ethylene concentrations inhibit the polar transport of auxin down the petiole, the abscission layer forms (Fig. 10.19) and the leaf soon abscises. An ethylene-releasing chemical is under investigation for promoting abscission of fruits for mechanical harvesting. Under certain conditions, it is successful for some fruits without producing too many undesirable side-effects.

Orchid flowers are in part valued for their longevity in cultivation as well as their beauty. In cultivation they are isolated from their natural pollinators and this contributes to their longevity. The packets of pollen grains, pollinia, normally carried by pollinators, are very rich in auxin, and when a pollinium or auxin is applied to the orchid stigma a rapid production of ethylene ensues which triggers the further development of the flower and the rapid senescence of the petals. This can be most disappointing to the orchid fancier.

Ethylene and Fruit Maturation

During the normal development of a flower, on through to a ripened fruit with mature seeds, there is a period which requires extensive growth and development of the embryo in the seeds. During this time, dispersal would be premature. There is in many fruits a programmed pattern of changes that convert the fruit from a seed-manufacturing to a seed-dispersal organ. One change frequently is a color change from green to yellow, orange, red, or blue; this increases its visibility to potential animal dispersal agents. Chlorophylls are broken down and other pigments such as the red and blue anthocyanins are synthesized. Another change is the conversion of starch and organic acids to sugar so that the fruit becomes sweet. Some of the materials in the cell walls are broken down so that the cells become more loosely bound to each other and the fruit is softer. Volatile flavor components are synthesized, and these contribute much of the flavor we value. One important link in this grand plan is the rise in ethylene to a critical concentration within the fruit. When this critical concentration is reached, the whole pattern of changes we call **ripening** is initiated. In many fruits, ripening can be hastened or at least made to occur uniformly by applying ethylene to fruits which are fully enough developed but have not attained the critical internal level of ethylene to trigger the ripening process.

The whole battery of changes requires extensive synthetic activities by the fruit. This requires the expenditure of considerable metabolic energy, as is shown by a large increase in respiration while these changes are being made. This rise is called the **climacteric rise** in respiration. It can be measured as CO_2 production, O_2 consumption, and a sudden rise in ethylene production well over the rise that triggered the process initially.

It is standard commercial practice to pick bananas green in the tropics, transport them to storage rooms near the point of use and then treat them with ethylene gas to ensure that all the bananas in the bunch ripen at the same rate. They will ripen in about 5 days.

Ethylene may cause ripening when it is not wanted. In fruit held in cold storage, elaborate precautions are used to prevent ethylene from accumulating in the atmosphere to a point where it would trigger ripening. When a fruit goes through the phase of rapid ripening, the production of ethylene may increase many times over the rate that was initially necessary to reach the critical triggering level. Thus, "one bad apple in a barrel" producing ethylene may trigger the ripening of the rest. Or the ethylene produced by the mold on an orange may

Abscisic Acid

Abscisic Acid and Stomatal Closure

The role of abscisic acid (ABA) (Fig. 20.6) in regulating abscission was outlined in Chapter 10. There are other roles played by ABA that make it a part of a rapidly responding system sensitive to water stress. When transpiration exceeds water absorption, the water content of leaves decreases. Turgor is lost and wilting ensues. It has been shown that in less than half an hour, under water stress, the leaf may begin to make and accumulate abscisic acid. Within a few hours, a manyfold increase in ABA concentration has been observed in a number of species. It has also been shown that ABA fed to leaves with open stomata can start stomatal closure in times as short as 5 minutes. Abscisic acid appears to act by interfering with the uptake or retention of potassium (or sodium) in guard cells. Since the high guard cell turgor required to maintain open stomata cannot be maintained without potassium ions, the stomata close. When water loss by transpiration is thus slowed, turgor in the leaf as a whole is regained, and the leaf recovers from wilting. The high level of ABA is metabolized away in a day or two and stomata begin to open normally.

Abscisic Acid as an Inhibitor

Another role for abscisic acid probably accounts for its frequent occurrence in high concentration in fleshy fruits. Abscisic acid is a potent inhibitor of seed germination, without being injurious to the seed. It probably functions in preventing germination in the fruit. It may carry over in the seed after release from the fruit and serve as part of the mechanism which prevents premature germination. It may be removed from seeds by the leaching action of rains.

Abscisic acid is a potent inhibitor of growth in many systems, but it differs from many other inhibitors in being nontoxic and that its inhibitory action is easily reversible with GA or auxin, depending upon the system.

The Use of Growth Regulators in the Control of Growth

We have discussed some specific characteristics of five types of plant hormones. This knowledge of hormone action has been very effectively applied to the practical control of plant growth. A very few of many possible examples in which growth-regulating chemicals are successfully used are: the selective killing of dandelions in a grass lawn; the inhibition of the growth of tree branches growing under city power lines; the thinning of flowers and fruit; the prevention of premature drop of fruit; the hastening of fruit ripening; the stimulation of growth of larger fruit. In addition to the examples we have discussed already, we will consider a few more examples in detail.

Control of Sprouting

We have seen how auxin can suppress the growth of lateral buds. This property may be used to advantage in controlling sprouting. Undesirable bud development may be delayed or inhibited by the application of certain auxins. For example, potato tubers exposed to the vapors of a synthetic growth regulator do not sprout in storage.

Rooting of Cuttings

Cuttings of certain species normally root slowly or produce a very small number of roots. It is now possible in some of these species to secure vigorous root production (Fig. 20.15) by treating the cuttings with various auxins, such as indoleacetic acid, indolebutyric acid, β-naphthoxyacetic acid, and numerous substituted phenoxy and benzoic acids. These growth regulators in proper concentration initiate cell divisions and stimulate the production of root primordia in the cuttings. The bases of the cuttings are immersed in the growth-regulator solution for a period of 12–24 hr, the strength of the solution ranging from 1 part of the substance in 5000–100,000 parts of water. The length of treatment and strength of solution that give optimum rooting vary with the species.

Root formation in cuttings of *Hibiscus* was found to depend upon a combination of indoleacetic acid and sugars and nitrogenous substances. The sugars and nitrogenous substances are contributed by the leaves, which explains the fact that rooting is very slight, even when cuttings are treated with a growth regulator, unless leaves are present.

Of significance is the fact that certain growth regulators affect the course of differentiation. For example, short segments of the tap root of dandelion normally develop buds from the upper cut surface, and callus and roots from the lower cut surface. If the auxin, indolebutyric acid, for example, is applied to these segments, roots differentiate from both cut surfaces; or, if the growth regulator in tissues is decreased by proper treatments, leaves can be caused to differentiate from both ends of the segments.

Figure 20.15 The promotion of root initiation by the synthetic auxin, indolebutyric acid, on American holly (*Ilex opaca*) cuttings. Row *A*, cuttings that stood in an aqueous solution of 0.01% indolebutyric acid for 17 hr before being placed in sandy rooting medium. Photographed after 21 days. Row *B*, untreated controls. Courtesy of U.S. Department of Agriculture.)

Control of Fruit Development

The development of fruits, particularly tomatoes, without pollination has been induced by application of certain growth regulators. A high percentage of seedless fruit results when the flowers are treated before pollination; but flowers treated after pollination set fruit, most of which have seed. Substances employed for this purpose include the auxins indoleacetic acid, β-naphthoxyacetic acid, and 4-chlorophenoxyacetic acid. They are readily available synthetic compounds.

Control of Abscission

Certain substances are effective in retarding the preharvest drop of fruit such as apples, pears, and citrus. After treatment with these chemicals, the fruits cling to the trees for a number of days longer than normal and may attain more satisfactory color and maturity. Growth regulators effective for this purpose are the auxins, naphthaleneacetic acid and 2,4-dichlorophenoxyacetic acid (2,4-D). The growth regulator apparently retards the processes that result in the formation of the abscission zone at the base of the petiole (pp. 181–182).

Growth Regulators as Herbicides

Certain synthetic growth regulators, although stimulative in extremely small quantities, seriously disturb physiological processes in plants when added in larger amounts. Some of these compounds have turned out to be very potent weed-killers. Although not all growth regulators can be used to kill weeds, the so-called "phenoxy" compounds are particularly effective. Because these compounds are selective, killing broad-leaved plants and leaving the grasses relatively unharmed, they are particularly useful for destroying many kinds of weeds growing in combination with various grasses or crops of grain. Thus, it is possible to spray a lawn with 2,4-D (2,4-dichlorophenoxyacetic acid) and, without injuring the grass, kill dandelions growing in it. The use of this and a related compound, 2,4,5-T (2,4,5-trichlorophenoxyacetic acid) in defoliation is being studied because of possible harmful side-effects to humans. It appears that it is an impurity in some samples of these chemicals, rather than the chemicals themselves, that is harmful to man.

The fact that growth regulators are translocated throughout the plant in the phloem makes these herbicides particularly effective in killing deep-rooted perennial plants. Studies of the factors affecting their movement in plants have been greatly facilitated through the use of radioactive isotopes. Fig. 20.16 shows the movement of amitrol (ATA) containing radioactive carbon through a plant.

There are numbers of mechanisms through which herbicides kill plants or parts of plants. A few mechanisms can be described in some detail, while others cannot. Effective doses of auxin-type herbicides, such as 2,4-D and 2,4,5-T, produce growth changes that indicate they are doing many of the same things natural auxin can do. This includes the stimulation of abnormally high rates of ethylene synthesis and/or stimulation of cell division in the phloem region with blockage of phloem transport.

Other herbicides may block specific metabolic reactions. The phenyl urea type of compounds reach a specific site in the electron transport chain, within chloroplasts, resulting in a blockage of the oxygen-evolving reactions of photosynthesis. Other nonselective herbicides, such as Paraquat or Diquat, are in themselves not very damaging, but when they get

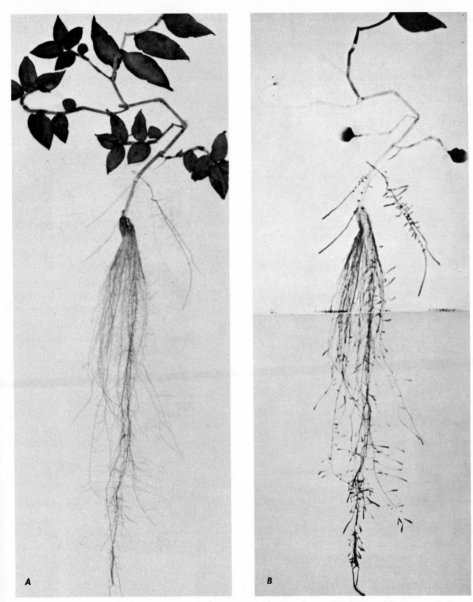

Figure 20.16   *A*, silhouette of the entire *Zebrina* plant; *B*, autoradiogram of the
Zebrina plant showing the path of movement of ATA (amitrol) containing
radioactive carbon. ATA was applied to the leaf at the upper right. (Courtesy of
A. S. Crafts.)

into chloroplasts they are reduced and the resulting free radicals do great damage to cell constituents. These compounds require that the plant be exposed to sunlight to exert their effect.

There are many bases of selectivity whereby one plant is killed and another survives application of herbicides. Again, some modes of resistance are well-understood while many more are not. The most important method of achieving selectivity is proba-bly by applying precisely controlled amounts of herbicide that have been found, by trial and error, to be effective in killing some, and not other, plants. The reasons it takes less of a compound to kill one plant than it does another may have to do with the amounts sticking on the surface of two plants, the amounts penetrating the cuticle and entering cells, the amount and pattern of translocation, the relative rates of breakdown or detoxification in the plants, and the abilities of different plants to recover after portions of the plant have been severely injured. Successful use of herbicides requires highly trained people, and it is most effective in crop situations, where one is concerned only with the enhanced survival of one species.

Vitamins and the Regulation of Growth

Extensive research in the field of animal nutrition has shown that animals, including man, may suffer from certain so-called vitamin deficiencies. The diet may contain water, mineral salts, carbohydrates, fats, and proteins in sufficient quantities to supply their energy requirements, but something may be lacking which is necessary for normal development and good health. If this something, which we call a **vitamin,** is added, even in the minutest quantities, the individual recovers from the deficiency symptoms.

Today, a number of different vitamins have been discovered, each with specific physiological roles. Not only do vitamins function in animal metabolism, but also, many of these same chemicals have been found to regulate physiological processes in plants as well. One of the chief difficulties in studying the functions of vitamins in plants is that, in contrast to animals that are unable to synthesize vitamins, green plants in general produce their own supply. Thus, it is difficult for a plant physiologist to produce a vitamin deficiency in a plant. If the plant manufactures the vitamin, the physiologists cannot easily deprive it of the vitamin.

Certain organs in plants, however, are unable to produce all the vitamins they need. The culture of these organs isolated from the plant has led to a better understanding of the role that vitamins have in plant growth. As an example, let us look at the influence of vitamins on root growth.

Being devoid of chlorophyll, roots are unable to synthesize sugar, and hence are dependent upon the green organs of the plant. We may ask: Do the leaves produce substances, other than sugar, essential for the normal growth of roots? If growing root tips are cut off and placed in a medium containing water, essential mineral salts, and sugar, the isolated roots will not continue to grow normally. They will grow for a prolonged period of time, however, if an extract of yeast cells is added to this medium. The growth substances in yeast extract are shown to be **thiamine, nicotinic acid,** and **pyridoxine,** all part of the vitamin B complex. The living cells of the roots of some species of plants cannot synthesize thiamine, nicotinic acid, or pyridoxine. These substances are manufactured in the leaves under the influence of light

and move from them to the roots. Roots of other kinds of plants can produce pyridoxine but not nicotinic acid or thiamine.

Thiamine is widely distributed in the plant kingdom. It has been found in many kinds of bacteria, in yeasts and other fungi, in both fresh-water and marine algae, and in mosses, liverworts, ferns, and the higher plants. Some bacteria, yeasts, and other fungi are able to synthesize thiamine, whereas others are not. Almost all the thiamine in a wheat grain is in the bran. It is now common practice to enrich white flour with thiamine.

Young plant tissues of higher plants, principally the leaves, are particularly active in synthesizing thiamine. From these tissues it moves to other parts of the plant. That thiamine is produced in leaves and translocated through the phloem to roots may be shown by removing a girdle of bark from the stem of a plant. It will be found that thiamine accumulates above the girdle. If the girdle is made on the petiole between the leaf and stem, thiamine accumulates in the leaf. As stated, thiamine is not produced in roots (at least in most plants that have been studied), but it is essential for root growth. Thus, thiamine has some of the characteristics of plant hormones in that it is produced in one part of a plant, and translocated to another part, where in small amounts it is very effective in influencing growth. However, it differs from the hormones discussed above in that thiamine, nicotinic acid, and pyridoxine all are cofactors (p. 62) of known enzyme reactions. Thus, thiamine may be classed as a root-growth hormone; it plays a role in the normal formation of roots in many seed plants.

Carefully controlled experiments lead us to believe that most ordinary plants growing with their roots in a soil containing the required mineral nutrients, and with their leaves in light, synthesize sufficient quantities of thiamine in the leaves to meet all requirements of the plant. Seeds usually are relatively rich in thiamine. Thus, the seedling probably has an adequate supply of the vitamin to support normal growth until such time as newly formed leaves are synthesizing the substance. Cuttings, however, may be deficient in thiamine. A few leaves on stem cuttings promote root growth, not only because they contribute carbohydrates to the roots, but also because they may contribute other growth-influencing factors.

Light and the Control of Development

It is not particularly surprising that green plants should have evolved mechanisms to ensure placement of their leaves to maximize light capture. What is remarkable, however, is the variety of developmental processes into which are woven light-mediated controls. These controls are sensitive to the presence or absence, direction, daily duration, and spectral composition of the light in the plants' environment. Any light-mediated event requires a pigment to absorb the light and to be altered by that light absorption long enough to react in some biochemical system in the cell. Only a very small fraction of the kinds of chemicals making up a living plant cell absorb visible light. These light-absorbing chemicals are called pigments. Photosynthetic pigments are present in high concentrations and contribute color to tissue, but growth-controlling pigments are low in concentration and do not impart a noticeable color to tissues. There are two well-characterized growth-controlling pigments, **phytochrome** and **protochlorophyll,** plus a third, a **phototropic pigment** that is not chemically identified, and possibly several other distinct but less well-known pigments. Most of the remaining pages in this chapter will be devoted to phytochrome. We will discuss protochlorophyll in relation to the greening process at the end of the chapter.

Phytochrome is a remarkable pigment that seems essentially universal in vascular plants. It is a protein that exists in two stable forms, active and inactive, and the two forms are different in color. The inactive form absorbs red light and is called P_{Red} or P_R, and the active form is called $P_{Far-red}$ or P_{FR}, because it absorbs far-red light just at the extreme of human visual sensitivity. When either form absorbs light, it is converted to the other form. Since P_R absorbs red light most efficiently, red light (660 nm wavelength) is most efficient in converting P_R to P_{FR}. Conversely, far-red (730 nm) light converts P_{FR} to P_R.

$$P_R \xrightleftharpoons[\text{far-red}]{\text{red}} P_{FR} \quad \text{(active)}$$

(dark reversion in some plants)

When chlorophyll absorbs light, the change induced in the molecule lasts about 1 billionth of a second (10^{-9} sec). When phytochrome absorbs light, the ultimate effect is long-lasting. P_R is stable indefinitely in the dark, and P_{FR} may exist and exert its effect for many hours, even in darkness, before being inactivated or spontaneously converted back to P_R. Thus, a little bit of red or far-red light can essentially switch phytochrome activity on or off. In contrast to chlorophylls and phytochrome, pigments of another class, the red- or blue-colored **anthocyanins** do not do anything with the light energy they absorb other than convert it to heat.

Phytochrome absorbs the various wavelengths of sunlight to different extents, the net result being that in sunlight a mixture of P_R and P_{FR} is reached. When exposed to artificial light sources, incandescent or fluorescent lamps, the relative amounts of P_R and P_{FR} are quite different from that achieved by sunlight. Growth of plants is somewhat different in artificial light unless a mixture of incandescent and fluorescent light that approximates sunlight is used to achieve a mixture of P_R and P_{FR} similar to that in sunlight.

As mentioned earlier, some seeds do not germinate in darkness. After seeds are soaked, a brief exposure to sunlight or, more particularly, red light will activate their germination. Red light acts by converting phytochrome to the active P_{FR} form. P_{FR} relieves some as yet unknown block in metabolism, and the seed can go on to germinate. If far-red light is given immediately following the red treatment, P_{FR} is converted back to P_R before P_{FR} has enough time to act. Now the seeds do not germinate. The far-red reverses the red treatment. If, after the red treatment, a delay is introduced prior to the far-red reversing treatment, the P_{FR} is given more time to act, and it may do enough in a short time to enable germination to proceed. In that case, the far-red reversal treatment is ineffective as far as the physiological response goes, although it is still effective in converting the pigment back to P_R.

Thus, phytochrome functions in many seeds as an indicator of the presence or absence of light. Some seeds that are buried are prevented from germination by the accumulation in them of volatile metabolites as they sit in the restricted air supply of the soil environment. Though initially they may not have had a light requirement for germination, they can acquire one under these conditions. That light requirement for germination is only satisfied when the soil is disturbed naturally or by cultivation.

The main signal to a seedling that it has emerged from the soil is the formation of P_{FR} by light. This P_{FR} causes changes in many growth processes in the stem and leaf. In dicotyledonous plants, ethylene synthesis is shut down to a low level in the apical meristem or terminal bud and leaf enlargement starts and the hook opens (Fig. 20.3). Stem elongation is generally inhibited by P_{FR}, and changes in the differentiation of xylem tissue lead to added strength necessary to raise and support leaves above ground. In grasses, light promotes unrolling of the first leaf as it breaks out of the ensheathing coleoptile (Figs. 16.33, 20.4, 20.5). But the rolled-up leaf can attain considerable size in grasses in the dark, so that leaf expansion is not an important initial response in a typical grass. It appears that formation of P_{FR} rapidly leads to a release of active gibberellin from an inactive form and that increased GA is the more immediate cause of leaf unrolling. Applied GA can substitute for light in this particular response.

In some tissues, phytochrome-controlled differentiation can be followed at the biochemical level. For instance, irradiation of etiolated seedlings may induce the synthesis of the enzyme phenylalanine ammonia lyase that catalyzes an important step in the biosynthesis of lignin deposited in secondary cell walls. P_{FR} may also stimulate the synthesis of anthocyanin pigments that give apples their red color.

When a plant finds itself almost continuously in the shade of other plants, the quality of light falling on the plant is changed in a way that can markedly affect the phytochrome status of the tissue. As light passes through successive layers of leaves, the red wavelengths are filtered out by chlorophyll, while the far-red wavelengths pass through. In the shade created by other plants, light rich in far-red tends to markedly lower the P_{FR} concentration. The result is more rapid stem elongation and an increased probability of growing out from under the shade of other plants. Even in light-grown plants, phytochrome is present, and experiments can be devised to show how it exerts subtle controls on stem elongation. In this example, a change in light quality can trigger a response that has obvious survival advantages.

While the mechanism of action of phytochrome has not yet been established, some work suggests that it may be attached to membranes and exert some control over membrane function.

Photoperiodism

It was mentioned earlier (p. 374) that photoperiodism frequently provides the timing signal for flowering and other activities that prepare a plant to survive a season unfavorable for growth. The formation of dormant buds, resistance to freezing temperatures, formation of tubers and bulbs may be responsive to photoperiodic stimuli.

Photoperiodic control of flowering involves the perception of day length in the leaves. Under the right conditions, a chemical stimulus is made in the leaves, and moved out through the phloem to terminal or axillary buds, where it induces a change in the kind of organs being initiated, a change from initiation of leaf primordia to the initiation of flower primordia. In the case of photoperiodic control of bud dormancy, the leaves produce something that moves to the shoot apex, causing it to stop elongation and to make bud scales and the beginnings of next season's shoot. Similarly, a factor from the leaves may initiate tuber formation.

Photoperiod and Flowering

Plants that generally are induced to flower during the lengthening days of spring respond to day lengths longer than a certain minimum or critical photoperiod. These are called **long-day plants** (Figs. 20.17, 20.18). A long-day plant might flower after having been exposed to day lengths longer than 11, 12, or 13 hr, depending on its critical photoperiod. Examination of Fig. 20.17 will show that to initiate flowering around May 1 requires progressively longer critical day lengths at higher latitudes. Conversely, if a plant is stimulated to flower by 13 hr days in its native habitat, that critical day length will come earlier in the year the farther away from the equator the plant is moved. Such a plant might be poorly adapted to the season at more northerly (or southerly) latitudes because of flowering prematurely.

Short-day plants are those whose flowering is initiated by days shorter than a critical length. They are generally induced by the shortening days of late summer or early fall. Short-day plants may be thought of as long-night plants, because it is the length of an uninterrupted dark period that is critical to timing (Fig. 20.19). Thus, a plant requiring days of less than 13 hr in reality is responding to nights of more than 11 hr.

The photoperiod may be either an absolute requirement or a partial requirement, causing flowering at an earlier time or with greater intensity. Many other plants indifferent to photoperiod are called **day-neutral plants.** Their flowering may be controlled either by other environmental factors such as temperature and water stress, or by a genetically determined pattern of development.

Phytochrome and the "Biological Clock"

By what mechanism does a plant measure time? First of all, the time-measuring system can be very accurate and quite independent of temperature. Some plants commonly respond quite differently to day (or night) lengths differing by half an hour and even to differences as small as 10 minutes. This precision enables response to the progression of the seasons occurring in 1 week. Another characteristic of the timing mechanism is that it involves phytochrome as the photoreceptor that indicates whether it is light (day) or dark (night). A long dark period that would otherwise be inductive to short-day plants can be rendered ineffective by forming phytochrome P_{FR} with a short light treatment (a night interruption) in the middle of the dark period (Fig. 20.19). Evidently, phytochrome P_{FR} present at the end of the day disappears in the early hours of dark-

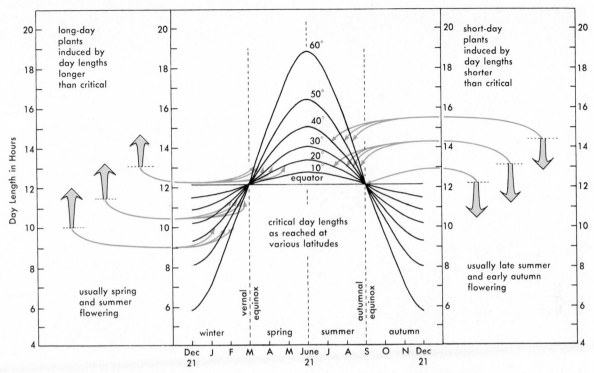

Figure 20.17 Diagram illustrating the response of photoperiodically sensitive plants to the seasonal progression of daily light and dark periods and how the day lengths vary with latitude. Approximate latitudes: Mexico City, 19°; New Orleans, 30°; San Francisco, 38°; New York, 41°; Chicago, 42°; Montreal, 45°; Paris, 49°; London, 51°; Oslo, 60°; Leningrad, 60°.

Figure 20.18 Effect of day length on behavior of the long-day plant henbane (*Hyoscyamus*). Plants were grown with 8 hr photoperiod until they were about a month old and were then subjected to 24 photoperiods of 10, 11, 12, 13, 14, or 16 hr. Initiation of flowers and accompanying stem elongation (bolting) occurs on days longer than 12 hr. (Courtesy of Plant Industry Station, Crops Research Division, Agricultural Research Service, U.S. Department of Agriculture.)

Figure 20.19   Effect of day length on behavior of *Chrysanthemum*, a short-day plant. *A*, plant that received light of natural short days of autumn and blossomed at the usual time. *B*, plant that received an hour of light near the middle of each night for several weeks beginning just before flower buds would normally have been initiated; thus, each long dark period was divided into two short ones. This interruption was sufficient to delay flowering. (Courtesy of Plant Industry Station, Crops Research Division, Agricultural Research Service, U.S. Department of Agriculture.)

a cycle once approximately every 24 hr. It serves to synchronize physiological, biochemical, and activity functions on a daily cycle. Biological clocks must be present in organisms ranging from single-celled algae to flowering plants and in animals including humans. It is clear that phytochrome and the "clock" interact to measure time, and under the proper circumstances the block to making the flowering stimulus is removed. The functioning of the biological clock can be observed independently of the flowering process in plants that link another process to the clock. Even under continuous light or continuous darkness, bean plants can maintain a daily cycle of "sleep" movements of their leaves. The leaves hang down during the "night" part of the cycle and are extended more horizontally during the "day" part of the cycle as determined by their internal clock.

Thus, many plants are able to time various activities on the annual cycle by virtue of accurately measuring the length of the day or night. Flowers are initiated in time to mature seeds prior to frost or drought. During favorable seasons, dormant buds are made capable of withstanding unfavorable cold weather that probably lies ahead.

Supplemental Factors Controlling Flowering

One might inquire why a short-day plant does not respond to the short days of late winter or early spring prior to the time the days get too long to be inductive. There are generally supplemental factors controlling the development of plants in their natural habitat that keep a short-day plant from blooming in the early spring when the days might otherwise be short enough. For instance, many annual plants are not sensitive to the photoperiod until they have reached a certain size. The normal season for seed germination may preclude the plants attaining the necessary maturity prior to the arrival of the long days of spring. Thus, the only season in which it would be capable of responding under normal conditions would be late summer or early fall.

The Nature of the Flowering Stimulus

We have spoken of the flowering stimulus without indicating more of its nature other than it is something that is made in the leaves and moves through the phloem to the buds. Some experiments strongly suggest that the same flowering stimulus serves long-day, short-day, and day-neutral plants. They apparently differ only in the conditions required to bring about its synthesis. Much work has gone into attempts to extract and identify such a univer-

ness as it spontaneously reverts to P_R. A short-day plant requires more than a minimum dark period without the active phytochrome P_{FR}. Conversely, a long-day plant is prevented from flowering if it has less than a minimum daily period with P_{FR} (light). So phytochrome P_{FR} is inhibitory to flowering during the night in short-day plants and is promotory to flowering in long-day plants.

The least well-known component of the time-measuring system is the "biological clock." The biological clock is a metabolic system that goes through

sal substance without substantial success. Empirical testing has, however, uncovered procedures whereby known plant hormones and metabolites can, in some plants, be used to cause flowering under noninductive photoperiods or to prevent flowering under otherwise inductive conditions.

Gibberellins applied to many long-day rosette plants will produce flowering under short days and would bring about a response similar to long days in the plants shown in Fig. 20.16. Gibberellins are, on the other hand, inhibitory to flower initiation in many other plants, especially woody plants. In woody plants, it is often possible to stimulate flowering by treating with growth retardants that inhibit gibberellin synthesis. Recall also that auxin or ethylene are effective in initiating flowering in pineapple. In some plants, abscisic acid application has induced flowering. The potential practical importance of being able to control flowering time in more crops cannot be underestimated.

Vernalization and Flowering

Another mechanism for timing flowering makes the passage of winter a prerequisite to flowering. Plants having this type of control include both winter annuals and biennials. Winter annuals typically germinate just prior to or during the winter so that the young plant passes through a period of weeks of exposure to low temperatures prior to the warmer weather of spring and summer when they flower.

Biennials generally have a full season of vegetative growth the first year, exposure to cold weather in a winter and then flower the second spring and summer. They need to achieve a larger size before the cold treatment is effective. Without exposure to the low temperatures, neither group will flower, or flowering will be delayed (Fig. 20.20). Gibberellins can sometimes substitute for the cold requirement. This promotion of flowering in response to cold treatment is called **vernalization.** The cold treatment generally requires temperatures of 10°C or less for a period of weeks. Frequently, the cold requirement is followed by a long-day photoperiod requirement that prevents the flowering from occurring too early in the spring. In vernalization, the growing tip may be the site of perception of the cold stimulus, and no transportable factor·is involved.

"Winter" cereals, wheat and rye, are normally planted in the fall so that they germinate prior to winter and go on to flower promptly the next spring and summer after making a minimum number of leaves. If they are planted in the spring and are not vernalized, flowering is delayed until many more leaves are formed. "Spring" cereal strains are nor-

Figure 20.20 Substitution of gibberellic acid for the cold requirement (winter) in the flowering of the biennial carrot (*Dacus carota*). *A*, control plant under long days only; *B*, long days plus gibberellic acid, no cold treatment; *C*, long days plus cold treatment, no gibberellic acid. (Courtesy of Lang.)

mally planted in the spring and flower promptly in response to long days. Winter cereal strains can be artificially vernalized prior to a spring planting date by moistening them and holding them at about 1°C for a period of weeks.

Dormancy, Photoperiodism, and Temperature

In the discussion of photoperiodism, we indicated that many woody perennial plants use the photoperiod to induce a dormant condition in the bud prior to the onset of unfavorable weather. The bud typically goes through a period of increasing dormancy until it can no longer be rapidly reactivated by favorable environmental conditions or a shock treatment such as defoliation. By fall it may be in a state of rest wherein no treatment will activate it other than the passage of time at low temperature, a process called **chilling.** The satisfaction of a chilling requirement requires temperatures a little above freezing but generally below 10°C. Normally, the requirement is met by midwinter and at that time a branch brought into the greenhouse will have buds bursting in a few weeks. The buds are no longer in a condition of rest. Their chilling requirement has been satisfied. They are able to respond to the warming temperatures of spring with renewed growth. The photoperiod induced the dormant condition and the chilling process signaled that it was safe to respond to spring-like weather.

The chilling requirement for many deciduous fruit trees has been extensively studied. Typical varieties may require 250–1000 hr of exposure to chilling temperatures to produce a rapid uniform burst of flower and leaf buds in the spring. A variety planted in an area of insufficient chilling will have limited and sporadic bud opening and will set a poor crop of fruit. The lack of sufficient cold weather sets the southern limit for crops such as peaches, apricots, cherries, plums, and apples, and for effective flowering of ornamentals such as lilac (*Syringa vulgaris*).

Another temperature-related phenomenon is the development of **frost hardiness.** In the summertime, a few degrees below freezing will kill leaves and buds. During the fall, there can be (if the genotype provides the capacity to so react) development of resistance to temperatures below freezing. Short photoperiods and exposure to temperatures just above freezing lower the temperature at which buds and even leaves are irreversibly damaged. This cold resistance may progressively increase from −5°C early in the fall to −30° or even −50°C in the buds and twigs by late fall. This protects the buds and stems from the extreme cold and drying effects of winter weather.

Olive trees are evergreen and live in Mediterranean climates with mild winters. They make use of a chilling requirement for a different purpose, the initiation of flowers, rather than for the termination of rest. Thus, one environmental signal may be linked to the control of a variety of aspects of development. The pattern of linkage of signals and responses is one of the characteristics of an ecotype (Chap. 18) or population.

The Greening Process

The process of greening, the formation of chlorophyll in leaves and young stems, etc., is precisely controlled to prevent the use of food reserves and energy in constructing the photosynthetic apparatus unless light is available for photosynthesis. Angio-spermous plants do not make chlorophyll in the dark. Each molecule of chlorophyll is made by a reaction requiring the absorption of light by protochlorophyll hooked up to a special catalytic protein, the photochlorophyllide holochrome. In darkness, a small amount of this photoreceptor system is maintained in plastids with a minimum of structural and enzymatic differentiation. Without light there is very little development of this etioplast (Fig. 4.10B). When light is absorbed by the protochlorophyllide holochrome, chlorophyll is formed and some new protochlorophyll is made to replace the protochlorophyll that was used up. In a period of 36 or 48 hr an etiolated leaf can be converted to a fully green leaf (Fig. 20.3, 20.4). New membranes are formed in the plastid and the enzymes involved in the electron transport and carbon cycles (Chaps. 14, 15) accumulate. Phytochrome frequently participates in this light-activated greening process, too. In a few hours after the start of illumination, photosynthesis can be observed.

Summary

1. Plants are able to respond to a wide variety of environmental situations because they have systems for monitoring important environmental factors and they have programmed patterns of response to these stimuli.

2. The influence of one part of a plant on the growth and development of another part is important in keeping the plant growing as a coordinated unit. The correlation of activities in one part with another may be through food, mineral nutrient or water supplies, or by the production and distribution of specific hormones.

3. Plant hormones are natural compounds, usually transported from a site of synthesis to a site of action, which act in small quantities to regulate growth and development in many ways.

4. Auxin, gibberellins, cytokinins, abscisic acid, and ethylene are the important plant growth hormones.

5. Auxins promote cell enlargement by altering cell walls, making them more extensible. Growth is controlled by controlling the synthesis, transport, and inactivation of auxin. Phototropic and geotropic curvatures in the shoot are regulated by changing the transport of auxin to the growing cells.

6. Auxins also help control other developmental processes such as apical dominance, cell division, and differentiation of specific cell types in the xylem.

7. Gibberellins also control stem elongation through effects on both cell elongation and on division in the subapical region. Other developmental effects of gibberellins include promotion of the synthesis of specific enzymes.

8. Cytokinins are active in control of cell division in specific tissues, in maintenance of active nucleic acid and protein metabolism in leaves, and in correlating root and shoot growth.

9. Ethylene is important in controlling stem elongation along with auxin, in maintaining the etiolated growth habit in darkness, and in regulating the ripening process in many fruits.

10. Abscisic acid is important in regulating the leaf's response to water stress through stomatal closure, in control of seed germination, and in the development of the abscission layer.

11. Light is absorbed by specific growth-active photoreceptors in addition to the photosynthetic pigments. These receptor systems include phytochrome, the phototropic pigment, protochlorophyll, and others.

12. Phytochrome exists in active and inactive forms that are interconvertible by red and far-red light. It is used to sense whether or not light is present, and when activated by light it is involved in starting the development of the above-ground mode of growth. It may inhibit stem elongation, promote leaf expansion, promote the germination of light-sensitive seeds, and many other developmental events.

13. Flowering is frequently precisely timed with the seasons through photoperiodism. A system in the leaf can measure the length of the day or night and regulate the synthesis of a flowering stimulus. The flowering stimulus moves from the leaf to receptive lateral or terminal buds, where it causes the meristem to make flowers rather than leaves.

14. There are two main types of photoperiodically controlled plants, long-day and short-day plants, responding respectively to day lengths longer than critical or to night lengths longer than critical. Phytochrome is the photoreceptor involved in sensing day or night. Many plants are photoperiodically unresponsive, or day-neutral.

15. Buds may be induced to go dormant by a photoperiodic system and only activated again by favorable conditions after passing through the winter season, a prolonged period of low temperatures, the chilling period.

16. A prolonged period of low temperatures may vernalize certain plants, that is, remove a metabolic block to the formation of flowers by the apex. This serves as a signal denoting the passage of winter.

17. Light is necessary in flowering plants for the conversion of protochlorophyll to chlorophyll. This requirement for light prevents the development of the photosynthetic apparatus in plant parts not exposed to light.

21

Bacteria

The bacteria are organisms which obtain energy and nitrogen by a variety of metabolic pathways. All species of the division to which the bacteria belong are prokaryotes. This division is known as the Schizomycophyta. The great majority of bacteria are heterotrophic forms obtaining energy and nitrogen from organic matter. Those utilizing dead organic matter are **saprophytes,** while others are **parasites** obtaining food from living organisms. Some may obtain food from both sources. Many living naturally as parasites may be cultivated on a synthetic medium. As saprophytes, they are responsible for releasing to the atmosphere an estimated 3.5 billion tons of carbon monoxide each year.

Some are aerobic; others live only in the absence of oxygen and are strictly anaerobic. A number of species may live in either environment. For example, *Clostridium tetani* is a parasite in the absence of oxygen and a saprophyte in its presence. The autotrophic species are largely anaerobic forms obtaining energy in a variety of ways. Some oxidize NO_2^- to NO_3^-, gaining energy from the process; others oxidize sulfur or iron compounds. Three species of bacteria are photosynthetic and anaerobic. These possess a pigment similar to chlorophyll *a* and utilize the energy of sunlight. In fact, in past ages, these photosynthetic forms, as suppliers of oxygen, may have played a highly significant role in the evolution of our present atmosphere.

See Table 21.1 for a tabulation of the various ways in which bacteria obtain energy and nitrogen.

General Characteristics and Distribution

The Schizomycophyta are of world-wide distribution. They may conveniently be divided into two classes, the Schizomycetes and the Microtatobiotes. There are several orders in the Schizomycetes, the largest and most important is the Eubacteriales, or the true bacteria. We shall devote most of our attention to this order. The class Microtatobiotes, although comprising organisms known for many years, has been but recently established as a formally recognized class, and the name may be changed. We shall consider briefly only two genera in two different groups of the Microtatobiotes: *Myxoplasma* and *Rickettsia*. They are of great interest because some of them have only recently been recognized as infectious agents of plants and their recognition as such has explained numerous puzzling diseases and suggests more workable methods for controlling these diseases.

Table 21.1 Nutritional and Energy Requirements of the Schizomycophyta

Bacteria	Source of nitrogen	Source of energy	Photosynthetic pigments present	Heterotrophic or autotrophic
1. Many bacteria	Living organisms (parasitic)	Organic compounds	None	Heterotrophic
2. Many bacteria	Nonliving organic matter (sapro-phytic)	Organic compounds	None	Heterotrophic
3. Some bacteria	Free nitrogen of atmosphere	Organic compounds	None	Heterotrophic
4. Soil bacteria	Ammonia and nitrates	Inorganic salts	None	Autotrophic
5. Soil bacteria	Ammonia and nitrates	Organic compounds	None	Heterotrophic
6. Purple bacteria	Organic or in-organic nitrogen	Radiant energy	A purple pigment and chlorophyll	Autotrophic

Rickettsia

Rickettsias have been known for some time and have long been feared as the infectious agents of rocky mountain spotted fever (*Rickettsia rickettssi*) and scrub typhus (*Rickettsia tsutsugmushi*). They are of wide occurrence, and their presence in insects which suck plant juices, e.g., aphids, suggested that they might be found in plants. A rickettsia-like organism was isolated in 1970 from dodder, an angiospermous parasite, and from the myxomycete *Didymium*. Pierce's disease has long been a very serious and mysterious disease of grapes. Pierce first described it in 1880 and actually noted bacteria in relation to diseased tissue. He was unable to cultivate bacteria from his diseased grapes, and the causative organism has eluded many investigators for close to 100 years. In 1972, Coheen, working at the University of California in Davis, discovered with the electron microscope the presence of rickettsia-like organisms in grape vines having the symptoms of Pierce's disease (Figs. 21.1, 21.2A).

Rickettsia cells are coccoid, rod-shaped, or pleomorphic bacteria about 2.0 by 0.08 μ in size. They are bounded by a cytoplasmic membrane and a well-developed cell wall which is frequently rippled. Internally, ribosomes and areas with DNA strands have been identified. So far it has not been possible to culture them in artificial media. They will grow within cells and intracellularly in tissue cultures.

Myxoplasma

Myxoplasmas are highly variable, plastic prokaryotes bounded by a unit plasma membrane and lacking a cell wall. They contain DNA, ribosomal RNA, and transfer RNA. They are spherical bodies ranging from 50 to 1000 nm in diameter (Fig. 21.3). They multiply by budding or fission and may sometimes produce short filaments. They are disseminated by leaf hoppers or by grafting. Disease symptoms range from yellowing, growth abnormalities like witches broom, and phloem necrosis (Fig. 21.2B; see Color Plate 11). Although bovine pleuropneumonia, a disease now known to be caused by a myxoplasma, was described in 1898, *Myxomplasma*, as a genus of bacteria, was not formally recognized by bacteriologists until 1966.

Myxoplasmas apparently have a rather broad distribution; they have been isolated from sewage and decaying matter as well as from dogs, rats, mice, and other animals. They will grow on an enriched artificial medium containing a beef heart infusion with serum or other similar additions.

Virus yellows in plants were long described as virus diseases of plants. Recently, over 30 of these virus yellows diseases have been shown to be caused by a myxoplasma. Other long-baffling plant diseases once thought to be due to a virus infection but now known to be caused by myxoplasmas are pear decline (Fig. 21.2D; see Color Plate 11), citrus stubborn

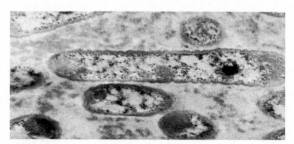

Figure 21.1 Rickettsia-like organism from grape vines infected with Pierce's disease, × 20,000. (Courtesy of G. Nyland.)

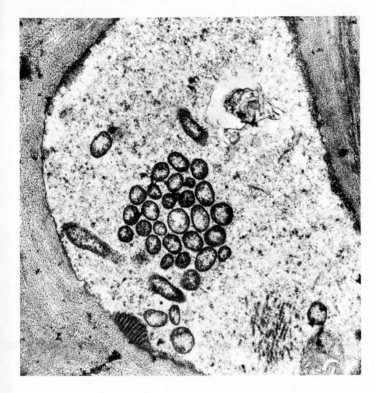

Figure 21.3 Myxoplasma organisms in a sieve tube of a peach tree showing symptoms of peach-x disease, ×10,000. (Courtesy of G. Nyland.)

disease (Fig. 21.2C), and Western X disease of peaches and cherries. Correct knowledge of the true causative organisms is a great step to devising effective means of controlling these baffling plant diseases.

Schizomycetes

Bacteria are of worldwide distribution, comprising a large number of species, mostly single-celled or simple filamentous forms between 0.5 and 5 μ in diameter and up to 15–20 μ long. Multiplication results from the simple division of single cells, a process known as **fission.** Many members of the Schizomycophyta form spores, but the spores are more a means of carrying the species over periods unfavorable to growth than a means of multiplication. While the bacteria are not to be classified as fungi, certain types of bacteria have characteristics that resemble those of fungi. Some bacteria resemble slime molds, while others are filamentous.

Bacteria play an important role in the decomposition of organic matter, are indispensable in maintaining soil fertility, affect the quality of the water and milk we drink and of the food we eat, and cause diseases of animals and plants. We are likely to gain the impression that all bacteria are harmful, because we associate them with such dread human diseases as typhoid fever, tuberculosis, and many other types of infections. But many kinds of bacteria are beneficial, in fact, indispensable in the life of the world.

Discovery of Bacteria

Our knowledge of the shapes and morphology of bacteria depends largely on the type of microscopes with which they are studied. Antony van Leeuwenhoek, of the Netherlands, first saw them in the summer of 1676. He was curious as to what made pepper hot, so he set some peppercorns aside in a cup of water and several days later examined this water under a simple single-lens microscope at a magnification of about 400 times. To his amazement he saw many actively moving, very small organisms which he called **animalcules.** He did not prepare any drawings of these small organisms until several years later, and these drawings, when finally prepared, showed organisms taken from his own mouth. These drawings, which are shown in Fig. 21.4, clearly indicate that he was describing bacteria for the first time.

While van Leeuwenhoek suggested that these small organisms might be related to fermentation, decay, and disease, his suggestion was not taken seriously. Their relation to these processes was finally discovered quite accidentally. Pasteur, a young French chemist, was retained in 1860 by the French government to improve the quality of French beers. A chemist was considered the proper person for the

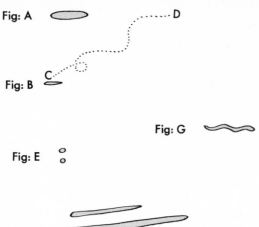

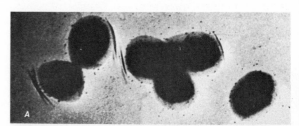

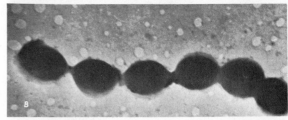

Figure 21.4 A copy of van Leeuwenhoek's drawings of the bacteria he obtained from his mouth. He described them thus: "I then most always saw, with great wonder, that in said matter there were many very little animalcules very prettily a-moving." (From *Arcana Natura Detecta Delphis Batavorum*, 1695.)

task because it was then thought that fermentation was a purely chemical process. Pasteur, however, firmly established the part played by living organisms in fermentation. He definitely showed for the first time that yeasts were necessary for the production of wine from grapes. Pasteurization, the flash heating of liquids to kill undesirable organisms, was first used to preserve French wines destined for export. During the 1890s the French Academy of Sciences, of which Pasteur was a member, was hotly debating whether or not life could arise spontaneously. The outcome of this debate was the definitive demonstration by Pasteur that microorganisms always arise from preexisting organisms. Living organisms, whether bacteria or insect larvae, never arose *de novo* from decaying matter. A long series of experiments in Pasteur's laboratory succeeded in establishing the very definite relationship between bacteria (and yeasts) and the processes of fermentation, decay, and disease. Burrill, in 1878, first recognized a bacterium as a plant pathogen.

Forms of Bacteria

Bacteria are of three general shapes: **spherical, rod-shaped,** and **spiral.** Spherical bacteria are called **cocci** (**coccus,** singular) (Figs. 21.5*A*, 21.5*B*, 21.5*C*), rod-shaped ones are called **bacilli** (**bacillus,** singular) (Fig.

21.5*D*), and spiral forms are known as **spirilla** (**spirillum,** singular). There is some intergrading between these forms. It is sometimes difficult, for instance, to distinguish between a very short rod and a coccus that has elongated in preparation for cell division. Furthermore, shape depends to a certain extent upon the age of the bacteria and upon the environment. Bacterial plant pathogens, except for the species of *Streptomyces,* are nonspore-forming rods.

Bacteria of the coccus form, in chains resembling strings of pearls, are called **streptococci** (Fig. 21.5*B*). Several species of the genus *Streptococcus* are very important. Some of the diseases they cause are scar-

Figure 21.5 First electron micrographs of bacteria. *A, Pneumococcus; B, Streptococcus; C, Staphylococcus aureus; D, Bacillus anthrax.* All ×10,000. (*A,* courtesy of Charles Pfizer and Company; *B, C, D,* courtesy of J. Hillier.)

let fever, erysipelas, mastitis of cows, and sinus infections. *Streptococcus lactis* is an agent in souring milk. Spherical bacteria may sometimes remain associated in irregular flattened masses; these forms are frequently referred to as **staphylococci** (Fig. 21.5C). One species of *Staphylococcus* is the usual causal agent of boils and abscesses. Some species of spherical bacteria remain associated in cubical packets of 8, 64, or even more cells; these species belong to the genus *Sarcina*.

Size of Bacteria

Bacteria are among the smallest living organisms. Some of them are close to the limit of visibility with the most powerful light microscopes. Several million bacteria may be present in a cubic centimeter of soil, or milk, and not be crowded in the least bit. If milk containing 10 million bacteria in each cubic centimeter is examined under a lens magnifying 440

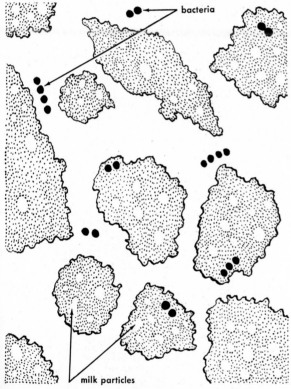

Figure 21.6 Spatial relationship of bacteria and milk particles in a cubic centimeter of milk containing approximately 10 million bacteria. (Redrawn from F. Lohnis and E. B. Fred, *Textbook of Agricultural Bacteriology*. Copyright 1923 by McGraw-Hill Book Company, New York. Used with permission of McGraw-Hill Book Company.)

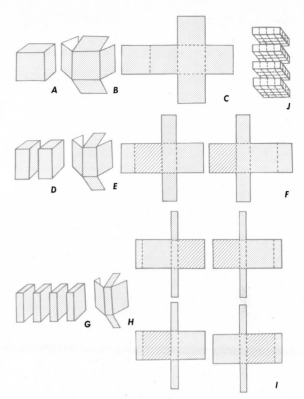

Figure 21.7 Diagram showing relationship between volume and surface.

times, the relationship between bacteria and milk particles would approximate that shown in Fig. 21.6.

Most bacteria average from 1 to $2\,\mu$ in diameter. If the cube *A* in Fig. 21.7 is 10 mm along one edge, there are 100 sq mm on each face and 1000 cubic mm in the cube. Ten thousand cocci, each $1\,\mu$ in diameter, could be placed along one edge of this cube. There would be room for 100 million of them on one face; 1000 billion would fit nicely into the cube. Thus, 1 million bacteria per cubic centimeter of soil, milk, or blood is really a small number.

The very smallness of bacteria is a factor of major importance in their life. All the nourishment, all gases and inorganic salts that a bacterium requires must be taken in through its cell wall. All its waste products, all toxins or poisons, certain enzymes and other special substances that are associated with bacterial activity must pass outward through the surface of the bacterial cell. Therefore, the more surface a bacterium possesses in relation to its volume, the more readily will these substances diffuse.

Distribution of Bacteria

These minute organisms are ever-present about us. They occur in the air, water, and soil. They exist on the surfaces of all animal and plant bodies and on the surfaces of almost everything we touch. Live bacteria are present in milk and in all foods that have not been sterilized. Even foods that have undergone the usual process of sterilization will sometimes contain viable or live spores of bacteria. In 1939, viable spores of certain heat-resistant bacteria were found in canned roasted veal that had traveled with the explorer Parry to the Arctic Circle in 1824. Bacteria occur naturally in the digestive tracts of animals.

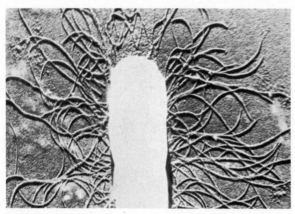

Figure 21.8 *Caryophanon latum* showing many flagella, ×5000. (Courtesy of C. F. Robinow.)

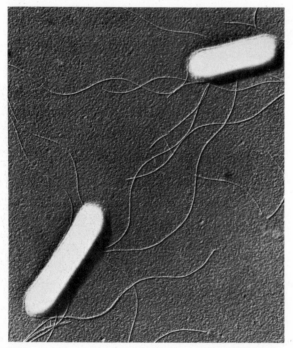

Figure 21.9 Diphtheroid bacteria with associated flagella, ×8000. (Courtesy of R. C. Williams and M. P. Starr.)

Nitrogen-fixing bacteria are closely associated with the living cells in the roots of legumes.

After a rain or snow, the dust is washed from the air and carried to the ground, and then the air may be practically free of bacteria. Generally, the number of bacteria in the air decreases with altitude. The air at altitudes at which commercial airliners fly is bacteria-free.

Water from deep, cold wells or springs is usually devoid of bacteria. Whereas the surface few inches of most soils teem with bacteria, the number decreases as the depth increases.

Although bacteria are almost universally distributed, it is possible, by taking great precautions, to keep rooms relatively free of these organisms. Such precautions are observed in hospital operating rooms and in certain types of biological and bacteriological laboratories. Surfaces are washed in disinfecting fluid (0.133% alkyl dimethylbenzl ammonium sodium hypochlorite or Chlorox), sterilized gowns are worn, and only instruments that have been placed in boiling water are used. Even the air may be purified by steam or ultraviolet radiation. This condition of freedom from bacteria is known as **asepsis. Sterilization** is the destruction of all living forms, including bacteria, that may be within or upon a particular object.

Movement

Many bacteria move. Some do so by simply flexing their bodies, others rotate and have a corkscrew type of motion. Motility may be obtained by a flexing of the protoplast or by the action of two sorts of exterior motor organelles. The bacterial surface may be provided with many short **cilia** (*Caryophanon latum*, Fig. 21.8) or with one, two, or several long whip-like filaments or **flagella** (Fig. 21.9 and *Escherichia coli*). However, neither of the bacterial motor organelles, cilia or flagella, have an ultrastructure similar to that of the flagella of eukaryotic organisms. In the latter case, the flagella are provided with a central complex of 9 + 2 strands (Fig. 22.25), while only a single central filament, about 12 nm in diameter, is present in the bacterial cilia or flagella (Fig. 21.10). Ciliated bacteria may travel at the rate of some 50 μ per second, whereas those lacking motor organelles move at a much slower rate, about 5 μ per second. In proportion to its body size, a speed of 50 μ per second is equal to a speed of 90 mph by a 6 ft long antelope.

Nutrition and Energy

Bacteria, like all other living organisms, need food for growth and multiplication. Food is any organic

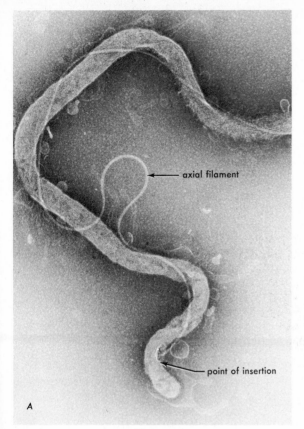
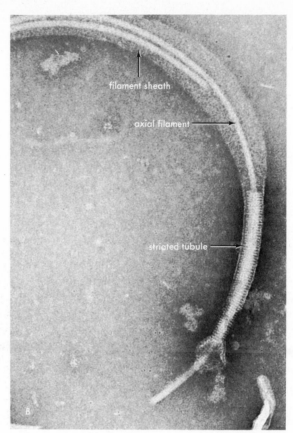

Figure 21.10 Axial filaments of *Spirochaeta stenostrepta*. *A*, the axial filament is attached to an end of the spirochete at an indentation of the protoplasmic layer, ×43,000. *B*, segments of the axial filament are enclosed in a cross-striated tubule which in turn is surrounded by an outer filament sheath, ×120,000. (Courtesy of S. C. Holt and E. Canale-Parola, *J. Bacteriol.* **96**, 822, 1968.)

substance that supplies energy and basic organic materials for use in metabolism. Most bacteria utilize nonliving organic matter as a source of food. In other words, most bacteria are saprophytes. They may use relatively simple organic compounds, such as alcohol, sugars, and fats, or very complex organic compounds, such as proteins. For example, the bacteria that form acetic acid utilize ethyl alcohol as a source of food, and they secure from its oxidation energy for their own purposes. These bacteria are common in cider, beer, and wine. The reaction may be represented as follows:

alcohol + oxygen → acetic acid + water + energy

Many bacteria depend on sugars as a source of food. The lactic acid bacteria use simple sugars in milk, and the souring of milk is caused by these bacteria. Many species of bacteria rely on proteins as foods, and in the process of decomposition proteins are broken down to simple products, many of

which have offensive odors. Ammonia is one of the simplest products formed from proteins by bacteria.

In addition to the saprophytic bacteria, there are those that obtain their food from other living organisms. Such bacteria are parasites. Among them are the disease-causing bacteria of plants and animals. Sometimes the distinction between parasite and saprophyte is not clear-cut. For example, certain types of bacteria thrive as saprophytes in the soil. When they become embedded in wounded flesh, however, they may become parasites. Such are the bacteria causing tetanus and gaseous gangrene in man, and blackleg in cattle, and bacteria causing soft rot of fruit and vegetables.

Some bacteria are able to obtain energy from simple inorganic salts, such as sodium nitrate, ammonia, and hydrogen sulfide. They bring about the oxidation of these simple substances and utilize the energy so released to synthesize organic materials from which enzymes, protoplasm, cell walls, and other cellular materials are formed. This synthesis is carried out in a manner comparable to that by which green plants manufacture foods.

We may classify bacteria on the basis of their method of obtaining energy, as follows: (a) Most bacteria secure energy from the oxidation or breakdown of either nonliving or living organic matter. (b) A few obtain energy from the oxidation of inorganic materials and use this energy to synthesize their own foods. (c) One group of bacteria synthesizes bacteriochlorophyll, and thus can utilize the energy of sunlight for the manufacture of food.

Structure of Bacteria

The Light Microscope

Most bacteria are single-celled organisms with a very primitive or simple organization. With the light microscope, one sees little internal structure other than a central region which responds to nuclear stains and is known as the **chromatin body** (Fig. 21.11). We may consider the bacteria to possess a **cell wall,** which surrounds a **protoplast.** Many secrete an external mucilaginous matrix. Though characteristic of the bacteria they are associated with, these materials are not necessarily part of the living organism, for they may frequently be removed without in any way damaging the living cell. Some bacteria are surrounded by a thick capsular material; others possess a diffuse layer of loose slime or are entangled in a mesh or loose cellulose fibrils. For bacteria

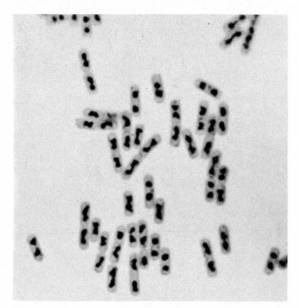

Figure 21.11 Dividing chromatin bodies in *Bacillus cereus.* (Courtesy of C. F. Robinow.)

lacking the capsules, slime, or other dispensable surface components, the cell wall constitutes the outermost structure. While this cell wall may be completely separated from a bacterium and studied chemically in a test tube, its composition is not yet completely understood. Indeed, it apparently varies greatly from one organism to another. In most cases, it seems to consist of a fat-like substance, a lipid, a polysaccharide, and an amino acid associated with a sugar. Experimental evidence indicates that penicillin owes its antibacterial effects, at least in part, to its interference in the synthesis of new cell wall materials. Cell walls of bacteria grown in the presence of penicillin become thin and eventually burst open, allowing the cell to separate into two hemispherical fragments. In some experiments, when *Escherichia coli* is incubated in a penicillium medium, the organism is converted in about 2 hr to spherical protoplasts. This appears to arise because the penicillin interferes with the normal stiffening component of the cell wall.

The protoplast is surrounded by a plasma membrane which, as in other cells, is differentially permeable. That such a membrane is present is borne out by the fact that the bacterial cell may be plasmolyzed. For example, when foods are preserved in sugar, sugar syrup, dry salt, or brine, the bacteria are unable to grow because of the withdrawal of most of the water in them by osmosis.

The protoplast can be isolated and studied intact by enzymic removal of the cell wall. Its cytoplasm, like that of the higher plants, is probably fluid. The enveloping plasma membrane, upon puncture, allows a rapid outpouring of the protoplasm. It is even thought that there actually may be some protoplasmic streaming in these extremely minute organisms. These isolated protoplasts frequently take the shape of spheres, a fact which furnishes additional evidence of the fluid nature of the protoplasm and indicates that it is the cell wall which determines the shape of the individual bacterial cell.

The Ultrastructure of Bacteria

Heterotrophs. Bacteria are prokaryotes, and the structure of the prokaryotic cell was discussed in Chapter 2. All bacteria possess a cell wall, of which the ultrastructure and composition is complex and varies with the genus. While a definite plasmalemma surrounds the protoplast, it has not always been easy to distinguish it with the electron microscope (Fig. 21.12). Internally the chromatin body consists of a tangle of filaments about 4.5 nm in diameter (Fig. 21.12). Since digestion with the enzyme DNAase removes them, they are considered to be DNA fibrils.

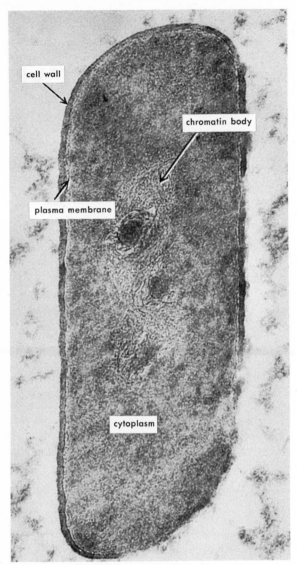

cell wall

chromatin body

plasma membrane

cytoplasm

Figure 21.12 A longitudinal section through a cell of *Bacillus subtilis*, ×80,000. (Courtesy of A. M. Glauert, E. M. Brieger, and J. M. Allen, *Exptl. Cell Res.* **22**, 73, 1961. Copyright Academic Press.)

The fibrils may be isolated and spread on the surface of clean water (a Langmuir trough). They are then picked up on a grid and shadowed with platinum while the grid is rotated. DNA fibrils prepared by this technique are shown in Fig. 21.13. Such preparations demonstrate the presence of a single circular strand of DNA, or bacterial chromosome, embedded in the matrix of the chromatin body. The chromatin body is not separated from the cytoplasm by a membrane of any sort. Several types of granules may occur in the cytoplasm, some of which are ribo-

somes. It is interesting that the sedimentation constant of the bacterial ribosomes is similar to that of mitochondrial and chloroplast ribosomes. The only internal membrane known in bacteria is an array of concentric membranes, the **mesosome,** which appears to arise from an invagination of the plasmalemma, and which seems to have an association with the chromatin body (Fig. 21.14). Some evidence indicates that the respiratory enzymes may be bound to the mesosome or to the plasmalemma.

Autotrophs. There are three families of photosynthetic bacteria, the purple sulfur bacteria, or Thiorhodaceae; the purple nonsulfur bacteria, or Athiorhodaceae; and the green sulfur bacteria, or Chlorobacteriaceae. They are all provided with membrane-bound pigments related to chlorophyll. The membranes of the purple bacteria which bind bacteriochlorophyll bear a resemblance to the grana of chloroplasts (Fig. 21.15), while the photosynthetic membranes of the green bacteria carrying chlorobium chlorophyll seem to be simply spherical bodies called **chromatophores** (Fig. 21.16). Bacteriochlorophyll resembles chlorophyll *a;* however the maximum absorption spectrum of bacteriochlorophyll is in the infrared. Photosynthesis may thus proceed in the absence of visible light.

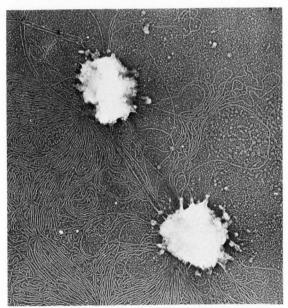

Figure 21.13 Fibrils of bacterial DNA produced by bursting a bacterial protoplast with distilled water and allowing the DNA fibrils to expand on the surface, ×100,000. (Courtesy of A. K. Kleinschmidt and D. Lang, *Fifth International Congress of Electron Microscopy*, 1962, Academic Press.)

Structure of Bacteria **409**

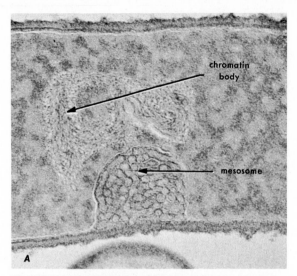

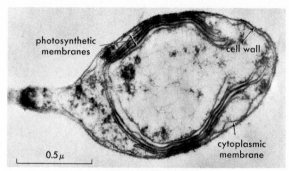

Figure 21.15 Section of *Rhodomicrobium vannielii* showing array of peripheral photosynthetic membranes, ×60,000. (Courtesy of E. S. Boatman and H. C. Douglas, *J. Cell Biol.* **11**, 469, 1961.)

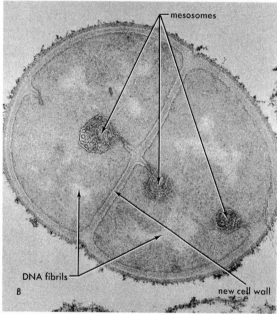

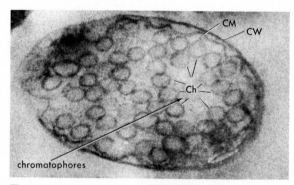

Figure 21.16 Section of *Rhodospirillum rubrum* showing the chromatophores (CH). The cell wall (CW) and plasma membrane (CM) are also apparent, ×60,000. (Courtesy of E. S. Boatman and H. C. Douglas, *J. Cell Biol.* **11**, 469, 1961.)

Figure 21.14 Mesosomes. *A*, the array of coiled membranes in the cytoplasm of *Bacillus subtilis* constitutes a mesosome. *B*, thin section of a vegetative cell of *Sarcina ureae*. Note the distinct association of the mesosomes with the areas of cross-wall formation. Nuclear material is present in the less dense areas throughout the cytoplasm, ×80,000. (*A*, courtesy of A. Ryter, *Bact. Rev.* **32**, 45, 1969; *B*, courtesy of S. C. Holt and E. R. Leadbetter, *Bact. Rev.* **32**, 346, 1969.)

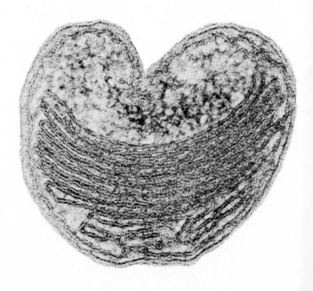

Figure 21.17 Right. The chemotroph, *Nitrosocystis*, oxidizing NO_2^- to NO_3^- anaerobically, has an array of parallel membranes, ×90,000. (Courtesy of R. G. E. Murray.)

The purple bacteria reduce sulfur compounds, frequently H_2S, in the photosynthetic process. In the Thiorhodaceae, the sulfur is deposited within the bacterial cell, while in the green sulfur bacteria, it is deposited outside the cell. The purple sulfur bacteria and the green sulfur bacteria are anaerobic. The purple nonsulfur bacteria are photosynthetic and anaerobic in the light but heterotrophic and aerobic in the dark.

It is also of interest that the anaerobic chemotroph, *Nitrosocystis*, obtains its energy through the oxidation of NO_2^- to NO_3^- and has an array of membranes resembling the grana of chloroplasts (Fig. 21.17).

Reproduction of Bacteria

Asexual Reproduction

Fission. Most bacteria divide by simple fission; the cell is "pinched" in two—much as a sausage might be cut by the tightening of a slip knot. Photographs taken with the electron microscope show the process (Figs. 21.11, 21.18). The slime layer surrounding bacteria is distinctly demonstrated (Fig. 21.18A). Near the middle of the central cell (Fig. 21.18B) a band of some denser material may be noted. Adjacent to this band, a notch in the cell is apparent. A deeper notch occurs in Fig. 21.18C. As the process nears completion, the two daughter cells pull apart, stretching the cell wall and plasma membrane (Fig. 21.18D). This thin, stretched portion finally breaks and the ends of the two daughter cells round up.

The chromatin body apparently divides before the division of the cell (Fig. 21.11), although it appears from this figure that chromatin merely pinches in two.

Under optimum conditions, a bacterium may divide and the daughters grow to full size, ready for another division, in 30 minutes. If such conditions were to prevail for $1\frac{1}{2}$ days, 200 trucks of 5 ton capacity would be required to haul away the progeny of a single cell. Obviously, these conditions have never prevailed for so long a period. Nevertheless, the reproductive capacity of bacteria is enormous. Bacterial growth is everywhere held in check. Probably the most important checks are (a) lack of food; (b) the production of substances unfavorable to growth by the bacteria themselves, by other bacteria, by fungi, by higher plants, or by the host in which the bacteria may be living; and (c) competition for oxygen. The rate of increase of a bacterial population is, in general, very similar to that of other organisms, including yeasts, fruit flies, and man. There is a short initial period during which increase in numbers is very slow; then the bacteria increase very rapidly and the population reaches a peak at which it remains constant for a time; and after the peak the number of bacteria decreases rather slowly (Fig. 21.19).

Spores. Several kinds of bacteria are able to form resistant cells, known as **spores,** at the end of their period of active growth (Figs. 21.20, 21.21, 21.22). Each spore is protected by several sets of membranes and probably contains a somewhat dehydrated cytoplasm. Spores may remain alive for many years under conditions quite unfavorable to growth. They resist desiccation, are not killed by long exposures to temperatures as high as 80°C, and may be suspended in boiling water for 20 minutes or longer. Their resistance to heat, coupled with their ability to become active immediately upon being placed in a suitable environment, makes them of great interest to food technologists, doctors, and also students of living cells. Most of the spore-forming bacteria are

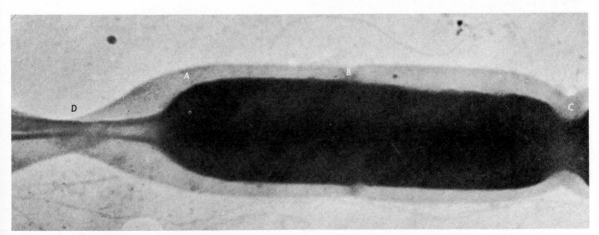

Figure 21.18 Changes in the surface of *Bacillus cereus* accompanying cell division, ×60,000. (Courtesy of F. H. Johnson, *J. Bacteriol.* **47,** 551, 1944.)

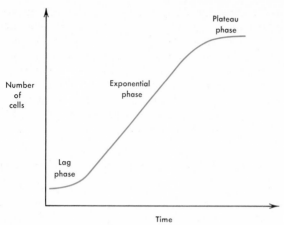

Figure 21.19 A growth curve showing an initial slow rate of growth followed by a rapid logarithmic rate of growth, and finally by the establishment of a colony with little or no growth.

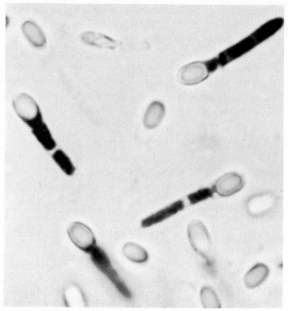

Figure 21.21 Sporulating cells and free spores of *Clostridium pectinovorum*. (Courtesy of C. F. Robinow, *The Bacteria*, Vol. 1, eds. I. C. Gunsalus and R. Y. Stanier, Academic Press, New York.)

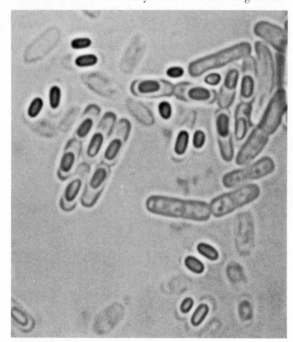

Figure 21.20 Vegatative forms, sporulating cells, and free spores of *Bacillus cereus*. (Courtesy of C. F. Robinow, *The Bacteria*, Vol. 1, eds. I. C. Gunsalus and R. Y. Stanier, Academic Press, New York.)

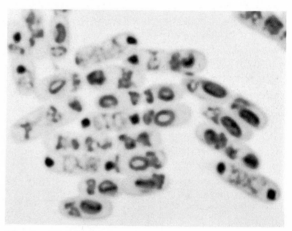

Figure 21.22 Arrangement of the chromatin at different stages in the development of spores of *Bacillus cereus*. (Courtesy of C. F. Robinow, *The Bacteria*, Vol. 1, eds. I. C. Gunsalus and R. Y. Stanier, Academic Press, New York.)

rod-shaped forms belonging to the genera *Bacillus* (Fig. 21.20) and *Clostridium* (Fig. 21.21). Only a few cocci and spirilla are able to form spores. While most bacteria form but one spore per cell, a few are known which may form two or more spores per cell. Such spores occur very commonly in soil bacteria. Fortunately, few disease-producing **virulent** (or **pathogenic**) bacteria form spores. Plant pathogenic

bacteria are nonspore formers. The form taken by the chromatin body in spores is shown in Fig. 21.22.

Sexual Reproduction

Sexual reproduction is definitively indicated by crossing over or gene recombination. It was through gene recombination that sexual reproduction was first demonstrated in the Schizomycophyta. The first suggestion of gene recombination occurred quite unexpectedly in studies on antibody formation and virulent and nonvirulent colonies of *Diplococcus pneumoniae*. Here genetic information was passed from one living bacterium to another through the medium of dead bacteria. The process was called **transformation.** Its discovery stimulated a search for gene recombination in bacteria through a more normal sexual process.

Transformation. Pneumonia is caused by *Diplococcus pneumoniae,* and the virulent bacteria in the sputum of a patient are surrounded by large capsules formed by a carbohydrate. When grown on agar these virulent bacteria form large smooth colonies. It has been shown that the carbohydrates of the capsules are able to stimulate the formation of antibodies in the host, and that four strains of bacteria occur as indicated by the type of carbohydrate present in the capsule and its antibody response. These four types are known simply as Type I, II, III, and IV. Avirulent bacteria producing rough colonies on agar occur as frequent mutations of the virulent smooth form. Once obtained, these mutants do not revert to the virulent smooth form but may be carried unchanged through hundreds of culture transfers and tens of thousands of generations.

We have thus established for study three traits of *Diplococcus pneumoniae:* (a) virulence, (b) type of colony formed, and (c) type of carbohydrate in the capsule. Each of these traits may mutate, virulent to nonvirulent, rough to smooth colony form, and one type of capsule carbohydrate to another. We may study these characters just as we did the various characters of peas (Chap. 17). Thus, avirulent bacteria producing rough colonies on agar frequently mutate to a virulent, smooth form.

In some studies on the activity of these bacteria, *living,* avirulent, rough colony bacteria with Type I capsules were injected into a mouse with a large number of *heat-killed,* virulent, smooth colony, Type II bacteria.

These injections should have been harmless because only *living* virulent bacteria can induce the disease. However, the mouse died, and *living,* virulent Type II cells were recovered from its dead body, although *no living* virulent cells of *D. pneumoniae* were injected into the mouse. We know that the dead bacteria cannot cause disease, and we do not believe that dead bacteria can become alive. The only other explanation is that somehow or other genetic information must have passed from the dead, virulent, smooth, Type II bacteria into the living, avirulent, rough, Type I bacteria. This hypothesis was confirmed when it was shown that an extract

taken from the dead, virulent form could indeed transform the living, avirulent bacteria into a lethal culture. When this extract was purified, DNA turned out to be the material which brought about the change or the **transforming principle.** This change of the genetic information in a bacterial cell by the addition of DNA from another culture of bacteria is known as **transformation.** This experiment may be outlined as follows:

living bacteria	dead bacteria	living bacteria
avirulent rough colony Type I capsule	+ virulent smooth colony Type II capsule →	virulent smooth colony Type II capsule

living bacteria	DNA from	living bacteria
avirulent rough colony Type I capsule	+ virulent smooth colony Type II capsule →	virulent smooth colony Type II capsule

Gene Recombination in Bacteria. Transformation indicated that gene recombination (Chap. 17) occurred in bacterial cells in spite of the absence of the usual nuclear apparatus. It further stimulated a search for the presence of naturally occurring gene recombination in bacterial cells. Evidence for gene recombination was first obtained by Lederberg in a brilliant, yet simply designed, experiment. *Escherichia coli,* the common bacillus of the large intestine, needs no specific growth factors. It is able to synthesize all its protoplasm from a simple broth of meat extract. Lederberg obtained two mutant strains that were unable to synthesize from the meat extract certain compounds that were required for their metabolism. In order to obtain growth, biotin (B) and methionine (M) had to be added to the broth of one strain and threonine (T) and leucine (L) had to be added to the broth of the second strain. Since each strain could synthesize the metabolite that the other could not, *gene recombination, if it occurred,* should result in progeny that would grow on the minimal broth medium, thus:

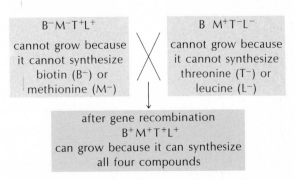

$B^-M^-T^+L^+$
cannot grow because it cannot synthesize biotin (B^-) or methionine (M^-)

$B\ M^+T^-L^-$
cannot grow because it cannot synthesize threonine (T^-) or leucine (L^-)

after gene recombination
$B^+M^+T^+L^+$
can grow because it can synthesize all four compounds

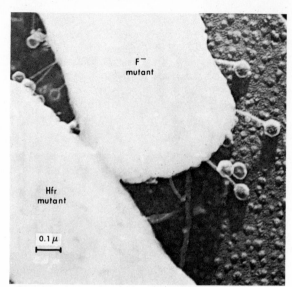

Figure 21.23 Mating between mutant strains of *Escherichia coli* (Hfr and F⁻). Note the attachment of phage particles to the F⁻ mutant (upper bacterium) and their absence from the lower, Hfr mutant. The DNA from the phage heads has been injected into the bacterial protoplast, ×95,000. (Courtesy of T. F. Anderson, E. L. Wollman, and F. Jacob, *Ann. Inst. Pasteur* **93**, 450, 1957.)

A bacterium that now has all the plus genes should give rise to a colony in the minimal broth culture to which these compounds have not been added, because it will be able to synthesize them from the meat extract in the broth. When this experiment was set up, a strain that was able to synthesize all four compounds was actually found. This demonstrated that recombination of genes did actually occur in bacteria. As it took place only once in about a million cells, its occurrence was too infrequent to allow for detailed study.

Eventually, a strain of *E. coli* was found in which recombinations were sufficiently frequent to make a genetic analysis of inheritance in bacteria possible. This strain is known as the Hfr strain, and Fig. 21.23 shows a mating involving the Hfr strain. Note that the contact is rather tenuous, occurring at a small spot on the cell surface. This attachment is of a relatively short duration, and when the cells separate, each is able to form a colony in a normal fashion. With the knowledge of this Hfr strain and the 17 distinctive characters it possesses, it has been possible to prepare a genetic map for the *E. coli* chromosome just as for corn or *Drosophila*, although the methods are of necessity somewhat different from those used for higher forms.

It must first be demonstrated that a true mating has occurred and that a transforming principle is not involved. This is easily accomplished by placing a fine sintered glass disk in the base of a U-tube and growing the different strains of the bacteria in the opposite arms (Fig. 21.24). A transforming principle can diffuse through the sintered glass disk, but bacteria cannot pass it. If a recombination occurs when strains are mixed in the same culture medium but does not occur when they are kept separated by a filter, the recombination must be the result of a mating. This experiment proves that a transforming principle was not present and that a mating occurred.

As with higher plants, it is possible to observe certain phenotypic changes. Hfr bacteria genetically marked for motility and shape may easily be distinguished in a mixed culture from a strain (F⁻) possessing the opposite characters (Fig. 21.25). The attachment between Hfr and F⁻ cells, which lasts only an hour, may be observed. When the cells separate, each produces a clone, but only the F⁻ progeny show any gene recombination. The progeny of the Hfr bacterium resemble the parent cell. Thus, transfer of genetic material is only from the donor Hfr cell to the receptor F⁻ cell (Fig. 21.25). Such an experiment tells us two things: (a) The transfer of genetic material goes only in one direction from Hfr to F⁻. (b)

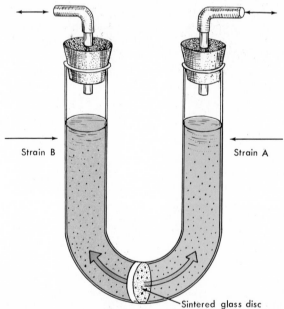

Figure 21.24 Diagram of an experiment to test whether a bacterial mating has taken place. If genetic information is not transferred across the sintered glass disk, a mating is necessary to bring about gene recombination.

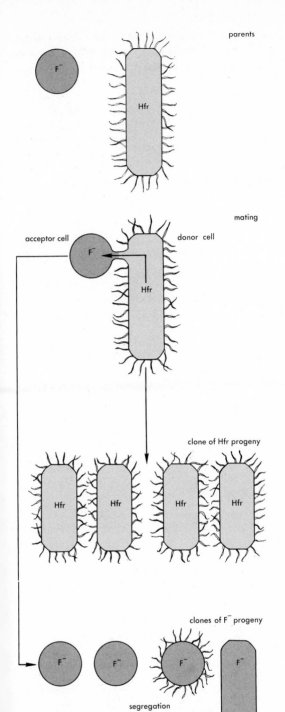

parents

mating

acceptor cell

donor cell

clone of Hfr progeny

clones of F⁻ progeny

segregation

Figure 21.25 Diagram to show how gene recombination may be traced in bacterial matings. Experiment starts with mating of Hfr and F⁻ mutant strains of *E. coli*. After mating, these two bacteria are allowed to produce clones; the appearance of the individuals of the clones is shown.

Since the Hfr bacterium still survives and reproduces in its original form, it cannot have donated all its genes to the receptor cell. Two factors may help to explain this survival: (a) There are probably at least two sets of genes in *E. coli* in separate DNA-containing bodies and (b) the Hfr cell does not donate a complete genetic complement.

Genetic evidence for these conclusions will not be discussed here, but separating the conjugations at successively longer times during the 60 minute period supplies interesting information. If the mating period is interrupted after 10 minutes, one marker gene may have entered the F⁻ cell; after 20 minutes, two more marker genes have entered; after 30 minutes, another two genes have entered, and so on. Even after the full 60 minutes, when the mating is naturally ended, all genes have not been transferred. As explained in Chapter 17, genes are linked in a linear arrangement on a chromosome. The passage of genes takes time—60 minutes for only partial passage—and the bridge between the mating bacteria is narrow (Fig. 21.23). The suggestion is that the DNA- or gene-bearing body is long and slender.

This method of breaking the mating period allows easy mapping of the genes, and the order of the genes so determined corresponds with that found by the conventional system of the percentage of recombinations occurring in a population (p. 319). We can only analyze the Hfr chromosome; we know nothing about the chromosomes in other types of bacterial cells. Special conditions are necessary for the analysis of the F⁻ chromosome, but when they are obtained, it is possible to study them by the method of interrupted mating. This F⁻ chromosome turns out to have the gene order identical to the Hfr chromosome. Different clones obtained after periods of interrupted mating did not have the same genes, although those that were present were in the correct order. Thus, one clone would have ABC, another CDE, and a third, EAB. This may best be explained by assuming that the F⁻ chromosome is circular and that it may break at any point for injection by the donor cell into the receptor cell. Thus, gene recombination may be demonstrated to occur in certain bacterial cells. The mechanism by which it occurs remains to be elucidated.

Respiration of Bacteria

Bacteria, like all living organisms, require energy to carry on living processes. Bacteria that generally use atmospheric oxygen in respiration are called aerobes. Many bacteria, however, are able to respire without atmospheric oxygen and without the subsequent oxidation of food materials to carbon dioxide and water (see Chap. 14); they are able to obtain sufficient energy from glycolysis, the oxidation of food to organic acids. They do not absolutely require the greater amount of energy that is released when oxygen combines with alcohol, pyruvic acid, or some other simple organic compound to form carbon dioxide and water. Indeed, a few species of bacteria cannot grow in the presence of oxygen. Bacteria that grow in the absence of oxygen are called anaerobes. If small amounts of oxygen hinder their growth, they are known as obligate anaerobes. The enzymes responsible for cellular oxidation appear to be membrane-bound, as in the eukaryotes. This membrane system is known as a mesosome.

Influence of Environmental Factors on Bacteria

The ability of bacteria to adapt to changing environments varies greatly from species to species. Some bacteria can grow only in certain very specific and constant surroundings. Others are able to adjust themselves to a wide variety of conditions. We have already noted that some bacteria require oxygen, some do not grow in its presence, and some may grow either with or without it. Factors besides food supply and oxygen that influence the rate of bacterial growth are water, temperature, sunlight, and chemicals.

Water

All bacteria require water to grow. They will not grow in foods containing much less than 15–20% water. Strong salt and sugar solutions inhibit their growth; hence, salt and sugar may be used as preservatives. Most disease bacteria are unable to live if they become dried out in the air. However, when quick-frozen and dried in a vacuum, bacteria may remain inactive but viable for years.

Temperature

Bacteria may grow at temperatures ranging from about 0°F to around 160°F, but individual species usually have a narrow range. Spores of certain bacteria in a dry atmosphere may resist temperatures of 240°F for some minutes, or they may withstand boiling water (212°F) for hours. Most disease-producing bacteria do not form spores and are easily killed at 140°F after about 10 minutes' exposure. Disease-producing bacteria in milk may be killed by heating the milk to 160°F and holding it at that temperature for 15 seconds. This process is known as **pasteur-**

ization. On the other hand, many vegetables must be heated to high temperatures for 20–40 minutes to kill the spores of soil bacteria that might cause spoilage.

Sunlight

Bacteria are rapidly killed by direct exposure to sunlight, ultraviolet rays being particularly effective.

Chemicals

Exposure of bacteria to a bactericide can induce development of tolerant strains. The use of streptomycin hydrochloride to control fire blight of pears, caused by *Erwinia amyboreora,* has resulted in streptomycin-resistant *Erwinia.*

Classification of Bacteria

Because of their small size and similar structure, bacteria are very difficult to classify. For instance, two species of bacteria are known which are almost indistinguishable from each other. Both grow in association with the roots of plants. One, *Rhizobium leguminosarum,* does not harm the host but instead makes nitrogen of the air available to it. The other species, *Agrobacterium tumefaciens,* rapidly kills the host plant. Thus, the two species, although very similar both morphologically and physiologically, differ in one very important physiological process that results, in one instance, in a beneficial relationship and, in the other, in a very harmful relationship (Figs. 21.26, 21.27).

In order to name bacteria correctly, one must have complete data as to their size, shape, motility, and spore formation. In addition, many physiological characters must be studied. For example, it may be necessary to know how the bacterium grows in milk, on agar, and on potato, what sugars it is able to ferment, what pigments it produces, and a number of other physiological reactions.

The Schizomycophyta are separated into seven orders on the basis of obvious morphological and physiological differences. Some have characters that suggest molds or algae or one-celled animals. The largest order—the Eubacteriales, or true bacteria—have characteristics described in the previous pages. It is the only order that will be discussed here.

Order Eubacteriales

The Eubacteriales is divided into two suborders: (*a*) spore-forming bacteria, with a single large family, the Bacillaceae; (*b*) non-spore-forming bacteria, with eleven families.

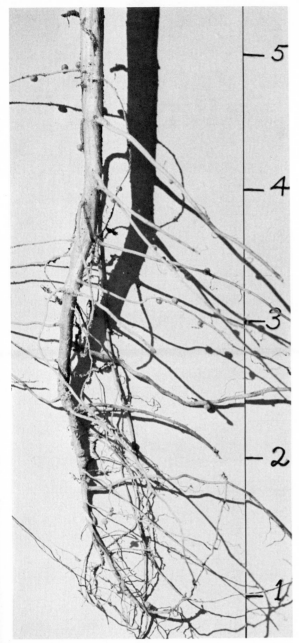

Figure 21.26 Nodules produced by *Rhizobium leguminosarum* on the roots of a legume. Scale is in inches. (Courtesy of Francis Smith.)

The Spore-Forming Eubacteriales. The family Bacillaceae contains several species of great economic importance. Most of them are normal inhabitants of the soil or of animal wastes. Many of them will grow well in the absence of oxygen, and some are able to secrete powerful poisons or toxins. *Clostridium botulinum* is common in garden and farm soils. It

Figure 21.27 Crown gall on almond. (Courtesy of Ark.)

not infrequently occurs on vegetables, such as corn and beans. If *Clostridium botulinum* is not killed in the process of canning, it grows and multiplies in the canned product, producing one of the most powerful poisons known. Food poisoning from **botulinus toxin,** however, has become quite rare. Commercial canning processes are adequate to kill all spores, and home canners have been so well-educated that home-canned food is as unlikely to contain viable *Clostridium botulinum* as is commercially canned food.

Clostridium tetani, like *C. botulinum,* is also a normal inhabitant of well-manured soils. When introduced into deep wounds, it may produce a very potent toxin. Tetanus infections were common during wartime. The organism is almost invariably present in dirty wounds, where it multiplies. The toxin secreted by it causes lockjaw. Fortunately, both soldiers and horses may be injected with a substance that gives them almost complete protection against the toxin of *C. tetani.*

The Non-Spore-Forming Eubacteriales. Although this suborder contains eleven distinct families, it is divided into three natural groups on the basis of food requirements.

1. One family (Nitrobacteriaceae) includes bacteria that make their own food. These bacteria, like green plants, are autotrophic, although they do not possess chlorophyll.
2. Four families are heterotrophic but obtain the nitrogen they require from inorganic salts, such as sodium nitrate, or from atmospheric nitrogen.
3. The remaining six families are also heterotrophic but require organic substances as a source of energy and nitrogen. They may utilize dead plant or animal wastes; that is, they are saprophytes. Or they may require living cells for their source of nitrogen, in which event they are parasites.

Summary of Bacteria

1. Bacteria are prokaryotes; they lack membrane-bound organelles and a nucleus. The light microscope shows them to possess a cell wall enclosing a protoplast. They contain a central chromatin body, so-called because it stains with certain nuclear stains and pinches in two before cell division. The electron microscope shows the presence of DNA fibrils in the chromatin body, and a special technique indicates only a single, circular strand of DNA in each chromatin body.
2. There are heterotrophic and autotrophic bacteria. The heterotrophic forms may be anaerobic or aerobic, but the autotrophic bacteria are anaerobic. The autotrophic bacteria may be chemotrophs oxidizing such substances as NO_2^- to NO_3^-, or phototrophic, utilizing the energy of sunlight.
3. Respiratory enzymes may be bound to the mesosome or to the plasmalemma, or to both membranes.
4. The photosynthetic pigment of the photosynthetic bacteria resembles chlorophyll a and is membrane-bound. The photosynthetic membranes of the green sulfur bacteria are spherical chromatophores, while those of the purple bacteria resemble grana of chloroplasts. In both these bacteria, photosynthesis is carried on anaerobically.
5. Transformation is the transfer of genetic information from cell to cell without cellular contact.
6. Gene recombination in bacteria results from the sexual process.
7. The chromosome of a bacterium consists of a single, circular DNA fibril.
8. Bacteria cause diseases of plants and animals.

22

Algae

Algae are autotrophic, aquatic, or semiaquatic plants. They are common in most semipermanent pools; in ponds, lakes, and streams; along ocean shores; and in surface waters of oceans. Terrestrial forms grow on many moist surfaces, including the surface of, or in the top few inches of, moist soil. Although they occur all about us, only the common seaweeds (Figs. 2.7C, 22.1H, see Color Plates 2, 12; 22.7) along rocky coasts are large enough to impress themselves upon us. They range from microscopic forms (Figs. 21.1A, B) to plants many meters in length (Fig. 22.7C). They all contain chlorophyll a, yet they are so distinctly colored by other pigments, that pigmentation plays an important role in their classification.

Probably the most commonly noticed natural urban habitats for algae are the sides of glass fish tanks. They may also generally be found around leaking faucets and in garden or park pools that are not kept "pure" with chemicals. The "bloom" occurring during the summer on many lakes or the scum found on ponds are the result of algal activity. Microscopic forms occur in most natural waters, including the top 75 meters of the ocean. Here they constitute a primary food source for marine animals of all types and account for about 70% of the oxygen released into the atmosphere through photosynthesis.

Economic Importance of Algae

The direct economic importance of algae is minimal, although they are of minor importance in the economy of many countries. The middle lamella and primary walls of many brown algae contain derivatives of alginic acid. These compounds, like the pectins found in the cell walls of higher plants, are nontoxic, highly viscous, and readily form gels. One tablespoon of alginic acid will give a quart of water the viscosity of honey. Alginates are used in many ways: They may find their way into marshmallows, ice cream, processed cheese, cosmetics, printing inks (for textiles), pharmaceuticals, and paints. As a food for man, some species of blue-green, brown, and red algae are useful to the populations of islands and coasts of the eastern Pacific Ocean. Algae find some use in sewage disposal, where they may greatly facilitate oxygenation. They may become a nuisance in drinking waters, where they may impart an unpleasant taste or odor as well as clog filters. They are also an indication of pollution. For instance, the water of Lake Tahoe high in the Sierra Nevada has been crystal clear because its mineral deficiency prevented

419

any extensive algal growth. With the increase in human population in the Tahoe basin, algae are appearing in Lake Tahoe waters. This indicates seepage of waste products into the lake.

Algal Classification

Grouping the aquatic, autotrophic, nonvascular plants under the general classification of algae is highly artificial but thoroughly practical (although, as we shall see, algae have much in common). There does exist a far greater diversity of form and metabolism among them than among the mosses and the vascular plants. The algae have chlorophyll *a* as do all other autotrophic plants. Some have only chlorophyll *a*, others have *a* and *b*, while two divisions are characterized by having *a* and *c*. Some alage have, in addition to chlorophyll, accessory pigments not occurring in the higher plants. Algae, including the blue-greens, require oxygen for respiration and evolve oxygen in photosynthesis, thus differing from the anaerobic photosynthetic bacteria and resembling higher plants.

Although large algae grow in the oceans (Fig. 22.7), there are no large terrestrial forms. Consequently, supporting sclerenchymatous tissue is lacking in the algae. Their aquatic environment lacked survival pressures for the evolution of sclerenchyma and specialized absorptive organs such as roots. With the probable exception of giant kelps, all the cells of an alga can carry on photosynthesis or are close to cells that can. So a food-conducting, phloem-like, tissue has evolved only in the long stipes of the kelps (Figs. 22.1G, 22.7). Algae lack roots, stems, and leaves, and their plant body is therefore referred to as a **thallus.** All plants with such a plant body are sometimes called **thallophytes**.

Although it is generally easy to visually distinguish an alga from a mature moss, it does turn out to be difficult to find definitive traits that are common to all algae yet absent from all other divisions. Possibly the only structure possessed by all higher plants and rarely occurring in the algae is a protective jacket of sterile cells surrounding the gametes.

Separation of the algae into divisions is based largely upon differences in pigments, food reserves, the ultrastructure of the chloroplast, type of flagellation, and the chemical constitution of the cell walls. Thus, the algal divisions are largely set up on biochemical and structural differences associated with photosynthesis. Detailed differences in reproductive cycles and vegetative structure, although associated with the biochemical differences, are of secondary importance in determining algal divisions, because

parallel reproductive and vegetative evolution has occurred in several divisions.

The electron microscope, coupled with a more precise knowledge of algal metabolism, is providing information leading to a revision of the classification of the algal groups. For instance, the Cyanophyta are prokaryotic cells (Figs. 22.2, 22.20), and as such have recently been thought to form a natural group with the bacteria, particularly with the anaerobic photosynthetic bacteria (p. 409). However, the internal membranes differ in the two groups, and the metabolic processes occurring in the prokaryotic Cyanophyta more closely resemble those occurring in the eukaryotic algae and in the higher plants than they do the metabolic processes occurring in the prokaryotic bacteria. Photosynthesis and aerobic respiration definitely establish the blue-green algae as a separate algal division, the Cyanophyta. It stands as an intermediate group between the prokaryotic bacteria and the eukaryotic algae. The algae thus consist of one prokaryotic division, the Cyanophyta, and eight eukaryotic divisions, the Rhodophyta, Pyrrophyta, Chrysophyta, Bacillariophyta, Phaeophyta, Xanthophyta, Euglenophyta, and Chlorophyta. Of these, our discussion will mainly concern the Cyanophyta, Rhodophyta, Bacillariophyta, Phaeophyta, and Chlorophyta.

Cyanophyta

These are the blue-green algae. They are ubiquitous. Characteristically, they may be found on moist barren soil, footpaths, tidal flats, cliffs, and sidewalks kept wet by rains, spray from falling water, or leaking faucets. They occur in hot springs and in intimate association with other plants or animals. Their plant bodies consist of single cells, short filaments (**trichomes**) and simple colonies of various forms. The cells are almost always encased in a mucilaginous envelope (Figs. 22.1A, see Color Plate 12; 22.2). They contain chlorophyll a, β-carotene, and two water-soluble pigments (phycobiliproteins), c-phycocyanin, and c-phycoerythrin. As a reserve, they accumulate cyanophyte starch and certain characteristic proteins. No flagellated cells occur and sexual reproduction has not been detected. Examples are *Nostoc*, *Synechococcus*, *Oscillatoria*, *Lynghya*, and *Spirulina* (Fig. 22.2).

Rhodophyta

These are the red algae (Figs. 22.1D, I, 22.10). They occur mostly as feathery plants growing most abundantly on rocky coasts of warm oceans. There are a number of fresh-water Rhodophyta (Fig. 22.3D).

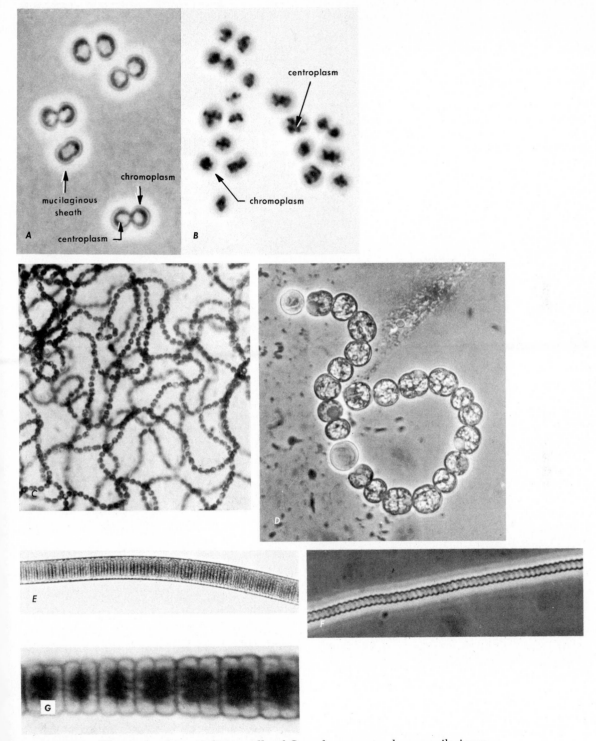

Figure 22.2 Blue-green algae. *A,* living cells of *Synechococcus* to show mucilaginous sheath and general location of photosynthetic pigments. *B,* cells killed and stained to show location of nuclear material, ×1000. *C* and *D, Nostoc, C,* ×800, *D,* ×1500. *E, Lyngbya,* ×1200. *F, Spirulina versicolor,* ×1250. *G, Oscillatoria,* ×1500. (*A* and *B,* courtesy of W. A. Cassel and W. G. Hutchinson, *Exptl. Cell Res.* **6,** 134, Academic Press; *G,* courtesy of R. Norris.)

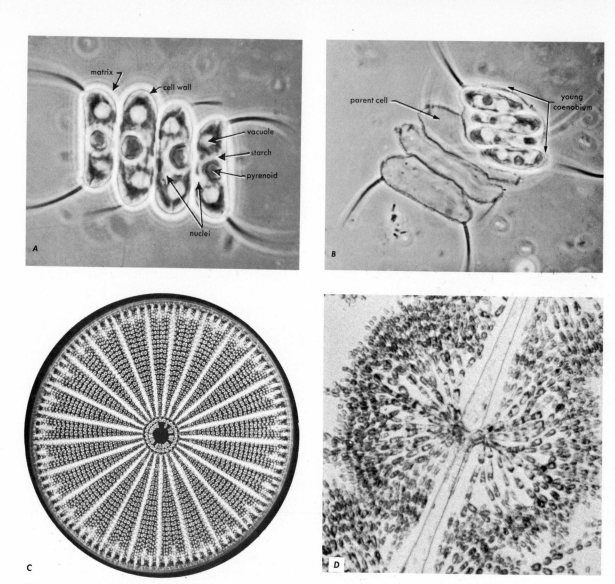

Figure 22.3   Examples of fresh-water algae. *A*, *Scenedesmus quadricauda*, green alga; *B*, young coenobium of *Scenedesmus* leaving the parent cell; *C*, the diatom, *Arachnoidiscus ehrenbergii*; *D*, a fresh-water red alga, *Batrachospermum moniliforme*. (*A*, *B*, courtesy of T. Bisalputra; *C*, courtesy of CCM: General Biological, Inc.; *D*, courtesy of David Brown.)

They are frequently small and must be searched for. Red algal plant bodies range from single unicellular (*Porphyridium*) forms through simple filaments to leaf-like thalli (Fig. 22.10) or to delicate (*Polysiphonia*) feathery plants (Fig. 22.1*l*). They contain only chlorophyll *a* and carotene and the water-soluble phycobilins (b-, c-, r-phycocyanin and b-, c-, r-phycoerythrin). As a reserve they accumulate outside the chloroplast, a carbohydrate, floridean starch. It turns red, not blue, upon straining with iodine. No flagellated cells are present. *Batrachospermum* is a freshwater species. *Porphyridium* is a unicell of con-

siderable laboratory interest because the phycobiliproteins can be seen with the electron microscope; they show a close relationship to the photosynthetic membranes. The genus *Polysiphonia* shows considerable evolutionary advancement and has a complex reproductive cycle (Fig. 22.19).

Bacillariophyta

The algae belonging to this division are the **diatoms.** They are widespread, especially in colder waters. They occur in surface waters of temperate and northern oceans, in lakes, ponds, and semipermanent pools, on moist barren earth and in top layers of soil. They are either single-celled forms (**unicells**), or a simple aggregation of cells forming filaments. They may be free-floating, or attached to other algae, submerged roots, or other material. Their cell walls consist of an intricate layer of silica sandwiched between two delicate organic membranes. They contain chlorophylls a and c, plus δ- and ε-carotene. While many diatoms show movement, flagellated vegetative cells are not present. Their ability to metabolize silica and to deposit it in their cell walls results in a delicate wall sculpturing, sometimes of great beauty (Fig. 22.3C).

Phaeophyta

These are the brown algae, or seaweeds, that grow abundantly along the rocky coasts of all oceans, but particularly those of the temperate zones (Fig. 22.7). In certain tropical areas, the free-floating *Sargassum* (Fig. 22.7D) forms dense beds. Brown algae range from small, almost microscopic forms to giant **kelps** that may exceed 30 meters in length. No unicellular forms are known. They contain chlorophylls a and c, carotene, and a characteristic brown pigment, fucoxanthin. They store a carbohydrate known as laminarin. The reproductive cells of a number of species are flagellated.

Chlorophyta

These are the green algae (Figs. 22.1C, F, 22.4), which are common fresh-water algae found, frequently abundantly, in all fresh-water streams, lakes, ponds, and semipermanent pools. Three orders inhabit tropical oceans, and a few marine forms are scattered through other groups. They are grass-green because of chlorophylls a and b (Fig. 22.1F). They also contain β-carotene and store starch. They show great morphological diversity, both in vegetative structure and in reproduction. Flagella are lacking altogether in some genera. In others, only gametes are flagellated. In some primitive forms the vegetative cells also have flagella.

Green algal plant bodies consist of single cells, filaments, or delicate flattened blades. There are many common green algae; almost any microscope slide of native water will contain some examples. *Chlamydomonas* and associated genera are common motile, unicellular, or colonial forms (Fig. 22.4A). *Spirogyra, Ulothrix, Oedogonium,* and *Cladophora* are common filamentous forms (Figs. 22.1C, F, 22.4). Sea lettuce, or *Ulva,* is an example of a species whose thallus is flat and blade-like (Fig. 2.7B).

Phylogeny of Algae

Since photosynthetic pigments are important in algal classification, a comparison of pigments should reveal relationships between the divisions. Although such a comparison does supply some indication of evolutionary tendencies, the situation is complicated. Consider relationships between prokaryotic Cyanophyta (pp. 9–11) and eukaryotic Rhodophyta. In spite of great differences between them, they are thought by some to be related (a) because they both contain only chlorophyll a and the two phycobiliprotein pigments, phycocyanin and phycoerythrin, and (b) because flagellated cells are absent from both divisions. Otherwise, the great differences in cellular detail, vegetative form, and reproductive structures indicate a widely divergent evolution.

A different situation exists when pigments and form are compared among eukaryotic divisions, including the Rhodophyta. Here, while each division is distinguished by characteristic pigments, they all contain species with similar plant bodies and similar reproductive stages. Only the Euglenophyta consists solely of unicellular forms, and only the Phaeophyta lacks unicellular species. Otherwise, each division contains species ranging from unicells to filaments, branching filaments, aggregations of cells forming parenchymatous thalli, and intertwining filaments forming more complicated plant bodies. Reproductive structures have not been of great assistance in indicating relationships between divisions. A recent account of algal phylogeny postulates three lines of development from a supposed common ancestor. The Cyanophyta and Rhodophyta, for reasons already discussed, form one line of ascent. A second line is made up of the Chrysophyta, Pyrrophyta, Bacillariophyta, and Phaeophyta, while the third line is comprised of the Xanthophyta, Euglenophyta, and Chlorophyta. There is general agreement among botanists that ancient Chlorophyta were the ancestors of both higher nonvascular plants and vascular plants.

Diversity of Form within the Algae

Within divisions, evolutionary relationships are demonstrated by thallus morphology. It is widely accepted that unicells are the most primitive and that evolutionary advance proceeded, in general, along similar lines in all divisions. There were three tendencies in this evolutionary expression: (a) single cells cooperated to form rather complex colonies (the volvocine tendency, Fig. 22.5A); (b) single cells

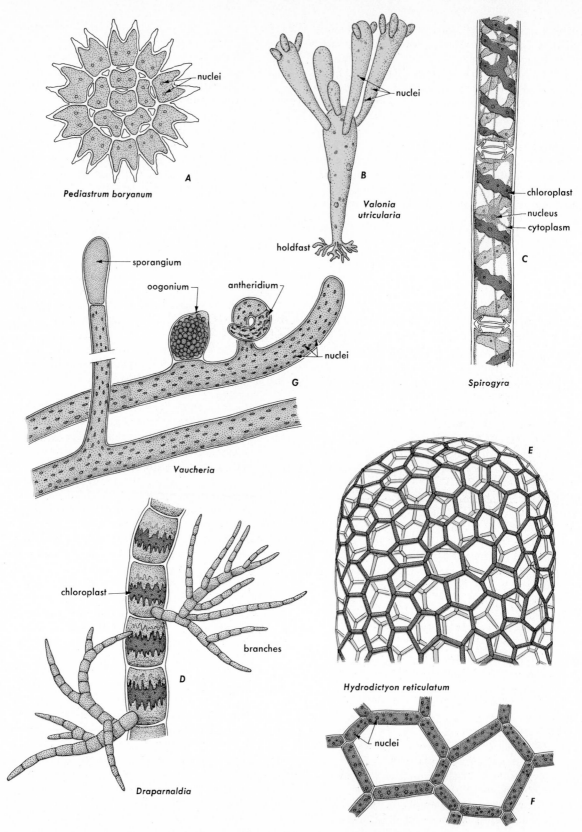

Figure 22.4 Fresh-water algae. *A* through *F*, Chlorophyta; *G*, Xanthophyta. *A*, *Pediastrum*; *B*, *Valonia*; *C*, *Spirogyra*; *D*, *Draparnaldia*; *E*, portion of net of *Hydrodictyon*; *F*, several multinucleate cells forming net of *Hydrodictyon*; *G*, *Vaucheria*.

became locally highly differentiated (the siphonous tendency, Fig. 22.5B); and (c) single cells remain attached together after division to form filaments or simple sheets of parenchymal-like tissue (the tetrasporine tendency, Fig. 22.5C). While parallel changes have apparently occurred in all divisions, they have not occurred to the same extent in all of them. For instance, there are only unicells in Euglenophyta, and unicellular forms are not known in the Phaeophyta. Further, thalli of Phaeophyta and Rhodophyta show a much higher degree of specialization than is found in the other divisions. We shall discuss the morphology of algal thalli, proceeding from less complex to more complex arrangements. The discussion will be illustrated by examples of Chlorophyta, Phaeophyta, and Rhodophyta.

Chlorophyta

There are three distinct evolutionary tendencies in the Chlorophyta. All three are thought to originate from a single-celled alga similar to *Chlamydomonas*. The difference in these three lines of development lies in the different ways in which cells can be associated to form a plant body. We can imagine a plant consisting of (a) a large number of individual cells (*Volvox*, Fig. 22.5A); (b) coenocytic filaments with thousands of nuclei sharing a common mass of cytoplasm lacking cross-walls (*Vaucheria*, Fig. 22.4G); or (c) a plant body in which adjacent cells share a common cell wall (*Ulothrix*, Fig. 22.14). The first of these lines is the **Volvocine** line, the second, the **Siphonous** line, and the third, the **Tetrasporine** line.

Volvocine Line. An example of this line is the order Volvocales in the Chlorophyta (Fig. 22.5A). Most individuals in the *Chlamydomonas* genus are oval or oblong. They have two flagella at their anterior end and a cup-shaped chloroplast. Most species have a single red body of carotene with a characteristic ultrastructure; this is the **eye spot** (Fig. 22.26). The nucleus is centrally located within the cup formed by the chloroplast. Sometimes some species go through a stage when the individuals lose their flagella and become embedded in a common mucilaginous matrix.

The permanent association of chlamydomonad-like cells into curved plates (*Gonium*, Fig. 22.5A) of four linear cells (*Scenedesmus*, Fig. 22.3A) may be considered the first step in advance in the volvocine line. Each new colony is formed within a parent cell and released from the parent cell as a small but fully formed new colony (Figs. 22.3A, B).

Further advance is occasioned by the formation of a hollow sphere of 32 similar individual cells. They

are connected to each other by delicate strands. Increase in the number of cells forming a colony and specialization of some cells for reproduction culminates in the genus *Volvox* (Fig. 22.5A). The colonies contain from 500 to 20,000 cells, only a few of which are reproductive cells confined to a small area. The vegetative cells of *Volvox* still resemble *Chlamydomonas*.

Siphonous Line. The outstanding characteristic of this line is the complete lack, or rare production, of cross-walls (**septa**). The cells of most siphonous species are thus multinucleate and coenocytic (Fig. 22.5B). In forms completely lacking septa, the peripheral cytoplasm, with embedded nuclei, plastids, and all other organelles, is continuous throughout the plant body. This also means that there exists a single central vacuole extending through all filaments. Such an evolution of form may be envisioned with a chlamydomonad-like cell as the primitive starting point. Evolution of the siphonous plant body occurred several distinct times in the Chlorophyta and has been paralleled at least once (*Vaucheria*) in the Xanthophyta. The siphonous habit did not develop in other divisions.

The most primitive forms like *Protosiphon* (Chlorophyta) or *Botrydium* (Xanthophyta) (Fig. 22.5B) consist only of an aerial green, multinucleate vesicle anchored to the soil by colorless branching rhizoids (*Botrydium*) or by a single rhizoid (*Protosiphon*). The irregular branching or budding of this vesicle would result in the irregular thallus of *Valonia* (Chlorophyta) (Fig. 22.4B). Simple filamentous forms such as *Vaucheria* (Fig. 22.4G) occur in the Xanthophyta. Two types of multinucleate forms occur in the Chlorophyta. Some, like *Cladophora* (Fig. 22.5B) and *Hydrodictyon* (Fig. 22.4E), have cross-walls separating multinucleate cells; others are strictly coenocytic. These coenocytic forms, belonging to the order Codiales, branch extensively and with a high degree of regularity, conferring considerable diversity of form on the adult thalli. The genus *Caulerpa*, which forms extensive mats on shallow tropical sea floors, is a good example of the degree of specialization attained by intertwining coenocytic filaments (Fig. 22.5B). Large tracts of shallow areas of the floor of the Mediterranean Sea are covered by *Caulerpa*. The plants of this genus consist of a cylindrical rhizome up to 3 meters in length. It is formed by a characteristic intertwining of filaments. This rhizome bears branching rhizoids growing downward into the ocean floor. It also sends upward a number of photosynthetic shoots that take a great variety of shapes. In some species, the shoots are feather-like, while in other species they are flat and leaf-like. They, like

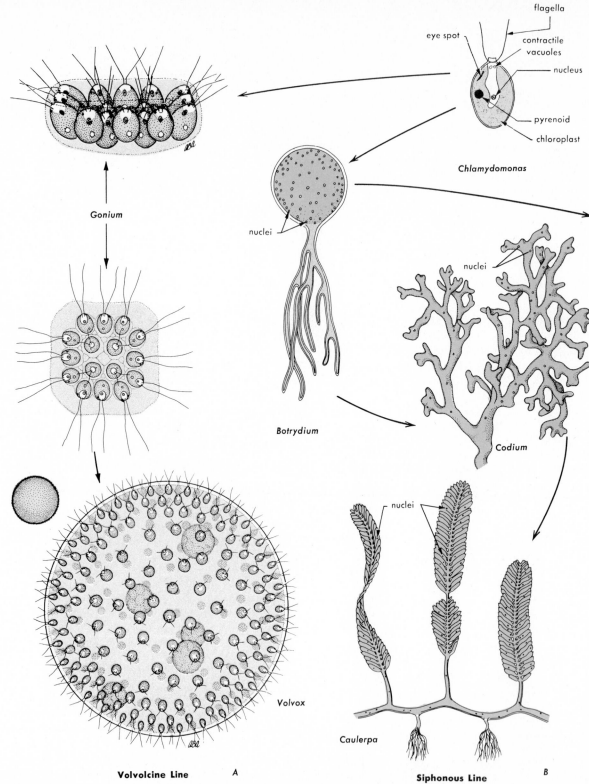

flagella

eye spot

contractile
vacuoles

nucleus

pyrenoid

chloroplast

Chlamydomonas

Gonium

nuclei

nuclei

Botrydium

Codium

nuclei

Volvox

Caulerpa

Volvocine Line   *A*

Siphonous Line   *B*

Figure 22.5   Diagrammatic representation of the three lines of algal evolution. *Botrydium* belongs to the division Xanthophyta; all others are Chlorophyta. All three lines are thought to take their origin from a type similar to *Chlamydomonas*. *A*, the volvocine line (light green); *B*, the siphonous line (medium green); *C*, the tetrasporine line (dark green). In the volvocine line, the cells retain their identity; in the siphonous line, mitosis is not followed by cytokinesis; in the tetrasporine line, mitosis and cytokinesis comprise a true cell division. Only plants with cell division successfully invaded the land, and Chlorophyta

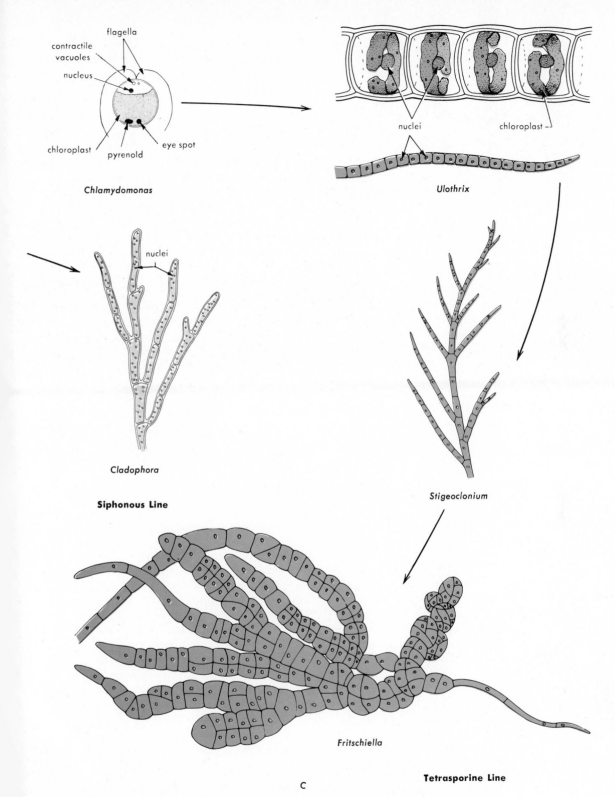

flagella

contractile
vacuoles

nucleus

chloroplast   eye spot
        pyrenoid

Chlamydomonas

nuclei   chloroplast

Ulothrix

nuclei

Cladophora

Siphonous Line

Stigeoclonium

Fritschiella

Tetrasporine Line

C

similar to *Fritschiella* may have been the progenitors of land plants. (*Codium* and *Caulerpa* redrawn after W. R. Taylor, *Marine Algae of the Eastern Tropical and Subtropical Coasts of the Americas*, University of Michigan Press. *Fritschiella* after M. O. P. Lyengar, *New Phytologist* **31,** 329. *Botrydium* redrawn after G. M. Smith, *Freshwater Algae of the United States.* Copyright 1933 by McGraw-Hill Book Company, New York. Used with permission of McGraw-Hill Book Company.)

the rhizome, are formed of branching coenocytic filaments arranged in a characteristic pattern. Growth in these forms is at the apex of each filament. where concentration of protoplasm also occurs. Note that here, two types of filaments are formed, a horizontal rhizome and upright photosynthetic branches.

In contrast to this extensive growth and differentiation of coenocytic filaments, stands the uninucleate single-celled *Acetabularia* (Fig. 4.17). Here, a high degree of specialization occurs within the limits of a single nucleated cell about 25 mm long and divided into three regions of specialization, (a) a branched rhizoid system, one branch of which contains the single nucleus; (b) an upright stalk; and (c) an expanded upper portion, the cap.

Tetrasporine Line. Although a *Volvox* colony in the volvocine line may consist of 20,000 cells with some degree of specialization, all these cells were formed within a single parent cell. The daughter cells provided their own cell wall at a later stage in development. In the siphonous line, mitoses occur, but cytokinesis does not, and the resulting nuclei are not set apart in individual protoplasts. They remain within the parent cell wall, which enlarges and may take on a complex shape. In the tetrasporine line, cell divisions occur; each mitosis is accompanied by cytokinesis. Now, the parent cell wall is retained on three sides as an essential part of the two daughter cells. A new common wall forms the fourth side. Protoplasts and enclosing cell walls become cooperating units in a tissue or plant body. The need to confine cell division to a single specific period of the life cycle has been eliminated. Cell divisions, mitosis and cytokinesis, may occur through all, or much, of the life of a plant. Size and complexity of form may now be attained by continued cell division and by controlling (a) polarity of cell divisions, (b) relative rates of cell divisions, and (c) relative rates of cell enlargement. As an example of the influence of polarity of cell divisions, if all mitotic spindles were aligned on a straight line, a simple unbranched filament would result (Fig. 22.6A).

Gametes of the green alga *Ulothrix* resemble mature chlamydomonad cells, as do the gametes of many other algae. When zoospores of *Ulothrix* germinate, the new cells adhere to each other and the cells divide with spindles aligned in a straight line, so that repeated division of all cells results in a filament formed by a single row of cells. This is a **uniseriate** filament. Two very important things have happened: (a) cytokinesis results in adherent, cooperating daughter cells, and (b) a definite polarity for all cells has been established for this simple multicellular plant. Adhering cells and genetic control of polarity of cell division are two basic requirements for attainment of great diversity of plant form (Figs. 22.6, 22.11).

In *Ulothrix* and other simple unbranched forms, all spindles are oriented in the same direction and all cells in the filament may divide. Upon establishing a second pair of poles at right angles to the axis of the first pair, a diversity of thalli will arise, depending upon the timing of cell divisions in the two planes. If all cells divided regularly at right angles to each other, but in the same plane, a thin tissue, one cell in thickness, would form a **monostromatic thallus** (Fig. 22.6C). This occurs in the green alga *Monostroma*. If, instead, every tenth cell divided, and continued to divide, at right angles to the line of divisions forming the filament, a flat filament with regular branches would result (Fig. 22.6B). If every twentieth cell could now divide in a plane at right angles to the plane of the flat branching filament, a branching filament with its branches at right angles to each other would result. If the rates and locations of cell divisions are confined to specific cells of the main filament, lateral branches will arise at definite points as in *Draparnaldia* (Fig. 22.6B). Note that the cells of lateral branches are much smaller than those of the central branch. If all cells early in the development of a monostromatic thallus divided but once in a plane at right angles to the plane of the monostromatic thallus, a thallus two cells in thickness (**distromatic thallus**) would result. Sea lettuce, or the marine green alga, *Ulva*, has a distromatic thallus (Figs. 2.7B, 22.6D). With division possible in three planes, form is determined by the relationship between the rates of division, the location of the divisions, the rates of cell expansion, and the kinds of differentiation that occur in the three planes. Cell divisions under these conditions are generally confined to specific locations or meristems. These generally occur at the apex of the thallus, or at definite intercalary points. In many brown algae the epidermal cells may also be meristematic.

While polarity of cell division is of importance in determining plant forms, environmental factors may also be involved. Nevertheless, it would be exciting to discover the genetic mechanism that determined that given cells were to divide at right angles to their neighbors. Just what type of change in base pairs brings about the change from a filament to the monostromatic thallus of *Monostroma* or the distromatic thallus of *Ulva*?

Heterotrichy

Caulerpa, of the Chlorophyta, produces a stout horizontal rhizome provided with rhizoids for an-

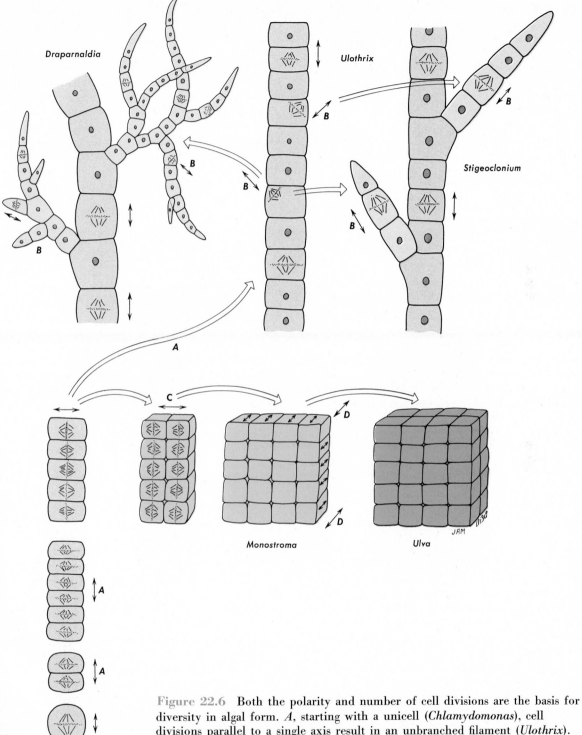

Figure 22.6   Both the polarity and number of cell divisions are the basis for diversity in algal form. *A*, starting with a unicell (*Chlamydomonas*), cell divisions parallel to a single axis result in an unbranched filament (*Ulothrix*). *B*, occasional changes in polarity of cell divisions give rise to various types of branched filaments (*Draparnaldia, Stigeoclonium*). *C*, all cells of a filament dividing at right angles to the axis of the filament, and in the same plane, give rise to a blade one cell in thickness (*Monostroma*). *D*, should all cells divide once at right angles to the plane of the blade, a two-celled lamina would result (*Ulva*).

chorage and complex arrangements of upright filaments for photosynthesis. Two different types of filaments are formed. This type of growth habit is called **heterotrichy.** It is a type of specialization that has occurred numerous times in the Chlorophyta, Rhodophyta, and Phaeophyta. It is thought to represent a definite evolutionary advance.

Phaeophyta and Rhodophyta

With the exception of a lack of unicells in the Phaeophyta and of motile cells in the Rhodophyta, a series of thalli paralleling the tetrasporine line, as exemplified by the Chlorophyta, is to be found in both Phaeophyta and Rhodophyta. However, both red and brown algae have developed much greater specialization than occurs in thalli of other algal divisions. This development has proceeded along separate pathways in these two divisions, and it will be necessary to consider the vegetative structure of the Phaeophyta and Rhodophyta separately.

Phaeophyta. The most primitive Phaeophyta, the Ectocarpales, are, as a group, small-branched filamentous forms (Figs. 22.1E, 22.7A). The heterotrichous habit is developed in most, if not all, of them. The prostrate system consists of a few filaments and serves to anchor an upright system of branching photosynthetic and reproductive filaments to other algae or to rocks. This prostrate system of filaments exhibits apical growth. The upright system appears as feathers or compact brownish tufts (Figs. 22.7A). Cell division is diffusely distributed along the branching filaments. It is rarely apical or confined to special regions.

The most advanced Phaeophyta are in the Laminariales. They are large plants with regions specialized for anchorage, conduction, support, and photosynthesis. A **holdfast** attaches the plant to the rocky ocean floor; photosynthesis occurs in a leaf-like blade or **lamina,** and a **stipe** connects the lamina with the **holdfast. Bladders** are found in many species and add to the buoyancy of the laminae. These features are well-exhibited by *Fucus* (Fig. 22.1H), *Laminaria* (Fig. 22.7B), and *Macrocystis* (Fig. 22.7C). This increase in size and complexity of brown algae over green algae is accompanied by an increase in the degree of cell specialization. Aggregates of cells form tissue systems such as those shown for the stipe of *Postelsia* (Fig. 22.8A). The outermost layer not only serves as a protective barrier around the stipe but is also meristematic, and cuts off new cells adding to the internal cortex. This outer layer of meristematic cells is appropriately called the **meristoderm.** The cells forming the cortex are parenchyma-like. Mucilage-secreting cells form definite canals throughout the cortex. The **medulla** or midrib contains mechanical or supporting cells as well as sieve cells. The latter have sieve plates, form callose (Fig. 22.1G), and strongly resemble sieve cells and sieve-tube members of the vascular plants. Work with ^{14}C indicates that the carbohydrate mannitol is translocated and that the mechanism of transport resembles that found in vascular plants.

The stipe of many Laminariales is perennial, but the meristems are not active during winter months. Because of this, annual rings may sometimes be observed in cross-sections of stipes of large kelps. Meristematic activity is also found in a transition region between the stipe and lamina. Here, cell formation adds new tissue to both stipe and lamina (Fig. 22.7B). Growth in larger Phaeophyta may be apical, arising from one or several apical cells.

The problems of polarity of cell division (Fig. 6.8) in the Phaeophyta have been studied experimentally in the germination of the zygote of *Fucus*. The first division of the zygote is preceded by formation of a small protuberance in the zygote cell wall. The plane of the first division is at right angles to this protuberance (Fig. 22.9A). This division is unequal; the larger of the two cells formed receives most of the chloroplasts and develops into a vegetative and reproductive thallus. The smaller cell, with fewer chloroplasts, first forms a slender rhizoid that later develops into a stripe and holdfast (Fig. 22.9A).

The protuberance in the zygote cell wall develops in response to blue light, lower pH, or warmth. The plane of the first cell division of the zygote appears to be determined by purely physical factors. It does not seem to be under genetic or nuclear control. Polarity does not seem to play a prominent part in the next few cell divisions, for the successors of the larger cell simply form a spherical cell mass. However, this sphere soon elongates into a pear-shaped mass of cells, thus indicating that a degree of polarity in cell divisions has been established (Fig. 22.9A). Soon a small depression appears in the top of the young pear-shaped thallus. A group of from one to eight apical cells forms at the bottom of this depression (Fig. 22.9B). The young thallus now becomes flat, and cells in the apical region follow each other in an orderly sequence to produce the flat strap-shaped dichotomously branching *Fucus* plant body (Fig. 22.1H). In spite of a fairly good knowledge of the sequence of events from zygote to the mature thallus, the basic questions of why and how cells come to divide in such precise planes still require explanation.

Rhodophyta. Early evolution of diversity of form in Rhodophyta proceeds along lines similar to that

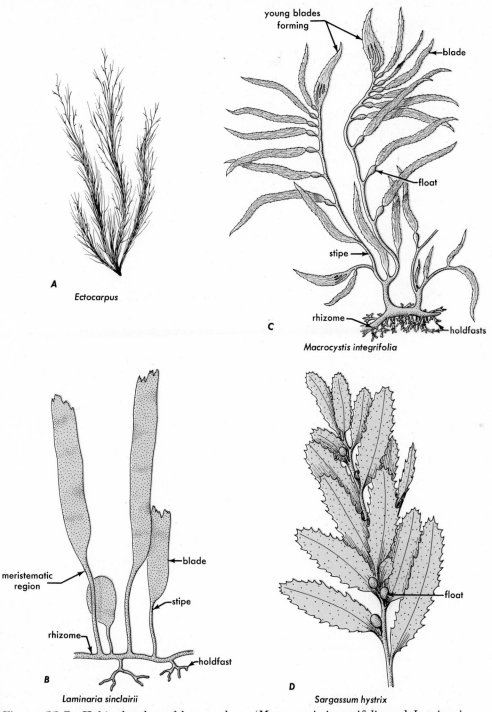

Figure 22.7 Habit sketches of brown algae. (*Macrocystis integrifolia* and *Laminaria sinclairii* redrawn after G. M. Smith, *Marine Algae of the Monterey Peninsula*, second edition, incorporating supplement by G. J. Hollenberg and I. A. Abbott. Copyright 1969 by Stanford University Press. *Ectocarpus* redrawn from W. R. Taylor, *Marine Algae of the Northeastern Coast of North America*, University of Michigan Press. *Sargassum hystrix* redrawn after W. R. Taylor, *Marine Algae of the Eastern Tropical and Subtropical Coasts of the Americas*, University of Michigan Press.)

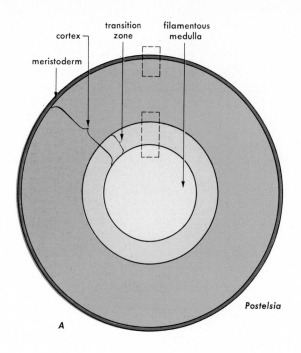

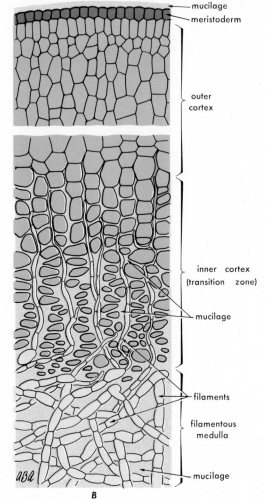

Figure 22.8 Cellular organization of blade and stipe of *Postelsia*. *A*, cross-section of stipe showing different tissue regions, ×25. *B*, enlarged portion of cross-section of stipe showing cellular detail, ×2500. *C*, cross-section of blade showing different tissue regions, ×50. *D*, enlarged portion of blade showing cellular detail, ×450.

already described for the Chlorophyta. The more primitive subclass (Bangiophycidae) contains unicellular species (*Porphyridium,* Fig. 22.1*D*), unbranched filamentous forms (*Erythrotrichia*), and branching filamentous forms (*Goniotrichum*). There are also species whose thalli consist of single layers of cells bearing several upright threads (*Erythrotrichia*), while in still other forms the thalli have two cell layers (*Porphyra variegata*) forming blades similar to *Ulva* (Fig. 2.7*B*). Several red algae are shown in Fig. 22.10.

Morphological diversity in the much larger and more complex subclass Florideophycidae is attained by an aggregation of filaments. The simplest algae in the subclass are much branched, small, rather delicate forms. They have developed the heterotrichous habit, as in the brown algae, but this type of growth disappears with increasing morphological

complexity. In more complex forms, filaments are compacted into a parenchyma-like mass of cells, and the plant body consists of a dissected blade, stipe, and holdfast such as exhibited by *Gigartina*.

Growth is almost always apical and, as in all other species, morphological form is related, among other things, to cellular polarity. This is well-demonstrated in *Polysiphonia* (Fig. 22.11). *Polysiphonia* has a single apical cell. Cell division with the mitotic spindle parallel to the long axis of the filament results in an increase in length of the filament (Fig. 22.11*A*). Cell division at right angles to the axis of the filament produces an array of five pericentral cells, precisely arranged around the central axial cell (Fig. 22.11*B*). This means that the planes of division giving rise to the pericentral cells are also precisely oriented to each other. Furthermore, the order of production of the new pericentral cells is the same in all plants:

Cells two and three are formed on opposite sides of cell one. Cell four forms beside two, and cell five forms between cells three and four (Fig. 22.11*B*).

Polarity at an angle to the main axis of the filament results in lateral branches (Fig. 22.11*A*). In this case, the daughter nuclei in the dividing apical cell end up in opposite corners of the cell, and the cell plate cuts the dividing cell at an angle to the axis of the filament. Normal polarity is restored in the apical cell. Polarity in the second cell is now established at an angle to the main filament, and a protuberance forms from this cell in line with this newly established polarity (Fig. 22.11*A*). Subsequent cell divi-

sions are at right angles to this polarity, cutting off successive cells to form a lateral branch.

Thus, the mature thallus of *Polysiphonia* (Fig. 22.1*I*) is a branching thallus formed by a line of central or axial cells surrounded by a regular number of pericentral cells (Fig. 22.11*B*). Axial and pericentral cells are the same length. A comparable structure occurs in many of the Florideophycidae. Morphological variation, which is considerable in the subclass, results from the number of cells forming the axis, variation in the size of cells, and the degrees of branching. Thus, the parenchyma-like body is basically filamentous in origin.

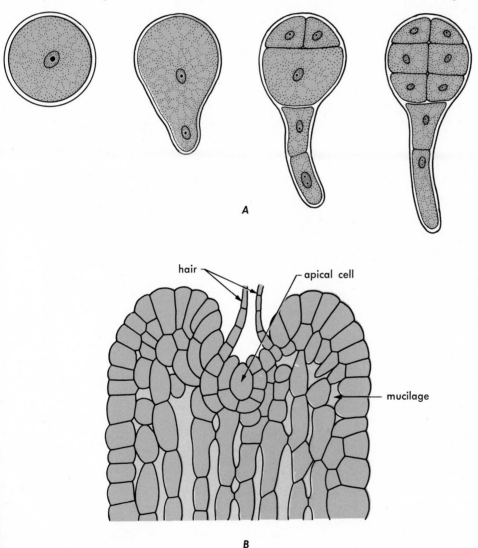

Figure 22.9   Growth in *Fucus*. *A*, germination of zygote. The polarity of the first division is determined by environmental factors such as light. *B*, apical region of blade showing notch with meristematic apical cell.

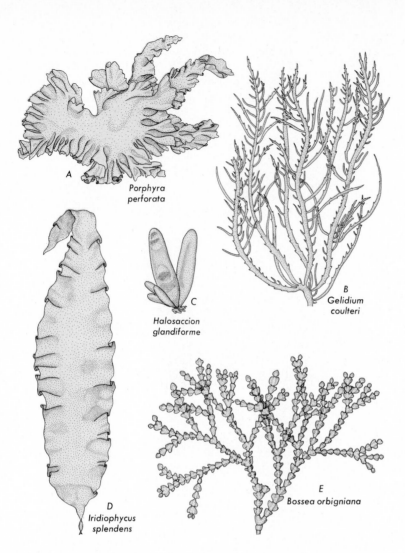

A
*Porphyra
perforata*

B
*Gelidium
coulteri*

C
*Halosaccion
glandiforme*

D
*Iridiophycus
splendens*

E
Bossea orbigniana

Figure 22.10 Habit sketches of
red algae. *A*, ×$\frac{7}{8}$; *B*, ×3; *C*, ×$\frac{7}{10}$;
D, ×$\frac{1}{3}$; *E*, ×2$\frac{1}{4}$. (Redrawn after
G. M. Smith, *Marine Algae of the
Monterey Peninsula*, second
edition, incorporating supplement
by G. J. Hollenberg and I. A.
Abbott. Copyright 1969 by
Stanford University Press.)

Summary of Morphological Diversity in the Algae

Morphological diversity, which is very considerable in the algae, is attained with thalli consisting of (*a*) single cells (sometimes motile), (*b*) simple colonies of single cells, (*c*) unbranched filaments, (*d*) branched filaments, (*e*) heterotrichous filaments, (*f*) intertwining filaments, and (*g*) a simple parenchymatous tissue. Most divisions exhibit all or significant portions of this range of diversity, and it may be repeated in several families. Examples of divisions not showing the full range of diversity include the Euglenophyta, which consists only of unicells, many of which are motile, the unicellular Rhodophyta, which are not motile, and the Phaeophyta, which contains no unicellular forms. The simplest filaments of the Phaeophyta have developed the heterotri-

chous habit, and the most advanced Rhodophyta show extensive intertwining of filaments resembling parenchyma. This latter tissue is, however, present in some of the primitive red algae.

Two conclusions can be drawn from these observations: (*a*) a high degree of morphological diversity can be obtained by different arrangements of single cells, filaments of cells, and sheets of cells; (*b*) the evolutionary pathway—single cells to parenchyma tissue with localized growth regions, or parts of this sequence—occurred numerous times in most of the algal divisions.

It is also important to note that divisions thought to have, at best, a very remote relationship with higher plants, followed evolutionary pathways leading to parenchyma tissue, apical growth, intercalary meristems, and meristems giving rise to increase in girth—all things found in present-day vascular

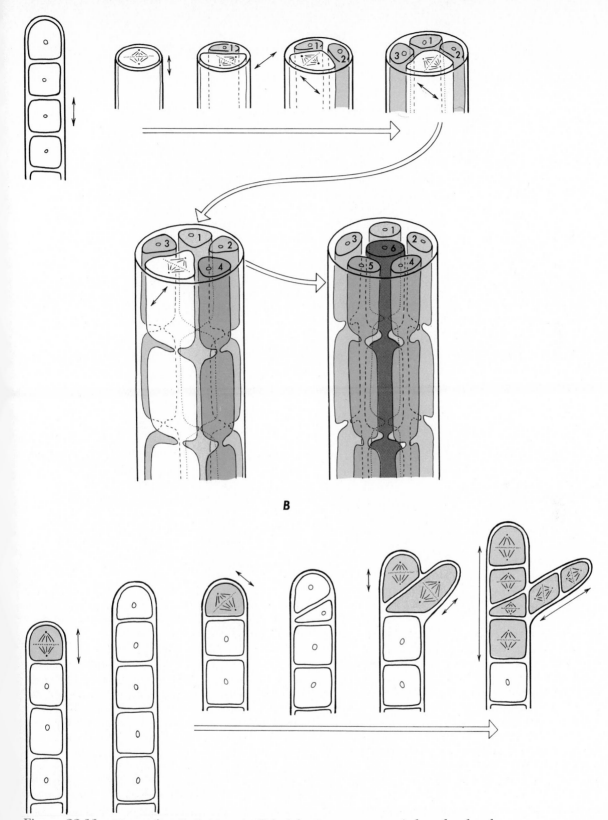

B

Figure 22.11 The early cell divisions in *Polysiphonia* are very precisely ordered and easily followed. They are responsible for the organization of the mature thallus. *A*, cell divisions only parallel to the axis of the filament would result in a uniseriate filament. Regularly spaced divisions at an angle to the axis produce regular branches. *B*, divisions at right angles to the axis of the filament, always in the order indicated, result in a thallus formed of a central cell surrounded by five pericentral cells. Arrows indicate polarity of cell divisions.

plants. Organs developed to anchor plants in place: specialized flattened laminae involved in photosynthesis, and specialized transport cells similar to sieve cells, developed in long stipes of kelps, whose holdfast is submerged and some distance from photosynthetic laminae.

Reproduction—Definition

Details of reproduction as it occurs in higher plants have been discussed in the chapters on the cell (Chap. 4), flower (Chap. 14), fruit (Chap. 15), and on inheritance (Chap. 17). In general, the higher plants reproduce (a) sexually through seeds and (b) vegetatively through cuttings, grafts, and other asexual processes. In fungi and algae, we may distinguish between three general types of reproductive processes, asexual reproduction (a) by vegetative means and (b) by specialized cells formed after mitosis (**mitospores**), and (c) sexual reproduction. In strictly vegetative reproduction, any portion, or specialized portions, of a vegetative plant may separate from the parent plant to form new individuals. We have seen examples of this in rooted cuttings (p. 385) and the nucellar buds in oranges (p. 282). As chromosomes pass unchanged from one individual to the next, succeeding generations have identical genotypes. Asexual reproduction, in fungi and algae, also occurs through the formation of spores or single cells specialized in the replication of new individuals, while preserving unchanged, the genetic makeup. Thus succeeding generations have identical genotypes. Both types of asexual reproduction commonly occur under conditions favorable for growth, and result in a rapid expansion of a given population. Sexual reproduction involves the union of two gametes, thus doubling the chromosome number. Meiosis must thus occur at least once in the life cycle of every sexually reproducing organism. The random distribution of chromosomes at anaphase I of meiosis and the recombination of genes in chromosomes during meiosis results in each individual of the progeny of a single mating having its own unique genotype. Sexual reproduction in the fungi and algae is commonly associated with conditions less favorable to growth, and in many instances the cells resulting from the union of gametes develop resistant walls that carry the species over periods unfavorable to growth. It will be recalled that, in the same way, seeds of most gymnosperms and angiosperms are adapted to remain viable for long periods of dryness, and that they too result from processes of sexual reproduction.

Knowledge of reproductive processes in the algae is incomplete and largely confined to a relatively few species in the Rhodophyta, Phaeophyta, and Chlorophyta. In some other divisions, detailed knowledge is known only for a very few species, genera, or families.

Asexual Reproduction: Vegetative

Vegetative reproduction is widespread in many forms and probably occurs in all divisions. The longitudinal splitting of a biflagellate motile *Euglena* into two, still motile, but uniflagellate daughter cells is simple vegetative reproduction. The diatoms may divide vegetatively for upwards of five years. In this case, the cell walls contain much silica. The cells are flat and their walls have two sections that fit into each other like the top and bottom of a petri dish. After division, a new wall forms within the old wall (Fig. 22.12A). This means that the new cell receiving the lower and smaller wall will be smaller than the parent cell. Eventually, a minimum size is reached and division stops. Full size can only be regained by sexual reproduction (Figs. 22.12B, C).

In still another type of vegetative reproduction exhibited by many Chlorophyta, a new, fully constituted young thallus is formed entirely within a parent cell. In *Scenedesmus* the parent plants consist of four flattened, linearly arranged cells (Figs. 22.3A, B). Two mitoses occur simultaneously in each of the four cells. Within each of these cells, a new four-celled colony is now formed. These fully formed young colonies, replete with an intricate surrounding net, are released simultaneously from each cell of the parent colony.

Vegetative reproduction in filamentous forms occurs by a simple fragmentation of the filament. In some species, specialized cells do seem to be associated with breaking of the thallus. In species of the Cyanophyta, notably *Oscillatoria* (Fig. 22.2G), the point at which fragmentation will occur is marked by a dead intercalary cell. Separation at these dead cells results in short filaments that have a gliding motion. This enables them to change their position in the mucilaginous mass in which the filaments grow, and to gradually enlarge the mass. Many filamentous species of Cyanophyta have occasional enlarged living cells (**heterocysts**). When located within a filament (*Nostoc*, Figs. 22.2C, D), they appear to be associated with the fragmentation of the filaments and nitrogen fixation.

Vegetative reproduction is probably common in the heterotrichous forms in which the prostrate system may behave in the manner of the stolons of higher plants (Fig. 7.26). In some species of Chlorophyta and Phaeophyta, specialized **propagules** consisting of small masses of meristematic cells serve

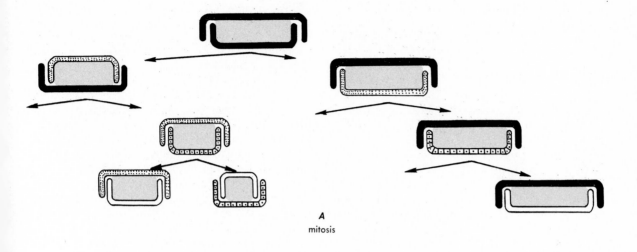

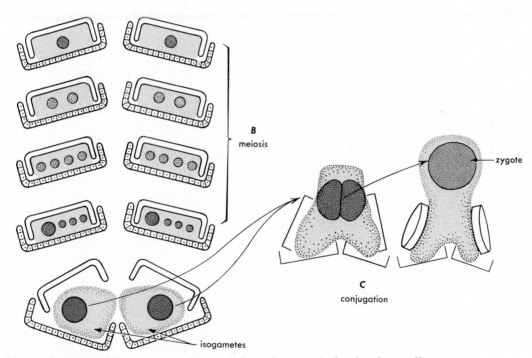

Figure 22.12 A gametic life cycle as represented by a diatom. *A*, the daughter cells continually become smaller if, after cell division, the new cell forms within the silica shell of the old cell. *B*, meiosis occurs; three of the nuclei with the *n* number of chromosomes degenerate, and the single remaining haploid cell becomes a gamete. *C*, fusion of the two gametes to form the zygote. (*A* and *C*, redrawn after G. M. Smith, *Freshwater Algae of the United States.* Copyright 1933 by McGraw-Hill Book Company, New York. Used with permission of McGraw-Hill Book Company.)

as a means of vegetative reproduction. One of the most remarkable cases of fragmentation is exhibited by members of the family Sargassaceae, particularly the genus *Sargassum* (Fig. 22.7*D*). This genus is a floating kelp covering thousands of acres in the tropical Sargasso Sea. Fragmentation of its large thallus is apparently its only means of reproduction.

Asexual Reproduction: Mitospores

Asexual spores are always preceded by mitosis and may be called **mitospores** (Figs. 22.14, 22.15) to distinguish them from spores formed after meiosis, or **meiospores** (Figs. 22.14, 22.16). The latter constitute a stage in sexual reproduction and will not be considered here. Since only mitotic divisions are involved, the genotypes of all asexual spores arising from one and the same parent are identical with each other and with the parent plant. They will produce a population of genetically identical individuals. When a population has the genotype best fitted to a given growing condition, or locality, it will multiply by asexual reproduction and, frequently explosively, populate that locality.

Mitospores may be either motile or nonmotile. Motile spores are called **zoospores** (Figs. 22.14*B*, 22.15), and nonmotile mitospores are frequently called **aplanospores** or, in the case of *Polysiphonia*, **carpospores** (Fig. 22.19). They are usually borne in specialized cells called **sporangia.** In many algae, the sporangia show little, if any, difference in appearance from ordinary vegetative cells (Figs. 22.14*B*, 22.15). The number of zoospores produced depends upon the species, being one for *Vaucheria* and several hundred for *Cladophora*. Most sporangia produce 16–64 zoospores. In *Ectocarpus* (Fig. 22.15), the mitospores are produced in characteristic sporangia. These sporangia are divided into many cells and each cell produces a single spore. Most of the Rhodophyta produce nonmotile spores. Here, the process is complex and variable. In general terms, short filaments arise from the zygote and, together with short filaments from nearby cells, form a structure known as a **cystocarp** (Fig. 22.19*D*). Mitospores or **carpospores** are cut off from the ends of filaments that take their origin from the zygote.

Most zoospores are pear-shaped or spherical (Figs. 22.13, 22.14*B*). Depending upon the species, they

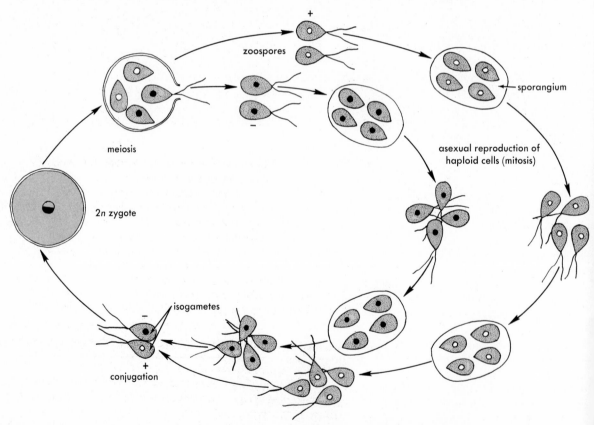

Figure 22.13   Zygotic life cycle as represented by the green alga *Chlamydomonas*. Meiosis occurs during the germination of the zygote; it is the only diploid, or sporophyte, cell. (Redrawn after F. Moewus.)

438   **Algae**

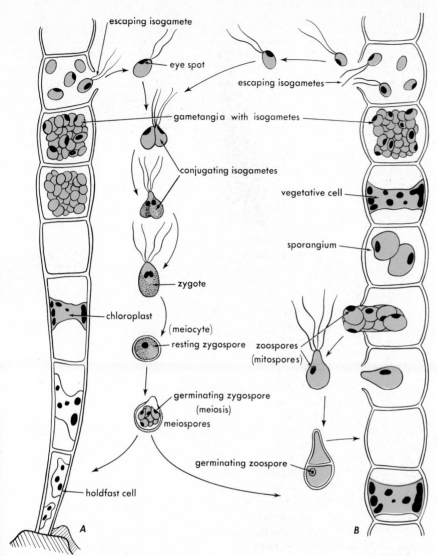

Labels in figure:
- escaping isogamete
- eye spot
- escaping isogametes
- gametangia with isogametes
- conjugating isogametes
- vegetative cell
- sporangium
- zygote
- chloroplast
- (meiocyte)
- resting zygospore
- zoospores (mitospores)
- germinating zygospore (meiosis)
- meiospores
- germinating zoospore
- holdfast cell
- A
- B

Figure 22.14 Zygotic life cycle as represented by the green alga *Ulothrix*. Meiosis occurs in the germination of the zygote; it is the only diploid, or sporophyte, cell.

have two, four, or whorls of many flagella. After liberation from the sporangium, zoospores are actively motile for as short a time as 4 minutes (*Pediastrum,* Fig. 22.4*A*) or for as long as 3 days (*Ulothrix,* Fig. 22.14*B*). Their movement and periods of activity are frequently affected by light. After their period of activity they settle to the bottom of a pond or culture tank, lose their flagella, and secrete a cell wall (Fig. 22.14*B*). Development of the new thallus is now initiated (Fig. 22.14*B*).

Sexual Reproduction

Sexual reproduction, involving as it does the segregation of genetic traits, complements asexual re-

production, in that sexual reproduction is responsible for variation within a population. Since sexual reproduction is frequently associated with resistant cells that carry the plants over seasons unfavorable for growth, the new plants formed in a given population by sexual reproduction will, upon the resumption of growth, all have different genotypes, and those best adapted for the new growing conditions will become established.

The cells that fuse in sexual reproduction are known as **gametes** (Fig. 22.17), and the single cell resulting from the fusion is a **zygote** (Fig. 22.17). Gametes have the **haploid** or *n* number of chromosomes, while the zygote has the **diploid** or 2*n* number. The number of chromosomes is returned to the haploid, or *n*, number through meiosis. The cells resulting from meiosis, with the *n* number of chromosomes, may be called **meiospores.** Gametes are produced in cells known as **gametangia** (Figs. 22.14*A*, 22.15*A*). Meiosis occurs at different places in the sexual cycle, and the cells in which it occurs have been called **meiocytes.** Plants that produce gametes are known as **gametophytes,** and those that produce meiospores are called **sporophytes.** It follows that

gametophytic plants are haploid and sporophytic plants are diploid (Figs. 22.12–22.19).

A gametangium, in the most unspecialized case, may be indistinguishable from an ordinary vegetative cell, except that several additional mitoses occur within it, giving rise to 16–32 flagellated cells (Figs. 22.13, 22.14*A*). In the unicellular, motile *Chlamydomonas*, these cells strongly resemble the vegetative cells, but they may be smaller (Fig. 22.13). In the

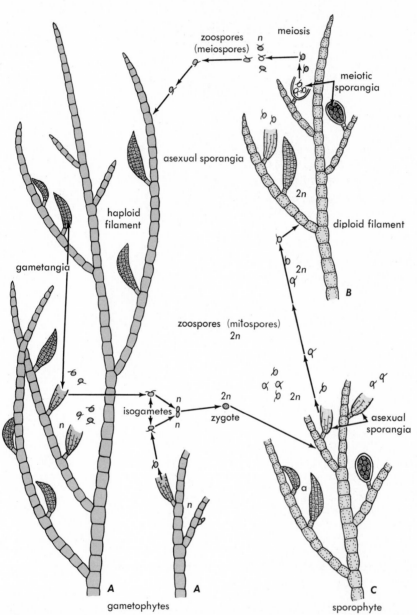

Figure 22.15 Sporic life cycle, alternation of isomorphic generations, differing only in having diploid and haploid chromosome numbers, as shown by *Ectocarpus*.

440 **Algae**

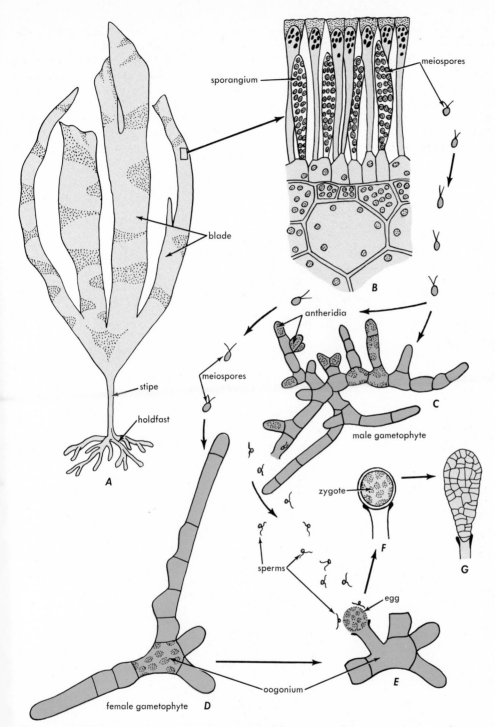

Figure 22.16 Sporic life cycle, alternation of heteromorphic generations, as shown by the brown alga *Laminaria*.

simple filamentous species *Ulothrix* (Fig. 22.14), an ordinary vegetative cell gives rise to 8–64 flagellated cells, again resembling the motile vegetative *Chlamydomonas* cells. When only 8 motile cells are formed, they germinate directly, acting simply as zoospores; and the cell in which they are formed is called a **zoosporangium.** However, when a greater number of cells is formed, the cells are smaller. These smaller

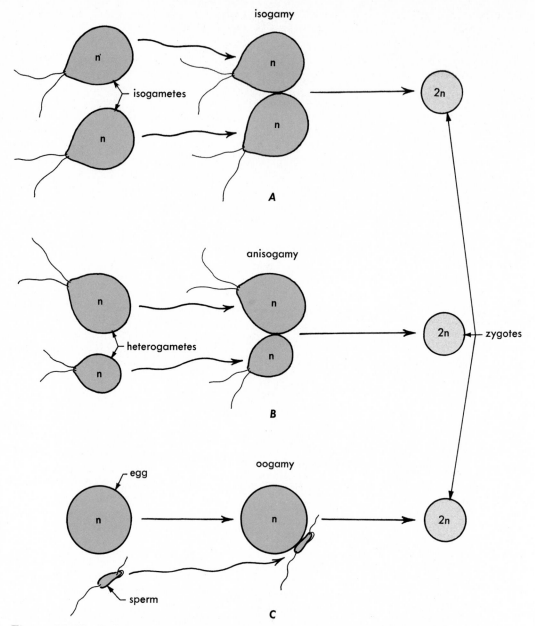

Figure 22.17 Differences in gametes. *A*, isogamy, the gametes are identical; *B*, anisogamy, the gametes are slightly different in size or activity; *C*, oogamy, the gametes are very different in size and activity—they are eggs and sperms.

cells behave as gametes and fuse to form a zygote. When gametes are alike in appearance, they are called **isogametes,** and the species producing them are said to be **isogamous** (Fig. 22.17A). Thus, *Chlamydomonas, Ulothrix,* and *Ectocarpus* exhibit isogamy. The union of isogametes is known as **conjugation.** In some species, the gametes differ in size, one being slightly smaller than the other. This situation occurs in two close relatives of *Chlamydomonas.* In

Gonium (Fig. 22.5), large gametes fuse with small gametes. Occasionally in *Gonium,* similar-sized gametes have been observed to fuse. Gametes having only a slight size difference are said to be **anisogametes** (heterogametes), and species expressing this state are **anisogamous species** or exhibit **anisogamy** (heterogamy) (Fig. 22.17B). In *Volvox,* chlamydomonad cells are joined in a large hollow sphere (Fig. 22.5A). A few of these cells undergo rapid divisions, producing up to 512 slender biflagellate sperms. A sperm-producing gametangium is known as an **antheridium** (Fig. 22.16). Another vegetative cell may become quite enlarged, sometimes becoming flask-shaped. The protoplast retracts slightly from the cell wall. It has become an **egg cell,** and the cell in which it is produced is known as an **oogonium** (Fig. 22.16E). The sperms are liberated from the antheridium and swim toward the oogonium. One sperm enters the oogonium, either through the funnel end or by a breaking down of the oogonial cell wall. Union of egg and sperm now ensues and the process is known as **fertilization** (Fig. 22.18Q). When gametes differ in size and activity as they do in *Laminaria* (Fig. 22.16), *Fucus* (Fig. 22.18), and *Polysiphonia* (Fig. 22.19), the plants exhibit **oogamy** (Fig. 22.17C).

How does a sperm cell find an egg cell? It could happen by chance, as wind-blown pollen lands on, among other things, stigmas of flowers of its own species. And how are gametes able to recognize each other, perceiving not simply that they are members of the same species, but also that they have been produced by different individuals or by different strains of the same species? There is considerable evidence that the female gametangia or gametes in plants produce chemical agents that seduce the sperms to the close proximity of egg cells. For instance, a drop of a homogenate prepared from the mature female filaments of the alga *Oedogonium* may be drawn up into a capillary tube. When the tube is placed in a suspension of sperms, the sperms will collect at the tip of the capillary tube and some will eventually enter the tube. This attraction seems to be species-specific for *Oedogonium*. This is important, because numerous general attractants for sperm cells are known. Most, if not all, cells probably have the ability to recognize each other and to accept or reject each other. Some pollen tubes will grow only in styles of their own species. Species, and even different strains, must be taken into account when budding or grafting. And as has been recently demonstrated, human tissues frequently reject transplanted organs from other humans.

Alternation of Generations

Fertilization constitutes one of the two critical steps of a complete sexual life cycle; meiosis is the second critical step. We may distinguish three types of life cycles, depending upon the relation of meiosis to fertilization: (a) In most animals, meiosis results in gamete formation, so the gametes are haploid. They are the equivalent of meiospores. In this case, the mature individuals are diploid gameto-

phytes and the haploid stage is limited to the gametes. This is a **gametic life cycle.** It occurs in the diatoms (Bacillariophyta, Fig. 22.12) and in the Codiales of the Chlorophyta. (b) In many of the more primitive algae, meiosis accompanies the germination of the zygote. In this case, the mature plants are haploid and the diploid stage is limited to the zygote. This is a **zygotic life cycle.** It occurs in many Chlorophyta (Figs. 22.13, 22.14) and few primitive Rhodophyta. (c) In most plants, meiosis and fertilization are separated by distinct generations; one is haploid and produces gametes; the other is diploid and produces meiospores. This is a **sporic life cycle** (Figs. 22.15, 22.16, 22.18, 22.19).

In each of these cases, haploid and diploid phases have alternated with each other. In a gametic life cycle, the dominant generation is diploid; meiosis precedes the formation of gametes, which are the only haploid cells in the life cycle. In a zygotic life cycle, the dominant generation is haploid; the zygote constitutes the diploid generation and meiosis accompanies its germination. In a sporic life cycle, mitoses occur in both haploid and diploid generations, giving rise to vegetative plants in both generations. The plants of the sporophyte and gametophyte generations may be identical in appearance. This is **alternation of isomorphic generations.** It occurs in some Chlorophyta, primitive Phaeophyta, and all higher Rhodophyta (Figs. 22.15, 22.19). In **alternation of heteromorphic generations,** the haploid plants of the gametophyte generation have different forms from the diploid plants of the sporophyte generation. This occurs in a few Chlorophyta and in all of the more advanced Phaeophyta (Figs. 22.16, 22.18).

Gametic Life Cycle. The diatoms (Figs. 22.3C, 22.12) are unicellular forms that may divide by cell division for periods as long as 5 years. With each division, one of the daughter cells is smaller than the parent cell (Fig. 22.12A). At the end of a period of mitotic divisions, meiosis occurs and four gametes are produced, not all of which may function (Fig. 22.12B). Depending upon the species, these most generally are isogametes, but they may be anisogametes or even eggs and sperm. After fusion, the zygote (Fig. 22.12C) increases greatly in size and, again depending on the species, may enter upon a rest period. The germination of the zygote is by mitotic cell division. Note that in the case of the diatoms, meiosis results, not in meiospores, but in gametes. The diploid generation may then be considered as a gametophyte and the gametes themselves are the only haploid cells.

Zygotic Life Cycle. *Chlamydomonas* and *Ulothrix* are examples of a zygotic life cycle (Figs. 22.13,

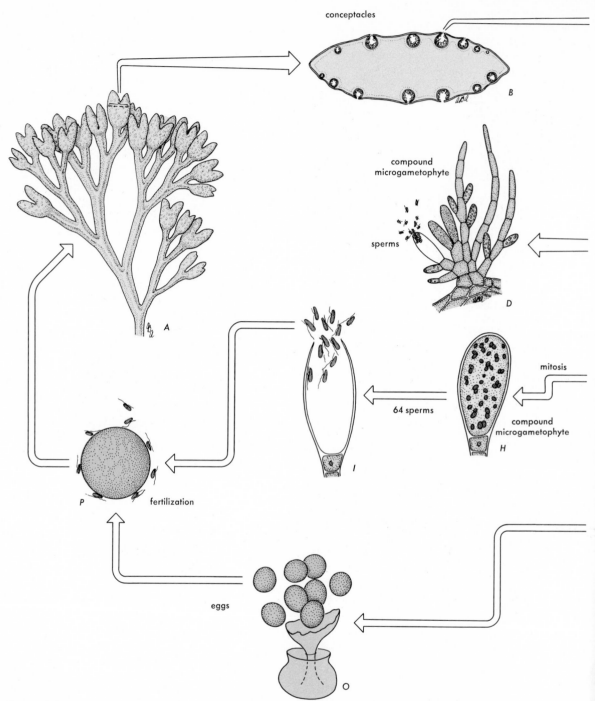

conceptacles

B

compound
microgametophyte

sperms

D

mitosis

64 sperms

compound
microgametophyte

H

I

P   fertilization

eggs

O

Figure 22.18 Sporic life cycle, as shown by *Fucus. A,* dichotymously branching blade
with swollen tips, ×1. *B,* cross-section of tip showing cavities, or conceptacles, ×8. *C,* the
conceptacles contain paraphyses, short, branching filaments bearing many small
(antheridial) meiocytes, and large (oogonial) meiocytes on short stalks, ×120. *D,* enlarged
view of filaments bearing sperm. *E,* enlarged view of the oogonial meiocyte after meiosis. *F*
and *G,* meiosis in male meiocyte. *H,* mitotic division resulting in 64 sperms. *I,* sperms
leaving antheridium. *J, K,* and *L,* meiosis in the oogonial meiocyte. *M,* a single mitosis

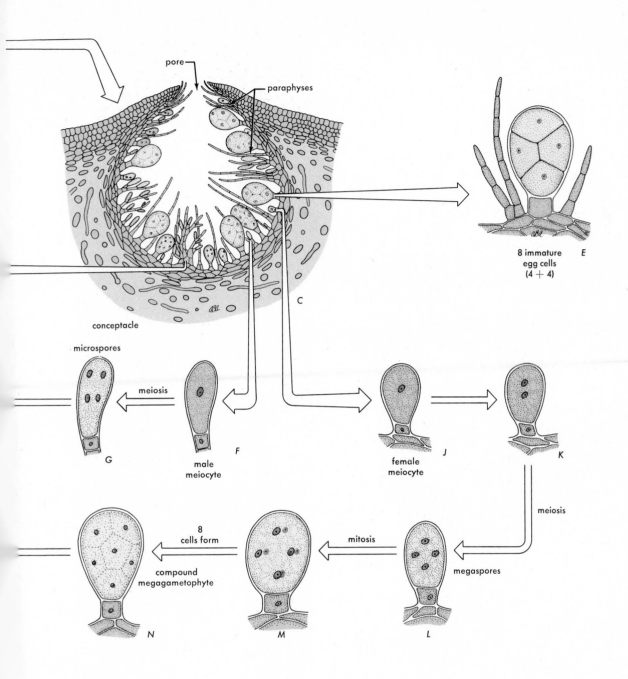

pore

paraphyses

conceptacle

C

8 immature
egg cells
(4 + 4)

E

microspores

meiosis

male
meiocyte

G

F

female
meiocyte

J

K

meiosis

megaspores

8
cells form

mitosis

compound
megagametophyte

N

M

L

following meiosis. *N* and 8 egg cells form. *O*, the egg cells leave the oogonium. *P*, eggs and sperms pass out into the sea water where fertilization takes place, ×2000. There is no free-living gametophyte generation. Magnification, *D–P*, inclusive, about ×850. (*F* through *P*, adapted from Robert F. Scagel et al., *An Evolutionary Survey of the Plant Kingdom.* Copyright 1965 by Wadsworth Publishing Company, Belmont, Ca. Reprinted with permission of the publisher.)

22.14). In each of these cases, two haploid gametes fuse to produce a diploid zygote. Meiosis occurs in the germination of the zygote, which is the only diploid cell.

Sporic Life Cycle: Alternation of Isomorphic Generations. Two genera, *Ulva* (Chlorophyta) (Fig. 2.7*B*) and *Ectocarpus* (Phaeophyta) (Fig. 22.7*A*), will serve to illustrate alternation of isomorphic generations. *Ulva*, allied to *Ulothrix,* has a distromatic thallus, while *Ectocarpus* is a branched filamentous form. Both are marine, growing in the intertidal zone.

Ulva. In *Ulva,* any cell of the haploid gametophyte thallus, except those near the base, may produce gametes. After conjugation of the gametes, the zygote grows into a sporophyte thallus identical to the gametophyte except for the diploid chromosome number. Again, meiosis may take place in any cell of the sporophyte thallus. In some *Ulva* species, the gametes, zygotes, and meiospores are similar in appearance, except that the gametes have two flagella, while the zygotes and meiospores have four flagella.

Ectocarpus. In *Ectocarpus* (Fig. 22.15), specialized reproductive cells are formed. The gametangia are grouped together, forming elongated, slightly curved structures. Each cell in this structure is a gametangium that has resulted from mitotic divisions and will produce one motile gamete with the haploid chromosome number. They are unable to germinate, but conjugate to form the usual diploid zygote (Fig. 22.15*A*). The zygote, upon germination, will grow into a diploid sporophyte (Fig. 22.15*C*) whose thallus is identical in appearance to that of the gametophyte thallus. Two different kinds of sporangia are formed on this sporophyte plant. One of these resembles exactly the group of gametangia found on the gametophyte plant. Like the gametangia, all sporangium cells have been formed through mitotic divisions, and each cell will produce a motile diploid zoospore that can germinate directly into a new diploid plant (Fig. 22.15*B*). The other type of sporangium found on the sporophyte plant is spherical; it will produce, after meiosis, numerous motile zoospores or meiospores with the haploid chromosome number (Fig. 22.15*B*). These meiospores grow into new gametophyte plants (Fig. 22.15*A*). Haploid gametes, halploid meiospores, and diploid zoospores are identical in appearance.

Polysiphonia. In the Rhodophyta, an alternation of isomorphic generations is complicated by a second, distinctive sporophyte phase. The delicate feathery thalli of male and female gametophytes and tetrasporophyte of *Polysiphonia* are identical in appearance and growth (Fig. 22.1*I*). No motile cells are produced. The nonmotile sperms (**spermatia**) are produced in great profusion near the tips of male gametophyte filaments (Fig. 22.19*C*). The oogonia (**carpogonia**) also form near the tips of the female gametophyte. The oogonia are flask-shaped cells with long slender necks (**trichogynes**) that protrude from a protecting envelope formed from the cells at the bases of the oogonia (Fig. 22.19*D*). Spermatia are carried by water currents to make contact with the necks of the oogonia. Nuclei enter the necks and pass downward to unite with the female nuclei.

Fertilization stimulates the development of an array of short diploid filaments (the **carposporophyte**), which will eventually produce diploid mitospores (**carpospores**). The original protective envelope of haploid filaments is also stimulated to grow. It enlarges and gives rise to a surrounding protective case formed of haploid filaments (**cystocarp,** Fig. 22.19*E*). Thus, the first sporophyte generation (carposporophyte) in *Polysiphonia* and many other Rhodophyta is composed of short diploid filaments that produce diploid spores. This sporophyte is protected by an envelope of gametophyte cells. When discharged, the diploid carpospores give rise to sporophyte thalli. (**tetrasporophyte**) identical in appearance to the two gametophyte plants. Meiosis occurs in sporangia in the sporophyte plant in cells formed between rows of axial and pericentral cells (Fig. 22.19*G*). Each sporangium produces four meiospores (**tetraspores**) which upon germination give rise to male or female gametophyte plants similar in appearance to the sporophyte plants.

Sporic Life Cycle: Alternation of Heteromorphic Generations

Kelp. Alternation of heteromorphic generations occurs in both the Chlorophyta and Phaeophyta but is most highly developed in the Phaeophyta, particularly in the kelps or Laminariales. The sporophyte generation of the kelps is the larger and better known. Its vegetative characteristics have been discussed in some detail earlier in the chapter (see also Fig. 22.7*B*). The sporangia arise from meristematic cells, or meristoderm, that form the outermost layer of cells on the lamina (Figs. 22.8*C,* 22.16*B*). The sporangia usually arise in groups or **sori** and are accompanied by an elongation of adjacent cells into paraphyses that serve to protect the developing sporangia (Fig. 22.16*B*). Depending on the species, each sporangium produces 8–64 motile meiospores.

The meiospores germinate, generally, into male or female gametophyte plants, both of which consist of a small branched filament (Figs. 22.16*C, D*). The antheridia are produced in large numbers, either

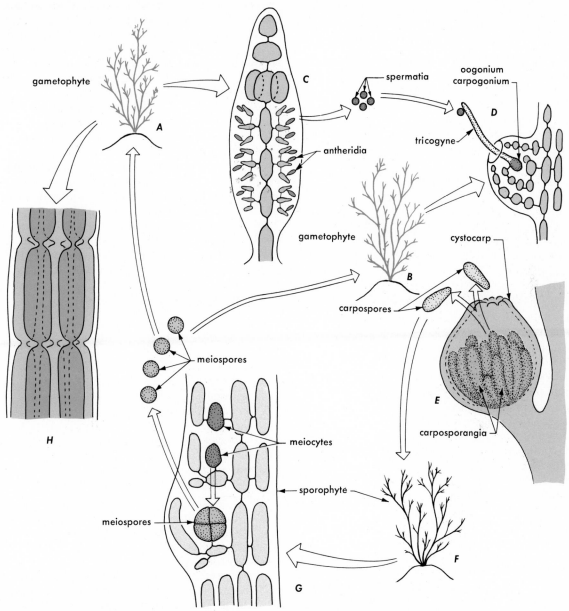

Figure 22.19 Sporic life cycle, alternation of isomorphic generations, as represented by the red alga *Polysiphonia*. *A*, male gametophyte plant. *B*, female gametophyte plant. *C*, production of spermatia at tip of male gametophyte. *D*, the carpogonium (oogonium) is protected by filaments, and spermatia become attached to the protruding tricogyne. *E*, the zygote produces a body of 2*n* carpospores. The gametophytic cystocarp enlarges and surrounds the carposporophyte. *F*, a sporophyte, identical in appearance to the gametophyte, arises from the carpospores. *G*, some pericentral cells give rise to meiocytes which, through meiosis, produce four meiospores, or tetraspores. *H*, the vegetative structure of both gametophytes and the sporophyte are identical. (Adapted from Robert F. Scagel et al., *An Evolutionary Survey of the Plant Kingdom.* Copyright 1965 by Wadsworth Publishing Company, Belmont, Ca. Reprinted with permission of the publisher.)

singly or in groups, at the ends of short branches. There appear to be no specialized oogonia, as eggs may be produced in any of the cells of the female gametophyte (Fig. 22.16*E*). The eggs are usually discharged from the oogonium before fertilization, but remain attached to it for some time. The egg is thus fertilized outside of the oogonium in open sea water.

Fucus. Alternation of heteromorphic generations has reached its highest stage of advancement in *Fucus* (Fig. 22.18). The diploid generation is the conspicuous one (Fig. 22.1*H*), and the haploid phase has been reduced to a few haploid cells. These do not develop into free-living vegetative gametophytes but become transformed into sperms and eggs and fuse to produce a new diploid generation (Fig. 22.18).

The ends of the diploid strap-shaped thallus of *Fucus* are swollen and notched. They bear small raised areas (Fig. 22.18*A*). A section through these areas shows them to be small cavities (**conceptacles**) that open by small conical pores to the sea water (Fig. 22.18*B*). The gametangia arise in these cavities. In Atlantic coast species of *Fucus*, the male and female gametangia occur on separate plants, but in the Pacific coast *Fucus*, they arise in the same cavities. The antheridia are formed profusely at the ends of short branching filaments and the oogonia are large spherical cells separated from the wall of the cavity by a single cell (Fig. 22.18*C*). Sterile hairs, or paraphyses, surround the gametangia, and may protrude outward through the pore.

The initial antheridial cell contains a single large nucleus (Fig. 22.18*F*). Meiosis takes place, and four haploid cells (microspores) are produced. This is followed by four rapid mitoses, giving rise to 64 cells that become sperms (Fig. 22.18*I*). They are discharged into the conceptacle. In the formation of the egg cell, meiosis is followed by a single mitotic division, so 8 haploid cells are formed, which will become mature eggs (Figs. 22.18*E, O*). They are discharged, unfertilized, from the oogonium and pass with the sperms into the sea water where fertilization takes place. Thus, in *Fucus*, the male gametophyte phase is reduced to four mitotic divisions and 64 sperm cells. Only a single mitotic division occurs in the female gametophyte generation which consists of 8 egg cells. Compare this with the development of pollen and the embryo sac in higher plants.

Summary of Reproduction in Algae

Reproduction in the prokaryotic Cyanophyta is vegetative insofar as it is known. In all other algae—which are eukaryotes—vegetative, asexual, and sexual reproduction are identical to reproductive processes of other eukaryotes. Reproduction serves the same purpose of either maintaining similarity of generations or of introducing variation between generations. There are broad variations across the algal divisions in the time interval between meiosis and fertilization. In many of the more primitive families, meiosis accompanies the germination of the zygote. In the Bacillariophyta, meiosis gives rise directly to gametes. In some Chlorophyta, most Rhodophyta, and all Phaeophyta, vegetative stages intervene between meiosis and fertilization on the one hand, and between fertilization and meiosis on the other. These intervening vegetative thalli may be similar in appearance or differ greatly in size and form. In *Fucus,* the gametophyte, or haploid, phase has been reduced to single haploid cells which, instead of growing into gametophyte plants, become gametes and fuse to produce new diploid plants.

Ultrastructure

The Prokaryotic Algal Cell

The cells of Cyanophyta resemble those of the bacteria in that they lack nuclei, mitochondria, dictyosomes, and endoplasmic reticulum. They are different from bacteria in that mesosomes (Fig. 21.14) have not yet been recognized. Of much greater significance, the blue-green algae carry on photosynthesis with the liberation of oxygen. The blue-green algal cell thus may be classified as a prokaryotic cell with regard to internal structure, but photosynthesis and respiration are aerobic, making it definitely algal-like.

Each individual protoplast is surrounded by a cell wall that may be similar to a bacterial cell wall. Under the electron microscope it may be resolved into at least four layers, but details of its ultrastructure have not been agreed upon (Fig. 22.20). A mucilaginous sheath surrounds individual cells, colonies, and trichomes. It varies from species to species in thickness, consistency, and pigmentation. This sheath pigment frequently is responsible for the color of the algal mass. Cells and trichomes with their sheaths may be embedded in this mucilaginous matrix.

Light microscopists early established that the photosynthetic pigment of the blue-green algal cell was confined to the periphery of the cell, and fluorescent studies showed it to be chlorophyll *a*. Figure 22.2*A* shows individual living cells of *Synechococcus* photographed by phase contrast. The mucilaginous sheath, cell wall, pigmented peripheral region, and

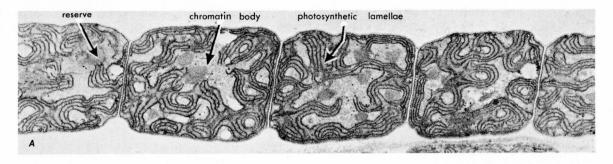

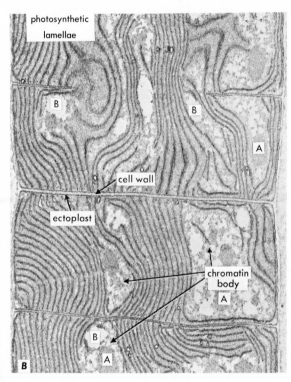

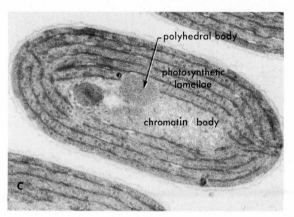

Figure 22.20   Ultrastructure of the blue-green
algae. *A, Nostoc,* ×15,000; *B, Oscillatoria,*
×37,000; *C, Spirulina,* ×40,000. (*A* and *B,*
courtesy of W. Menke; *C,* courtesy of N. J.
Lang.)

clear central region are evident. Feulgen staining of
Synechococcus not only demonstrates the presence
of DNA in the center of these cells but also shows
that it must extend outward into the peripheral
pigmented region. The presence of the chloroplast
pigments in the outer region and DNA in the interior
region has led to the designation of these regions
as the **chromatoplasm** and the **nucleoplasm.**

Electron microscopy shows the photosynthetic
membranes, or thylakoids, to be confined to the
peripheral regions of the cell (Fig. 22.20). Note that
the thylakoids are indeed individual flattened vesicles, or sacs, each composed of membranous sheets
enclosing a flattened space. The thylakoids are always closed at their margins. The thylakoids adjacent to the cell wall generally parallel the wall; the
more internal thylakoids, while still parallel to each

other, range internally toward the center of the cell.
The thylakoids, as a group, are not delimited by any
membrane from the remainder of the cell.

The nucleoplasm occupies the center of the cell
and also extends outward into spaces between the
thylakoids (Fig. 22.20). Mitochondria and dictyosomes are lacking. Numerous other particulates may
be observed in these cells, but their nature and
functions are not understood.

The Eukaryotic Algal Cell

All other algae have eukaryotic, compartmented
cells similar to those of vascular plants. The cells
contain one or more nuclei which divide by mitosis;
they contain mitochondria, dictyosomes, endoplasmic reticulum, ribosomes, and chloroplasts (Figs. 4.3,

22.21, 22.22, 22.23, 22.24). The filaments of the siphonous line lack septa, thus eliminating definite boundaries between otherwise typical eukaryotic cells. Unequal distribution of the other organelles occurs as in other divisions and is probably related to variations in metabolism. For instance, in some unicellular forms, the mitochondria have a peripheral location just under the plasmalemma. In some cases, they apparently form a very open reticulum.

Variations in flagellation are important in algal classification, but details will not be considered here. Of interest, however, is the remarkable similarity of the ultrastructure of flagella. The motile organelles of bacteria are supplied with a single axial filament (Fig. 21.10). The flagella of all other organisms from *Chlamydomonas* to man have a uniform internal arrangement of microtubules.

The flagella of *Chlamydomonas* will serve as an example. Here, there are two anterior flagella entering the cell at an angle. They end in **basal bodies** that are connected by a bridge of fibers (Fig. 22.24). A longitudinal section near the tip of the flagellum shows a unit plasma membrane enclosed by a flagellar sheath (Fig. 22.25*B*). Within the flagellum there

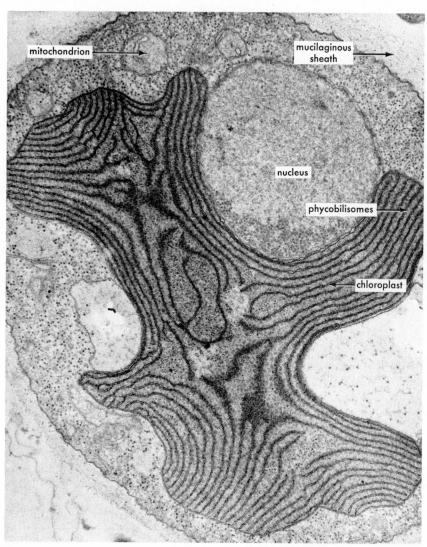

Figure 22.21 Electron micrograph of *Porphyridium purpureum*. Chloroplast membranes do not form grana. The accessory water-soluble pigments are localized in the phycobilisomes along the plastid membranes. Note the nuclear envelope with pores, ×52,000 (Courtesy of R. Chapman.)

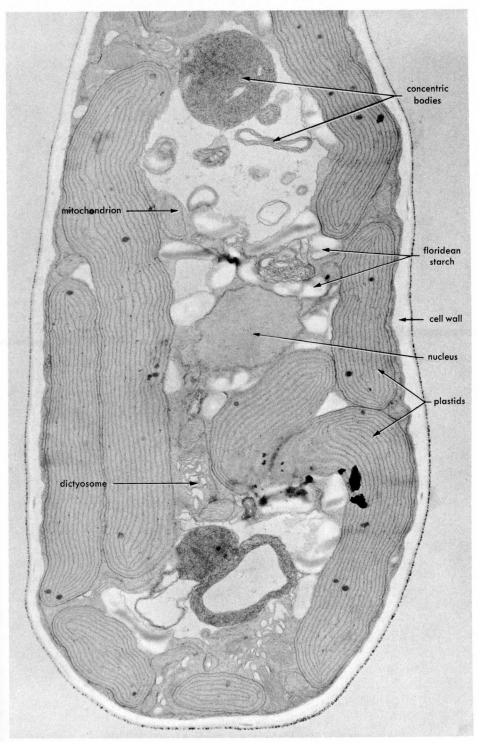

Figure 22.22 A vegetative cell from the red alga *Batrachospermum moniliforme.* Chloroplast membranes do not form grana, and the carbohydrate is stored free in the cytoplasm, ×20,000. (Courtesy of David Brown.)

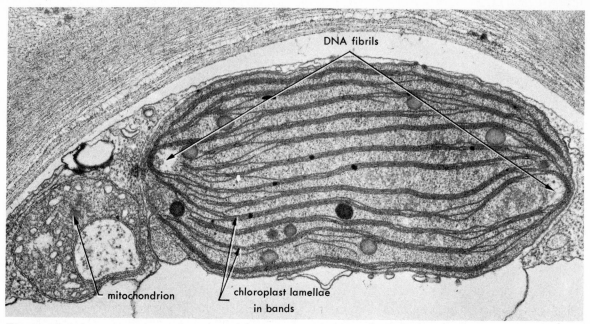

DNA fibrils

mitochondrion

chloroplast lamellae
in bands

Figure 22.23 A chloroplast and a mitochondrion from a cell of the brown alga
Egregia menziesii. Note clear areas at opposite ends with DNA fibrils and the
chloroplast membranes in parallel bands, ×26,000. (Courtesy of T. Bisalputra and A.
Bisalputra, *J. Cell. Biol.* **33,** 511.)

are three bands of microtubules parallel to the long
axis of the flagellum. The central band is longer. The
tubules are separated by a matrix of some sort. An
array of indefinite cross-connections may also be
seen. A cross-section of the flagellum at this level
shows the flagellar sheath and plasmalemma. It also
reveals that the outer circle of microtubules is composed of nine sets of double microtubules and that
a pair of single microtubules forms the central band
(Fig. 22.25A). Other cross-sections taken at different
levels show changes in the relationship of the microtubules as the sections approach the basal body.

Motility is gained by a flagellar motion similar to
the swimming breast stroke. At the beginning of the
stroke, the flagella are extended straight forward.
They sweep backwards without bending. The return
stroke is an upward wave motion and again resembles the return of the arms in the swimming breast
stroke.

Most motile forms, including adult unicells, ga-

metes, sperm cells, and zoospores are provided with
a red **eye spot.** This generally appears as two rows
of globules of carotene separated by a thylakoid at
the periphery of the chloroplast (Fig. 22.26).

There is a close relationship between pigments
and the ultrastructure of the plastids. The Cyanophyta and the Rhodophyta have only chlorophyll a.
The photosynthetic membranes of the Rhodophyta
resemble those of the Cyanophyta (Figs. 20.20, 22.21,
22.22) in being single flattened sacs or lamellae.
These lamellae may be more or less parallel, but they
never appear as appressed membranes forming grana
as occurs in the higher plants. In all Rhodophyta, the
photosynthetic lamellae are embedded in a stroma
and are separated from the cytoplasm of the cell by
a double plastid envelope. The floridean starch characteristically found in the large and more advanced
group of the Rhodophyta always occurs outside the
chloroplast and in the cytoplasm (Fig. 22.22). The
accessory phycobiliproteins may be present as gran-

Figure 22.24 Right. *Chlamydomonas reinhardi*. At the anterior end two flagella
are attached to a basal body. The single cup-shaped chloroplast occupies the
periphery of the cell. Starch grains are distributed throughout the chloroplast and
starch is prominent in the lower half of the cell, where it surrounds a darkly
staining pyrenoid. The chloroplast envelope separates the nucleus and surrounding
cytoplasm from the pyrenoid and inner plastid membranes. Ribosomes are
prominent, mitochondria stain diffusely, and dictyosomes are not apparent, ×19,000.
(Courtesy of D. Ringo, *J. Cell. Biol.* **33,** 543.)

452   Algae

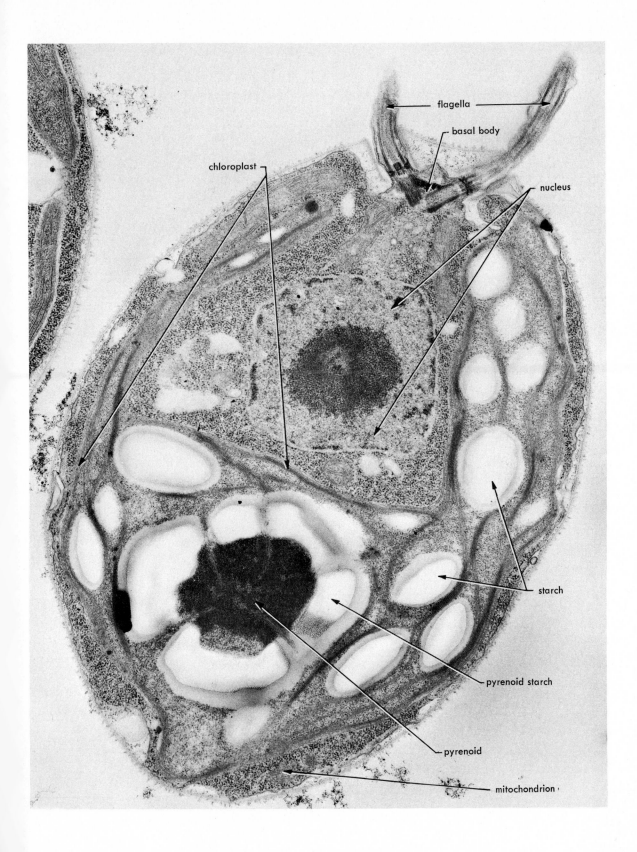

flagella

basal body

chloroplast

nucleus

starch

pyrenoid starch

pyrenoid

mitochondrion

ules aligned along the photosynthetic lamellae. These pigments are able to trap energy from light of low intensity and to pass the trapped energy on to chlorophyll, where an electron becomes excited in the usual manner (p. 219).

Most chloroplasts of the Rhodophyta show denser, roughly spherical areas embedded in stroma, but not delimited from it by a membrane. These are **pyrenoids** and they are thought to play some part in the formation of the starch.

The chloroplasts of the Phaeophyta contain chlorophyll c in addition to chlorophyll a. Here, too, the lamellae occur as bands of single parallel thylakoids (Fig. 22.23). They are never appressed to form grana, but they do always occur in groups of three to seven.

They are parallel to each other. The storage carbohydrate elaborated in these brown algal cells occurs as granules lying outside the chloroplast in the cytoplasm.

Pyrenoids in the Phaeophyta are limited to definite orders. They project from either the ends or sides of the chloroplast, and the photosynthetic lamellae do not enter them. The pyrenoids are frequently separated from the cytoplasm by a distinct cap.

The Chlorophyta and Euglenophyta possess both chlorophylls a and b in a ratio similar to that found in the higher plants. The thylakoids in these algae are appressed into grana just as they are in the chloroplasts of higher plants. Pyrenoids are present in many species of the Chlorophyta. The pyrenoids are

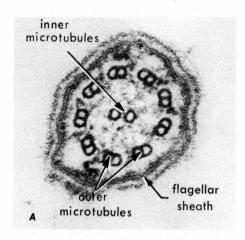

Figure 22.25 Flagella of *Chlamydomonas reinhardi*. *A*, cross-section, taken near tip of flagellum. There is an outer circle of nine paired microtubules (doublets) and two central single microtubules. The microtubules are embedded in a matrix which is surrounded by a flagellar membrane and a flagellar sheath, ×100,000. *B*, a longitudinal section of a flagellum at its tip end; the central microtubules are longer than those composing the outer circle of doublets, ×68,000. (Courtesy of D. Ringo, *J. Cell. Biol.* **33**, 543.)

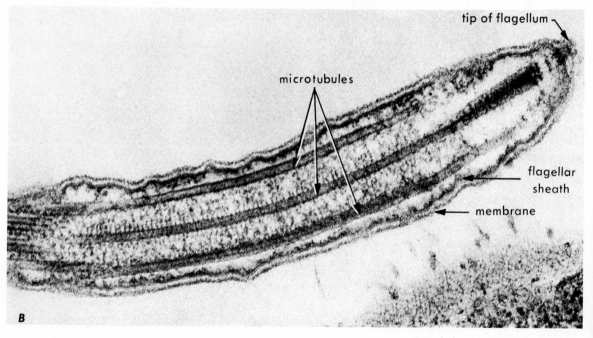

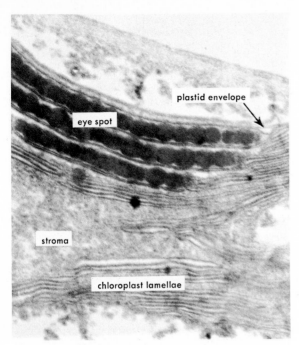

Figure 22.26 The eye spot in *Chlamydomonas* consists of several rows of granules containing carotene. It lies wholly within the chloroplast, ×25,000. (Courtesy of Carole Lembi.)

always embedded in the stroma and seem to have a close relationship with the starch grains, which are always within the chloroplast (Figs. 22.24, 22.27).

Summary of the Algal Cell

The cyanophyte cell is related to prokaryotic forms by the similarity of its genetic mechanism to that found in bacteria. It contains membrane-associated chlorophyll a, photosynthesizes a carbohydrate from carbon dioxide and water, and evolves oxygen. This relates its metabolism to that of higher plants. The cells of all other algae are typical eukaryotic cells. The ultrastructure of their chloroplasts is related to the photosynthetic pigment present. Grana, similar to those observed in the chloroplasts of higher plants, are found only in the Chlorophyta and Euglenophyta, which also have chlorophyll a and b in ratios close to those of higher plants. In cross-section, flagella show the typical arrangement of microtubules found in the motor organelles of all organisms exclusive of the bacteria.

Figure 22.27 Right. The ultrastructure of a portion of a cell of *Scenedesmus quadricauda*, ×15,000. (Courtesy of T. Bisalputra.)

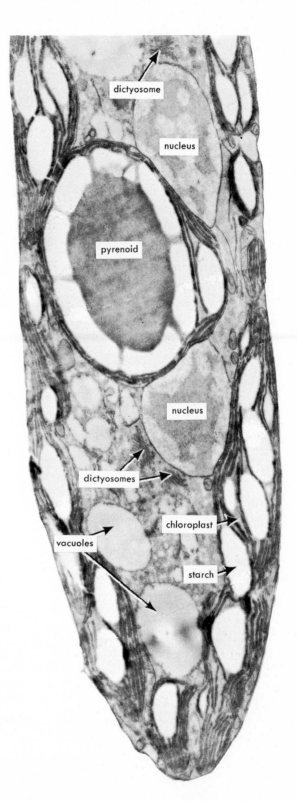

23

Lower Fungi

It is well-known that the ravages of human diseases have changed or modified the history of peoples. Similarly, epidemics of diseases that lay waste important food plants may influence greatly the course of events. From 1843 to 1847, a fungus known scientifically as *Phytophthora infestans* spread rapidly through the potato fields of Ireland. This fungus causes a severe disease of potatoes known as **late blight.** It not only kills the foliage but also infects the tubers, causing them to rot rapidly.

In the middle of the nineteenth century, Irish peasants depended chiefly on the potato as a source of food. Consequently, from 1843 to 1847, when weather conditions were just right for the rapid development of *Phytophthora infestans,* and when the potato disease that it caused attained epidemic proportions, the Irish experienced a disastrous food famine. During these years, over a million Irishmen moved across the ocean to New York City. They and their descendants have left an imprint upon American life. The potato famine in Ireland, produced by a fungus, was responsible for the death of a quarter of a million people and the mass movement of Irish people to the United States, chiefly in 1846 and 1847.

More than 100,000 species of fungi have been recognized. They play an indispensable role in the life of the world. They help to maintain soil fertility; decompose organic matter of both plants and animals that would soon make the surface of the earth uninhabitable were it not disposed of; cause diseases of plants and animals, including man; and are of great importance in various industries, such as in the making of cheese, manufacture of alcohol, and retting of flax.

Common examples of parasitic fungi are those that cause mildew of roses, late blight of potatoes, and rust of cereals. The common field mushrooms are good examples of saprophytic forms. A parasite may or may not be "disease-producing," that is, **pathogenic.** The fungus *Phytophthora infestans* is pathogenic; it produces a disease of the potato. On the other hand, certain bacteria subsist on the roots of alfalfa, beans, and other members of the legume family but do not cause a disease. The organism upon which the parasite lives is the **host.** For example, the potato plant is the host of the parasitic fungus *Phytophthora infestans,* and man is host to the fungi that are responsible for athlete's foot.

Some parasites, like those that cause rusts of cereals, are obliged to secure their nourishment from living tissues. They are known as **obligate parasites.** And certain saprophytes, like some mushrooms, thrive only on nonliving organic materials. They are

457

known as **obligate saprophytes.** On the other hand, some fungi have the faculty of growing as either a parasite or a saprophyte and are referred to as **facultative parasites** or **facultative saprophytes.** The most common plant pathogens are facultative parasites.

Classification of Fungi

The various ranks of classification in the fungi may be recognized by the terminal syllable or syllables of the name of the rank. The division name Mycota may be broken into two syllables; *Myc-* comes from the Greek *myketos,* which means fungus, and *-ota* is New Latin meaning "to have." So Mycota literally means "fungus having." The names of the subdivisions end in *-mycotina,* which is a combination of Greek and Old Latin and means "having a fungal likeness." Classes end in *-mycetes,* which is simply the classical Greek word for fungus. Thus, the name Ascomycetes may be broken down into *Ascos-* (in Greek *ascus* means sac) and *-mycetes,* or literally, the sac fungi.

There is a single fungal division, the Mycota, which may be divided into two subdivisions, the Myxomycotina and the Eumycotina. The first of these subdivisions comprises the slime molds. It is a small subdivision whose members have a vegetative plant body consisting of a naked mass of protoplasm. The prefix *myxo-* is a Greek combining form meaning slime. Thus Myxomycotina would be "slime having a fungal likeness." The other subdivision, the Eumycotina, is a large group containing thousands of species of diverse forms. Here, the prefix *Eu-* is also from Greek and means "true." The Eumycotina is divided into five classes, one being a **form class.** We shall consider the following:

Class 1.   The Oomycetes (egg fungi). The common water molds and the fungus that causes late blight of potatoes are examples.

Class 2.   The Zygomycetes (zygote fungi). Bread mold and fungi causing decay of sweet potatoes and peaches belong to this class.

Class 3.   The Ascomycetes (sac fungi). Representatives in this group include a common mold that grows on jellies and jams. Many sac fungi cause severe diseases of certain orchard trees, and one produces a destructive disease of elms.

Class 4.   The Basidiomycetes (club fungi). Well-known members of this group are mushrooms and the fungi that cause rusts and smuts of the cereals.

Class 5.   The Fungi Imperfecti (imperfect fungi). The fungi causing athlete's foot are members of this class. Also, many diseases of crop plants result from the attacks of representatives of this group.

Because of the very large numbers of fungi and their great importance in medicine, industry, and agriculture, several branches of science are concerned with them. For instance, one group of workers has studied the structure and activities of the true fungi. These scientists are not greatly concerned with the economic importance of fungi; rather, they are interested in the life histories, in modes of reproduction, and food requirements. They are called **mycologists,** and their science is **mycology. Pathology** deals with the diseases of man, other animals, and plants. A **plant pathologist** is one who specializes in a study of plant diseases. He studies the life histories of the organisms that cause disease, the environmental factors that influence the growth of the organisms, and methods of combating the diseases. He is also interested in plant diseases due to other causes. The annual reduction in yield of orchard, garden, and field crops due to diseases in the United States amounts to many millions of dollars. On the other hand, the development of control methods by plant pathologists have reduced these losses to a level where it is possible to grow a crop economically.

Subdivision Myxomycotina

General Characteristics and Distribution.   The slime fungi or slime molds, of which there are some 300 species, are of scientific interest because they seem to combine the characteristics of both plants and animals. The vegetative body consists of a slimy mass of naked protoplasm in which there are many nuclei without separating walls. The vegetative body is called a **plasmodium** (plural, **plasmodia**) (Fig. 23.1). This plasmodium has no definite shape; it creeps slowly by amoeboid movement, usually over shaded, rotting tree trunks, across leaves, or in crevices, engulfing solid particles of food as it goes. Most slime fungi are saprophytes. The absence of cell walls, amoeboid movement, and the ability to take solid food particles into the protoplasm are characteristics that are usually associated with animals. When slime fungi reproduce, however, they form spores with cellulose walls; thus, they have reproductive characteristics that are definitely those of plants.

Reproduction.   Prior to the reproductive stage, the plasmodium moves to a drier substratum. After a time, the plasmodium ceases moving and forms one or more spore-producing structures (**sporangia** or fructifications, Fig. 23.1). There is great variation in

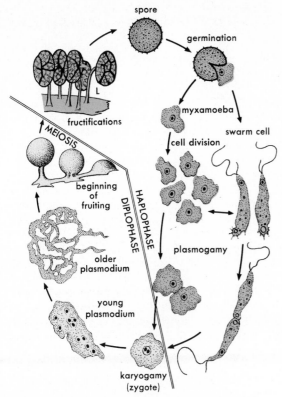

Figure 23.1 Life cycle of a myxomycete. (Redrawn from C. J. Alexopoulous, *Introductory Mycology*, John Wiley & Sons, New York.)

the form and color of the sporangia among the different species. Some have a very delicate structure and brilliant coloring. The spores formed within the sporangia are uninucleate and are surrounded by a cellulose wall. They are discharged from the sporangia and spread by wind. In the presence of water, they germinate; the wall is ruptured, and the contents escape in the form of a flagellated or amoeboid naked mass of protoplasm that may later multiply by fission. In some species of slime fungi, pairs of these bodies may fuse to form zygotes. Nuclear fusion occurs in the zygote. Each zygote may grow into a plasmodium, or zygotes may coalesce to form a plasmodium.

Subdivision Eumycotina

General Characteristics. With the exception of the simplest of the Eumycotina, the vegetative thallus, or plant body of the members of this subdivision, is composed of a mass of threads of filaments called **hyphae (hypha,** singular) (Figs. 23.2A, B). Three of the four different classes of this subdivision have distinctive types of hyphae. The vegetative body may be webby and delicate, as in common bread mold (Figs. 23.4, 23.17) and other molds (Fig. 23.5), or it may be quite hard and leathery, as in species that cause wood rot. The mass of hyphae forming the vegetative body is called the **mycelium** (Fig. 23.2C).

The cell walls of the hyphae may contain cellulose,

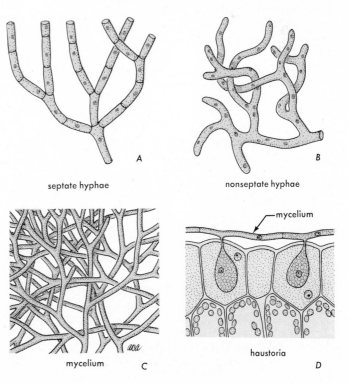

Figure 23.2 Various aspects of the fungal plant body.

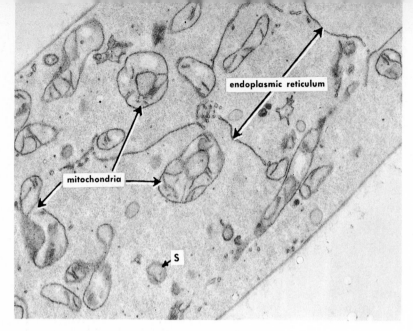

endoplasmic reticulum

mitochondria

S

Figure 23.3 Portion of hypha of *Rhizoctonia solani,* ×24,000. (Courtesy of C. E. Bracker.)

and also in some species, **chitin,** a substance identical with that found in the cuticle of insects and crustaceans. The nuclei divide normally by mitosis. Cell division does not always follow immediately after nuclear division. In certain instances, a half dozen or more nuclear divisions will occur before walls form between the daughter nuclei. In the Oomycetes and Zygomycetes, the vegetative hyphae usually have no cross-walls (**septa,** singular **septum**). They are multinucleate, and there is complete continuity of the cytoplasm throughout the whole vegetative thallus. Hyphae of this sort are said to be **coenocytic,** and since no cross-walls or septa are formed in such filaments (Fig. 23.2*B*), they are also called **nonseptate.** When cross-walls are present, the filaments are said to be **septate** (Fig. 23.2*A*).

The cytoplasm of fungal hyphae resembles that of other eukaryotic plant cells in having mitochondria and an endoplasmic reticulum. The mitochondria are small rings as may be seen in Fig. 23.3. The endoplasmic reticulum is in the form of slender paired membranes. The small profiles (S) may be similar to bodies seen in the cytoplasm of higher plants which have no generally accepted name, but are sometimes called **microbodies.**

Function of the Mycelium. The vegetative mycelium carries on the general activities of plant cells, such as absorption, digestion, respiration, and secretion, but not photosynthesis. Since it is incapable of synthesizing its own foods, it must obtain nourishment either from nonliving organic matter or from living plants or animals. The food must be rendered soluble so that it may diffuse through the walls of the hyphae and reach the protoplast. Some fungi obtain food from even the hardest of woods and from solids of many other sorts. These solid substances are liquefied or otherwise rendered diffusible by enzymes that are secreted by the hyphae. For example, certain enzymes are secreted by fungi that break down the complex carbohydrates of wood to simple soluble sugars. Many fungi actually carry on very active digestion of solid materials. Some fungi have specialized hyphae known as **haustoria** that can penetrate living cells and obtain nourishment therefrom (Fig. 23.2*D*).

Since the principal function of the mycelium is to obtain nutritive materials, it is usually found in close association with a source of food. It may be growing inside a living tree, in a dead stump, on or in a leaf, in aging cheese, in a manure pile, or in many other kinds of organic substances. The mycelium of most fungi is not adapted to withstand much drying; hence it seldom grows openly exposed to the atmosphere, unless the relative humidity is high.

Reproduction. The sporangia, spores, gametangia, and gametes of the fungi show a great range of structure and recall in some respects the reproductive structures found in algae. Simple cell division and fragmentation occur in some fungi. Both motile and nonmotile spores are produced in various types of sporangia. Isogamy is characteristic of several genera. Heterogamy, involving gametangia resembling those of *Vaucheria,* occurs in several families. In two classes of fungi, the Ascomycetes and Basidiomycetes, the sexual cycle is complicated by the failure of the two haploid nuclei to fuse immediately after the gametes unite. This characteristic results in two distinct types of hyphae: (*a*) those in which the cells are haploid, with one haploid nucleus in each cell, and (*b*) those in which the cells contain two haploid nuclei in each cell. In the latter type, the two haploid nuclei will ultimately fuse just before meiosis (see Fig. 24.3).

Although the vegetative mycelium of most fungi is the actively destructive portion, the reproductive structures are of more importance from the disease-control standpoint and as a basis of classification. Usually, fungi multiply and are disseminated by abundant spores; it is therefore important to prevent the dissemination of spores, or to stop their germination, or to kill the young hyphae after germination. It is seldom possible to rid the infected plant of a vigorous mycelium.

Sexual reproduction. As in the algae, sexual reproduction involves two fundamental steps: (*a*) the union of gametes, during which the chromosome number per nucleus is doubled; and (*b*) meiosis, during which the chromosome number is halved.

Asexual reproduction. As usual, asexual reproduction does not involve changes in the number of chromosomes. Asexual spores are frequently formed by one of the following methods.

1. Spores may be borne in specialized cells called **sporangia** (spore cases). The sporangia (sporangium, singular) may occur either on typical filaments on special upright hyphae, or on highly branched hyphae. Aerial sporangia are formed as in bread mold (Fig. 23.18). In other species, the sporangia are submerged and germinate in thin films of water. The spores produced by such sporangia are motile and are called **zoospores.** The hyphae that bear sporangia are called **sporangiophores** (Figs. 23.9, 23.18).

2. In other cases most of the cells of a septate filament separate from each other. Before separating, heavy walls are secreted around them. Reproductive bodies formed in this manner are known as **chlamydospores.** They are formed in enormous numbers by the various types of fungi.

3. End cells of a hypha may round up and be cut off from the remaining portion of the hypha. These rounded cells are light in weight, a condition that favors rapid dissemination of the fungus. Such cells are called **conidia.** If they germinate to form hyphae or new fungus plants, these conidia may be called **conidiospores.** In many other species, the conidia will become sporangia and produce zoospores, in which event they may be called **conidiosporangia.** The hyphae that bear conidia are called **conidiophores** (Figs. 23.9, 24.2).

Class Oomycetes. Many Oomycetes are severe plant pathogens. Some, however, may infect fish or insects or cause diseases in man. Many members of this class, like the common water mold, are normally

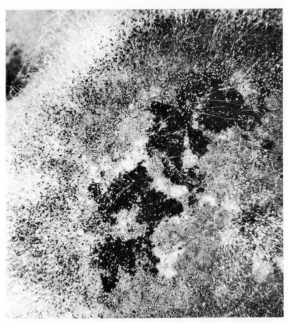

Figure 23.4 *Rhizopus* on peach, ×4. (Courtesy of R. N. Campbell.)

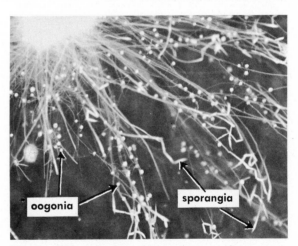

Figure 23.5 Water mold growing on a hemp seed, ×3.

saprophytes or weak parasites. Some are minute forms of one to several cells, and a few of these smaller sorts are reported to attack small aquatic animals. Most representatives of the Oomycetes develop a more or less extensive mycelium of indefinite form (Figs. 23.4, 23.5).

The hyphae of actively growing Oomycetes are generally **coenocytic** and **nonseptate.** Septa may form occasionally, however, in old hyphae of some species, and normally the reproductive cells are cut off from the vegetative hyphae by cross-walls.

The Oomycetes are divided into several orders. Specialists do not agree completely as to the best manner of classifying the plants in this class, and therefore, the number of recognized orders depends upon the authority one is consulting. We shall consider several forms distributed in the orders Saprolegniales and Peronosporales. Note that the names of orders end in -*ales*.

Saprolegniales. Fungi of this order live in fresh or salt water and are generally saprophytic. Even the species that attack the gills of fish grow on tissues that have been weakened or suffered injury. Because of the aquatic habitat of many members of this order, they are frequently called **water molds.** They may be easily cultivated by placing small pieces of meat, egg albumin, radish seeds, or dead flies in a dish of pond water (Fig. 23.5). After the mycelium has ramified through the dead fly of other substratum, hyphae grow outward into the water. A small ball of white hyphae, from $\frac{1}{2}$ to 1 inch in diameter, may thus be formed. Reproductive cells are produced by these hyphae.

Asexual reproduction. With ample food supply, the mycelium increases in size with but little tendency to produce reproductive cells. If, however, a well-developed mycelium is transferred to distilled water in which a food supply is lacking, sporangia will usually appear. Sporangia are formed at the ends of hyphae by a cross-wall cutting off the tip of a hypha from the rest of the mycelium as in *Saprolegnia* (Fig. 23.6A). The sporangium is a multinucleate structure. After a time, the protoplasmic contents of the sporangium divide into a large number of spores, each with one nucleus (Figs. 23.6B, C). Upon maturity, the spores are discharged from the sporangium. In *Saprolegnia,* each zoospore has two flagella attached to its anterior end which enable it to swim actively (Fig. 23.6D). After a time, the zoospores settle down, lose their flagella, develop a cellulose wall, and pass through a resting period. Upon resuming activity, they escape from the wall; the two newly developed flagella are now attached laterally, and the zoospores, thus equipped, swim about for a period. If they come to rest on a suitable substance, each sends out a tubular outgrowth that penetrates and infects this substance.

In a few species of water molds, one or even both zoospore stages are suppressed. When zoospores emerge from the sporangia, they may germinate directly into a new mycelium. In certain other species, the spores never leave the sporangia but germinate while still enclosed within it, and the germ tubes pierce the old sporangial wall. In still other species, spores are not even formed; the sporangia germinate directly into a coenocytic mycelium.

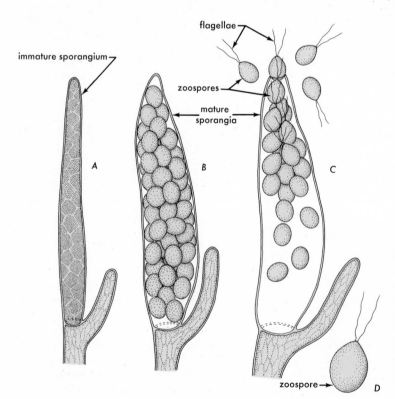

flagellae

immature sporangium

zoospores

mature sporangia

A

B

C

zoospore

D

Figure 23.6 Zoosporangia of *Saprolegnia. A,* immature sporangium; *B,* mature sporangium; *C,* discharge of zoospores; *D,* a zoospore. (Based on W. C. Coker, *The Saprolegniaceae,* University of North Carolina Press, 1923 [*Saprolegnia angiospora,* Plate 8].)

Sexual reproduction. The **oogonia** are spherical cells, formed at the tips of short side branches (Figs. 23.7A, 23.8). When mature, they are three to four times the diameter of ordinary hyphae. The cytoplasm in the swollen tip becomes denser than that in regular hyphae. From one to twenty eggs may be formed from the protoplasmic contents in each oogonium, depending upon the species. The eggs are spherical, dense bodies of protoplasm, each containing one nucleus.

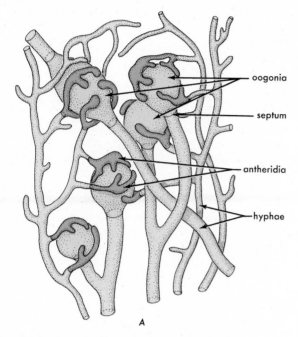

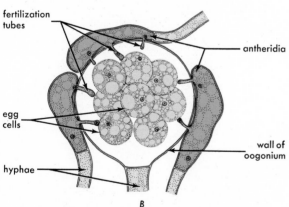

Figure 23.7 Gametangia in *Saprolegnia*. *A*, oogonia and antheridia with associated hyphae; *B*, antheridia, oogonium with many eggs, fertilization tubes in place. (Based on W. C. Coker, *The Saprolegniaceae*, University of North Carolina Press, 1923 [*Saprolegnia angiospora*, Plates 9 and 10].)

The **antheridia** are also formed at the tips of branches, in some species near the oogonia (Figs. 23.7A, 23.8). Each is separated from the main filament by a cross-wall. The antheridia are usually curved and not much greater in diameter than the hyphae from which they arise. The contents of a mature antheridium consist of several nonflagellated male gametes. The antheridium comes in contact with an oogonium (Figs. 23.7A, 23.8). In some instances, one oogonium may have several antheridia attached to it. A short slender hypha, the **fertilization tube,** grows from the side of an antheridium adjacent to an oogonium (Figs. 23.7B, 23.8), penetrates the oogonial wall, and comes in contact with one or more eggs. If the oogonium contains several eggs, the fertilization tube usually branches, sending a branch to each egg. Nuclei (male gametes) from the antheridium migrate into the tube and any branches present. One nucleus and possibly some cytoplasm pass into each egg, and fertilization ensues. The fertilized egg or zygote develops a heavy wall, becoming an **oospore** (Fig. 23.8), and usually will not germinate for several months, even under favorable conditions; hence, it is well-adapted to survive unfavorable conditions. Upon germination, the oospore sends out new hyphae (Fig. 23.8), which, if they find a source of food, rapidly grow into a typical mycelium. If food is scarce, the formation of zoospores follows soon after germination. Not uncommonly, eggs may develop into new hyphae without fertilization.

Peronosporales. Nearly all the Peronosporales are obligate parasites. In general, heavy-walled oospores carry them over unfavorable periods (Fig. 23.16), and various sorts of asexual spores (Figs. 23.9, 23.10) bring about rapid multiplication under suitable conditions. Because many of them form a downy growth on the surface of the host or substrate, they are sometimes called **downy mildews.**

Asexual reproduction. The asexual reproductive structures of the different genera are adapted to various habitats and show more variation than do the sexual reproductive bodies.

A few Peronosporales, for example, *Pythium,* may live in water or on moist soil as saprophytes; they may infect aquatic plants and are frequently responsible for the "damping off" of the seedlings of many farm and hothouse plants. In keeping with their moist habitat, these fungi are propagated by zoospores that strongly resemble those of the water molds. Sporangia are formed at the tips of the hyphae (Fig. 23.9A). When proper conditions prevail, the sporangia open and the zoospores emerge. The zoospores are small and are able to swim in the

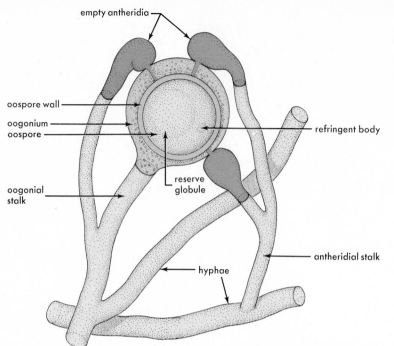

empty antheridia

oospore wall

oogonium
oospore

oogonial
stalk

reserve
globule

refringent body

antheridial stalk

hyphae

Figure 23.8 Sexual reproduction in beet water mold. (Based on J. F. Middleton, *Memoirs of the Torrey Botanical Club* **20,** 98; and V. D. Mathews, *Studies on the Genus Pythium,* (University of North Carolina Press, 1931.)

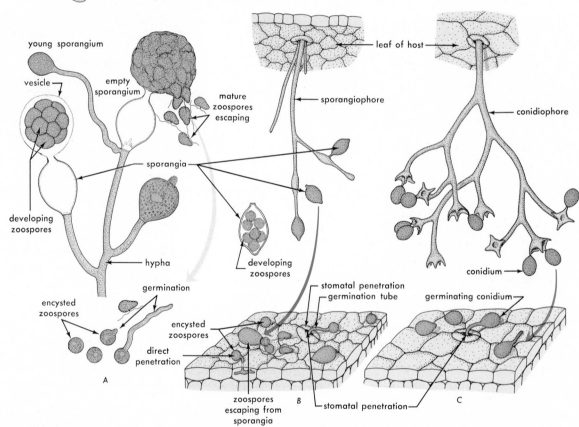

young sporangium

vesicle

empty sporangium

mature zoospores escaping

leaf of host

sporangiophore

conidiophore

developing zoospores

sporangia

developing zoospores

hypha

conidium

germination

stomatal penetration
germination tube

germinating conidium

encysted zoospores

encysted zoospores

direct penetration

zoospores escaping from sporangia

stomatal penetration

A

B

C

Figure 23.9 Asexual reproduction in the Peronosporales. *A, Phythium; B, Phytophthora; C, Bremia. A,* from C. Dreschler, *Phytopath.* **29,** 391, 1939 (*Pythium helicoides*); *B,* redrawn from L. R. Jones, N. J. Giddings, and B. F. Lutman, "Investigations of the Potato Fungus," Vermont Exp. Sta. Bull. **168,** 9, 1912; and J. Webster, *Introduction to Fungi,* Cambridge University Press, 1970; *C,* based on D. G. Milbraith, "Downy Mildew on Lettuce in California," *J. Agr. Res.* **23,** 989, 1923.)

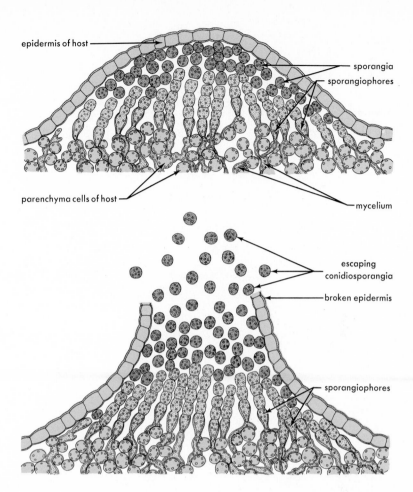

epidermis of host

sporangia
sporangiophores

parenchyma cells of host

mycelium

escaping
conidiosporangia

broken epidermis

sporangiophores

Figure 23.10 Pustules on shepherd's purse caused by conidiosporangia of *Albugo*. (Redrawn from G. M. Smith, *Cryptogamic Botany*, vol. I, *Algae and Fungi*, McGraw-Hill, New York, 1955.)

tually, they germinate and develop a new mycelium. Seedlings may be infected at or just below the soil surface and are quickly killed.

Other Peronosporales (*Phytophthora, Plasmorpara, Albugo*) require a moist habitat and grow best in rainy or humid weather, but they are strictly terrestrial. Many of them are parasites on flowering plants. The mycelium of downy mildews grows within a leaf or stem of a plant (Fig. 23.11) between the cells. Haustoria penetrate the cells.

Aerial sporangia are formed in one of two ways: (a) on long sporangiophores that extend out through the stomata of the infected plant (Figs. 23.9B, 23.11) (*Phytophthora, Plasmopara,* and others), or (b) in compact pustules beneath the epidermis of infected plants (*Albugo*, Fig. 23.10). In both structures, the sporangia are disjoined from the special hyphae producing them and are disseminated by air currents. Upon germination, they usually produce zoospores. Sporangia of this sort may also be called **conidiosporangia** (since they are pinched off), and the branched hyphae that bear them are known as either **sporangiophores** or **conidiophores.**

Some conidiosporangia eventually come to rest on the leaves of susceptible plants. When moisture on the leaf surface is sufficient, zoospores escape from them. The zoospores are very small and are able to move about in the film of water on the leaf surface. They soon send out small hyphae called **germ tubes** that penetrate the host tissues and bring about new infections (Figs. 23.9B, C). *Phytophthora infestans* reproduces asexually in this manner. One fact that the potato growers of England and Ireland learned from the great potato famine of 1846 was that the disease was most severe in the dampest areas and almost nonexistent in dry areas.

Downy mildew is produced by a fungus having asexual reproduction similar to that of *Phytophthora*. Downy mildews are very common infections on many cultivated and wild plants (Fig. 23.12). They are frequently found on hops, beans, grasses, melons, alfalfa, peas, sugar beets, and other plants. They are easily identified by the conidiophores that may, in severe cases, nearly cover the leaf. Brought into the laboratory, the sporangia can usually be induced to germinate by floating them on cool water.

Classification of Fungi **465**

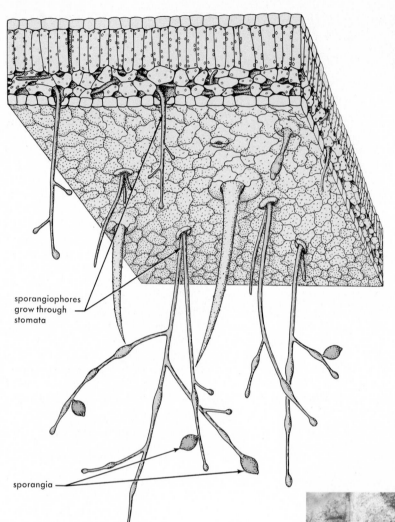

sporangiophores
grow through
stomata

sporangia

Figure 23.11 Sporangia of *Phytophthora infestans.* (Redrawn from L. R. Jones, N. J. Giddings, and B. F. Lutman, "Investigation of the Potato Fungus," *Vermont Exp. Sta. Bull.* **168,** 9, 1912.)

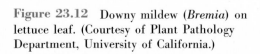

Figure 23.12 Downy mildew (*Bremia*) on lettuce leaf. (Courtesy of Plant Pathology Department, University of California.)

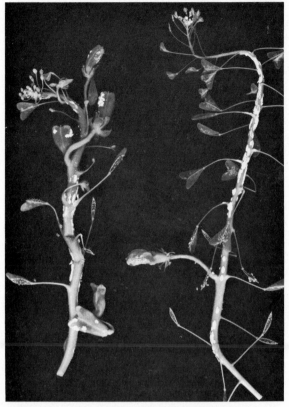

A small group of the Peronosporales, known as the "white rusts," develop sporangia beneath the epidermis of such crop plants as mustards and spinach. There is but one genus (*Albugo*) in this group. The conidiophores do not grow out of the stomata as in the downy mildews. Instead, they collect in pustules under the epidermis of the stem or leaf. Conidiosporangia are cut off in chains from the tips of the conidiophores. They accumulate in large numbers and finally rupture the epidermis, forming creamy-white pustules (Figs. 23.10, 23.13).

In some of the Peronosporales (*Peronospora*), the conidia are formed at the tips of hyphae and dispersed by air currents, but never give rise to zoospores. Instead, they germinate directly by sending out one to several germ tubes, which infect the host plant (Fig. 23.9C). A downy mildew causing a severe disease of lettuce behaves in this manner.

Summarizing, the types of asexual reproduction in Peronosporales may be characterized as follows: (*a*) Sporangia germinate while still attached to the parent hyphae, giving rise to zoospores (*Pythium*, Fig. 23.14). (*b*) Conidiosporangia are dispersed by wind to other hosts, where the zoospores are liberated (*Phytophthora*). (*c*) Conidia are identical in appearance, manner of formation, and method of dispersal to conidiosporangia, but, upon germination, give rise to germ tubes (*Peronospora*).

Figure 23.13 *Albugo* on shepherd's purse. (From R. M. Holman and W. W. Robbins, *A Textbook of General Botany*, John Wiley & Sons, New York.)

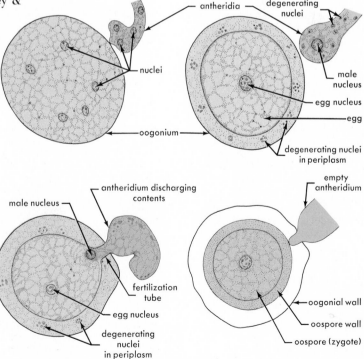

Figure 23.14 Fertilization in *Pythium*. (Redrawn from K. Miyake, "Fertilization of *Pythium de Baryanum*," *Ann. Botany* (*London*) **15**, 653, 1901.)

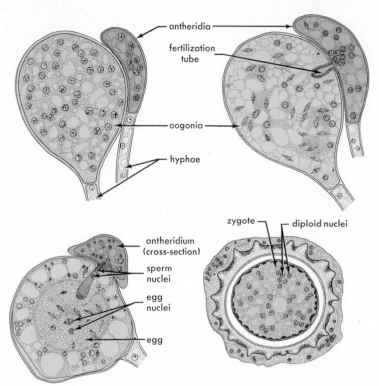

antheridia

fertilization
tube

oogonia

hyphae

zygote   diploid nuclei

antheridium
(cross-section)

sperm
nuclei

egg
nuclei

egg

Figure 23.15   Fertilization in *Albugo*. (From F. L. Stevens, "Gametogenesis and fertilization in *Albugo*," *Botan. Gaz.* **32,** 77. Copyright The University of Chicago Press.)

Sexual reproduction. Sexual reproduction in the Peronosporales is similar to that observed in the Saprolegniales. In many forms, it immediately precedes death of the host plant. Gametangia (oogonia and antheridia) are formed on short side branches, the ends of which have been cut off by cross-walls (Figs. 23.8, 23.15). Some end cells swell to form spherical oogonia, each of which contains one egg. Close by, the antheridia are formed from end cells of other side branches. One antheridium presses closely to the wall of an oogonium. A fertilization tube forms and penetrates the oogonial wall until it contacts the egg. Fertilization results by the fusing of one sperm nucleus with an egg nucleus. The resulting fertilized egg becomes an oospore by developing a thick cell wall that protects the protoplasm against adverse conditions (Fig. 23.16). When conditions are favorable for growth, the oospore germinates, forming, either immediately or after the development of a short hypha, a large number of zoospores, each of which may develop into a hypha.

Class Zygomycetes.   The name of this class refers to the production of a sexual resting spore which typically results from the fusion of two gametangia. No motile cells of any sort are found in this class; all its members produce typical aerial sporangia or conidia, and they regularly have a coenocytic mycelium of indefinite form. Most species are saprophytic, some are weak parasites of plants, a few are specialized parasites of animals, and some are actually obligate parasites on other Zygomycetes. We shall consider only the common bread mold.

The mycelium at first grows chiefly within the substrate, which may be composed of various kinds of organic matter. Eventually, aerial hyphae develop so that the surface of the substrate may become covered with a mass of hyphae (Fig. 23.17).

Common bread mold, *Rhizopus stolonifer,* is a member of the order Mucorales and is of worldwide distribution. It grows well on a large variety of organic substances. Although it is mainly a saprophyte, it does attack and considerably damage sweet potatoes, berries, and fruit while in transit or in storage. It grows very luxuriantly on sweet potatoes and bread, and these foods may be utilized as media for the growth of laboratory cultures.

Asexual reproduction. After the mycelium has become well-established upon and in a substrate, certain aerial hyphae, usually of larger diameter than

468   **Lower Fungi**

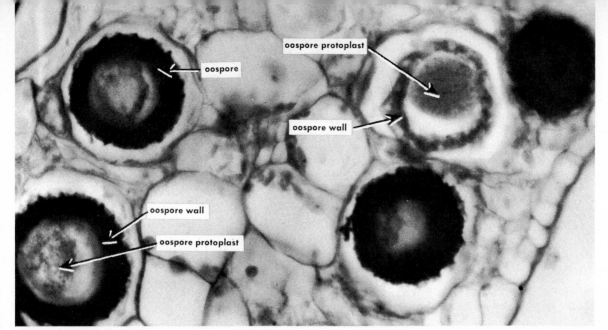

Figure 23.16 Oospores of *Albugo*, ×150.

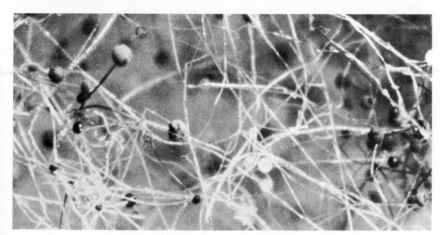

Figure 23.17 Hyphae, mature sporangia (black), and immature sporangia (white) of *Rhizopus stolonifer*, ×4.

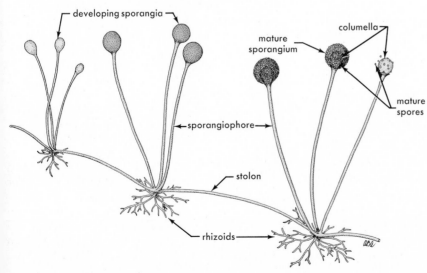

Figure 23.18 Asexual reproduction in *Rhizopus stolonifer*.

Figure 23.19 Stereoscan micrograph of spores of *Rhizopus stolonifer*, ×400. (Courtesy of S. Cook and D. Hess.)

those in the substrate, grow just above its surface for a short distance and then come in contact again with the substrate. They are known as **stolons.** At the point of contact with the substrate, new hyphae form. Hyphae are of three types: (a) stolons; (b) short, branched hyphae, called **rhizoids,** that penetrate the substrate and serve to anchor the mycelium and to absorb nutrients; and (c) hyphae that grow upright and produce sporangia at their tips and therefore are **sporangiophores** (Fig. 23.18).

The development of a sporangium is as follows: The tip of the sporangiophore swells, and a bulging wall cuts off an apical cell, which becomes the sporangium. The dome-shaped wall separating the sporangium from the parent hypha is called the **columella.** Numerous spores are formed from the protoplasm within the sporangium, and when the spores are ripe (Fig. 23.19), the outer wall of the sporangium falls apart and the spores are dispersed by air currents and insects. On a suitable substrate, the spores germinate and develop new hyphae. Immature sporangia are white, mature ones are black (Fig. 23.17).

Sexual reproduction. The characteristic type of sexual reproduction in the order Mucorales is conjugation. The first step is the chance contact of the tips of short club-shaped hyphal branches (Fig. 23.20). Once in contact, the ends of these two short hyphae swell and elongate slightly. A cross-wall forms back from the tip of each hypha, separating a terminal cell with many nuclei from the parent hypha. The two tip cells thus formed are gametangia. The walls of the gametangia that are in contact dissolve, permitting the two protoplasts to fuse. The zygote resulting from this union develops into a

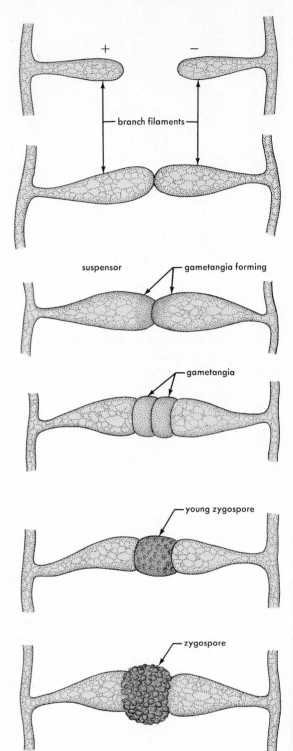

Figure 23.20 Sexual reproduction in *Rhizopus stolonifer.* (Based on G. M. Smith, *Cryptogamic Botany*, vol. I, *Algae and Fungi*, McGraw-Hill, New York, 1955.)

470   **Lower Fungi**

Figure 23.21 Zygospore with suspensors, *Rhizopus stolonifer*, ×300. (Courtesy of R. Riding.)

zygospore (Fig. 23.21), which has a thick wall and is quite resistant to unfavorable conditions.

There is some variation in the way different species of the Mucorales develop gametangia and zygospores. Let us assume that in a given community no bread mold has ever been found. By chance, one bread mold spore is brought in and grows, producing many millions of spores that spread the fungus throughout the community. We now have a strain of bread mold growing in this community that has arisen from a single spore. The probabilities are that sexual union will never occur, because many Mucorales, although morphologically similar, are physiologically differentiated into **sexual strains.** Their similar appearance and behavior make it impossible to designate them as male and female. Instead they are called, for convenience, **plus** (+) and **minus** (−) **strains.** The bread mold spore introduced under these conditions into this community would be either plus or minus. A plus strain will not conjugate with a plus strain, nor a minus strain with a minus strain, and hence no sexual reproduction will occur. Asexual reproduction, however, will be normal. If the opposite strain were now introduced and established, isogametes and zygospores would form. Species of fungi that are differentiated into plus and minus strains are said to be **heterothallic.**

The zygospores of some Mucorales germinate only with difficulty. This is true of common bread mold, *Rhizopus stolonifer*. In these plants, little is known about segregation of the plus and minus characters. In other species, the zygospores do germinate readily, and the segregation of the two sexual strains has been observed.

Significant Features of the Oomycetes and Zygomycetes

A. Vegetative mycelium
Saprolegniales, Peronosporales, Mucorales: a coenocytic mycelium; saprophytes, parasites and intermediate types; parasites mainly plant pathogens

B. Reproduction
Asexual
Saprolegniales: zoospores form in elongated sporangia
Peronosporales: zoospores from attached sporangia and conidiosporangia; germination of conidia by germ tubes
Mucorales: aerial sporangia; spores dispursed by air currents
Sexual
Saprolegniales: heterogamy, one to several eggs in each oogonium
Peronosporales: heterogamy, one egg in each oogonium
Mucorales: isogamy with coenogametes; heterothallic

Classification

Division	Mycota
Subdivision	Myxomycotina
Class	Myxomycetes
Subdivision	Eumycotina
Class	Oomycetes
Order	Saprolegniales
Genus	*Saprolegnia*
Order	Peronosporales
Genera	*Pythium*
	Phytophthora
	Plasmopara
	Albugo
	Bremia
	Peronospora
Class	Zygomycetes
Order	Mucorales
Genus and species	*Rhizopus stolonifer*

24

Higher Fungi

The higher fungi comprise three classes of the division Mycota: the Ascomycetes, Basidiomycetes, and Fungi Imperfecti. They are of great importance, being parasitic upon such crops as wheat, corn, stone fruits, and forest trees. Certain species are required for baking bread, brewing, and producing antibiotics.

Hyphae of ascomycetes are septate, and for most of their lives the cells are mononucleate. The septa are perforated by a central opening enabling the protoplasm to flow from cell to cell.

The hyphae of the Basidiomycetes, like those of the Ascomycetes, are septate with pores through the septa. Fig. 24.1 shows a septum in *Rhizoctonia solani*. The pore has a thickened doughnut-like margin much broader than the adjacent septum. Note the presence of the endoplasmic reticulum paralleling the wall and turning abruptly away from the pore, which contains a mitochondrion apparently on its way from one cell to another. There is reason to think that nuclei, too, may pass through these pores from one cell to another. This mycelium is the active vegetative portion of the plant ramifying through the host or substrate, causing disease or decay.

Class Ascomycetes

The fine flavors of some cheeses are due to Ascomycetes, and some of the edible mushrooms are members of this class. Ascomycetes include several severe, though uncommon, human pathogens, and certain other representatives are known to the medical profession mainly for the beneficial drugs derived from them. Ergot, a drug widely used to control bleeding, is derived from an Ascomycete that infects grasses. Yeasts, also members of this class, are a source of many vitamins and are of great importance in the production of alcohol and in bread-making.

Reproduction

Asexual Reproduction. Various types of asexual spores are chiefly responsible for the dissemination of Ascomycetes during their period of active growth. In mild climates asexual spores may survive the unfavorable seasons, but they are usually killed by cold or by very hot and dry weather.

In many Ascomycetes, the spores are formed from the end cells of specialized hyphae. In the process, the end cells round up, are cut off, and are subse-

473

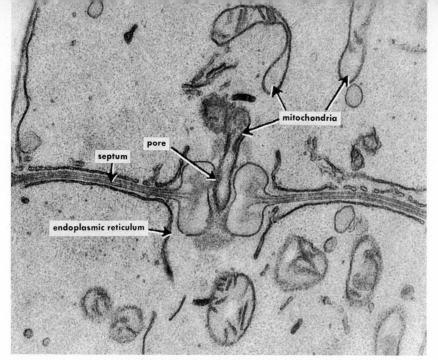

Figure 24.1 A small portion of a hypha of *Rhizoctonia solani* showing a septum, with pore, through which a mitochondrion appears to be passing, ×24,000. (Courtesy of C. E. Bracker and E. E. Butler.)

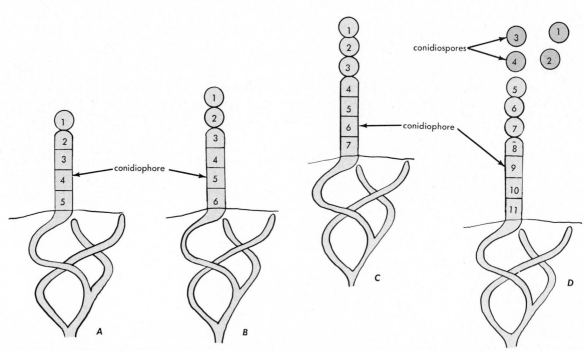

Figure 24.2 Diagram showing formation of conidiospores. *A*, tip of a filament, with the conidiophore divided into cells of similar sizes; first cell (1) has rounded up; *B*, second cell has rounded up, and another cell (6) has been added to the series of tip cells; *C*, cell (3) has rounded up, and cell (7) has been added to the series at the tip of the conidiophore; *D*, cells (1), (2), (3), and (4) have cut off as conidiospores, (5), (6), and (7) have rounded up, and (8), (9), (10), and (11) have been added to the series.

Higher Fungi

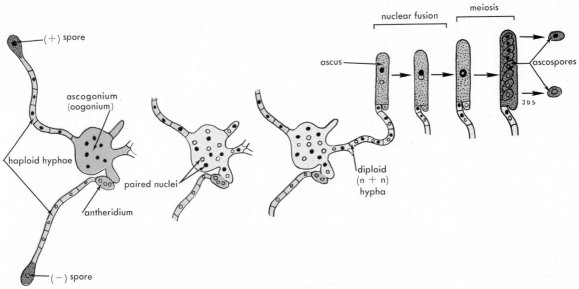

Figure 24.3 Diagram showing sexual life cycle of an Ascomycete.

Figure 24.4   Ascocarps. *A*, apothecia of *Peziza*, ×1. *B*, *C*, stereoscan micrographs; *B*, cleistothecia of powdery mildew (*Uncinula*) on grape, ×300; *C*, perithecia of apple canker (*Nectria*) on apple, ×500. (*A*, courtesy of John P. Roche, *B*, courtesy of D. Hess, *C*, courtesy of E. Butler.)

quently carried away by wind (Fig. 24.2). The spores so formed are called **conidiospores,** or frequently, simply **conidia,** and the specialized hyphae are called conidiophores (see p. 461). Frequently, the conidiospores remain attached to each other for some time, forming long chains. In many species, the conidiophores are associated in a characteristic manner.

Sexual Reproduction. Meiosis followed by mitosis in most Ascomycetes results in eight meiospores in a terminal cell (Fig. 24.3). The appearance of these spores suggests beans or marbles within a cellophane sac, thus the common name **sac fungi.** The Greek word for sac is *ascus* (plural, *asci*), from which is derived the class name, Ascomycetes. The meiospores are generally called **ascospores.** The **asci** are always the terminal cells of special hyphae. They are usually located in a reproductive structure, the **ascocarp** (Fig. 24.4), and have developed by a specialized modification of fertilization from the female

gametangium, the oogonium, or, as it is known in this group, the **ascogonium** (Fig. 24.3). The male gametangium is called an **antheridium,** as in preceding groups.

The ascocarp, composed of both vegetative and ascus-bearing hyphae, is characteristic of the species. It may be microscopic or as much as 6 inches in diameter. There are three general types of ascocarps:

1. **Cleistothecium:** hollow, completely closed sphere (Fig. 24.5A, B).
2. **Perithecium:** hollow, flask-shaped body with narrow opening (Fig. 24.5C, D).
3. **Apothecium:** open, cup-shaped body (Fig. 24.5 E, F).

The end cells (asci) of the ascus-bearing hyphae, in many forms, line the inner surface of the ascocarp. This surface layer is the **hymenium** or fertile layer. Sterile cells, called **paraphyses,** also arise in the

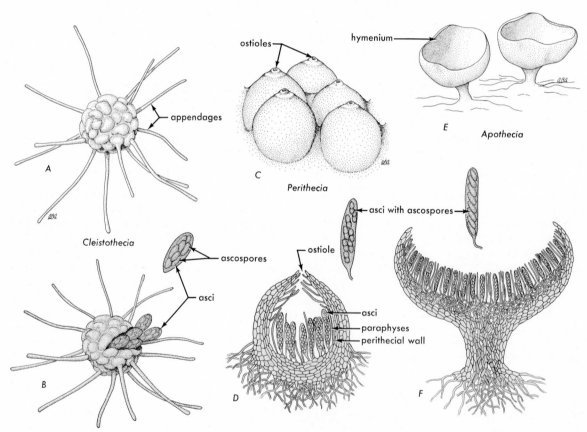

Figure 24.5 Diagram showing three types of ascocarps. *A* and *B*, cleistothecia, ×500; *C* and *D*, perithecia, ×150; *E* and *F*, apothecia, *E* ×¼, *F* ×10. (*D*, redrawn from E. A. Gäuman, *The Fungi. A Description of Their Morphological and Evolutionary Development,* Hafner Publishing Co., New York, 1952; and E. A. Gäuman, *Comparative Morphology of Fungi,* McGraw-Hill, New York, 1928.)

476 Higher Fungi

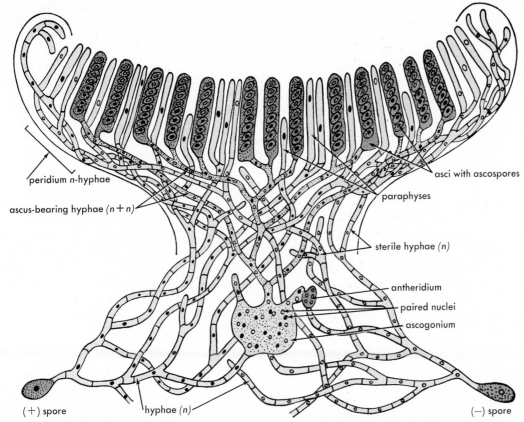

peridium n-hyphae

ascus-bearing hyphae (n + n)

asci with ascospores

paraphyses

sterile hyphae (n)

antheridium

paired nuclei

ascogonium

(+) spore

hyphae (n)

(−) spore

Figure 24.6 Diagram of a cross-section of an apothecium. (Redrawn from L. W. Sharp, *Fundamentals of Cytology.* Copyright 1943 by McGraw-Hill Book Company, New York. Used with permission of McGraw-Hill Book Company.)

hymenium (Fig. 24.6) and are more numerous and generally longer than asci.

In most Ascomycetes, the ascocarps are the direct result of cellular fusion. The description that follows is a generalized account of cellular fusion as it occurs in many Ascomycetes. The gametangia develop from the haploid mycelium growing within the host or substrate. The female gametangium, or oogonium, is a single cell with many nuclei (Fig. 24.3). Antheridia are elongated cells borne on short side branches of adjacent filaments. Both gametangia may have special accessory cells. After an antheridium establishes contact with an oogonium, the male nuclei pass into it. Male and female nuclei pair, but do not fuse as is normally the case. This process now stimulates the growth of hyphae from the oogonium and from the surrounding haploid mycelium. The cells forming from the oogonium are binucleate, each having one male and one female nucleus. This is not a true diploid condition; it may be designated by $n + n$ rather than $2n$, and the paired nuclei are frequently referred to as a **dikaryon.** Since it is from

these hyphae that the asci will eventually develop, they may be called **ascogenous hyphae.** The coordinated growth of the haploid and ascogenous hyphae results in the formation of the **ascocarp.** The terminal cells (Fig. 24.6) of the very much branched ascogenous hyphae, together with the ends of some haploid hyphae, form the **hymenial layer.** The hymenial layer is more or less surrounded on the outside and protected by the vegetative haploid hyphae which form a layer called the **peridium.** It is in the young ascus that the nuclei of the dikaryon finally fuse to form a true diploid cell. Meiosis occurs immediately after the nuclear fusion and is followed, in most species, by a mitotic division giving rise to eight ascospores.

The sexual phase of most Ascomycetes is thus complicated by the failure of the male and female nuclei to fuse immediately after the union of protoplasts from the antheridium and oogonium. This gives rise to a prolonged stage between the union of the sex protoplasts and the fusion of the sperm and egg nuclei. Fertilization in the Ascomycetes may

Class Ascomycetes   477

then be said to involve two steps: (a) the union of the two sex protoplasts, or **plasmogamy,** and (b) the union of the two nuclei, or **karyogamy.** These stages are shown diagrammatically in Figs. 24.3 and 24.6.

The gametangia of the Ascomycetes vary in structure and development. For instance, the simplest type of structure and development occurs in the yeasts, which are single-celled fungi. In some yeasts, two cells unite. Meiosis occurs immediately after conjugation, and ascospores develop (Fig. 24.9).

At the other extreme are filamentous forms in which the oogonium develops a beak and no antheridia are formed. Instead, bodies called **spermatia** are produced at some distance from the oogonium. The spermatia come in contact with the beak of the oogonium. A nucleus from the spermatium passes into the beak and migrates downward to the female nucleus in the inflated base of the oogonium, thus effecting fertilization.

In the Ascomycetes so far described, the development of the ascocarp has been preceded by cellular fusion. In some forms, an ascocarp may develop without cellular fusion.

Classification of the Ascomycetes

The fact that about 25,000 species of Ascomycetes exist, many of them severe pathogens, presents a formidable problem to the taxonomist. To group all these species into a system that will show their relationships and be convenient to use is no mean task. They are commonly grouped into two subclasses and seven orders.

Reproduction. Yeasts are normally single-celled Ascomycetes. In old cultures, the cells may remain attached, forming short, branched chains. Some yeasts divide by fission, as do bacteria. In the majority of yeasts, however, new cells grow out from the mother cell, much as a small bubble would form if a piece of thin rubber were made to expand through a small opening in some heavier material. The small "bubbles" formed from the mother yeast cell are called **buds.** They enlarge and finally separate from the parent cell. This process of vegetative reproduction is termed **budding** (Figs. 24.7A, B, C).

Most yeasts form asci, each containing from one to eight ascospores, the number being constant for a given species. As in other Ascomycetes, the production of asci usually is associated with a sexual cycle. In the formation of ascospores, the nucleus of a diploid cell divides into several nuclei, and each, with some associated cytoplasm, becomes delimited as a spore. The spores lie within the parent cell, which is essentially an **ascus** (Fig. 24.8). During the formation of ascospores, a reduction division occurs. Thus, in yeast, the parent cell is diploid, and the ascospores are haploid. In some yeasts, the nuclei of adjoining haploid vegetative cells fuse, the resulting zygote dividing to produce ascospores which multiply by vegetative division. In certain other yeasts, a fusion of haploid ascospores takes place.

The life cycle of *Saccharomyces cerevisiae,* the common yeast of commerce, is of considerable interest because it may reproduce asexually by budding in both haploid and diploid phases. Its life

Classification of Ascomycete Genera Discussed

Subclass	Order	Genus	Common name
Hemiascomycetidae	Endomycetales	*Saccharomyces*	Yeasts
	Taphrinales	*Taphrina*	Leaf-curl fungi
Euascomycetidae	Eurotiales	*Eurotium* (*Aspergillus*)	Green mold
		Talaromyces (*Penicillium*)	Blue mold
	Erysiphales	*Erysiphe*	Powdery mildews
	Clavicipitales	*Claviceps*	Ergot
	Helotiales	*Monilinia*	Brown rot fungus
	Pezizales	*Peziza*	Cup fungus

***Saccharomyces* (Yeasts).** Yeast cells may be spherical, ellipsoid, more or less rectangular, or, in vigorously growing cultures, sometimes hypha-like. Their rate of growth seems to influence their shape to some extent. Under ordinary microscopic magnification, living cells appear lacking in much structural detail: only an outer membrane, a protoplast with cytoplasm, vacuoles, and granules of reserve food may generally be seen. A nucleus is present but difficult to observe unless the slides are specially prepared (Fig. 24.7D).

history is shown in Fig. 24.9 and is of such a nature as to make this species particularly suited for studies of fundamental biological significance. The life cycle may be written simply as shown in the diagram.

Economic Importance of the Yeasts. The main sources of energy for many yeasts are sugars, which are oxidized in respiratory processes within the yeast cell. Oxygen is not required for the first steps of the oxidation. The sugar, nevertheless, is broken down to a simple organic acid (p. 239) and energy is re-

478 **Higher Fungi**

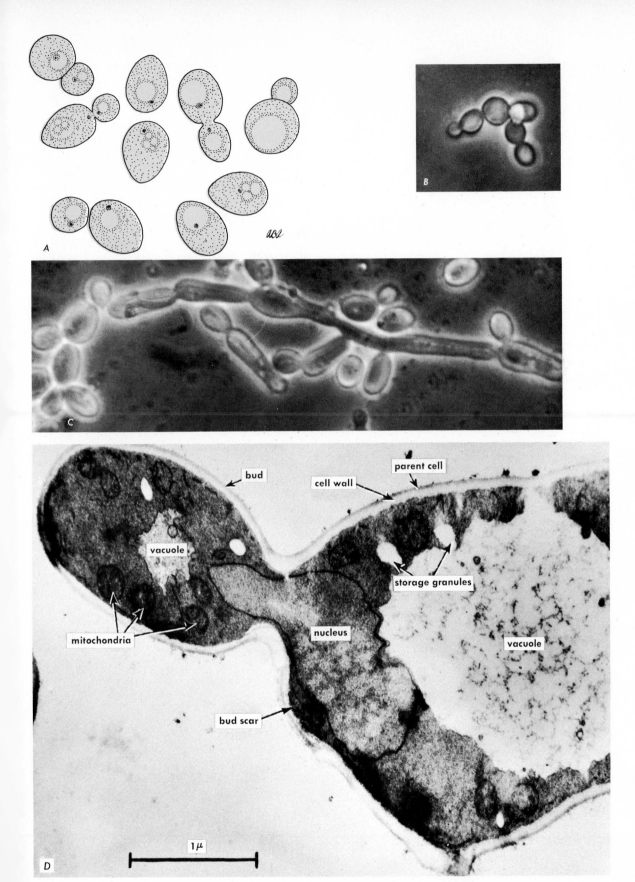

Figure 24.7 Yeasts in various stages of budding. *A, Saccharomyces cervisiae; B, Torulopsis apicola; C, Candida tropicalis; D, electron micrograph of budding in Saccharomyces cervisiae. A to C, about ×1000; D, ×25,000. (B, C, courtesy of H. Pfaff, D, courtesy of E. Vitols, R. J. North, and A. W. Linnane, J. Cell. Biol.* **9,** *689.)*

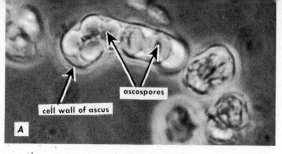

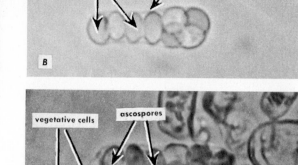

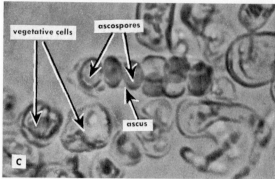

Figure 24.8 Ascospores of yeast (*Schizosaccharomyces octosporus*). *A* and *B*, as seen in ordinary light; *C*, as seen by phase contrast. (Material courtesy of Priscilla Stoner.)

leased in the process. If the oxygen supply is ample, this acid is oxidized, with the release of relatively large amounts of energy, to carbon dioxide and water. If the oxygen supply is deficient, the organic acid will be changed to carbon dioxide and *alcohol*. Thus, yeasts growing in a sugar solution well supplied with oxygen will produce carbon dioxide and water, and the yeast plants will multiply rapidly. On the other hand, yeasts growing in a sugar solution poorly supplied with oxygen will form carbon di-

oxide and alcohol and will multiply slowly. The oxidation of sugars to carbon dioxide and alcohol by yeasts without the presence of oxygen is known as **alcoholic fermentation.** This process is utilized in the production of industrial alcohol, wine-making, brewing, and bread-making.

Yeasts have played a very important part in vitamin research. Several vitamins are synthesized by yeast plants, which are an important commercial source of these highly valued substances.

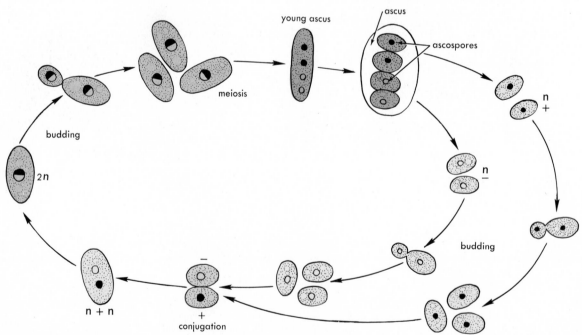

Figure 24.9 Life cycle of *Saccharomyces cerevisiae*.

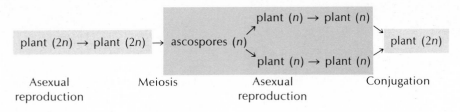

plant (2n) → plant (2n) → ascospores (n)
plant (n) → plant (n)
plant (n) → plant (n)
plant (2n)

Asexual reproduction   Meiosis   Asexual reproduction   Conjugation

Taphrina. All members of this genus are highly pathogenic to plants, especially fruit trees. They infect leaves, flowers, fruits, and young shoots. They induce unequal growth in leaf cells, causing diseases frequently referred to as **leaf curls.** The fungus causing peach leaf curl is well-known. The mycelium of this fungus is intercellular in the palisade tissue of leaves. Here, it stimulates some cells to more rapid growth, which results in wrinkling of the leaf. Eventually, the hyphae push their way between the epidermis and the cuticle, where the individual cells swell, round off and thicken their walls, and form a compact layer. These cells then elongate, rupturing the cuticle, and become asci, each one of which contains ascospores. Thus, the asci are produced at the surface of the infected organ. The opening buds as they start to grow in the spring are infected by spores that have overwintered between the bud scales. A copper fungicide applied just before the buds swell is a very effective control. Typical gametangia are not formed.

Form Genera *Penicillium* and *Aspergillus*. Since classification is based on the perfect stages, fungal forms in which the sexual stages have not been found cannot be accurately classified. These forms are grouped together in form genera based upon similarities of the asexual spores.

Asexual reproduction in *Penicillium* and *Aspergillus* is effected by means of conidiospores. In *Penicillium,* the spores are formed on profusely branched conidiophores (Figs. 24.10*A,* 24.11). In *Aspergillus,* the tip of the conidiophore swells and conidiospores form in long chains radiating from this swollen tip (Figs. 24.10*B, C*).

In only a few cases are sexual stages known for these forms, but even so they may be divided into five genera in the family Eurotiaceae. One of these may be briefly described. Ascogonia have been found to occur in the form species *Penicillium vermiculatum* which would place it in the true Ascomycete genus *Talaromyces. Penicillium vermiculatum* is thus also *Talaromyces vermiculatum* (Fig. 24.12).

In the yeasts, the ascospores are produced in typical vegetative cells; in *Taphrina,* special asci are formed, but they are not grouped in ascocarps. In

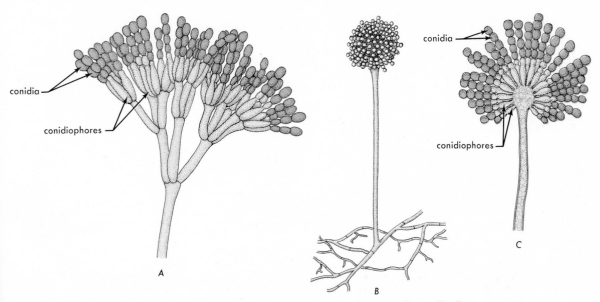

Figure 24.10 **Diagrams of different forms of conidiophores.** *A, Penicillium; B, C, Aspergillus; B,* three-dimensional appearance; *C,* optical section through *B.* (*A,* after Raper and Alexander, *J. Elisha Mitchell Sci. Soc.* **61,** 74–113.)

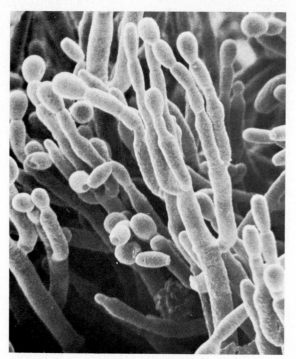

Figure 24.11 Stereoscan micrograph of conidiophores with conidia of *Penicillium digitatum*, ×3500. (Courtesy of E. Butler.)

Talaromyces (*Penicillium*), the asci are grouped in an ascocarp but distributed at random within it rather than being oriented in a hymenial layer. These ascocarps are of the cleistothecial type.

The ascogonium of *Talaromyces vermiculatum* is an elongated slender multinucleate cell, and the antheridium is simply a club-shaped, swollen hypha. Details of fertilization and meiosis are obscure, but eventually cleistothecia develop, containing scattered asci. The sexual cycle is shown in Fig. 24.12. Asexual reproduction is by means of conidia. It must be emphasized that the asexual stage is the usual and greatly predominant mode of reproduction, the sexual stage being rare indeed. The life cycle diagram thus gives much greater emphasis to the sexual stage than is warranted.

Penicillium and *Aspergillus* are probably the most widespread fungi. They are the common blue, green, and black molds which occur on citrus fruits, jellies, and preserves. Their conidia are everywhere in the air and soil, and in the biological laboratory they are frequent contaminations in culture media. Figure 24.13*A* shows a growth of *Penicillium* on an apple, and Fig. 24.13*B* shows it on an orange. Enzymes secreted by these fungi are particularly active in digesting starch and other carbohydrates. When purified, these enzymes are important industrial preparations. *Aspergillus oryzae* is used in the preparation of rice wine and soybean sauces. Several species of *Aspergillus* are important in cheese man-

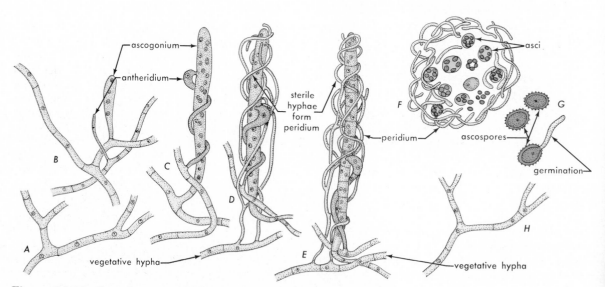

Figure 24.12 Sexual stages in *Talaromyces vermiculatum* (*Penicillium vermiculatum*). *A*, vegetative hypha; *B*, formation of antheridium and ascogonium; *C*, *D*, and *E*, development of antheridia and ascogonia with first stage of fertilization; *F*, cross-section of the loosely formed primitive ascocarp; *G*, ascospores; *H*, vegetative hyphae.

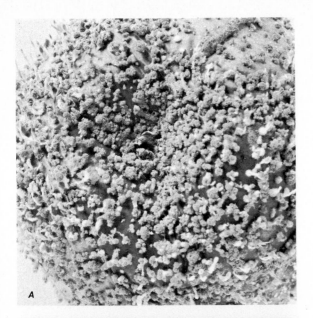

A

B

Figure 24.13   *A, Penicillium* on apple; *B, Penicillium* on orange. (*A,* courtesy of Plant Pathology Department, University of California.)

ufacture. A disease resembling tuberculosis is caused by *Aspergillus fumigatus,* and a number of other *Aspergillus* species cause diseases in plants. Strains of *Aspergillus flavus* form aflotoxins which are toxic to animals.

Erysiphe. *General Characteristics.* The powdery appearance of the surface of leaves infected with many members of *Erysiphe* and related genera sug-

gests the common name, **powdery mildews.** All the powdery mildews are obligate plant parasites (Figs. 24.14*B, C;* see Color Plate 13). Their food requirements are closely integrated with the metabolism of the host plants, and frequently the host plant is not killed. This relationship ensures the fungus a continued food supply. Many can live only on a special host, whereas others have a wide host range. They grow poorly or not at all in artificial culture media. These characters indicate a high degree of specialization and a longstanding relationship between parasite and host.

The mycelium is generally confined to the surface of the leaves, flowers, or fruits (Figs. 24.14*B, C*). Haustoria (Fig. 24.15) penetrate epidermal and parenchyma cells, from which they secure nourishment. At first the mycelium on the surface of the leaf appears like a delicate cobweb. Eventually, it assumes a white powdery or dusty appearance owing to the development of numerous conidiospores.

Asexual Reproduction. This type of reproduction is effected by conidiospores and usually accounts for the rapid propagation of the fungus during the growing season. The conidiophores are short filaments that stand outward from the mycelium on the surface of the host (Fig. 24.16). The spores themselves may remain attached to each other and form long characteristic chains. Sulfur dusted on the host plants at this stage of the fungal life cycle is an effective control to prevent infections from these spores.

Sexual Reproduction. The ascogonium is always located at the end of a hypha and may be slightly swollen. The antheridia are small cells, also occurring at the tips of hyphae. An antheridium becomes closely appressed to an ascogonium; an opening appears in the wall separating the gametangia; and the male nucleus migrates through it into the ascogonium. Binucleate hyphae develop from the binucleate zygote, and simultaneously adjacent haploid vegetative hyphae also develop. A small **cleistothecium** is formed and encloses one or several asci (Figs. 24.5, 24.14*A*). Nuclear fusion and reduction division take place in the ascus, and ascospores develop. The ascospores are usually discharged by force from the cleistothecium.

The cleistothecia are readily seen as small black specks on the surface of infected leaves. Appendages, characteristic of the genera, extend outward from the cleistothecia (Figs. 24.4*B,* 24.5*A, B*). Appendages may aid in the dispersal of the cleistothecia and in attaching them to a new host.

While the powdery mildews may not kill their host, they weaken it and greatly reduce the crop

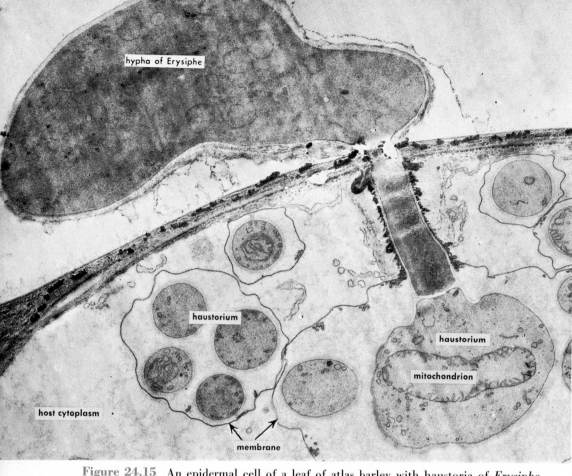

Figure 24.15 An epidermal cell of a leaf of atlas barley with haustoria of *Erysiphe graminis nordei.* Note that the haustoria are separated from the hyaloplasm of the host cell by a membrane and that the haustoria are thus in a vacuole-like space. Cellular organelles are prominent in the haustoria and absent from the host cell protoplast, ×11,200. (Courtesy of C. E. Bracker.)

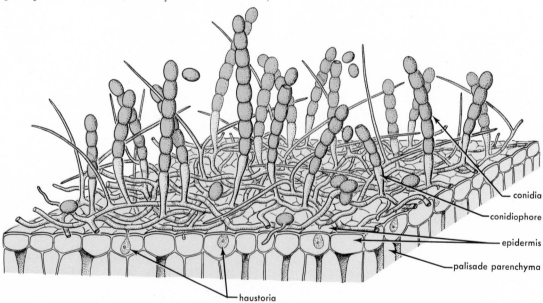

Figure 24.16 Powdery mildew on leaf surface. (Redrawn from *Selecta Fungorum Fungorum Carpologia of the Brothers Tulasne, L. R. and C.,* vol. I, 1861–65. Translated into English by W. B. Grove; A. H. R. Buller and C. L. Shear, eds., Oxford (Clarendon Press), 1931.)

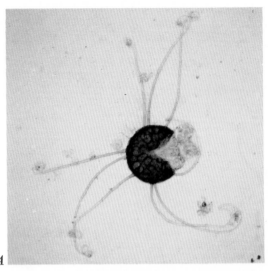

B

A

Figure 24.14 *A*, cleistothecium with asci of powdery mildew (*Uncinula*), ×5; *B*, whitish areas on grape leaf are mycelium of *Uncinula*, ×⅓; *C*, grapes heavily infected with powdery mildew, ×½. (Courtesy of Joe Ogawa.)

C

Color Plate 13

Various Fungi

A

B

Figure 24.18 *A*, apothecia of the brown rot fungus (*Monilinia fructicola*), ×½; *B*, brown rot on cherry compared with a healthy fruit, ×½; *C*, brown rot on peach, ×⅕. (Courtesy of Joe Ogawa.)

C

Figure 24.46 Various types of lichens. *A*, the conspicuous green foliose lichen is *Peltigera aphthosa*, and it is surrounded by the white branches of the fruticose lichen *Cladonia* sp.; *B*, crustose lichen *Acarospora flava* colors a rock on the skyline of a Sierra Nevada ridge; *C*, *Ramalina reticulata*, a fruticose lichen, clothes the branches of an oak in the foothills of the Coast Ranges along the Pacific Coast.

Figure 24.17 *Peziza. A*, apothecia quiescent; *B*, cloud of ascospores discharged upon stimulation of apothecia, $\times\frac{1}{3}$.

yield. Powdery mildews cause diseases of apples, grasses, grains, grapes (Fig. 24.14), cherries, and many other plants.

Monilinia. The ascocarps of the order to which *Monilinia fructicola* belongs are cup- or disk-shaped (apothecia) (Fig. 24.18*A;* see Color Plate 13). They may be as much as 6 inches in diameter, depending on the species, and are sometimes brilliantly colored. Members of one genus, *Morchella* (the morels), are edible. A saprophyte, *Peziza* (Fig. 24.17), is common and probably the best-known. Brown rot of stone fruits, a very severe disease of peaches, cherries, plums, apricots, and nectarines, is caused by a disk fungus (*Monilinia,* Fig. 24.18), which infects mainly blossoms and fruits (Fig. 24.18). Because of these infections, the orchardist may experience lower yields or spoilage of the ripe fruit on the tree or during shipment.

Life History of *Monilinia fructicola*

1. Ascospores formed during spring, or conidia formed on a mummy, infect blossoms (Fig. 24.19). Conidia form on blighted blossoms.
2. Conidiospores produced in blossoms rapidly infect healthy fruits (Fig. 24.19) under favorable weather conditions.
3. An extensive mycelium develops within the fruit, causing it to rot (Figs. 24.18, 24.19).
4. The rotted fruit dries, becoming a mummy. It may drop to the ground or remain on the tree (Fig. 24.19).
5. In early spring, gametangia form in the mummied fruits that have fallen to the ground.
6. Fertilization ensues.
7. Apothecia develop from ascogenous and vegetative hyphae (Figs. 24.18*A*).
8. The resulting asci shoot ascospores several inches into the air (Fig. 24.17) and reinfection occurs.
9. Bordeaux mixture or a benzimidazole fungicide applied as the buds swell, together with a destruction of the mummies, will help control blossom blight.
10. Preharvest fruit infection is prevented by sulfur, captan or benzimidazole fungicide applications.

Claviceps. The genus *Claviceps* is parasitic on grasses, including grains. A dormant mycelium that replaces the mature grain is known as **ergot.** It pos-

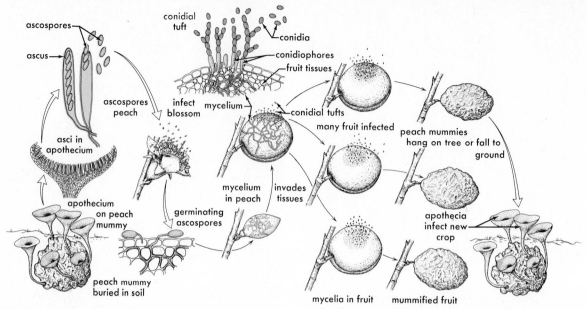

Figure 24.19 Diagram of life cycle of the brown rot fungus (*Monilinia fructicola*). (Based on G. N. Agrios, *Plant Pathology*. Academic Press, New York, 1969.)

sesses several alkaloids that have medicinal properties. Ergot constricts the blood vessels, particularly those that pass into the hands and feet, thus depriving the extremities of a normal blood supply. In humid summers in Central Europe, *Claviceps* may infect rye heavily. In centuries past, before the nature of the fungus was understood, ergot would be milled along with the grains of rye. The contami-

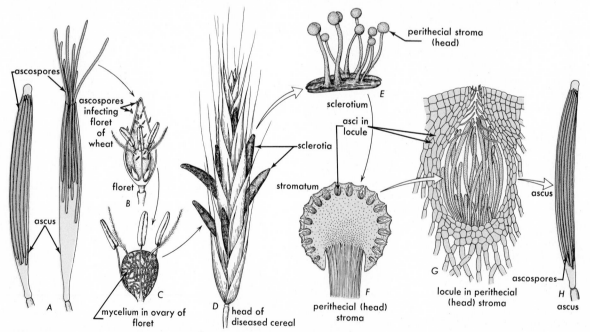

Figure 24.20 Life cycle of *Claviceps*. (*A, D, E, F,* and *H*, from E. A. Gäuman, *Comparative Morphology of Fungi*. McGraw-Hill, New York, 1928; *B*, From P. A. Munz, and D. D. Keck, University of California Press, Berkeley, 1963; *G*, G. M. Smith, 1955. *Cryptogamic Botany*, vol. I, *Algae and Fungi*, McGraw-Hill, New York, 1955.)

486 **Higher Fungi**

Figure 24.21   Ergoted head of a grass. *A*, whole head showing untidy appearance of head; *B*, enlargement of *A*, showing several spikelets with a large ergot protruding from the base of one, ×20.

nated flour, which might contain as much as 10% of powdered mycelium, would be baked in bread. A continued diet of bread from this flour resulted in much misery. Because of the contraction of the blood vessels and the limited supply of blood reaching feet or hands, gangrene set in. Hands, arms, and legs would die and finally drop off. The disease was known as "Holy Fire" because of a burning sensation in the extremities. Today, ergot is a valued drug used to control hemorrhage, particularly during childbirth.

With the knowledge of the poisonous nature of the ergot, diseased grain is no longer milled and *Claviceps* is not of concern in human diet. However, cattle may occasionally feed on infected grasses having ergots in place of seeds (Fig. 24.21) and thus be poisoned.

Life History of *Claviceps*

1. Ascospores are mature when the first flowers of rye, wheat, or other host grasses open. A given race of *Claviceps* infects certain species of cereals or grasses. The mature ascospores infect the young flowers (Fig. 24.20).

2. A mycelium develops throughout the ovary of the infected flower. Finally it completely replaces the ovary, assuming its general shape (Fig. 24.20).

3. Conidia appear on the surface of this mycelium, together with a sticky, sweet secretion. Insects collect the secretion and, in doing so, disperse the conidia.

4. The compact mycelium replacing the ovary grows upward and becomes a hard and horny body. It is somewhat longer than the mature grain and is purple in color. This dormant, elongated mycelium is the **ergot** (Fig. 24.21*B*). It is composed of a compact mass of tough hyphae that, in a dormant condition, survive periods unfavorable to growth (Fig. 24.20).

5. When conditions become suitable, the hyphae of the ergot produce ascogonia and antheridia. After fertilization the ascogenous hyphae, accompanied by adjacent vegetative hyphae, develop a short, upright stalk. The stalk supports a number of small flask-shaped perithecia (Fig. 24.20).

6. Meiosis takes place; asci and ascospores are formed within the perithecia (Fig. 24.20).

Class Ascomycetes   487

7. As the asci mature, they extend out of the perithecial opening. Pressure develops in the asci, shooting the ascospores into the air.

Significant Features of the Ascomycetes

General Characteristics

1. The mycelium is septate. The cells of the main vegetative hyphae have a single haploid nucleus. The ascogenous hyphae have two haploid nuclei ($n + n$) in each cell.
2. Many are severe plant pathogens, causing diseases of fruit and nut trees and of grains.
3. Some saprophytic forms, such as the yeast, are of considerable economic importance.
4. Yeasts are simple Ascomycetes, consisting of a single cell.

Reproduction

1. Asexual
   (a) Usually by means of conidiospores.
   (b) By budding (in the yeasts).
2. Sexual
   (a) Definite gametangia are not developed in the more primitive species; typical ascogonia and antheridia form in the more advanced types.
   (b) Fertilization generally involves:
      (1) The union of two protoplasts without the fusion of the nuclei (plasmogamy).
      (2) Stimulation to growth of ascogenous hyphae ($n + n$) and of haploid vegetative hyphae.
      (3) Union of two haploid nuclei in ascus (karyogamy), ($n + n \rightarrow$ **2n**).
   (c) An ascocarp develops as a result of the growth of ascogenous and haploid vegetative hyphae.
      (1) The inner layer (hymenium) of the ascocarp gives rise to asci.
      (2) The outer layer (peridium) of the ascocarp is composed of haploid vegetative filaments.
   (d) There are three types of ascocarps, (1) cleistothecia, (2) perithecia, and (3) apothecia.
   (e) Meiosis occurs in the formation of the ascospores in the ascus. Eight ascospores are generally produced in each ascus.

Class Basidiomycetes

Representatives of the Basidiomycetes include mushrooms, toadstools, bracket fungi, puffballs, and species causing such diseases as wheat rust and corn smut. Many of them are saprophytes, being particularly important in the decay of dead forest trees. Others are parasitic and cause considerable damage to forest and orchard trees, wheat, corn, onions, snapdragons, roses, and many other plants.

Reproduction and Classification

No specialized sex organs are formed by the Basidiomycetes; nearly all, however, reproduce sexually. Conjugation can occur (a) between ordinary hyphal cells, (b) between two special cells, or (c) between special sperm-like bodies and receptive hyphae. Some species are **heterothallic;** that is, only plus and minus strains conjugate. Other species are **homothallic,** conjugation taking place between any two cells of any two hyphae, or even between cells of the same hypha. In both types, prior to conjugation the cells of the vegetative mycelium contain a single nucleus; after conjugation the resulting cell contains two haploid nuclei ($n + n$). The two nuclei in each cell do not fuse until just before meiosis (Figs. 24.24A, B). Fusion of the two nuclei and subsequent meiosis take place in special cells known as **basidia** (Figs. 24.24C, D). The meiospores formed by these basidia are called **basidiospores.** The basidia may be either (a) a single club-shaped cell (Fig. 24.22), (b) a single short, filamentous cell, or (c) a short four-celled filament (Fig. 24.22). The club-shaped basidia occur on special spore-bearing structures, as in mushrooms, bracket fungi, and puffballs. This type of basidium characterizes the subclass Homobasidiomycetidae. When the basidium is a single short, filamentous cell or a short four-celled filament, it results from the germination of special resistant spores (Figs. 24.31, 24.38). Basidiomycetes with this type of basidia constitute the subclass Heterobasidiomycetidae. In both types, the basidiospores are attached to the basidia by short stalks, the **sterigmata** (singular, **sterigma**).

A generalized life history of a Basidiomycete is shown diagrammatically in Fig. 24.22.

Subclass Homobasidiomycetidae

General Characteristics and Distribution. The mushrooms, bracket fungi, and puffballs are Homobasidiomycetidae. Only a few are parasitic, but these few do extensive damage to forest trees (Figs. 24.29, 24.30), some orchard trees, and a small number of garden crops. Several Homobasidiomycetidae are always found growing in association with certain species of trees. It is believed that some of them may form **mycorrhizae:** a symbiotic relationship with tree roots. The fungus is able to make nitrogen and other nutrients more available to the roots. Certain Homo-

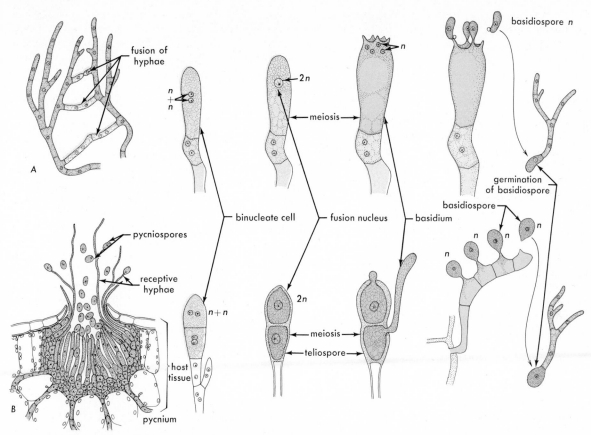

Figure 24.22   Generalized life cycles of Basidiomycetes showing fusion of protoplasts, fusion of nuclei, and meiosis (production of basidiospores). *A*, common mushroom; *B*, a fungus causing wheat rust. *A*, redrawn from G. M. Smith, *Cryptogamic Botany*, vol. I, *Algae and Fungi*, McGraw-Hill, New York, 1955, based on H. Kniep, and A. H. R. Buller, *Researches on Fungi*, vol. VII, *The Sexual Process in the Uredinales*. Royal Society of Canada, University of Toronto Press, 1922; *B*, after A. H. R. Buller (see *A*) and R. F. Allen, *Phytopath.* **23,** 574–586, 1933.)

basidiomycetidae obtain their carbohydrates from wood and may even cause extensive decay of timbers in mines and buildings. Specialized tissues and organs may be formed, some of the $n + n$ hyphae becoming specialized to transport food and water, and others becoming tough and woody.

Although both haploid and $n + n$ mycelia develop extensively, the $n + n$ phase is of special interest because it gives rise to the spore-bearing body, or **basidiocarp** (sporophore, Fig. 24.23), in which the basidia are developed. The hyphae form a tangled mass in the substrate or host and emerge to form a basidiocarp—the mushroom, puffball, or bracket fungus. As the basidiocarp grows, the ends of certain hyphae generally become aligned in a layer that will bear spores when mature. This is the **hymenium layer,** some cells of which gradually enlarge to become basidia.

The development of the basidiospore is shown in Fig. 24.24. The young basidium contains the $n + n$ nuclei, or the dikaryon (Fig. 24.24*A*). These two nuclei fuse to give the true diploid nucleus shown in Fig. 24.24*B*. Meiosis gives rise to the four haploid nuclei destined to pass into the basidiospores appear (Fig. 24.24*C*). When this stage is reached, the sterigmata have extended outward from the tip of the basidium and are ready to receive the basidiospore nuclei. One of the nuclei present in the basidium of Fig. 24.24*C* would have passed through the sterigma shown in the figure. The basidiospore of Fig. 24.24*D* is mature and ready for dispersal as the haploid nucleus has moved into it. Note that one nuclear division has been completed here.

There are five orders of Homobasidiomycetidae. We shall consider mainly the order Agaricales, of which common mushrooms and bracket fungi are

Class Basidiomycetes   **489**

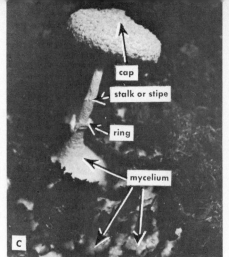

Figure 24.23 Stages in the development of a basidiocarp. A small button just appearing from the substrate shows at *A*; *B*, enlarging basidiocarp, gills still enclosed; *C*, cap open showing mature basidiocarp; *D*, additional basidiocarps grow around the first-formed one to give a cluster of fruiting structures, $\times \frac{1}{2}$.

members. Many of them are edible, and a few are violently poisonous. The common edible mushroom of commerce, *Agaricus campestris*, is a species of the order (Fig. 24.25). The mycelium, or, as it is known to the mushroom growers, the **spawn**, grows extensively in the substrate. It is common in many pastures and is cultivated in well-rotted organic matter.

The Basidiocarp. The basidiocarp of a typical mushroom such as *Agaricus campestris* (Fig. 24.25) consists of a short upright stalk or **stipe**, attached at its base to a mass of mycelium and expanding on top into a broad cap or **pileus.** Many mushrooms have a ring of tissue around the stipe rather close to the pileus. This ring occurs because when it is young, the pileus was attached to the stipe at this point. The underside of the pileus of *Agaricus* is formed by thin **gills** radiating outward from the stipe. These gills are lined with a hymenium or spore-bearing layer. If a young gill is sectioned, the majority of basidiospores will still be attached to the basidia. Four basidiospores are attached to each basidium by short stalks, the **sterigmata.** If the gill is very young, developing basidia on which basidiospores have not yet been formed may be observed (Fig. 24.26). Fig. 24.27 shows a section through three gills. Note two stages of development of basidia and the loose tangle of hyphae forming the center of the gill. When a basidiospore is discharged, it is shot horizontally from its position on the basidium straight into the space between two gills. A basidiospore is somewhat like a toy rubber balloon, having a large volume for its weight. It is shot from the basidium with a force sufficient to carry it about midway between the two gills. It then falls straight downward. It has been estimated that some basidiocarps may discharge as

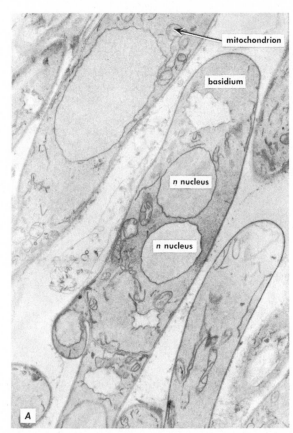

Figure 24.24 Basidiospore formation in *Schizophyllum commune.* Above. *A*, young basidium with $n + n$ nuclei, $\times 3000$. Right. *B*, basidium with $2n$ nucleus, $\times 5000$. *C*, four n-nuclei after meiosis, $\times 3000$. Note development of sterigma. *D*, basidiospore attached to sterigma; the n-nucleus is completing its first division. (Courtesy of K. Wells.)

490   Higher Fungi

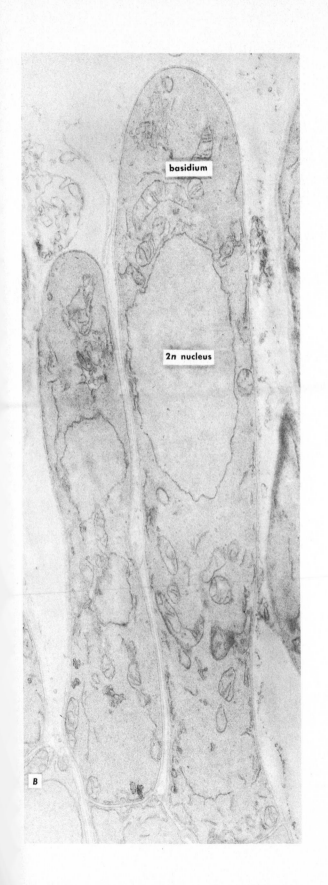

basidium

2n nucleus

B

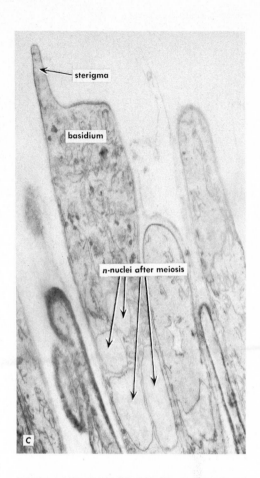

sterigma

basidium

n-nuclei after meiosis

C

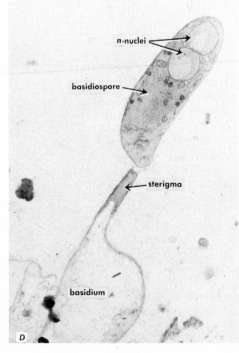

n-nuclei

basidiospore

sterigma

basidium

D

Figure 24.25 *Agaricus campestris*, the common edible mushroom, ×1.

many as a million spores a minute for several days.

If the cap of a mushroom is placed on a piece of paper and carefully protected from wind, the spores will fall straight downward and form a **spore print.** Spore prints are important in the identification of mushrooms, since they indicate the color of the spores and the pattern of the gills.

The spores of *Agaricus campestris* are brown, whereas those of other Agaricales may be white, black, red, yellow, or ochre. The spores of *Amanita muscaria,* the "Fly-Mushroom," are white. *Amanita muscaria* has other distinguishing features: When the basidiocarp is in the button stage, it is completely surrounded by a thin tissue known as the universal veil (Figs. 24.8A, B, C). As the basidiocarp expands, this veil is torn (Figs. 24.28D, E). Part of it remains attached to the cap (Figs. 24.28F, G), where ruptured fragments of it may be seen in the mature mushroom; the remaining part stays underground, forming a cup-shaped structure out of which the stalk grows (Figs. 24.28F, G). The enlarged basal structure is referred to as the **volva** or "death cup." The fragments of white on the reddish yellow cap are remnants of the universal veil. The white gills and ring are visible. The volva of *A. muscaria* tightly clasps the stipe; it may be seen here as the swollen, scaly base (Fig. 2.7A; see Color Plate 2).

Amanita phalloides is one of the most deadly poi-

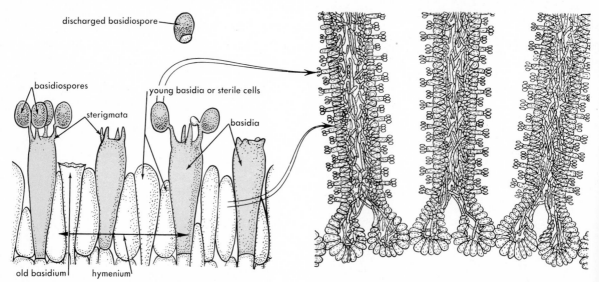

Figure 24.26 Section of gill cut parallel with plane of basidia. It shows the production of basidiospores. (Redrawn from A. H. R. Buller, *Researches on Fungi.* Used with permission of Longman Green Ltd.)

Figure 24.27 Section showing three gills of a mushroom. (Redrawn from A. H. R. Buller, *Researches on Fungi.* Used with permission of Longman Green Ltd.)

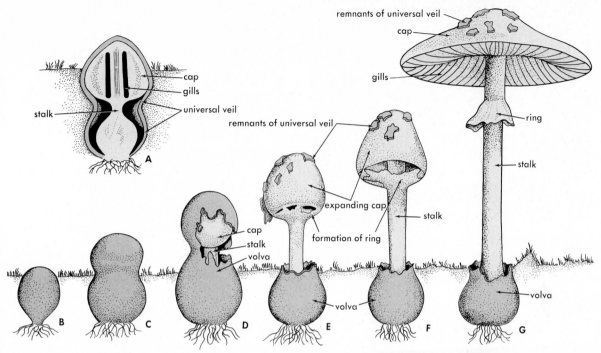

sonous of all mushrooms. It may be recognized by (a) its white spores; (b) the volva or "death cup" (one must be sure to dig deeply); (c) the ring on the stipe below the cap; (d) the color of cap, which is reddish in the center and fades to yellow toward the edges; and (e) the torn remains of the universal veil on top of the cap. As a general precaution, one should avoid all white-spored mushrooms with a death cup and a ring on the stipe. Many species of Agaricales, however, are excellent eating.

Not all Agaricales have gills. In some, the lower surface of the cap or bracket has many small pores that extend upward to form small tubules (Fig. 24.29). The hymenium lines the tubules. Many bracket fungi found on the trunks of forest trees are **pore fungi.**

Some bracket fungi cause serious damage to forest trees and lumber in mines and wooden buildings. During World War II, much green lumber had to be used, and destruction by fungi of improperly cured lumber resulted in much loss. One of the best-

Figure 24.29 The underside of the pore fungus *Polyporus brumalis*, $\times\frac{1}{3}$. (Courtesy of Roche.)

Figure 24.30 Basidiocarp of *Fomes applanatus*, ×⅓. (Courtesy of Brownell.)

known fungi in this group is *Fomes applanatus* (Fig. 24.30). It causes a disease of forest trees known as *white-mottled rot,* to which many hardwoods, such as poplar, birch, maple, basswood, oak, and elm are susceptible. True firs, Douglas fir, spruce, and hemlock may also suffer from this disease.

The basidiocarp of *Fomes applanatus* is a very elegant shelf-shaped bracket or **conk.** The conks are perennial and grow to a large size. They are gray on top, and the lower surface has millions of pores lined with a creamy-white hymenium.

Most of the bracket fungi of the forest are facultative parasites. They normally live in the dead heartwood, to which they gain entrance by wounds. They are, however, able to invade the living cells of sapwood, which they may destroy.

Other types of basidiocarps also occur, such as those in puffballs and bird's-nest fungi.

Subclass Heterobasidiomycetidae

General Characteristics. The basidium of the Heterobasidiomycetidae is a short, swollen filament developing directly from a specialized resistant spore (Figs. 24.31, 24.38). Most members of the subclass are parasites on vascular plants. Like the Homobasidiomycetidae, they possess, as far as is known, both haploid and $n + n$ phases, and reduction division takes place in the development of the basidia or basidiospores. Asexual reproduction may occur in either the haploid or $n + n$ phase or in both. Conjugation may take place between two nonspecialized cells, as in the Homobasidiomycetidae. In some forms, however, special cells conjugate or a sperm-like cell may conjugate with a receptive hypha.

Some of the Heterobasidiomycetidae may have

two hosts; that is, separate phases of their life history are passed on different plants. Such forms are **heteroecious.** If only one host is required to complete their life history, they are said to be **autoecious.**

Uredinales. The Uredinales, or rust fungi, are of worldwide distribution. They are known to infect a very large number of higher plants, and all are obligate parasites. Black or red spores develop in pustules (**sori**) on the leaves of many grasses, roses, and mallows, and they (particularly the red spores), suggested the name of **rust.**

There are many types of rust fungi and they show marked variation in their life histories. In nearly all of them, however, basidiospores are produced on short four-celled basidia that usually develop from an overwintering spore, known as a **teliospore** (Figs. 24.31, 24.36). Fusion of the two haploid nuclei and reduction division occur during the development of the basidia and basidiospores. The details of conjugation are obscure in many species.

Common wheat rust, caused by *Puccinia graminis,* is of great economic importance, and its life history is well-known. It produces all the different types of spores common to the order. We shall follow its life history in detail.

Life History of Puccinia graminis. In the early spring, the overwintering teliospores germinate on the soil or stems of wheat, in fact any place where there is moisture.

Teliospores of *Puccinia graminis* are two-celled spores. A short four-celled basidium develops from each cell; only one is shown in Fig. 24.31. A single haploid basidiospore is formed from each cell, so that four basidiospores eventually develop from

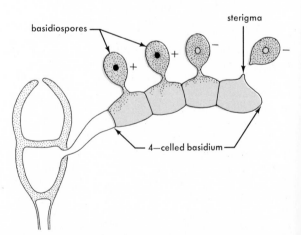

Figure 24.31 One stage in the life cycle of *Puccinia graminis.* Teliospore with basidium and basidiospores.

each cell of the teliospore. Two of these four basidio-spores are plus strain (+), two are minus strain (−).

The basidiospores are carried by air currents and can infect only the young leaves, fruit, or twigs of common barberry bushes. They germinate chiefly on barberry leaf surfaces and produce a germ tube, which penetrates the epidermis. Soon a mycelium develops in the tissues of the host. Then, after a time, pustules, called **spermagonia,** appear on the upper surface of leaves. In section a spermagonium is pear-shaped and has a small pore opening through the upper epidermis to the exterior of the leaf (Figs. 24.32, 24.33D). Some hyphae extend upward through the pore; others, lining the interior of the spermago-nium, cut off large numbers of sperm-like cells, called **spermatia.** These are forced out of the sper-magonia through the pores.

Puccinia graminis is heterothallic; both plus and minus spermagonia and spermatia appear on bar-berry leaves. These develop from plus or minus ba-sidiospores, both of which are haploid. Spermatia are transferred by wind or insects to adjacent sperma-gonia, where they come in contact with the **re-ceptive hyphae,** which are essentially female game-tangia. A plus (+) spermatium fuses with a cell of a minus (−) receptive hypha, and vice versa. An *n* + *n* diploid mycelium results; its cells each con-tain two haploid nuclei.

Shortly after fusion, a mass of *n* + *n* hyphae ap-pears in the spongy parenchyma of the barberry leaf close to the lower epidermis. Here, a second series of pustules called **aecia** are formed (Figs. 24.32, 24.33A, B). The aecia face downward, and the upper side toward the leaf mesophyll becomes lined with *n* + *n* hyphae. The growth of the aecia eventually brings about the rupture of the lower epidermis, and binucleate **aeciospores,** cut off from the ends of the numerous hyphae lining the base of the aecium, are exposed to the atmosphere (Figs. 24.32, 24.33C, D, E). The aeciospores thus released are disseminated by wind and they can infect only *wheat* plants.

The hyphae of germinating aeciospores gain en-trance to wheat plants by growing through the sto-mata. A delicate mycelium develops and the hyphae form haustoria that draw nourishment from sur-rounding cells of the wheat plant but do not kill them. About 10 days after infection, red binucleate spores, **uredospores,** are produced by the mycelium. The epidermis of the wheat plant is ruptured by the mass of spores, forming open sori, the **uredia** (Figs. 24.32, 24.34A, B, C). The uredospores thus produced are carried by wind currents to other wheat plants. The fungus is spread throughout a grain field during the growing season by the red uredospores. The progress of the disease throughout a wheat-growing

area may be so rapid as to attain epidemic propor-tions.

As the wheat begins to mature, the production of red uredospores by the infecting mycelium gives way to the production by the same mycelium of heavy-walled two-celled overwintering spores—**teliospores.** The sori producing teliospores are called **telia** (Figs. 24.32, 24.34D, E).

Each cell of the teliospore is, at the outset, binu-cleate; the two haploid nuclei, one (+) and one (−), are still present and distinct (Figs. 24.32, 24.34E). These two haploid nuclei now fuse **(karyogamy),** the true diploid condition resulting. As in the Ascomy-cetes and Homobasidiomycetidae, meiosis follows directly upon nuclear fusion. In the rusts, a short basidium grows from each cell of the teliospore, the four haploid nuclei resulting from meiosis migrate into it, and cell walls develop. Basidiospores now form, one from each cell of the basidium (Figs. 24.35, 24.36), and the cycle starts again.

Puccinia graminis is **heteroecious;** it requires two hosts to complete its life cycle. Certain other rusts are **autoecious,** needing but one host to complete their life cycle. Not all rusts develop all the spore types found in *Puccinia graminis,* nor do they all have an identical life history.

Wheat rust may be controlled in cold climates by eradicating all barberry bushes (the alternate host) near wheat fields. Barberry will not be present for new infection in the spring, and if uredospores are killed by the winter cold, the rust will die. In regions with warm winters, the uredospores may overwinter, wheat may be infected in the spring, and wheat rust may therefore be more difficult to control.

Different varieties of *Puccinia graminis* are able to attack other grains. Variety *avenae* attacks oats; vari-ety *tritici* attacks wheat. Breeding work with wheat, however, has shown that many **biological strains** exist even within the variety *tritici.* Indeed, new rust strains seem to appear from time to time, and so it becomes necessary to continue the breeding of wheat varieties that will be relatively resistant to them.

Significant Steps in the Life History of
Puccinia graminis

1. Two hosts, wheat and barberry, are necessary to complete the life history. In other words, *Puc-cinia graminis* is heteroecious.
2. Uredospores (one-celled, red) and teliospores (two-celled, black) are produced on the wheat plant. Uredospores can infect other wheat plants; they are killed by vigorous winters. Telio-spores cannot infect any plant; they are over-

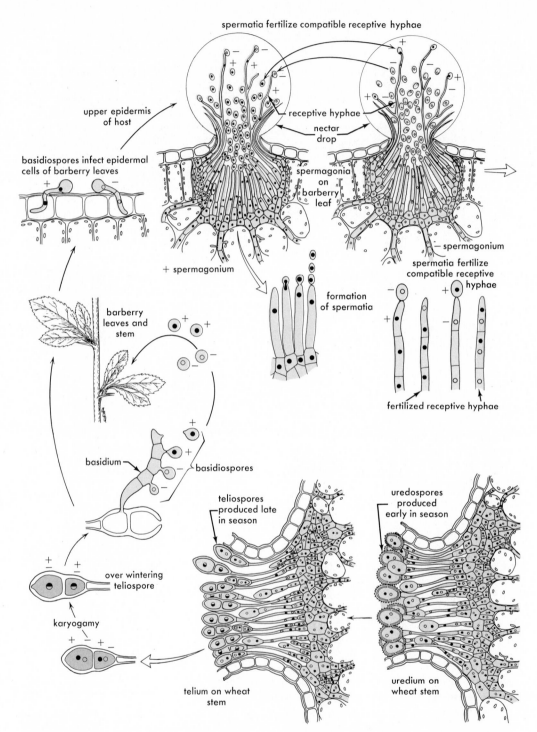

spermatia fertilize compatible receptive hyphae

upper epidermis of host

basidiospores infect epidermal cells of barberry leaves

receptive hyphae

nectar drop

spermagonia on barberry leaf

spermagonium

+ spermagonium

spermatia fertilize compatible receptive hyphae

barberry leaves and stem

formation of spermatia

fertilized receptive hyphae

basidium

basidiospores

teliospores produced late in season

uredospores produced early in season

over wintering teliospore

karyogamy

telium on wheat stem

uredium on wheat stem

Figure 24.32 Diagram of the life cycle of *Puccinia graminis*. (Based on A. H. R. Buller, *Researches on Fungi.* vol. VII, *The Sexual Process in the Uredinales*, Royal Society of Canada, University of Toronto Press, 1922; R. F. Allen, *Phytopath.* **23**, 574–586, 1933; and E. A. Bessey, *Morphology and Taxonomy of Fungi*, The Blakiston Co., Philadelphia, 1950.)

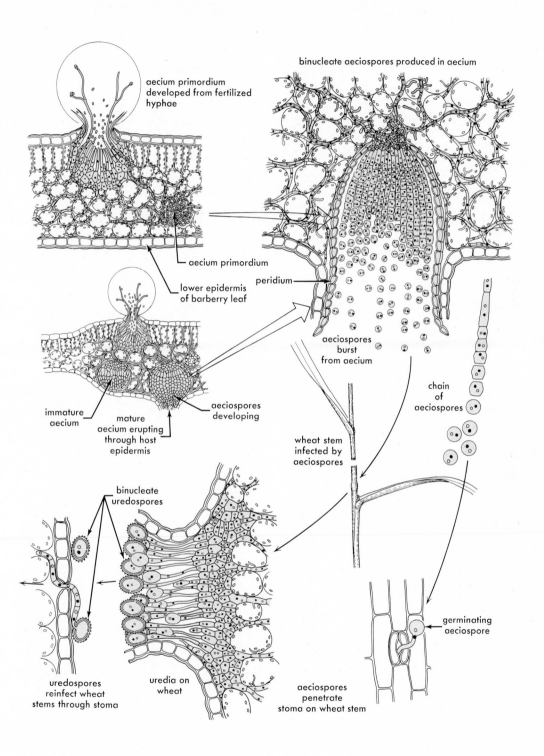

binucleate aeciospores produced in aecium

aecium primordium developed from fertilized hyphae

aecium primordium

lower epidermis of barberry leaf

peridium

aeciospores burst from aecium

chain of aeciospores

immature aecium

mature aecium erupting through host epidermis

aeciospores developing

wheat stem infected by aeciospores

germinating aeciospore

binucleate uredospores

uredospores reinfect wheat stems through stoma

uredia on wheat

aeciospores penetrate stoma on wheat stem

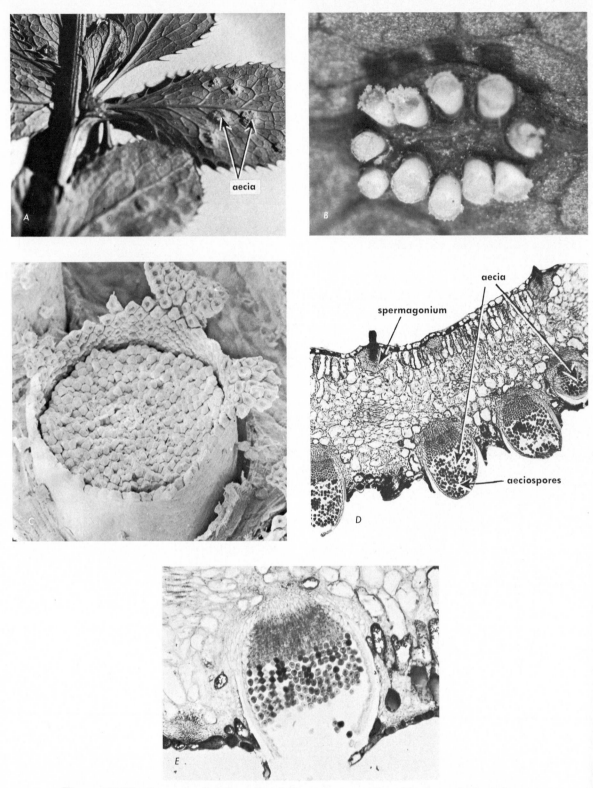

Figure 24.33 Stages in the life cycle of *Puccinia graminis* on barberry leaf. *A*, lower surface of barberry leaf (*Berberis vulagaris*) showing groups of aecia, ×1; *B*, enlarged view of groups of aecia, ×150; *C*, stereoscan view of aecium with aeciospores, ×250; *D*, cross-section of barberry leaf showing spermogonium on upper surface, aecia on lower surface, ×150; *E*, section of an aecium, ×1000.

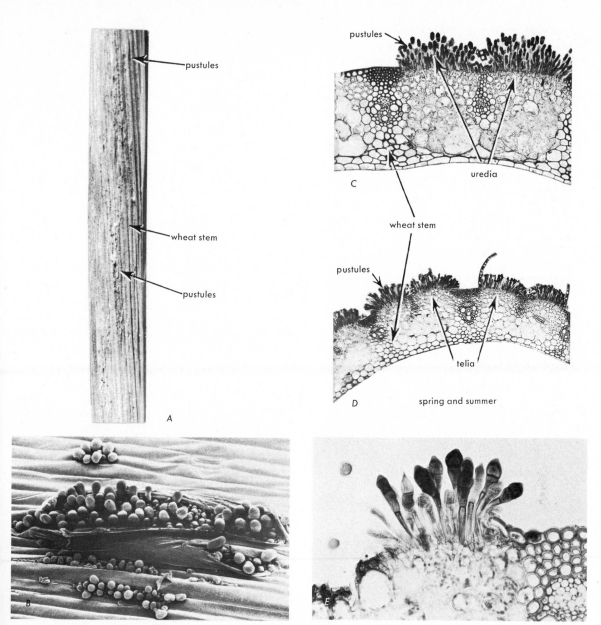

Figure 24.34 Stages in life cycle of *Puccinia graminis* on stem of wheat. *A*, sori on wheat stem, ×1; *B*, stereoscan view of a uredial sorus on wheat, ×500; *C*, cross-section of uredia, ×100; *D*, cross-section of telia, ×100; *E*, enlarged view of telium showing teliospores, ×1000. (*B*, courtesy of E. Butler.)

wintering spores, germinating in the spring on the soil or wherever else they find moisture.

3. A short four-celled basidium develops from each of the two cells of a teliospore. Meiosis takes place in this development. One basidiospore (haploid) forms from each of the four cells of the basidium. Basidiospores, either plus or minus, infect young tissues of the barberry.

4. Plus or minus spermagonia (both kinds, haploid) develop in the upper portion of the barberry leaf.

The spermagonia produce spermatia and receptive hyphae.

5. Fertilization results from union of a spermatium with a receptive hypha of the opposite strain.

6. The resulting $n + n$ mycelium forms an aecium that discharges binucleate aeciospores from openings in the lower surface of the leaf.

7. Aeciospores infect young wheat plants, producing an $n + n$ mycelium which eventually produces uredospores and teliospores.

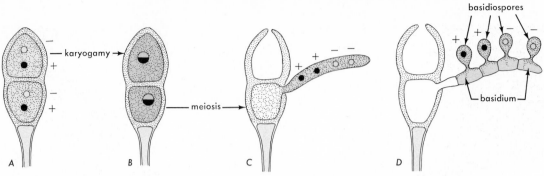

Figure 24.35 Stages in the life cycle of *Puccinia graminis*. Teliospore; *A*, binucleate cells; *B*, 2*n* cells; *C*, step in formation of basidium and basidiospores; *D*, basidium with basidiospores. For convenience, only one of the teliospore cells is followed.

Ustilaginales. All members of the Ustilaginales are plant parasites causing a group of diseases known generally as the **smuts.** Two families are generally recognized: Ustilaginaceae and Tilletiaceae. The life histories of the fungi causing cereal smuts vary, particularly in the manner of infection of the hosts: there may be **local infection, blossom infection,** or **seedling infection.** We shall discuss several genera under these headings.

Local Infection. *Ustilago maydis,* a member of the Ustilaginaceae, causes common corn smut (Figs. 24.37*A*, *B*). Infection by spores of U. *maydis* is said to be **local,** meaning that infections remain localized, producing pustules or large tumors. Perhaps the most noticeable tumors are those that occur in the corn ear. The kernels become much enlarged due to the development within them of an extensive mycelium that eventually gives rise to a mass of black, heavy-walled, overwintering spores. Since haploid nuclei fuse and meiosis occurs within them, they are teliospores. These spores may lie dormant in the soil, on old corn stalks, or in manure.

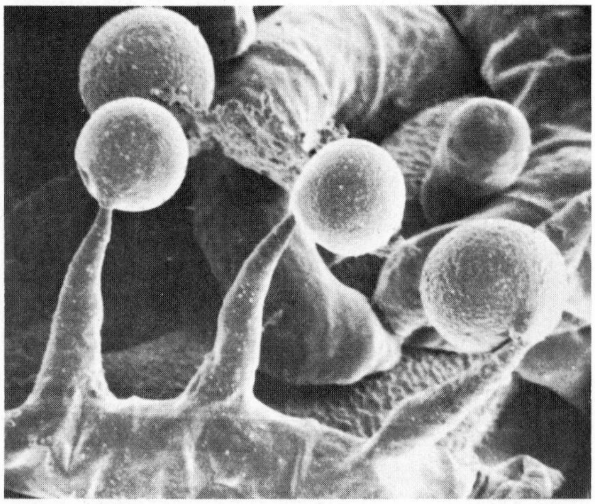

A four-celled basidium develops when the teliospore germinates. Two plus and two minus basidiospores are formed, one from each of the four cells of the basidium (Fig. 24.38). The basidiospores may infect corn plants directly or they may produce **sporidia,** which may multiply by budding and which are able to bring about infection. At any rate, a small haploid mycelium, either plus or minus, develops soon after infection (Fig. 24.39). Plus and minus mycelia develop closely together, and conjugation occurs by the fusion of cells from small haploid mycelia of opposite strains. After conjugation, the *n* + *n* mycelium, depending upon its location in the corn plant, may form a small pustule or develop into a tumor as large as a baseball. Most of the hyphal cells in the pustule or tumor change into teliospores.

Blossom Infection. Spores of *Ustilago tritici* and *Ustilago avenae* can infect *only the pistil* of wheat and oat flowers, respectively. They are known as **blossom-infecting smuts,** and the disease they produce is called the loose smut of wheat or oats. Teliospores that mature in heads of diseased plants at the time healthy plants are in blossom are transferred to healthy blossoms by air currents (Fig. 24.40). The young pistil is directly invaded and a small mycelium eventually develops in the embryo. The infection has no injurious effect on the developing grain, which matures normally. The mycelium within the embryo remains small and dormant, thus carrying the fungus over seasons unfavorable to growth. When the grain germinates, the mycelium within resumes activity. It grows best in or near meristematic tissue, and thus keeps pace with the development of the wheat or barley plant. The presence of

Figure 24.36 Stereoscan view of several basidia, the foremost one shows three of the four cells, with sterigmata and attached basidiospores, ×1000. (Courtesy of E. Butler.)

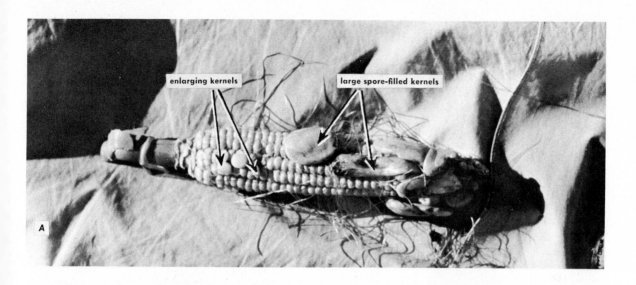

enlarging kernels

large spore-filled kernels

Figure 24.37   Smuts (Ustilago). *A*, ear of corn with smutted kernels; *B*, smut of corn leaf; *C*, healthy (left) and smutted (right) panicles of oats.

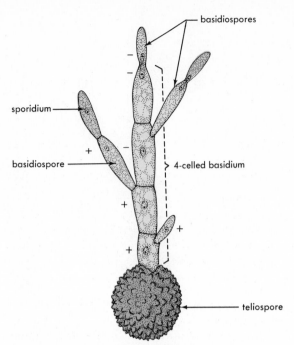

basidiospores

sporidium

basidiospore

4-celled basidium

teliospore

Figure 24.38 Teliospore of *Ustilago zeae* with basidium. (Redrawn from G. M. Smith, *Cryptogamic Botany*, vol. I, *Algae and Fungi*, McGraw-Hill, New York, 1955.)

the mycelium close to the meristematic tissue has an accelerating influence on the growth of the host, which matures rapidly, producing flower heads. The mycelium invades the developing ovary, eventually replacing the host tissue completely. The entire head of grain becomes a black, loose mass of teliospores (Fig. 24.37C), and when the spores blow away nothing is left but the axis of the spike.

Meanwhile, healthy plants have been growing more slowly and developing normal florets. Teliospores, and florets ready for pollination, occur at the same time. Teliospores are carried by wind to the healthy florets, where the ovaries are eventually invaded. (Fig. 24.40).

Seedling Infection. Tilletia tritici, belonging to the family Tilletiaceae, causes **bunt** or **stinking smut** of wheat. Only young seedlings can be infected. The smuts of this family are consequently referred to as the **seedling-infecting smuts.**

As in the blossom-infecting smuts, masses of smut spores (teliospores) replace the mature grain of wheat (Fig. 24.41). In contrast to loose smuts, however, the grain coats remain intact, and infected grains resemble normal ones. At harvest time, the infected grains are broken and the teliospores are thoroughly mixed with the healthy grains, to which they firmly adhere. As the name stinking smut indicates, the teliospores have a very unpleasant odor. Grain mixed with them is unfit for flour or animal feed. The presence of large numbers of smut spores as dust during threshing greatly increases the fire hazard; explosions and serious fires were not infrequent when heavily smutted grain was threshed.

The teliospores carry the fungus over the season unfavorable for growth. Adhering to the grain, they are sown with it. Meiosis occurs in the teliospore, and frequently, several mitotic divisions follow, producing several haploid nuclei. Upon germination of the teliospore, a short haploid mycelium is formed that gives rise to numerous spores called **sporidia.** The sporidia conjugate by means of conjugation tubes, either while still attached to the haploid mycelium or after falling away. After conjugation, a short $n + n$ mycelium is formed which produces conidia. Seedlings may be infected by these conidia, or a second short mycelium may form that produces another crop of conidia. At any rate, infection of seedling wheat plants is brought about by conidia. The young mycelium grows best in, or adjacent to, meristematic tissue. It keeps pace with the developing wheat plant. When the grain starts to head out, the mycelium enters the young ovaries of developing

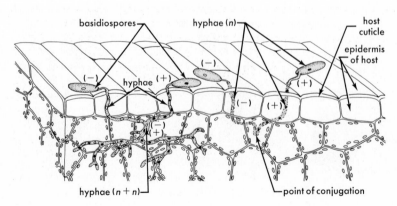

basidiospores — hyphae (n) — host cuticle

hyphae — epidermis of host

hyphae (n + n) — point of conjugation

Figure 24.39 The upper epidermis of a corn leaf (*Zea mays*) showing diagrammatically the formation of an $n + n$ mycelium of *Ustilago zeae*. (Original leaf detail from G. M. Smith, E. M. Gilbert, G. S. Ryan, R. I. Evans, and J. F. Stauffer, *A Textbook of General Botany*, Macmillan, New York, 1953.)

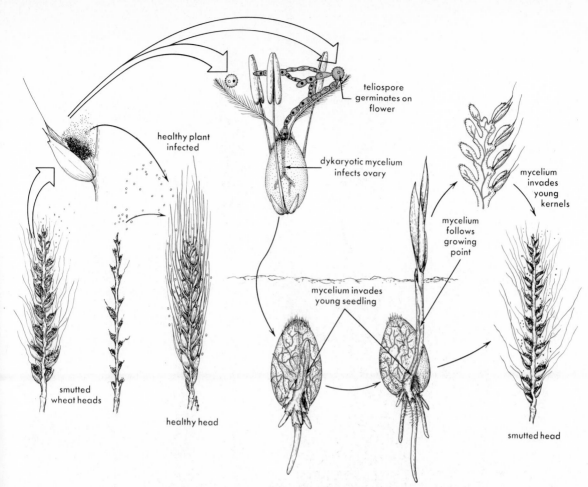

Figure 24.40 Life cycle of *Ustilago tritici*, a blossom-infecting smut. (Parts based on G. N. Agrios, 1969. *Plant Pathology*. Academic Press, New York, 1969.)

flowers. Hyphal filaments may eventually replace all normal cells of the grain, except the coats. Teliospores are formed by these hyphal cells, and the cycle is repeated (Fig. 24.41).

Comparison of Seedling- and Blossom-Infecting Smuts. In both types, teliospores are formed in the inflorescence of cereals. In seedling infection, the teliospores overwinter in the soil or *on the grain*. Upon germination, they produce sporidia that conjugate and develop a small mycelium from which conidia are cut off. Conidia infect only seedling grain plants. In blossom-infecting smuts, teliospores lead to the infection of the pistils of nearby flowers, and the rudimentary mycelium remains dormant *in the grain* over winter. In both types of infection the young seedling wheat plant has within its cells an actively growing smut mycelium that keeps pace with the development of the cereal. This mycelium invades young ovaries and fills the ovary with hyphal cells, which become teliospores.

Significant Features of the Basidiomycetes

A. *Vegetative characteristics.*
   1. Hyphae septate; extensive haploid mycelium, one haploid nucleus in each cell; extensive $n + n$ mycelium, two haploid nuclei in each cell.
   2. Homobasidiomycetidae: mostly saprophytic forms, a few species parasitic on trees.
   3. Heterobasidiomycetidae: parasitic on plants; some, such as the heteroecious rusts with a complicated life cycle; in a few species true basidiospores not produced.

B. *Reproductive characteristics.*
   1. Asexual: asexual spores not formed by the Homobasidiomycetidae; various sorts formed by different species of Heterobasidiomycetidae; uredospores and aeciospores formed by the rusts, sporidia formed by some smuts.
   2. Sexual: no motile sex cells or gametangia.

Class Basidiomycetes 503

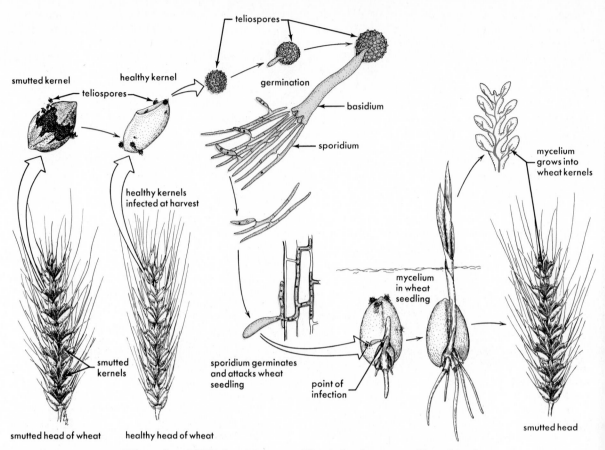

teliospores

smutted kernel

teliospores

healthy kernel

germination

basidium

sporidium

mycelium grows into wheat kernels

healthy kernels infected at harvest

smutted kernels

healthy kernels infected at harvest

sporidium germinates and attacks wheat seedling

mycelium in wheat seedling

point of infection

smutted head

smutted head of wheat

healthy head of wheat

Figure 24.41 Life cycle of *Tilletia tritici*, a seedling-infecting smut. (Parts based on G. N. Agrios, 1969. *Plant Pathology.* Academic Press, New York, 1969.)

(a) Homobasidiomycetidae: (*1*) union of two haploid hyphal cells; (*2*) *n + n* mycelium with two haploid nuclei in each cell developed from the union of the two haploid hyphal cells; (*3*) basidiocarp (mushroom, puffballs, bracket fungi) produced by the *n + n* mycelium; (*4*) hymenium consisting of club-shaped basidia developed on the basidiocarp; (*5*) two haploid nuclei fuse in the basidia; (*6*) meiosis results in four basidiospores.

(*b*) Heterobasidiomycetidae: (*1*) basidia develop from resistant teliospores; may be short filamentous cells or short four-celled filaments; meiosis occurs in formation of basidiospores; (*2*) fertilization with eventual fusion of two haploid nuclei varies considerably from species to species.

Class Fungi Imperfecti

About 24,000 species of fungi, in some 1200 genera, are known only by their asexual stages. The class

name, Fungi Imperfecti, arises from the custom of calling the sexual stages of fungi **perfect stages** and the asexual stages **imperfect stages.** Since only the imperfect stages of this large group of fungi are known, they are called the Fungi Imperfecti. Obviously, the classification is an artificial one.

In general, the structure of the hyphae, which are septate, and of spores, suggests that many imperfect fungi may be Ascomycetes; others may be Basidiomycetes. The Fungi Imperfecti may be considered to be Ascomycetes or Basidiomycetes whose sexual stages have not been observed or no longer exist. The classification of these fungi is fraught with great difficulties, largely because a natural classification is based upon sexual stages and the morphology of sexual and asexual stages is by no means coordinated. For instance, two forms may have very similar conidial stages but very different sexual stages. The various categories of classification in this class are designated as **form genera** or **form families** to show that the members of the genera or families do not necessarily have a natural or family relationship. For

504 **Higher Fungi**

instance, some species have been named simply on the basis of the host upon which they were found, a procedure that resulted in naming hundreds of literally nonexistent "species."

Only a single form group will be mentioned, the form order Moniliales; it is the largest in the class and has over 10,000 form species. Most of the fungal pathogens of man belong here. Some of them have been demonstrated to be causal agents of various types of allergy. *Penicillium* and *Aspergillus* are genera frequently placed in this order because there are many species belonging to these genera whose sexual stages are unknown. This means that the *Penicillium* of Roquefort cheese and the forms from which penicillin is obtained are Fungi Imperfecti because the formation of asci has never been observed, even though tons of mycelium of these species have been grown.

Penicillium notatum has won deserved fame because **penicillin,** a drug derived from it, will inhibit the growth of bacteria without injuring human tissue. Substances such as penicillin, formed by one organism which inhibits the growth of other organisms, are called **antibiotics** (Fig. 24.42). Our knowledge of them dates back to 1870, but their significance was not fully appreciated until the discovery of penicillin.

In 1928, Alexander Fleming investigated extracts of *Penicillium* and after demonstrating their antibiotic properties gave the antibiotic the name penicillin.

It was not until after 1940 that a concentrated effort was made to produce penicillin on a commercial scale.

Many thousands of pounds of the mycelium of *Penicillium notatum* have been grown, and the species has been subjected to intensive study, yet no sexual stages have ever been observed. The absence of sexual reproduction presents the geneticists who are studying *Penicillium notatum* with a difficult breeding problem. Nevertheless, they have succeeded in growing strains that produce many times more penicillin per pound than do the original strains. Fig. 24.43 illustrates several steps in the commercial production of penicillin.

Several forms in the Moniliales are adapted to capture and destroy microscopic animals. A few actually trap nematode worms by forming small rings which quickly constrict when stimulated by the contact of a nematode crawling through them. A nematode-trapping form and several types of conidiospores are shown in Fig. 24.44.

The Parasexual Cycle

Since the sexual process has not been observed to occur in the Fungi Imperfecti and genetic variation is not thought to be a part of asexual reproduction, the question may be asked: Is genetic variation possible in the Fungi Imperfecti? The answer is, apparently, yes, at least in the laboratory. Careful

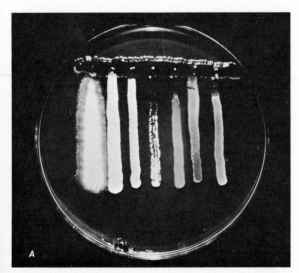
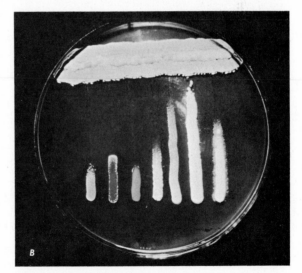

Figure 24.42 Testing a bacterium for its ability to prevent or retard the growth of other bacteria. *A,* seven different cultures have grown from parallel streaks made on an agar plate. A culture of the bacterium to be tested is streaked across the top of the plate. *B,* the upper streak has grown and an antibiotic from this colony has inhibited the growth of the bacteria in the parallel streaks.

Figure 24.43 Commercial production of an antibiotic. *A*, fermentation tanks in which *Penicillium* or another organism is grown; *B*, equipment for the extraction of the antibiotic from the fungal mycelium. (Courtesy of Charles Pfizer and Company.)

genetic analysis has produced evidence for two possible sources of variation. One is very similar to heterosis; a selective advantage appears to result from certain stable combinations of nuclei in a heterokaryon. The second observation yields results similar to those observed in a true sexual life cycle. The two nuclei of a dikaryon fuse to form a diploid nucleus. Recombination of genes takes place as it does in true meiosis. Eventually, in the fungal filaments a somatic reduction takes place, the haploid condition is again attained, and the crossing over of selected traits may be observed. This process is called the parasexual cycle. It is well-established in laboratory cultures where it may account for the great variability of some Fungi Imperfecti. It has not been established that the process is of significance in wild populations.

The Lichens

General Characteristics and Distribution

The lichens are composite plants composed of algae and fungi. The algal components are generally single-celled forms belonging to either the Chlorophyta or the Cyanophyta. When free from the fungus, the algae may exist normally.

The fungal component is generally an Ascomycete, but algal associations with Basidiomycetes are also known. The fungal component of the lichen cannot live separated from the alga unless supplied with a special nutritive medium.

The association of a fungus and an alga in the lichen is generally believed to be **symbiotic;** that is, both the fungus and the alga derive benefit from the association. The alga furnishes food for the fungus, which supplies moisture and shelter from high light intensity for the alga.

A section through a lichen thallus reveals four distinct layers. The top and bottom layers consist of a compact mass of intertwining fungal filaments. The algal cells form a green layer beneath the top mass of fungal filaments, and a loose layer of hyphae lies directly below the algal cells (Fig. 24.45).

Lichens are widespread. They form luxuriant growths on the frozen, northern tundras, where they supply feed ("reindeer moss") for deer. In our own country they frequently cover rocks, trees, and boards exposed to sun and wind (Fig. 24.46; see Color Plate 14). They are slow but efficient soil formers; rocks are disintegrated slowly by their action. Lichens are pioneer plants, appearing before any other plants on recent lava flows. Lichens and mosses accumulate soil and organic matter sufficient

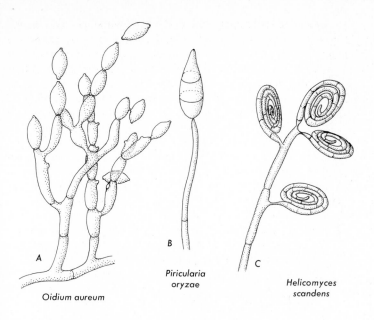

A

Oidium aureum

B

Piricularia
oryzae

C

Helicomyces
scandens

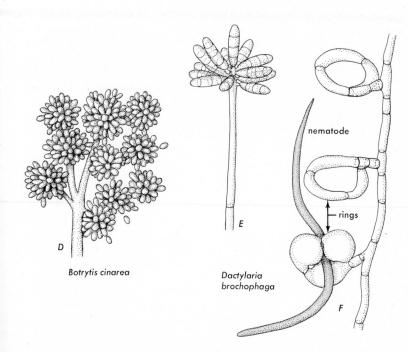

D

Botrytis cinarea

E

Dactylaria
brochophaga

nematode

rings

F

Figure 24.44 *A–E*, Conidia of various Fungi Imperfecti, *F*, nematode trapping loops of *Dactylaria*. (*A, B, C, E, F*, redrawn from E. A. Gäuman, *Fungi, A Description of Their Morphological Features and Morphological Development*, Hafner Publishing Co., New York, 1950; *D*, redrawn from E. A. Bessey, *Morphology and Taxonomy of Fungi*, The Blakiston Co., Philadelphia, 1950.)

to allow herbs and later shrubs and trees to become established.

For convenience in classifying, lichens may be divided into three groups simply on the basis of their vegetative body. This is a highly artificial grouping. Probably the most conspicuous lichens are members of the **fruticose** type. They are formed by small tubules or branches. Fruticose lichens may make ex-

tensive and attractive growths on oaks or the dead branches of certain pines (Fig. 24.46). Some lichens have a leaf-like plant body. They constitute the **foliose** type and are common in moister climates, where they make extensive growths on tree trunks or on fallen logs (Fig. 24.46A). However, the **crustose** lichens are the most common type, as they grow extensively on rocks, bark, poles, boards, and other

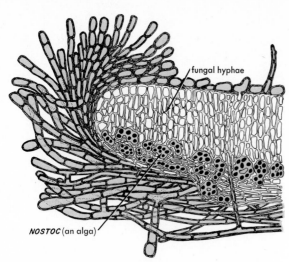

Figure 24.45 Cross-section through a lichen thallus. (Redrawn from Darbishire.)

hard surfaces. Indeed any bright coloring on the dark rocks (Fig. 24.46*B*) and gray granites of our northern mountains, on unpainted wood, or on bark of trees is quite likely to be due to a growth of crustose lichens.

Asexual Reproduction. Lichens may multiply simply by the distribution of small pieces of the vegetative thallus. Many lichens produce special powdery bodies, composed of both a fungus and an alga, that serve as a means of vegetative reproduction. These bodies are called **soredia** (singular, **soredium**).

Both the algal and the fungal components may produce spores. The algae may grow independently, but if, upon germination of the fungal spores, the young hyphae do not find the required algal cells, they soon die.

Sexual Reproduction. Sexual reproduction is characteristic of the type of fungus present in the lichen. Since the fungal component is, in the great majority of lichens, an Ascomycete, ascocarps with asci and ascospores are formed. Upon the germination of the ascospore, the germ tube must find its proper algal associate in order to survive.

Mycorrhizae

Mycorrhizae (singular, mycorrhiza) are fungal associations with the roots of higher plants. In some cases, the fungi are unicellular and live within individual root cells, but in many cases, the fungi have typical hyphae that cover the root tips in a thick mat and penetrate between cortical cells. These hyphae do not form haustoria, but contact with root cells

is nevertheless very close and metabolites are transferred in both directions. Radioactive carbon in carbon dioxide, for example, can be fixed in photosynthesis by the higher plant and later be detected in the fungus. Radioactive isotopes of phosphorus, calcium, and potassium can also be shown to be taken up in greater amounts by plants with mycorrhizae than by plants without them.

Mycorrhizal associations occur throughout the plant kingdom. They have been found on roots of many trees, shrubs, and herbs of flowering plants; on conifers, ferns, mosses, and lower vascular plants. Possibly more than 80% of all angiosperms exhibit them. In some cases, the presence of mycorrhizae is essential to normal plant development. Seedlings of orchids and heaths fail to survive if the fungi are absent, and plantings of pines in areas where they do not normally grow (such as Australia) do poorly unless the mycorrhizal fungi are introduced to the soil. These fungi may also contribute to plant development by secreting hormones, particularly IAA.

Fungi of all types are involved in these associations, but Basidiomycetes are by far the most common. Those which cannot be identified are given a trinomial name beginning with the binomial *Mycelium radicans*. The mycorrhizal fungus of scotch pine (*Pinus sylvestris*), for example, has been isolated but never identified with a known, free-living fungus; it has been given the temporary name *Mycelium radicans sylvestris*.

Classification

Division	Mycota
Class	Ascomycetes
Subclass	Hemiascomycetidae
Order	Endomycetales
Genus and species	*Saccharomyces cervisiae*
Order	Taphrinales
Genus	*Taphrina*
Subclass	Euascomycetidae
Order	Eurotiales
Genera	*Eurotium* (*Aspergillus*)
	Talaromyces (*Penicillium*)
Order	Erysiphales
Genus	*Erysiphe*
Order	Clavicipitales
Genus	*Claviceps*
Order	Helotiales
Genus	*Monilinia*
Order	Pezizales
Genus	*Peziza*

Class	Basidiomycetes	Family	Ustilaginaceae
Subclass	Homobasidiomycetidae	Genera and	*Ustilago maydis*
Genera and		species	*Ustilago tritici*
species			*Ustilago nuda*
	Agaricus campestris	Family	Tilletiaceae
	Amanita muscaria	Genus and	*Tilletia tritici*
	Amanita phalloides	species	
	Fomes applanatus	Class	Fungi Imperfecti
Subclass	Heterobasidiomycetidae	Form order	Moniliales
Order	Uredinales	Form genera	*Penicillium notatum*
Genus, species,	*Puccinia graminis tritici*	and species	*Aspergillus*
and strain			*Dactylaria brachophaga*
Order	Ustilaginales		

25

Viruses

Toward the close of the nineteenth century, Iwanowski, a Russian plant scientist, carefully pressed some sap from a stunted and diseased tobacco plant. With a small amount of this sap he inoculated a healthy tobacco plant. The healthy plant soon developed signs of a disease similar to that of the plant from which the sap was taken. The experiment was repeated, but this time the sap from the diseased plant was forced through a filter with pores so small that bacteria could not pass through them. The healthy plants inoculated with this filtered sap soon contracted the disease of the original plant. This was a new discovery—a disease could be transmitted from one plant to another by a filtered sap that was free of visible living bodies even when viewed by the most powerful light microscopes. The sap contained something (a **virus**) that caused an infectious disease.

Characteristics and Distribution

This experiment of Iwanowski's has been confirmed many times. The infective principle has been isolated and studied. We have gained from these studies much information about viruses that infect plants, animals, and even bacteria. A healthy tobacco plant compared with a plant of the same age but infected with tobacco virus is shown in Fig. 25.1. Diseases transmissible from one plant to another by an agent too small to be distinguished even when enlarged some 3000 times are called **virus diseases.** Such diseases occur in both plants and animals. The name **virus** is given to the agent or substance that causes the disease.

Smallpox, poliomyelitis, and measles are examples of virus diseases of man. Dog distemper is caused by a virus. The so-called mosaic diseases of plants, characterized by a mottling of the leaves, are also virus diseases. Even though the actual agent causing these virus diseases was long unknown, much has been learned of methods to combat the diseases.

For instance, a disease known as curly top had practically eliminated sugar beet fields in California by 1920. Careful study of the disease showed that it was caused by a virus and that it was transmitted by leaf hoppers (Fig. 25.16). The leaf hopper overwinters on weedy plants adjacent to sugar beet fields and rapidly infects the young beet seedlings in the spring. It is still impossible to cure a virus infection once it has become established in a host, and it is not practical to develop a vaccine to inoculate sugar

Figure 25.1 *A*, healthy tobacco plant; *B*, tobacco plant infected with curly top virus. (Courtesy of K. Esau.)

beets against the virus. Two pathways are open: Develop sugar beet plants resistant to the virus or control the population of leaf hoppers. Sugar beets resistant to the virus have been developed and the leaf hoppers are controlled by insecticides and herbicides that keep down the overwintering weeds. Today, the sugar beet industry in California is worth approximately $100 million.

Recall from Chapter 21 that some diseases of plants long thought to be virus-induced (e.g., some "virus" yellow diseases) are now known to be caused by bacteria (myxoplasmas).

Culture of Viruses

The study of viruses is hampered because it is only possible to culture them in living cells. Rabbit skin, tobacco leaves, bacteria, or cultures of living tissue are necessary for multiplication of viruses. Tobacco virus is, for instance, increased in quantity by grow-

ing it in tobacco leaves; cowpox virus may be grown on rabbit skin or obtained from cattle.

The virus of poliomyelitis is grown on a culture of cells derived from human cancer or from living cells obtained from the kidneys of the rhesus monkey.

The Nature of a Virus

Viruses are small particles, less than $0.2\,\mu$ in one dimension and thus below the resolving power of the light microscope. They are all, as far as known, unable to multiply outside a host. While these characteristics serve to distinguish viruses as a group, they do not separate all bacteria from all viruses. Smallpox virus is large enough to be seen with the light microscope, and there are some very small parasitic bacteria. The metabolic relationship between the virus and the host cell, however, differs from the relationship between a bacterium and its host cell.

The bacterium is a true cell comprised of a cell wall and an internal organized protoplast, and supplied with enzyme systems responsible for energy conversion; these factors make possible synthesis of cellular products needed for growth and cell division. A bacterium lives as an individual living within the host cell; it synthesizes its own characteristic proteins, catalyzed by its own enzyme system and utilizing energy produced through respiration in its own prototoplast. Its raw materials are absorbed from the host cell. Bacteria give off waste products which may be harmful to the host cell. In so doing, bacteria may change the metabolism of the host cell, but so far as is known, the host cell never elaborates bacterial components.

On the other hand, viruses appear to become a part of the host cell. They do not possess energy-yielding or synthetic enzyme systems. Viral constituents, particularly their nucleic acids, are similar to those of their hosts and induce the host cell to elaborate viral, rather than host, materials. Therefore, viruses cannot multiply outside cells. Due to this close association with living cells, we know viruses only because of their direct effect upon either bacteria, animal, or plant cells. Indeed, viruses may be more logically compared to cellular constituents than to whole cells. Viruses, as cell constituents, offer a logical explanation of their infectious nature. The similar cell and viral constituents are the nucleic acids which carry genetic information to specify cell functions. Since the host cell cannot distinguish viral nucleic acid from its own, it utilizes these new genetic messages to obediently synthesize virus constituents. Some of the new virus-specified constituents are enzymatically active. These and host enzymes synthesize new virus and utilize the energy providing pathways of the host cell in these new syntheses.

Viruses, as particles, exist outside cells, where they may be considered nonliving chemicals. They gain entrance into cells in one way or another from the outside, and once inside they multiply with the assistance of host cells, doing harm to the host cells in the process. Viruses, then, are true infectious agents. Inside the host cell they subvert its metabolism for their own uses and may be transmitted from one cell generation to the next or from one cell to another in the host organism. They are in a vegetative stage.

Host cell and viral constituents differ in that host cell constituents never exist as such outside the cell, while the viral constituents may so exist. Host cell constituents are active strictly in host cell metabolism, while viral constituents induce the host cell to elaborate viral products.

All viruses are small infectious agents. Most of them can be seen only with the electron microscope, and the study of them with it has yielded much information on their structure and morphology. We may catalog them as rods (Fig. 25.7), spherical particles, or more complicated particles, but we cannot properly classify them by their morphology alone as we do many living organisms. Table 25.1 gives a list of several plant viruses with shapes and sizes. For convenience, the present procedure is to group them as bacterial, plant, or animal viruses. This is an artificial classification, and we may find that some viruses are infectious for a wide range of host organisms.

Bacterial Viruses

Bacterial viruses were first known as **bacteriophages,** a term which has been shortened to **phage** and is now used to designate a bacterial virus. Phages were discovered independently by two investigators, Twort (1915) and d'Herelle (1917), both of whom were curious over the sudden dissolution, or **lysis,** of colonies or portions of colonies of the bacteria they were studying. These spots of lysed bacteria are known as plaques. They were able to show that the destruction of the bacteria was due to an infectious agent (a) too small to be seen with the light microscope, (b) capable of passing through a fine filter, and (c) capable of being transmitted in bacterial cells. Its characteristics thus conformed to the definition of a virus.

The discovery of an infectious agent that could rapidly destroy the bacterial cell explained the failure of transmission of bacterial-induced human diseases in such places as the religious rivers of India where large numbers of sick bathe. The pathogenic bacteria discharged in such places are rapidly destroyed by phages.

Perhaps the best-known phages are those infecting strains of *Escherichia coli,* for which several phage strains are known. These phage strains are designated as T1, T2, T7, and so on.

The Replication of Bacterial Viruses

Virulent Phage. When phage particles are added to a thinly spread culture of bacteria on an agar plate, they become attached by their tails on the cell walls of the susceptible bacteria (Fig. 25.2). Note that the phage particle consists of the aforementioned tail and a head which in end view has six sides. The tail ends in an attachment region. In Fig. 25.3, this attachment can be seen to consist of spread protein fibrils. The tail is a hollow protein rod with, it is thought, a pin at the bottom end, and a sheath which surrounds the rod just below the head. The head consists of a protein coat and a DNA core. The head is about 70nm in diameter. Viral DNA does not contain the base cytosine (p. 87), as does bacterial DNA, but another base, hydroxymethylcytosine (HMC). This is a convenient circumstance, as it makes it possible for an investigator to distinguish viral and bacterial DNAs. Shortly after the attachment of the phage particle to the bacterium, phage DNA moves down the hollow core of the tail and is injected into the bacterial protoplast. The protein

TABLE 25.1 Some Plant Viruses with Their Dimensions and Vectors

Virus	Size, nanometers	Vector
Spherical viruses		
Tobacco necrosis	20	Fungus
Bean mosaic	30	Aphids
Squash mosaic	30	Beetle
Potato yellow dwarf	110	Leafhopper
Short-rod viruses		
Hydrangia ringspot	44 × 16	Unknown
Tobacco mosaic	280 × 15	Unknown
Long-rod viruses		
Bean yellow mosaic	750 × 12	Aphids
Beet yellows	1250 × 10	Aphids

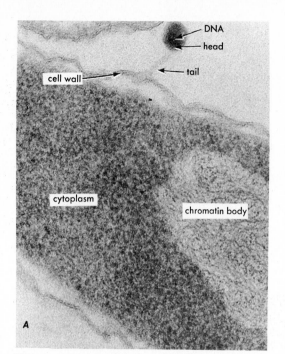

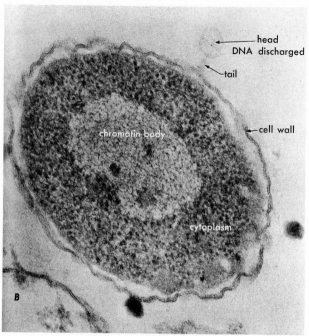

Figure 25.2   Phage T2 attached to the surface of an *E. coli* cell. *A*, before discharge of DNA from the head, ×20,000; *B*, after discharge of DNA from head, ×22,000. (Courtesy of E. H. Cota-Robles and M. D. Coffman, *J. Bacteriol.* **63**, 266.)

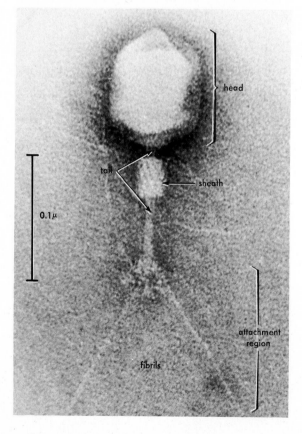

coat of the head remains outside the bacterium, and may be seen in Fig. 25.2*B*. Immediately upon the entry of the viral DNA, replication of bacterial DNA stops. Synthesis of new bacterial enzymes also stops, but those present continue to function for a time. Under the influence of viral DNA, bacterial enzyme systems either synthesize, or are themselves converted into systems capable of elaborating necessary phage constituents. New viral RNA synthesis is necessary to initiate and direct the synthesis of new virus constituents. Phage DNA is now replicated in the bacterial cell. This activity results in the elaboration of 100 or more new phage particles, complete with protein fibrils, tails, and head coating, the latter enclosing a viral DNA core. The bacterium now disintegrates or undergoes lysis and viral particles are released to infect more bacteria (Fig. 25.4). The whole process takes about 1 hour and is shown diagrammatically in Fig. 25.5. Phages which bring about such a rapid lysis of the host cell are called **virulent phages.**

Temperate Phage.   There are many phages known that do not immediately kill bacteria they invade.

Figure 25.3   Left. An electron micrograph of a T-even phage particle, ×60,000. (Courtesy of S. Brenner, R. W. Horne, et al., *J. Mol. Biol.* 1959. Copyright Academic Press.)

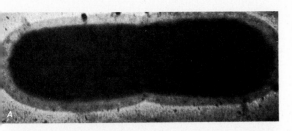

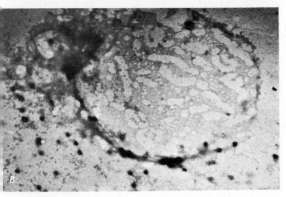

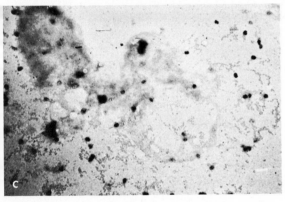

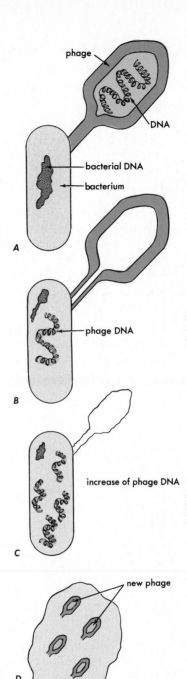

Figure 25.4 Disintegration of a bacterium and multiplication of virus particles. (Courtesy of S. E. Luria, M. Delbrück, and T. F. Anderson, *J. Bacteriol.* **46**, 57.

The bacterial hosts are called **lysogenic** bacteria, and phages that infect them are known as **temperate** phages. Metabolism of these temperate phages is so subtly linked to that of the host bacterium that both may exist through many generations without apparent injury to the host. In this state, the phage is known as a **prophage** and it confers immunity upon its host against invasion by other similar phage particles. The prophage changes to an active phage when its DNA becomes separated from that of the host. At this juncture, phage DNA directs the bacterial cell to elaborate phage constituents; the phage multiplies and eventually matures; infectious, temperate

Figure 25.5 Diagram of a life cycle of a virulent phage.

phage particles are released from lysing bacterial hosts.

During replication of phage, bacterial DNA may be broken down and the products incorporated into phage DNA. A temperate phage particle can excise a segment of bacterial DNA and carry it into another bacterium; the new host may be influenced by bacterial DNA carried by the phage. This process is known as **transduction** and occurs as described below.

Transduction. A strain of *Salmonella typhinurium* is known which is resistant to streptomycin; it may be designated as str-r. Phages may be reproduced in bacteria resistant to streptomycin. A phage becomes attached to the bacterial cell wall and phage DNA enters the bacterial protoplast (Fig. 25.6A). The bacterial DNA disintegrates and there is a multiplication of the vegetative virus (Fig. 25.6B). Disintegration of bacterial DNA does not destroy the individual bits of genetic information but seems to distribute it at random in the bacterial cell (Fig. 25.6C). When phage DNA becomes incorporated in protein heads of newly forming phage particles, replicated phage DNA carries with it some original bacterial DNA (Fig. 25.6D). This original bacterial DNA is known as the **transducing fragment.** Upon lysis, phage particles are released into the medium, where they may be collected and transferred to a culture of streptomycin-sensitive bacteria (str-s). If this culture is now grown on a medium containing streptomycin, about one in a million bacterial cells will be str-r. What has happened is this: When phage DNA is injected into the bacterium, the tranducing fragment enters also (Fig. 25.6D) and is replicated along with phage DNA. Since this is a temperate phage, lysis will not occur immediately, and upon division of the bacterial cell the str-r gene in the transducing fragment somehow takes the place of the str-s gene in the bacterial DNA (Fig. 25.6F). Note that in this case the transduced fragment merely goes along with the viral DNA. There is also evidence that the host str-r gene may become incorporated in the viral DNA by gene recombination. **Transduction** is the transfer of host DNA by means of a phage to another host where it is incorporated into the DNA of the new host. It might be compared to transformation; in transduction, bacterial DNA is transferred by a virus, while in transformation, bacterial DNA is transferred by agents other than phage.

Lysogenic Conversion. When lysogenic bacteria are infected with a phage, lysis does not occur. The phage becomes a **prophage;** it takes up residence on the bacterial chromosome and its replication is synchronized with replication of bacterial DNA so that bacteria and phage may exist together in a type

Figure 25.6 Right. Diagram to show probable steps in the transduction of genetic information from one bacterium to another by way of a transducing virus. (Redrawn after R. Sager and F. J. Ryan, *Cell Heredity*, John Wiley & Sons, New York.)

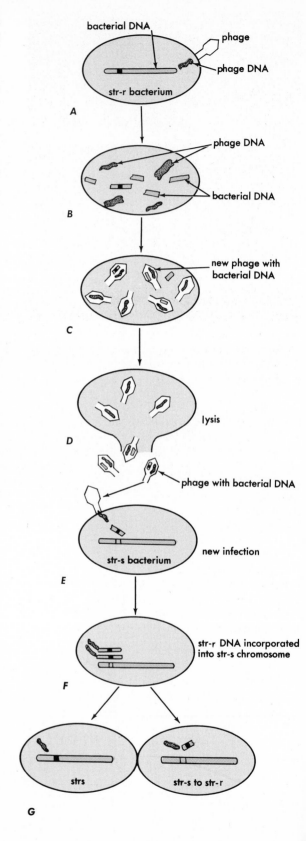

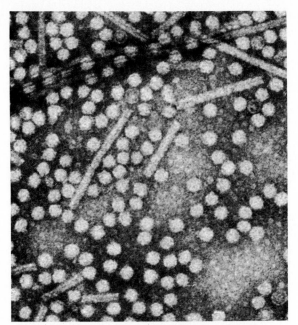

Figure 25.7 Rod-shaped particles of tobacco mosaic virus and isodiametric particles of grape fan leaf virus. TMV virus particle is 18 nm wide. (Courtesy of Robert Taylor.)

of symbiosis. Under these conditions, phage DNA may directly influence the bacterial DNA. This is known as **lysogenic conversion.**

Summary of Phage

At this point, to summarize, we mention that it has been rigorously demonstrated that DNA of bacterial cells may be changed by incorporation of DNA from other outside sources. In other words, DNA in these forms does not have a strict genetic continuity; it may fail to pass unchanged from one cell generation to another, and instead be influenced by DNA introduced from sources outside strictly hereditary lines. We do not know whether this happens in higher forms. There have been, for intsance, suggestions that cancer might be due to a virus.

Plant Viruses

Plant viruses occur as either spheres, bacilliform particles, or elongated rods (Fig. 25.7). In each case, there is some sort of RNA or DNA core surrounded by a protein sheath. In the case of the spherical bushy stunt virus, the center of the sphere is the RNA core, which is surrounded by a protein sheath (Fig. 25.8). The rod forms, as illustrated by tobacco mosaic, seem to be more complicated; for (although

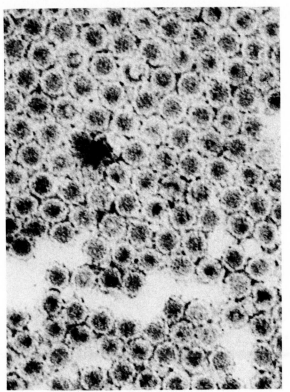

Figure 25.8 Bushy stunt virus stained to show the central core, which is composed of RNA. The clear area around the core is a layer of protein, and the dark circles separating the virus particles are formed by the precipitation of uranyl nitrate absorbed on the protein sheath, ×150,000. (Courtesy of H. E. Huxley and G. Zubay, *J. Biophys. Biochem. Cytol.* **11**, 273.)

there is still a central core of RNA) this elongated rod is twisted into a uniform spiral to form what appears, in gross form, to be a hollow rod (Fig. 25.9). The presence of virus particles may be detected in the chloroplasts of tobacco plants infected with tobacco mosaic virus (Fig. 25.10). Under certain conditions, these particles may become arranged in a precise crystalline array (Fig. 25.11).

Classification of Viruses and Virus Diseases

Many angiosperms are susceptible to virus infection. It is also known that bacteria may be attacked by a virus-like agent. So far, no virus diseases of mosses has been reported. Tobacco itself may suffer from 15 different viruses, and potatoes from 18. As early as 1935, virus diseases were known for 1100 species of angiosperms. Furthermore, a single virus

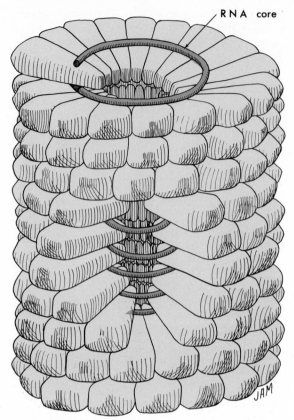
RNA core

Figure 25.9 A diagrammatic interpretation of the structure of a tobacco mosaic rod.

may exist as several distinct strains. Fifty-four strains have been described for the common tobacco mosaic. A system for the classification of this large number of viruses is now being devised by a committee called the International Committee for the Nomenclature of Viruses.

Symptoms of Virus Diseases of Plants

One of the most common symptoms of virus infection in plants is the occurrence of light-green or yellow areas on leaves (Fig. 25.12B). These areas range from small circles to large irregular patches. Uniform stripes or irregular bands of the normal green color may appear in the leaf. In some instances, the entire diseased plant may be a much lighter green than the healthy one. Many variegated horticultural varieties of plants undoubtedly are infected with some mild-acting virus. Diseases characterized by mottling or variegation of leaves are referred to as "mosaics," and almost all are regarded as virus diseases. We have, for instance, tobacco mosaic, potato mosaic, and sugar cane mosaic.

The yellow or lighter areas of diseased leaves are usually thinner in cross-section than are those of normal healthy leaves. Chloroplasts may be absent or very reduced in size (Fig. 25.12C, D) and in number, or they may be yellow in color. This reduction in green pigment, and therefore in the color of the leaf, is called **chlorosis.** A chlorotic leaf is one that has light green or yellow areas. Not all chlorotic leaves, however, harbor a mosaic virus; chlorosis may be due to other factors, including mineral deficiencies.

In addition to being chlorotic, leaves of virus-infected plants may be irregular in shape with a rough or puckered surface, may be rolled in along the edges, may be reduced in width, or may turn yellow and drop prematurely (Fig. 25.13).

Flowers of infected plants are frequently dwarfed, mottled, or streaked. Streaking of flowers occasionally produces a highly prized horticultural variety; many hundreds of dollars have been paid for tulip bulbs infected with a virus that causes a streaking of the flowers (Fig. 25.14). Fruits may be small, distorted, and mottled. Tomatoes, peppers, melons, and beans infected with certain viruses show a distinct mottling; tomatoes and cucumbers may be distorted and much reduced in size. Stems may show discoloration of various sorts; frequently, they are dwarfed and contain cankers or dead areas.

Viruses belonging to the mosaic group induce changes primarily in parenchyma tissue. Vascular tissues, however, also may be injured as the result of a primary injury to the parenchyma tissue of leaves and stems. Other viruses primarily harm the conducting tissue, particularly the phloem. The sieve tubes are injured and frequently killed, forming a mass of dead cells on the outer edge of the vascular bundle. This mass impairs the movements of food in the plant and results in injury to other plant tissues.

In many plants, viruses seem to interfere with the tissues responsible for production, as well as translocation, of food (Fig. 25.15). This damage may result in a stunting of all organs of the plant (Fig. 25.1). Viral diseases of plants claim a large toll from worldwide agriculture. Swollen shoot virus infects cacao trees in West Africa; over 1 million cacao trees were lost in one western province of the Gold Coast in 1950, and in one 5 year period 74% of the 30–40 year old trees were killed. In the 1950s, another virus disease, tristeyo caused the loss of 7 million citrus trees in São Paulo State, Brazil. Although the curly top disease of sugar beets is under control, another virus disease of sugar beets, virus yellows, probably causes a 35% reduction in sugar yields over large areas of the Western United States each year. Unfortunately, mant plant virus diseases, particularly of the mosaic type, are more insidious. They induce less conspicuous symptoms and cause less dramatic losses, and hence are tolerated more willingly by growers. Over a period of years, their cumulative effect represents an enormous loss of food production. In developing countries of the world these losses are frequently of crucial importance.

Some viruses are so highly infectious that they may be carried from one plant to another by almost any sort of contact. Workers in the tobacco field may carry tobacco virus from one plant to another on their hands or clothing. The tobacco virus is not destroyed in the manufacture of smoking tobacco. Thus, should cigarettes made from virus-infected leaves be smoked by a field worker, he may easily

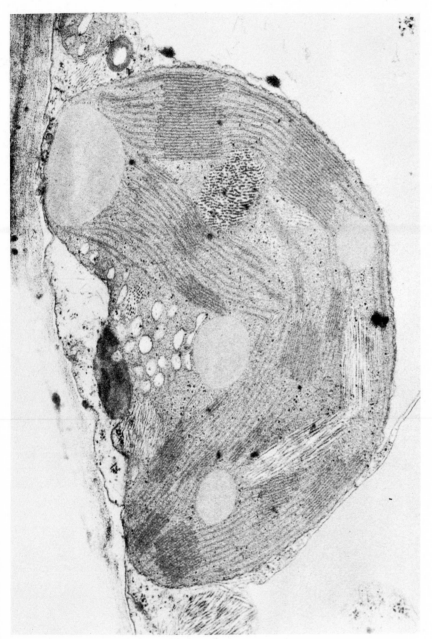

Figure 25.10 Chloroplast from a tobacco plant infected with tobacco mosaic virus (U-5 strain). Note array of crystalline-like particles, ×25,000. (Courtesy of T. A. Shalla.)

Figure 25.11 A crystalline particle of the tobacco necrotic virus, ×84,000. (Courtesy of L. W. Lablaw, *J. Ultrastruct. Res.* **2,** 177. Copyright Academic Press.)

Figure 25.12 *A*, healthy sugar beet leaf; *B*, sugar beet leaf infected with mosaic virus; *C*, cells from healthy leaf; *D*, cells from diseased leaf. (Courtesy of K. Esau and C. W. Bennett.)

carry the virus on his hands to healthy plants. Tobacco virus may also cause a disease of tomato plants.

A few of the more common virus diseases of plants are alfalfa mosaic, beet yellows, and beet mosaic. Mosaic of legumes and potato mosaic are caused by several discrete viruses. Wheat mosaic is transmitted by root-inhibiting fungi common to many soils. Tobacco ring spot and similar viruses are transmitted by soil-inhibiting nematodes which feed on plant roots.

The Reproductive Cycle of a Plant Virus

Transmission of Viruses. Any organism that carries a disease-causing agent from one living thing to another is called a **vector.** Insects, particularly those that feed on plant juices, are the most important vectors of viruses of plants. Certain types of aphids, thrips, and leafhoppers account for the transmission of viruses from one plant to another. The aphids feed by injecting a long stylet into plant leaves or stems and sucking out plant juices. They obtain sap mostly from phloem tissues, from parenchyma cells of the cortex of stems, and from the mesophyll of leaves. The stylet may wind its way between cells of the

Figure 25.13 Healthy plants compared with those infected with a virus. *A,* healthy bean; *B,* diseased bean; *C,* healthy grape; *D,* diseased grape. (Courtesy of Plant Pathology Department, University of California.)

Plant Viruses 521

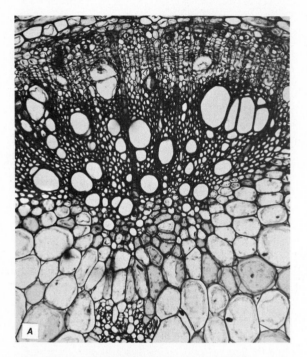

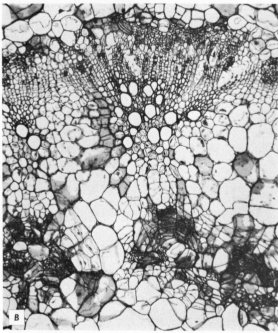

Figure 25.15 Healthy and diseased vascular tissues of tobacco. *A*, healthy; *B*, infected with curly top mosaic. (Courtesy of K. Esau.)

cortex until it reaches the phloem, or it may pass directly through all cells in its path (Fig. 25.16). Thrips have large mandibles with which they chisel their way into epidermal cells. They then suck out the cell contents.

When insects feed on plants infected with virus diseases, they ingest virus along with cell sap. The virus passes into the intestine of the aphid or thrip, is absorbed into the bloodstream, and is carried to the salivary glands. Saliva secreted by feeding insects then carry virus, and plants upon which aphids or thrips feed may be come infected with virus. There is some reason to believe that virus may undergo development in the insect. With certain diseases, the insect cannot infect a healthy plant immediately after feeding upon a diseased plant; an incubation period (within the insect) of several days must intervene before the virus is capable of causing infection when given off with saliva. Some plant viruses have a more casual relationship with their insect vectors. Though virus may be ingested by the vector, this is irrelevant, as only a small amount of virus which adheres to the stylet of the insect participates in transmission.

Some viruses can be transmitted from one host plant to another only through specific insects. Often, however, the relationship between insect, virus, and plant is not specific. One insect may carry several virus diseases, several different insects may carry the same virus, and several species of plants may be infected with the same virus. Cucumber mosaic virus is transmissible by more than 60 species of aphids, and the virus has the capacity to infest plants in more than 40 dicotyledonous and monocotyledonous families.

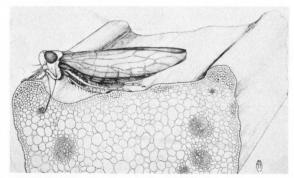

Figure 25.16 Leafhopper feeding on beet petiole. (Courtesy of C. W. Bennett.)

Replication of Viral RNA. In the plant cell, an invading virus particle soon sheds its protein coat to release the viral RNA. In the case of viruses that have a chromosome consisting of single-stranded RNA, the viral nuclei acid combines directly with host ribosomes which now synthesize virus-specified proteins. The viral RNA behaves like *m*RNA. Some of these newly synthesized viral proteins are enzymatically active, and they set to work replicating the virus nucleic acid. This occurs in several stages. In the first step, a complementary strand of viral nucleic acid, the plus strand, is synthesized using the invading viral RNA as a template. This is **transcription.** This new strand is composed of the matching complementary nucleotide bases. This new strand and the template from which it was formed combine spontaneously to produce a double-stranded helical molecule with many of the properties of DNA. Additional virus-elicited enzymes, or perhaps the same enzyme which constructed the new nucleic acid strand, now proceed to copy both strands simultaneously to produce 100–200 of the double-helical DNA-like molecules. These new molecules are a replication form of the virus and can be isolated from the infected plant. These replication molecules now instigate the replication of many new plus RNA strands identical to the original invading strand. They possess the same property of combining the host ribosomes to produce new virus-specified proteins. Late in the replication process, some virus-directed enzymes activate host cell metabolism to synthesize viral coat proteins. These proteins now combine with the plus strands of RNA to produce mature virus particles. This is **reconstitution.**

Summary of the Reproductive Cycle of a Plant Virus

1. Infection of the host cell.
2. Separation of the single RNA strand from the protein coat.
3. Combination of viral RNA with host ribosomes.
4. Synthesis of viral-directed proteins, some of which are enzymatically active.
5. Transcription of a complementary RNA strand.
6. Combination of the two complementary RNA strands, the original invading strand and the newly synthesized strand, to form a double-helix DNA-like molecule.
7. Replication of 100–200 DNA-like molecules.
8. Synthesis by the replicated molecules of many RNA strands identical to the invading strand.
9. Late in the replication process, the host cell metabolism is directed to synthesize viral protein coat.
10. Union of protein coat with replicated invading strands to form new virus particles.

Some plant viruses are now known to contain DNA. In this case, the DNA is double-stranded, and the replication of this virus is undoubtedly quite different from the version discussed for single-stranded RNA viruses. One difference, for example, that would be expected is that the double-stranded DNA would be transcribed into single-stranded RNA which would function as *m*RNA with host ribosomes to produce unique virus-induced proteins. The latter would then participate in the enzymatic synthesis of virus nucleic acid and coat protein.

26

Bryophytes

The division Bryophyta includes hornworts, mosses, and liverworts. Although all bryophytes are morphologically distinct from thallophytes, it is difficult to separate them from algae in general on the basis of any single character. Perhaps the two greatest differences between algae and bryophytes are the general structure of the plant bodies and gametangia. It will be recalled that the plant body of thallophytes is generally composed either of single cells (or filaments of cells), or of intertwining filaments, resulting in a more or less complex body. Layers of parenchyma tissue occur in some forms. With the exception of one stage in the life history of mosses, the bryophyte plant body is never filamentous. It is composed of blocks or sheets of cells forming a parenchymatous tissue.

General Characteristics

Gametangia of thallophytes are unicellular or, if multicellular as in *Ectocarpus,* lack a protective jacket of sterile cells. Gametangia of bryophytes, on the other hand, are multicellular and always have a protective jacket of sterile cells. Each and every cell of the multicellular structure of *Ectocarpus* produces gametes. However, multicellular female gametangia characteristic of all members of the Bryophyta produce but a single gamete, the egg, which is surrounded by a layer of protecting cells. Male gametangia of both *Ectocarpus* and the bryophytes produce numerous sperms, but sperm-producing cells of the bryophytes are always surrounded by a protective layer of parenchyma cells. These differences are shown in Table 26.1 at the end of this chapter.

Water is required by most algae and bryophytes to effect fertilization. A definite alternation of generations occurs in all members of the Bryophyta, and the sporophyte (diploid) generation is always more or less dependent on the gametophyte (haploid) generation. In all Bryophyta, an **embryo sporophyte,** consisting of a spherical or elongated mass of tissue, is formed directly after the germination of the zygote (Figs. 26.2D, 26.3C).

Asexual spores are not formed in Bryophyta. Members of this division reproduce asexually either by fragmentation of gametophyte or by special bodies known as **gemmae** (Fig. 26.4).

Bryophytes occur on soil, trunks of trees, and rocks, and many are truly aquatic.

Alternation of Generations

As just mentioned, all Bryophyta have a distinct alternation of generations. A haploid plant, produc-

525

ing gametes, alternates with a diploid plant; meiosis occurs in the diploid plant, and haploid spores containing *n* sets of chromosomes in each nucleus are formed. Thus, a haploid gamete-producing plant—the **gametophyte**—regularly alternates with a diploid spore-producing plant—the **sporophyte,** as shown in the diagram. Thus, the sporophyte and gametophyte

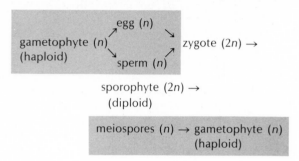

plants are but two phases in a complete sexual life history.

Classification

The Bryophyta are divided into three classes, Hepaticae (liverworts), Anthocerotae (hornworts), and Musci (mosses). The Hepaticae are the most primitive bryophytes. The simplest liverworts consist of a flat, ribbon-like, green thallus which produces gametes and a meiospore-producing capsule or sporangium embedded within the gametophyte thallus. The name liverwort is very old, having been used in the ninth century. It was probably applied to these

plants because of their fancied resemblance to the liver and the belief that plants resembling human organs would cure diseases of the organs they resembled. At any rate, a prescription for a liver complaint in the 1500s called for "liverworts soaked in wine."

Less primitive Hepaticae have simple "stems" and "leaves," and sporophytes are borne above green gametophytes. Mosses are probably the best-known class of Bryophyta. They have either an upright or prostrate leafy gametophyte, and may form extensive mats on moist, shady soil. The sporophyte is raised above the leafy gametophyte. *Anthoceros* has the simplest gametophyte of the group but its sporophyte has a meristematic region, which is a type of tissue more characteristic of higher forms than of Bryophyta.

Class Hepaticae

General Characteristics and Distribution

Hepaticae are commonly known as liverworts. There are some 8500 species of liverworts. The great majority of them grow in moist, shady localities. The gametophyte is the prominent plant, and the sporophyte is usually partially dependent on the gametophyte. The gametophyte is green and may grow either as a flat ribbon or as a leafy shoot. In either event, the plant body is frequently called a thallus, even though its internal structure does not corre-

Figure 26.1 Habit of, *A, Ricciocarpus,* and *B, Riccia,* ×1. (Courtesy of W. Russell.)

spond to that of the thallophytes. Of the four orders in the Hepaticae, we shall consider briefly the characteristics of two genera, *Riccia* and *Marchantia,* both belonging to the order Marchantiales.

Gametophytes of the Marchantiales are small, green, ribbon-shaped plants (Fig. 26.1). They branch regularly by a simple forking at the growing tip, resulting in a number of Y-shaped branches. In some species, a rosette may be formed. The upper surface of the thallus is composed of cells adapted for photosynthesis and arranged to form air chambers that open to the surface by a pore (Fig. 26.2A). Several types of storage cells generally make up the lower surface. **Rhizoids,** specialized elongated cells performing the functions of roots (anchorage, and absorption of water and mineral nutrients), extend downward from the lowermost layer of cells. The thallus is thus differentiated into distinct upper and lower portions. Scales, frequently brown or red, are formed on the lower surface (Fig. 26.2A).

Riccia

Riccia is a widely distributed genus, and, although it requires water for active growth, most species will withstand considerable drought. Several species are aquatic, growing either on mud or on the surface of small ponds.

Gametophyte. The gametophyte is a small green thallus, frequently forming a rosette (Fig. 26.1). Tissue on the lower side is composed of colorless cells that may contain starch. Tissue on the upper side consists of vertical rows of chlorophyll-bearing cells, between which are air chambers (Fig. 26.2A). The gametangia are embedded in deep, lengthwise depressions or furrows, in the dorsal surface of the thallus. Antheridia and archegonia are usually found on the same gametophyte.

Antheridia. Antheridia of different representatives of the Bryophyta are similar in structure, though they may vary somewhat in shape. In *Riccia,* they are pear-shaped and composed of two types of cells, (a) fertile and (b) sterile (Figs. 26.2B, 26.3A). Fertile cells give rise to sperms, which are relatively numerous, small, and dense with protoplasm. Sterile cells form a protective jacket, one cell in thickness, around the fertile cells. Antheridia of *Riccia,* and of other Bryophyta, are thus composed of a mass of fertile cells, all of which develop into sperms, and a protective jacket of sterile cells. The antheridia are connected to the gametophyte by a short stalk.

Mature sperms consist mainly of an elongated nucleus with two long flagella. They may be shot from the mature antheridium with considerable force, or they may be extruded slowly in a single mucilagenous mass. In any event, they do not leave the antheridium until enough moisture is present to allow them to swim about.

Archegonia. The archegonium of *Riccia* is a flask-shaped structure consisting of two parts, (a) an expanded basal portion, the **venter,** and (b) an elongated **neck** (Fig. 26.2C). Four **cover cells** are located at the top of the neck. Each archegonium contains a single egg cell, which is located in the venter. A short stalk attaches the archegonium to the gametophyte. Archegonia of a similar structure are found in other Bryophyta.

Shortly before the egg cell is mature, the cover cells separate. At the same time, the cells in the center of the neck dissolve, so that an open canal connects the venter with moisture outside the archegonium. Free-swimming sperms move toward certain chemical substances formed by the archegonium, for they swim to it and then down the canal opened by dissolution of the neck canal cells. This response to chemical stimuli is known as a **chemotactic response.** Several sperms may enter the archegonium, but only one fertilizes the egg in the venter.

Sporophyte. As a result of fusion of sperm and egg nuclei, the zygote contains the diploid, or 2n, set of chromosomes. Mitosis now proceeds in a more or less orderly manner until a spherical mass of some 30 or more similar cells is formed (Figs. 26.2D, 26.3C). These cells are partially dependent upon the gametophyte. This mass of undifferentiated, cells comprising the young sporophyte is an **embryo.** It is located within the venter of the archegonium.

Further development of the embryo involves differentiation of an outer layer of cells to form a protective jacket of sterile tissue surrounding a mass of cells that are capable of forming spores. This fertile or spore-forming tissue is called **sporogenous tissue.** The sporogenous cells continue to divide by ordinary cell division until many have formed. All these cells are **spore mother cells,** and will undergo meiosis, producing spores with the haploid number of chromosomes (Fig. 26.2E, F, G, H).

Meiosis consists of two cell divisions, during which the number of chromosomes is halved. Since two divisions are involved, four spores are formed from each spore mother cell. Each group of four spores, formed from a single spore mother cell, is usually referred to as a **tetrad** or quartet of spores.

The mature sporophyte of *Riccia* (Figs. 26.2G, 26.3E) is called a **sporangium** or **capsule.** It is composed of a jacket of sterile cells enclosing a mass of spores. The sporophyte wall breaks down when the spores are mature. These spores remain embedded in the gametophyte thallus and are not released until after the death and decay of the gametophyte.

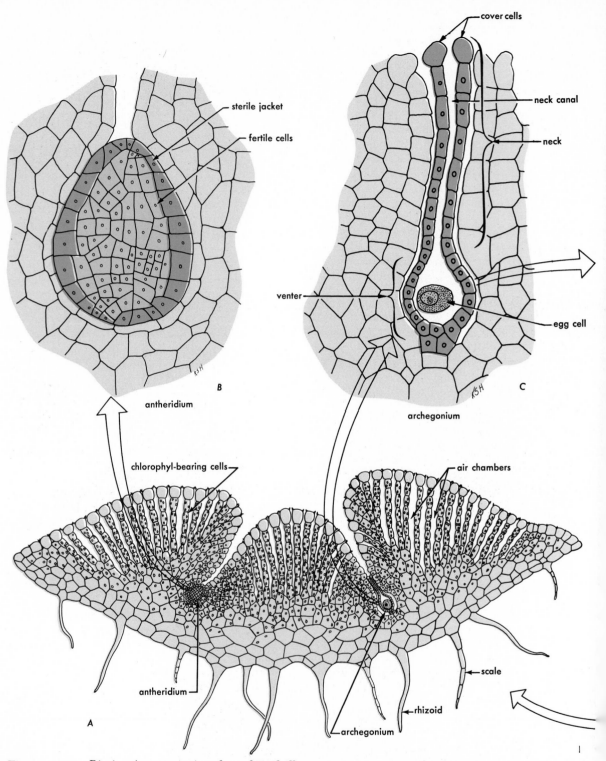

Figure 26.2 *Riccia. A,* cross-section through a thallus; gametangia are on the upper surface of the thallus between the photosynthetic filaments or in a notch in the thallus, ×8. *B,* an antheridium; note jacket of sterile cells surrounding the developing sperm cells. *C,* archegonium on thallus, neck canal open, egg cell in venter. *D,* young sporophyte developing in venter of old archegonium. *E,* sporophyte with spore mother cells still enclosed in venter of old archegonium. *F,* meiosis occurs within the sporophyte. *G,* the venter now contains fragments of the old spore case (sterile cells of the sporophyte) and spores. *H,* a spore. *B* through *G,* about ×50; *H,* ×100.

Labels in figure:

B — sterile jacket, fertile cells, antheridium

C — cover cells, neck canal, neck, egg cell, venter, archegonium

A — chlorophyl-bearing cells, air chambers, antheridium, archegonium, scale, rhizoid

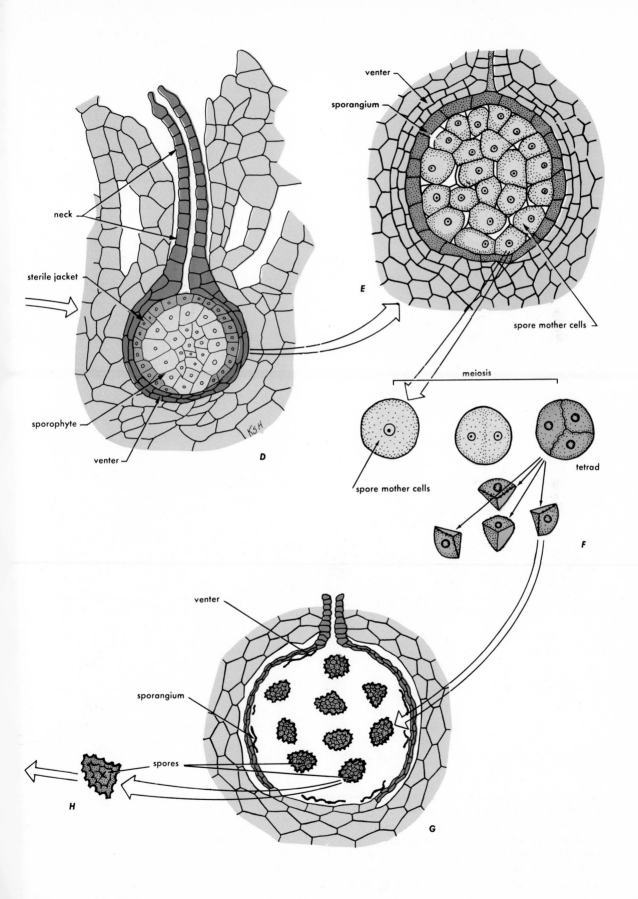

neck

sterile jacket

sporophyte

venter

D

venter

sporangium

spore mother cells

E

meiosis

spore mother cells

tetrad

F

venter

sporangium

spores

G

H

When spores germinate, each grows into a typical gametophyte plant.

The complete life history of *Riccia* may be diagrammed as shown.

Significant Steps in the Life History of *Riccia*

1. The gametophyte, a small, green, flat plant, absorbs water and mineral salts from the soil and carbon dioxide from the air. It contains chlorophyll and can synthesize food. All gametophytic

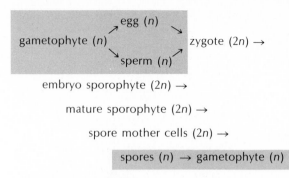

$$\text{gametophyte } (n) \nearrow \text{egg } (n) \searrow \text{zygote } (2n) \rightarrow$$
$$\searrow \text{sperm } (n) \nearrow$$

embryo sporophyte $(2n) \rightarrow$

mature sporophyte $(2n) \rightarrow$

spore mother cells $(2n) \rightarrow$

spores $(n) \rightarrow$ gametophyte (n)

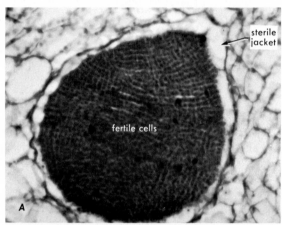

A — fertile cells, sterile jacket

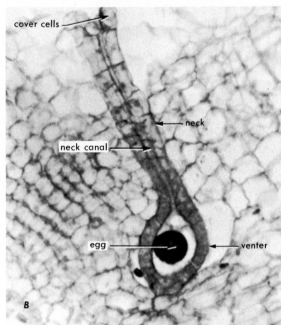

B — cover cells, neck, neck canal, egg, venter

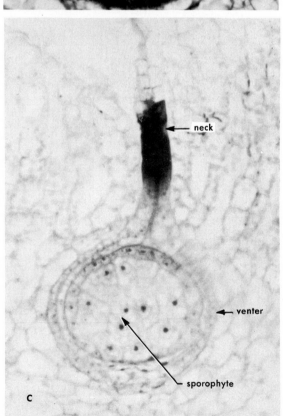

C — neck, venter, sporophyte

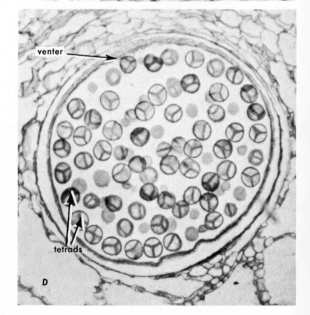

D — venter, tetrads

nuclei are haploid (contain *n* sets of chromosomes).

2. Gametangia (antheridia and archegonia) develop in a furrow on the upper surface of the gametophyte.

3. Each antheridium produces thousands of sperms.

4. A single egg is formed in the venter of each archegonium.

5. Gametes (eggs and sperms) contain *n* sets of chromosomes.

6. When sufficient moisture is present, mature sperms are extruded from antheridia. The neck of the mature archegonium is opened and sperms swim down to the egg in the venter.

7. Each egg is fertilized by a single sperm.

8. The zygote, as a result of the union of two haploid gametic nuclei, is diploid.

9. An embryo sporophyte, consisting of a spherical mass of undifferentiated cells, partially dependent upon the gametophyte, develops from the zygote. Each nucleus of the embryo sporophyte is diploid ($2n$).

10. The cells of the embryo differentiate into a sporangium consisting of a jacket of sterile cells enclosing a mass of fertile cells.

11. Spore mother cells, each with $2n$ chromosomes, form in sporangia.

12. Spore mother cells undergo meiosis. A quartet of spores forms from each spore mother cell. All spore mother cells contain $2n$ chromosomes, and spores contain *n* chromosomes.

13. Upon death and decay of old gametophytes, spores are liberated and new gametophytes are developed from them.

Marchantia

Marchantia may grow in large mats on moist rocks and soil in shady locations. It is widely distributed and fairly common on banks of cool streams. It is somewhat better adapted to growing on land than

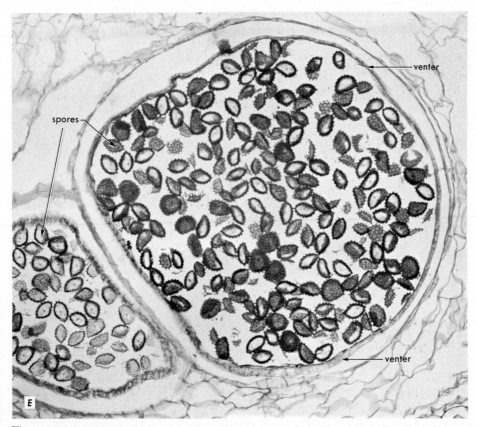

Figure 26.3 *Riccia.* Left. *A,* antheridium, note jacket of sterile cells. *B,* archegonium in notch. *C,* developing sporophyte within venter of old archegonium. *D,* nearly mature sporophyte; meiosis has occurred but spores have not separated. Above. *E,* mature spores in venter of old archegonium, all about ×100.

Class Hepaticae   531

is *Riccia,* but considerable moisture is still required for active growth and for fertilization.

Gametophyte. The gametophyte of *Marchantia,* and of other members of the family to which it belongs, differs from that of *Riccia* in that gametangia are on special disks raised some distance above the vegetative thallus (Figs. 26.4C, D; see Color Plate 15). The thallus of *Marchantia* is similar to that of *Riccia,* but much coarser. It is strap-shaped, with a prominent midrib. It shows dichotomous branching. The tips of the branches are notched, indicating growth by one or several apical cells. An individual thallus may be $\frac{1}{2}$-$\frac{3}{4}$ inch broad. Their length ($\frac{1}{2}$-$\frac{3}{4}$ inch) and the degree of branching depend upon growth conditions. The thalli are frequently crowded and overlap each other to some extent. It is usual for new growth to overlap the older decaying ends of adjacent thalli. On its upper surface are polygonal areas, with a small but conspicuous pore in the center. These areas, with their air pores, mark the outlines of air chambers, each filled with short filaments of cells containing chloroplasts. As in *Riccia,* the lower surface of the thallus is composed of colorless cells, some of which are modified for storage. Rhizoids, which anchor the thallus and serve as organs for absorption, grow from the cells covering the lower surface. Several rows of scales are also attached to this lower surface.

Asexual Reproduction. Marchantia reproduces asexually in two ways: (*a*) Older parts of the thallus die and younger portions, no longer attached, develop into new individual plants; and (*b*) small cups, known as **gemmae cups,** form on the upper surface, and small disks of green tissue, called **gemmae,** grow from the bottom of these cups (Fig. 26.4B). Gemmae, when mature, break off from the thallus and are distributed into nearby areas. Raindrops are often the agents that break off gemmae and scatter them away from the thallus. New gametophyte plants grow from gemmae.

Sexual Reproduction. The gametangia are very similar in structure to those of *Riccia.* Antheridia are

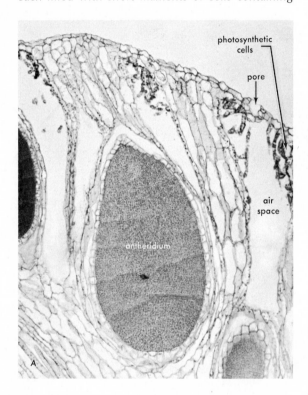

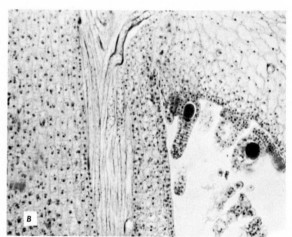

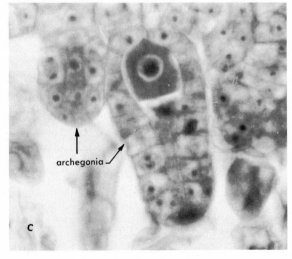

Figure 26.5 Light micrographs of gametangia of *Marchantia. A,* antheridium in antheridial head, ×100; *B,* archegium suspended from lower surface of archegonial head, ×50; *C,* details of archegonia, ×200.

Color Plate 15

Marchantia

Figure 26.4 *A*, habit, shows dichotomously branching thallus with prominent midrib, gemmae cups, antheridial heads, and archegonial heads with mature (yellow) sporophytes, $\times \frac{1}{6}$. *B*, gemmae cups, $\times 1$. *C*, antheridial heads, $\times 2$. *D*, archegonial heads, $\times 1.5$. *E*, under surface of an archegonial head; the sporophytes are enveloped by a protective perianth. Immature sporophytes are green; ripening sporophytes show yellow, $\times 3$. *F*, *G*, and *H*, time sequence showing opening of sporophyte capsule, (arrows), $\times 3$. *I*, mass of spores being extruded from sporophyte capsule, $\times 10$.

A

B

Figure 27.11 Above: Two species of *Selaginella*. *A, Selaginella watsonii*, a montane species growing in crevices in granite, ×1; *B, Selaginella emmeliana*, a species native to tropical America and requiring a moist warm climate, ×$\frac{1}{3}$.

Figure 27.13 Left: Several strobili of *Selaginella watsonii*.

pear-shaped bodies composed of a jacket of sterile tissue surrounding sperms (Fig. 26.5A). They are borne on disks raised above the thallus on slender stalks, called **antheridiophores** (Figs. 26.4C, 26.5A). Antheridiophores are modified portions of the thallus having furrows, rhizoids, and air chambers. Antheridia develop in cavities on the upper surface of the disk, the youngest antheridia being close to the outer margin of the disk. Mature sperms are extruded in a mucilagenous mass.

Archegonia are flask-shaped and have the same structures as those of *Riccia*—venter, neck, and cover cells (Fig. 26.5C).

Antheridia and archegonia of *Marchantia* are heterothallic; that is, gametophytes are either male or female. Archegonia are borne on specialized branches called **archegoniophores.** The disk at the top of the archegoniophore is frequently an eight-lobed structure. In development, the lobes bend downward and then grow inward toward the stalk. Thus, the youngest portion of an older disk bearing archegonia is on the underside and close to the stalk (Fig. 26.5B). Archegonia develop in the lower surface of the disk but in tissue similar to that found on the upper surface of the thallus. Striking finger-like processes grow out from between the lobes. The archegonia mature, and fertilization takes place when the disk of the archegoniophore is but slightly elevated above the thallus.

Sporophyte. Zygotes develop, as in *Riccia,* into embryos, which are spherical masses of undifferentiated, colorless, dependent tissue (Fig. 26.6A). Embryos are diploid, the nuclei of all cells containing $2n$ chromosomes. Subsequent growth, however, is more complicated than in *Riccia.* Some cells of the embryo divide to form a mushroom-like growth that becomes embedded in gametophyte tissue. This growth is called the **foot.** Cells in the central portion of the embryo form a stalk or **seta.** As the seta elongates, the lower cells develop into a **sporangium** (Fig. 26.6B). Lengthening of the seta suspends the sporangium below the disk. While this development is taking place, the stalk of the archegoniophore elongates, lifting the disk with its archegonia well above the thallus (Figs. 26.4D, E).

Fertilization stimulates enlargement of the old archegonium, which keeps pace with the enlargement of the developing sporophyte. As a result, the sporophyte is continually enclosed within the archegonium, which, because of its increase in size and change in function and shape, is now known as the **calyptra.** In addition, the surrounding gametophyte tissue produces two other envelopes that protect the sporophyte.

The mature sporophyte (Fig. 26.4E) is composed of three parts, **foot, seta,** and **sporangium** (Figs. 26.4E, H, 26.6B). The foot is an absorbing organ. The seta serves to lower the sporangium away from the archegonial disk and thus to facilitate distribution of spores. Before meiosis, the sporangium or capsule is composed of a jacket of sterile cells surrounding a mass of spore mother cells, among which are a number of sterile elongated cells (Fig. 26.4E). Four spores, each containing n chromosomes, develop from each spore mother cell. The elongated sterile cells are transformed into spiral elements that change shape under the influence of varying moisture conditions. These cells are called **elaters** and they aid in dispersal of spores (Fig. 26.6D).

Dissemination of spores is aided by two structural features not found in *Riccia:* (a) the sporangium of the sporophyte hangs from the lower side of the raised archegonial disk (Fig. 26.6C), and (b) elaters help to empty the spore case. Spores develop immediately into new gametophytes if they land in an appropriate habitat.

The life history of *Marchantia* can be represented as shown.

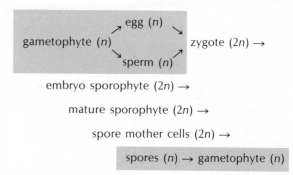

egg (n)

gametophyte (n)   zygote ($2n$) →

sperm (n)

embryo sporophyte ($2n$) →

mature sporophyte ($2n$) →

spore mother cells ($2n$) →

spores (n) → gametophyte (n)

Significant Steps in the Life History of *Marchantia*

1. The gametophyte thallus absorbs water and mineral salts from soil and carbon dioxide from air. It is green and can synthesize foods. All nuclei of the gametophyte are haploid (n).
2. Gametangia (antheridia and archegonia) are borne on upright branches called antheridiophores and archegoniophores.
3. The zygote is diploid ($2n$).
4. The embryo sporophyte, consisting of a spherical mass of undifferentiated cells, is formed from the zygote.
5. Cells of the embryo sporophyte differentiate into foot, seta, and sporangium.
6. The foot is an absorbing organ. It is embedded in the gametophyte, from which it receives nourishment.
7. The sporangium consists of a jacket of sterile cells surrounding a mass of sporogenous tissue interspersed with elongated sterile cells.

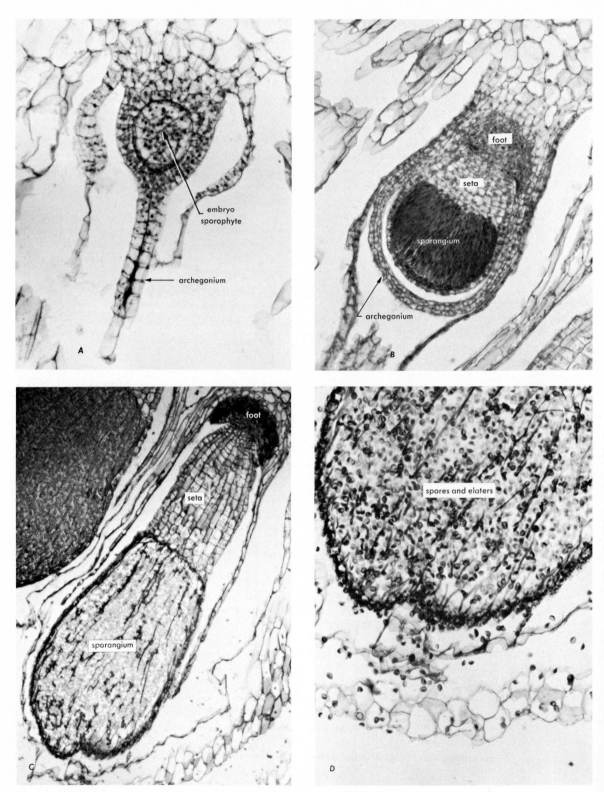

Figure 26.6 Light micrographs of developing sporophyte of *Marchantia*. *A*, embryo sporophyte; *B*, elongating sporophyte showing three regions; *C*, mature sporophyte; *D*, detail of lower end of the capsule.

8. The seta is a stalk that, in lengthening, lowers the sporangium below the surface of the archegonial disk.
9. Repeated mitotic cell divisions in sporogenous tissue result in spore mother cells, each containing 2n chromosomes.
10. Spore mother cells divide by meiosis into quartets of spores, each containing n chromosomes.
11. The elongated sterile cells are transformed into spiral elaters.
12. The presence of elaters and the hanging position of the sporophyte aid in disseminating spores.
13. Spores germinate immediately into new gametophyte plants.

Class Anthocerotae

The Anthocerotae have the simplest gametophytes of the Bryophyta. They are small, green thallus plants with little internal differentiation of vegetative tissues. They are slightly lobed, with numerous rhizoids growing from the lower surface (Fig. 26.7). The antheridia are similar in structure to those encountered among the Hepaticae. They are located in roofed chambers in the upper portion of the thallus. The archegonia are embedded within the thallus and in *direct contact* with the vegetative cells surrounding them (compare Figs. 26.2C and 26.8).

The sporophyte (Figs. 26.7, 26.9) of the Anthocerotae is in striking contrast to those of the Hepaticae. The subepidermal cells contain chloroplasts, and typical stomata are found in the epidermis. A foot

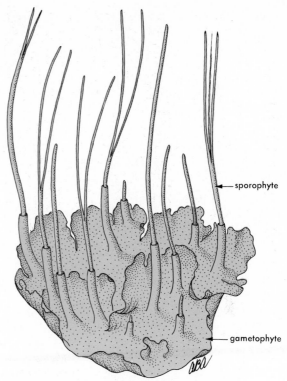

Figure 26.7 *Anthoceros* gametophyte with upright, dependent sporophyte, ×3.

embedded in the thallus serves as an absorbing organ. The sporangium is an upright elongated structure. Sporogenous tissue forms a cylinder parallel with the elongated axis of the sporangium. Spores

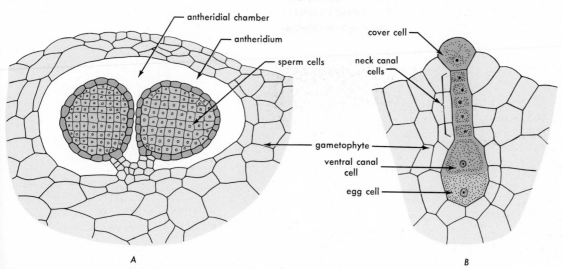

Figure 26.8 Section through an antheridium (*A*) and an archegonium (*B*) of *Anthoceros*, ×50. (Redrawn from G. M. Smith, F. M. Gilbert, G. S. Bryan, R. I. Evans, and J. F. Stauffer. *A Textbook of General Botany*, Macmillan, New York, 1953, p. 419.)

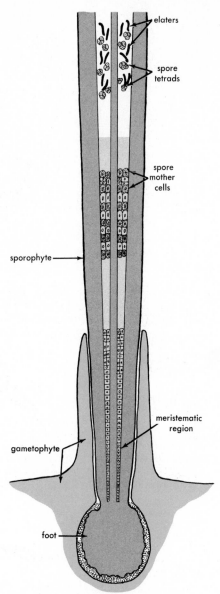

Figure 26.9 Longitudinal median section through sporophyte of *Anthoceros*. (Redrawn from R. M. Holman and W. W. Robbins, *A Textbook of General Botany*, John Wiley & Sons, New York.)

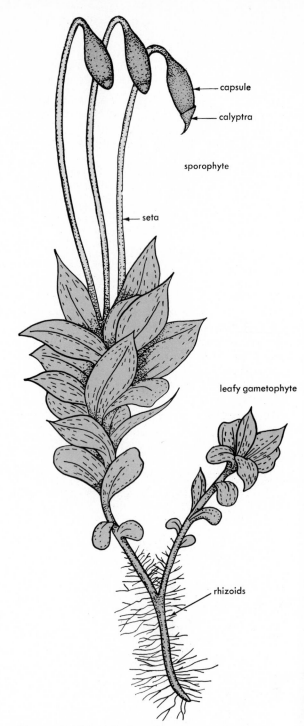

Figure 26.10 *Mnium*, showing gametophytes with attached sporophytes.

mature in progression from top down. A meristematic region lies just above the foot and continually adds new cells to the base of the sporangium. Its presence means that spores may be produced over long periods.

Under exceptionally favorable growing conditions, the sporophyte may lengthen greatly. Some sporogenous tissue at the base of the sporangium may be replaced by a conspicuous conducting strand. The foot enlarges and, through decay of the gameto-

phyte, comes into more or less direct contact with the soil. Such sporophytes are capable of maintaining themselves independently for some time.

Class Musci

General Characteristics and Distribution

We have seen that the liverworts are small and inconspicuous plants; they are not generally noticed where they grow on some stream bank or moist roadside. Although the mosses are small plants, they are nevertheless conspicuous. They frequently cover rather large areas of stream banks. They grow on rocks, on trees, and sometimes submerged in streams. Some are able to resist considerable drought, but, like the liverworts, all require moisture for active growth and reproduction.

Funaria and *Polytrichum*

Mosses as a class show great structural uniformity. *Funaria* and *Polytrichum* may serve as examples. Gametophytes of nearly all species have two growth stages: (a) a creeping, filamentous stage (the **proto-**

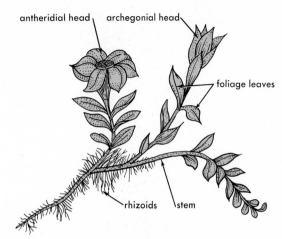

Figure 26.11 *Mnium*, showing location of gametangia, ×1.

nema), from which is developed (b) the moss plant with an upright or horizontal stem bearing small, spirally arranged green leaves. Rhizoids are found at the base of the stem. Gametangia occur at the tips

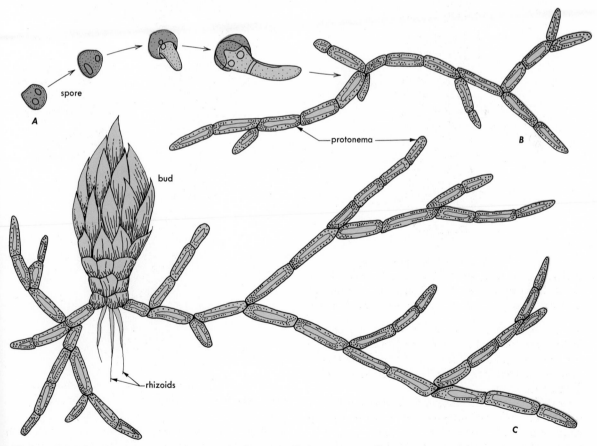

Figure 26.12 *Funaria. A*, germination of spores; *B*, protonema; *C*, protonema with bud. (Redrawn from W. P. Schimper, *Recherches sur les mousses*, Strasbourg, 1848.)

of either the main or lateral branches (Figs. 26.10, 26.11).

On the protonema, leafy shoots arise from buds (Fig. 26.12), and a protonema derived from a single spore may give rise to many moss shoots, each with numerous rhizoids at its base. The rhizoids, unlike those found in most of the liverworts, are composed of filaments of cells. They absorb water and solutes, anchor the leafy shoot to soil and may give rise to protonemata under favorable conditions.

The mature sporophyte, like that of some of the Hepaticae, is composed of a foot, seta, and sporangium. Unlike the Hepaticae sporophyte, it has a greater amount of sterile tissue and the capsule is considerably more complex (Fig. 26.19).

Although there are three orders of mosses, their structural uniformity is such that a single life history is sufficient to illustrate the general features of development in this class.

Gametophyte. Germinating spores of mosses do not develop directly into a leafy gametophyte but first become a filamentous structure. This early stage of gametophyte is called the protonema (Fig. 26.12). The protonema is not a permanent structure, although it may branch considerably under favorable conditions and cover a rather large area of soil, sometimes forming a green coating resembling algal growth. Cells composing the protonema contain numerous chloroplasts (Fig. 13.5).

The mature stem of leafy gametophytes is differentiated into a **central cylinder, cortex,** and **epidermis.** Recent observations indicate that certain cells of the central cylinder (Fig. 26.13A) may function in upward movement of water and solutes. There are, however, no vessels or tracheids such as occur in vascular plants. Some cells are strikingly similar to sieve elements (Fig. 26.13B).

Sexual Reproduction. The gametophytes of many mosses are **monoecious** (homothallic); that is, both antheridia and archegonia are produced by the same gametophyte (Fig. 26.11). Some mosses are **dioecious** (heterothallic); in other words, antheridia and archegonia are produced on separate gametophytes. Gametangia of *Funaria* are formed at the summit of a leafy shoot. First, antheridia develop, and then the gametophyte may branch, with archegonia forming at tips of the branches. In *Mnium* and *Polytrichum*, shoots bearing antheridia are easily recognized, for the leaves surrounding them are spread somewhat like petals of a flower. The group of antheridia appears as an orange spot in the center of the terminal cluster of leaves (Figs. 26.11, 26.14).

Antheridia of most true mosses (Fig. 26.15) are essentially similar to those of liverworts. Cells form-

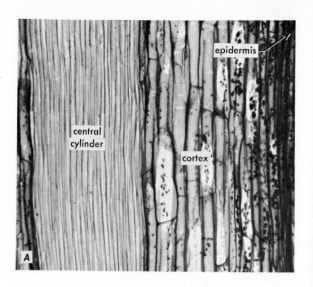

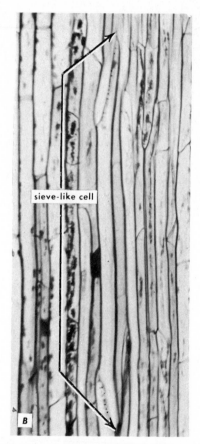

Figure 26.13 Longitudinal sections through gametophyte stems of mosses. *A*, stem of *Mnium* showing central strand of sclerenchyma-like cells, ×100; *B*, stem of *Polytrichum* with phloem-like cells, ×100. (*B*, courtesy of E. M. Gifford, Jr.)

Figure 26.14 Antheridial heads of *Polytrichum*, ×1.

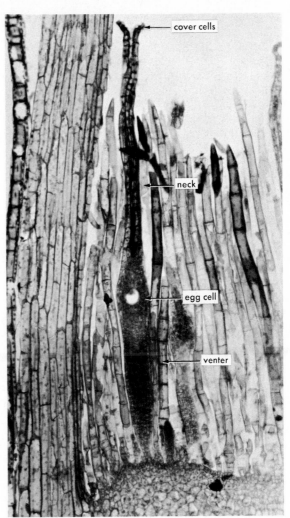

Figure 26.16 Archegonium surrounded by paraphyses (*Mnium*), ×50.

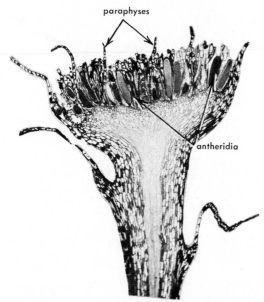

Figure 26.15 Antheridial head of *Mnium*, ×35.

ing the sterile jacket of the antheridium contain chloroplasts that become orange-red when the antheridium ripens. As in liverworts, the sperms consist mainly of an elongated nucleus, and each has two flagella. The antheridia are surrounded by club-shaped, multicellular, sterile hairs with conspicuous chloroplasts. They are called **paraphyses.**

The archegonia differ from those of the Hepaticae in having a longer neck and a longer stalk between venter and tip of the gametophyte (Fig. 26.16).

Mosses, like liverworts, are dependent upon the presence of water to effect fertilization as well as opening of the antheridia and archegonia.

Sporophyte. Soom after fertilization, the zygote begins to develop into a spindle-shaped embryo that differentiates into a sporophyte consisting of a foot,

sporophyte

leafy gametophytes

Figure 26.17 *Polytrichum* showing leafy gametophytes with sporophytes.

seta, and sporangium or capsule. The foot penetrates the base of the venter and grows into the apex of the leafy shoot. It absorbs water and nourishment from the gametophyte for growth and development of the sporophyte. The seta elongates rapidly, raising the sporangium $\frac{1}{2}$ inch or more above the top of the leafy gametophyte (Figs. 26.10, 26.17). The old archegonium increases in size as the sporophyte enlarges. When the seta elongates, the top of the expanded archegonium is torn from its point of attachment to the gametophyte and elevated with the sporangium. The old archegonium, now known as the calyptra, remains for a time as a covering for the sporangium.

The sporangium of mosses is larger and more complex than that of the Hepaticae (Fig. 26.18). It may measure $\frac{1}{16}-\frac{1}{8}$ inch in diameter and twice that in length. It is surrounded by an epidermal layer composed of cells similar to those in the epidermis of higher plants. Stomata occur in the epidermis covering the lower half of the sporangium. The sterile tissue forming the inner portion of the sporangium may conveniently be divided into three regions, each of which may be recognized by the type

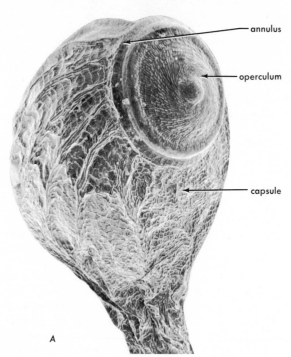

annulus

operculum

capsule

A

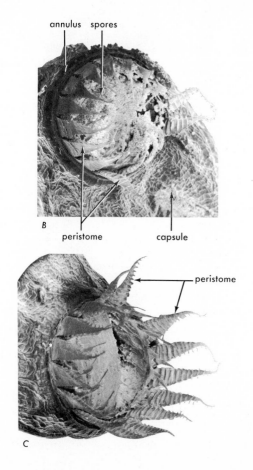

annulus   spores

B

peristome   capsule

peristome

C

Figure 26.18 Stereoscan micrograph of a capsule of *Funaria*. *A*, capsule with operculum in place; *B*, operculum has been shed, peristome teeth partially open, spores attached to them; *C*, peristome teeth fully open, capsule empty, ×300. (Courtesy of W. Russell and D. Hess.)

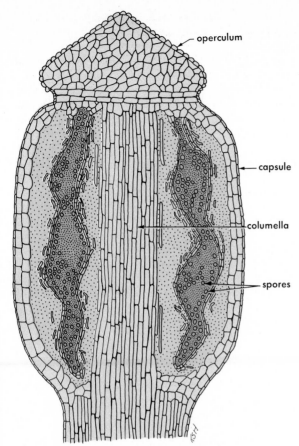

Figure 26.19 Median section through a mature but unopened capsule of *Mnium*, ×25.

(Figs. 26.18, 26.19). The cells immediately beneath the operculum form a double row of long, triangular teeth called the **peristome.** The broad bases of the teeth are attached to the thick-walled deciduous cells forming the **annulus** around the upper end of the sporangium. When the sporangium matures and becomes dry, thin-walled cells holding the operculum in place break down, allowing the operculum to fall away and thus expose the teeth of the peristome. By this time, most of the thin-walled cells within the columella have collapsed and the cavity thus formed is filled with a loose mass of spores. Peristome teeth are rough and are very sensitive to the amount of moisture in the air (Fig. 26.18*B*). When they are wet or the atmospheric humidity is very high, they bend into the cavity of the sporangium; when dry, they straighten and lift out some of the spores, which are then disseminated by air movements. If a spore comes to rest on moist soil and if illumination and temperature are favorable, it germinates and grows into a protonema.

The life history of *Funaria* may be represented as shown in the diagram.

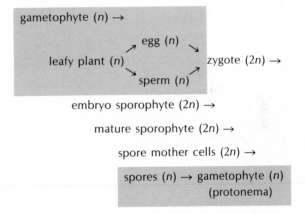

Significant Steps in the Life History of a Moss

1. The gametophyte consists of (a) a filamentous, branched, algal-like growth called a protonema, and (b) leafy shoots that develop from buds on the protonema. The shoots consist of a stalk bearing rhizoids at its lower end and leaves throughout its length. The gametophyte is green and able to synthesize food. All gametophyte nuclei are haploid.
2. Antheridia and archegonia develop at the apex of the leafy gametophyte.
3. Each antheridium produces hundreds of sperms.
4. A single egg is formed in the venter of each archegonium.
5. Gametes (eggs and sperms) each contain *n* chromosomes.

of cells comprising it. A fourth region is formed from fertile sporogenous cells (Fig. 26.19). The sterile regions may be briefly characterized as follows: (a) cells comprising the base of the sporangium contain chloroplasts and are fairly compact; (b) cells forming the outer region of the upper half of the sporangium also contain chloroplasts, but these cells are rather loosely associated, large air chambers being regularly formed; and (c) cells forming a core in this upper portion are devoid of chloroplasts and are compact, **columella.** The fourth region, composed of sporogenous cells, forms a layer around the columella. As in the liverworts, the cells of the sporogenous tissue may increase in number by mitotic cell division. Spore mother cells, each containing the diploid number of chromosomes, develop from the sporogenous cells. Meiosis takes place in spore mother cells, and haploid spores containing *n* chromosomes result. They are the first cells of the gametophyte generation.

The columella projects upward, forming a small dome above the main mass of the sporangium. The four or five outer layers of cells of this dome differentiate into a dry, brittle cap called the **operculum**

6. When sufficient moisture is present, mature sperms are extruded from the antheridium. The neck of the mature archegonium is open and sperms swim down it to the egg in the venter of the archegonium.
7. The egg is fertilized by one sperm.
8. The zygote, as a result of the union of gametic nuclei, is diploid.
9. An embryo sporophyte, consisting of a spindle-shaped mass of undifferentiated cells, partially dependent on the gametophyte, develops from the zygote.
10. The embryo develops into a sporophyte consisting of a foot, seta, and sporangium. The sporangium of most mosses is more complex than those found in the Hepaticae and the Anthocerotae. In addition to sporogenous tissue, it contains several types of sterile tissues, among which may be mentioned operculum, annulus, and peristome.
11. Spore mother cells, each containing $2n$ chromosomes, form from sporogenous cells.
12. Spore mother cells undergo meiosis. Four spores result from each mother cell.
13. When spores germinate, they form the protonema.

Classification

Division   Bryophyta
  Class   Hepaticae
    Order   Marchantiales
      Genera   *Riccia*
                 Ricciocarpus
                 Marchantia
  Class   Anthocerotae
    Genus   *Anthoceros*
  Class   Musci
    Genera   *Funaria*
                 Mnium
                 Polytrichum
                 Sphagnum

TABLE 26.1   **Tabulation of Differences between Thallophytes and Bryophytes**

	Thallophytes		Bryophytes
1.	Mostly aquatic	1.	Mostly terrestrial
2.	A few of the kelps have sieve-like elements	2.	Some mosses have cells suggesting sieve cells; otherwise food conducted in relatively undifferentiated cells
3.	None have water-conducting elements	3.	Simple water-conducting cells (not vessels or tracheids)
4.	In general, no specialized water- and nutrient-absorbing tissue; rhizoids and haustoria may occur in some fungi	4.	Rhizoids anchor and absorb water and mineral nutrients
5.	Mostly filamentous or a lacework of intertwining filaments; parenchyma in a few	5.	Only one stage of mosses if filamentous; all others are formed of parenchyma cells
6.	A definite alternation of generations in many forms	6.	All have an alternation of heteromorphic generations
7.	Both sporophytes and gametophytes independent	7.	Gametophyte independent; sporophyte dependent
8.	Sporophyte infrequently large and complex	8.	Sporophyte small and relatively simple
9.	Gametangia are either single cells or groups of single cells not accompanied by a jacket of sterile vegetable cells	9.	Gametangia are always composed of gamete-producing cells protected by a jacket of sterile vegetative cells
10.	Spores often water-dispersed	10.	Spores wind-dispersed

27

Lower Vascular Plants

In contrast to algae, fungi, liverworts, and mosses, vascular plants possess a well-developed vascular system that serves for conduction of water, mineral salts, and foods. Most of them are land plants, but some require free water for fertilization. The sporophyte is the dominant generation; except for the youngest stages in the formation of an embryo, it is independent of the gametophyte. Gametophytes are small, and in conifers and flowering plants they are dependent upon the sporophyte. The plant body of the sporophyte is, in its simplest form, an axis, with meristematic tissue terminating the opposite ends, which form shoots or roots.

Classification

There are ten divisions of vascular plants, six are seed-bearing and four do not bear seeds. The latter four divisions of lower vascular plants, are the **Psilophyta, Lycophyta, Sphenophyta,** * and **Pterophyta.** The number of families and genera in these divisions is small and their members do not form a very conspicuous part of the current land flora. The Pterophyta is currently the largest of the four divisions; it comprises the ferns, which are prominent in certain cool moist habitats. However, plant fossils indicate that there was a period in the earth's history when the members of the Lycophyta and Sphenophyta formed a large and dominant flora.

That changes have occurred in vegetation of the earth's surface is a well-established fact; beautifully preserved portions or imprints of plants are found buried in many types of sedimentary rocks, coal deposits, and peat bogs. Such plant remains are called fossils (Fig. 2.6). Sometimes fossils appear to be remains of plants almost identical with some living today. For instance, although the coast redwoods of California are now confined to a very narrow fog belt along the coast of northern California and southern Oregon (Fig. 18.1), fossil redwoods show that they were once prevalent in the Northern Hemisphere (Fig. 27.1). Fossil vascular plants differ in greater or lesser degree from present-day forms. All of them, however, can be classified, and it is possible to draw conclusions regarding their relationship with existing plants. Some of these fossils will be discussed under the headings of their appropriate divisions; others will be discussed in Chapter 30.

*As noted in Chapter 2, we choose to use the old designation, Sphenophyta, for the division to which the horsetails belong, to avoid confusion. The new proposed designation, Arthrophyta, might be confused with Anthophyta or Arthropoda.

543

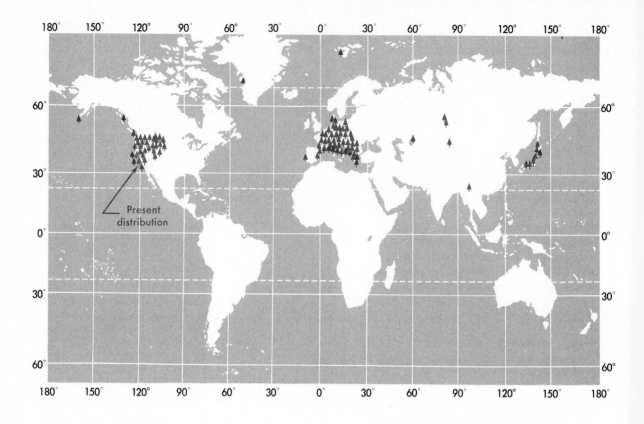

Present
distribution

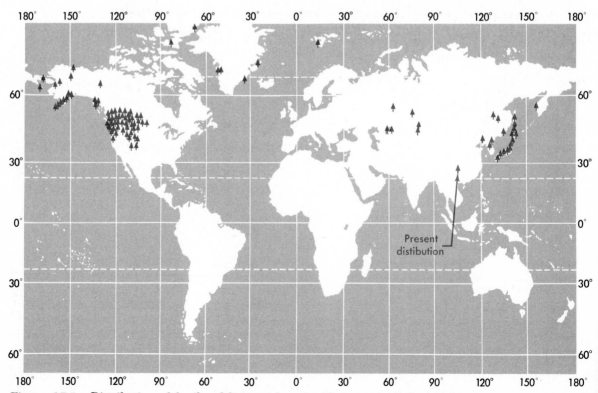

Present
distibution

Figure 27.1   Distribution of fossil and living redwoods, *Metasequoia* (below) and
Sequoia (above). (From A. Florin, Acta Horti Bergiani, **20.**)

Division Psilophyta

This division is represented in the existing flora by a single family, the Psilotaceae, comprised of two genera, *Psilotum* and *Tmesipteris*. They are rare plants found mainly in the tropics, one form of *Psilotum* growing as far north as Florida. Fig. 27.2 shows the simple plant body of *Psilotum nudum*. Roots are not present and leaves are small and scale-like. These are not "true" leaves (megaphylls, see Chap. 30); they are instead more like epidermal outgrowths. The branching, upright stem is slightly flattened and contains chlorophyll. A fungus is always associated with a branched rhizome, which is clothed with many rhizoids. The vascular tissue is simple (Fig. 27.3), consisting of poorly developed phloem and xylem in a radial arrangement. An endodermis, with a conspicuous Casparian strip, separates the central vascular tissue from the outer cortex. The xylem is composed of spiral and scalariform tracheids. Sporangia are borne in axils of some of the scale-like leaves (Fig. 27.2*B*).

Fossil Psilophyta are quite well-known. They are found in the older sedimentary rocks, indicating that the plants flourished well over 300 million years ago. In spite of their great age, their structure may be beautifully preserved. The presumed habit of one species, *Rhynia major,* and a cross-section of its stem, are shown in Figs. 27.4 and 27.5. Note the remarkable preservation of cellular structure. The plant consisted of a slender rhizome upon which were borne erect, cylindrical stems that branched sparingly. True leaves were absent. Sporangia developed at the tips of some branches, and typical quartets of spores suggesting normal meiosis have been found. Rhizoids growing from the rhizome absorbed moisture and nutrients and anchored the plant to the soil.

Division Lycophyta

The Lycophyta are well-represented in the northern hemisphere by three families, Lycopodiaceae, Selaginellaceae, and Isoetaceae. There is one living genus from each family in our modern flora: *Lyco-*

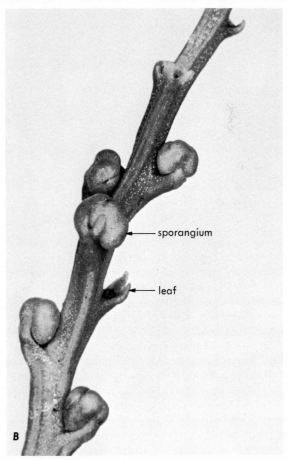

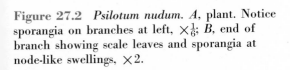
Figure 27.2 *Psilotum nudum. A*, plant. Notice sporangia on branches at left, $\times\frac{1}{6}$; *B*, end of branch showing scale leaves and sporangia at node-like swellings, $\times 2$.

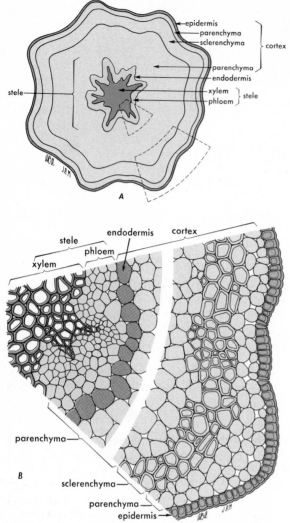

Figure 27.3 Cross-section of *Psilotum* stem. *A*, diagram to show arrangement of tissues, ×8; *B*, enlarged sector showing cellular detail, ×450.

podium, Selaginella, and *Isoetes*. We shall consider two genera, *Lycopodium* and *Selaginella*.

There are many fossil Lycophyta belonging to several orders, Lepidodendrales being one of the best known. Now-extinct representatives of this order were forest trees, some of which bore seeds (Chap. 30).

The leaves of the living members of this division are generally small and usually arranged in a spiral. In flowering plants, leaf gaps occur at the juncture of leaf and stem (Fig. 8.16). Pay particular attention to Fig. 9.13E, noting the gaps formed by the departure of the leaves as compared with the solid core of xylem in the root. Leaf gaps are never formed at the junction of stem and leaf in the Lycophyta, so

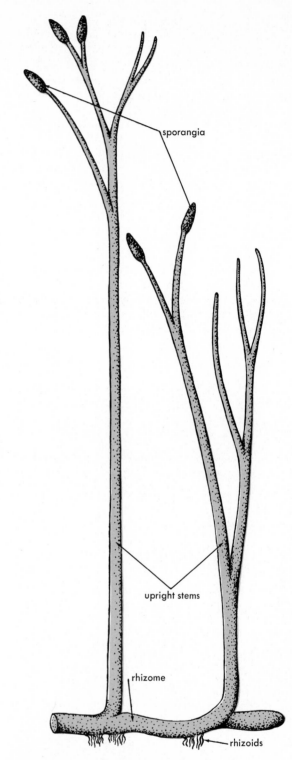

Figure 27.4 Presumed habit of *Rhynia*, a plant that lived over 300 million years ago. (Redrawn from R. Kidston and W. H. Lang, *Trans. Roy. Soc. Edinburgh* **52**, 831.)

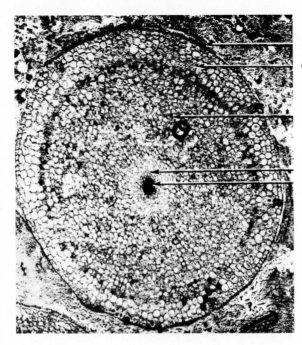

epidermis

outer cortex

inner cortex

phloem

xylem

Figure 27.5 Cross-section of *Rhynia* stem. (From R. Kidston and W. H. Lang, *Trans. Roy. Soc. Edinburgh* **52**, 603.)

Figure 27.6 *Lycopodium annotinum* in its natural habitat in the deep shade of a cool forest.

that a living stem of a member of this division has, like many angiosperm roots, a core of solid xylem.

Lycopodium

Approximately 400 species are in this genus. Most are trailing plants, many forming short upright branches that sometimes recall pine seedlings (Figs. 2.1, 27.6). They are frequently called "ground pine" or "club moss." Widely distributed, they are found in largest numbers in subtropical and tropical forests. They cannot grow in arid habitats. Several species grow in the eastern and northwestern United States, but none occur in the more arid states of the southwest. Some eastern species are in danger of extinction because of their ready sale in florist shops for Christmas decorations.

Mature Sporophyte. The main stem branches freely and is prostrate. Upright stems, approximately 8 inches in height, grow from the horizontal stem (Figs. 27.7A, B). Both types of stems are sheathed with small green leaves. Small but well-developed adventitious roots rise irregularly from the underside of the horizontal stem (Fig. 27.10A). As in many higher plants, the primary root, which grows from the embryo, is not long-lived.

The xylem in a *Lycopodium* stem occurs as a complex system of anastomosing strands, so that its distribution in the stem varies greatly in different

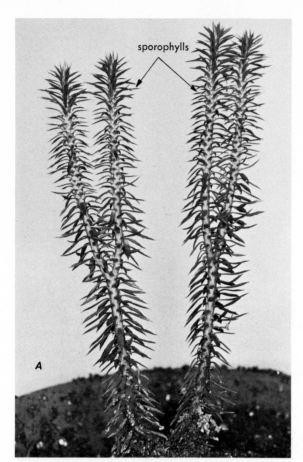

Figure 27.7 Sporophylls in *Lycopodium*. *A*, sporophylls similar to vegetative leaves, *Lycopodium selago*; *B*, sporophylls different from vegetative leaves and grouped in strobili, *Lycopodium obscurum*, ×1.

transections of the stem. The phloem is present between the strands of xylem and thus, too, forms a complex system of anastomosing strands. The irregular arrangement of xylem and phloem is shown in Fig. 27.8. The xylem is composed of tracheids, whereas the phloem contains sieve cells and some parenchyma cells. An endodermis encircles the vascular cylinder. *Lycopodium* stems also lack pith, as do roots of most higher vascular plants.

Reproduction. *Lycopodium* may reproduce asexually or through an alternation of generations involving, as usual, gametophytic and sporophytic generations. Like some of the more advanced algae, and unlike the bryophytes, both generations are independent of each other. Furthermore, the sporophyte generation is the dominant generation. Meiosporangia are borne in sporangia in, or close to, axils of the small leaves (Figs. 27.9, 27.10*B*), sheathing the stem. Spores (Fig. 27.10*C*) and sporangia of a given

species are all alike. Leaves bearing sporangia are called spore-bearing leaves or **sporophylls** (Fig. 27.9). In some species, they closely resemble ordinary nonspore-bearing leaves in their structure and appearance (Fig. 27.7*A*). In other species, they differ in size, shape, position, and color from sterile leaves. Such modified sporophylls are grouped together closely at the ends of stems, forming a cone or **strobilus** (Figs. 2.1, 27.6, 27.7*B*).

Gametophyte. The spores do not germinate readily. Gametophytes in some species grow above ground and are green; in other species, gametophytes are subterranean and lack chlorophyll. The gametophytes are always associated with a fungus. Underground gametophytes (Fig. 27.10*D*) are hard to find and, being difficult to grow in culture, are not well-known. Those that have been studied are monoecious (homothallic), and the gametangia resemble those of *Anthoceros*. Gametangia are borne on the

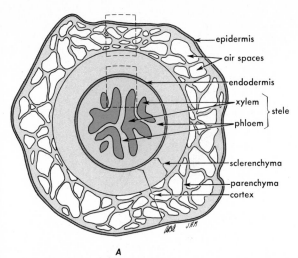

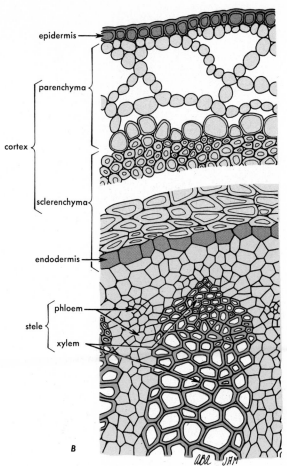

Figure 27.8 Cross-section of *Lycopodium* stem. *A*, showing irregular distribution of xylem and phloem, ×65; *B*, enlarged sector showing cellular detail, ×260.

upper portion of the gametophyte (Fig. 27.10*E*); cells of this portion are free of fungal filaments. Archegonia consist of a neck projecting slightly from the gametophyte surface, and a venter with one enclosed egg (Fig. 27.10*H*). The antheridia are composed of a mass of fertile cells buried shallowly in the gametophyte (Fig. 27.10*F*) and surrounded by a jacket of sterile cells. The sperms have two flagella.

Fertilization occurs when sufficient free water is present to allow sperms to swim to mature archegonia.

Sporophyte Embryo. The embryo sporophyte possesses (*a*) a well-developed foot, (*b*) rudiments of a short primary root, (*c*) leaf primordia, and (*d*) a short shoot apex (Fig. 27.10*I*). The embryo grows directly into the mature sporophyte plant (Figs. 27.10*A, J*).

In some species, small masses of tissue, called **bulbils,** are formed, drop from the parent plant, and grow directly into new young sporophyte plants.

Selaginella

Selaginella species resemble those of *Lycopodium* in their general appearance but are smaller. They number more than 500 and, although they are widely

Figure 27.9 The sporangia of *Lycopodium selago* are borne in the axils of sporophylls resembling vegetative leaves, ×3.

Division Lycophyta **549**

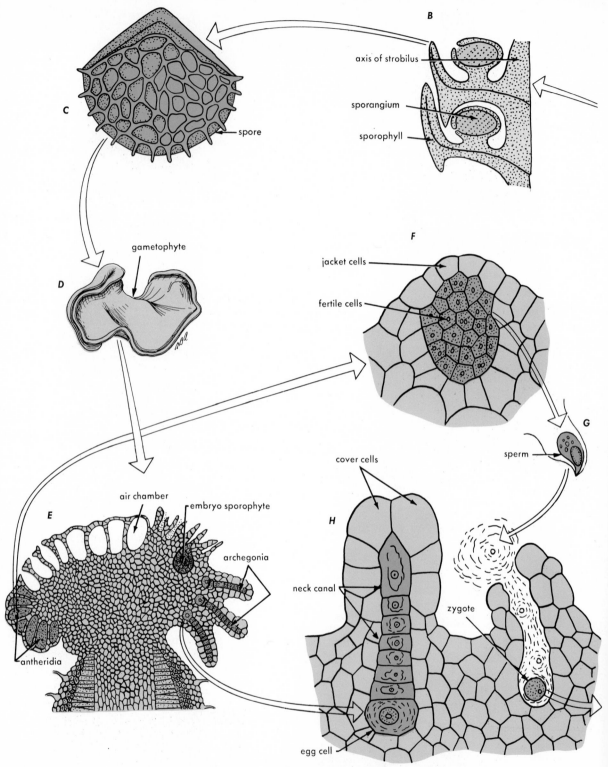

Figure 27.10 Stages in the life cycle of *Lycopodium*. *A, Lycopodium annotinum*, ×1; *B,* section of sporophyll showing location of sporangia, ×3; *C,* spore, ×50; *D,* gametophyte, ×1; *E,* section of gametophyte showing location of gametangia, ×25; *F,* antheridium, ×100; *G,* sperm, ×200; *H,* archegonium, ×100; *I,* section of embryo,

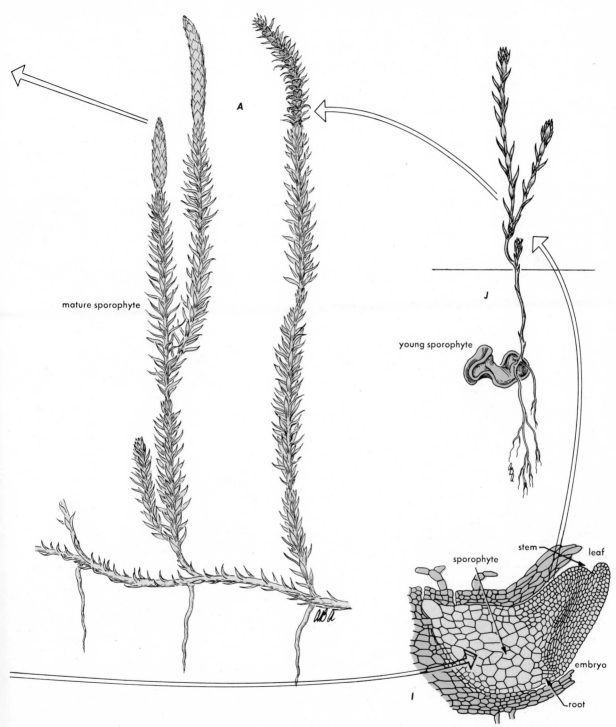

mature sporophyte

A

J

young sporophyte

stem

leaf

sporophyte

embryo

root

I

×50; *J*, underground gametophyte with sporeling, ×3. (*B*, after M. G. Sykes, *Ann. Botany* (*London*) **22**, 63; *C*, after E. Pritzel, in A. Engel and K. Prantl, *Die näturlichen Pflanzinfamilien; E, G, H*, after H. Bruchmann, *Flora* **101**, 220; *J*, material supplied by A. J. Eames.)

distributed, most of them are tropical; a few grow in temperate zones. Some species are adapted to withstand periods of drought and hence may grow in relatively dry localities.

Mature Sporophyte. As in *Lycopodium,* the sporophyte of *Selaginella* generally consists of a branched, prostrate stem with short, upright branches, usually only a few inches high. In some species, the stem is upright and slender and may climb 50 ft in height. Both horizontal and upright stems are sheathed with small leaves in four longitudinal rows or ranks (Figs. 27.11, 27.13; see Color Plate 16).

Two species of *Selaginella* are shown in Fig. 27.11. One of them, *Selaginella watsonii,* grows in exposed rock crevices in the higher Sierra Nevada and is able to withstand periods of drought. The more delicate *Selaginella emmeliana* grows best in a humid environment; it is frequently grown as an ornamental plant.

The vascular system in stems of *Selaginella* consists of a **stele** of one to several branching strands, each having a central core of xylem surrounded by phloem. The cross-sections of Fig. 27.12 show a single strand in a stem of *Selaginella.* It occurs in a large air space and is supported by strands of radially arranged endodermal cells. Vessels are present in the xylem of several species of *Selaginella.* Parenchyma and sclerenchyma tissue form a cortex, which is bounded externally by an epidermis.

Reproduction. As in *Lycopodium,* spores are borne in sporangia, which grow in or near axils of sporophylls. Although sporophylls do not differ greatly in appearance from sterile leaves, they are always grouped to form cones, or strobili, at the ends of upright branches (Fig. 27.13, see Color Plate 16, Fig. 27.14A).

Two types of sporangia are formed: **megasporangia** (Gr., *mega,* large) and **microsporangia** (Gr., *micro,* small). As the names indicate, megasporangia produce larger spores. A single strobilus usually contains both types of sporangia, in some species microsporangia are borne above megasporangia (Figs. 27.14B, C, D).

Within a developing megasporangium, all but one of the spore mother cells degenerate. This remaining spore mother cell, nourished in part by fluid resulting from the degenerating spores and in part by **tapetal** cells (layer of cells surrounding spore cavity), increases greatly in size. During meiosis, four large spores, called **megaspores,** are formed (Figs. 27.14C, F). Each megaspore may germinate and give rise to a female gametophyte (Fig. 27.14M).

Only a few spore mother cells within the develop-

ing microsporangium degenerate. The 250 or so that remain undergo meiosis, each forming four small spores, **microspores** (Figs. 27.14E, H).

The production by a given species of two distinct types of meiospores—megaspores and microspores—is called **heterospory.** Thus, *Selaginella* is heterosporous; *Lycopodium,* which produces but one spore type, is said to be **homosporous.**

Female Gametophyte. Repeated cell divisions within the megaspore result in the female gametophyte, which is contained within the megaspore until it nears maturity. Its increase in size eventually ruptures the megaspore wall, and a small cushion of colorless gametophytic tissue protrudes from the megaspore along the lines of rupture (Fig. 27.14M). Archegonia develop on the protruding cushion. Although similar in structure to archegonia of liverworts, they are much reduced in size. Archegonia of *Selaginella* are sunken within the gametophyte tissue; only two short cells of the neck protrude (Fig. 27.14N).

The megaspore (containing the female gametophyte) may be shed from the cone or strobilus at almost any stage of the development of the gametophyte. In some species, it may be retained in the strobilus until well after fertilization. In any event, fertilization occurs only when sufficient water, either rain or dew, is present, allowing sperms to swim to archegonia.

Male Gametophyte. Upon germination, microspores divide into two cells. One of them, the **prothallial cell,** is small and does not divide further; it represents the vegetative portion of the male gametophyte. The other cell, by repeated divisions, develops into an antheridium composed of a jacket of sterile cells enclosing 128 or 256 biflagellated sperms (Figs. 27.14I, J, K, L). This development occurs *within* the microspore wall. Microspores are shed from the

$$\text{sporophyte (2n)} \begin{array}{l} \nearrow \text{megaspore mother} \rightarrow \\ \quad \text{cell (2n)} \\ \\ \searrow \text{microspore mother} \rightarrow \\ \quad \text{cell (2n)} \end{array}$$

megaspore (n) → female gametophyte (n) → egg

microspore (n) → male gametophyte (n) → sperm

zygote (2n) → embryo (2n)

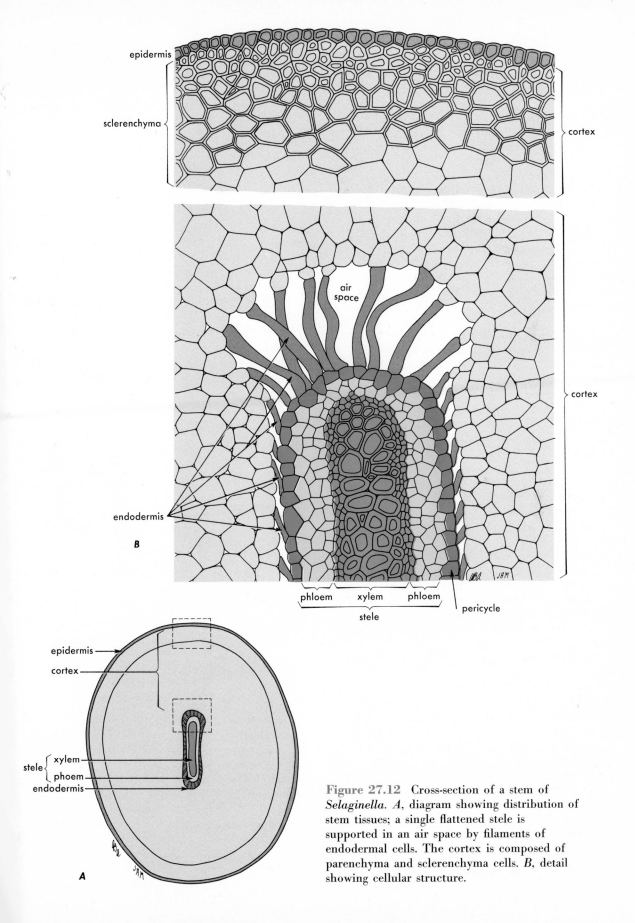

epidermis

sclerenchyma

cortex

air space

cortex

endodermis

B

phloem xylem phloem

stele

pericycle

epidermis

cortex

stele { xylem
phoem
endodermis

A

Figure 27.12 Cross-section of a stem of *Selaginella*. *A*, diagram showing distribution of stem tissues; a single flattened stele is supported in an air space by filaments of endodermal cells. The cortex is composed of parenchyma and sclerenchyma cells. *B*, detail showing cellular structure.

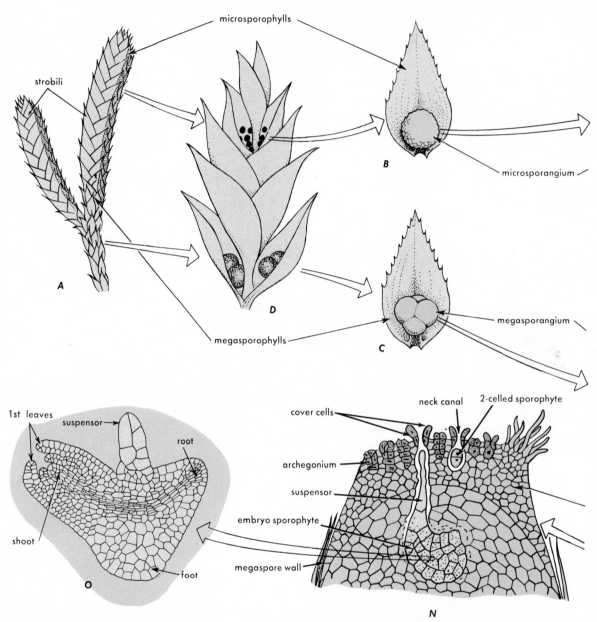

Figure 27.14 Stages in the life cycle of *Selaginella*. *A*, shoot of *Selaginella watsonii* showing two strobili, ×2; *B*, microsporophyll with microsporangium, ×5; *C*, megasporophyll with megasporangium, ×5; *D*, strobili showing sporophylls with sporangia, ×2; *E*, microsporangium discharging microspores, ×50; *F*, megasporangium discharging megaspores, ×50; *G*, microspores sift downward and lodge in axils of megasporophylls close to megaspores; *H*, a microspore, ×100; *I*, section of a microspore, ×150; *J*, male gametophyte within microspore, ×150; *K*, antheridium with developing sperm cells; *L*, mature sperm cells, ×100; *M*, female gametophyte within megaspore, ×100; *N*, section of female gametophyte showing location of archegonia in a young embryo pushed, by its suspensor, into the vegetative tissue of the female gametophyte, ×100; *O*, the young embryo sporophyte generally remains embedded in vegetative gametophyte tissue, ×100. (*E, H, I, J, K, L*, after R. A. Slagg, *Am. J. Botany* **19,** 106; *I, J, O*, after H. Bruchmann, *Flora* **104,** 180.)

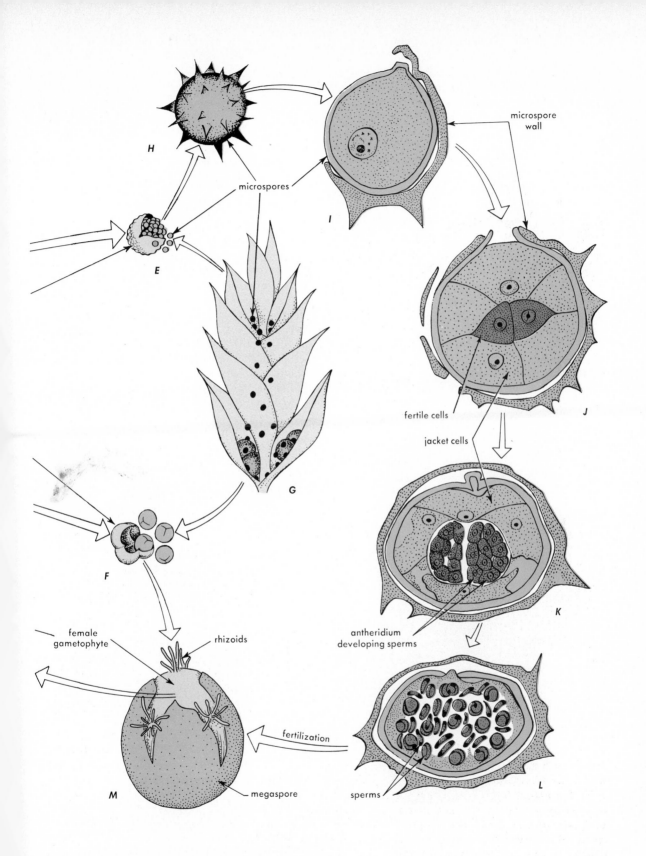

microspores

microspore wall

H

I

E

G

J

fertile cells

jacket cells

K

F

antheridium developing sperms

female gametophyte

rhizoids

fertilization

M

megaspore

sperms

L

microsporangium midway in the development of the male gametophyte and grow to maturity without direct connection with parent sporophyte or soil. Usually, microspores sift down to bases of the megasporophylls (Fig. 27.14*D*). In this position, they are close to the developing female gametophytes. The sperms escape when the microspore wall ruptures.

Sperms swim to the archegonia, which grow on that portion of the female gametophyte protruding from the megaspore (Fig. 27.14*N*). Fertilization ensues, and the resulting zygote initiates the diploid or sporophyte generation.

Sporophyte Embryo.   Of the two cells formed by the first division of the zygote, only one develops into an embryo. The other cell grows into an elongated structure, the **suspensor,** which pushes the developing embryo into gametophytic tissue, where there is a food supply (Fig. 27.14*N*).

The embryo is a structure with (*a*) a foot, (*b*) a root, (*c*) two embryonic leaves, and (*d*) a shoot (Fig. 27.14*O*). In certain species of *Selaginella,* the embryo is held by the megaspore and retained within the strobilus. It does not pass into a dormant state, as

do seeds, but continues to grow. Young sporophytes may be found extending from the strobilus of parent sporophytes. Should the developing embryo pass into a period of dormancy while being held by the mother sporophyte, a condition would arise that approaches the seed habit.

The life history of *Selaginella* is shown in the diagram.

Lepidodendrales

The Lepidodendrales is an important fossil order of the division Lycophyta. Many representatives were heterosporous, and true seeds were formed by some species. No members of the order are in existence today. It was, however, a dominant order during the period when the great deposits of coal were being formed. If one can imagine a lush forest of *Selaginella*-like trees, 150 ft tall and 6 ft in diameter, a fair picture of this forest may be obtained. There were also smaller, shrubby and herbaceous Lepidodendrales. Since two spore types, and seeds, were formed by these ancient plants, we must believe that both heterospory and the seed habit are very old plant characters and not characters confined to present-day plants (see also Chap. 30).

Significant Features of the Division Lycophyta

1. Spores are borne on or near sporophylls, grouped at ends of upright branches to form cones or strobili.
2. Both homospory and heterospory occur.
3. Gametophytes of some species of *Lycopodium* are colorless and subterranean, and are associated with a fungus; in other species, they are autotrophic.
4. Gametophytes of *Selaginella* develop within spores.
5. Spores of some species of *Selaginella* are retained in the strobili until the gametophytes have matured and fertilization has occurred.
6. Members of the fossil order Lepidodendrales were heterosporous and produced seeds.
7. The vascular system in stems of the Lycophyta resembles that found in roots of Angiospermae.

Division Sphenophyta

Members of this division once grew very abundantly, as is evidenced by their good representation in the fossil record. Today, the division is represented by a single family (Equisetaceae) with but one genus (*Equisetum*) of about 25 species (Fig. 27.15). Many

Figure 27.15 *Equisetum telmateia,* showing much-branched vegetative shoots and straight unbranched fertile shoots with terminal strobili, $\times\frac{1}{4}$.

species inhabit cool, moist places, but *Equisetum arvense* grows in dry habitats. Most species are characterized by the presence of silica in the epidermis of stems. Because of this trait, they were used in colonial days to scour pots and pans and hence were called scouring rushes. The genus is also commonly known as the horsetails.

Mature Sporophyte

The sporophyte of *Equisetum* is the dominant phase of its life history. In one tropical species, the sporophyte is vine-like and may reach 25 ft in length. Usually, however, 5 ft represents the maximum upright growth. All species are perennials and have a branched rhizome from which upright stems arise. Stems, depending upon the species, may branch either profusely or sparingly. In either event, they are straight and marked by ridges and distinct nodes (Fig. 27.16). Tissue just above nodes remains meristematic and structurally weak so that the stems may be easily pulled apart at these points. Bases of nodes are sheathed by whorls of small simple leaves that are fused laterally (Figs. 27.15, 27.16). When branch-

ing occurs, the branches arise at nodes immediately below leaves and, since there are frequently many leaves in a whorl, many branches may form at each node. The leaves are much reduced in size, nongreen and, in many species, short-lived. The stems are green and are, therefore, the organs of food manufacture. Roots occur only at the nodes of the rhizomes or bases of upright stems. Leaves of plants of this division are generally arranged in a whorl, and alternate with branches at nodes. Leaf and branch gaps are present as in flowering plants, but their arrangement is different. The ribbed stem of Sphenophyta is such a characteristic feature that fossil stems showing ribs are regularly placed in this division.

Except at nodes, stems of *Equisetum* are hollow and the ribs make prominent markings on their outer circumferences (Fig. 27.16). The ridges, formed of sclerenchyma tissue, not only project outward, but extend inward almost to the small vascular bundles (Fig. 27.17). There are air spaces, or canals, between vascular bundles. Photosynthetic tissue lies between these air canals and the thin outer layer of sclerenchyma tissue and epidermis. Vascular strands are also marked by a canal (Fig. 27.17). Arms of xylem extend

Figure 27.16 *Equisetum hyemale*, vegetative shoot. *A*, upper six nodes showing the telescoping of the youngest internodes and their rapid elongation; ridges and whorls of leaves are evident, ×2. *B*, enlarged view of stem showing three nodes. Note that the leaves match the ridges below them, but alternate with the ridges above them, ×6. *C*, the reproductive shoot ends in a strobilus, ×1.

outward from this canal, and phloem tissue lying outward from the canal is present between radial arms of xylem. An endodermis is present, but its location varies from species to species; it may surround each vascular bundle, or encircle a ring of vascular bundles. In some species, there is also an inner endodermis.

Reproduction

In all species of *Equisetum,* the sporangium-bearing organs, **sporangiophores,** are specialized structures very different from ordinary leaves. They are grouped together in **strobili** (Fig. 27.18) at the summit of main upright branches and occasionally on lateral branches (Figs. 27.15, 27.18*A*). In most species, cones or strobili are borne on ordinary vegetative shoots; in a few species, they are formed only on special fertile shoots (Figs. 27.15, 27.18*A*).

The sporangiophores are stalked, shield-shaped structures borne at right angles to the main axis of the cone. The cone may be compared to a pole to which open umbrellas have been fastened, the handle of the umbrella being at right angles to the pole (Fig. 27.18*C*). Sporangia are attached to the underside of the shield, close to its edge. They extend horizontally inward, toward the axis of the cone.

Meiosis is normal, and four meiospores with the haploid number of chromosomes are formed from

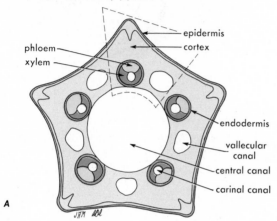

Figure 27.17 Cross-section of *Equisetum* stem. *A,* diagram showing arrangement of stem tissues, ×70; *B,* enlarged view showing details of cellular structure, ×400.

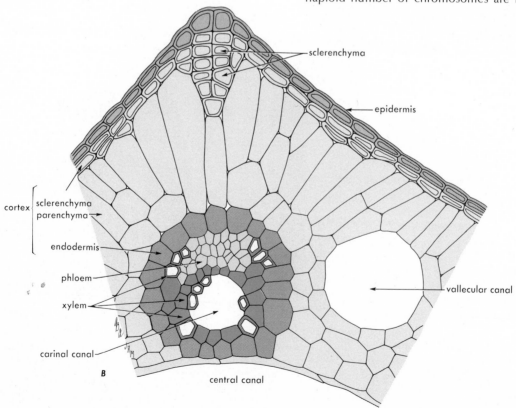

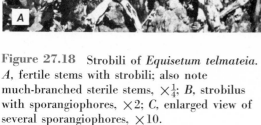

Figure 27.18 Strobili of *Equisetum telmateia*. *A*, fertile stems with strobili; also note much-branched sterile stems, $\times \frac{1}{4}$; *B*, strobilus with sporangiophores, $\times 2$; *C*, enlarged view of several sporangiophores, $\times 10$.

each spore mother cell. All meiospores are morphologically alike. *Equisetum*, therefore, is homosporous. Upon maturity the meiospores are discharged from the sporangia. The spores have a thick, double wall, and, when they are mature, the outer half of the wall unfolds to form narrow hygroscopic ribbons that move under the influence of moisture changes and aid in distribution of spores (Fig. 27.19). The spores are fragile and normally live but a few days.

Gametophytes. Gametophytes are small green bodies about the size of a pinhead and consist of a cushion-like base with many erect, delicate lobes (Fig. 27.20). They may be easily cultured.

Gametangia are similar to those of *Lycopodium*.

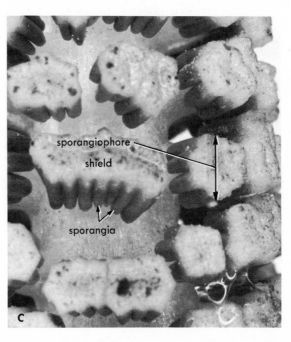

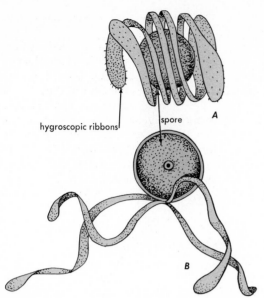

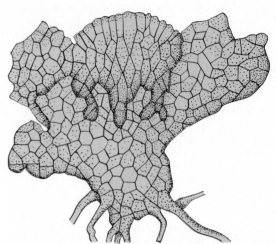

Figure 27.19 Spores of *Equisetum* showing the hygroscopic ribbons, ×100. (Redrawn from R. von Wettstein.)

Figure 27.20 Gametophyte of *Equisetum*, ×20. (Redrawn from E. R. Walker, *Botan. Gaz.* **92**, 1. Copyright the University of Chicago Press.)

Fertilization takes place only when there is free water. Sperms are spiral, multiciliate cells.

The zygote develops directly into an embryo. No suspensor is formed. The embryo is similar to those already described, except that the foot is small or lacking.

Significant Features of the Division Sphenophyta

1. The genus *Equisetum,* consisting of about 25 species in the family Equisetaceae, is the only living group of this once widely distributed division.
2. The stem of *Equisetum* has distinct nodes and internodes. Meristematic tissue occurs just above the nodes, which are sheathed by whorls of small, brownish, short-lived leaves. Photosynthesis is carried on by green stems.
3. The strobili are composed of specialized structures, sporangiophores consisting of shield-shaped disks, supported by short stalks growing at right angles to the stem.
4. The autotrophic gametophyte is a single thallus with many lobes.
5. Present-day species of *Equisetum* are homosporous; certain extinct species of Sphenophyta were heterosporous.

Division Pterophyta

The division Pterophyta comprises the ferns, most of which are shade-loving plants of small size, their upright leaves, or **fronds,** generally being their most prominent feature. All species native to temperate zones have an underground rhizome with leaves and roots at nodes. In this respect they resemble the growth form of many monocotyledons (Fig. 7.27A). Some ferns of tropical regions do grow into fairly large trees (Fig. 27.21). All Pterophyta have a definite alternation of generations, with both sporophyte and gametophyte being autotrophic plants.

The Pterophyta comprise four orders, the largest of which, the Filicales, includes the **true ferns.** The Filicales are divided into eleven or more families, of which the largest and best-known in the United States is the Polypodiaceae. Most of the following discussion will deal with representatives of this family. There are also several fossil orders. One of them, the Coenopteridales, will be briefly discussed because of its characteristic leaf structure.

Mature Sporophyte

The sporophyte, which is the dominant generation of all ferns, possesses an underground stem or rhizome from which leaves and adventitious roots arise (Fig. 27.22). Leaves are the most prominent part of the fern plant and vary greatly in size and form. They differ from leaves of flowering plants in two important respects: (a) they have an apical meristem that usually continues active for some time (Fig. 27.23B), and (b) spores frequently are borne on their lower surface (Fig. 27.26). Young fern leaves are rolled in tight spirals and consist chiefly of meristematic tissue (Fig. 27.23). As the leaf matures, it unwinds from the base upwards. The upright, expanded portion is mature, but the leaf continues to grow by cell division at its coiled tip. The uncoiled, fully expanded fern leaf lacks meristematic tissue, as the cells of all such tissue have differentiated into permanent cells. In certain species, the apical meristem may remain active for years, resulting in leaves nearly 10 ft long.

Most fern leaves are compound, although simple types exist in all groups. The most common type of fern leaf has a stout or rigid petiole, which is pro-

longed to form a rachis from which leaflets arise (Figs. 27.22, 27.24).

The vascular tissue of ferns is organized into one or more vascular strands, each having a core of xylem, surrounded by phloem and separated from the ground parenchyma by an endodermis. These vascular bundles are interconnecting, so, as in *Lycopodium*, the precise arrangement will vary from section to section. However, there are arrangements characteristic of various genera. For instance, in *Polypodium* there is a ring of small bundles placed well out from the center of the rhizome (Fig. 27.25B). Each strand is formed of a central core of xylem,

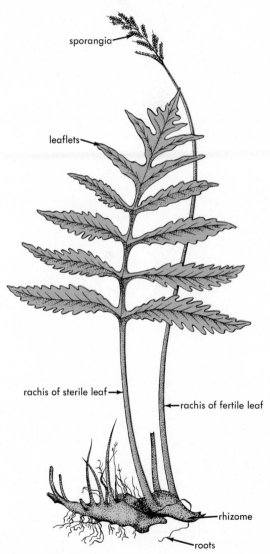

Figure 27.22　The sensitive fern (*Onoclea sensibilis*), $\times\frac{1}{20}$. (Redrawn from L. Diels, in A. Engler and K. Prantl, *Die näturlichen Pflanzenfamilien.*)

Figure 27.21　A tree fern (*Cyathea*) growing in Golden Gate Park, San Francisco, $\times\frac{1}{50}$.

Figure 27.23 Fern fronds retain meristematic tissue at their tightly rolled tips. As differentiation proceeds, the tips uncoil. *A*, tightly rolled tip of a young frond of the tree fern, *Cyathea*, ×¼. *B*, maturing frond of the bracken fern (*Pteris aquilina*); the tips are still tightly rolled, ×½. *C*, uncoiling tip of a frond of the sword fern (*Nephrolepsis exaltata*), ×2.

Figure 27.24 Fern fronds. *A*, simple entire frond of hart's-tongue fern (*Phyllitis scolopendrium*), $\times\frac{1}{10}$; *B*, pinnate frond of *Polypodium*, $\times\frac{1}{2}$; *C*, bipinnate frond of *Pellaea*; *D*, bipinnate frond, with serrate leaflets, of the rock brake fern, *Cryptogramma acrostichoides*, $\times1$; *E*, fertile leaf of *Cryptogramma*, $\times1$; *F*, much-dissected frond of *Asplenium*, young plants form at meristematic leaf tips, $\times\frac{1}{2}$.

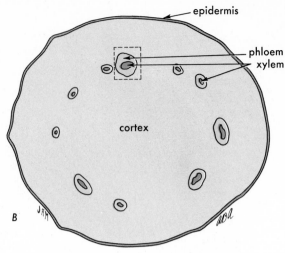

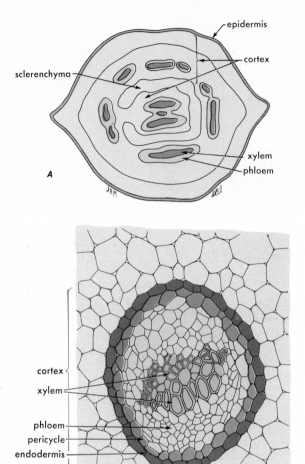

Figure 27.25   Cross-sections of fern rhizomes. *A*, Rhizome of *Pteris*, showing relationship of stem tissues, ×37; *B*, *Polypodium*, showing relationship of stem tissues, ×45; *C*, a higher magnification of *B*, showing cellular detail, ×450.

surrounded by phloem, and the whole has a bounding endodermis (Fig. 27.25C). In *Pteris*, there is an outer circle of somewhat flattened vascular bundles with a central set of interconnected bundles. Each vascular bundle has a bounding endodermal layer (Fig. 27.25A). The xylem consists largely of tracheids, vessels being known in only two genera. Sieve cells occur in the phloem.

Reproduction

Vegetative Reproduction.   This type of reproduction may occur in one of two ways: (a) by death and decay of the older portions of the rhizome and the subsequent separation of the younger growing ends; and (b) by the formation of deciduous leaf-borne buds, which become detached and grow into new plants. Such buds occur in only a few genera.

Sexual Reproduction.   In the sexual life cycle, independent sporophyte and gametophyte generations

alternate with each other. The vegetative structure of the sporophyte has been described above. Spores are borne in sporangia, which ordinarily develop on the lower surface or margins of fronds (Fig. 27.26). Not all leaves are fertile, that is, spore-producing; and fertile leaves are not, in all species, similar in structure to sterile (non-spore-producing) ones (Figs. 27.22, 27.24D, E). The distribution of sporangia on the leaf surface varies considerably in different genera and species (Fig. 27.26). Sporangia may (a) cover much of the lower surface, (b) be grouped in **sori** (singular, **sorus**) and grow in a definite relationship with veins, or (c) grow only along margins or edges of the leaf.

When sporangia are grouped together in sori, a structure called the **indusium** (Figs. 27.26B, D, 27.28), which is sometimes umbrella-like, may be present, thereby protecting young and developing sporangia. Frequently, marginal sporangia are protected by the curled edge of the leaf, which forms a **false indusium**.

564   **Lower Vascular Plants**

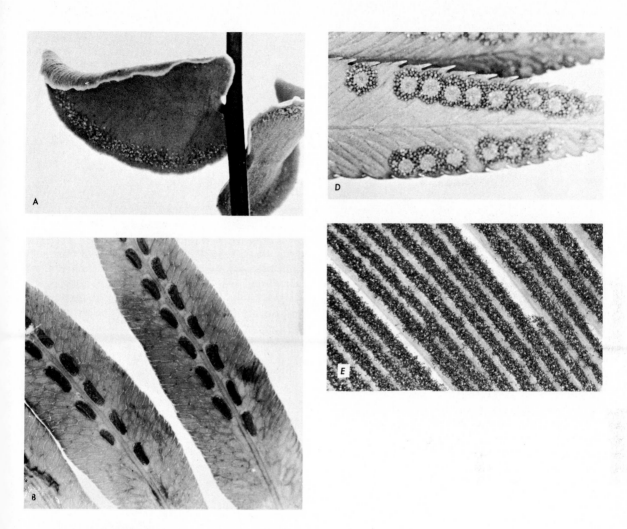

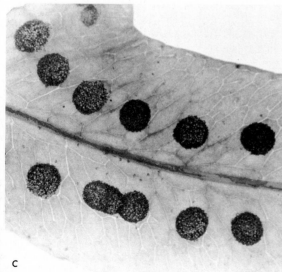

Figure 27.26 Sori. *A*, sporangia in a band close to margin of leaflet, *Pellaea*; *B*, sporangia in elongated sori parallel to midrib of leaflet with indusia, *Woodwardia*; *C*, sporangia in round sori, *Polypodium*; *D*, sporangia grouped in round sori; centrally placed indusia are present, *Polystichum*; *E*, sporangia in long rows on veins, *Asplenum nidus*. All about ×2.

Figure 27.27 Fern gametophytes growing on a flower pot, ×2. (Courtesy of E. M. Gifford.)

The sporangium is a delicate watch-shaped case, consisting of a single layer of epidermal cells, only one row of which possesses heavy walls. This row, which nearly encircles the sporangium (in the Polypodiaceae), is the **annulus;** it functions in opening dried mature sporangia and aids in dispersal of ripe spores (Figs. 27.28, 27.29*B*).

The young sporangium is filled with sporogenous cells, which eventually give rise to spore mother cells. Meiosis in the spore mother cell results, as always, in spores with a reduced number of chromosomes. Ferns, with the exception of two families, the Marsileaceae and the Salviniaceae, are homosporous. These two families, though widely distributed, are small, uncommon, and unfern-like in appearance.

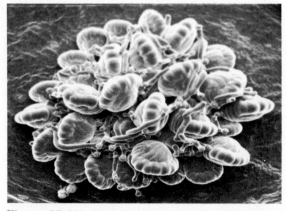

Figure 27.28 Stereoscan micrograph of sorus of *Polypodium heterophylla*, ×500. (Courtesy of E. M. Gifford and R. Falk.)

Gametophyte. Gametophytes, or **prothallia,** of the Polypodiaceae are small, flat, green, heart-shaped structures (Fig. 27.27) with rhizoids on their lower surface (Fig. 27.29*D*). In most species, they apparently mature rapidly and are not long-lived. Antheridia and archegonia are borne on the same prothallium. Antheridia are formed when the prothallium is very young and are scattered over its lower surface. Antheridia are small and have a jacket formed from only three or four sterile cells (Fig. 27.29*E, F*). Antheridia project only slightly from the surface of the gametophyte. Normally, 32 sperms develop within each antheridium (Fig. 27.29*F*).

Archegonia form later than antheridia and are usually clustered close to the notch, also on the undersurface of the gametophyte (Figs. 27.29*D, H*). They, too, are small but typical, consisting of a short neck and a venter that encloses the egg cell. Neck cells protrude slightly from the lower surface of the gametophyte.

Fertilization occurs when moisture is present and sperms are thus able to swim to the neck of the archgonium. The resulting zygote is diploid and rapidly develops into an embryo sporophyte comprising a foot, root, stem, and leaf (Figs. 27.29*I, J, K*). The embryo develops directly into a young sporophyte (Fig. 27.29*L*).

Significant Features of the Division Pterophyta

1. The sporophyte plant, in most species, consists of a rhizome bearing adventitious roots and upright leaves.
2. A well-developed vascular system is present.
3. Usually, the leaves are dissected into leaflets, arranged pinnately on a stout midrib or rachis. They are condensed branch systems and, when young, have meristematic tissue at their tips.
4. Sporangia are borne on the undersurface of leaves. They may be grouped into sori, which, in some species, are protected by an indusium.
5. All but two families of Pterophyta are homosporous.
6. Gametophytes (prothallia) are independent of sporophyte. They are small, green, heart-shaped, and short-lived plants. Both antheridia and archegonia are borne on the same gametophyte.

Order Coenopteridales (Fossil Ferns)

The fossil ferns as represented by some of the Coenopteridales, though different in external ap-

TABLE 27.1 Comparison of Psilophyta, Lycophyta, Sphenophyta, and Pterophyta

Generation	Psilotum	Lycopodium	Selaginella	Equisetum	Ferns
Sporophyte	A branching stem, scale leaves, no roots	Prostrate branching stem with upright branches, roots, and leaves	Prostrate branching stem with upright branches, roots, and leaves, also scales	Rhizome, upright, jointed stem, small leaves, stems carry on photosynthesis	Rhizome, roots, and leaves Herbs (mostly) Stem: several strands of vascular tissue with xylem of each surrounded by phloem Vessels occur in two genera Sieve cells, but no companion cells in phloem
	Simple vascular system	Simple vascular system	Simple vascular system, but vessels are present	Simple vascular system	
	One type of sporangium Homospory No sporophylls	One type of sporangium Homospory Sporophylls present	Megasporangia Microsporangia Heterospory Sporophylls all similar	One type of sporangium Homospory Sporangiophores	One type of sporangium (mostly) Homospory (mostly) Leaves bear spores in sporangia
	No cone	A cone in some species	Cone	Cone composed of sporangiophores	No cones
	Embryo develops within gametophyte	Embryo develops within gametophyte	Embryo develops within female gametophyte on sporophyll	Embryo develops within gametophyte	Embryo attached to gametophyte
	No suspensor	Suspensor	Suspensor	No suspensor	No suspensor
Gametophyte	Irregular subterranean structure associated with fungus	Irregular to tapered subterranean structure associated with fungus	Male gametophyte, one prothallial cell, and an antheridium within microspore	A single green thallus resembling *Anthoceros*	Heart-shaped, small, green, completely independent thallus
	Gametangia embedded in thallus	Gametangia embedded in thallus	Female gametophyte, small amount of tissue within megaspore, cushion protrudes in which gametangia are embedded	Gametangia embedded in thallus, neck of archegonium protruding	Antheridium of approximately 6–10 cells Archegonia partially embedded in gametophyte
	Motile sperms	Motile sperms	Motile sperms	Motile sperms	Motile sperms

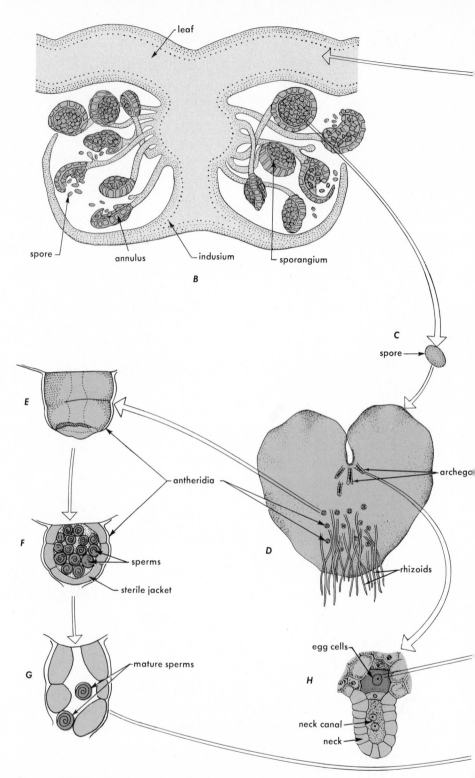

Figure 27.29 Stages in the life cycle of a fern. *A*, a frond of a fern, ×1; *B*, section through a sorus, showing indusium, sporangia, and spores, ×50; *C*, spore, ×100; *D*, gametophyte, ×5; *E*, antheridium, ×100; *F*, section of antheridium showing sperm, ×100; *G*, open antheridium discharging sperm, ×100; *H*, archegonium, ×100; *I*, archegonium receptive to sperm, ×100; *J*, section of gametophyte showing zygote within the venter of the

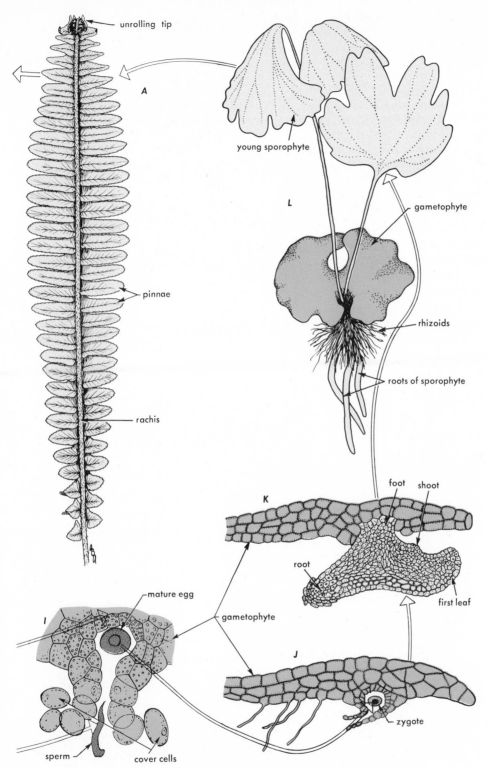

A

young sporophyte

L

gametophyte

rhizoids

roots of sporophyte

pinnae

rachis

foot   shoot

K

root

first leaf

mature egg

gametophyte

I

gametophyte

J

zygote

sperm   cover cells

archegonium, ×100; *K*, section of gametophyte and embryo, ×100; *L*, gametophyte with attached young sporophyte, ×10. (*E, F, G*, after M. E. Hartmann, *Botan. Gaz.* **91**, 252. Copyright The University of Chicago Press. *H, I*, after D. H. Campbell, *Mosses and Ferns*, Macmillan, New York, 1905. *J, K*, after R. M. Holman and W. W. Robbins, *A Textbook of General Botany*, John Wiley & Sons, New York.)

pearance from present-day ferns, possessed sporangia, spores, and a vascular system that resembled the corresponding structures of present-day ferns. They were small plants, either erect, creeping, or climbing. They are of interest because of their primitive leaves, which differed only slightly from the stem. As the shoot developed, it branched; one division continued to grow, whereas the other soon ceased growth. This short, determinate branch of the main axis was the leaf. Structurally, it closely resembled the stem. This primitive type of fern leaf forms one basis for the belief that the leaves of flowering plants are in reality condensed branch systems (see also Chap. 30).

A Summary of the Morphological Traits of the Lower Vascular Plants

These divisions of more primitive vascular plants have striking similarities, as well as differences. They can be arranged in a sequence suggesting the evolution of traits better adapting them to survive in an environment becoming increasingly drier. These traits are compared in Table 27.1.

Classification

Division	Psilophyta
Family	Psilotaceae
Genera	*Tmesipteris*
	Psilotum nudum
Family	Rhyniaceae
Genus	*Rhynia*
Division	Lycophyta
Family	Lycopodiaceae
Genus	*Lycopodium*
Family	Selaginellaceae
Genus	*Selaginella*
Family	Isoetaceae
Genus	*Isoetes*
Division	Sphenophyta
Family	Equisetaceae
Genus	*Equisetum*
Division	Pterophyta
Orders	Coenopteridales
	Filicales
Family	Polypodiaceae
Genera	*Polypodium*
	Pteris
Family	Cyatheaceae
Genus	*Cyathea*

28

Gymnosperms

This group of plants is represented by such common trees as pines (*Pinus*), spruces (*Picea*), firs (*Abies*), and cedars (*Cedrus*), all of which possess well-developed cones in which seeds are borne (Fig. 28.1; see Color Plate 17). Other gymnosperms, such as *Ginkgo*, yews (*Taxus*), and Mexican tea or joint pine (*Ephedra*), do not bear cones. However, a trait common to all gymnospermous plants is the absence of a protecting case, like the ovary wall, around the seeds. In yews, seeds are partially surrounded by a red berry-like structure (Fig. 28.20). In *Ephedra*, ovules are borne in axils of short bracts (Fig. 28.3) and in *Ginkgo*, naked ovules are attached to ends of short branches (Fig. 28.8*B*). By contrast, in angiosperms, the seed of peach, for example, is surrounded by a matured ovary wall, which is in part fleshy and in part hard and stony. In pines and other cone-bearers, seeds are borne on the surface of scales that comprise the cone, and, though well-protected by the scales, they are not surrounded by floral parts. Such seeds, lacking protection of an ovary wall, are said to be "naked" (Fig. 28.2)—thus the name gymnosperm, derived from two Greek words *gymnos* (naked) and *sperma* (seed). It will be recalled that fossil Lepidodendrales, belonging to the Lycophyta, bore seeds, thus indicating that the seed habit is not of recent geological origin.

General Characteristics and Classification

The external morphology of gymnosperms shows considerable variation, as may be seen from the following examples: (*a*) Fossil gymnosperms so resemble ferns that they are commonly called "seed ferns." (*b*) Some modern gymnosperms of Mexico and other tropical areas are palm-like in appearance. (*c*) Most north temperate gymnosperms have needle or scale leaves and are mainly evergreen. (*d*) Some gymnosperms of temperate regions, particularly of Australia, are broad-leaved. (*e*) Certain gymnosperms are shrubby and thrive under very arid conditions.

Vascular tissue of the class has been discussed at some length in Chapters 7 and 8. It should be recalled that vessels are absent from the xylem in all but a few species, and that companion cells do not occur in the phloem, where there are sieve cells rather than sieve tube members. Pith is present in stems but not in roots.

There are five divisions of gymnosperms encompassing 650 species grouped into many genera and

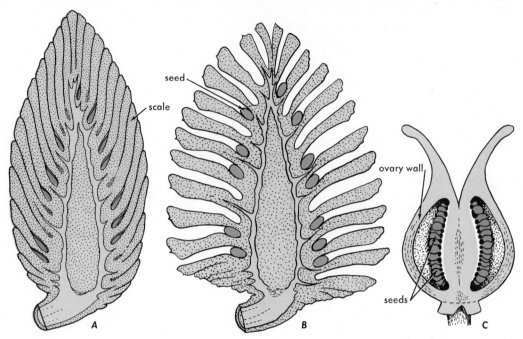

Figure 28.2 *A* and *B*, median longitudinal sections of cones of a gymnosperm. *A*, immature cone tightly closed to protect ovules; *B*, open to disseminate seeds; *C*, angiosperm fruit showing seeds enclosed in an ovary.

families. Of these divisions, we shall only briefly consider (a) the Cycadophyta, or cycads; (b) the Pteridospermophyta, or seed ferns; and (c) the Ginkgophyta, composed of only a single species, the ginkgo or maiden-hair tree. We will discuss a fourth division, the Coniferophyta, in much more detail. The remaining division, the Gnetophyta, contains about 71 species in three genera and three orders, and all are tropical or subtropical. Only one genus, *Ephedra*, occurs in the United States; it is a rather small shrub of the southwestern deserts, is almost leafless, and bears small naked seeds (Figs. 28.1*E*, 28.3).

Division Pteridospermophyta

The Pteridospermophyta are extinct, but they are well-known as fossil plants. They grew over much of the earth when coal deposits were being formed. Their fossils, consisting of beautifully preserved stems, roots, leaves, seeds, and even pollen grains, have been found encased in rock mined as waste material in coal mines (Fig. 28.4). They have a distinct fern-like appearance (Fig. 28.5) and for many years were regarded as ferns. When, however, seeds were found attached to leaves, it became necessary to

revise their classification. Since the seeds were naked, these plants were regarded as gymnosperms. The structure of these seeds (Fig. 28.4*D*) resembles such modern gymnosperms as cycads (p. 574) and *Ginkgo* (p. 575). Furthermore, pollen grains have been found associated with these fossil seeds in such a manner as to suggest strongly a life history similar in its essential steps to that of living gymnosperms. Because of these findings, Pteridospermophyta are now considered to be the most primitive gymnosperms.

Figure 28.3 Naked seeds of *Ephedra*; the only genus of the division Gnetophyta living in the United States. See Fig. 28.1*E*, ×2.

A

B

C

Figure 28.1   Gymnosperms. *A* through *D*, Coniferophyta. *A*, cones and branches of *Cedrus deodar; B, C,* young and mature ovulate cones of *Pinus ponderosa; D,* staminate cones of *Juniperus occidentalis; E, Ephedra viridis,* an evolutionarily advanced gymnosperm belonging to the division Gnetophyta.

D

E

Figure 30.1 Monument Valley buttes, Arizona.

Figure 30.2 The Grand Canyon, Arizona.

Figure 30.3 Towers of the Virgin River, Utah.

A

C

Figure 30.23 Reconstruction of vegetation in eastern Oregon during the past 50 million years. *A*, montane rainforest of Eocene; *B*, mixed conifer-hardwood forest of Oligocene; *C*, deciduous forest of Miocene; *D*, sagebrush desert scrub of today.

D

B

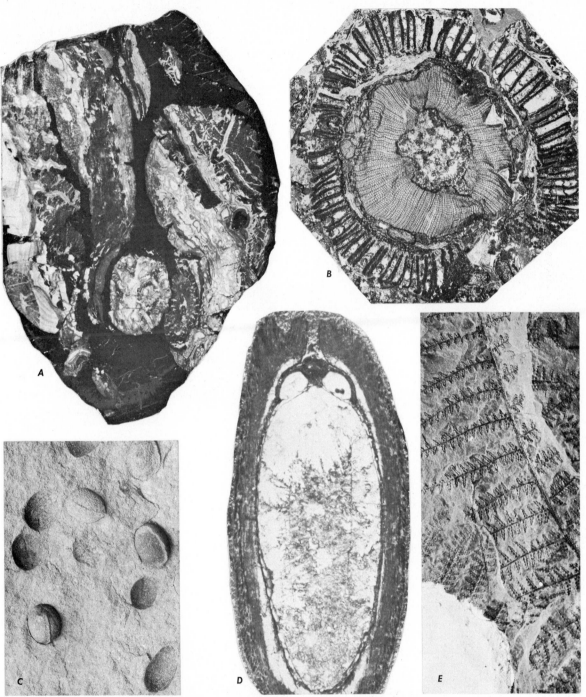

Figure 28.4 Fossil seed ferns. *A*, section through a mass of coal showing embedded plant material; *B*, cross-section of fossilized stem of *Lyginopteris*, an ancient plant that grew in England; *C*, seeds of a large seed fern that grew in Illinois; *D*, median longitudinal section of the seed of a seed fern; *E*, impressions in rock of leaves of *Lyginopteris*, from Scotland. (*A*, *B*, *D*, *E*, courtesy of H. N. Andrews, Jr.; *C*, courtesy of Field Museum of Natural History.)

Figure 28.5 Probable appearance of a seed fern. (From H. N. Andrews, Jr.)

Division Cycadophyta

Some 200 million years ago members of the Cycadophyta formed an extensive portion of the earth's flora, probably constituting the food supply for at least some of the herbivorous dinosaurs. Today, there remain some nine well-defined genera of cycads with about 100 species growing in widely separated areas of the earth's surface and largely confined to the tropics. Only one genus, *Zamia* (Fig. 28.7), occurs naturally in the continental United States; it is found in southern Florida. Various genera are, however, under cultivation, outdoors in warmer regions, and in large greenhouses in colder regions. Two cycads, *Cycas revoluta* and *Dioön spinulosum*, are shown in Figs. 28.6*A*, *B*. Both were grown in the conservatory at Golden Gate Park. These trees are palm-like in appearance; in fact, *Cycas revoluta* is known as Sago palm. In *Zamia,* the genus native to Florida, the trunk is largely subterranean (Fig. 28.7), but in *Cycas* it forms a single straight bole. These

trees are very slow-growing, a 6 ft high specimen being perhaps as much as 1000 years of age.

That these trees are really gymnosperms and not palms is readily evident from the forms of their cones. The large, upright cone shown in Fig. 28.6*A* bears pollen, while that shown in Fig. 28.6*B* is a

microsporangiate cone

ovulate cone

Figure 28.6 *A, Cycas revoluta* showing a microsporangiate cone; *B, Dioön spinulosum* showing ovulate cone.

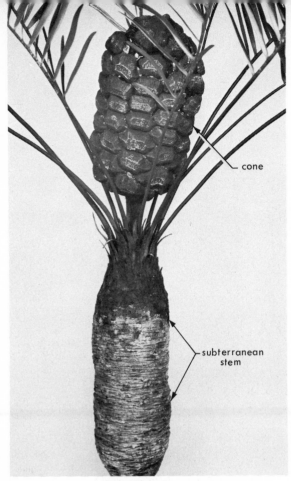

Figure 28.7 *Zamia*, showing ovulate cone and subterranean stem. (Courtesy of Field Museum of Natural History.)

seed-bearing, or ovulate, cone. While we shall not discuss the nature of these seeds, they are not like those of angiosperms but consist of erect unprotected ovules with their micropyles turned outward, forming striking examples of naked ovules. An additional primitive feature is the presence of flagellated sperm. Since *Ginkgo* is cultivated widely in the United States and, like cycads, possesses both naked ovules and flagellated sperm, these features are described in somewhat more detail in the following section.

Division Ginkgophyta

Only one living representative, the maiden-hair tree (*Ginkgo biloba*), remains of this very ancient division of plants. It has been reported as growing wild today in forests of remote western China. It has, however, been grown for centuries on Chinese and Japanese temple grounds and is now cultivated in many countries. It is a large tree with characteristic small, fan-shaped leaves (Figs. 10.5*B*, 28.8), which are divided into two lobes. The trees are dioecious.

Figure 28.8 *Ginkgo biloba*. *A*, growing on the campus of the University of California at Berkeley; *B*, branch showing seeds. (*B*, courtesy of John H. Gerard from National Audubon Society.)

Figure 28.9 Mature ovulate cones. *A, Cedrus deodara; B, Picea sitchensis; C, Sequoia sempervirens; D, Metasequoia glyptostroboides; E, Sequoiadendron gigantea; F, Pinus monophylla; G, Pinus contorta; H, Pinus jeffreyi; I,* base of cone of *Pinus jeffreyi.*

F

G

H

I

Reproduction

With the exception of a few algae and some fungi, all living plants so far studied have been characterized by motile sperms, and free water has been necessary for fertilization. Moreover, gametophytes have been mostly independent of the sporophyte. It will be recalled that *Selaginella* has these characteristics, except that female gametophytes are occasionally retained by the sporophyte until well after fertilization. The sexual cycle in *Ginkgo* is somewhat comparable to that of *Selaginella;* however, female gametophytes are almost completely surrounded by sporophyte tissue, and, though motile sperms are formed, the fluid in which they swim is produced by the parent sporophyte.

Female gametophytes of *Ginkgo* are buried deep in the ovule, where small, reduced archegonia are formed. The archegonia have two neck cells and no neck-canal cells except the ventral canal cell. Entrance to the ovule is through a small opening, the micropyle, at one end of the ovule. Microspores develop in catkin-like strobili on male trees. When mature, microspores are shed and are carried by air currents to mature ovules on female trees. The microspores are drawn into the micropyle and come to lie in contact with the nucellus. In this position they develop into male gametophytes, forked pollen tubes, that grow parisitically into the upper portion of the nucellus. When the end of a pollen tube nears an archegonium, two motile sperms are released into a cavity just above the archegonium. One sperm eventually passes into an archegonium and fuses with the egg nucleus. An embryo develops from the zygote. Integuments become the seed coat and surround the dormant embryo and female gametophyte, thus forming a true seed.

The important feature of this life history is the development of a method to *ensure fertilization without the necessity of free water.* Such a development has made possible the evolution of an extensive dry land flora.

Division Coniferophyta

Classification

Without exception, the well-known and economically important gymnosperms of temperate zones belong to the Coniferophyta. There are nine families and numerous genera. They comprise pines, hemlocks, firs, spruces, junipers, yews, redwoods, and many others. Not all of them bear cones, yet the cone is such a conspicuous feature of a large number of them that the division has been named the Conif-

erophyta or cone-bearers. True cones are formed by the majority of them (Fig. 28.1, see Color Plate 17; Fig. 28.9). Juniper berries are in reality cones with fleshy adhering scales (Fig. 28.17). In yews, seeds are surrounded at the base by a more or less pulpy, berry-like body and are not borne in cones (Fig. 28.20). In either event, the seeds are naked, not being surrounded by an ovary wall.

The following brief descriptions of several common genera will serve as illustrations of the more important conifers.

Family Pinaceae

Pinus (*Pines*). Pines are usually large trees. The trees may be either pyramidal (Fig. 28.10) or flat-topped. The leaves are needlelike, two or more growing together (except in *Pinus monophylla*) in a **fascicle** or group, which is sheathed at the base (Fig. 28.11). The cones vary greatly in size and shape and are very characteristic of the species to which they belong (Fig. 28.9*F, G, H*).

Abies (*Firs*). Firs are stately trees of a symmetrical cylindrical or pyramidal shape. The leaves are flat and linear; in cross-section they are relatively broad, without marked angles. The cones are erect on the branches and shatter at maturity (Fig. 28.12).

Picea (*Spruces*). These trees closely resemble the firs, from which they can be distinguished by the position of the leaves on the branchlets (Fig. 28.13) and by the angular appearance of the leaves in cross-section. The cones are pendent and do not shatter at maturity (Figs. 28.9*B,* 28.14).

Tsuga (*Hemlocks*). The trees of this genus are pyramidal with slender horizontal branches. The leaves are usually two-ranked, linear, flat, and with a short petiole. They resemble the leaves of firs but are much shorter. The cones are small (Fig. 28.15).

Pseudotsuga (*Douglas Fir*). Only two species occur in the United States. One of them, the Douglas fir (*Pseudotsuga menziesii*), is the most heavily cut timber tree of the United States. It may grow to a height of 200 ft, with a diameter of 12 ft. Its leaves are flat, like those of the true firs, but have white lines on either margin and a groove along the upper surface. The cones are 2–4½ inches long, pendulous, and easily recognized by bracts extending outward below each scale (Fig. 28.16).

Juniperus (*Junipers*). This genus is composed of both trees and shrubs. The trees are inclined to be somewhat irregular in shape and frequently have more than one trunk. The leaves are of two sorts,

Figure 28.10 *Pinus jeffreyi*, showing the trunk of one tree and the crown of a second tree. Ponderosa pine (*P. ponderosa*) has a similar appearance.

Figure 28.11 Fascicle of Jeffrey pine needles, ×1.

(*a*) in some species, spreading and needle-like and (*b*) in other species, scale-like and closely pressed to the twigs. Junipers are dioecious; ovulate and staminate cones being borne on different trees. The cone of the junipers is usually called a "berry;" it is a fleshy structure enclosing several seeds. Morphologically, it is a modified cone, composed of fleshy adhering scales that completely enclose the seeds (Fig. 28.17).

Larix (*Larches*). Trees belonging to this genus grow in the cooler portions of the Northern Hemisphere. They differ from most other members of the Pina-

ceae in that they are deciduous. The American larch, or tamarack, is a tall tree frequently found in bogs. The needles are short, linear, and grouped in crowded clusters on short spurs. On leading shoots, however, the needles are arranged spirally. The cones are small and persistent (Fig. 28.18).

Family Taxaceae (Yews)

With two exceptions, ovulate cones are not borne by members of the yew family. The seed of this family so strongly resembles a drupe or a nut that

Figure 28.12 *Abies. A*, a mature tree. *B*, top of a young tree showing ovulate cones. *C*, ovulate cone and branch. Note arrangement of needles. *D*, staminate cones.

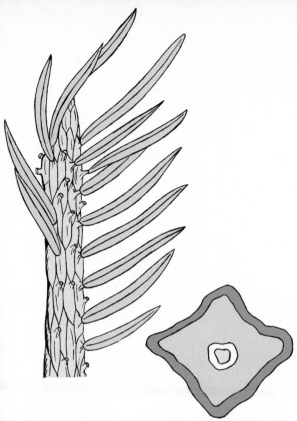

Figure 28.13 Leaves of *Picea*. (Redrawn from L. H. Bailey, *The Cultivated Conifers in North America*, Macmillan, New York.)

Figure 28.14 Branch of *Picea alba* with ovulate cones, $\times\frac{1}{2}$. (Courtesy of American Museum of Natural History.)

Figure 28.15 *Tsuga mertensiana. A,* mature tree, $\times\frac{1}{2}$; *B,* ovulate cones. Note arrangement of needles on branch.

Figure 28.16 Branch of *Pseudotsuga menziesii* (Douglas fir) with ovulate cone, $\times\frac{1}{2}$.

Figure 28.17 *A, Juniperus occidentalis* growing at 8000 ft in the Sierra Nevada; *B*, branch with staminate cones; *C*, branch with ovulate cones. *B* and *C*, $\times 1$.

Figure 28.18 *Larix lyalli* (larch). *A*, mature trees; *B*, branch with needles; *C*, mature ovulate cones. *B* and *C*, ×1.

Figure 28.19 A branch of *Torreya californica* with seed, ×1.

it is usually referred to as a fruit. Fig. 28.19 shows a seed of *Torreya californica,* or the California nutmeg, a rare forest tree. The embryo is protected by an outer flesh called an **aril,** and an inner hard pit. Both these tissues may be derived from the integuments of the ovule; hence, the structure is morphologically a naked seed. The seed of ground pine (*Taxus canadensis*), which forms a shrubby ground cover in forests of North Atlantic states, is similar (Fig. 28.20). The outer red aril almost encloses the seed and drops away when the seed is mature.

Yews are not common in the United States. Two examples have been mentioned; two other species occur—western yew and the stinking cedar of southern states. Possibly the best-known yew is the English yew, which, because of the excellent bows that were made from its wood, is closely linked with English history and folklore.

The leaves of yews, as illustrated in Fig. 28.20, are flat and linear, not unlike those of firs.

Life History of a Conifer

The following outline of a gymnosperm life cycle is based largely on the genus *Pinus.* It differs from life cycles of other genera, mainly in requiring three seasons for completion. Significant stages of pollination, fertilization, and maturation of the seed occur within a few inches of each other on a given branch, thus making it an ideal subject for class study.

All conifers produce two kinds of spores—**microspores** and **megaspores**—borne in cones that are morphologically distinct. The two types of cones are known, respectively, as **staminate** (Figs. 28.21, 28.23D) and **ovulate cones** (Figs. 28.1B, C, see Color Plate 17; Fig. 28.9).

Staminate Cone

Staminate cones average $\frac{1}{2}$ inch, or less, in length by $\frac{1}{4}$ inch in diameter. They are borne in groups, usually on lower branches of trees (Fig. 28.21). Each cone is composed of a large number of small scales (microsporophylls) arranged spirally on the axis of the cone. Two microsporangia develop on the under surface of each scale (Figs. 28.23E, F).

Stages of microspore development are quite similar to those of *Selaginella* and to spores of mosses and ferns. Sporogenous cells give rise to spore mother cells, which are surrounded by a nutritive cell layer, the **tapetum.** Each spore mother cell undergoes meiosis, and four microspores result. As usual, microspore contains the haploid number of chromosomes. The nucleus within the newly formed microspore divides several times by mitosis, forming a **pollen grain** that contains two viable haploid nu-

Figure 28.20 Branches of *Taxus* with seeds; note berry-like aril surrounding seed in *A*, and seed without aril in *B*, ×1.

Figure 28.21 End of a branch of *Pinus* with a cluster of staminate cones holding ripe pollen, ×1.

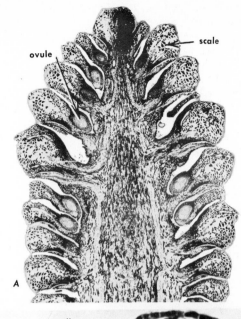

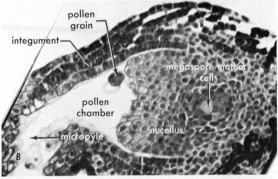

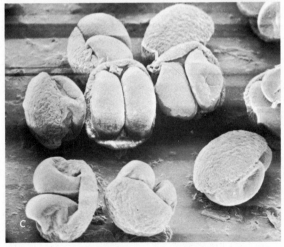

Figure 28.22 Ovulate cone of *Pinus* at time of pollination. *A*, median longitudinal section of cone, ×6; *B*, median section of an ovule; *C*, stereoscan micrograph of pollen grains (probably shrunken from treatment in SEM).

clei and vestiges of several vegetative cells (Figs. 28.23*G*, *H*). The pollen grain is finally shed from the microsporangium. Enormous numbers of pollen grains are formed. They are light in weight and bear two **wings** which facilitate dispersal by wind.

Ovulate Cone

The ovulate cone, when mature, is the well-known cone of pines, firs and other genera of conifers (Fig. 28.9). Each cone is composed of an axis, upon which are borne, in a spiral fashion (Figs. 28.9*I*, 28.23*A*), a number of woody scales. Young ovulate cones appear at tips of young branches in early spring (Fig. 28.1*B*; see Color Plate 17; Fig. 28.23*A*). Two ovules, each enclosing a single megasporangium, develop on the upper surface of **ovuliferous scales** (Figs. 28.23*B*, *C*). A longitudinal section through a young cone is shown in Fig. 28.22.

The ovules first appear as small protuberances on the upper surface of this scale, close to the axis of the cone. A protective layer of cells, the **integument,** develops early on the outer surface of the ovule. In

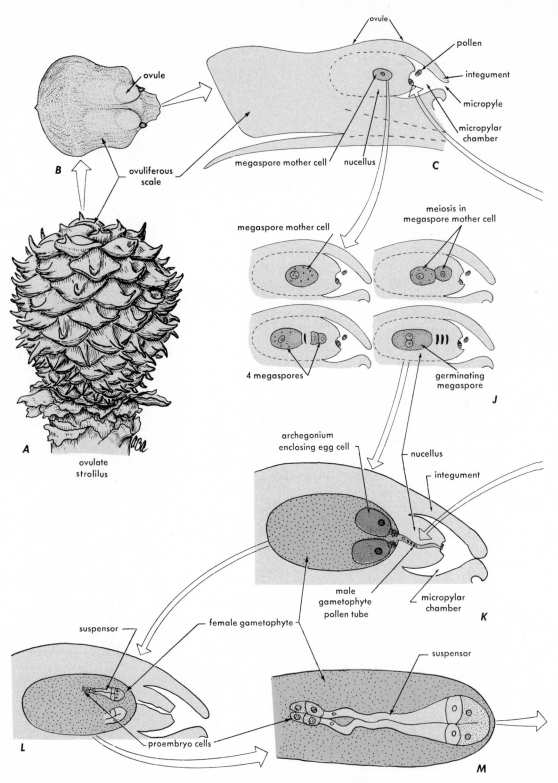

Figure 28.23 Stages in the life cycle of a pine. *A*, young ovulate cone, ×2; *B*, scale from ovulate cone showing two ovules, ×2; *C*, section of ovulate cone showing megaspore mother cell and pollen in pollen chamber, ×50; *D*, staminate cone, ×2; *E*, scale from staminate cone, ×3; *F*, cross-section of staminate scale, ×50; *G*, winged pollen grain, ×75; *H*, development of male gametophyte in pollen grain; *I*, pollen

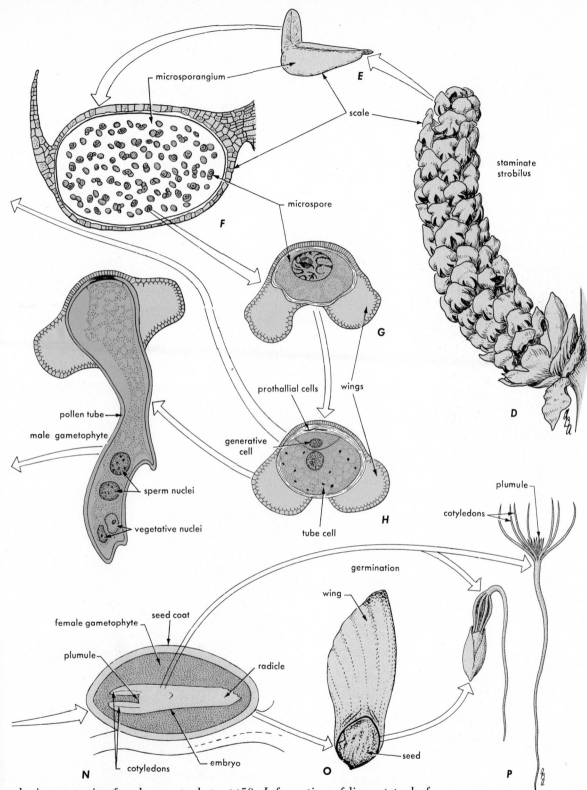

tube is penetrating female gametophyte, ×50; *J*, formation of linear tetrad of megaspores, ×75; *K*, female gametophyte with egg cell and pollen tube, ×75; *L*, suspensor and proembryo sporophyte within female gametophyte, ×75; *M*, proembryo, ×100; *N*, section of seed showing seed coats, female gametophyte, and embryo sporophyte, ×50; *O*, winged seed, ×2; *P*, seedlings, ×1. (*G, H, I*, after Coulter and Chamberlain, *Morphology of the Gymnosperms*, University of Chicago Press.)

the end of the ovule, nearest the axis of the cone, there is a small opening, the **micropyle,** through which pollen grains may enter.

In pine, one megaspore mother cell lies in the center of each ovule (Fig. 28.23*J*); several are contained in ovules of redwoods and cypresses. Like microspore mother cells, the megaspore mother cell is surrounded by a nutritive tissue, here called the **nucellus.** (Fig. 28.23*J*). It is in reality a megasporangium because it is within this tissue that megaspores arise.

Female Gametophyte. The megaspore mother cell soon divides by meiosis. Four megaspores, usually arranged in a single row of four cells, or a linear tetrad, (Fig. 28.23*J*), result from the meiotic division of each spore mother cell. The nucleus within each megaspore has the haploid number of chromosomes. Generally, only one of the four megaspores develops into a female gametophyte; the other three degenerate. Germination of megaspores and growth of female gametophytes progress very slowly. Several months are required in most conifers, and 13 months are required in pine.

The development of the female gametophyte takes place entirely within the ovule. There are approximately eleven mitotic divisions of the megaspore before cell walls begin to appear between the newly formed nuclei. Walls gradually form, however, resulting in a small mass of gametophytic tissue, completely enclosed by the diploid cells of the ovule. The sporophytic tissue of the ovule adjacent to the female gametophyte is the nucellus. While cell walls are being laid down in the developing female gametophyte, two or more archegonia differentiate at its micropylar end. Reference to Figs. 28.23*K*, 28.24*B* shows that the ovule at this stage consists of integuments, nucellus, and female gametophyte, which contains several archegonia, each with its enclosed egg. Directly beneath the micropyle is a space, the **micropylar chamber.** Note the nucellar tissue lying between the micropylar chamber and the archegonia.

Pollination

It will be recalled that pollination, in typical flowers, is the transfer of pollen from anther to stigma; in conifers, it is the transfer of pollen from staminate cone to ovulate cone. Conifer pollen is wind-blown. Since, in many species, ovulate cones are borne on higher branches of the tree and staminate cones are concentrated on lower branches, and since pollen does not blow vertically upward, cross-pollination is usual. Pollination occurs in most conifers in early spring soon after the ovulate cone emerges from the

dormant bud (Fig. 28.1*B*; see Color Plate 17)—at about the time of meiosis. At this age, scales of young cones turn slightly away from the axis so that pollen grains sift down to the axis of the cone. Here, they come in contact with a sticky substance secreted by the ovule. As this material dries it draws some pollen grains through the micropyle into the micropylar chamber where they lodge close to the developing female gametophyte (Fig. 28.23*C*). The pollen grain develops slowly into a male gametophyte (pollen tube, Figs. 28.23*I, K*). Thus, male and female gametophytes of conifers develop to maturity in close proximity within the ovule. They are both dependent upon nucellar tissue for nourishment and protection.

Male Gametophyte. A layer of nucellar tissue separates the female gametophyte from the micropylar chamber. The pollen grain lodged in the micropylar chamber sends out a tube, the **pollen tube,** which grows into nucellar tissue. The tube may branch slightly, and it apparently secretes enzymes that digest the nucellar tissue (Figs. 28.23*K*, 28.24*B*). Several

A

nuclear divisions occur in the tube, but no cell walls are formed. Two of the last-formed nuclei are **sperm nuclei.** This branched pollen tube, containing two sperm nuclei and several vegetative nuclei, is the **male gametophyte** (Fig. 28.23*I*).

Development of male and female gametophytes is so coordinated that the egg is formed and ready for fertilization when the pollen tube, containing two sperm nuclei, has reached the archegonium. In pine, this development requires about 1 year. At the time of fertilization, pine cones are generally green and the scales are tightly closed, showing the spiral pattern of arrangement (Fig. 28.24). The sperm nuclei, together with other protoplasmic contents of the pollen tube, are discharged directly into the egg cell. Sperm nuclei do not possess cilia and hence are not actively motile. One sperm nucleus comes in contact with the egg nucleus and unites with it. The nonfunctioning sperm nucleus and the other protoplasmic material discharged into the egg cells soon undergo disorganization.

The formation of the embryo is preceded by the development of a relatively elaborate **proembryo.**

This structure, in pines consists of four tires of four cells each (Figs. 28.23*L, M*). The four cells farthest from the micropylar end of the proembryo may each develop into an embryo. The intermediate cells are the suspensor cells; they elongate greatly and push the embryo cells deep into the female gametophyte (Fig. 28.23*M*). While this development is taking place, the female gametophyte continues to grow. It enlarges and becomes packed with food to be used not only for the growth of the embryo but also as a reserve in the seed.

It will be recalled that the female gametophytes usually contain two or more archegonia apiece. Since the egg in each archegonium may be fertilized and since each of the four embryo-forming cells may give rise to an embryo, a number of embryos may develop in every seed. Normally, however, only one embryo survives, but seeds with two well-formed embryos are not rare.

The mature embryo consists of several **cotyledons** or seed leaves, **epicotyl, hypocotyl,** and **radicle** or rudimentary root (Fig. 28.23*N*). The embryo is embedded, as previously mentioned, in the enlarged

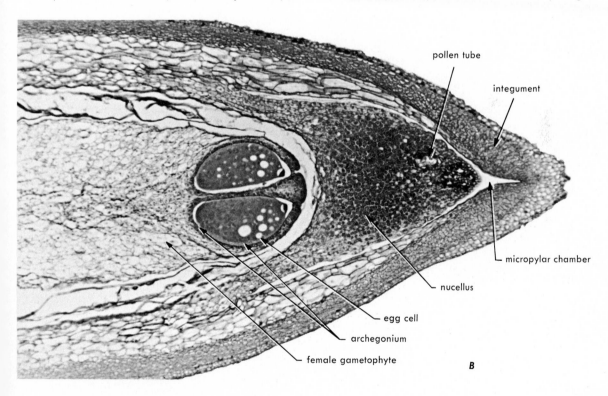

Figure 28.24 Ovulate cones of *Pinus*. Left. *A*, about 13 months old, ready for fertilization. Above. *B*, section through tip of ovule after pollination. (*B*, slide courtesy of Triarch Products.)

female gametophyte. Both embryo and female gametophyte are embedded in a papery shell and a hard protective **seed coat,** both formed from the integument of the ovule. The whole structure is the **seed** (Fig. 28.23O). In many pines, the seeds are winged. The development of the seed is summarized in Table 28.1.

Normally, in pines, the seed matures some 12 months after fertilization, and since fertilization is effected about 13 months after pollination, 2 years

TABLE 28.1   Comparison of Ovule and Seed of Pine

Time of pollination	Time of fertilization	Mature seed
Ovule	Ovule	Seed
Megaspore mother cell	Female gametophyte	Female gametophyte (stored food)
Pollen grain	Egg cell → Zygote	Embryo
	Pollen tube sperm nucleus	
Megasporangium	Megasporangium or nucellus	Perisperm
Integument	Integument	Seed coat
Micropyle	Micropyle	Micropyle

TABLE 28.2   A Generalized Life History of a Pine

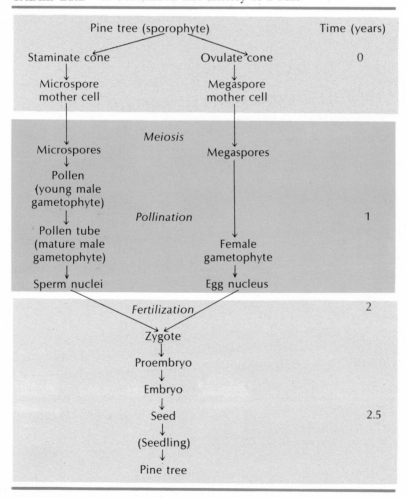

intervene between initiation of ovules and formation of seeds. Young pine seedlings (Fig. 28.23P) have several cotyledons. A generalized life history of a pine is shown in Table 28.2.

Gymnosperm seeds may remain dormant for many years, and some may remain embedded in the old mature cones for 6 years or more. Heat causes the cones of some species to open and release the seeds. This behavior has considerable survival value, since large numbers of seeds are released after fires, and injured trees are thus replaced by the young ones. In many species, however, seeds are shed soon after they are mature.

Significant Features of the Gymnosperms

1. All forms in this group bear seeds which are not surrounded, or protected, by an ovary wall; a true fruit is never formed.
2. All are trees or shrubs of varying form; many have needle or scale leaves, and most of these forms are evergreen. Some have broad leaves, and in others the leaves are palm-like.
3. With few exceptions, vessels are lacking in the xylem, and companion cells are absent from the phloem.
4. The best-known gymnosperms of the Northern Hemisphere bear cones of one sort or another. In yews, the seed is surrounded by a fleshy aril.
5. Sexual reproduction involves:
   (a) Production of microspores (through meiosis) in staminate cones and their transformation, by several nuclear divisions in the microspore, into pollen grains.
   (b) Pollination by wind.
   (c) Development of an ovule on the upper surface of scales of the ovulate cone.
   (d) Production of megaspores through meiosis within the ovule.
   (e) Development of a female gametophyte within the ovule from one megaspore. Several reduced archegonia form within the female gametophyte.
   (f) Penetration of a pollen tube or male gametophyte through the nucellus and discharge of sperm cells into the egg cells. A few gymnosperms have flagellated sperm cells.
   (g) Formation of a proembryo from the zygote, one end of which gives rise to the embryo.
   (h) Enlargement of the female gametophyte, which stores food.
   (i) The seed, which consists of a hard seed coat, the perisperm, the food-storing female gametophyte, and the embryo.

Classification

Division	Pteridospermophyta
Division	Cycadophyta
Genera	*Cycas*
	Zamia
	Dioön
Division	Ginkgophyta
Genus and species	*Ginkgo biloba*
Division	Coniferophyta
Family	Pinaceae
Genus and species	*Pinus monophylla*
Genus	*Abies*
Genus	*Picea*
Genus	*Tsuga*
Genus and species	*Pseudotsuga menziesii*
Genus	*Juniperus*
Genus	*Larix*
Family	Taxaceae
Genus and species	*Taxus brevifolia*
	Taxus canadensis
Genus and species	*Torreya claifornica*
	Toreya taxifolia
Division	Gnetophyta
Genus	*Ephedra*

29

Angiosperms

The angiosperms, or flowering plants, are the dominant plants of the world today. They include nearly all the crop plants of orchard, garden, and field. Hardwood forests, shrublands, grasslands, and deserts are composed chiefly of flowering plants. In fact, except for coniferous forests and waters of the world, the conspicuous and dominating vegetation wherever we go is mainly flowering plants. They show great variation in form, from simple stemless, free-floating duckweed (*Lemna*) through a whole series of herbaceous types to shrubs of varying sizes and finally to trees such as oaks and beeches. Structurally they are adapted to a land habitat, and certain forms, as exemplified by cacti (Fig. 18.38), are able to live and grow in very dry deserts. We have already studied the structure and physiology of the angiosperm plant body in considerable detail. Angiosperms constitute the Anthophyta which is divided into two classes, Monocotyledonae and Dicotyledonae.

Flower, Fruit, and Reproduction

The outstanding and unique structures of angiosperms are the *flower* (Color Plate 6; Fig. 29.1) and the *fruit* (Fig. 16.1), which develops from the ovary of the flower (Figs. 16.1, 16.2), with its enclosed seeds (Fig. 16.2). The flower is a shoot bearing floral leaves. In a complete flower (Figs. 15.1, 29.1), the floral leaves are sepals, petals, stamens, and carpels. Some flowers have only the essential reproductive structures (Fig. 15.12A), stamens and carpels. Other flowers are unisexual (Fig. 15.13), that is, have either stamens or carpels, not both. Other variations in flower structure are described on pp. 261–269.

Free water is not needed for fertilization in angiosperms; motile, ciliated sperms are not produced by any members of this division. Sexual reproduction is made possible by production of pollen (Fig. 29.1C), pollination, and growth of a pollen tube (Fig. 29.1D). This results in the union of the egg and sperm (Figs. 29.1A, B, C).

All flowering plants produce seeds (Fig. 2.14) except some that have been modified by man so that they are now seedless. In angiosperms, seeds are borne within a closed structure, they ovary, which eventually becomes the fruit (Fig. 16.1). In gymnosperms, the seeds are not surrounded completely, but are borne on the upper surface of a scale (Fig. 28.2).

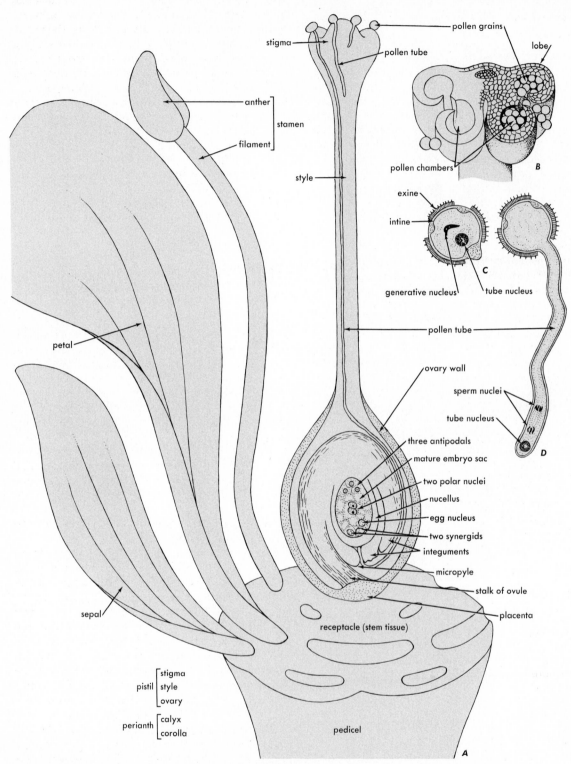

Figure 29.1 *A*, diagram of a flower; *B*, cross-section of an anther; *C*, mature pollen grain; *D*, germinating pollen grain. (*C* and *D*, redrawn from Bonnier and Sablon, *Cours de Botanique*, Librairie Generale de l'Enseignement.)

Flower Structure Compared with Reproductive Structures of More Primitive Forms

Since the flower is the angiosperm structure adapted for sexual reproduction, its parts should, to some degree, be comparable to the reproductive structures of more primitive forms. Let us examine the flower with this in mind. Petals and sepals, the showy and protective whorls of floral leaves, are not directly represented in any of the lower forms. Meiosis occurs in the anthers of stamens and microspores are produced in pollen sacs (Fig. 29.1*B*). The stamen, then, is a microsporophyll and a pollen sac is a microsporangium. The pollen grain itself, containing, as it does, two nuclei or cells (Fig. 29.1*C*), would be the first stage in development of a male gametophyte. The pollen tube would constitute a mature male gametophyte. Meiosis takes place within the ovule. The tissue immediately surrounding the developing megaspore is the nucellus, which is borne within the ovule (Fig. 29.1*A*). It will be recalled that ovules, in many instances, are attached to margins of carpels, which are floral leaves (Fig. 16.2). Thus, the carpel, since it bears megaspores, is a megasporophyll, and the nucellus is the megasporangium. Generally, one of the four megaspores gives rise to an embryo sac or female gametophyte (Fig. 29.1*A*). These two sets of terms are summarized in Table 29.1.

Life Cycle

With these comparisons in mind, let us proceed with an outline of the essential steps in the life cycle of an angiosperm. The dominant vegetative plant is the sporophyte and, as already emphasized, the flower is the organ concerned with sexual reproduction.

TABLE 29.1 Comparative Terminology of Reproductive Parts

Anthophyta	Other divisions
Stamen	Microsporophyll
Pollen sac	Microsporangium
Microspore	Microspore
Pollen grain	Germinated microspore or young male gametophyte
Pollen tube	Mature male gametophyte
Carpel	Megasporophyll
Nucellus	Megasporangium
Megaspore	Megaspore
Embryo sac	Female gametophyte
Ovule	Ovule

Male Gametophyte. The anther is the particular part of the stamen responsible for production of microspores. These are the products of meiosis. The microspore mother cell forms within the pollen sac (Fig. 15.4). Each microspore contains one haploid nucleus, which soon divides into generative and tube nuclei (Fig. 29.1*C*). While this mitosis is taking place, a heavy, sculptured wall forms about the microspore and a mature pollen grain results. The pollen is now shed and conveyed in one fashion or another (Chap. 15) to a stigma where it germinates. A pollen tube penetrates the stigma (Figs. 29.1*A, D*) and grows through the style, down to the ovule within the ovary. The tube nucleus, depending upon the species of plant, may degenerate early or may persist until the pollen tube is well-developed. In any event, the generative nucleus usually divides within the pollen tube into two sperm nuclei (Fig. 29.1*D*). Each sperm nucleus, with its associated cytoplasm, is a sperm cell. The pollen tube with sperm cells and tube nucleus, if present, constitute the mature male gametophyte.

Female Gametophyte. An enlarged cell within nucellar tissue of a young ovule undergoes meiosis and forms four megaspores arranged in a row. While this process is occurring, integuments form about nucellar tissue, resulting in formation of a typical ovule (Fig. 15.10). The three megaspores closest to the micropyle generally disintegrate. The remaining megaspore undergoes three successive mitotic divisions, resulting in eight nuclei of the embryo sac or female gametophyte: one egg cell, two synergid cells, one endosperm mother cell with two polar nuclei, and three antipodal cells (Fig. 29.1*A*).

Fertilization and Seed Development. With penetration of the pollen tube to the mature embryo sac, the stage is set for fertilization. In all previous forms we have studied, this involves only the union of egg cell and sperm. In angiosperms, fertilization involves not only the union of the egg cell with the sperm cell but, in addition, the union of a second sperm cell with the endosperm mother cell to form the primary endosperm cell (Fig. 29.2). Since there are two nuclei in the endosperm mother cell, the nucleus of the primary endosperm cell will have three sets of chromosomes. The fusions of two sperm cells, one with the egg, the second with the endosperm mother cell, is **double fertilization.**

The resulting zygote generally develops directly into a small proembryo, from one end of which a typical embryo (seed) develops. The fate of the primary endosperm cell depends upon the species. Probably in all forms it develops into an endosperm

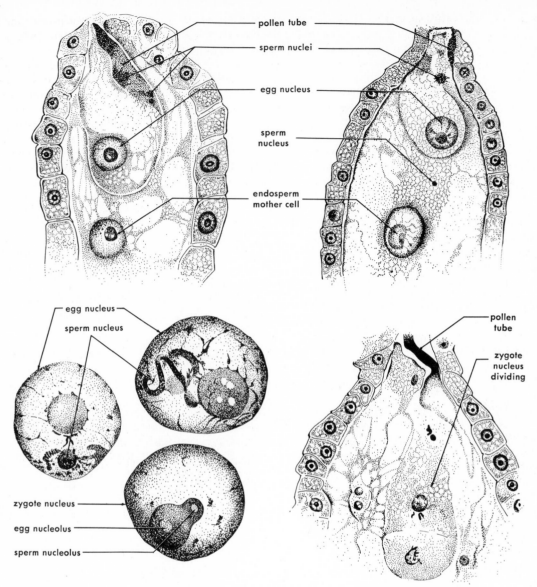

with labels: pollen tube, sperm nuclei, egg nucleus, sperm nucleus, endosperm mother cell, egg nucleus, sperm nucleus, zygote nucleus, egg nucleolus, sperm nucleolus, pollen tube, zygote nucleus dividing

Figure 29.2 Fertilization and embryo development in *Crepis*. (Redrawn from H. Gerassimova, *La Cellule* **42**, 103.)

which plays some role in nourishing the developing embryo (Fig. 16.21). In a few genera, notably the grains (Fig. 16.14), the endosperm enlarges and persists in the seed to form a main source of nourishment for the young seedling. There are four general types of seeds, depending upon the various combinations of cotyledon number and location of stored food (Figs. 16.22, 16.24, 16.26). A seed always includes a dormant embryo and a food supply to enable the young seedling to establish itself. Food is generally stored either in cotyledons of the embryo or in endosperm.

With the germination of a seed and the development of the seedling into a flowering plant, the life cycle of an angiosperm is completed. The essential steps are shown in Table 29.2.

Comparative Reproductive Structures

Certain comparative details between structures and life histories of members of the gymnosperms and angiosperms are given in Table 29.3. The pine seed and the corn grain are compared in Table 29.4. Details concerning the Psilophyta, Lycophyta,

Sphenophyta, and Pterophyta are given in Table 27.1.

In all Pterophyta the sporophyte is the dominant generation, and both gametophyte and sporophyte generations are independent. In gymnosperms and angiosperms, there occurs a further reduction in size and complexity of the gametophyte to a condition of complete dependency on the parent sporophyte. This is accompanied by an overall increase in the general complexity of the sporophyte. It can be stated that, as a guiding principle, the more advanced forms in these divisions are more highly adapted to grow and flourish in a dry land habitat than are the more primitive forms. The most obvious developments designed to accomplish this adaptation are (a) those which remove any dependence of the plants on free moisture for fertilization, and (b) adaptations, both vegetative and reproductive, which enable the plant to grow and reproduce on dry land.

Protection of the Female Gametophyte. The gynoecium of the angiosperm flower is essentially a modified leaf, or leaves, enclosing an ovule, or ovules, in which the female gametophyte is to be found. The ovary, as we know, eventually develops into a fruit. This envelopment of ovules, and subsequently seeds, by an ovary wall is a new development which sets angiosperms apart from all other groups of plants. Recall that in ferns the egg cells have little protection from their immediate environment. The gametophyte thallus in ferns is a delicate structure, and the venter of the archegonium opens directly, by way of the ventral canal, to external moisture (Figs. 27.29I, J). The female gametophyte of *Selaginella* (Figs. 27.14M, N) may be afforded some protection by the parent sporophyte, but motile sperms must reach eggs under their own power, and the archegonium opens to the external environment by a very short neck canal. In pine, the female gametophyte has gained the protection of integuments in the ovule and the scales of the ovulate cone, as well as being retained within the megasporangium. Pollen must still, however, be deposited very close to the female gametophyte in order to effect fertilization. Pine no longer requires free water for movement of sperms and, except for a few primitive gymnosperms, flagellated sperms are absent from these plants.

The female gametophyte (embryo sac) of angiosperms has gained, in the ovary wall, an added protective barrier. However, this has necessitated further adaptations to enable sperms to pass this added protection: (a) development of a receptive stigma and (b) growth of a pollen tube through the style.

Size of Gametophytes. Reduction in size of gametophytes has thus proceeded to a point in which the male gametophyte, the pollen tube, consists of one vegetative nucleus and two sperm cells (Fig. 29.1D). The female gametophyte is a seven-celled structure, one of whose cells is an egg cell. A second cell, the endosperm mother cell, is an innovation in that it, too, is receptive to a sperm cell. The remaining five cells are vegetative cells (Fig. 29.1A). It should be pointed out that embryo sacs of some angiosperms have more cells, while a few have less (Fig. 19.10). Should the gametophyte stages be completely eliminated, gametes, instead of meiospores, would result from meiosis. This actually occurs in animals and in

TABLE 29.2 Generalized Life History of an Angiosperm

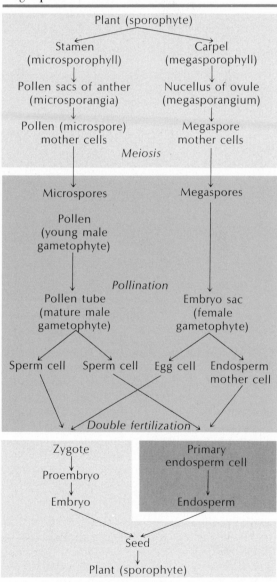

TABLE 29.3 Comparison of the Conifers and the Seed Plants

Generation	Gymnosperms	Angiosperms
Sporophyte	Roots, stems, and leaves present	Roots, stems, and leaves present
	Trees and shrubs, no herbs	Trees, shrubs, and herbs
	Stem: one core of xylem surrounded by phloem, vessels in one shrubby genus only	Stem: one core of xylem surrounded by phloem, or several vascular strands with phloem exterior
	Sieve cells without companion cells	Companion cells and sieve-tube members
	Pith present in stems, not in roots	Pith in stems, not in roots, of dicots; in some monocot roots
	Cambium develops in all species	Cambium in perennial forms; absent generally in annuals
	Staminate and ovulate cones, generally, with staminate and ovulate scales	Flowers, stamens, and carpels
	Heterospory	Heterospory
	Microspores in microsporangia	Microspores in anther sacs
	Megaspores in megasporangia or nucellus	Megaspores in nucellus
	Ovules present, exposed on scales	Ovules present, enclosed within carpel
	Embryo develops within a seed	Embryo develops within a seed
Fertilization	Double fertilization absent; no endosperm	Double fertilization, resulting in zygote and primary endosperm cell
Gametophyte	Pollen tube: male gametophyte	Pollen tube: male gametophyte
	Female gametophyte within ovule	Female gametophyte (embryo sac) within ovule
	Antheridium absent	Antheridium absent
	Sperm cells formed in pollen tube	Sperm cells formed in pollen tube
	Motile sperms in several primitive genera, but free water not needed for fertilization	
	Nonmotile sperms in all other forms	Nonmotile sperms
	A very greatly reduced archegonium completely embedded in female gametophyte	Archegonium only suggested by synergid cells of the embryo sac
	Embryo within female gametophyte	Embryo within seed coats, remains of nucellus, and endosperm; female gametophyte absent

some diatoms (p. 437). The haploid phase in these forms is reduced to the sperms and egg cells.

Nourishment for the Developing Embryo. Double fertilization is still another difference between the life cycle of the angiosperms and more primitive forms. In the latter forms, nourishment for the young developing sporophyte is generally provided by the female gametophyte (Figs. 27.14*M, N,* 28.23*L*) or, in the case of many ferns, by the simple monoecious (homothallic) gametophyte (Fig. 27.29*J*). In angiosperms, food for the developing seedling is stored either in endosperm (Fig. 16.14) or in the cotyledons of the embryo itself (Fig. 16.24).

Fruit. Fertilization in angiosperms, as in all other forms, stimulates certain tissues surrounding the zygote to begin further development. In angiosperms, the ovary wall is stimulated to produce a fruit, and the fruit is a development originating with,

and characteristic of, angiosperms. The protection of the seed from its surroundings by the ovary wall is accompanied by various devices for releasing the seed from its enclosure or otherwise bringing it under conditions favorable for germination. Among these adaptations may be mentioned the different sorts of dehiscence (Figs. 16.10, 16.11), various kinds of fleshy fruits (Figs. 16.7, 16.20), and the reduction of the ovary wall so that it becomes indistinguishable from the seed coats (Fig. 16.14).

Evolution in the Angiosperms

If, as assumed, angiosperms represent the culmination of an extensive evolutionary development, it would appear likely that, within the division itself, evolution has been, and probably still is, active. In other words, families of angiosperms must be related and it should be possible to arrange the families in

TABLE 29.4 Comparison of Pine Seed and Corn Grain

Pine seed	Corn grain
Seed lacks pericarp	Seed coats fused with pericarp
Numerous cotyledons	One cotyledon
Food stored in female gametophyte	Food stored in endosperm
Radicle unprotected	Radicle protected by coleorhiza
Plumule protected by cotyledons	Plumule protected by coleoptile

an order that would give some indication of their relationship. Some families will be found to be more primitive than others; and from the more primitive types the more advanced and specialized forms have presumably arisen (Chap. 19). What is the primitive floral type? What have been the trends in the evolution of the flower? What system of classifying flowering plants will be most truly natural?

The Besseyan System. Many systems of classifying plants have been proposed. Two are in use in the United States. One was proposed by Engler and is the basis for the arrangement of plants in many books. A second system that has found much favor with American botanists is described here. This system (Besseyan) regards the order Ranales as being the most primitive. In this order, the Magnoliaceae is one of the more primitive families, with the Ranunculaceae somewhat more advanced. The Christmas rose (Fig. 15.2), magnolias (Figs. 15.3A, 16.10A), and buttercups (Fig. 29.3) are representatives of these two families. While these all have flowers with many characteristics in common, the fact that magnolias are trees or shrubs and most Ranunculaceae are herbaceous forms, and evidence from fossil remains suggest that trees appeared before herbs (Table 30.1), indicates that the Magnoliaceae is the more primitive family. In both magnolias and buttercups, the flower parts are arranged as follows:

1. The floral axis is elongated.
2. All, or at least some, of the flower parts are spirally arranged.
3. Stamens and carpels are numerous and separate.
4. Sepals and petals are numerous and separate.
5. Flowers are perfect and complete.
6. Floral symmetry is regular.
7. Perianth segments and stamens are attached to the receptacle *below* the points of attachment of the carpels (i.e., the flowers are hypogynous).

These, then, are a group of characteristics indicative of primitive flowers; changes in any of them indicate more advanced forms. The principal tendencies in evolution of the flower, according to the Besseyan system, are changes in these characteristics and they may be enumerated as follows:

1. From an elongated to a shortened floral axis.
2. From a spiral to a whorled condition of floral parts.
3. From numerous and separate stamens and carpels to few and coalesced stamens and carpels.
4. From numerous and separate sepals and petals to few and coalesced sepals and petals.
5. From complete and perfect flowers to incomplete and imperfect flowers.
6. From regular to irregular flowers.
7. From hypogyny to epigyny.

According to the Besseyan view, there were at least

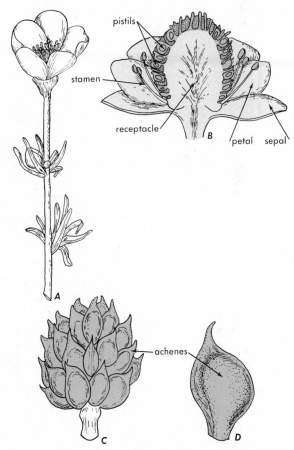

Figure 29.3 Ranunculaceae. Flower and fruit of buttercup (*Ranunculus*). *A*, flower stalk; *B*, lengthwise section of flower; *C*, floral diagram; *D*, cluster of fruits; *E*, a single fruit. (*A*, *B*, and *D*, redrawn from Korsmo.)

Flower, Fruit, and Reproduction 599

three main lines of advance from the primitive Ranalian type of flower as exemplified by magnolias and buttercups. These lines culminated in (a) mints (Labiatae), (b) asters (Compositae), and (c) orchids (Orchidaceae). In comparing plants with evolution in mind, it must be emphasized that not only do certain characteristics have more weight than others, but groups of characteristics must be given thoughtful consideration. For example, epigyny is thought to be an advanced condition. However, it should be noted that all flowers in one of the three main lines of ascent are hypogynous. This line culminates in mints and places this family at a level of evolutionary advance comparable to that held by the Compositae and Orchidaceae.

Selected Families of Angiosperms

Let us now proceed to examine selected families of angiosperms with two objectives in mind: (a) to learn the characteristics of some important angiosperm families and (b) to discover how these families fit into the Besseyan system of angiosperm classification. In our discussion of these matters so far, floral characteristics have been emphasized. The single exception is that of the tree form of the magnolias versus the herbaceous habit of the buttercups. The emphasis is placed on floral parts because their form and structure are little influenced by the external environment. They also are structurally more constant from generation to generation than are vegetative parts. Because of this constancy in structure, floral parts have a much greater value in both plant identification and in tracing evolutionary relationships than do vegetative characters. However, the latter cannot be neglected altogether, and many of the characteristics discussed in detail in Chapters 7, 8, 10, and 16 are used in identification and in evolutionary studies. It might be well to review briefly and independently the morphological details presented in these chapters.

Since the magnolia, tulip tree, and buttercups are thought to be primitive forms, we shall begin our discussion with these families, then proceed to follow through a line of ascent ending in the mint family. We shall consider the line culminating in the composites next and complete our discussion with the line of monocotyledonous families. It should be further noted that lines of development are not straight but branched and tree-like. The arrangement of the families discussed here as they occur in the Besseyan system is shown in Fig. 29.4.

From Ranales to Labiatae

In this line of ascent from Ranales, no change has occurred from the condition of hypogyny; all flowers, in all families in the line are hypogynous. Syncarpy occurred very early, as did a reduction in number of floral parts. Other types of coalescence and the change to irregular flowers appeared somewhat later but still early in the line of ascent. The families involved here are: Magnoliaceae, Ranunculaceae, Malvaceae, Cruciferae, Solanaceae, Salicaceae, and Labiatae.

The Malvaceae is a clearly marked family with relatively primitive characters as shown by its regular, perfect, and hypogynous flowers. There are numerous stamens with coalesced anthers, and the five or more carpels are more or less firmly united. In Cruciferae, flower parts are reduced in numbers and only carpels are coalesced. In Solanaceae, the flowers, while perfect and regular, show coalescence of sepals, petals, and carpels, with stamens adnate to the corolla tube. The willows (Salicaceae) are an example of a family in which reduction in both number and size of floral parts has resulted in an advanced condition. The flowers are imperfect and grouped in catkins, and both calyx and corolla are lacking. The pistillate flowers are hypogynous and, in staminate flowers, the number of stamens has been reduced to one or two. The willows are a part of a branch from the main line in which wind, rather than insect, pollination is the rule. The Labiatae, or mints, represent the most advanced family in this line of ascent. The flowers, even in this advanced family, are hypogynous; they are, however, irregular; there is a reduction in number of parts to four or two; and sepals, petals, and carpels are coalescent. The stamens are not only adnate to the corolla tube but are also highly specialized for insect pollination. Let us consider each family in more detail.

Magnoliaceae. This group shares with the Ranunculaceae the distinction of being very ancient. The tulip tree, *Liriodendron*, as shown by fossil remains, once had a very wide distribution. Most species of the Magnoliaceae are trees or fairly large shrubs. Magnolias themselves are magnificent trees, *Magnolia grandiflora* growing to 100 ft or more with its stiff, large, evergreen leaves and large white blossoms 7–8 inches across. The magnolias are largely warm-climate species, but tulip trees will stand northern winters. There are some 10 genera and 100 species in the family. Leaves are alternate, mostly entire, and pinnately veined, with large stipules generally enclosing buds.

Flowers are perfect, regular, and with distinct,

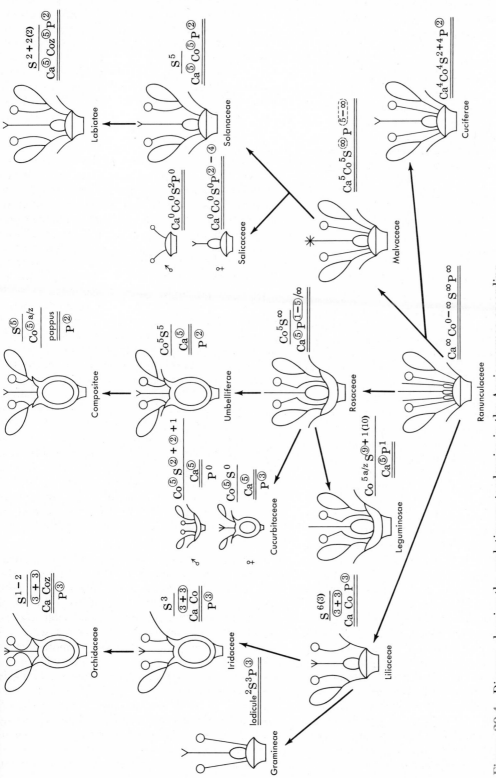

Figure 29.4 Diagram showing the evolutionary tendencies in the Angiospermae according to the Besseyan system. Key to symbols is in Laboratory Manual, exercise on variation in floral structure. (Courtesy of McMinn.)

separate parts. There are six or more petals. There are generally numerous stamens and carpels all arranged spirally on the receptacle (Fig. 15.3A). Fruits (Fig. 16.10A) are follicles or samaras and are frequently grouped to form a cone-like body.

Ranunculaceae. Most of these are herbs, but there are a few small shrubs and woody climbers. Leaves are characteristically very much divided, a feature which has led to the common name of Crowfoot Family. There are about 35 genera and approximately 2000 species growing largely in temperate and arctic regions. Many of our common garden flowers belong to this family. Among these are *Delphinium, Aquilegia, Paeonia, Ranunculus, Anemone,* and *Clematis.* The marsh marigold and hepatica, common spring flowers, are members of this family. One species, *Aconitum napellus,* yields the drug aconite, a cardiac and respiratory sedative, and some of the wild *Delphinium* are poisonous to livestock.

Floral parts are quite similar in structure and arrangement to those of Magnoliaceae. They indicate a primitive angiosperm family (Fig. 29.3). There are numerous separate parts to each whorl, and they are arranged in a spiral order on the receptacle. Flowers are generally regular and hypogynous. The fruit is a follicle, an achene, or sometimes a berry. There are exceptions to this arrangement of parts; for instance, flowers of *Delphinium* are very irregular.

Cruciferae. This is a large, distinct family of many cultivated forms, as well as some that are noxious weeds. Cabbage, cauliflower, broccoli, Brussel sprouts, kohlrabi, and kale are all varieties of a single species, *Brassica oleracea.* There is reason to think that kale was the original form. Wild mustard, another *Brassica* species, is a noxious weed in grain fields, though sometimes used as cover crop in orchards. Radishes and stocks are other members of this family, as are many garden herbs. The family has a worldwide distribution in temperate and subarctic zones; all are herbs. There are about 350 genera and 3000 species.

The flowers and fruits are very characteristic of the family. Flowers (Fig. 29.5A) are small but conspicuous because of their large numbers. They are complete, regular, and hypogynous. There are four sepals and four petals, usually somewhat sharply angled close to their midline and bending backward. There are six stamens, four longer than the other two. The single pistil is composed of two carpels and has two locules. The arrangement of petals generally suggests a cross, hence the family name, Cruciferae or "cross-bearing." The fruit is a silique (Fig. 16.12),

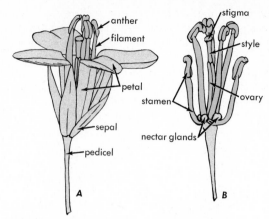

Figure 29.5 Cruciferae. Mustard flower (*Brassica*). *A,* all flower parts intact; *B,* sepals and petals removed. Note four long and two short stamens and the single pistil.

generally dehiscing in the typical fashion. The silique of radish (*Raphanus*) does not dehisce.

Malvaceae. This is a large family of 80 genera and 1500 species. There are no major food-producing species (although okra is in this family), and ornamentals in the family, such as *Hibiscus* and *Abutilon,* are not as conspicuous as roses, irises, and orchids. Cotton, however, taken from seed coats of various members of the genus *Gossypium,* make this family of plants as important from the viewpoint of politics, agriculture, and industry as grasses, legumes, and roses. Herbs, shrubs, and trees are in the family; their leaves are mostly palmately veined and lobed, with small deciduous stipules.

Flowers (Figs. 15.16B, 29.6) are regular, generally perfect, with five sepals usually united, and a tubular, five-lobed corolla. The many stamens are united by their filaments to form a tube surrounding the pistil. This tube is frequently adnate at its base to the five petals. The ovary is superior, generally with two to many locules, each containing one to several seeds. In most species, the number of carpels can be told from the number of styles and stigmas. The fruit of most species is a capsule, although in a few it is a berry. The cotton of commerce occurs as long hair or fuzz on seeds, which are borne in large capsules.

Solanaceae. A large family with many tropical forms, it is also well-represented in temperate regions. There are about 85 genera and over 2300 species, some 1200 of the latter being in a single genus, *Solanum.* To this family belong tomatoes, potatoes, tobacco, eggplant, peppers, *Petunias,* and *Salpiglossis.* While the family is of world-wide distribution, most of the cultivated forms were brought under domestication in the Western Hemisphere. There are many poisonous and drug plants, such as belladonna and atropine, in the family. Even such a common plant as the tomato was long supposed to be poisonous, as appears to be indicated by its scientific name *Lycopersicon,* which means "wolf peach." There are many erect and climbing herbs in the family, also some shrubs and small trees. The alternate leaves are mostly entire, although in a few species they may be variously dissected.

The perfect flowers (Fig. 29.7) are mostly regular; calyx and corolla are coalesced and generally five-lobed. The number of adnate, hypogynous stamens (five) corresponds to the number of corolla lobes with which they alternate. Sometimes one or more

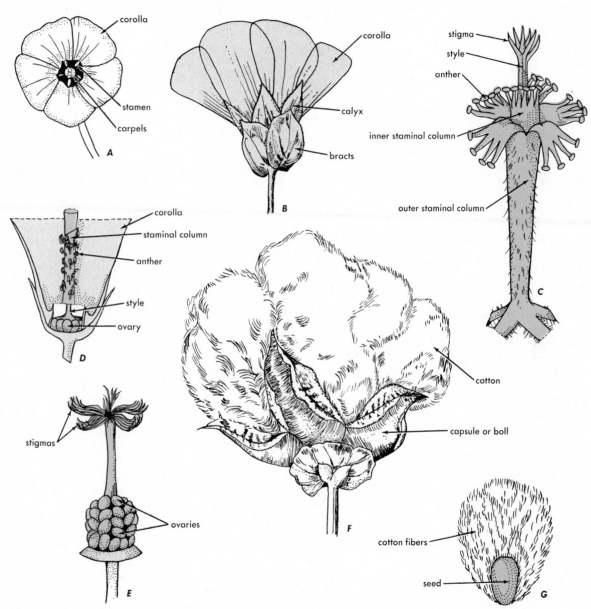

Figure 29.6   Malvaceae. *A* and *B,* regular flowers of *Anoda* and *Malope; C,* double staminal column of *Sidalcia; D,* ovary and staminal column of *Anoda; E,* gynoecium of *Malope; F,* dehisced capsule or boll of *Gossypium* with cotton fibers; *G,* seed of *Gossypium* with attached cotton fibers. (Redrawn from L. H. Bailey, *Manual of Cultivated Plants,* Macmillan, New York.)

Selected Families of Angiosperms   603

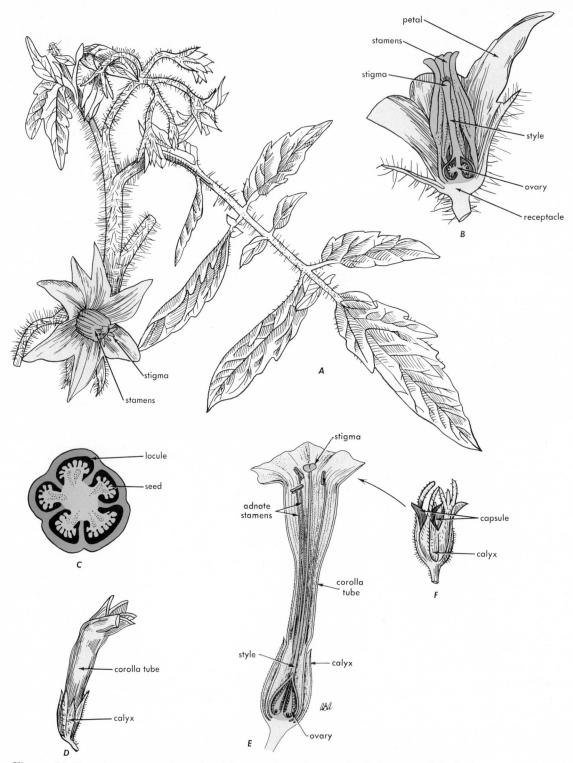

Figure 29.7 Solanaceae. *A*, branch of *Lycopersicon* showing leaf, flower, and buds; *B*, longitudinal section of flower of *Lycopersicon*; *C*, cross-section of tomato fruit; *D*, unopened flower of *Nicotiana*; *E*, longitudinal section of *Nicotiana* flower; *F*, capsule of *Nicotiana*, calyx is persistent. (*A, B, C, D, F,* redrawn from L. H. Bailey, *Manual of Cultivated Plants*, Macmillan, New York.)

of the stamens may be sterile. The superior ovary normally has two locules (five in tomato), with a single style and a two-lobed stigma. The fruit is a berry or, less frequently, a capsule with many seeds.

Salicaceae. The willow family is comprised of two genera (*Salix*, willows; *Populus*, poplars), with about 350 species. They are mostly trees, but some shrubby forms are known. They are very abundant in the Northern Hemisphere, mostly in temperate zones. Baskets are woven from branches of the basket willow, and paper pulp is obtained from trunks of one species. Willows are common along water courses, frequently overhanging or even choking mountain streams, to the ill comfort of fishermen. All species have alternate simple stipulate leaves.

The trees are generally dioecious, and flowers are borne in catkins in spring in advance of leaves. The perianth is lacking; the one or more stamens and a single pistil are found in axils of bracts. Most often, there are two stigmas (Fig. 29.8). Fruits are small capsules having many seeds with long down or hairs in a single locule. Willows have one or two stamens per flower, while poplar flowers have between eight and ten. Pussy willow is a familiar example of this family.

Labiatae. Mints, like orchids and composites, represent the supposed highest advance of one of the three lines of angiosperm evolution. It is a large family of considerable economic importance, largely because of volatile oils produced by certain of its members. Peppermint, spearmint, thyme, sage, and lavender are examples. There are about 200 genera with 3200 species well-distributed over the surface of the earth. There are herbs and shrubs in the family, generally with characteristic square stems. Leaves are simple and opposite or whorled.

Flowers (Fig. 29.9) may be showy, are generally irregular, and are usually complete, although some may be imperfect. The calyx is regular or two-lipped, partly coalesced and commonly five-toothed. Four or five parts of the corolla are coalesced characteristically to form two distinct lips, from which the name of the family is derived. There are four stamens, of two sizes, two shorter than the other two. The superior ovary comprised of four carpels is four-parted, each carpel containing a seed. The mature fruit consists of four single-seeded nutlets.

From Ranales to Compositae

The second line of ascent from the Ranales involves the Leguminosae, Rosaceae, Umbelliferae, and Compositae, with an offshoot family, the Cucur-

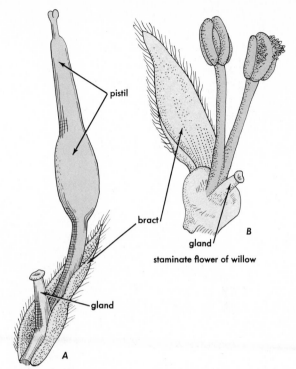

pistillate flower of willow

Figure 29.8 Salicaceae. *A*, pistillate flower of willow (*Salix*); *B*, staminate flower. (Redrawn from A. M. Johnson, *Taxonomy of the Flowering Plants.* Copyright 1931, 1959 by Meredith Corporation. Used with permission of Appleton-Century-Crofts.)

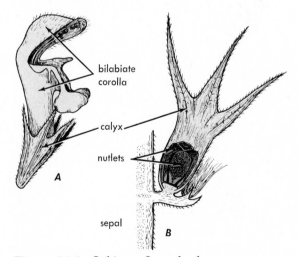

Figure 29.9 Labiatae. Irregular hypogynous flower of mint, showing union of sepals and petals.

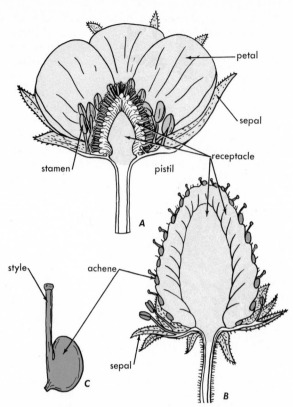

labels on figure: petal, sepal, receptacle, stamen, pistil, A, style, achene, sepal, C, B

Figure 29.10 Rosaceae. Flower and aggregate fruit of strawberry (*Fragaria*). *A*, median lengthwise section of a flower; *B*, median lengthwise section of a fruit; *C*, single achene. (Redrawn from W. W. Robbins, *Botany of Crop Plants.* Copyright 1931 by Blakiston. Used with permission of McGraw-Hill Book Company.)

bitaceae. This large group of families is characterized by an early change from hypogyny to epigyny. Following this there is a division into two sublines, one being marked by a lack of coalescence and a retention of regular flowers. In the other subline, coalescence and irregular flowers both occur in advanced forms.

The Rosaceae (rose family) is one of the more primitive families in this line of evolutionary development. There exists a considerable variety of flower structure within the family. The more primitive members of the family are regular, perfect, with numerous parts and with such a slight degree of perigyny as to be recognizable only with careful observation. This is exemplified by the strawberry (Figs. 16.6, 29.10). Changes within the family involve a reduction in the number of floral parts, a shift from slight perigyny to true perigyny (Fig. 15.17B, see Color Plate 7; Fig. 29.10) and to epigyny (Fig. 16.20). Coal-

escence also occurs (Fig. 16.20). Members of the family generally have regular flowers.

Leguminosae (pea family) are considered to be a more advanced family than Rosaceae, even though their flowers are hypogynous. There is a reduction in number of parts, the gynoecium having only one carpel; coalescence of stamens and of some petals occurs, and the corolla parts are irregular. Umbelliferae represent a further advance over Rosaceae in that the flowers are epigynous and exhibit syncarpy (Fig. 29.13). This line of ascent finds its climax in the Compositae, whose flowers all possess advanced characters, such as reduction in number of parts, modification of parts (pappus), irregular flowers, some imperfect flowers, coalescence in each whorl, and adnation.

The Cucurbitaceae culminate a branch line from the rose order. Flowers of Cucurbitaceae have such advanced characters as loss of floral parts, imperfect flowers, epigyny, and coalescence.

Rosaceae. Whereas grasses and legumes supply bread and vegetables of basic importance, the rose family supplies fruit for dessert and roses for decoration. There are better than 3000 species in this family and over 100 genera, not counting the almost numberless cultivated forms of roses, peaches, apples, cherries, strawberries, almonds, and so on. The family is of world-wide distribution and somewhat heterogeneous. Many of the principles of angiosperm evolution can be demonstrated with its members. There are trees, shrubs, and herbs in the family. Leaves are usually alternate and bear stipules.

While flowers of different genera show much variation, they are generally perfect and regular. The parts are borne on a floral disk which, as it develops after fertilization, becomes responsible for certain fruits characteristic of the family. In rose the disk or calyx tube is hollow and surrounds numerous pistils; the numerous stamens and corolla are adnate to this tube, the flower being perigynous. Almonds (Fig. 15.17B; see Color Plate 7), plums, peaches, and apricots have a similar structure except that the number of parts of the gynoecium is reduced to one or occasionally two. In strawberry (Figs. 16.6, 29.10), the reverse situation obtains, in that instead of a hollow structure bearing the parts of the gynoecium, the carpels are borne on the outside of a raised disk. In apple (Fig. 16.20) and quince, the calyx tube or floral disk grows attached to the ovary and expands to become the edible portion of the fruit. A simple classification of flowers of the rose family is not possible. There is much variation in the type of fruits. In strawberry, the matured ovary is an achene; in apples and pears, it is a pome; in peaches, apricots, blackberries, and cherries, it is a drupe; capsules occur in some genera and follicles in others.

Cucurbitaceae. The gourd or melon family is of world-wide distribution in at least warmer regions of the world, and various peoples have selected different forms for domestication. It is also probably the only group in which fruits are highly prized for ornamental purposes and for use as vessels of various types. Pumpkins and squashes are thought to be of American origin. Cucumbers and melons have probably come from Africa and central Asia. Other species appear to have been first cultivated in the tropics of Asia, Polynesia, and India. They are frequently tendril-climbing, annual, or perennial vines. Most are rapid-growing and frost-tender. There are about 100 genera with 850 species. Stems are usually soft and hairy or prickly. The generally simple leaves are large and sometimes deeply cut. Lateral tendrils are often present and they may be simple or branched.

Flowers (Figs. 15.13C, D, 29.11) are imperfect and plants may be either dioecious or monoecious. In pistillate flowers, the five-lobed calyx is adnate to the ovary. There may be five petals or they may be coalesced into a five-lobed corolla. The inferior

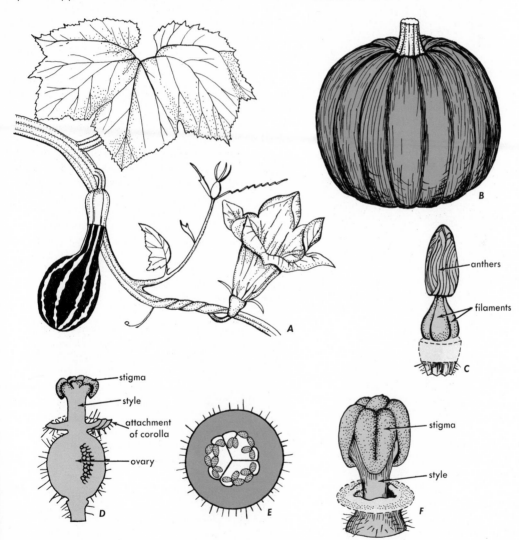

Figure 29.11 Cucurbitaceae. *A, Cucurbita pepo,* flowering and fruiting stem; *B, C. pepo,* fruit; *C, C. maxima,* staminal column; *D, C. maxima* stigma, longitudinal section through pistil, corolla tube removed; *E, C. maxima,* cross-section through ovary; *F, C. maxima,* style and stigma. (Redrawn from L. H. Bailey, *Manual of Cultivated Plants,* Macmillan, New York.)

ovary is composed commonly of three carpels with styles and stigmas also coalesced, but the latter may be lobed. Fruit is the pepo or berry so characteristic of the family. The perianth of staminate flowers is similar to that of the pistillate flower. There are five stamens, two of which may be fused. Anthers twist so that the stamens appear to form a solid central column.

Leguminosae. The legume family has such distinctive characteristics that its members can frequently be recognized with but little experience. It is one of the three largest families, with 550 genera and 13,000 species. All sorts of plant bodies are represented: herbs, both annuals and perennials, shrubs, vines, and trees. It is of world-wide distribution. While less heterogeneous than the rose family, considerable variation is found among its various members. It is a family of considerable importance in supplying food for man and his animals. Many legumes are used for ornaments, from shade trees to cut flowers. Some lumber is obtained from the black locust, and some of the tropical species furnish wood for fine cabinet work. Association of the nitrogen-fixing bacteria with roots of legumes places this family in a unique position relative to maintenance of soil fertility. Peas, beans, peanuts, clovers, and lupines are common herbaceous legumes; wisteria is a vine, brooms and redbuds are shrubs or low trees, and locusts and acacias are tree types. The *Mimosa* genus alone has some 300 species, including the sensitive plant of the florist shop; there are others of varying habit from tall trees to low herbs. Leaves of legumes are prevailingly pinnately compound and quite generally with stipules. Sometimes a leaflet will be modified to a tendril.

Flowers of most genera are of the well-known sweet pea or "butterfly" type (Figs. 15.5A, 29.12). They are thus irregular, the five-parted corolla forming a standard, two wings, and a keel. Sepals are present. Nine of the ten stamens are united to form a tube around the ovary; the tenth stamen is free. The gynoecium is composed of a single carpel. In some genera, flowers, while quite irregular, are not of the typical sweet pea type, and in *Mimosa* and allied genera the flowers are regular and small. In this group, some species have numerous, free stamens. The legume (Figs. 16.2, 16.9) or pod is the characteristic fruit of the family. It forms from a single carpel, and has one locule and numerous seeds. In many genera, it dehisces at maturity along both sutures; in a few genera, it is indehiscent.

Umbelliferae. The parsley family forms a very distinctive group of plants, many of which can be recognized with little experience. There are over 200 genera and 3000 species growing in all regions of the world, though confined to mountains in the tropic zone. The name is derived from the typical umbel type of inflorescence (Fig. 15.27). The family contains many crop plants, as well as some that produce drugs and a number of kitchen herbs. Poison hemlock, which is famous because of its use by the Greeks as a means of carrying out the death sentence and which was given as such to Socrates, is a member of this family. The genus is widespread, constituting a hazard to livestock not only in pastures of Greece, but on open ranges of California. Carrots, celery, and parsnips as well as parsley are members of this family. They are mostly herbs, rarely small shrubs with generally hollow stems. Leaves are alternate, mostly compound, and frequently much dissected. The petioles expand at the base and may somewhat sheath the stem.

Flowers are small, being borne in umbels, which are in turn grouped in umbels, so a large inflorescence is formed. Frequently, umbels at all levels are subtended by bracts, forming a characteristic involucre. Flowers (Fig. 29.13) are regular or the outer flowers irregular, always epigynous and perfect, although they may not always be complete. There are five stamens alternating with the same number of petals. The gynoecium is composed of two carpels, each with but a single seed. Fruits are indehiscent, although carpels separate from each other when mature.

Compositae. The composite family forms the largest family of angiosperms. Only a few are woody, and except for rare instances they are not designated as trees or shrubs. In this family of herbaceous plants there are around 950 genera and about 19,000 known species. The family is not only of world-wide distribution but in most places is relatively abundant. The family is not noted for its food plants; endive, artichokes, chicory, lettuce, and sunflower, form most of its contribution in this respect. Neither is it famous as a producer of drugs or other commercial products. One species, safflower, is now being grown for oil and another has been considered as a possible source of latex for rubber. But paging through a seed catalog or a visit to a florist shop will show members of this family in their full grandeur. *Dahlia, Chrysanthemum, Aster, Zinnia, Tagetes, Gaillardia, Ageratum,* and many others are representatives of this family. Dandelions color lawns, and various wild species add to fall color of meadows and fields. There are 200 species of the sagebrush, genus *Artemisia*, that cover arid portions of the world and supply, among other things, tarragon for fancy vinegar and nectar for desert honey. Leaves are of various shapes.

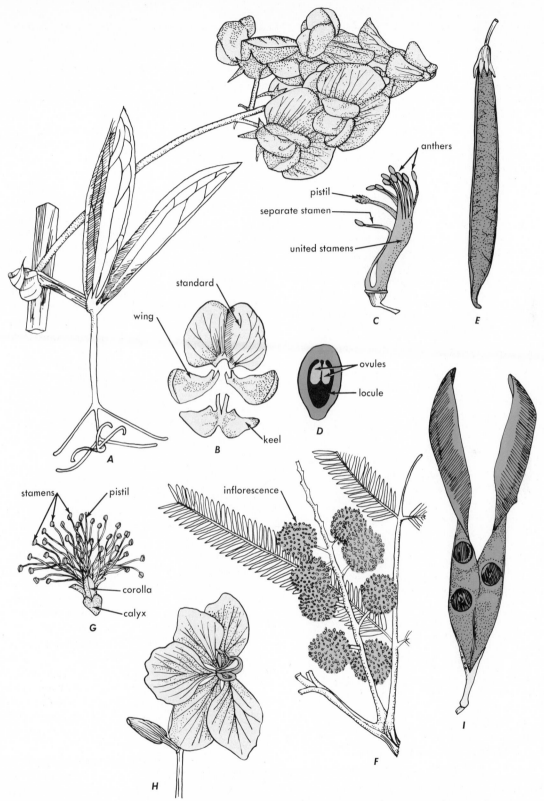

Figure 29.12 Leguminosae. *A,* inflorescence and leaf of *Lathyrus; B,* exploded view of *Lathyrus* corolla; *C,* essential organs of *Lathyrus* flower; *D,* cross-section of ovary; *E,* pod of *Lathyrus; F,* branch showing inflorescences and leaves of *Acacia; G,* a single *Acacia* flower; *H,* flower of *Bauhinia; I,* dehisced pod of *Bauhinia.* (Redrawn from L. H. Bailey, *Manual of Cultivated Plants,* Macmillan, New York.)

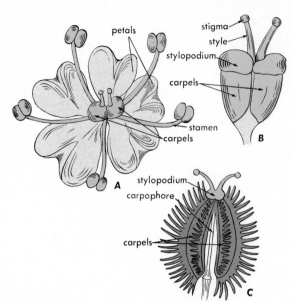

Figure 29.13 Umbelliferae. Flower and fruit of carrot (*Daucus carota*). *A*, flower; *B*, two carpels; *C*, mature fruit.

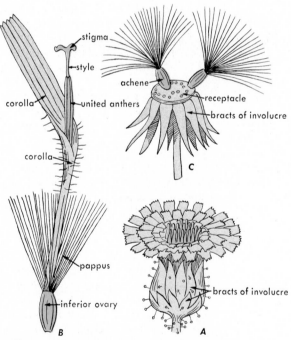

Figure 29.14 Composite flower and fruit of sow thistle (*Sonchus*). *A*, flowering head; *B*, single ray flower; *C*, mature achenes. (Redrawn from Korsmo.)

Flowers and fruits of the family have been described in detail on pp. 273–274, 296 and are shown in Fig. 29.14. They will not be discussed further here. They are, however, so characteristic of the family that any composite should be recognizable with a minimum of experience.

Monocotyledonous Line from Ranales to Orchids

There are many obvious differences between the Ranales and the monocotyledonous line of evolution. These have been discussed and may be briefly reviewed in Table 29.5.

However, some primitive monocotyledons, particularly certain water weeds such as arrowheads and water plantains, have much in common with buttercups and marsh marigolds. The families selected to represent this line of ascent are Liliaceae, Gramineae, Iridaceae, and Orchidaceae. Lilies differ from Ranales in all distinctive monocot characters. In addition, they show a reduction in number of floral parts and a coalescence of carpels. Irises show a single advance over lilies in that they are epigynous. Many irises are regular, perfect, and with separate perianth parts. Stamens show no coalescence. Syncarpy occurs as it does in lilies. Orchids show evolutionary advance in that they have irregular flowers very highly specialized for insect pollination. Stamens have been reduced in number to one or two.

The order to which grasses belong is thought to have arisen from the order to which lilies belong. The primitive characters of grasses include hypogyny, regular flowers, and only coalescence of carpels. Loss of perianth parts and reduction in number of stamens and carpels marks them as a more advanced family than the Liliaceae.

Liliaceae. Unlike grasses, lilies are grown largely for ornamental purposes, although two genera, onions (*Allium*) and *Asparagus,* are grown extensively for food. Some species yield drugs, and others have poisonous properties that may cause trouble in pastures or on ranges of western states. There are about 4200 species in some 240 genera. Most lilies grow from a bulb or bulb-like organ and flower in a single growing season after which the shoot dies down. A few, such as Joshua trees (Fig. 7.29C), are woody perennials. Tulips, hyacinths, day lilies, and aloes are other examples of the lily genera.

Members of the lily family may be generally quite easily recognized because of the typical showy flowers (Fig. 15.1C; see Color Plate 6). The regular perianth has six parts, usually separate; but if united, they are lobed, and corolla and calyx are generally not dis-

TABLE 29.5 Tabulation of Differences between Monocotyledons and Dicotyledons

Monocotyledons

1. One cotyledon or seed leaf.
2. Generally marked parallel leaf venation.
3. Flower parts typically in groups of three or multiples of three.
4. Vascular bundles of stems scattered throughout a cylindrical mass of ground tissue.
5. Vascular cambium lacking in most forms.

Dicotyledons

1. Two cotyledons or seed leaves.
2. Generally marked netted venation of leaves.
3. Flower parts typically in groups of four or five.
4. Vascular bundles of stems usually arranged in the form of a cylinder.
5. Vascular cambium present in forms having secondary growth.

tinct from each other. There are six stamens and one pistil. The superior ovary is single and three-celled; there are three stigmas. The fruit is mostly many-seeded and may be a capsule (Fig. 16.5) or a berry.

Gramineae. Man and other mammals have been associated with the grass family for far more years than those of recorded history. It is probably not an overstatement to say that without the grass family and man's ability to learn to exploit it, civilization, as we know it, could not have developed. Different grasses were domesticated by the three great civilizations: wheat, rye, barley, and oats by Mediterranean people; corn (*Zea mays*) by Aztecs, Mayans, and Incas and other natives of the Western Hemisphere; and rice and millet by the Chinese. Grasses are grown for human and animal consumption, for ornament, and for uses in arts and industry. There are over 500 genera and about 8000 species. Most grasses are herbaceous, either annuals or perennials; a few bamboos are climbers and some bamboos are woody. Grasses of one kind or another grow in all kinds of soil and situations. Grasses may be conveniently divided into five groups according to their use. (a) Bamboos, mostly evergreen, stout perennials, are used for construction in many parts of the Orient, and some have been introduced into warmer parts of the United States for ornamental purposes. (b) Cereals, supply grain and forage for both man and animals. (c) Sugar-producing species, such as sugar cane, are strong, upright perennials growing only in the tropics. (d) Sod-forming grasses, perennials that cover many square miles of the earth's surface, are used in lawns and meadows. (e) There is a large group of grasses grown for ornamental purposes, such as pampas grass. Flowers and vegetative characteristics of grasses have been described in some detail in Chapter 15 and shown in Figs. 15.19 and 15.20. The leaf is characteristically divided into a sheath, which surrounds the stem and expands to a linear blade (Fig. 10.4).

Flowers, known as florets, are small and usually grouped in spikelets. They may be arranged in a variety of ways, frequently in spikes or heads. The spikelet normally has two glumes at its base and a number of florets attached to the rachilla. The floret itself has two bracts, lemma and palea, from which may extend an awn, protecting a single pistil, with its feathery stigmas, and three stamens. Lodicles may also be present. The fruit is the typical grain or caryopsis, with a single seed whose seed coats have become attached to the ovary wall.

Iridaceae. The iris family, contains about 60 genera and some 1500 species. They are all herbaceous, largely perennial forms, usually with rhizomes, bulbs, or corms, as in lilies. Leaves and flowering stalks last for only one season. Some of the choicest florists' plants occur in this family—*Iris, Gladiolus, Freesia, Crocus, Watsonia,* and so forth.

The regular perianth, like that of lily, has six parts, united in some genera and separated in others (Fig. 29.15). In some genera, flowers are irregular. Three parts of the perianth may differ from the other three parts, or all may be similar. Flowers are showy and usually emerge from a spathe consisting of a pair, sometimes more, of herbaceous or scaly bracts. The single ovary is inferior and sets this family definitely apart from the Liliaceae. The ovary is three-celled and has three stigmas. The fruit is always a many-seeded capsule.

Orchidaceae. It is agreed by all that orchids represent the most advanced family in the Monocotyledoneae. They are all herbaceous plants, circling the world largely in the tropics, but with a few genera extending into colder temperate regions. There are some 15,000 recognized species in several hundred genera, making this one of the three largest families of angiosperms. All forms are perennial, with tuberous, bulbous, or otherwise thickened roots. They may be erect, prostrate, or climbing. A few are saprophytic and lack chlorophyll; others are epiphytes, growing on trees without benefit of soil.

Flowers are very showy and highly specialized for insect pollination. A typical flower is shown in Fig. 29.16. Sepals are present, the corolla is irregular, and the petals are united. The ovary is epigynous, one-

Selected Families of Angiosperms 611

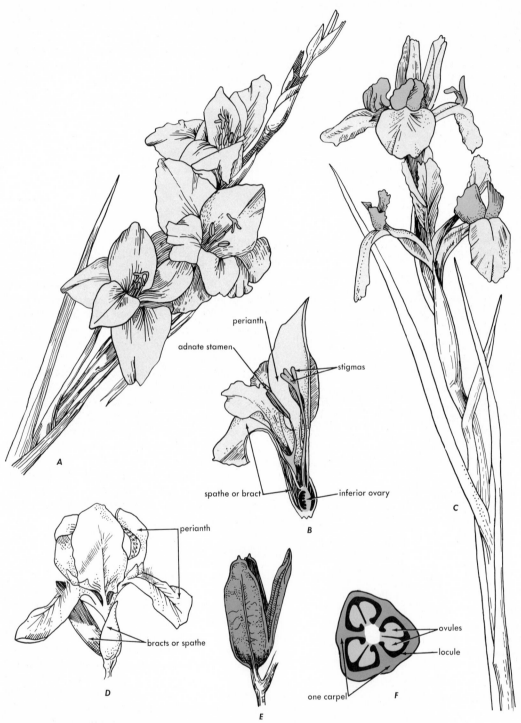

Figure 29.15 Iridaceae. *A, Gladiolus* inflorescence; *B,* longitudinal section of a gladiolus flower; *C, Iris* inflorescence; *D, Iris germanica; E,* capsule of *Iris sibirica; F,* cross-section of ovary of *Freesia.* (Redrawn from L. H. Bailey, *Manual of Cultivated Plants,* Macmillan, New York.)

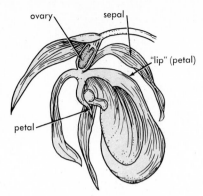

Figure 29.16 Orchidaceae. Irregular epignyous flower of orchid, longitudinal section.

or three-celled. Stamens unite with the pistil to form a structure sometimes called the **column.** There may be only one fertile stamen, sometimes with a number of **staminodes** or infertile stamen-like structures. Pollen hangs together in masses known as **pollinia.** Insects lift such a pollinium and place it on the beak or projection of the stigma, which may stand at the base of anthers. The fruit is a dehiscent capsule having many minute seeds.

Summary of Angiosperm Innovations

In our short discussion of the life history of an angiosperm, several references were made to structures which were new to, and characteristic of, angiosperms. This combination of structures has adapted angiosperms to the present climatic conditions and resulted in their great variety and extensive vegetative growth. While this discussion has considered only reproductive structures, various vegetative adaptations have also played their part in making the angiosperms such successful competitors for their ecological niche. We shall not discuss these vegetative adaptations here. The reproductive innovations may, however, be summarized as follows:

1. Development of the carpel, involving (a) ovary, (b) style, (c) stigma.
2. Enclosure of ovule within ovary, affording added protection to ovule and female gametophyte within it.
3. Provision, by the style and stigma, of a precise pathway guiding sperm cells in pollen tube to egg cell in embryo sac.
4. Reduction of female gametophyte to a seven-celled embryo sac.
5. Various highly successful systems of pollination, made possible by the stamen, which (although not an innovation of comparable rank to that

of the carpel) has proved to be a very adaptable part.
6. Reduction of male gametophyte to a small pollen tube and complete elimination of motile sperms.
7. Elimination of female gametophyte as the source of nourishment for young embryo and seedling.
8. Presence of double fertilization, resulting in formation of an endosperm as a source of nutrient, at least for the young embryo and in some cases for the seedling.
9. Food storage within the embryo itself.
10. Development of fruit from ovary wall.

Significant Features of the Angiosperm Life Cycle

1. Microspores are produced by meiosis in microsporangia or pollen sacs of anthers; the latter are a part of the microsporophylls or stamens.
2. Each pollen grain, developing from a microspore by mitosis, has a generative and tube nucleus and is provided with a thick, sculptured cell wall.
3. After pollination the pollen tube or male gametophyte grows down the style to the ovule.
4. Meiosis in the megaspore mother cell within the nucellus or megasporangium of the ovule results in four megaspores aligned in a row.
5. One megaspore commonly develops into a seven-celled embryo sac or female gametophyte.
6. Two sperms are discharged into the embryo sac, and double fertilization occurs.
7. The zygote develops an embryo.
8. Primary endosperm cell forms an endosperm that may not persist into the seed.
9. After fertilization, the ovule becomes the seed.

Summary of Angiosperm Evolution

1. Angiosperm families, according to the Besseyan system of classification, are all derived from the primitive Ranales, to which order buttercups and magnolias belong.
2. Important changes in floral evolution, according to the Besseyan system, are as follows: (a) from a spiral to a whorled arrangement of floral parts; (b) from many parts to few or even to a loss of parts; (c) from separate floral parts to coalesced parts; (d) from a regular flower to an irregular flower; (e) from hypogyny to epigyny.
3. According to the Besseyan system of classification there are three branched lines of ascent from the Ranales.
4. The first line of ascent following the Ranales

613

Bessey's system	Engler's system	
DICOTYLEDONAE	MONOCOTYLEDONAE	Primitive
Magnoliaceae	Gramineae	
Ranunculaceae	Liliaceae	
Malvaceae	Iridaceae	
Cruciferae	Orchidaceae	
Solanaceae	DICOTYLEDONAE	
Labiatae	Salicaceae	
Rosaceae	Ranunculaceae	
Leguminosae	Magnoliaceae	
Cucurbitaceae	Cruciferae	
Salicaceae	Rosaceae	
Umbelliferae	Leguminosae	
Compositae	Malvaceae	
MONOCOTYLEDONAE	Umbelliferae	
Liliaceae	Labiatae	
Gramineae	Solanaceae	
Iridaceae	Cucurbitaceae	
Orchidaceae	Compositae	Advanced

includes: Malvaceae (mallow family), Cruciferae (mustard family), Solanaceae (potato family), Salicaceae (willow family), and Labiatae (mint family).

5. The second line comprises: Rosaceae (rose family), Leguminosae (pea family), Umbelliferae (carrot family), and Compositae, with Cucurbitaceae (melon family) as an offshoot from Rosaceae.

6. The third line comprises the Monocotyledoneae in the following order: Liliaceae (lily family), Iridiaceae (iris family), and Orchidaceae (orchid family). The Gramineae (grass family) are an offshoot from the Liliaceae.

Classification

In our survey of families of flowering plants, we have generally been following the Besseyan system of classification. Many taxonomists, however, follow other systems of classification, the system of Engler being a favorite alternative. For the purpose of comparison, the families studied are listed here according to the systems of both Bessey and Engler. For each system, classes and families considered most primitive are given first, advanced are given last. Note that Bessey considered monocotyledons more advanced while Engler considered them to be more primitive.

30

Evolution

In a broad sense the term **evolution** refers to a process involving gradual changes. It is well-known that neither animals, nor plants, nor cities, nor states remain the same. We may speak of the evolution of means of transportation, the evolution of human clothing, the evolution of mountains and valleys, and the evolution of many other nonorganisms. **Organic evolution** pertains to the gradual changes that have taken place in living organisms. Coal, for instance, is formed from the remains of plants that were different in appearance from those growing today. The student of organic evolution may be interested in, among other things, accounting for the disappearance from the earth's flora of the Coal Age plants and the appearance of the modern flowering plants.

Organic evolution is closely allied to genetics and plant breeding. We have seen that the factors which determine the characteristics and activities of plants are associated with the cell nuclei. Genes are transmitted from generation to generation by the chromosomes during the processes of fertilization and meiosis. If the origin of new plants is to be understood, experiments on genetics will furnish important information. Many such experiments have been carried out, and they have achieved two ends: (a) they have made possible clearer insight into some of the mechanisms of evolution, and (b) they have greatly increased the yield and improved the quality of agricultural crops.

The Geologic History of Plants

Since plants cannot be dissociated from their environment, some knowledge of the principal changes in the earth's crust is essential to an understanding of plant evolution. Broadly speaking, two different types of rock are found in the earth's crust. One of them, like lava, was formed from molten material. The oldest known rocks are of this nature; granite is an example. Because of their molten origin, no fossils are ever found in these or in other **igneous** rocks.

The second type of rock (**sedimentary** rocks) is formed by the deposition of large amounts of sediment in bodies of water. As layer upon layer of sediment accumulates, the pressure upon the lower layers becomes enormous and they are turned to rock. Shale, limestone, and sandstone are examples of sedimentary rock. If plants or animals are buried in the sediment, and if decay does not occur, the buried individuals are preserved as fossils.

However, because great changes have taken place

in the earth's crust over time, sedimentary and igneous rock are not always found where you would expect to find them.

We Live on a Restless Earth

The billions of years of earth's history have been characterized by great changes in the atmosphere and the crust. During this time, entire floras and faunas have evolved, been almost completely eliminated, and new forms arisen in response to these geophysical changes. Perhaps the most dramatic evidence for change is displayed in the desert canyon country of southern Utah and northern Arizona. Monument Valley, in northeastern Arizona, lies between 5000 and 7000 ft above sea level, yet it consists entirely of sedimentary rocks which were deposited on the bottom of some primeval sea! The sedimentary sandstone must have been thrust up thousands of feet from sea level, and now the rock lies exposed to the erosive action of rain and wind. The buttes are mute evidence that already several hundred feet of sandstone have been eroded from the surface of Monument Valley (Fig. 30.1; see Color Plate 18).

About 150 miles to the west, the Colorado River has cut a gorge over a mile deep—the Grand Canyon (Fig. 30.2, see Color Plate 18)—and all the rock exposed is sedimentary. We can surmise that a similar mile-deep deposit of sandstone and shale underlies Monument Valley, all deposited by an open ocean. Eighty miles north of the Grand Canyon, the Towers of the Virgin River in Zion National Park, Utah, rise a mile into the air—that is, a mile higher than the buttes in Monument Valley. These towers (Fig. 30.3; see Color Plate 18) are also made of sedimentary rock. The inference is that a layer of sedimentary rock, 2 miles thick, covers the entire region; the bottom of a primeval sea has been lifted thousands of feet and its eroded remains form much of what we today call Utah and Arizona.

But that is not all. Less than 1000 years ago, Sunset Crater, 100 miles south of Monument Valley, was formed by a burst of lava boiling through the marine sediments. This was the latest of many volcanic eruptions in the region. The San Francisco Mountains just north of Flagstaff were formed from cubic miles of lava. In Monument Valley itself, Mt. Agatha and some smaller adjacent peaks are lava plugs, once the throats of volcanoes. The volcanoes themselves and the lava they spread out in the Monument Valley have since eroded away. All this sedimentation, eruption, deposition, uplift, and erosion has taken place in the past 350 million years. And trapped within the sediments are fossil remains of many plants and animals which existed during that period of time.

How Fossils Are Formed

A **fossil** may be a shell or bone, little changed from its original aspect; it may be the microscopic cell wall of a spore or pollen grain, which are also very resistant. But most plant parts are not hard, and plant fossils seldom consist of unchanged structures. They often are simply the impressions of leaf or stem fragments that were trapped in mud—mud that later was compressed into sedimentary rock. To yield a good fossil, plant parts must settle in quiet waters, then be buried under silt or volcanic ash to lie trapped in conditions unfavorable to decay. They are seldom entombed where shed, but are more typically carried by water or wind for some distance, first. If the plant part is large enough, for example, a tree trunk, the protoplast or lumen of the cell may be replaced by silica quartz crystals. The cellulose walls remain, and thus anatomical details are faithfully preserved (Fig. 30.4). In addition, degradation products from tissue decay may remain trapped as "chemical fossils." The porphyrins from chlorophyll degradation, for example, are very inert and long-lasting. Their presence in sediments indicate that photosynthetic plants were present at the time the sediments were laid down.

It is important to note that plant fossils consist almost entirely of plant remains that grew in or near water or were carried by streams into lakes, bays, or other sites of deposition. The remains of plants growing in arid or mountainous regions are rarely deposited where conditions are favorable for their preservation as fossils. Either it is too dry or the site is undergoing erosion and the deposits are of very short duration. This means that the geological record or plants must probably remain incomplete and that the records we do have usually represent only the floras of lowlands and moist habitats. Considerable evidence indicates that the great environmental extremes prevalent in arid or semiarid mountainous areas are especially favorable to rapid evolution. This factor, we shall see, may be very important in connection with the problem of determining the origin of the angiosperms and may serve to explain peculiarities that surround the first records of the angiosperms.

When Did the World Begin?

When did the world begin? Earth scientists have no definite clues, but the calculations of astronomers and physicists, coupled with known age of

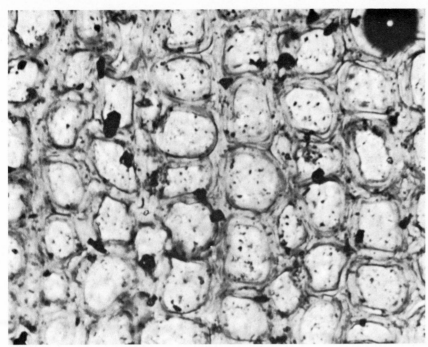

Figure 30.4 A thin section of petrified wood, as seen through a light microscope. Tracheid lumens are filled with silica stone.

meteorite fragments and of rocks from the moon, lead us to estimate that the surface of the earth cooled to a crust about 4.5–5.0 billion years ago. This primordeal crust probably contained uplands, ocean basins, plateaus, and mountainous volcanoes. Volcanic eruptions filled the atmosphere with carbon dioxide, methane, toxic carbon monoxide, nitrogen, hydrogen, and strong-smelling ammonia and hydrogen sulfide. Free oxygen was absent. The first rains washed, dissolved, and eroded the uplands, carrying many nutrients to the young oceans.

Life, or some sort of half-way form of life, probably originated very early. It may have consisted of aquatic, small, self-replicating, heterotrophic cells that lived on the rich, anaerobic ocean soup around them. However, our present methods of detecting remains of ancient life require that the fossils be preserved in rocks. So far, the oldest rocks found on earth are 3.4 billion years old. This means that we have no record of the first 1–1.5 billion years of earth's history (Fig. 30.5).

Aquatic life did exist 3 billion years ago, for we have fossil evidence of it. Some of the oldest rocks known lie exposed near the gold-mining town of Barberton, on the border between the Republic of South Africa and Swaziland. About 3.2 billion years ago, this area was a shallow, warm sea or embay-

ment. Living things existed in thin sheets at the bottom of a silica-rich sea. Apparently, conditions were perfect for preservation of the organisms. They were preserved in the sediment in a siliceous solution that later crystallized into rock called chert, much as a modern biological specimen is preserved by being embedded in plastic. There was no distortion in the soft bodies of these early organisms because the silica matrix of chert is incompressible. Today, these beds of chert, up to 400 ft thick, are exposed. When thin sections are placed on a microscope slide, the perfectly preserved organisms **(microfossils)** can be seen. The chert itself has not been dated, but rock layers above and below it have. Its age is estimated to be 3.2 billion years.

Many unusually shaped structures, which may or may not be organisms, have been seen in this chert (called the Fig Tree Formation), but only two are seen frequently enough and in great enough detail to leave no doubt of their once-living nature. One is a rod-shaped bacterium-like cell, called *Eobacterium isolatum* (Fig. 30.6), the other is possibly a blue-green alga, *Archaeosphaeroides barbertonensis*. Organic residues were also found in the chert. Analysis of them showed the presence of certain hydrocarbons which can most reasonably be regarded as breakdown products of chlorophyll, and that the

The Geologic History of Plants   617

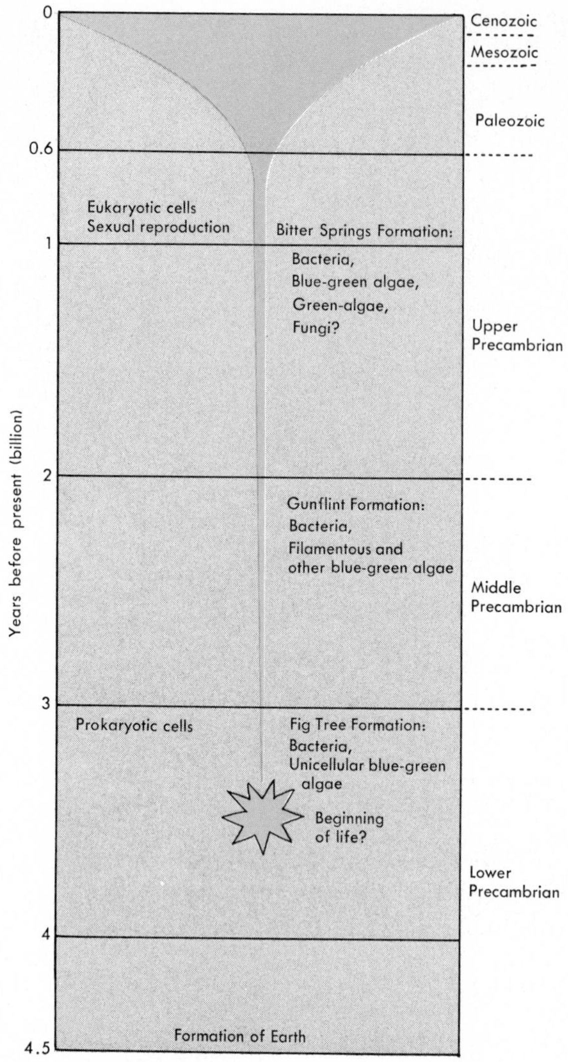
0
0.6
1
2
3
4
4.5
Years before present (billion)
Cenozoic
Mesozoic
Paleozoic
Upper Precambrian
Middle Precambrian
Lower Precambrian
Eukaryotic cells
Sexual reproduction
Bitter Springs Formation:
Bacteria,
Blue-green algae,
Green-algae,
Fungi?
Gunflint Formation:
Bacteria,
Filamentous and other blue-green algae
Prokaryotic cells
Fig Tree Formation:
Bacteria,
Unicellular blue-green algae
Beginning of life?
Formation of Earth

Figure 30.5 The diversity of life through geologic time. Increasing width of the green band indicates increasing diversity.

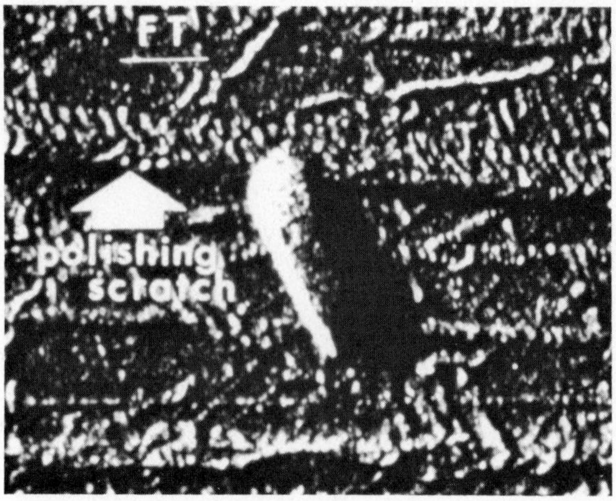
FT
polishing scratch

Figure 30.6 Oldest known bacterium. *Eobacterium isolatum* from the Fig Tree Formation, about 3.2 billion years old. (Courtesy of E. S. Barghoorn.)

ratio of carbon-13 to carbon-12, two natural isotopes of carbon, was lower than the ratio in today's atmosphere. When plants photosynthesize, they have preference for carbon-12, so their tissue and fossil residues have a low carbon-13 to carbon-12 ratio, corresponding to the ratio in the chert. This combination of direct and indirect evidence strongly suggests that plant life existed 3.2 billion years ago.

A similar, but richer, collection of microfossils has been preserved in another chert deposit, the Gunflint Formation, along the Ontario shore of Lake Superior. By dating layers above and below it, this deposit is thought to be 2 billion years old. Filaments are most abundant, and some resemble modern blue-green algae (Fig. 30.7). Small spheres that might be colonial blue-green algae or spores are also common. Analysis of the organic residue and determination of the carbon isotope ratio show that photosynthetic organisms were present; however, they were still prokaryotic organisms.

The first eukaryotic plants did not appear until 1 billion years ago; their fossils appear in the Bitter Springs Formation, about 40 miles northeast of Alice Springs, in the heart of Australia. These fossils are still richer and more varied, and they indicate that at least four groups of organisms existed at that time: (*a*) prokaryotic filamentous blue-green algae, akin to modern *Oscillatoria* and *Nostoc;* (*b*) bacteria; (*c*) eukaryotic fungi (Fig. 30.8); and (*d*) eukaryotic green algae (Fig. 30.9). Cytological details are preserved so well, that stages of cell division can even be detected (Fig. 30.10).

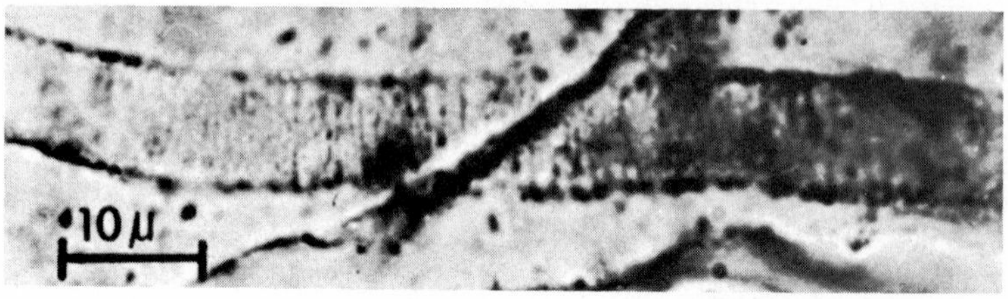
10 μ

Figure 30.7 The filamentous blue-green alga, *Animikiea septata*, from the Gunflint Formation, 2 billion years old. (Courtesy of J. W. Schopf.)

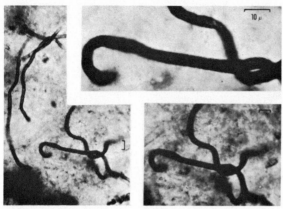

Figure 30.8 Hyphae of a fungus, *Eomycetopsis robusta*, from the Bitter Springs Formation, *1 billion years* old. (Courtesy of J. W. Schopf.)

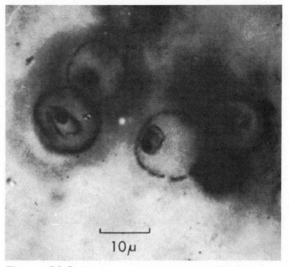

Figure 30.9 Eukaryotic green algal cells from the Bitter Springs Formation. (From J. W. Schopf, *J. Paleontol.* **42,** 651–688, 1968.)

This early appearance of eukaryotes is impressive, because the differences between prokaryotes and eukaryotes are relatively enormous. These changes, as you may recall from earlier chapters, involve structure of DNA, organization of a nucleus, endoplasmic reticulum, ribosomes, chloroplasts, mitochondria, development of new chlorophylls, the elaboration of sexual reproduction (with meiosis, chromosome recombination, and even mitosis itself), and the formation of many new enzymes in new pathways of metabolism. The evolution of eukaryotes from prokaryotes, even in the long time span of 2 billion years, was a considerable accomplishment.

The evolution of more modern plants during the past 1 billion years may seem rapid by comparison,

but it is based on the permutations and combinations of life processes which were already set during the first 2 billion years of life.

Sexual reproduction brings with it variation in offspring and a more rapid rate of evolution. It made possible a greater diversity of forms to develop and proliferate. A spurt of evolution resulted. By the end of the **Proterozoic** era (Table 30.1), the seas had become crowded with life, including many forms of animals and plants. The rate of oxygen formation had increased, and by the end of the Proterozoic era the level of oxygen in the air may have been 1% of its

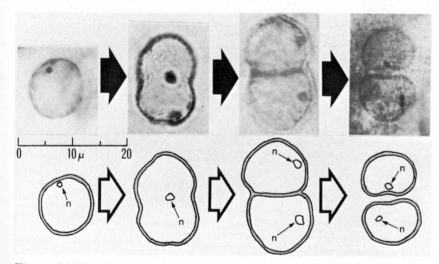

Figure 30.10 Sequence of fossil impressions from the Bitter Springs Formation arranged to show cell division. (From J. W. Schopf, *J. Paleontol.* **42,** 651–688, 1968.)

TABLE 30.1   Geologic Time and the Dominance of Different Plant Groups through Time[a]

Era	Period	Epoch or part	Began (millions of years ago)	Dominants
Cenozoic	Quarternary	Recent	Last 5000 years	Flowering plants
		Pleistocene	2.5	
	Tertiary	Pliocene	7	
		Miocene	26	
		Oligocene	38	
		Eocene	54	
		Paleocene	65	
Mesozoic	Cretaceous	Upper	90	
		Lower	136	Gymnosperms
	Jurassic	Upper	166	
		Lower	190	
	Triassic	Upper	200	
		Lower	225	
Paleozoic	Permian	Upper	260	
		Lower	280	Lower vascular plants
	Carboniferous	Pennsylvanian	325	
		Mississippian	345	
	Devonian	Upper	360	
		Middle	370	
		Lower	395	Algae
	Silurian		430	
	Ordovician		500	
	Cambrian		570	
Proterozoic	Precambrian		4,500–5,000	

[a]Shaded sections were times of great evolutionary changes in the plant kingdom. The time scale is not drawn to scale. All named epochs do not appear on this chart.

present level. Oxygen not only supports life, it also screens out ultraviolet radiation from the sun. The radiation can be damaging to cell activities. This amount of oxygen—1% of present level—produced a sufficient filter of radiation to permit life in all but the top inch of water. Life on land, except in sheltered places, would still have been impossible.

Algal Diversity

The "sudden" evolutionary spurt 600 million years ago marked the end of the Proterozoic era and the beginning of the long **Paleozoic** era, a time when the land was low-lying, and inland seas moderated the climate so that seasonal and latitudinal effects

620   Evolution

were minor. It was a time when a climate much like that in the tropics today dominated the entire world. It was a time of enormous evolutionary change in the plant kingdom. However, the rate of evolution did not proceed at a constant pace during the entire Paleozoic era. Most of the new and successful experiments in plant form appeared in only 25–50 million years, during the upper-Silurian to mid-Devonian periods. It was then that plants came to dominate the land; it was then that the oxygen level of the air may have reached 10% of the present level.

Through the Cambrian, Ordovician, and most of the Silurian periods, algal forms continued to dominate the plant kingdom. Blue-green and green algae were most abundant, but lime-encrusted reds and some browns resembling *Laminaria* and *Fucus* were also present. The family Codiaceae in the Chlorophyta was especially prominent, and the early forms closely resemble living members of that family. Many of these showed branching, coenocytic filaments; larger forms had a central axis with radiating branches whose ends united to form an enveloping net. The lime-encrusted forms secreted calcium carbonate just as modern forms do. In the Devonian period, many algal fossils can be directly related to living genera of desmids, browns, and reds.

Of particular interest are fossils from this time of *Chara*-like algae (Fig. 30.11), for *Chara* is today only known from fresh-water habitats. The fossil forms had a very thick-walled resting zygospore that would

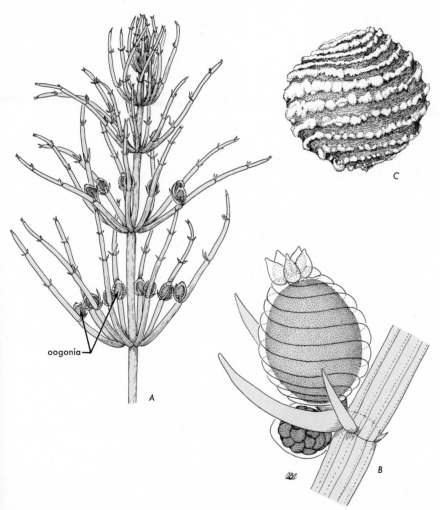

oogonia

A

C

B

Figure 30.11   Living and fossil *Chara*, a green alga. *A*, Living *Chara*, showing whorled branches and location of oogonia, ×70; *B*, one oogonium, ×700; *C*, a fossil oogonium or zygospore from the lower Devonian, ×400; (*C*, redrawn from H. P. Banks, *Evolution and Plants of the Past*, Wadsworth, Belmont, Ca.)

The Geologic History of Plants   **621**

have been resistant to drought. Such spores are rare in marine algae. In addition, there is evidence that some other algae may have been terrestrial, forming thick, moss-like mats on wet ground. So, in the early Paleozoic era, the invasion of the land began. It should be recalled from earlier chapters that the green algae, particularly the genus *Fritchiella*, is thought to be a logical starting point for the evolution of a terrestrial, vascular flora.

Invasion of the Land: Silurian and Devonian Periods

What are some of the adaptations required for plant life on dry land? In water, plants need no complex structures for support nor for the uptake of nutrients. The surrounding water buoys them up and bathes them with soluble nutrients. In contrast, land plants must not only develop roots for the absorption of nutrients and the elaboration of stiff tissue for support, they must provide pathways for water and nutrient transport—xylem and phloem. Reproductive cells on land must be carried by agents other than water, and they must be thick-walled and cutinized to avoid desiccation. All external surfaces of land plants, in fact, must be adapted to reduce water loss; these adaptations can take the form of stomata and cuticle.

Despite tempting fragments of evidence, such as cutinized spores and bits of xylem dating back to the Cambrian period, the first undisputed fossils of terrestrial, vascular plants do not appear until the upper-Silurian period. But what appeared at that time was not only one type of vascular plant. Many types suddenly occur in the fossil record: representatives of the Psilophyta, Lycophyta, Spenophyta, and pre-ferns. So we can surely believe that the first vascular plant evolved many millions of years before that. Did they all evolve from something like *Fritchiella*, or did they evolve from several separate lines? At this time there is not enough evidence to rule out either possibility.

The early vascular plants consisted of slender dichotomously branching stems, with or without leaf-life appendages, and with sporangia either at the tips of the branches or along their sides. Perhaps the most primitive example is *Cooksonia*, discovered in 395 million-year-old fossil beds in Czechoslovakia and Wales, from the end of the Silurian period (Fig. 30.12). It was a plant probably less than 4 inches tall, made up of dichotomizing branches less than $\frac{1}{8}$ inch in diameter that terminated with sporangia. The spores had a waxy cuticle, indicating that they were adapted for dissemination on land. The xylem in the stem was a solid, central core; there was no pith.

We know nothing of the root system. Another early vascular plant was *Rhynia* (pp. 545–547), and its excellent state of preservation in the fossil record allows for detailed reconstructions that have made it the best-known primitive vascular plant. The plant consisted of a slender rhizome upon which were borne erect, cylindrical stems that branch dichotomously and sparingly. True leaves and roots were absent; the stem tissue was probably photosynthetic, and rhizoids growing from the rhizome absorbed moisture and nutrients from the soil. Modern *Psilotum* shares many of these characteristics with *Rhynia*.

From this primitive, herbaceous start, evolution progressed rapidly to the first trees only 25 million years later. *Pseudosporochnus nodosus* was one of the first forest trees of the world, though comparison with trees of today would make us call it a sapling at best (Fig. 30.13). It had a main trunk up to 3 inches in diameter and several feet tall, topped with a crown of finely branched limbs. There were no leaves. Xylem in the branches no longer was in a central bundle (as with *Rhynia* and *Cooksonia*); instead it exhibited a more complicated, dissected pattern.

There were several other tree species in the mid- to late-Devonian period, but they were much larger. *Aneurophyton*, for example, was up to 40 ft tall. *Archeopteris* was more than 100 ft tall and up to 5 ft in diameter at the base (Fig. 30.14). Both had evolved pith, cambium, and the capacity to produce considerable secondary xylem. Annual rings are absent from petrified wood samples, however, so it is not possible to determine their age.

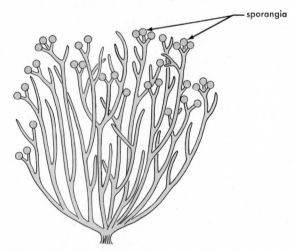

Figure 30.12 A reconstruction of *Cooksonia*, one of the first vascular plants. The above-ground portion, shown here, was several inches tall. (Redrawn from H. P. Banks, *Evolution and Plants of the Past*, Wadsworth, Belmont, Ca.)

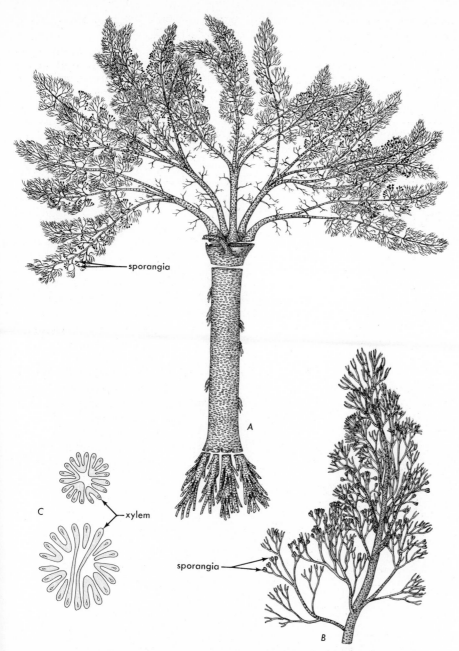

Figure 30.13 *Pseudosporochnus nodosus*, one of the first trees. *A*,
reconstruction of the entire plant, about 3 ft tall; *B*, detail of a terminal
branch with sporangia; *C*, xylem pattern in cross-sections of small and large
branches. (Redrawn from S. Leclerq and H. B. Banks, *Palaeontographica*
110(B), 1–34, 1962.)

Leaves had also evolved, changing from very small epidermal outgrowths of the primitive vascular plants, to larger true leaves having a vascular system of their own, as found in *Aneurophyton* and *Archeopteris*. The "true" leaves are called **megaphylls;** the epidermal outgrowths are called **microphylls.** Some botanists theorize that megaphylls evolved as modifications of branch systems: starting with a system of bifurcating branches, one section of branches first becomes reduced in size (it is overtopped by the other portions); then the branches come to lie all in one plane; then epidermal outgrowths connect the branches, much like webbing on a duck's foot. The result is a leaf blade with many veins (Fig. 30.15). Megaphylls, then, are considerably different in form and evolution from the superficial microphylls that were on the first vascular plants and which still exist on modern *Psilotum* and other lower vascular plants. Megaphylls today are only found in ferns and seed plants.

The evolution of large plants created another problem for paleobotanists. They rarely find a whole plant preserved as a unit. Leaf impressions may be found in one location in 1910, petrified trunk fragments somewhere else in 1932, and spores or roots still elsewhere at other times. Until the fragments can be connected by collections that show the parts attached to each other, each piece gets its own taxonomic name: a **form genus.** For example, *Archeopteris* at first was only known by fossil fronds, and this name was the form genus for that part of the unknown whole plant. *Callixylon* was the form genus for petrified wood fossils which all had the same anatomical details. Only in 1960 was a fossil

A

B

Figure 30.14 Reconstructions of large Devonian trees. *A, Aneurophyton,* 25–40 ft tall; *B, Archeopteris,* about 100 ft tall. (*A,* redrawn from W. Goldring, *Sci. Monthly* **24,** 515–529, 1927; *B,* redrawn from C. B. Beck *Amer. J. Botany* **49,** 373–382, 1962.)

624   **Evolution**

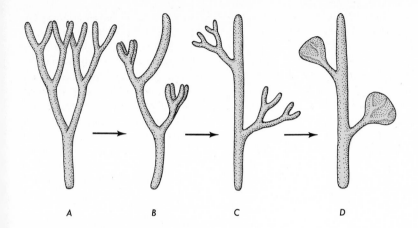

Figure 30.15 Hypothetical origin of true leaves (megaphylls). *A*, bifurcating branches; *B*, overtopping; *C*, the small branches lie in a plane; *D*, webbing occurs. (Redrawn from A. S. Foster and E. M. Gifford, *Comparative Morphology of Vascular Plants*, W. H. Freeman, San Francisco.)

A   B   C   D

frond found attached to a fossil stem and then it was possible to put the two together. *Archeopteris* was the name adopted for the entire plant.

Coal Age Forests and Seed Plants

Plant development during the Devonian period not only modified the vegetative size of plants, it affected their reproduction as well. Early vascular plants were homosporous, but *Archeopteris*, in the upper-Devonian period, was heterosporous. Two sizes of meiospores were formed: many small ones which produced the male gametes, and a few large ones which produced the female gametes. With four additional steps, heterospory can lead to the development of seeds: (*a*) the number of functional female meiospores (megaspores) produced inside the megasporangium is reduced to one; (*b*) the single megaspore is not shed from the megasporangium, but is retained within it; (*c*) the fertilized egg, or some later stage in embryo development becomes dormant; (*d*) protective structures (integuments) grow around the old megasporangium (with its megaspore or later, the embryo).

The first seeds did develop in the upper-Devonian period. A reconstruction of *Archeosperma* (a form genus for the seed) appears in Fig. 30.16. Actually, it is on the way to being an angiosperm, with enclosed seeds, because outer branches are beginning to fuse and form a protective layer around the several seeds. Unfortunately, we do not know what kind of plant produced this seed. It is not until the Coal Age (Mississippian and Pennsylvanian epochs, also known as the Carboniferous period) that seed plants become better known.

Extremely lush, swamp forests dominated the **Coal Age** landscape (Fig. 30.17). Because the land was low, minor changes in sea level successively inundated, buried, then supported one forest after another. The buried organic remains have become compressed and changed through time. Today, they form the coal, gas, and oil reserves of the world. These sources of energy are the chemically changed, fossil remains of Coal Age forests, and for that reason they are referred to as **fossil fuels**. Burial of plants and transformation into fossil fuel has taken place all through time, but not to the extent it did during the Coal Age.

These forests were dominated by Lycophyta. In the Devonian period, this division evolved rapidly into both herbaceous and woody types; in the Coal Age, the woody types reached tree stature. One example is *Lepidodendron* (Fig. 30.18), up to 120 ft tall and with a trunk 3 ft across. Strap-like leaves and sporangia occurred near the ends of the branches. A cross-section of the trunk reveals pith, primary and secondary xylem, cambium, and an enormous amount of cortex and cork. Very little of the stem area served for conduction or strengthening. The roots were dichotomously branched.

Second in abundance were giant horsetails, such as *Calamites* (Fig. 30.19). *Calamites* was smaller than *Lepidodendron,* but still was 30 ft tall and had a trunk about 1 ft in diameter. Whorls of branches, and smaller branches from those, and leaves from those, arose at nodes. The upright stems developed from a horizontal rhizome system.

Third in abundance—not tall, but dominating the forest floor—were ferns and seed ferns. Seed ferns were fern-like, but in addition possessed seeds on their fronds instead of spores (pp. 572–574). Exactly what the ferns evolved from is not clear, for they appear rather suddenly in the fossil record in the Coal Age. Some botanists think *Archeopteris* of the Devonian period is a likely ancestor of the ferns because of its fern-like leaves, but other evidence tends

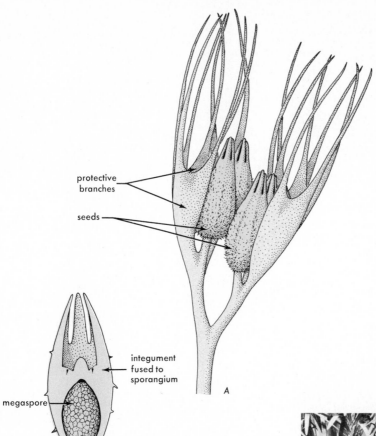

protective
branches

seeds

integument
fused to
sporangium

megaspore

A

B

Figure 30.16   Reconstruction of
the first seed, *Archeosperma*. *A*,
cluster of seeds within protective
branches; *B*, section of one seed.
(Redrawn from H. P. Banks,
Evolution and Plants of the Past,
Wadsworth, Belmont, Ca.)

Figure 30.17   Reconstruction of a Coal Age
forest. Note the seeds attached to fern-like leaf in
left center. (Courtesy of Field Museum of Natural
History.)

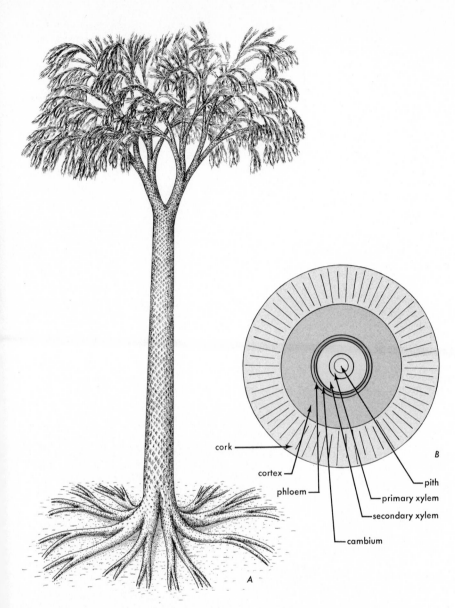

cork

cortex

phloem

pith

primary xylem

secondary xylem

cambium

B

A

Figure 30.18 *Lepidodendron*, a dominant of the Coal Age forest. *A,* reconstruction of the entire plant, about 100 ft tall; *B,* a cross-section of the trunk. (Redrawn from M. Hirmer, *Handbuch der Paläobotanik* **1,** 182–232, 1927.)

to show that *Archeopteris* is more closely related to gymnosperms. Perhaps ferns and gymnosperms both came from the same stock?

Gymnosperms were fourth in abundance. But many groups make up the gymnosperm category. Two groups of gymnosperms evolved in the Coal Age, precursors of the Cycadophyta and Coniferophyta (Fig. 30.20). Gymnosperms did not become abundant until 100 million years later.

Herbaceous forms of Lycophyta and Sphenophyta were common, but members of the Psilophyta—direct descendants of the first, weak, vascular plants to climb onto land only 70 million years earlier—were rare. The lower vascular plants dominated world vegetation during the Coal Age just as thoroughly as the algae dominated the world in previous ages. The reign of the lower vascular plants was short, however, and great extinctions lay ahead.

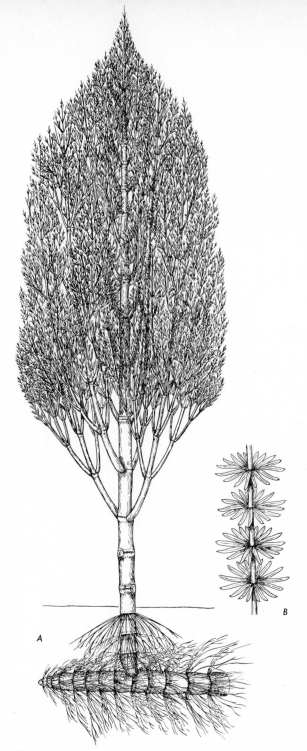

Figure 30.19 *Calamites*, another Coal Age forest tree. *A*, reconstruction of an entire plant, including rhizome and roots, of *C. carinatus*. *B*, detail of whorled leaves of the form species *Annularia radiata*, which are very similar to the leaves of this species of *Calamites*. The leaves are about $\frac{1}{2}$ inch long. (Redrawn from M. Hirmer, *Handbuch der Paläobotanik* **1**, 409–452, 1927.)

Permian Extinctions and on to the Mesozoic Era

The 375 million-year-long Paleozoic era was marked with two concentrated periods of evolution: a 25 million year period of creativity in the lower-Devonian period during which weak vascular herbs became forest trees; and a 50 million year period of major extinctions in the Permian period. The creativity was associated with increasing oxygen in the air; the extinction may have been caused by a cooling and drying climate and uplifting of the land (Fig. 30.21). In many ways, the sudden rise and fall of the Coal Age forests is just as striking and mysterious as the rise and fall of the dinosaurs many years later. The Permian extinctions marked the start of the **Mesozoic** era.

The plants which replaced the lower vascular plants in dominance were gymnosperms: the true conifers (Coniferophyta), Cycadophyta, and Ginkgophyta. Most of the Mesozoic era was the age of the gymnosperms. However, the group which dominates the world today, the flowering plants, must have been evolving all through the Mesozoic era. Pieces of wood, leaf impressions, and pollen scattered through the geologic record as far back as the Triassic period seem to be angiosperm in nature. But the first uncontestable appearance of fossil angiosperms is in the Cretaceous period.

The Cretaceous period was a third interval of rapid evolution, undoubtedly because of major climatic and geologic changes that took place. According to the best evidence we have today—and much of it has only been collected in the past 10 years—the world's continents were once connected into a single land mass near the equator. During the Triassic period, this primeval mass, called Pangaea, began to separate. North and South America split from Europe and Africa, and the Atlantic Ocean formed; Australia and India and Antarctica separated, and India rammed into the Asian continent; Saudi Arabia split from Africa, and the Red Sea was formed. These movements did not happen all at once, but at the rate of only a few centimeters each year, and they continue today (Fig. 30.22). During the early Cretaceous period, then, land areas became distributed over all lines of latitude, so plants were exposed to different climatic regimes. In addition, the once low-lying continents with great inland seas began to rise and form large, dry continental areas. The climate cooled.

Two major biological changes that accompanied these great continental and climatic shifts during the close of the Cretaceous period were (a) a spread of the flowering plants, and (b) an extinction of the

Figure 30.20 Reconstructions of some now-extinct gymnosperms. *A*, member of the Cordaitales, 30–100 ft tall; *B* and *C*, members of the Bennetitales (*C* was about 6 ft tall and *B* is drawn to the same scale as *C*). (Redrawn from A. S. Foster and E. M. Gifford, *Comparative Morphology of Vascular Plants*, W. H. Freeman, San Francisco; also from A. and C. Black, *Studies in Fossil Botany*, London Press; also from C. A. Arnold, *An Introduction to Paleobotany*, McGraw-Hill, New York.

dinosaurs. A number of other changes in the plant world accompanied all this (for example, the extinction or near-extinction of several gymnosperm groups), but the rise of the angiosperms was the greatest change. Where had they come from, and why did they evolve so quickly? We may never know the answers. At present, one leading theory is that flowering plants evolved over a long period of time. This took place in tropical uplands where fossil preservation in sediments is rare. These early angio-

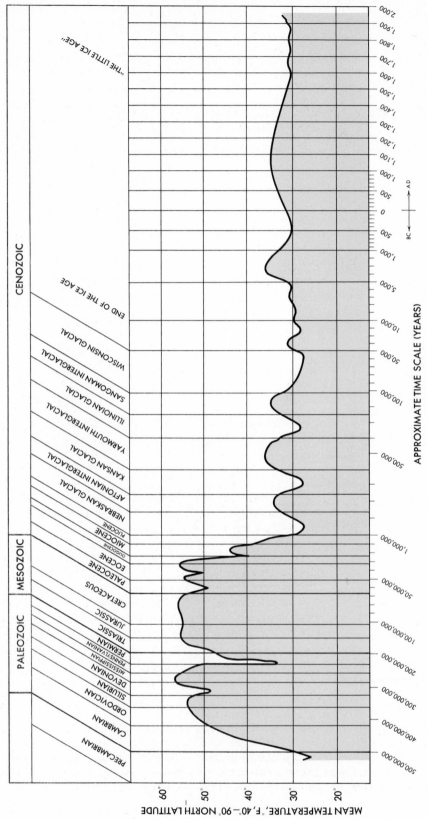

Figure 30.21 Average temperature during geologic time for the earth at 40–90° north latitude. The time scale is distorted to show more detail for the last 1 million years. (Redrawn from E. Dorf, *Amer. Sci.* **48**, 341–364, 1960.)

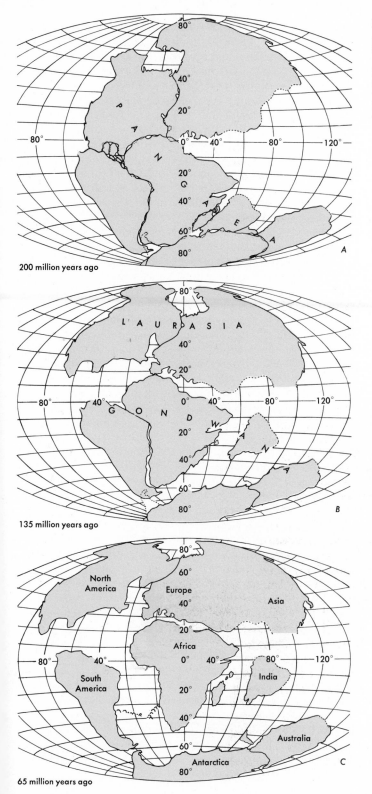

200 million years ago

135 million years ago

65 million years ago

Figure 30.22 Continental drift. *A*, position of the continents in the lower Triassic period; *B*, lower Cretaceous period; *C*, Paleocene epoch. (Redrawn from R. S. Dietz and J. C. Holden, *in* J. T. Wilson, *Continents Adrift*, W. H. Freeman, San Francisco.)

sperms may have occupied warm, seasonally wet, rocky slopes with great variations in microhabitats due to differences in exposure, elevation, drainage, and soil type. This environmental variability could have been a stimulus to evolution. As the land lifted and Cretaceous seas withdrew, new lowlands could have been rapidly invaded by diverse forms of flowering plants which had evolved on mountain slopes. By the time the **Cenozoic** era began, 65 million years ago, the conifers dominated only the cold temperate and polar regions, and to the angiosperms went everything else.

Climatic Change in the Cenozoic Era

You have already seen in Chapter 18 how fossil assemblages of plants can be used to estimate past climate. Quite simply, the theory is that the present is the key to the past. Where one finds modern relatives of fossil plants today, one finds a climate which must have been similar to the one at the time and place where the fossils were deposited. The most complete record of fossil plants for western North America during the Cenozoic is in the John Day Basin of eastern Oregon. If the present is indeed the key to the past, this fossil record shows a cooling and drying climatic trend that has continued to the present.

In the Eocene epoch, the plant community of the John Day area contained cinnamon, fig, cycads, and tropical ferns now found in the mountain forests of Central America, with a rainfall of 1500 mm or more and no frost (Fig. 30.23A). Leaves were large, with entire margins. Twenty million years later, in the Oligocene epoch, there was a change to mixed conifer-hardwood forest with birch, alder, oak, redwood, elm, sycamore, beech, and others. Leaves had become smaller, with dentate or convoluted margins. We do not find quite such a mixture anywhere today, but close approximations exist along the cool, wet California coast or in the Smoky Mountains of Tennessee (Fig. 30.23B). Rainfall was still high at that time, about 1250 mm, but the climate had grown cooler. Ten million years later, in the Miocene epoch, there was a shift to deciduous trees, such as oak, hickory, and maple. This indicates a climate like modern Ohio, with 1000 mm of rainfall and prolonged freezing temperatures in winter (Fig. 20.23C). By the Pliocene epoch, the climate became so cold that conifers (fir, spruce, pine) were present along with the deciduous hardwoods. Some leaves were quite small and hard. Today, the area is dominated by sagebrush. Trees are absent except along watercourses, and rainfall is about 250 mm per year (Fig. 30.23D).

The change in temperature with latitude that we have today apparently did not exist at the beginning of the Cenozoic era. Looking at fossil deposits of marine plankton, it is possible to make estimates of ocean surface temperatures from the equator to the north pole at past times. In the mid-Cretaceous period, the difference from equator to pole was only 11°C, but by the end of the Cretaceous period, it had grown to 15°C. The difference continued to grow during the Cenozoic era. We can guess that similar temperature changes occurred on land. In the past, the climate over the globe was broadly zoned and without temperature extremes, but today many climatic zones exist between the pole and the equator. Fig. 30.24 shows how these zones may have formed and shifted during the last 60 million years. During Pleistocene glaciation, the temperature gradient was at a peak.

The Pleistocene **Ice Age**, which ended only a moment in geologic time ago—12,000 years or so—produced another period of extinctions. Glaciated areas were scraped clean of plants, and they are still today being slowly revegetated. The pattern of change in the north is not uniform, however. There is evidence from central Canada, for example, that timberline there is actually being pushed south and the tundra is advancing at a rate of 250 miles in the past 900 years; but in other areas of the world, glaciers are still retreating from a "minor ice age" in the fifteenth century. Pollen diagrams, such as the one shown in Chapter 18, are good illustrations of climatic and vegetation changes that have continued through the past 12,000 years.

Many large mammals disappeared from North America during glaciation. They include the mastodon, mammoth, giant ground sloth, llama, peccary, giant armadillo, sabertooth tiger, ox, yak, and horse. Some of these survived in other parts of the world, but many became extinct. A debate has raged over the cause of the extinctions, one side attributing them to climatic change and stress, another side attributing them to the hunting prowess of emerging man. The climatic theory may be the better choice because equally massive extinctions occurred in Permian and Cretaceous times without the help of man.

In the Recent period, conifers have become more and more restricted in the land area they dominate. The flowering plants, however, continue to evolve, especially in harshly cold or dry environments. Today, only two groups of vascular plants appear to be expanding in terms of diversity and abundance: ferns and angiosperms. Both have survived great physical changes of the earth, yet their evolution seems to have been stimulated by these stresses, while many other forms have become extinct or survive as tenuous remnants.

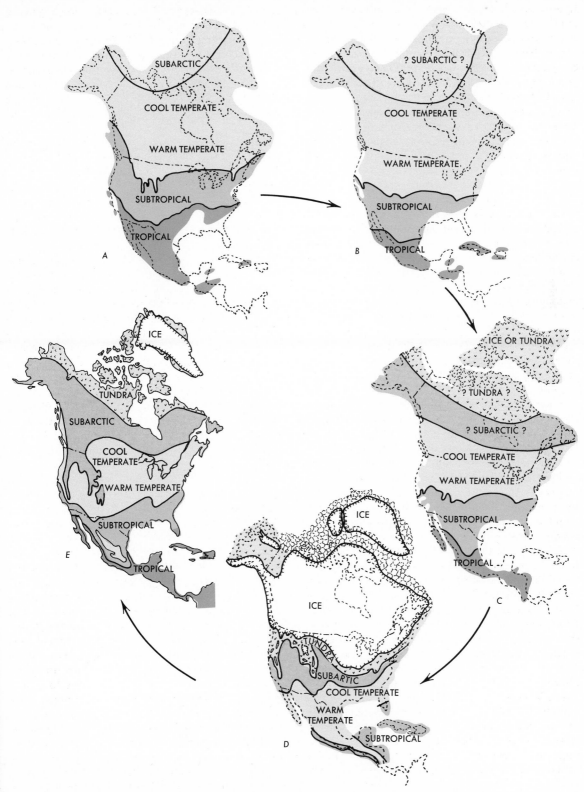

Figure 30.24 Climatic and vegetation zones in North America during the past 40 million years. *A*, Oligocene; *B*, Miocene; *C*, Pliocene; *D*, at the peak of the Pleistocene Ice Age; *E*, present. (Redrawn from E. Dorf, *Amer. Sci.* **48**, 341–364, 1960.)

Man himself began to play a role in the evolution and distribution of plants by the end of the Ice Age. Although the earliest archeological evidence for seed agriculture (cultivation of annuals such as grain, beans, squash) takes us back to 9000 BC, man may have been cultivating root crops (perennials propagated by cuttings and harvested for starch in the "roots" of cassava, sweet potato, and taro) in southern Asia as long ago as 13,000 BC. Some of these root crops have been asexually propagated for so many years by man that they have lost or nearly lost the capacity for sexual reproduction, and it is doubtful that they could survive in nature. Annuals have also been selected for productivity, rather than natural survival, and the result is they have become so changed from their wild relatives that it is difficult to determine where, and from what stock, they were first domesticated. Man has also accidentally domesticated and favored the evolution of certain other plants, the weeds. These are plants which grow well in disturbed or trampled soil, in waste areas rich in nitrogen, or in cropland. Some have evolved seeds which imitate the size of crop seeds, so they may not be separated during the threshing or sieving of crop seeds. When the next season's crop is sown, the weeds are sown inadvertently right along with it.

In our travels, we have taken these domesticated plants with us. Some strains of crop plants are so distinctive and peculiar to a given group of people, that their past migrations can be traced by determining the present location of the crop plants. This sort of evidence for ti plant, taro, and candlenut has been used to trace the movement of Polynesians from island to island in the South Pacific.

Weeds, of course, also are taken along inadvertently by moving people. In modern times, Europeans coming to the New World brought along both crops and weeds. As much as a third of the flora of some states is composed of weeds from various parts of the world, mainly Europe. The accelerated pace of land disturbance, which opens the way for weeds, combined with the planting of the same crops and ornamentals everywhere, are making the floras of the world increasingly similar.

Charles Darwin and Evolution

Many biologists since the time of the Greek philosophers have attempted to explain the mechanism of evolution. Great progress has been made during the nineteenth and twentieth centuries. Our better understanding was made possible chiefly because the work of Charles Darwin (see later pages in this chapter) and Gregor Mendel provided a basis that enabled biologists to approach the problem of evolution in an entirely new light.

Darwin, as a result of many years of careful study and observation of a large number of plants and animals, laid emphasis upon the following points:

1. Numbers of plants or animals may increase in a geometric ratio. For example, a given plant may produce 1000 seeds. If each seed grows into a new plant and each new plant produces 1000 seeds, 1 million new plants could result. A third generation from the one original plant would result in 1 billion plants. It may be seen from this example that plants (and animals also) potentially could increase in numbers at a tremendous rate.
2. Actually, the number of individuals of a given plant or animal remains fairly constant. There is no such tremendous increase in the number of individuals as seed production seemingly makes possible.
3. No two individual plants or animals are identical; there is variation.

Reasoning from these observations, Darwin arrived at these conclusions:

1. Any given population is usually able to reproduce many more young individuals than can adequately be raised in the region it occupies. Therefore, a struggle for existence occurs among the individuals.
2. In the struggle for existence, only those individuals survive which, because of their particular variation, are best adapted to their immediate environment. Thus, a natural selection takes place; the unfit do not survive.
3. These selected variations may be inherited, that is passed from one generation to another, and thus may gradually give rise to new species.

The Sources of Variation

Since variation plays such an important part in plant evolution, it will be well to examine the sources of variation a little more closely.

One source of variation is **mutation,** the spontaneous transformation of a gene (Fig. 30.25). Mutation does not create new genetic material (new DNA), it simply changes the arrangement of the existing material so that enzymes and other proteins coded for by the DNA are no longer quite the same. A mutated gene can back-mutate to the original condition, as well. A number of environmental factors, such as radiation, can cause a mutation; but most mutations seem to result from cellular "mistakes" in copying the DNA molecule during cell division.

Figure 30.25 Examples of mutations. *A*, common wild sunflower; *B*, mutant sunflower known as sun gold; *C*, variation in the seed coats of beans. (*A* and *B*, courtesy of C. Heiser; *C*, courtesy of F. Smith.)

Mutations are a universal fact of life. They are known to occur in every plant and animal that has been studied. This does not mean that a given gene is relatively unstable and is likely to mutate very often. Rates of mutation for an individual gene, or locus, probably average one for every 100,000 cells. The mutation rate does vary from gene to gene, however, as shown for corn (Table 30.2), where one gene may have a mutation rate 500 times that of another.

However, one mutation in every 100,000 copies is still a small input of variation. Is it significant enough

to account for major evolutionary change through time? Is it significant enough to create a new, quite different species, from a preexisting one? If we do a little multiplying, the answer is, yes. Consider that, although a single gene may mutate only once every 100,000 copies, there are probably 10,000 genes in one plant. Each one can mutate independently of the others, so the chance of one gamete, or sex cell, containing any mutant gene is reduced to one in ten (1/100,000 genes mutated × 10,000 genes). Most mutations are not beneficial and might be so detrimental that they would not be passed on to future

TABLE 30.2   Mutation Rates of Different Genes in Corn (*Zea*)

Gene	Number of mutations per 1,000,000 gametes
Seed color, not purple	492
Seed color inhibitor	106
Purple seed color	11
Sugary seed	2.4
Shrunken seed	1.2
Waxy seed	less than 0.1

generations. Let us assume that only one mutation out of a thousand is beneficial. This means that one gamete in 10,000 has a beneficial mutation (1/10 gametes with any mutation × 1/1000 beneficial gene mutations).

How many gametes, or plants in a species? Let us give one gamete to each plant, and estimate an average species size of 1,000,000 plants. The gametes fuse and produce a new generation. This new generation will then have 100 new, useful mutations scattered among its members (1/10,000 beneficial mutants per gamete × 1,000,000 gametes, or parents). A few mutations will be identical and a few will appear in a single individual, but most will be unique and scattered through the population. The rate of mutation will be repeated each generation. In 100 generations, the total number of new mutations

could be as high as 10,000—the same number of mutations as original genes. Of course, some mutants will back-mutate, and some mutations will be identical, so there will be somewhat fewer than 10,000 separate mutations. How long a time is 100 generations? For an annual plant, like sunflower, it is only 100 years; for a large tree, it might be 30,000 years. Even 30,000 years is well within the time limits of species evolution as revealed in the fossil record.

But can mutation provide enough variation to account for the great differences between larger taxonomic groups? Great discontinuities seem to exist between families, classes, and divisions of plants, and some—like the flowering plants—seem to have appeared in the fossil record very suddenly. Are there two kinds of evolution: (*a*) **microevolution,** providing minor changes from one species to another within the same genus; and (*b*) **megaevolution,** providing major, rapid change between quite unlike forms (Fig. 30.26)? The answer may be, no. There is only one path of evolution—microevolution—and it alone is sufficient to account for all the diversity present on earth today. Discontinuities are more apparent than real. The existence of several classification schemes for the plant kingdom supports this view.

As we pointed out in Chapter 2, our classification schemes of the plant kingdom still remain rather arbitrary. That is, boundaries between divisions, classes, and even families are artificial, and exactly where the lines are drawn depends on the botanist. If all the categories were "natural" and distinct, then every investigator would come to the same conclusions and would classify the plant kingdom the same way. Instead, we have a great number of classification schemes. The more study that is done, the more intermediate forms are found and the fuzzier the distinctions become between major groups. Only 20 years ago, primitive plants were found in southeast Asia with reproductive organs half-way between leaves and pistils or stamens. (Fig. 30.27). The fossil record also reveals intermediate forms. The extinction of the intermediates leaves us with apparent genetic discontinuities today, but these gaps are only artifacts of time. As the fossil record becomes better-known, "sudden" appearances of taxa will become less sudden. Angiosperm fossils, for example, are being found from earlier and earlier periods as collecting goes on.

Other sources of variation have been discussed in Chapter 17. These include chromosome aberrations such as deletions, duplications, inversions, translocations, and crossing-over. A final powerful source of variation is **recombination,** the reshuffling of chromosomes during sexual reproduction.

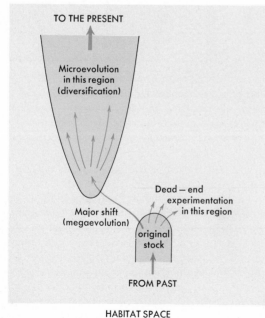

Figure 30.26   Megaevolution and microevolution.

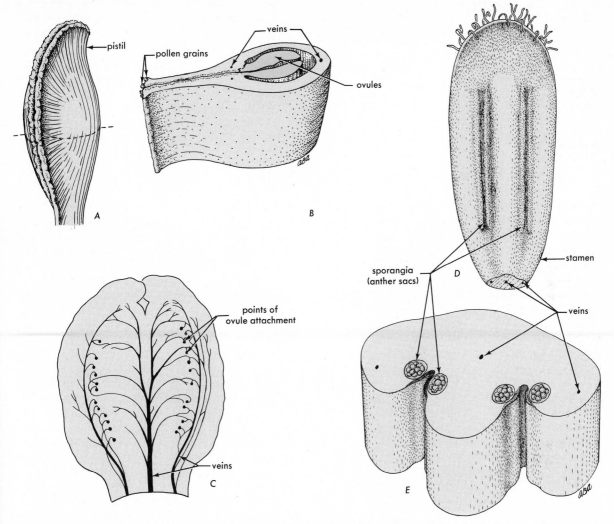

Figure 30.27 Primitive pistil and stamen from living plants in southeast Asia. *A*, pistil of *Drimys piperita*; *B*, cross-section of pistil (dotted line); *C*, pistil laid open; *D*, stamen of *Degeneria vitiensis*; *E*, cross-section of stamen. (Redrawn from A. S. Foster and E. M. Gifford, *Comparative Morphology of Vascular Plants*, W. H. Freeman, San Francisco; also from I. W. Bailey and A. C. Smith, *J. Arnold Arboretum* **23**, 256–265, 1942; also from J. E. Canright, *Amer. J. Botany* **39**, 484–497, 1952.)

The Role of Natural Selection

Mutation and recombination are the raw materials for evolution. They provide the variation among *individuals*. However, for an entire population of individuals in a species to progressively change into something new, this raw material is not enough. Some force, some pressure, must be exerted on the *population* so that the abundance of certain mutations becomes higher and higher, until all or most members of the species possess it. How does this happen? The presently accepted answer had its own

evolution, over the course of 150 years of debate and experimentation.

At the beginning of the nineteenth century, the French naturalist Jean Baptiste Lamarck (1744–1829) was a liberal—for his day—in his thoughts on evolution. Lamarck did not believe that all species had the same age, that all were created together at one time. Neither did he believe that species were constant; new ones were forming all the time as a result of changing environments. This stand was quite heretical, considering that Linnaeus 50 years earlier had practically founded "modern" taxonomy on the principles that there was only one time of creation, and that species from that time on were fixed and constant. Lamarck thought that new traits and new species could evolve from old ones if the species were placed under stress. For example, if a tall plant at sea level were transplanted to a severe, timberline habitat, the climate would stunt it. This stunted plant would shed seeds and the new seedlings would also be stunted, even if grown back at sea level. Charac-

ters acquired during the lifetime of one plant would be passed on to succeeeding generations.

Lamarck himself did no experimentation to bolster his theory. Other botanists did, however. Bonnier in France made reciprocal transplants of alpine and sea-level species. He grew a number of plants of each species to a convenient size at Paris, then cloned them (split them into pieces) and planted them in plots in the Alps, the Pyrenees, and Paris.

Figure 30.28   The dwarfing effect of high altitudes. *A, Prunella vulgaris* (heal-all) as it grows near sea level; *B*, the same species as it grows near 6000 ft elevation.

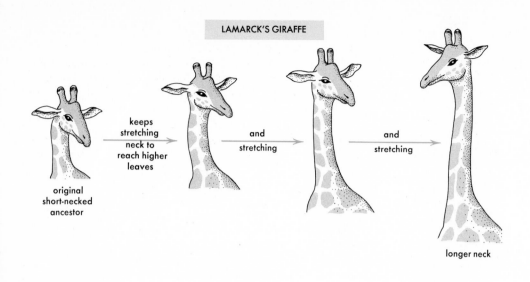

LAMARCK'S GIRAFFE

original
short-necked
ancestor

keeps
stretching
neck to
reach higher
leaves

and
stretching

and
stretching

longer neck

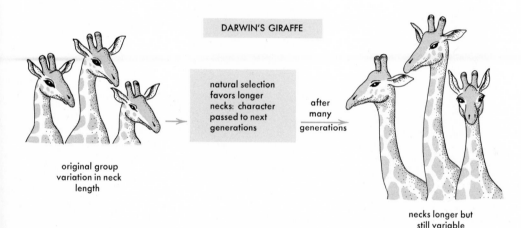

DARWIN'S GIRAFFE

original group
variation in neck
length

natural selection
favors longer
necks: character
passed to next
generations

after
many
generations

necks longer but
still variable

Figure 30.29 Comparison of Lamarck's and Darwin's concepts of evolution. (Redrawn from J. M. Savage, *Evolution*, Holt, Rinehart and Winston, New York.)

They were not planted in gardens, but were put in among natural vegetation. The plots were sometimes fenced, sometimes not; watering and fertilization were not practiced. The plots were periodically visited for a number of years (1884–1920) and any changes in the plants noted. He concluded that a number of lowland species were transformed into related alpine or subalpine species during the years of the study (Fig. 30.28). Also around 1920, the American ecologist Frederic Clements established a similar series of "gardens" from Pike's Peak to the California shore. He too concluded that some lowland species were transformed to related high-elevation species, and vice versa.

However, the most carefully documented and controlled transplant experiments were conducted last, between 1920 and 1940, by Clausen, Keck, and Hiesey in California (Chap. 19). Their detailed study

of some 60 taxonomically diverse species showed not one case of a lowland species becoming an alpine species, or the reverse. A lowland species at timberline might become dwarf or prostrate or stunted, but if its seeds were collected and sown back at sea level, normal tall plants would result. There had been no genetic change—only a temporary, plastic response to a harsh environment. Such temporary changes are called **phenotypic** changes, in contrast to **genotypic** changes which are genetically fixed and are passed on to offspring.

Darwin's concept of evolution, expressed over a hundred years ago, is the one accepted today; not Lamarck's. Its differences with Lamarck's are illustrated in Fig. 30.29. Basically, Darwin's concept is this: Variation exists in the initial population; an environmental stress places certain individuals at an advantage; because those individuals survive or re-

Charles Darwin and Evolution 639

produce more successfully, they leave more offspring which carry the same genetic traits; the abundance of the advantageous traits in this way increases in every generation, but variation still persists. This process of directed change is called **natural selection.**

How Species Remain Distinct

Evolution proceeds most rapidly in a population of organisms if it is "isolated" from other populations. But we mean a very special kind of isolation: **reproductive isolation.** If plants in one isolated population cannot breed with similar plants in other populations, then natural selection will produce a

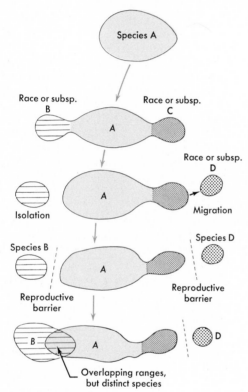

Figure 30.30 Diagram showing the sequence of events which leads to the formation of new races, subspecies, and species. One species (green, at top) extends into new habitats, and populations at the extremes become modified into races or subspecies. If the races or subspecies become isolated, they may evolve further from the original species and become incompatible with it; at this point, they are recognized as distinct species. The reproductive barrier will keep the species distinct, even if later migration brings them back into contact. (Redrawn from G. L. Stebbins, *Processes of Organic Evolution,* Prentice-Hall, Englewood Cliffs, N.J.)

TABLE 30.3   Summary of Some Important Reproductive Isolating Mechanisms

A. *Prezygotic mechanisms: prevention of pollination or fertilization*

1. Separation in space: two populations live far from each other so that pollen cannot be transferred, or they live close but are separated by barriers (mountains, bodies of water), or they occupy different habitats (lowlands versus uplands), or drifting continents move them apart

2. Separation in time: two populations flower at different times of the year, or if they flower at the same time, the pollen is shed before the stigmas are receptive

3. Biological separation: the pollinating insect or animal is different for each population, so that it is unlikely cross-pollination will occur, even if the plants are close to each other; or the flowers may open at different times of the day (morning versus evening)

4. Physiological differences: the pollen of one is unable to grow through the style of the other, or it grows more slowly than pollen of the other

B. *Postzygotic mechanisms: prevention of normal offspring development*

1. Incompatibility of zygotic or embryonic tissue with that of the mother plant produces seed abortion

2. Hybrids (F_1) are completely inviable, or considerably weakened

3. Hybrids are vigorous but sterile

4. F_1 is vigorous, but successive generations (F_2, etc.) are weak or sterile

distinctive species in the shortest amount of time. This species will remain distinct from all others, once it has formed (Fig. 30.30). Without isolation, crossbreeding would dilute the abundance of new genes that are of most value in one particular environment; all members of a species would continue to be more or less alike and not as well-adapted to extremes within their range, as they could be if isolated. Mediocrity—a population of generalists—would result. One large population of generalists is apparently not as successful a strategy for survival as several small populations of specialists.

How can reproductive isolation be established? Table 30.3 lists the most important isolating mechanisms. The prezygotic mechanisms have to do with preventing pollen of one population from reaching the stigmas of another. This can be accomplished by separating the populations in space (isolated valleys or separating continents), time (different seasons of flowering), or biology (different pollinating insects, which do not visit both flowers). On a world scale, continental drift may have been the most important isolating mechanism. Postzygotic mechanisms prevent normal offspring from developing, even though pollination occurs. The hybrids may be weak or sterile, for example, or the developing embryos can abort.

Hybrids are often weak in nature, or at least they are less well-adapted to a given habitat than either parent. So the most common type of hybrid in nature is not a "pure" hybrid, half-way between each parent. Instead, the successful hybrids are often the result of back-crossing between the original pure hybrid and either or both parental types (Fig. 30.31). The few pure hybrids in this way produce an entire series of partial hybrids between themselves and original parental types. This is called **introgressive hybridization,** and it creates great diversity (hybrid swarms) which may spell success for the group as a whole. It additionally creates headaches for taxonomists.

TABLE 30.4 Increasing Incidence of Polyploidy with Increasing Harshness of the Environment

Region	Polyploid species (% of all vascular plants)
West Africa (tropical)	26
Mediterranean shore	38
Great Britain	53
Iceland	66
Southern Greenland	71
Northern Greenland (edge of the tundra)	86

Hybridization may be especially difficult to achieve if two populations of a species have come to differ in chromosome number. Polyploidy, as explained in Chapter 17, is a multiplication of the chromosome number. It appears to be a common phenomenon among plants, and it seems to increase their tolerance of climatic extremes. Perhaps it is no mere coincidence that the two vascular groups showing most rapid evolution today, flowering plants and ferns, also show the highest incidence of polyploidy. Exactly how or why polyploidy increases tolerance of environmental extremes is not known, but the high incidence of polyploids in arctic and desert habitats has been well-documented (Table 30.4). It has been estimated that 25–33% of the angiosperms are polyploid. Polyploids include some important commercial crops (cotton, wheat, oats, potato, tobacco, alfalfa, many pasture grasses, apples, olives) and ornamentals (poplars, violets, asters, and many ferns).

Summary

A modern theory of the mechanism of evolution can be outlined as follows:

1. Genes, contained in the chromosomes, are largely responsible for the development, structure, and metabolism of plants and animals.
2. The complement of genes does not remain absolutely constant. Mutations (changes in chromosomes and genes) occur, which modify the structure and metabolism of the individuals containing them.
3. Mutations causing considerable phenotypic change are likely to kill or to greatly weaken the plant because they upset the delicate equilibrium existing between the plant and its environment.
4. Hybrids, because of the resulting new combina-

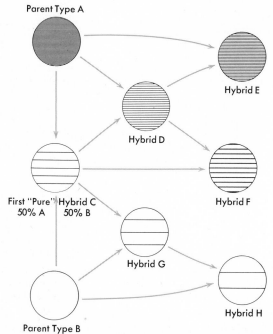

Figure 30.31 Introgressive hybridization. The degree of shading (horizontal lines in the circles) shows the degree of similarity of hybrids to parental types A and B. Introgressive hybrids D, E, F, G, and H would be more numerous in nature than "pure" hybrid C.

tions of genes, differ from their parents. Hybrid swarms, resulting from introgressive hybridization, are common.

5. If the variants produced by mutations or hybridization are better adapted to the environment than the parent plants, the parent plants may be replaced by the new forms. Many complicated factors are involved in this replacement.

6. As mountains are elevated or eroded, as glaciers advance or recede, as the climate of the earth becomes more or less arid, or as the habitats of plants are changed in other ways, the plants best adapted to the new environment replace the forms less well-adapted.

7. Evolution thus results from slow changes in the earth's surface, variation in plants and animals, and an adjustment between the changes in the earth and changes in the living organisms.

8. In summarizing evolutionary mechanisms, Dobzhansky, writing in *Scientific American*, has said, "Evolution is a creative response of the living matter to the challenges of the environment. Evolution is due neither to chance nor to design; it is due to a natural creative process."

Glossary

Abbreviation	Meaning
A.S.	Anglo-Saxon
D.	Dutch
dim.	diminutive
F.	French
Gr.	Greek
It.	Italian
L.	Latin
Lapp.	Lappland
M.E.	Medieval English
M.L.	Medieval Latin
N.L.	New Latin
O.F.	Old French
O.E.	Old English
R.	Russian
Sp.	Spanish

The glossary serves two purposes, (a) to define terms used in the text and (b) to give their derivation. It should be noted that in carrying out the first purpose, with few exceptions, only the meanings actually used in the text are given. Reference to dictionaries will give additional meanings, while other textbooks may define the same words in slightly different ways.

Scientific language and slang have much in common. They are both vital, growing, and changing phases of modern English. The same words will have different meanings in different parts of the country. Both slang and scientific vocabularies contribute their full share of standard everyday English words. And finally, both vocabularies are largely unintelligible to those not fully initiated into their intricacies.

Scientific language grows in a number of ways. Some botanical terms, such as "seed," are used by scientist and layman alike and can be traced back to Anglo-Saxon days, when the word *sed* was used to designate anything sown in the ground, or, in our terms, a seed. In other instances, words from Greek or Latin have been taken into the scientific language with their identical original meaning; for instance, the Greeks called wood *xylon,* and we use this term, changed to xylem, to designate the woody conducting tissue of plants. Other words have a more complex history. *Metabolos* is the Greek word "to change." When the scientists of Europe wrote in Latin they took this Greek word, made a Medieval Latin word from it, and used it to describe the changes undergone by some insects. It was used in 1639 to mean changes in health, and in 1845 the

German scientist who elaborated the cell doctrine made a German word of it and first used it as we use it here, designating the changes taking place within a cell. Perhaps the newest scientific word in this text is *eukaryote*. Look it up.

In using the glossary, pay as much attention to derivation as you do to the definition. To memorize the definition alone is to learn, parrot-fashion, only one word. To learn the derivation is to understand the word and possibly to introduce yourself to a whole family of new words. Pay particular attention to such combining forms as *hetero, auto, micro, phyll, angio, plast* or *plasm, spore,* and the like.

A, abbreviation for an **angstrom,** 0.0001 of a micron; there are 10,000,000 angstroms in a millimeter, 10 in a millimicron, and 10 in a nanometer

ABA, abscisic acid

Abscisic acid, a hormone variously inducing abscission, dormancy, stomatal closure, growth inhibition, and other responses in plants

Abscission zone (L. *abscissus,* cut off), zone of delicate thin-walled cells extending across the base of a petiole, the breakdown of which disjoins the leaf or fruit from the stem

Absorb (L. *ab,* away + *sorbere,* to suck in), to suck up, to drink up, or to take in; in plant cells materials are taken in (absorbed) in solution

Absorption spectrum, a graph relating the ability of a substance to absorb light of various wavelengths

Accessory (M. L. *accessorius,* an additional appendage), something aiding or contributing in a secondary way, such as buds in addition to the main axillary bud

Achene, simple, dry, one-seeded indehiscent fruit, with seed attached to ovary wall at one point only

Acid (F. *acide,* from L. *acidus,* sharp), a substance belonging to a class of which the commonest and most typical members are sour, and are compounds of hydrogen with another element or elements; oxygen is generally the third element; examples of inorganic acids are sulfuric acid, (H_2SO_4) and hydrochloric acid (HCl); examples of organic acids are acetic acid (vinegar) and pyruvic acid

Actinomorphic (Gr. *aktis,* ray + *morphe,* form), said of flowers of a regular or star pattern, capable of bisection in two or more planes into similar halves

Action spectrum (F. *acte,* a thing done), a graph relating the degree of physiological response (e.g., phototropism, photosynthesis) caused by different wavelengths of light

Active solute absorption, the intake of dissolved materials by cells against a concentration gradient and requiring an expenditure of energy

Adaptation (L. *ad,* to + *aptare,* to fit), adjustment of an organism to the environment

Adenine, a purine base present in DNA, RNA, and nucleotides such as ADP and ATP

Adhesion (L. *adhaerere,* to stick to), a sticking together of unlike things or materials

Adnation (L. *adnasci,* to grow to), in flowers, the growing together of two or more whorls to a greater or less extent; compare adhesion

ADP, adenosine diphosphate

Adsorption (L. *ad,* to + *sorbere,* to suck in), the concentration of molecules or ions of a substance at a surface or an interface (boundary) between two substances

Advanced (M.E. *advaunce,* to forward), said of a taxonomic trait thought to have evolved late in time from some more primitive trait

Adventitious (L. *adventicius,* not properly belonging to), referring to a structure arising from an unusual place: buds at other places than leaf axils, roots growing from stems or leaves

Aeciospore (Gr. *aikia,* injury + spore), one of the binucleate asexual spores of rust fungi formed as a result of the sexual fusion of cells but not of nuclei

Aecium (plural, **aecia**) (Gr. *aikia,* injury), in rust, a sorus that produces aeciospores

Aerate, to supply or impregnate with common air, such as by bubbling air through a culture solution

Aerobe (Gr. *aer,* air + *bios,* life), an organism living in the presence of oxygen

Aerobic respiration, respiration involving molecular oxygen

Agar (Malay *agaragar*), a gelatinous substance obtained mainly from certain species of red algae

Aggregate fruit (L. *ad,* to + *gregare,* to collect; to bring together), a fruit developing from the several separate carpels of a single flower; e.g., a strawberry.

Alcohol (M.L. from Arabic *al-kuhl,* a powder for painting eyelids; later applied, in Europe, to distilled spirits which were unknown in Arabia), a product of the distillation of wine or malt; any one of a class of compounds analogous to common alcohol; the ending **ol** designates a member of this class of compounds

Aleurone (Gr. *aleuron,* flour), proteinaceous material, usually in the form of small grains, occurring in the outermost cell layer of the endosperm of wheat and other grains

Alga (plural, **algae**) (L. *alga,* seaweed), a member of the large group of thallus plants containing chlorophyll and thus able to synthesize carbohydrates

Algin, a long-chain polymer of mannuronic acid found in the cell walls of the brown algae

Alkali (Arabic *alqili,* the ashes of the plant saltwort), a substance with marked basic properties

Allele (Gr. *allelon,* of one another, mutually each other), either of a pair of contrasting Mendelian characters; multiple-, several alleles at a single locus; pseudo-, factors that recombine within a single locus

Allopolyploidy (Gr. *allos,* other + polyploidy), ploidy in which sets of chromosomes are derived from different parents, generally not closely related

Alpine (L. *Alpes,* the Alps Mountains), meadow-like vegetation at high elevation, above tree line

Alternate, referring to bud or leaf arrangement in which there is one bud or one leaf at a node

Amensalism (L. *a,* not + *mensa,* table), a form of biological interaction in which one organism is inhibited by another, but the other is neither inhibited nor stimulated

Amino acid (Gr. *Ammon,* from the Egyptian sun god, in N.L. used in connection with ammonium salts), an acid containing the group $-NH_2$; one of the building blocks of a protein

Ammonification (*Ammon,* Egyptian sun god, near whose temple ammonium salts were first prepared from camel dung + L. *facere,* to make), decomposition of amino acids, resulting in the production of ammonia

Anaerobe (Gr. *a,* without + *aer,* air + *bios,* life), an organism able to respire in the absence of free oxygen, or in greatly reduced concentrations of free oxygen

Anaphase (Gr. *ana,* up + *phais,* appearance), that stage in mitosis in which half chromosomes or sister chromatids move to opposite poles of the cell

Androecium (Gr. *andros,* man + *oikos,* house), the aggregate of stamens in the flower of a seed plant

Aneuploid (Gr. *aneu,* without + *ploid*), the condition in which the number of chromosomes differs from the normal by less than a full set (*n*), e.g., $3n + 1$, $2n - 1$, $2n + 4$; compare Polyploid

Angiosperm (Gr. *angion,* a vessel + *sperma* from *speirein,* to sow, hence a seed or germ), literally a seed borne in a vessel, thus a group of plants whose seeds are borne within a matured ovary

Angstrom (after A. J. Angstrom, a Swiss physicist, 1814–1874), a unit of length equal to 0.0001 of a micron (0.1 nanometer), 10,000 angstroms = 1 micron = 0.001 millimeter

Anisogamy (Gr. *an,* prefix meaning not + *isos,* equal + *gamete,* spouse), the condition in which the gametes, though similar in appearance, are not identical

Annual (L. *annualis,* within year), a plant that com-pletes its life cycle within 1 year and then dies

Annular vessels (L. *annularis,* a ring), vessels with lignified rings

Annulus (L. *anulus* or *annulus,* a ring), in ferns, a row of specialized cells in a sporangium, of importance in opening of the sporangium; in mosses, thick-walled cells along the rim of the sporangium to which the peristome is attached

Anther (M.L. *anthera*—from the Gr. *anthros,* meaning flower—a medicine extracted from the internal whorls of flowers by medieval pharmacists; confined to pollen-producing parts by herbalists in 1700), pollen-bearing portion of stamen

Antheridium (anther + *-idion,* a Gr. dim, ending, thus a little anther), male gametangium or sperm-bearing organ of plants other than seed plants

Anthocyanin (Gr. *anthros,* a flower + *kyanos,* dark blue), a blue, purple, or red vacuolar pigment

Antibiotic (Gr. *anti,* against or opposite + *biotikos,* pertaining to life), a natural organic substance which retards or prevents the growth of organisms; generally used to designate substances formed by microorganisms that prevent growth of other microorganisms

Antibody (Gr. *anti,* against + body), a protein produced in an organism, in response by the organism to a contact with a foreign substance, and having the ability of specifically reacting with the foreign substance

Antipodal (Gr. *anti,* opposite + *pous,* foot), generally, referring to one object situated opposite another object. Specifically, referring to cells or nuclei at the end of the embryo sac opposite that of the egg apparatus

Apex (L. *apex,* a tip, point, or extremity), the tip, point, or angular summit of anything: the tip of a leaf; that portion of a root or shoot containing apical and primary meristems

Apical dominance, the inhibition of lateral buds or meristems by the apical meristem

Apical meristem, a mass of meristematic cells at the very tip of a shoot or root

Apomixis (Gr. *apo,* away from + *mixis,* a mingling), the production of offspring in the usual sexual structures without the mingling and segregation of chromosomes

Apothecium (Gr. *apotheke,* a storehouse), a cup-shaped or saucer-shaped open ascocarp

Archegonium (L. dim. of Gr. *archegonos,* literally a little founder of a race), female gametangium or egg-bearing organ, in which the egg is protected by a jacket of sterile cells

Aril (M.L. *arillus,* a wrapper for a seed), an accessory seed-covering formed by an outgrowth at the base of the ovule in *Taxus*

Ascocarp (Gr. *askos,* a bag + *karpos,* fruit), a fruiting body of the Ascomycetes, generally either an open cup, a vessel, or closed sphere lined with special cells called asci (*see* Ascus)

Ascogenous hyphae, hyphae arising from the ascogonium, after the formation of *n* + *n* paired nuclei; the hymenial layer of the ascocarp develops from the ascogenous hyphae

Ascogonium, the oogonium or female gametangium of the Ascomycetes

Ascomycetes (Gr. *askos,* a bag + *mykes,* fungus), a large group of true fungi with septate hyphae producing large numbers of asexual conidiospores and meiospores called ascospores, the latter in asci

Ascospore (Gr. *askos,* a bag + spore), meiospore produced within an ascus

Ascus (plural, **asci**) (Gr. *askos,* a bag), a specialized cell, characteristic of the Ascomycetes, in which two haploid nuclei fuse, immediately after which three (generally) divisions occur, two of which constitute meiosis, resulting in eight ascospores still contained within the ascus

Asexual (Gr. *a,* without + L. *sexualis,* sexual), any type of reproduction not involving the union of gametes or meiosis

Aspect (L. *aspectus,* appearance), the direction of slope of a surface, as a hillside with a south-facing aspect

Assimilation (L. *assimilare,* to make like), the transformation of food into protoplasm

Atoms (F. *atome,* from the Gr. *atomos,* indivisible), the smallest particles in which the elements combine either with themselves or with other elements, and thus the smallest quantity of matter known to possess the properties of a particular element; Atoms were first postulated by the Greek philosopher–scientists Leucippus and Democritus about 450 BC

ATP, adenosine triphosphate, a high-energy organic phosphate of great importance in energy transfer in cellular reactions

Auricles (L. *auricula,* dim. of *auris,* ear), ear-like structures; in grasses, small projections that grow out from the opposite side of the sheath at its upper end where it joins the blade

Autoecious (Gr. *auto,* self + *oikia,* dwelling), having a complete life cycle on the same host

Autoradiograph (Gr. *auto,* self + L. *radiolus,* a ray + Gr. *graphe,* a painting), a photographic print made by a radioactive substance acting upon a sensitive photographic film

Autotetraploidy, the condition in which the doubling of the chromosome number occurs in one cell or between cells on the same plant

Autotrophic (Gr. *auto,* self + *trophein,* to nourish with food), pertaining to a plant that is able to manufacture its own food

Auxin (Gr. *auxein,* to increase), a plant growth-regulating substance regulating cell elongation

Axil (Gr. *axilla,* armpit), the upper angle between a petiole of a leaf and the stem from which it grows

Axillary bud, a bud formed in the axil of a leaf

Bacillus (L. *baculum,* a stick), a rod-shaped bacterium

Bacteria (Gr. *bakterion,* a stick), common name for the class Schizomycetes

Bacteriology (bacteria + Gr. *logos,* discourse), the science of bacteria

Bacteriophage (bacteria + Gr. *phagein,* to eat), literally, an eater of bacteria; a virus that infects specific bacteria, multiples therein, and usually dissolves the bacterial cells

Banner (M.L. *bandum,* a standard), large, broad, and conspicuous petal of legume type of flower

Bark (Swedish *bark,* rind), the external group of tissues, from the cambium outward, of a woody stem or root

Base pair, the nitrogen bases that pair in the DNA molecule, adenine with thymine, and guanine with cytosine

Basidiomycetes (M.L. *basidium,* a little pedestal + Gr. *mykes,* fungus), group of fungi characterized by the production of meiospores on special cells, the basidia

Basidiospore (M.L. *basidium,* a little pedestal + spore), type of meiospore borne by basidia in the Basidiomycetes

Basidium (plural, **basidia**) (M.L. *basidium,* a little pedestal), a specialized reproductive cell of the Basidiomycetes in which nuclei fuse and meiosis occurs. It may be a special club-shaped cell, a short filamentous cell, or a short four-celled filament

Benthon (Gr. *benthos,* the depths of the sea), attached aquatic plants and animals, collectively

Berry, a simple fleshy fruit, the ovary wall fleshy and including one or more carpels and seeds

Biennial (L. *biennium,* a period of 2 years), a plant which completes its life cycle within 2 years and then dies

Bifacial leaf (L. *bis,* twice + *facies,* face), a leaf having upper and lower surfaces distinctly different

Binomial (L. *binominis,* two names), two-named; in biology each species is generally indicated by two names, the genus to which it belongs and its own species name

Bioassay (Gr. *bios,* life + L. *exagere,* to weigh or test), to test for the presence or quantity of a substance by using an organism's response as an indicator

Biology (Gr. *bios*, life + *logos*, word, speech, discourse), the science that deals with living things

Biosystematics (Gr. *bios*, life + *synistanai*, to place together), a field of taxonomy which emphasizes breeding behavior and chromosome characteristics

Biotic (Gr. *biokitos*, relating to life), relating to life

Biotin, a vitamin of the B complex

Bordered pit, a pit in tracheid or other cell of secondary xylem having a distinct rim of the cell wall overarching the pit membrane

Botany (Gr. *botane*, plant, herb), the science dealing with plant life

Bract (L. *bractea*, a thin plate of precious metal), a modified leaf, from the axil of which arises a flower or an inflorescence

Bud (M.E. *budde*, bud), an undeveloped shoot, largely meristematic tissue, generally protected by modified scale-leaves

Bud scale, a modified protective leaf of a bud

Bud scar, a scar left on a twig when the bud or bud scales fall away

Bulb (L. *bulbus*, a modified bud, usually underground), a short, flattened, or disk-shaped underground stem, with many fleshy scale-leaves filled with stored food

Bundle scar, scar left where conducting strands passing out of the stem into the leaf stalk were broken off when the leaf fell

C$_3$ cycle, the Calvin–Benson cycle of photosynthesis, in which the first products after CO_2 fixation are three-carbon molecules

C$_4$ cycle, the Hatch–Slack cycle of photosynthesis, in which the first products after CO_2 fixation are four-carbon molecules

Cactus (Gr. *kaktos*, a kind of prickly plant), any plant of the cactus family, famous plants of the desert

Callose (L. *callum*, thick skin + *ose*, a suffix indicating a carbohydrate), an amorphous, hardened carbohydrate constituent of cell walls, commonly developing upon injury

Callus (L. *callum*, thick skin), mass of large, thin-walled cells, usually developed as the result of wounding

Calyptra (Gr. *kalyptra*, a veil, covering), in bryophytes, an envelope covering the developing sporophyte, formed by growth of the venter of the archegonium

Calyx (Gr. *kalyx*, a husk, cup), sepals collectively; outermost flower whorl

Cambium (L. *cambium*, one of the alimentary body fluids supposed to nourish the body organs), a layer, usually regarded as one or two cells thick, of persistently meristematic tissue, giving rise to secondary tissues, resulting in growth in diameter

Canopy (Gr. *kanopeion*, a cover over a bed to keep off gnats), the leafy portion of a tree or shrub

Capillaries (L. *capillus*, hair), very small spaces, or very fine bores in a tube

Caprification, an artificial method of pollinating certain cultivated varieties of figs

Capsule (L. *capsula*, dim. of *capsa*, a case), a simple, dry, dehiscent fruit, with two or more carpels

Carbohydrate (chemical combining forms, *carbo*, carbon + *hydrate*, containing water), a food composed of carbon, hydrogen, and oxygen, with the hydrogen and oxygen frequently in a 2:1 ratio, as in water, H_2O

Carboxydismutase (F. *carbone*, carbon + Gr. *oxys*, acidic + *dismutation*, reduction of one metabolite by another), the enzyme responsible for the fixation of inorganic CO_2 into organic compounds in the dark reactions of the C$_3$ photosynthesis cycle; also called ribulose diphosphate carboxylase

Carotene (L. *carota*, carrot), a reddish-orange plastid pigment

Carpel (Gr. *karpos*, fruit), a floral leaf bearing ovules along the margins

Carpogonium (Gr. *karpos*, fruit + *gonos*, producing), female gametangium (in red algae)

Carpospore (Gr. *karpos*, fruit + spore), one of the spores produced in a carpogonium

Caruncle (L. *caruncula*, dim. of *caro*, flesh, wart), a spongy outgrowth of the seed coat, especially prominent in the castor bean seed

Caryopsis (Gr. *karyon*, a nut + *opsis*, appearance), a simple, dry, one-seeded, indehiscent fruit, with pericarp firmly united all around to the seed coat

Catalyst (Gr. *katelyein*, to dissolve), a substance that accelerates a chemical reaction but which is not used up in the reaction

Cation exchange (Gr. *kata*, downward), the replacement of one positive ion (cation) by another, as on a negatively charged clay particle

Catkin (literally a kitten, apparently first used in 1578 to describe the inflorescence of the pussy willow), a type of inflorescence, really a spike, generally bearing only pistillate flowers or only staminate flowers, which eventually fall from the plant entire

Caulescent (Gr. *kaulos*, a plant stem), a plant whose stem bears leaves separated by visibly elongated internodes, as opposed to a rosette plant

Cell (L. *cella*, small room), a structural and physiological unit composing living organisms, in which take place the majority of complicated reactions characteristic of life

Cellulose (cell + *ose*, a suffix indicating a carbohydrate), a complex carbohydrate occurring in the cell walls of the majority of plants; cotton is largely cellulose; it is composed of hundreds of simple sugar molecules, glucose, linked together in a characteristic manner

Cenozoic (Gr. *kainos,* recent + *zoe,* life), the geologic era extending from 65 million years ago to the present

Chalaza (Gr. *chalaza,* small tubercle), the region on a seed at the upper end of the raphe where the funiculus spreads out and unites with the base of the ovule

Chernozem (R. *cherny,* black + *zem,* earth), a soil characteristic of some grassland vegetation in warm areas with moderate rainfall; dark in color because of a high content of organic matter

Chiasma (Gr. *chiasma,* two lines placed crosswise), the cross formed by breaking, during prophase I of meiosis, of two nonsister chromatids of homologous chromosomes and the rejoining of the broken ends of different chromatids

Chitin (Gr. *chiton,* a coat of mail), a horny substance forming the outer coat of insects and crustaceans and also found in the cell walls of many fungi

Chlamydospore (Gr. *chlamys,* a horseman's or young man's coat + spore), a heavy-walled resting asexual spore

Chlorenchyma (Gr. *chloros,* green + *-enchyma,* a suffix meaning tissue), tissue possessing chloroplasts

Chlorophyll (Gr. *chloros,* green + *phyllon,* leaf), the green pigment found in the chloroplast, important in the absorption of light energy in photosynthesis

Chloroplast (Gr. *chloros,* green + *plastos,* formed), specialized cytoplasmic body, containing chlorophyll, in which occur important reactions of starch or sugar synthesis

Chlorosis (Gr. *chloros,* green + *osis,* diseased state), failure of chlorophyll development, because of a nutritional disturbance or because of an infection of virus, bacteria, or fungus

Chromatid (chromosome + *-id,* L. suffix meaning daughters of), the half-chromosome during prophase and metaphase of mitosis, and between prophase I and anaphase II of meiosis

Chromatin (Gr. *chroma,* color), substance in the nucleus which readily takes artificial staining; also, that portion which bears the determiners of hereditary characters.

Chromatin bodies, bodies in bacteria that give some of the histochemical reactions that are associated with the chromosomes of higher organisms

Chromatophores, (Gr. *chromo,* color + *phorus,* a bearer), in algae, bodies bearing chlorophyll; in bacteria, small bodies, about 100 nanometers in diameter, containing chlorophyll, protein, and a carbohydrate

Chromoplast (Gr. *chroma,* color + *plastos,* formed), specialized plastid containing yellow or orange pigments

Chromosome (Gr. *chroma,* color + *soma,* body), a group of nuclear bodies containing genes, and largely responsible for the differentiation and activity of a cell, and undergoing characteristic division stages; or, one of the bodies into which the nucleus resolves itself at the beginning of mitosis and from which it is derived at the end of mitosis

Cilia (singular, **cilium**) (Fr. *cil,* an eyelash), protoplasmic hairs which, by a whip-like motion, propel certain types of unicellular organisms, gametes, and zoospores through water

Cisterna (plural, **cisternae**) (L. *cistern,* a reservoir), generally referring to sections of the endoplasmic reticulum that appear as parallel membranes, each about 5 nanometers thick bounding a space about 40 nanometers wide

Cistron (L. preposition *cis,* on this side of, relating to a genetical test of gene location + *tron* or *ton,* from the last syllable of proton, electron, and other such words indicating an elementary particle), a complex unit of the gene, probably consisting of from a hundred to thousands of nucleotide pairs

Cladode (Gr. *kladodes,* having many shoots), a cladophyll

Cladophyll (Gr. *klados,* a shoot + *phyllon,* leaf), a branch resembling a foliage leaf

Clay, soil particles less than 2 microns in diameter, composed mainly of aluminum (Al), oxygen (O), and silicon (S)

Cleistothecium (plural, **cleistothecia**) (Gr. *kleistos,* closed + *thekion,* a small receptacle), the closed, spherical ascocarp of the powdery mildews

Climax community, the last stage of a natural succession; a community capable of maintaining itself as long as the climate does not change

Clone (Gr. *klon,* a twig or slip), the aggregate of individual organisms produced asexually from one sexually produced individual

Closed bundle, a vascular bundle lacking cambium

Coal Age, the Carboniferous period, beginning 345 million years ago and ending 280 million years ago

Coalescence (L. *coalescere,* to grow together), a condition in which there is union of separate parts of any one whorl of flower parts

Coccus (plural, **cocci**) (Gr. *kokkos,* a berry), a spherical bacterium

Coenobium (Gr. *koinois,* common + Gr. *bios,* life), a colony of unicellular organisms surrounded by a common membrane

Coenocyte (Gr. *koinos*, shared in common + *kytos*, a vessel), a plant or filament whose protoplasm is continuous and multinucleate and without any division by walls into separate protoplasts

Coenogamete (Gr. *koinos*, shared in common + *gamete*), a multinucleate gamete, lacking cross-walls

Coenzyme, a substance, usually nonprotein and of low molecular weight, necessary for the action of some enzymes

Cohesion (L. *cohaerere*, to stick together), union or holding together of parts of the same materials; the union of floral parts of the same whorl, as petals to petals

Coleoptile (Gr. *koleos*, sheath + *ptilon*, down, feather), the first leaf in germination of monocotyledons, which sheaths the succeeding leaves

Coleorhiza (Gr. *koleos*, sheath + *rhiza*, root), sheath which surrounds the radicle of the grass embryo and through which the young root bursts

Collenchyma (Gr. *kolla*, glue + -*enchyma*, a suffix, derived from parenchyma and denoting a type of cell tissue), a stem tissue composed of cells which fit rather closely together and with walls thickened at the angles of the cells

Colloid (Gr. *kolla*, glue + *eidos*, form), referring to matter composed of particles, ranging in size from 0.0001 to 0.000001 millimeter, dispersed in some medium. Milk and mayonnaise are examples

Community (L. *communitas*, a fellowship), all the populations within a given habitat; usually the populations are thought of as being interdependent

Companion cells, cell associated with sieve-tube members

Competition (L. *competere*, to strive together), a form of biological interaction in which both organisms (at least initially) decline in growth or success because of the insufficient supply of some necessary factor(s)

Complete flower, a flower having four whorls of floral leaves: sepals, petals, stamens, and carpels

Compound leaf, a leaf whose blade is divided into several distinct leaflets

Conceptacle (L. *conceptaculum*, a receptacle), a cavity or chamber of a frond (of *Fucus*, for example) in which gametangia are borne

Cone (Gr. *konos*, a pine cone), a fruiting structure composed of modified leaves or branches, which bear sporangia (microsporangia, megasporangia, pollen sacs, or ovules), and frequently arranged in a spiral or four-ranked order; for example, a pine cone

Conidia (singular, **conidium**) (Gr. *konis*, dust), asexual reproductive cells of fungi, arising by fragmen-

tation of hyphae, by the cutting off of terminal or lateral cells of special hyphae, or by being pushed out from a flask-shaped cell

Conidiophore (conidia + Gr. *phoros*, bearing), conidium-bearing branch of hypha

Conidiosporangium (Gr. *konis*, dust + sporangium), sporangium formed by being cut off from the end of a terminal or lateral hypha

Conidiospore (conidia + spore), spore formed as described for conidia

Conifer (cone + L. *ferre*, to carry), a cone-bearing tree

Conservative (L. *conservare*, to keep), said of a taxonomic trait whose expression is not modified to any great extent by the external environment; a trait that is constant unless its genetic base is changed

Convergent evolution, process of successive progeny, originally of quite distinct parents, coming to appear more and more alike through time because of selection pressure

Conjugation (L. *conjugatus*, united), process of sexual reproduction involving the fusion of isogametes

Cork (L. *quercus*, oak), an external, secondary tissue impermeable to water and gases

Cork cambium, the cambium from which cork develops

Corm (Gr. *kormos*, a trunk), a short, solid, vertical, enlarged underground stem in which food is stored

Corolla (L. *corolla*, dim. of *corona*, a wreath, crown), petals, collectively; usually the conspicuous colored flower whorl

Cortex (L. *cortex*, bark), primary tissue of a stem or root bounded externally by the epidermis and internally in the stem by the phloem and in the root by the pericycle

Cotyledon (Gr. *kotyledon*, a cup-shaped hollow), seed leaf; two, generally storing food in dicotyledons; one, generally a digestive organ in the monocotyledons

Cristae (L. *crista*, a crest), crests or ridges, used here to designate the infoldings of the inner mitochondrial membrane

Cross-pollination, the transfer of pollen from a stamen to the stigma of a flower on another plant, except in clones

Crossing-over, the exchange of corresponding segments between chromatids of homologous chromosomes

Cuticle L. *cuticula*, dim. of *cutis*, the skin), waxy layer on outer wall of epidermal cells

Cutin (L. *cutis*, the skin), waxy substance which is

but slightly permeable to water, water vapor, and gases

Cutinization, impregnation of cell wall with a substance called cutin

Cyme (Gr. *kyma,* a wave, a swelling), a type of inflorescence in which the apex of the main stalk or the axis of the inflorescence ceases to grow quite early, relative to the laterals

Cystocarp (Gr. *kystos,* bladder + *karpos,* fruit), a peculiar diploid spore-bearing structure formed after fertilization in certain red algae

Cytochrome (Gr. *kytos,* a receptacle or cell + *chroma,* color), a class of several electron-transport proteins serving as carriers in mitochondrial oxidations and in photosynthetic electron transport

Cytokinesis (Gr. *kytos,* a hollow vessel + *kinesis,* motion), division of cytoplasmic constituents at cell division

Cytokinin (Gr. *kytos,* a receptacle or cell + *kinetos,* to move), a class of growth hormones important in the regulation of nucleic acid and protein metabolism, in cell division, delaying senescence, and organ initiation

Cytology (Gr. *kytos,* a hollow vessel + *logos,* word, speech, discourse), the science dealing with the cell

Cytoplasm (Gr. *kytos,* a hollow vessel + *plasma,* form), all the protoplasm of a protoplast outside the nucleus

Cytosine, a pyrimidine base found in DNA and RNA

Deciduous (L. *deciduus,* falling), referring to trees and shrubs that lose their leaves in the fall

Decomposer, (L. *de,* from + *componere,* to put together), an organism which obtains food by breaking down dead organic matter into simpler molecules

Decomposition (L. *de,* to denote an act undone + *componere,* to put together), a separation or dissolving into simpler compounds; rotting or decaying

Dehiscent (L. *dehiscere,* to split open), opening spontaneously when ripe, splitting into definite parts

Deletion (L. *deletus,* to destroy, to wipe out), used here to designate an area, or region, lacking from a chromosome

Dendrogram (Gr. *dendron,* tree + *gramme,* what is written or drawn), a graph showing relationship between things at different levels of similarity; the graph resembles the limbs of a tree

Denitrification (L. *de,* to denote an act undone + *nitrum,* nitro, a combining form indicating the presence of nitrogen + *facare,* to make), conversion of nitrates into nitrites, or into gaseous oxides of nitrogen, or even into free nitrogen

Deoxyribose nucleic acid (DNA), hereditary material; the DNA molecule carries hereditary information

Deuterium, or heavy hydrogen; a hydrogen atom, the nucleus of which contains one proton and one neutron; it is written as ^2H; the common nucleus of hydrogen consists only of one proton

Development (F. *developper,* to unfold), changes in form in the plant body brought about by the appearance of qualitative differences between cells, tissues, and organs

Diastase (Gr. *diastasis,* a separation), a complex of enzymes that brings about the hydrolysis of starch with the formation of sugar

Diatom (Gr. *diatomos,* cut in two), member of a group of golden brown algae with silicious cell walls fitting together much as do the halves of a pill box

Dicotyledon (Gr. *dis,* twice + *kotyledon,* a cup-shaped hollow), a plant whose embryo has two cotyledons

Dictyosome (Gr. *diktyon,* a net + *soma,* body), one of the component parts of the Golgi apparatus; in plant cells a complex of flattened double lamellae

Differentially permeable, referring to a membrane through which different substances diffuse at different rates; some substances may be unable to diffuse through such a membrane

Differentiation (L. *differre,* to carry different ways), development from one cell to many cells, accompanied by a modification of the new cells for the performance of particular functions

Diffusion (L. *diffusus,* spread out), the movement of molecules, and thus a substance, from a region of higher concentration of those molecules to a region of lower concentration

Digestion (L. *digestio,* dividing, or tearing to pieces, an orderly distribution), the processes of rendering food available for metabolism by breaking it down into simpler compounds, chiefly through actions of enzymes

Dihybrid cross (Gr. *dis,* twice + *hybrida,* the offspring of a tame sow and a wild boar, a mongrel), a cross between organisms differing in two characters

Dikaryon (Gr. *di,* two + *karyon,* nut), the $n + n$ paired nuclei, each usually derived from a different parent one male, one female

Dioecious (Gr. *dis,* twice + *oikos,* house), unisexual; having the male and female elements in different individuals

Diploid (Gr. *diploos,* double + *oides,* like), having a double set of chromosomes, or referring to an individual containing a double set of chromosomes per cell; usually a sporophyte generation

Disease (L. *dis,* a prefix signifying the opposite + M. E. *aise,* comfort, literally the opposite of ease), any alteration from state of metabolism necessary for the normal development and functioning of an organism

Distromatic (Gr. *di* a prefix meaning two + *stroma,* a bed, currently meaning a supporting framework), referring to a thallus, two cells in thickness

Divergent evolution, process of successive progeny, originally of quite similar parents, coming to appear more and more different through time because of isolation and selection pressure

Division, a major portion of the plant kingdom; equivalent to phylum

DNA, deoxyribose nucleic acid

Dominant (L. *dominari,* to rule), referring, in ecology, to species of a community that receive the full force of the macroenvironment; usually the most abundant of such species, and not all of them; referring, in heredity, to that gene (or the expression of the character it influences) which, when present in a hybrid with a contrasting gene, completely dominates in the development of the character; in peas, tall is dominant over dwarf

Donor, one who gives

Dormant (L. *dormire,* to sleep), being in a state of reduced physiological activity such as occurs in seeds, buds, etc.

Dorsiventral (L. *dorsum,* the back + *venter,* the belly), having upper and lower surfaces distinctly different, as a leaf does

DPN, diphosphopyridine nucleotide, or more recently, NAD—nicotinamide adenine dinucleotide

DPNH, reduced diphosphopyridine nucleotide or, more recently, NADH—reduced nicotinamide adenine dinucleotide

Drupe (L. *drupa,* an overripe olive), a simple, fleshy fruit, derived from a single carpel, usually one-seeded, in which the exocarp is thin, the mesocarp fleshy, and the endocarp stony

Ecology (Gr. *oikos,* home + *logos,* discourse), the study of plant life in relation to environment

Ecosystem (Gr. *oikos,* house + *synistanai,* to place together), an inclusive term for a living community and all the factors of its nonliving environment

Ecotype (Gr. *oikos,* house + *typos,* the mark of a blow), genetic variant within a species which is adapted to a particular environment yet remains interfertile with all other members of the species

Ectoplast (Gr. *ektos,* outside + *plastos,* formed), cytoplasmic membrane surrounding the outside of the protoplast

Edaphic (Gr. *edaphos,* soil), pertaining to soil conditions that influence plant growth

Egg (A. S. *aeg,* egg), a female gamate

Elater (Gr. *elater,* driver), an elongated, spindle-shaped, sterile, hygroscopic cell in the sporangium of a liverwort sporophyte

Electron (Gr. *elektron,* gleaming in the sun, by way of L. *electrum,* a bright alloy of gold and silver, and finally amber, from which the first electricity was produced by friction), the elementary charge of negative electricity, equal to 4.80×10^{-10} electrostatic units; normally, the number of orbital electrons balances the positive charge on the nucleus and renders the atom electrically neutral

Electrophoresis (Gr. *electron,* amber + *phora,* motion + *esis,* drive), the process of causing charged molecules (e.g., proteins) to move between positively and negatively charged poles

Element (L. *elementa,* the first principles; according to one system of medieval chemistry as recent as 1700, there were four elements composing all material bodies: earth, water, air, and fire), in modern chemistry, a substance that cannot be divided or reduced by any known chemical means to a simpler substance; 92 natural elements are known, of which gold, carbon, oxygen, and iron are examples; several, including plutonium, have been formed in atomic piles

Embryo (Gr. *en,* in + *bryein,* to swell), a young sporophytic plant, while still retained in the gametophyte or in the seed

Embryo sac, the female gametophyte of the angiosperms; generally a seven-celled structure; the seven cells are two synergids, one egg cell, three antipodal cells (each with a single haploid nucleus), and one endosperm mother cell with two haploid nuclei

Emulsion (L. *emulgere,* to milk out), a suspension of fine particles of a liquid in a liquid

Endocarp (Gr. *endon,* within + *karpos,* fruit), inner layer of fruit wall (pericarp)

Endodermis (Gr. *endon,* within + *derma,* skin), the layer of living cells, with various characteristically thickened walls and no intercellular spaces, which surrounds the vascular tissue of certain plants and occurs in nearly all roots and certain stems and leaves

Endogenous (Gr. *endon,* within + *genos,* race, kind), developed or added from outside the cell

Endoplasmic reticulum (Gr. *endon,* within + *plasma,* anything formed or molded; L. *reticulum,* a small net), originally, a cytoplasmic net adjacent to the nucleus, made visible by the electron microscope; now, any system of paired membranes within the cytoplasm; frequently abbreviated to ER

Endosperm (Gr. *endon,* within + *sperma,* seed), the nutritive tissue formed within the embryo sac of seed plants; it is often consumed as the seed matures, but remains in the seeds of corn and other cereals

Endosperm mother cell, one of the seven cells of the mature embryo sac, containing the two polar nuclei and, after reception of a sperm cell, giving rise to the primary endosperm cell from which the endosperm develops

Enzyme (Gr. *en,* in + *zyme,* yeast), a protein of complex chemical constitution produced in living cells, which, even in very low concentration, speeds up certain chemical reactions but is not used up in the reaction

Epiblast (Gr. *epi,* upon + *blastos,* a sprout or shoot), a small appendage in front of the plumule in the embryo of many grasses

Epicotyl (Gr. *epi,* upon + *kotyledon,* a cup-shaped hollow), the upper portion of the axis of embryo or seedling, above the cotyledons

Epidermis (Gr. *epi,* upon + *derma,* skin), a superficial layer of cells occurring on all parts of the primary plant body: stems, leaves, roots, flowers, fruits, and seeds; it is absent from the root cap and not differentiated on the apical meristems

Epigyny (Gr. *epi,* upon + *gyne,* woman), the arrangement of floral parts in which the ovary is embedded in the receptacle so that the other parts appear to arise from the top of the ovary

Epiphyte (L. *epi,* upon + *phyton,* a plant), a plant that grows upon another plant, yet is not parasitic

Equatorial rain forest, vegetation with several tree strata; characteristic of warm, wet regions; synonymous with tropical rain forest of other texts

ER, endoplasmic reticulum

Ergot (F. *argot,* a spur), a fungus disease of cereals and wild grasses in which the grain is replaced by dense masses of purplish hyphae, the ergot

Erosion (L. *e,* out + *rodere,* to gnaw), the wearing away of land, generally by the action of water

Ethylene, C_2H_4, a growth hormone regulating fruit ripening, various aspects of vegetative growth, and also important in the abscission process

Etiolation (F. *etioler,* to blanch), a condition involving increased stem elongation, poor leaf development, and lack of chlorophyll, found in plants growing in the absence, or in a greatly reduced amount, of light

Eukaryote (L. *eu,* true + *karyon,* a nut referring in modern biology to the nucleus), any organism characterized by having the cellular organelles, including the nucleus, bounded by membranes

Evapotranspiration (L. *evaporare, e,* out of + *vapor,* vapor + F. *transpirer,* to perspire), the process of water loss in vapor form from a unit surface of land both directly and from leaf surfaces

Evolution (L. *evolutio,* an unrolling), the development of a race, species, genus, or other larger group of plants or animals

Exine (L. *exterus,* outside), outer coat of pollen

Exocarp (Gr. *exo,* without, outside + *karpos,* fruit), outermost layer of fruit wall (pericarp)

Exogenous (Gr. *exe,* out, beyond + *benos,* race, kind), produced outside of, originating from, or due to external causes

F_1, first filial generation in a cross between any two parents

F_2, second filial generation, obtained by crossing two members of the F_1, or by self-pollinating the F_1

Facultative (L. *facultas,* capability), referring to an organism having the power to live under a number of certain specific conditions, e.g., a facultative parasite may be either parasitic or saprophytic

Family (L. *familia,* family), in plant taxonomy, a group of genera; families are grouped in orders

Fascicle (L. *fasciculus,* a small bundle), a bundle of pine or other needle-leaves of gymnosperms

Fascicular cambium, cambium within vascular bundles

Fermentation (L. *fermentum,* a drink made from fermented barley, beer), an oxidative process in foods in which molecular oxygen is not involved, such as the production of alcohol from sugar by yeasts

Ferredoxin, an electron-transferring protein containing iron, involved in photosynthesis and in the biological production and consumption of hydrogen gas

Fertilization (L. *fertilis,* capable of producing fruit), that state of a sexual life cycle involving the union of egg and sperm and hence the doubling of chromosome numbers

Feulgen, a histochemical test for DNA first published in 1932 by R. Feulgen, a German histologist

Fiber (L. *fibra,* a fiber or filament), an elongated, tapering, thick-walled strengthening cell occurring in various parts of plant bodies

Fiber-tracheid, xylem elements found in pine that are structurally intermediate between tracheids and fibers

Filament (L. *filum,* a thread), stalk of stamen bearing the anther at its tip; also, a slender row of cells (certain algae)

Fission (L. *fissilis,* easily split), asexual reproduction involving the division of a single-celled individual into two new single-celled individuals of equal size

Flagellum (plural, **flagella**) (L. *flagellum,* a whip), a long, slender whip of protoplasm

Flora (L. *floris,* a flower), an enumeration of all the species that grow in a region; also, the collective term for all the species that grow in a region

Floret (F. *fleurette,* a dim. of *fleur,* flower), one of the small flowers that make up the composite flower or the spike of the grasses

Flower (F. *fleur,* L. *flos,* a flower), floral leaves grouped together on a stem and adapted for sexual reproduction in the angiosperms

Follicle (L. *folliculus,* dim. of *follis,* bag), a simple, dry, dehiscent fruit, with one carpel, splitting along one suture

Food (A. S. *fōda*), any organic substance that furnishes energy and building materials directly for vital processes

Food chain, the path along which caloric energy is transferred within a community (from producers to consumers to decomposers)

Form genus, a scientific name given to an organism from the fossil record, when only a portion of the entire plant has been recovered and is known

Fossil (L. *fossio,* a digging), any impression, natural or impregnated remains, or other trace of an animal or plant of past geological ages which has been preserved in the earth's crust

Fruit (L. *fructus,* that which is enjoyed, hence product of the soil, trees, cattle, etc.), a matured ovary; in some seed plants other parts of the flower may be included; also applied, as **fruiting body,** to reproductive structures of other groups of plants

Fucoxanthin (Gr. *phykos,* seaweed + *xanthos,* yellowish brown), a brown pigment found in brown algae

Fungus (plural, **fungi**) (L. *fungus,* a mushroom), a thallus plant unable to make its own food, exclusive of the bacteria

Funiculus (L. *funiculus,* dim. of *funis,* rope or small cord), a stalk of the ovule, containing vascular tissue

Galen, Claudius Galenus, 131–210(?), the last of the great Greek biologists, he probably wrote 256 treatises, of which 131 were of a medical nature; his medical studies, as expressed in his book *On the Natural Faculties,* had a very great influence on biological and medical thought during the middle ages

Gametangium (Gr. *gametes,* a husband, *gamete,* a wife + *angeion,* a vessel), organ bearing gametes

Gamete (Gr. *gametes,* a husband, *gamete,* a wife), a protoplast that fuses with another protoplast to form the zygote in the process of sexual reproduction

Gametophyte (gamete + Gr. *phyton,* a plant), the gamete-producing plant

Gel (L. *gelare,* to freeze), jelly-like collidal mass

Gemma (plural, **gemmae**) (L. *gemma,* a bud), a small mass of vegetative tissue; an outgrowth of the thallus

Gene (Gr. *genos,* race, offspring), a group of base pairs in the DNA molecule in the chromosome that determines or conditions one or more hereditary characters

Gene recombination, the appearance of gene combinations in the progeny different from the combinations present in the parents

Genetics (Gr. *genesis,* origin), the science of heredity

Genotype (gene + type), the assemblage of genes in an organism

Genus (plural, **genera**) (Gr. *genos,* race, stock), a group of structurally or phylogenetically related species

Geotropism (Gr. *ge,* earth + *tropos,* turning), a growth curvature induced by gravity

Germination (L. *germinare,* to sprout), the beginning of growth of a seed, spore, bud, or other structure

Gibberellins, a group of growth hormones (not identical with auxins), the most characteristic effect of which is to increase the elongation of stems in a number of kinds of higher plants

Glucose (Gr. *glykys,* sweet + -*ose,* a suffix indicating a carbohydrate), a simple sugar, grape sugar, $C_6H_{12}O_6$

Glume (L. *gluma,* husk), an outer and lowermost bract of a grass spikelet

Glycogen (Gr. *glykys,* sweet + *gen,* of a kind), a carbohydrate related to starch but found generally in the liver of animals

Glycolysis (Gr. *glykys,* sweet + *lysis,* a loosening), decomposition of sugar compounds without involving free oxygen; early steps of respiration

Golgi body or zone (Italian cytologist Camillo Golgi 1844–1926, who first described the organelle), in animal cells, a complex perinuclear region thought to be associated with secretion; in plant cells a series of flattened plates, more properly called dictyosomes

Grana (singular, **granum**) (L. *granum* a seed), structures within chloroplasts, seen as green granules with the light microscope and as a series of parallel lamellae with the electron microscope

Ground cover, the area of ground covered by a plant when its canopy edge is projected perpendicularly down

Ground meristem (Gr. *meristos,* divisible), a primary meristem that gives rise to cortex, pit rays, and pith

Grow (A. S. *growan,* probably from Old Teutonic *gro,* from which grass is also derived), of living bodies generally: to increase gradually in size by natural development

Growth retardant, a chemical (such as cycocel, CCC) that selectively interferes with normal hormonal promotion of growth—but without appreciable toxic effects

Guanine, a purine base found in DNA and RNA

Guttation (L. *gutta,* drop, exudation of drops), exudation of water from plants, in liquid form

Gynoecium (Gr. *gyne,* woman + *oikos,* house), the aggregate of carpels in the flower of a seed plant

Haploid (Gr. *haploos,* single + *oides,* like), having a single complete set of chromosomes, or referring to an individual or generation containing such a single set of chromosomes per cell; usually a gametophyte generation

Haustorium (plural, **haustoria**) (M. L. *haustrum,* a pump), a projection of hyphae that acts as a penetrating and absorbing organ

Head, in inflorescence, typical of the composite family, in which flowers are grouped closely on a receptacle

Helix (Gr. *helix,* anything twisted), anything having a spiral form; here, quite generally refers to the double spiral of the DNA molecule

Hemicellulose (Gr. *hemi,* half + cellulose), any one of a group of carbohydrates resembling cellulose

Herb (L. *herba,* grass, green blades), a seed plant that does not develop woody tissues

Herbaceous (L. *herbaceus,* grassy), referring to plants having the characteristics of herbs

Herbal (L. *herba,* grass), a book that contains the names and descriptions of plants, especially those which are thought to have medicinal uses

Herbarium (L. *herba,* grass), a collection of dried and pressed plant specimens

Herbicide (L. *herba,* grass or herb + *cidere,* to kill), a chemical used to kill plants, frequently chemically related to a hormone (as the herbicide 2,4-D is related to the hormone IAA); an herbicide may have narrow or wide selectivity (range of target organisms)

Heredity (L. *herditas,* being a heir), the transmission of morphological and physiological characters of parents to their offspring

Hermaphrodite flower (Gr. *hermaphroditos,* a person having the attributes of both sexes, represented by Hermes and Aphrodite), a flower having both stamens and pistils

Heterobasidiomycetidae (Gr. *heteros,* other + Basidiomycete), a subclass of Basidiomycetes with variable basidia, never club-shaped cells

Heterocyst (Gr. *heteros,* different + *cytis,* a bag), an enlarged colorless cell that may occur in the filaments of certain blue-green algae

Heteroecious (Gr. *heteros,* different + *oikos,* house), referring to fungi that cannot carry through their complete life cycle unless two different host species are present

Heterogametes (Gr. *heteros,* different + gamete), gametes dissimilar from each other in size and behavior, like egg and sperm

Heterogamy (Gr. *heteros,* different + *gamos,* union or reproduction), reproduction involving two types of gametes

Heterophyte (Gr. *heteros,* different + *phyton,* a plant), a plant which must secure its food readymade

Heterosis (Gr. *heteros,* different + *-osis,* suffix indicating a state of), the state of a genotype having a large degree of heterozygosity

Heterospory (Gr. *heteros,* different + spore), the condition of producing microspores and megaspores

Heterothallic (Gr. *heteros,* different + thallus), referring to species in which male gametangia and female gamentangia are produced in different filaments or by different individual plant bodies

Heterotrichy (Gr. *heteros,* different + *trichos,* a hair), in the algae, the occurrence of two types of filaments, erect and prostrate

Heterotrophic (Gr. *heteros,* different + *trophein,* to nourish with food), referring to a plant obtaining nourishment from outside sources

Heterozygous (Gr. *heteros,* different + *zygon,* yoke), having different genes of a Mendelian pair present in the same cell or organism; for instance, a tall pea plant with genes for tallness (*T*) and dwarfness (*t*)

Hexose (Gr. *hexa,* six + *-ose,* suffix indicating, in this usage, carbohydrate), a carbohydrate with six carbon atoms

Hfr gene, gene in *Escherichia coli* including high frequency of gene recombination

Hilum (L. *hilum,* a trifle), scar on seed, which marks the place where the seed broke from the stalk

Histology (Gr. *histos,* cloth, tissue + *logos,* discourse), science that deals with the microscopic structure of animal and vegetable tissues

Homobasidiomycetidae (Gr. *homo,* the same + Basidiomycete), a subclass of Basidiomycetes with a typical club-shaped cell as a basidium

Homologous chromosomes (Gr. *homologos,* the same), members of a chromosome pair; they may be heterozygous or homozygous

Homospory (Gr. *homos,* one and the same + spore), the condition of producing one sort of spore only

Homothallic (Gr. *homos,* one and the same + thallus), referring to species in which male gametangia and female gamentangia are produced in the same filament or by the same individual plant body

Homozygous (Gr. *homos,* one and the same + *zygon,* yoke), having similar genes of a Mendelian pair present in the same cell or organism; for instance, a tall pea plant with genes for tallness (*TT*) only

Hormogonia (singular, **homogonium**) (Gr. *hormos,* necklace + *gonos,* offspring), short filaments, the result of a breaking apart of filaments of certain blue-green algae at the heterocysts

Hormone (Gr. *hormaein,* to excite), a specific organic product, produced in one part of a plant or animal body, and transported to another part where, effective in small amounts, it controls or stimulates another and different process

Humidity, relative (L. *humidus,* moist), the ratio of the weight of water vapor in a given quantity of air, to the total weight of water vapor that quantity of air is capable of holding at the temperature in question, expressed as percent

Humus (L. *humus,* the ground), decomposing organic matter in the soil

Hyaloplasm (Gr. *hyalinos,* glossy + *plasma,* anything formed), the clear background portion of the cytoplasm, forming the continuous substance in which other protoplasmic bodies are embedded

Hybrid (L. *hybrida,* offspring of a tame sow and a wild boar, a mongrel), the offspring of two plants or animals differing in at least one Mendelian character; or the offspring of plants or animals differing in many characters

Hydathode (Gr. *hydro,* water + O.E. *thoden,* stem or *thyddan,* to thrust), a structure, usually on leaves, which releases liquid water during guttation

Hydrogen acceptor, a substance capable of accepting hydrogen atoms or electrons in the oxidation–reduction reactions of metabolism

Hydrolysis (Gr. *hydro,* water + *lysis,* loosening), union of a compound with water, attended by decomposition into less complex compounds; usually controlled by enzymes.

Hymenium (Gr. *hymen,* a membrane), spore-bearing tissue in various fungi

Hypantheum (L. *hypo,* under + Gr. *anthos,* flower), fusion of calyx and corolla part way up their length to form a cup, as in many members of the rose family

Hypertrophy (Gr. *hyper,* over + *trophein,* to nourish with food), a condition of overgrowth or excessive development of an organ or part

Hypha (plural, **hyphae**) (Gr. *hyphe,* a web), a fungal thread or filament

Hypocotyl (Gr. *hypo,* under + *kotyledon,* a cup-shaped hollow), that portion of an embryo or seedling between the cotyledons and the radicle or young root

Hypogyny (Gr. *hypo,* under + *gyne,* female), a condition in which the receptable if convex or conical, and the flower parts are situated one above another in the following order, beginning with the lowest: sepals, petals, stamens, carpels

Hypothesis (Gr. *hypothesis,* foundation), a tentative theory or supposition provisionally adopted to explain certain facts and to guide in the investigation of other facts

Hypotrophy (Gr. *hypo,* under + *trophein,* to nourish with food), an underdevelopment of an organ or part

IAA, indoleacetic acid

Imbibition (L. *imbibere,* to drink), the absorption of liquids or vapors into the ultramicroscopic spaces or pores found in such materials as cellulose or a block of gelatine; an adsorption phenomenon

Imperfect flower, a flower lacking either stamens or pistils

Imperfect fungi, fungi reproducing only by asexual means

Inclusion body, a body found in the cells of organisms with a virus infection

Incomplete flower, a flower lacking one or more of the four kinds of flower parts

Indehiscent (L.*in,* not + *dehiscere,* to divide), not opening by valves or along regular lines

Indicator species, (a species that has a narrow range of tolerance for one or more environmental factors so that, from its occurrence at a site, one can predict these factors at that site (e.g., nutrient availability or summer temperatures)

Indoleacetic acid, a naturally occurring growth regulator, an auxin

Indusium (plural, **indusia**) (L. *indusium,* a woman's undergarment), membranous growth of the epidermis of a fern leaf that covers a sorus

Infect (L. *infectus,* to put into, to taint with morbid matter), specifically to produce disease by such agents as bacteria or viruses

Inferior ovary, an ovary more or less, or even completely, attached to or united with the calyx

Inflorescence (L. *inflorescere,* to begin to bloom), a flower cluster

Inheritance (O. F. *enheritance,* inheritance), the reception or acquisition of characters or qualities by transmission of parent to offspring

Inorganic, referring in chemistry to compounds that do not contain carbon

Integument (L. *integumentum,* covering), coat of ovule

Inter, a prefix, from the Latin preposition *inter,* meaning between, in between, in the midst of

Intercalary (L. *intercalare,* to insert), descriptive or meristematic tissue or growth not restricted to the apex of an organ, i.e., growth at nodes

Intercellular (L. *inter,* between + cells), lying between cells

Interfascicular cambium (L. *inter,* between + *fasciculus,* small bundle), cambium that develops between vascular bundles

Internode (L. *inter,* between + *nodus,* a knot), the region of a stem between two successive nodes

Intine (L. *intus,* within), the innermost coat of a pollen grain

Intra, a prefix from the Latin preposition *intra* meaning on the inside, within

Intracellular (L. *intra,* within + cell), lying within cells

Introgressive hybridization (L. *intro,* to the inside + *gress,* walk + *hybrida,* halfbreed), backcrossing between complete or partial hybrids and the original parental stock

Involucre (L. *involucrum,* a wrapper), a whorl or rosette of bracts surrounding an inflorescence

Ion (Gr. *ienai,* to go), an electrified particle formed by the breakdown of substances able to conduct an electric current

Irregular flower, a flower in which one or more members of at least one whorl are of different form from other members of the same whorl; zygomorphic flower

Isobilateral leaf (Gr. *isos,* equal + L. *bis,* twice, twofold + *lateralis,* pertaining to the side), a leaf having the upper and lower surfaces essentially similar

Isodiametric (Gr. *isos,* equal + diameter), having diameters equal in all directions, as a ball

Isogametes (Gr. *isos,* equal + gametes), gametes similar in size and behavior

Isogamy (Gr. *isos,* equal + *gamete,* spouse), the condition in which the gametes are identical

Isomers (Gr. *isos,* equal + *meros,* part), two or more compounds having the same molecular formula, for example, glucose and fructose are both $C_6H_{12}O_6$

Jansen, Zacharias, sixteenth century Dutch spectacle maker who made the first compound microscopes with magnifications of about ten times

K selection, natural selection which favors long-lived, late-maturing individuals which devote a small fraction of their resources into reproduction; tree species are K strategists

Karyogamy (Gr. *karyon,* nut + *gamos,* marriage), the fusion of two nuclei

Karyolymph (Gr. *karyon,* nut + *lympha,* water), the ground substance of a nucleus

Karyon (Gr. *karyon,* a nut), term used in conjunction with the nuclei in cells of Ascomycetes and Basidiomycetes; **dikaryon,** two nuclei per cell, each derived from a different parent, *n +n;* **heterokaryon,** the situation in which members of a dikaryon pair carry different alleles

Keel (A.S. *ceol,* ship), a structure of the legume type of flower, made up of two petals loosely united along their edges

Kelp (M.E. *culp,* seaweed), a collective name for any of the large brown algae

Kinetochore (Gr. *kinein,* to move + *chorein,* to move apart), specialized body of a chromosome, which seems to direct its movement

Lamella (plural, **lamellae**) (Gr. *lamin,* a thin blade), cellular membranes, frequently those seen in chloroplasts

Lamina (L. *lamina,* a thin plate), blade or expanded part of a leaf

Lateral bud, a bud that grows out of the side of a stem

Laterite (L. *later,* a brick), a soil characteristic of rain forest vegetation; color is red from oxidized iron in the A horizon

Latex (L. *latex,* juice), a milky secretion

Leaf axil, angle formed by the leaf stalk and the stem

Leaflet, separate part of the blade of a compound leaf

Leaf primordium (L. *primordium,* a beginning), a lateral outgrowth from the apical meristem, which will become a leaf

Legume (L. *legumen,* any leguminous plant, particularly bean), a simple, dry dehiscent fruit with one carpel, splitting along two sutures

Lemma (Gr. *lemma,* a husk), lower bract that subtends a grass flower

Lenticel (M.L. *lenticella,* a small lens), a structure of the bark that permits the passage of gas inward and outward

Leucine (Gr. *leukos,* white), a white, crystalline amino acid

Leucoplast (Gr. *leukos,* white + *plastos,* formed), a colorless plastid

Liana (F. *liane* from *lier,* to bind), a plant that climbs upon other plants, depending upon them for mechanical support; a plant with climbing shoots

Lichen (Gr. *leichen,* thallus plants growing on rocks and trees), a composite plant consisting of a fungus living symbiotically with an alga

Lignification (L. *lignum,* wood + *facere,* to make), impregnation of a cell wall with lignin

Lignin (L. *lignum,* wood), an organic substance or

group of substances impregnating the cellulose framework of certain plant cell walls

Ligule (L. *ligula,* dim. of *lingua,* tongue), in grass leaves, an outgrowth from the upper and inner side of the leaf blade where it joins the sheath

Linkage, the grouping of genes on the same chromosome

Linked characters, characters of a plant or animal controlled by genes grouped together on the same chromosome

Lipase (Gr. *lipos,* fat + *-ase,* suffix indicating an enzyme), any enzyme that breaks fats into glycerin and fatty acids

Lipid (Gr. *lipos,* fat + L. *ides,* suffix meaning son of; now used in sense of having the quality of), any of a group of fats or fat-like compounds insoluble in water and soluble in fat solvents

Liverwort (liver + M.E. *wort,* a plant; literally, a liver plant, so named in medieval times because of its fancied resemblance to the lobes of the liver), common name for the Class Hepaticae of the Bryophyta

Loam (O.E. *lam* or Old Teutonic *lai,* to be sticky, clayey), a particular soil texture class, referring to a soil having 30–50% sand, 30–40% silt, and 10–25% clay

Lobed leaf (Gr. *lobos,* lower part of the ear), a leaf divided by clefts or sinuses

Locule (L. *loculus,* dim. of *locus,* a place), a cavity of the ovary in which ovules occur

Lodicules (L. *lodicula,* a small coverlet), two scale-like structures that lie at the base of the ovary of a grass flower

Longevity (L. *longaevus,* long-lived), length of life

Lumen (L. *lumen,* light, an opening for light), the cavity of the cell within the cell walls

Lysis (Gr. *lysis,* a loosening), a process of disintegration of cell destruction

Lysogen (from *lysin,* a substance which brings about lysis, + N.L. *genic,* to give rise to), a substance, virus, or bacteria that stimulates the formation of lysins

Lysogenic bacteria (lysis + N.L. *genic,* combining form meaning to give rise to), here used to indicate bacteria carrying phage which eventually disrupt the bacteria

Lysogenic conversion, the influence of prophage DNA on the host bacterium, including colony morphology, and on the presence of particular enzymes and other compounds

Macroenvironment (Gr. *makros,* large + O.F. *environ,* about), the environment due to the general, regional climate; traditionally measured some 4 ft

above the ground and away from large obstructions

Macronutrient (Gr. *makros,* large + L. *nutrire,* to nourish), and essential element required by plants in relatively large quantities

Map distance on a chromosome, the distance in crossover units between designated genes

Megaphyll (Gr. *megas,* great + *phyllon,* leaf), a leaf whose trace is marked with a gap in the stem's vascular system; megaphylls are thought to represent modified branch systems

Megasporangium (Gr. *megas,* large + sporangium), sporangium that bears megaspores

Megaspore (Gr. *megas,* large + spore), the meiospore of vascular plants, which gives rise to a female gametophyte

Megaspore mother cell, a diploid cell in which meiosis will occur, resulting in four megaspores

Megasporophyll (Gr. *megas,* large + spore + Gr. *phyllon,* leaf), a leaf bearing megasporangia

Meiocyte (meiosis + Gr. *kytos,* currently meaning a cell), any cell in which meiosis occurs

Meiosis (Gr. *meioun,* to make smaller), two special cell divisions occurring once in the life cycle of every sexually reproducing plant and animal, halving the chromosome number and effecting a segregation of genetic determiners

Meiospore (meiosis + spore), any spore resulting from the meiotic divisions

Meristem (Gr. *meristos,* divisible), undifferentiated tissue, the cells of which are capable of active cell division and differentiation into specialized tissues

Meristoderm (meristem + epidermis), the outer meristematic cell layer (epidermis) of some Phaeophyta

Mesocarp (Gr. *mesos,* middle + *karpos,* fruit), middle layer of fruit wall (pericarp)

Mesophyll (Gr. *mesos,* middle + *phyllon,* leaf), parenchyma tissue of leaf between epidermal layers

Mesophyte (Gr. *mesos,* middle + *phyton,* a plant), a plant avoiding both extremes of moisture and drought

Mesosome (Gr. *mesos,* middle + *soma,* body), one of a series of paired membranes occurring in many bacteria

Mesozoic (Gr. *mesos,* middle + *zoe,* life), a geologic era beginning 225 million years ago and ending 65 million years ago

Metabolism (M.L. from the Gr. *metabolos,* to change), the process, in an organism or a single cell, by which nutritive material is built up into living matter, or aids in building living matter, or by which protoplasm is broken down into simple substances to perform special functions

Metabolite (Gr. *metabolos,* changeable + *ites,* one

of a group), a chemical that is a normal cell constituent capable of entering into the biochemical transformations within living cells

Metamorphic rock (Gr. *meta,* change + *morphe,* shape or form), one of three major categories of rock; rocks whose original structure or mineral composition has been changed by pressures or temperatures in the earth's crust

Metaphase (Gr. *meta,* after + *phasis,* appearance), stage of mitosis during which the chromosomes, or at least the kinetochores, lie in the central plane of the spindle

Methionine, a sulfur-containing amino acid

Microbody (Gr. *mikros,* small + body), a cellular organelle, always bound by a single membrane, frequently spherical, from 20 to 60 nanometers in diameter, containing a variety of enzymes

Microcapillary space, exceedingly small spaces, such as those found between microfibrils of cellulose

Microenvironment (Gr. *mikros,* small + O.F. *environ,* about), the environment close enough to the surface of a living or nonliving object to be influenced by it

Microfibrils (Gr. *mikros,* small + fibrils, dim. of fiber; literally, small little fibers), the translation of the name expresses the concept very well; microfibrils are exceedingly small fibers visible only with the high magnifications of the electron microscope

Micrometer (Gr. *mikros,* small + *metron,* measure), one millionth (10^{-6}) of a meter, or 0.001 millimeter; also called a micron, and abbreviated μ (Greek letter mu)

Micron (Gr. *mikros,* small), a unit of distance, 0.001 millimeter or 0.000039 inch; symbol μ (Greek letter mu)

Micronutrient (Gr. *mikros,* small + L. *nutrire,* to nourish), an essential element required by plants in relatively small quantities

Microphyll (Gr. *mikros,* small + *phyllon,* leaf), a leaf whose trace is not marked with a gap in the stem's vascular system; microphylls are thought to represent epidermal outgrowths

Micropyle (Gr. *mikros,* small + *pyle,* orifice, gate), a pore leading from the outer surface of the ovule between the edges of the two integuments down to the surface of the nucellus

Microsporangium (plural, **microsporangia**) (Gr. *mikros,* little + sporangium), a sporangium that bears microspores

Microspore (Gr. *mikros,* small + spore), a spore which, in vascular plants, gives rise to a male gametophyte

Microspore mother cell, a cell in which meiosis will occur, resulting in four microspores

Microsporophyll (Gr. *mikros,* little + spore + Gr. *phyllon,* leaf), a leaf bearing microsporangia

Microtubule (Gr. *mikros,* small + *tubule,* dim. of tube), a tubule 25 nm in diameter and of indefinite length, occurring in the cytoplasm of many types of cells

Micurgical dissections (Gr. *mikros,* small + surgical), surgical experiments done on individual cells or groups of cells, such as a shoot apex

Middle lamella (L. *lamella,* a thin plate or scale). original thin membrane separating two adjacent protoplasts and remaining as a distinct cementing layer between adjacent cell walls

Millimeter, the 0.001 part of a meter, equal to 0.0394 inch

Mitochondrion (plural, **mitochondria**) (Gr. *mitos,* thread + *chondrion,* a grain), a small cytoplasmic particle associated with intracellular respiration

Mitosis (plural, **mitoses**) (Gr. *mitos,* a thread), nuclear division, involving appearance of chromosomes, their longitudinal duplication, and equal distribution of newly formed parts to daughter nuclei

Mitospore (mitosis + spore), a spore forming after mitosis

Mixed bud, a bud containing both rudimentary leaves and flowers

Molecular biology, a field of biology which emphasizes the interaction of biochemistry and genetics in the life of an organism

Molecule (F. *môle,* mass + *cule,* a dim.; literally, a little mass), a unit of matter, the smallest portion of an element or a compound that retains chemical identity with the substance in mass; the molecule usually consists of a union of two or more atoms, some organic molecules containing a very large number of atoms

Monocotyledon (Gr. *monos,* solitary + *kotyledon,* a cup-shaped hollow), a plant whose embryo has one cotyledon

Monoecious (Gr. *monos,* solitary + *oikos,* house), having the reproductive organs in separate structures, but borne on the same individual

Monohybrid (Gr. *monos,* solitary + L. *hybrida,* a mongrel), a cross involving one pair of contrasting characters

Monophyletic (Gr. *mono,* single + *phyle,* tribe), said of organisms having a common (but sometimes quite ancient) ancestor

Monostromatic (Gr. *monos,* single, solitary + *stroma,* a bed, currently meaning a supporting framework), referring to a thallus, one cell in thickness

Morphogenesis (Gr. *morphe,* form + L. *genitus,* to produce), the structural and physiological events involved in the development of an entire organism or part of an organism

Morphology (Gr. *morphe,* form + *logos,* discourse), the study of form and its development

Moss (L. *muscus,* moss), a bryophytic plant

Mu, μ, Greek symbol used to indicate a micron; there are 1000 microns in 1 millimeter

Multiciliate (L. *multus,* many + F. *cil,* an eyelash), having many cilia present on a sperm or spore or other type of ciliated cell

Multiple fruit, a cluster of matured ovaries produced by separate flowers; e.g., a pineapple

Mutation (L. *mutare,* to change), a sudden, heritable change appearing in an individual as the result of a change in genes or chromosomes

Muton (mutation + *tron,* or *on,* from the last syllable of proton, electron, and other such words indicating an elementary particle), the smallest element, the alteration of which can cause a mutation; about 0.08 map unit, possibly a single nucleotide pair

Mutualism (L. *mutuus,* reciprocal), a form of biological interaction in which both organisms must associate together for continued success of both

Mycelium (Gr. *mykes,* mushroom), the mass of hyphae forming the body of the fungus

Mycology (Gr. *mykes,* mushroom + *logos,* discourse), the branch of botany dealing with fungi

Mycorrhiza (Gr. *mykos,* fungus + *riza,* root), a symbiotic association between a fungus and usually the root of a higher plant

Myxomycophyta (Gr. *myxa,* mucus + *mykes,* mushroom + *phyton,* plant), a division comprising the "slime fungi"

NAD, nicotineamide adenine dinucleotide, a coenzyme capable of being reduced; also called DPN, diphosphopyridine nucleotide

NADH, reduced NAD

NADP, nicotineamide adenine dinucleotide phosphate, a coenzyme capable of being reduced; also called TPN, triphosphopyridine nucleotide

NADPH, reduced NADP

Naked bud, a bud not protected by bud scales

Nanometer (Gr. *nanos,* small), one millionth (10^{-6}) of a millimeter, similar to a millimicron; equals 10 angstroms

Natural selection, the effect of the environment in channeling the genetic variation of organisms down particular pathways

Nectar (Gr. *nektar,* drink of the gods), a fluid rich in sugars secreted by nectaries, which are often located near or in flowers

Nectar guide, a mark of contrasting color or texture that may serve to guide pollinators to nectaries within the flower

Nectary (Gr. *nektar,* the drink of the gods), a nectar-secreting gland

Net productivity, the arithmetic difference between calories produced in photosynthesis and calories lost in respiration

Net radiation, the arithmetic difference between incoming solar radiation and outgoing terrestrial radiation

Net venation, veins of leaf blade visible to the unaided eye, branching frequently and joining again, forming a network

Neutron (L. *neuter,* neither), an uncharged particle found in the atomic nucleus of all elements except hydrogen; the helium nucleus has two protons and two neutrons; mass of a neutron is equal to 1.67×10^{-24} gram

Niche (It. *nicchia,* a recess in a wall), the functional position of an organism in its ecosystem

Nitrification (L. *nitrum,* nitro, a combining form indicating the presence of nitrogen + *facere,* to make), change of ammonium salts into nitrates through the activities of certain bacteria

Nitrogen base, the nitrogen-containing, positively charged, components of nucleic acids and phospholipids

Node (L. *nodus,* a knot), slightly enlarged portion of the stem where leaves and buds arise, and where branches originate

Nonseptate, descriptive of hyphae lacking cross-walls

Nucellus (L. *nucella,* a small nut), tissue composing the chief part of the young ovule, in which the embryo sac develops; megasporangium

Nucleic acid, an acid found in all nuclei, first isolated as part of a protein complex in 1871 and separated from the protein moiety in 1889; all known nucleic acids fall into two classes, DNA and RNA; they differ from each other in the sugar, in one of the nitrogen bases, in many physical properties, and in function

Nucleolus (L. *nucleolus,* a small nucleus), dense protoplasmic body in the nucleus

Nucleosides, components of nucleic acids consisting of a nitrogen base and a sugar; in DNA, the sugar is deoxyribose, and in RNA, ribose; adenine, guanine, and cytosine occur in both DNA and RNA, thymine occurs in DNA, and uracil occurs in RNA

Nucleotides, components of nucleic acid: nucleoside (nitrogen base + sugar) + phosphoric acid

Nucleus (L. *nucleus,* kernel of a nut), a dense protoplasmic body essential in cellular synthetic and developmental activities; present in all eukaryotic cells except mature sieve-tube members

Numerical taxonomy, a field of taxonomy that does not place subjective weight on any particular type of evidence that shows relationships between taxa

Nut (L. *nux,* nut), a dry, indehiscent, hard, one-seeded fruit, generally produced from a compound ovary

Obligate anaerobe, an organism obliged to live in the absence of oxygen

Obligate parasite, an organism obliged to live strictly as a parasite

Obligate saprophyte, an organism obliged to live strictly as a saprophyte

Oogamy (Gr. *oion,* egg + *gamete,* spouse), the condition in which the gametes are different in form and activity, i.e., sperms and eggs

Oogonium (L. dim. of Gr. *oogonos,* literally, a little egg layer), female gametangium of egg-bearing organ not protected by a jacket of sterile cells, characteristic of the thallophytes

Oospore (Gr. *oion,* an egg + spore), a resistant spore developing from a zygote resulting from the fusion of heterogametes

Open bundle, a vascular bundle with cambium

Operculum (L. *operculum,* a lid), in mosses, cap of sporangium

Opposite, referring to bud or leaf arrangement in which there are two buds or two leaves at a node

Organ (L. *organum,* an instrument or engine of any kind, musical, military, etc.), a part or member of an animal or plant body or cell adapted by its structure for a particular function

Organelle, a membrane-bound specialized region within a cell such as the mitochondrion or dictyosome

Organic, referring in chemistry to the carbon compounds, many of which have been in some manner associated with living organisms

Osmosis (Gr. *osmos,* a pushing), diffusion of a solvent through a differentially permeable membrane

Ovary (L. *ovum,* an egg), enlarged basal portion of the pistil, which becomes the fruit

Ovulate, referring to a cone, scale, or other structure bearing ovules

Ovule (F. *ovule,* from L. *ovulum,* dim. of *ovum,* egg), a rudimentary seed, containing, before fertilization, the female gametophyte, with egg cell, all being surrounded by the nucellus and one or two integuments

Ovuliferous (ovule + L. *ferre,* to bear), referring to a scale or sporophyll bearing ovules

P$_{FR}$ and **P$_R$,** abbreviations for the far-red (FR) or red (R) absorbing form of phytochrome (P)

Palea (or **palet**) (L. *palea,* chaff), upper bract which subtends a grass flower

Paleoecology (Gr. *palaios,* ancient), a field of ecology that reconstructs past vegetation and climate from fossil evidence

Paleozoic (Gr. *palaios,* ancient + *zoe,* life), a geologic era beginning 570 million years ago and ending 225 million years ago

Palisade parenchyma, elongated cells, containing many chloroplasts, found just beneath the upper epidermis of leaves

Palmately veined (L. *palma,* palm of the hand), descriptive of a leaf blade with several principal veins spreading out from the upper end of the petiole

Panicle (L. *panicula,* a tuft), an inflorescence, the main axis of which is branched, and whose branches bear loose racemose flower clusters

Pappus (L. *pappus,* woolly, hairy seed or fruit of certain plants), scales or bristles representing a reduced calyx in composite flowers

Parasexual cycle (Gr. *para,* beside), a sexual cycle involving changes in chromosome number differing in time and place from the usual sexual cycle; occurring in fungi in which the normal cycle is suppressed or apparently absent

Parallel venation, type of venation in which veins of a leaf blade that are clearly visible to the unaided eye are parallel to each other

Paraphysis (plural, **paraphyses**) (Gr. *para,* beside + *physis,* growth), a slender, multicellular hair (*Fucus,* etc.); one of the sterile branches or hyphae growing beside fertile cells in the fruiting body of certain fungi

Parasite (Gr. *parasitos,* one who eats at the table of another), an organism deriving its food from the living body of another plant or an animal

Parenchyma (Gr. *parenchein,* an ancient Greek medical term meaning to pour beside and expressing the ancient concept that the liver and other internal organs were formed by blood diffusing though the blood vessels and coagulating, thus designating ground tissue) a tissue composed of cells which usually have thin walls of cellulose, and which often fit rather loosely together, leaving intercellular spaces

Parent material, the original rock or depositional matter from which the soil of a region has been formed

Parietal (F. *pariétal,* attached to the wall, from L. *paries,* wall), belonging to, connected with, or attached to the wall of a hollow organ or structure, especially of the ovary or cell

Parietal placentation, a type of placentation in which placentae are on the ovary wall

Parthenocarpy (Gr. *parthenos,* virgin + *karpos,* fruit), the development of fruit without fertilization

Parthenogenesis (Gr. *parthenos,* virgin + *genesis,*

origin), the development of a gamete into a new individual without fertilization

Passive absorption of water, the absorption of water by roots due to forces of transpiration in leaves

Passive solute absorption absorption due only to forces of simple diffusion

Pathogen (Gr. *pathos,* suffering + *genesis,* beginning), an organism that causes a disease

Pathology (GR. *pathos,* suffering + *logos,* account), the study of diseases, their effects on plants or animals, and their treatment

Peat (M.E. *pete,* of Celtic origin, a piece of turf used as fuel), any mass of semicarbonized vegetable tissue formed by partial decomposition in water

Pectin (Gr. *pektos,* congealed), a white amorphous substance which, when combined with acid and sugar, yields a jelly; the substance cementing cells together; the middle lamella

Pedicel (L. *pediculus,* a little foot), stalk or stem of the individual flowers of an inflorescence

Peduncle (L. *pedunculus,* a late form of *pediculus,* a little foot), stalk or stem of a flower that is borne singly; or the main stem of an inflorescence

Penicillin, an antibiotic derived from the mold *Penicillium*

Pentose (Gr. *pente,* five, + *-ose,* a suffix indicating a carbohydrate), a five-carbon-atom sugar

Perennial (L. *perennis,* lasting the whole year through), a plant that lives from year to year

Perfect flower, a flower having both stamens and pistils; hermaphroditic flower

Perianth (Gr. *peri,* around + *anthos,* flower), the petals and sepals taken together

Pericarp (Gr. *peri,* around + *karpos,* fruit), fruit wall, developed from ovary wall

Pericycle (Gr. *peri,* around + *kyklos,* circle), tissue, generally of root, bound externally by the endodermis and internally by the phloem

Peridium (plural, **peridia**) (Gr. *peridion,* a little pouch), external covering of the hymenium of certain fungi; in Myxomycetes, the hardened envelope that covers the sporangium

Perigyny (Gr. *peri,* about + *gyne,* a female), a condition in which the receptacle is more or less concave, at the margin of which the sepals, petals, and stamens have their origin, so that these parts seem to be attached around the ovary; also called half-inferior

Peristome (Gr. *peri,* about + *stoma,* a mouth), in mosses, a fringe of teeth about the opening of the sporangium

Perithecium (Gr. *peri,* around + *theke,* a box), a spherical or flask-shaped ascocarp having a small opening

Permafrost (L. *permanere,* to remain + A.S. *freosan,* to freeze), soil that is permanently frozen; usually found some distance below a surface layer that thaws during warm weather

Permeable (L. *permeabilis,* that which can be penetrated), said of a membrane, cell, or cell system through which substances may diffuse

Peroxysome, an organelle of the microbody class which contains enzymes capable of making and destroying hydrogen peroxide, glycolic oxidase, and catalase

Petal (Gr. *petalon,* a flower leaf), one of the flower parts, usually conspicuously colored

Petiole (L. *petiolus,* a little foot or leg), stalk of leaf

PGA, phosphoglyceric acid

Phage (Gr. *phago* to eat), a virus infecting bacteria; originally bacteriophage

Phelloderm (Gr. *phellos,* cork + *derma,* skin), a layer of cells formed in the stems of some plants from the inner cells of the cork cambium

Phellogen (Gr. *phellos,* cork + *genesis,* birth), cork cambium, a cambium giving rise externally to cork and in some plants internally to phelloderm

Phenotype (Gr. *phaneros,* showing + type), the external visible appearance of an organism

Phloem (Gr. *phloos,* bark), food-conducting tissue, consisting of sieve tubes with companion cells or sieve cells, phloem parenchyma, and fibers

Phosphoenolpyruvate carboxylase, the enzyme responsible for the fixation of inorganic CO_2 into oxaloacetic acid in a dark reaction of the C_4 photosynthesis cycle

Phosphoglyceric acid (PGA), a three-carbon compound formed by the interaction of carbon dioxide (CO_2) and a five-carbon compound, ribulose diphosphate; the reaction yields two molecules of PGA for each molecule of the ribulose diphosphate; the first step in the C_3 carbon cycle of photosynthesis

Phosphorylation, a reaction in which phosphate is added to a compound, e.g., the formation of ATP from ADP and inorganic phosphate

Photon, a quantum of light; the energy of a photon is proportional to its frequency; $E = \hbar\nu$, where E is energy; $\hbar$, Planck's constant, 6.62×10^{-27} erg-second; and ν is the frequency

Photoperiod (Gr. *photos,* light + period), the optimum length of day or period of daily illumination required for the normal growth and maturity of a plant

Photophosphorylation, a reaction in which light energy is converted into chemical energy in the form of ATP produced from ADP and inorganic phosphate

Photoreceptor (Gr. *photos,* light + L. *receptor,* a receiver), a light-absorbing molecule involved in converting light into some metabolic (chemical energy) form, *e.g.*, chlorophyll and phytochrome

Photosynthesis (Gr. *photos,* light + *syn,* together + *tithenai,* to place), a process in which carbon dioxide and water are brought together chemically to form a carbohydrate, the energy for the process being radiant energy

Phototropism (Gr. *photos,* light + *tropos,* turning), a growth curvature in which light is the stimulus

Phycobiliproteins, pigments found in the red and blue-green algae, similar to bile pigments and always associated with proteins

Phycocyanin (Gr. *phykos,* seaweed + *kyanos,* blue), a blue phycobilin pigment occurring in blue-green algae

Phycoerythrin (Gr. *phykos,* seaweed + *erythros,* red), a red phycobilin pigment occurring in red algae

Phycomycetes (Gr. *phykos,* seaweed + *mykes,* mushroom or fungus), a class of fungi which approaches the algae in some characters

Phylogeny (Gr. *phylon,* race or tribe + *genesis,* beginning), the evolution of a group of related individuals

Phylum (Gr. *phylon,* race or tribe), a primary division of the animal or plant kingdom

Physiology (Gr. *physis,* nature + *logos,* discourse), the science of the functions and activities of living organisms

Phytobenthon (Gr. *phyton,* a plant + *benthos,* depths of the sea), attached aquatic plants, collectively

Phytochrome, a reversible pigment system of protein nature found in the cytoplasm of green plants; it is associated with the absorption of light that affects growth, development, and differentiation of a plant, independent of photosynthesis, *e.g.*, in the photoperiodic response

Phytoplankton (Gr. *phyton,* a plant + *planktos,* wandering), free-floating plants, collectively

Pi, an abbreviation for inorganic orthophosphate (a mixture of $H_2PO_4^-$ and HPO_4^{2-}, depending on the pH)

Pileus (L. *pileus,* a cap), umbrella-shaped cap of fleshy fungi

Pinna (plural, *pinnae*) (L. *pinna,* a feather), leaflet or division of a compound leaf (frond)

Pinnately veined (L. *pinna,* a feather + *vena,* a vein), descriptive of a leaf blade with single midrib from which smaller veins branch off, somewhat like the divisions of a feather

Pioneer community, the first stage of a succession

Pistil (L. *pistillum,* a pestle), central organ of the flower, typically consisting of ovary, style, and stigma

Pistillate flower, a flower having pistils but no stamens

Pit, a minute thin area of a cell wall

Pith, the parenchymatous tissue occupying the central portion of a stem

Placenta (plural, **placentae**) (L. *placenta,* a cake), the tissue within the ovary to which the ovules are attached

Placentation (L. *placenta,* a cake + *-tion,* state of), manner in which the placentae are distributed in the ovary

Plankton (Gr. *planktos,* wandering), free-floating aquatic plants and animals, collectively

Plaque (D. *plak,* flat piece of wood), here, the clear area in a plate culture of bacteria caused by the lysis of the bacteria

Plasmodesma (plural, **plasmodesmata**) (Gr. *plasma,* something formed + *desmos,* a bond, a band), fine protoplasmic thread passing through the wall that separates two protoplasts

Plasmodium (Gr. *plasma,* something formed + mod. L. *odium,* something of the nature of), in Myxomycetes, a slimy mass of protoplasm, with no surrounding wall and with numerous free nuclei distributed throughout

Plasmogamy (Gr. *plasma,* anything molded or formed + *gamos,* marriage), the fusion of protoplasts, not accompanied by nuclear fusion

Plasmalemma (Gr. *plasma,* anything formed + *lemma,* a husk or shell of a fruit), a delicate cytoplasmic membrane found on the outside of the protoplast adjacent to the cell wall

Plasmolysis (Gr. *plasma,* something formed + *lysis,* a loosening), the separation of the cytoplasm from the cell wall due to removal of water from the protoplast

Plastid (Gr. *plastis,* a builder), the cellular organelle in which carbohydrate metabolism is localized

Plastoquinone, a quinone, one of a group of compounds involved in the transport of electrons during photosynthesis in chloroplasts

Plumule (L. *plumula,* a small feather), the first bud of an embryo or that portion of the young shoot above the cotyledons

Podzol (R. *pod,* under + *zola,* ashes), a soil characteristic of taiga vegetation; color of the A horizon is gray because of excessive leaching

Polar transport, the directed movement within plants of compounds (usually hormones) predominantly in one direction; polar transport overcomes the tendency for diffusion in all directions

Polarity (Gr. *pol,* an axis), the observed differentiation of an organism, tissue, or cell into parts having opposed or contrasted properties or form

Pollen (L. *pollen,* fine flour), the germinated microspores or partially developed male gametophytes of seed plants

Pollen mother cell, the cell in which meiosis will occur, resulting in the formation of four pollen grains

Pollen profile, a diagrammatic summary of the sequence and abundance of pollen types that have been chronologically trapped in sediments

Pollination, the transfer of pollen from a stamen or staminate cone to a stigma or ovulate cone

Pollinium (L. *pollentis,* powerful or *pollinis,* fine flour + *ium,* group), a mass of pollen which sticks together and is transported by pollinators as a mass; present in orchids and milkweeds

Polygenes, many genes influencing the development of a single trait; results in continuous variability; compare allele

Polymerization, the chemical union of monomers such as glucose, or nucleotides to form starch or nucleic acid

Polynomial (Gr. *polys,* many + L. *nomen,* name), scientific name for an organism composed of more than two words (as in binomial)

Polynucleotides (Gr. *polys,* much, many), long-chain molecules composed of units (monomers) called nucleotides; nucleic acid is a polynucleotide

Polyphyletic (Gr. *polys,* many + *phyle,* tribe), referring to organisms having more than one common ancestor

Polyploid (Gr. *polys,* many + *ploos,* fold), referring to a plant, tissue, or cell with more than two complete sets of chromosomes; e.g., 4n, 6n

Polyribosome (Gr. *polys,* many + ribosomes), an aggregation of ribosomes; frequently simply *polysome*

Polysaccharides (Gr. *polys,* much, many + *sakcharon,* sugar), long-chain molecules composed of units (monomers) of a sugar; starch and cellulose are polysaccharides

Pome (G. *pomme,* apple), a simple fleshy fruit, the outer portion of which is formed by the floral parts that surround the ovary

Population (L. *populus,* people), a group of closely related, interbreeding organisms

Prairie (L. *pratum,* meadow), grassland vegetation, with trees essentially absent; often considered to have more rainfall than does the steppe

Predation (L. *predatio,* plundering), a form of biological interaction in which one organism is destroyed (by ingestion); parasitism is a form of predation

Primary (L. *primus,* first), first in order of time or development

Primary endosperm cell, a cell of the embryo sac after fertilization, generally containing a nucleus resulting from fusion of the two polar nuclei with a sperm nucleus; the endosperm develops from this cell

Primary meristems, meristems of the shoot or root tip giving rise to the primary plant body

Primitive (L. *primus,* first), referring to a taxonomic trait thought to have evolved early in time

Primordium (L. *primus,* first + *ordiri,* to begin to weave; literally beginning to weave, or to put things in order), the beginning or origin of any part of an organ

Procambium (L. *pro,* before + cambium), a primary meristem that gives rise to primary vascular tissues and, in most woody plants, to the vascular cambium

Producer (L. *producere,* to draw forward), an organism which produces organic matter for itself and other organisms (consumers and decomposers) by photosynthesis

Proembryo (L. *pro,* before + *embryon,* embryo), a group of cells arising from the division of the fertilized egg cell before those cells which are to become the embryo are recognizable

Prokaryotes (L. *pro,* before + Gr. *karyon,* a nut, referring in modern biology to the nucleus), primitive organisms, bacteria, and blue-green algae, which do not have the DNA separated from the cytoplasm by an envelope

Prophage (Gr. *pro,* prefix meaning before), a noninfectious phage unit which multiplies with the growing and dividing bacteria but does not bring about lysis of the bacteria; prophage is a stage in the life cycle of a temperate phage

Prophase (Gr. *pro,* before + *phasis,* appearance), an early stage in nuclear division, characterized by the shortening and thickening of the chromosomes and their movement to the metaphase plate

Proplastid (Gr. *pro,* before + plastid), a type of plastid, occurring generally in meristematic cells, which will develop into a chloroplast

Protease (protein + *-ase,* a suffix indicating an enzyme), an enzyme breaking down a protein

Protein (Gr. *proteios,* holding first place), naturally occurring complex organic substances (egg albumen, meat) composed of amino acids, which are associated to form submicroscopic chains, spirals, or plates

Proterozoic (Gr. *protero,* before in time + *zoe* life), the earliest geologic era, beginning about 4.5–5 billion years ago and ending 570 million years ago; also called Precambrian era

Prothallium (Gr. *pro,* before + *thallos,* a sprout), in ferns, the haploid gametophyte generation

Protochlorophyll (Gr. *protos,* first + *chloros,* green + *phyllos,* leaf), one of the precursors of chlorophyll; it accumulates in dark-grown and potentially green tissue

Protochlorophyllide holochrome, a light-sensitive compound or complex composed of protochlorophyll and a protein; absorption of light converts the protochlorophyll part to chlorophyll

Protoderm (Gr. *protos*, first + *derma*, skin), a primary meristem that gives rise to epidermis

Proton (Gr. *proton*, first), the nucleus of a hydrogen atom is a single positively charged particle, the proton; the nucleus of all other elements consists of protons and neutrons; the mass of a proton is 1.67×10^{-24} gram

Protonema (plural, **protonemata**) (Gr. *protos*, first + *nema*, a thread), an algal-like filamentous growth; an early stage in development of the gametophyte of mosses

Protoplasm (Gr. *protos*, first + *plasma*, something formed), living substance

Protoplast (Gr. *protoplastos*, formed first), the organized living unit of a single cell

Pseudopodium (Gr. *pseudes*, false + *podion*, a foot), in Myxomycetes, an arm-like projection from the body by which the plant creeps over the surface

Purine, a group of nitrogen bases having a double-ring structure, one five-carbon, the other six-carbon

Pyrenoid (Gr. *pyren*, the stone of a fruit + L. *oïdes*, like), a denser body occurring within the chloroplasts of certain algae and liverworts and apparently associated with starch deposition

Pyrimidine, a nitrogen base having a single-ring structural formula

Quadrat (L. *quadrus*, a square), a frame of any shape which, when placed over vegetation, defines a unit sample area within which the plants may be counted or measured

Quantum (L. *quantum*, how much), an elemental unit of energy; its energy value is $h\nu$, where $\hbar$, Planck's constant, is 6.62×10^{-27} erg-second and ν is the frequency of the vibrations or waves with which the energy is associated

R selection, natural selection which favors short-lived, early-maturing individuals which devote a large fraction of their resources into reproduction; annual herbs are R strategists

Raceme (L. *racemus*, a bunch of grapes), an inflorescence in which the main axis is elongated but the flowers are born on pedicels that are about equal in length

Rachilla (Gr. *rhachis*, a backbone + L. dim. ending *-illa*), shortened axis of spikelet

Rachis (Gr. *rhachis*, a backbone), main axis of spike; axis of fern leaf (frond) from which pinnae arise; in compound leaves, the extension of the petiole corresponding to the midrib of an entire leaf

Radicle (L. *radix*, root), portion of the plant embryo which develops into the primary or seed root

Random plant distribution, a distribution of a plant species within an area such that the probability of finding an individual at one point is the same for all points

Raphe (Gr. *rhaphe*, seam), ridge on seeds, formed by the stalk of the ovule, in those seeds in which the funiculus is sharply bent at the base of the ovule

Raphides (Gr. *rhaphis*, a needle), fine, sharp, needle-like crystals

Receptacle (L. *receptaculum*, a reservoir), enlarged end of the pedicel or peduncle to which other flower parts are attached

Recessive character, that member of a pair of Mendelian characters which, when both members of the pair are present, is subordinated or suppressed by the other, dominant character

Recombination (L. *re*, repeatedly + *combinatus*, joined), the mixing of genotypes that results from sexual reproduction

Recon (from recombination and the ending *on* currently indicating an elementary particle), the smallest unit of gene recombination, approximately 0.02 map unit; possible one nucleotide pair

Reduction (F. *reduction*, L. *reductio*, a bringing back), originally "bringing back" a metal from its oxide, i.e., iron from iron rust or ore; any chemical reaction involving the removal of oxygen from or the addition of hydrogen or an electron to a substance; energy is required and may be stored in the process as in photosynthesis

Regular flower, a flower in which the corolla is made up of similarly shaped petals equally spaced and radiating from the center of the flower; star-shaped flower; actinomorphic flower

Replication, the production of a facsimile or a very close copy; here used to indicate the production of a second molecule of DNA exactly like the first molecule

Reproduction (L. *re*, repeatedly + *producere*, to give birth to), the process by which plants and animals give rise to offspring

Reproductive isolation, the separation of populations in time or space so that genetic flow between them is cut off

Respiration (L. *re*, repeatedly + *spirare*, to breathe), a chemical oxidation controlled and catalyzed by enzymes which in living protoplasm break down carbohydrate and fats, thus releasing energy to be used by the organism in doing work

Reticulum (L. *reticulum*, a small net), a small net

Rhizoid (Gr. *rhiza*, root + L. *oïdes*, like), one of the cellular filaments that perform the functions of roots

Rhizome (Gr. *rhiza*, root), an elongated, underground, horizontal stem

Rhizophores (Gr. *rhiza*, root + *phoros*, bearing), leafless branches that grow downward from the leafy stems of certain Lycophyta and give rise to roots when they come into contact with the soil

Ribose, a five-carbon sugar, a component of RNA

Ribose nucleic acid, a nucleic acid containing the sugar ribose, phosphorus, and the bases adenine, guanine, cytosine, and uracil; present in all cells and concerned with protein synthesis in the cell

Ribosomes (*ribo*, from RNA + Gr. *somatos*, body), small particles 10–20 nanometers in diameter, containing RNA

Ripening (A.S. *rifi*, perhaps related to reap), changes in a fruit that follow seed maturation and which prepare the fruit for its function of seed dispersal

RNA, ribose nucleic acid

Root (A.S. *rōt*), the descending axis of a plant, normally below ground, serving to anchor the plant and absorb and conduct water and mineral nutrients

Root cap, A thimble-like mass of cells covering and protecting the apical meristems of a root

Root pressure, pressure developed in the root as the result of osmosis and inducing bleeding in stem wounds

Rootstock, an elongated, underground, horizontal stem

Runner, a stem that grows horizontally along the ground surface

Samara (L. *samara*, the fruit of the elm), simple, dry, one- or two-seeded indehiscent fruit with pericarp bearing a wing-like outgrowth

Sand, soil particles between 50 and 2000 microns in diameter

Saprophyte (Gr. *sapros*, rotten + *phyton*, a plant), an organism deriving its food from the dead body or the nonliving products of another plant or animal

Savannah (Sp. *sabana*, a large plain), vegetation of scattered trees in a grassland matrix

Scalariform vessel (L. *scala*, ladder + form), a vessel with secondary thickening resembling a ladder

Schizocarp (Gr. *schizein*, to split + *karpos*, fruit), dry fruit with two or more united carpels which split apart at maturity

Sclereids (Gr. *skleros*, hard), more or less isodiametric cells having heavily lignified cell walls

Sclerenchyma (Gr. *skleros*, hard + *-echyma*, a suffix denoting tissue), a strengthening tissue composed of cells with heavily lignified cell walls

Scrub (A.S. *scrob*, a shrub), vegetation dominated by shrubs; described as thorn forest in areas with moderate rainfall, or as chaparral or desert in areas with low rainfall

Scutellum (L. *scutella*, a dim. of *scutum*, shield), single cotyledon of grass embryo

Sedimentary rock (L. *sedere*, to sit), rock formed from material deposited as sediment, then physically or chemically changed by compaction and hardening while buried in the earth crust

Seed (A.S. *sed*, anything which may be sown), popularly as originally used, anything which may be sown; i.e., "seed" potatoes, "seeds" of corn, sunflower, etc.; botanically, a seed is the matured ovule without accessory parts

Self-pollination, transfer of pollen from the stamens to the stigma of either the same flower or flowers on the same plant

Seminal root, the root or roots forming from primordia present in the seed

Sepals (M. L. *sepalum*, a covering; "sepalum" is a modern word formed by analogy with petalum [Gr. *petalon*]), outermost flower structures which usually enclose the other flower parts in the bud

Septate (L. *septum*, fence), divided by cross-walls into cells or compartments

Septicidal dehiscence (L. *septum*, fence + *caedere*, to cut; *dehiscere*, to split open), the splitting open of a capsule along the line of union carpels

Septum (L. *septum*, fence), any dividing wall or partition; frequently a cross-wall in a fungal or algal filament

Serpentine (L. *serpens*, a serpent), referring to soil derived from metamorphic parent material characterized (among other things) by low calcium (Ca), high magnesium (Mg), and a greenish-gray color

Sessile (L. *sessilis*, low, dwarf, from *sedere*, to sit), sitting, referring to a leaf lacking a petiole or a flower or fruit lacking a pedicel

Seta (plural, **setae**) (L. *seta*, a bristle), in bryophytes, a short stalk of the sporophyte, which connects the foot and the capsule

Sexual reproduction, reproduction that requires meiosis and fertilization for a complete life cycle

Sheath, part of leaf that wraps around the stem, as in grasses

Shoot (derivation uncertain, but early referring to new plant growth; 1450, "Take a feyr schoyt of blake thorne"), a young branch that shoots out from the main stock of a tree, or the young main portion of a plant growing above ground

Shoot tip, portion of the shoot containing apical and

primary meristems and early stages of differentiation

Sibling species, species morphologically nearly identical but incapable of producing fertile hybrids

Sieve cell, a long and slender sieve element without a companion cell, with relatively unspecialized sieve areas, and with tapering end walls that lack sieve plates

Sieve plate, perforated wall area in a sieve-tube member through which pass strands connecting sieve-tube protoplasts

Sieve tube, a series of sieve-tube members forming a long cellular tube specialized for the conduction of food materials

Sieve-tube members, portion of a sieve tube comprised of a single protoplast and separated from other sieve-tube members by sieve plates

Silique (L. *siliqua,* pod), the fruit characteristic of Cruciferae (mustards); two-celled, the valves splitting from the bottom and leaving the placentae with the false partition stretched between

Silt, soil particles between 2 and 50 microns in diameter

Simple pit, pit not surrounded by an overarching border; in contrast to bordered pit

Siphonous line, a line of evolutionary development in the algae in which mitosis is not followed by cytokinesis; this results in an elongated multinucleate, coenocytic filament

Soil (L. *solum,* soil, solid), the uppermost stratum of the earth's crust, which has been modified by weathering and organic activity into (typically) three horizons: an upper A horizon which is leached, a middle B horizon in which the leached material accumulates, and a lower C horizon which is unweathered parent material

Soil texture, refers to the amounts of sand, silt, and clay in a soil, as a sandy loam, loam, or clay texture

Solute (L. *solutus,* from *solvere,* to loosen), a dissolved substance

Solution (M.E. *solucion,* from O.F. *solucion,* to loosen), a homogeneous mixture, the molecules of the dissolved substance (e.g., sugar), the solute, being dispersed between the molecules of the solvent (e.g., water)

Solvent (L. *solvere,* to loosen), a substance, usually a liquid, having the properties of dissolving other substances

Soredium (plural, **soredia**) (Gr. *soros,* a heap), asexual reproductive body of lichens, consisting of a few algal cells surrounded by fungous hyphae

Sorus (plural, **sori**) (Gr. *soros,* a heap), a cluster of sporangia

Species (L. *species,* appearance, form, kind), a class of individuals usually interbreeding freely and having many characteristics in common

Sperm (Gr. *sperma,* the generative substance or seed of a male animal), a male gamete

Spermagonium (plural, **spermagonia**) (Gr. *sperma,* sperm + *gonos,* offspring), flask-shaped structure characteristic of the sexual phase of the rust fungi; bearing receptive hyphae and spermatia

Spermatium (plural, **spermatia**) (Gr. *sperma,* sperm), in rust fungi, a cell borne at the tip of the hyphae which line the interior of spermagonia (on barberry leaves)

Spermatophyte (Gr. *sperma,* seed + *phyton,* plant), a seed plant

Spike (L. *spica,* an ear of grain), an inflorescence in which the main axis is elongated and the flowers are sessile

Spikelet (L. *spica,* an ear of grain + dim. ending -*let*), the unit of inflorescence in grasses; a small group of grass flowers

Spindle (A.S. *spinel,* an instrument employed in spinning thread by hand), referring in mitosis and meiosis to the spindle-shaped intracellular structure in which the chromosomes move

Sporangiophore (sporangium + Gr. -*phore,* a root of *phorein,* to bear), a branch bearing one or more sporangia

Sporangium (spore + Gr. *angeion,* a vessel), spore case

Spore (Gr. *spora,* seed), a reproductive cell that develops into a plant without union with other cells; some spores such as meiospores occur at a critical stage in the sexual cycle, but others are asexual in nature

Sporidium (dim. meaning a little spore), the basidiospore of smut fungi

Sporophore (spore + Gr. *phorein,* to bear), the fruiting body of fleshy and woody fungi, which produces spores

Sporophyll (spore + Gr. *phyllon,* leaf), a spore-bearing leaf

Sporophyte (spore + Gr. *phyton,* a plant), in alternation of generations, the plant in which meiosis occurs and which thus produces meiospores

Stamen (L. *stamen,* the standing-up things or a tuft of thready things), flower structure made up of an anther (pollen-bearing portion) and a stalk or filament

Staminate flower, a flower having stamens but no pistils

Starch (M.E. *sterchen,* to stiffen), a complex insoluble carbohydrate, the chief food storage substance of plants, which is composed of several hundred hexose sugar units and which easily breaks down on hydrolysis into these separate units

Statolith (Gr. *statos,* standing + *lithos,* stone), an organelle that moves to its position in a cell as a result of gravity, thus providing an initial sensing of, or orientation to, gravity by a cell

Stele (Gr. *stele,* a post), the central cylinder, inside the cortex, of roots and stems of vascular plants

Stem (O.E. *stemn*), the main body of the portion above ground of tree, shrub, herb, or other plant; the ascending axis, whether above or below ground, of a plant, in contradistinction to the descending axis or root

Steppe (R. *step,* a lowland), an arid grassland vegetation

Sterigma (plural, **sterigmata**) (Gr. *sterigma,* a prop), a slender, pointed protuberance at the end of a basidium, which bears a basidiospore

Stigma (L. *stigma,* a prick, a spot, a mark), receptive portion of the style to which pollen adheres

Stipule (L. *stipula,* dim. of *stipes,* a stock or trunk), a leaf-like structure from either side of the leaf base

Stolon (L. *stolo,* a shoot), a stem that grows horizontally along the ground surface

Stoma (plural, **stomata**) (Gr. *stoma,* mouth), a minute opening, bordered by guard cells in the epidermis of leaves and stems, through which gases pass

Strobilus (Gr. *strobilos,* a cone), a number of modified leaves (sporophylls) or ovule-bearing scales grouped together on an axis

Stroma (Gr. *stroma,* a bed or covering), a mass of protecting vegetative filaments; the background substance of chloroplasts, probably the location of the carbon cycle of photosynthesis

Style (Gr. *stylos,* a column), slender column of tissue that arises from the top of the ovary and through which the pollen tube grows

Suberin (L. *suber,* the cork oak), a waxy material found in the cell walls of cork tissue

Succession (L. *successio,* a coming into the place of another), a sequence of changes in time of the species which inhabit an area, from an initial pioneer community to a final climax community

Succulent (L. *sucus,* juice), a plant having juicy or watery tissues

Sucrose (F. *sucre,* sugar + *-ose,* termination designating a sugar), cane sugar ($C_{12}H_{22}O_{11}$)

Superior ovary, an ovary completely separate and free from the calyx

Suspensor (L. *suspendere,* to hang), a cell or chain of cells developed from a zygote whose function is to place the embryo cells in an advantageous position to receive food

Suture (L. *sutura,* a sewing together; originally the sewing together of flesh or bone wounds), the junction, or line of junction, of contiguous parts

Symbiosis (Gr. *syn,* with + *bios,* life), an association of two different kinds of living organisms involving benefit to both

Sympetaly (Gr. *syn,* with + *petalon,* leaf), a condition in which petals are united

Synandry (Gr. *syn,* with + *andros,* a man), a condition in which stamens are united

Syncarpy (Gr. *syn,* with + *karpos,* fruit), a condition in which carpels are united

Synergids (Gr. *synergos,* toiling together), the two nuclei at the upper end of the embryo sac, which, with the third (the egg), constitute the egg apparatus

Synsepaly (Gr. *syn,* with + sepals), a condition in which sepals are united

Taiga (Teleut *taiga,* rocky mountainous terrain), a broad northern belt of vegetation dominated by conifers; also, a similar belt in mountains just below alpine vegetation

Tannin, a substance that has an astringent, bitter taste

Tapetum (Gr. *tapes,* a carpet), nutritive tissue in the sporangium, particularly an anther

Taxon (plural, **taxa**) (Gr. *taxis,* order), a general term for any taxonomic rank, from subspecific to divisional

Taxonomy (Gr. *taxis,* arrangement + *nomos,* law), systematic botany; the science dealing with the describing, naming, and classifying of plants

Teliospore (Gr. *telos,* completion + spore), resistant spore characteristic of the Heterobasidiomycetidae, in which karyogamy and meiosis occur and from which a basidium develops

Telium (plural, **telia**) (Gr. *telos,* completion), a sorus of teliospores

Telophase (Gr. *telos,* completion + phase), the last stage of mitosis, in which daughter nuclei are reorganized

Temperate phage, a phage that does not necessarily cause lysis of the bacterial cells in which it reproduces

Tendril (L. *tendere,* to stretch out, to extend), a slender coiling organ that aids in the support of stems

Terminal bud, a bud at the end of a stem

Testa (L. *testa,* brick, shell), the outer coat of the seed

Tetrad (Gr. *tetradeion,* a set of four), a group of four, usually referring to the meiospores immediately after meiosis

Tetraploid (Gr. *tetra,* four + *ploos,* fold), having four sets of chromosomes per nucleus

Tetraspores (Gr. *tetra,* four + spores), four spores formed by division of the spore mother cell, used particularly for meiospores in certain red algae

Tetrasporine line (tetraspore + L. suffix *-ine,* like), a line of evolutionary development in the algae in which mitosis is directly followed by cytokinesis, resulting in a filament, thallus, or complex plant body of varied form

Thallophytes (Gr. *thallos,* a sprout + *phyton,* plant), a division of plants whose body is a thallus, i.e., lacking roots, stems, and leaves

Thallus (Gr. *thallos,* a sprout), plant body without true roots, stems, or leaves

Threonin, an amino acid

Thylakoid (Gr. *thylakos,* sack + N.L. *oid,* a thing that is like), one of membranes in the chloroplasts

Thymidine, a nucleoside incorporated in DNA, but not in RNA

Thymidine ^{3}**H**, tritiated or radioactive thymidine

Thymine, a pyrimidine occurring in DNA, but not in RNA

Tiller (O.E. *telga,* a branch), a grass stem arising from a lateral bud at a basal node; tillering is the process of tiller formation

Tissue, a group of cells of similar structure which performs a special function

Tonoplast (Gr. *tonos,* stretching tension + *plastos,* molded, formed), the cytoplasmic membrane bordering the vacuole; so-called by de Vries, as he thought it regulated the pressure exerted by the cell sap

Toxin (L. *toxicum,* poison), a poisonous secretion of a plant or animal

TPN, triphosphopyridine nucleotide, or, more recently, NADP, nicotinamide adenine dinucleotide phosphate

TPNH, reduced triphosphopyridine nucleotide or, more recently, NADPH, reduced nicotinamide adenine dinucleotide phosphate

Tracheid (Gr. *tracheia,* windpipe), an elongated, tapering xylem cell, with lignified pitted walls, adapted for conduction and support

Tracheophytes (Gr. *tracheia,* windpipe + *phyton,* plant), vascular plants

Trait, a distinctive definable characteristic; a mark of individuality

Transduction, gene transfer in bacteria, in which bacterial genes are carried from one bacteria to another by way of a temperate phage

Transformation, mutation or change in genes of bacteria by the direct intervention of extracellular DNA

Translocation (L. *trans,* across + *locare,* to place), the transfer of food materials or products of metabolism

Transmit, to pass or convey something from one person, organism, or place to another person, organism, or place

Transpiration (F. *transpirer,* to perspire), the giving off of water vapor from the surface of leaves

Trichogyne (Gr. *trichos,* a hair + *gyne,* female), receptive hair-like extension of the female gametangium in the Rhodophyta and Ascomycetes

Trichome (Gr. *trichoma,* a growth of hair), a short filament of cells

Triose (Gr. *treis,* three + *-ose,* suffix indicating a carbohydrate), any three-carbon sugar

Trisomic (Gr. *treis,* three + *soma,* body), a plant containing one additional chromosome; $2n + 1$

Tritium, a hydrogen atom, the nucleus of which contains one proton and two neutrons; it is written as ^{3}H; the more common hydrogen nucleus consists only of a proton

Tropism (Gr. *trope,* a turning), movement of curvature due to an external stimulus that determines the direction of movement

Tuber (L. *tuber,* a bump, swelling), a much-enlarged, short, fleshy underground stem

Tundra (Lapp. *tundar,* hill), meadow-like vegetation at low elevation in cold regions that do not experience a single month with average daily maximum temperatures above 50°F

Turgid (L. *turgidus,* swollen, inflated), swollen, distended; referring to a cell that is firm due to water uptake

Turgor pressure (L. *turgor,* a swelling), the pressure within the cell resulting from the absorption of water into the vacuole and the imbibition of water by the protoplasm

Tylosis (plural, **tyloses**) (Gr. *tylos,* a lump or knot), a growth of one cell into the cavity of another

Type specimen, the herbarium specimen selected by a taxonomist to serve as a basis for the naming and descriptions of a new species

Umbel (L. *umbella,* a sunshade), an inflorescence, the individual pedicels of which all arise from the apex of the peduncle

Unavailable water, water held by the soil so strongly that root hairs cannot readily absorb it

Unicell (L. *unus,* one + cell), an organism consisting of a single cell; generally used in describing algae

Uniseriate (L. *unus,* one + M.L. *seriatus,* to arrange in a series), said of a filament having a single row of cells

Universal veil, a membrane completely surrounding the sporophyte of a homobasidiomycete at the button stage

Uracil, a pyrimidine found in RNA but not in DNA

Uredium (plural, **uredinia**) (L. *uredo,* a blight), a sorus of uredospores

Uredospore (L. *uredo,* a blight + spore), a red, one-celled summer spore in the life cycle of the rust fungi

Vaccination, the injection of vaccine

Vaccine (L. *vacca,* cow), a suspension of weakened or dead bacteria or other pathogens injected into the body to immunize against the same species of pathogen or their toxins

Vacuole (L. dim. of *vacuus,* empty), a watery solution of various substances forming a portion of the protoplast distinct from the protoplasm

Van Leeuwenhoek, Antony (1632–1723), a native of Delft, Holland, a dealer in dry goods, and a skilled lens maker; made many observations of biological materials at magnifications of about 100×; famous for his early observations on bacteria

Vascular (L. *vasculum,* a small vessel), referring to any plant tissue or region consisting of or giving rise to conducting tissue, e.g., bundle, cambium, ray

Vascular bundle, a strand of tissue containing primary xylem and primary phloem (and procambium if present) and frequently enclosed by a bundle sheath of parenchyma or fibers

Vascular cambium, cambium giving rise to secondary phloem and secondary xylem

Vector (L. *vehere,* to carry), an organism, usually an insect, that carries and transmits disease-causing organisms

Vegetation (L. *vegetare,* to quicken), the plant cover that clothes a region; it is formed of the species that make up the flora, but is characterized by the abundance and life form (tree, shrub, herb, evergreen, deciduous plant, etc.) of certain of them

Venation (L. *vena,* a vein), arrangement of veins in leaf blade

Venter (L. *venter,* the belly), enlarged basal portion of an archegonium in which the egg cell is borne

Ventral canal cell, the cell just above the egg cell in the archegonium

Ventral suture (L. *ventralis,* pertaining to the belly), the line of union of the two edges of a carpel

Vernalization (L. *vernalis,* belonging to spring + *izare,* to make), the promotion of flowering by naturally or artificially applied periods of extended low temperature; seeds, bulbs, or entire plants may be so treated

Vessel (L. *vasculum,* a small vessel), a series of xylem elements whose function it is to conduct water and mineral nutrients

Vessel element, a portion of a vessel derived from a single cell of the vascular cambium or procambium

Virulence (L. *virulentia,* a stench), the relative infectiousness of a bacteria or virus, or its ability to overcome the resistance of the host metabolism

Virus (L. *virus,* a poisonous or slimy liquid), a disease principle that can be cultivated only in living tissues, or in freshly prepared tissue brei

Vitamins (L. *vita,* life + amine), naturally occurring organic substances, akin to enzymes, necessary in small amounts for the normal metabolism of plants and animals

Volva (L. *volva,* a wrapper), cup at base of stipe or stalk of fleshy fungi

Volvocine line (*Volvox* + L. suffix *-ine,* like), a line of evolutionary development in the algae in which the cells remain separate, as in *Volvox,* and are never connected to form a filament or flattened thallus

Water potential, refers to the difference between the activity of water molecules in pure distilled water at atmospheric pressure and 30°C (standard conditions) and the activity of water molecules in any other system; the activity of these water molecules may be greater (positive) or less (negative) than the activity of the water molecules under standard conditions

Weed (A.S. *wēod,* used at least since 888 in its present meaning), generally a herbaceous plant or shrub not valued for use or beauty, growing where unwanted, and regarded as using ground or hindering the growth of more desirable plants

Whorl, a circle of flower parts, or of leaves

Whorled, referring to bud or leaf arrangement in which there are three or more buds or three or more leaves at a node

Wild-type, in genetics, the gene normally occurring in the wild population, usually dominant

Wings, lateral petals of legume type of flower

Wood (M.E., *wode, wude,* a tree), a dense growth of trees, or a piece of a tree, generally the xylem

Xanthophyll (Gr. *xanthos,* yellowish brown + *phyllon,* leaf), a yellow chloroplast pigment

Xerophyte (Gr. *xeros,* dry + *phyton,* a plant), a plant very resistant to drought, or that lives in very dry places

Xylem (Gr. *xylon,* wood), a plant tissue consisting of tracheids, vessels, parenchyma cells, and fibers; wood

Zoology (Gr. *zoon,* an animal + *logos,* speech), the science having to do with animal life

Zoosporangium (Gr. *zoon,* an animal + sporangium), a sporangium bearing zoospores

Zoospore (Gr. *zoon,* an animal + spore), a motile spore

Zygomorphic (Gr. *zygo,* yoke, pair + *morphe,* form), referring to bilateral symmetry; said of organisms, or a flower, capable of being divided into two symmetrical halves only by a single longitudinal plane passing through the axis

Zygospore (Gr. *zygon,* a yoke + spore), a thick-walled resistant spore developing from a zygote resulting from the fusion of isogametes

Zygote (Gr. *zygon,* a yoke), a protoplast resulting from the fusion of gametes (either isogametes or heterogametes)

Zymase (Gr. *zyme,* leaven), an intracellular, sugar-fermenting complex of enzymes

Table of Metric Equivalents

Exponents

Very large or very small numbers are often written in exponential form to save space. Thus, 0.1 can be written exponentially as 1×10^{-1}, or simply as 10^{-1}; $0.0000035 = 3.5 \times 10^{-6}$; $1000 = 1 \times 10^3$ or 10^3.

Common Prefixes

nano $= 10^{-9}$ (one billionth)
micro $= 10^{-6}$ (one millionth)
milli $= 10^{-3}$ (one thousandth)
kilo $= 10^3$ (one thousand times)
mega $= 10^6$ (one million times)

I. Units of Length

kilometer (km) $= 10^3$ m $= 0.62$ mile
meter (m) $= 1.09$ yd $= 3.28$ ft
decimeter (dm) $= 10^{-1}$ m $= 3.90$ in.
centimeter (cm) $= 10^{-2}$ m $= 0.39$ in.
millimeter (mm) $= 10^{-3}$ m $= 0.04$ in.
micron, micrometer (μ) $= 10^{-6}$ m
millimicron, nanometer (mμ or nm) $= 10^{-9}$ m
angstrom (Å) $= 10^{-10}$ m

II. Units of Weight

metric ton $= 10^3$ kg $= 10^6$ g $= 1.10$ short tons $= 0.98$ long ton $= 2200$ lb
kilogram (kg) $= 10^3$ g $= 2.20$ lb
gram (g) $= 0.035$ oz
milligram (mg) $= 10^{-3}$ g $= 0.015$ grain

III. Units of Area

are (a) $= 100$ m^2 $= 1075.84$ ft^2
hectare (ha) $= 100$ are $= 2.47$ acres

IV. Units of Volume

liter (l) $= 10^3$ ml $= 1.06$ U.S. liquid qt $= 0.91$ U.S. dry qt $= 0.88$ British qt
milliliter, cubic centimeter (ml or cm^3) $= 0.03$ fluid oz $= 0.06$ in.^{3}

V. Miscellany; Units of Pressure, Light Intensity, Energy, Work, Force

bar $= 0.99$ atm of pressure $= 14.54$ lb/in.^{2}
lux (L) $= 0.09$ ft-c of light intensity
joule (j) $= 0.735$ ft-lb of work $= 9.5 \times 10^{-4}$ British Thermal Units (BTU)
erg $= 10^{-7}$ joule
watt (w) $= 1.3 \times 10^3$ hp
calorie, gram-calorie (c) $=$ amount of heat needed

to raise the temperature of 1 g of water 1°C from 14.5 to 15.5°C

kilocalorie, kilogram-calorie (kc) = 1000 c = 3.97 BTU (1 BTU is the amount of heat needed to raise 1 lb of water 1°F; in 1970, the per capita energy consumption per year was 300 million BTU)

solar constant (radiation at Earth's outer atmosphere edge) = 2 cal/cm^2/min = 13,400 ft-c = 4.06 × 10^7 BTU/m^2/yr

dyne = force required to accelerate 1 g at the rate of

1 cm/sec/sec = 1 g-cm/sec^2

newton = 10^5 dyne = 1 kg-m/sec^2

VI. Units of Temperature

one degree Celsius, Centigrade (1°C) = 1°K = 1.8°F

absolute zero = −273°C = 0°K = −460°F

boiling point of water = 100°C = 373°K = 212°F

freezing point of water = 0°C = 273°K = 32°F

to convert from °F to °C: (°F − 32) × 0.56 = °C

Subject Index

Index to Genera

Illustrations are indicated by boldface type. Text discussion often appears on adjacent pages. Families are given for the Anthophyta, classes for the Mycota, and divisions for all other genera. Fossil plants are indicated by (f).

Common names are given for the vascular plants and for some of the bryophytes and thallophytes. The common names selected are those preferred by Bailey (*Manual of Cultivated Plants*) or Jepson (*Manual of the Flowering Plants of California*) for either the genus or a common species of the genus. It is probable that different common names will be used for the same species in different sections of the country.